MARINE BIOLOGY

Editor-in-Chief

John H. Steele

Marine Policy Center, Woods Hole Oceanographic Institution, Woods Hole,
Massachusetts, USA

Editors

Steve A. Thorpe

National Oceanography Centre, University of Southampton,
Southampton, UK
School of Ocean Sciences, Bangor University, Menai Bridge, Anglesey, UK

Karl K. Turekian

Yale University, Department of Geology and Geophysics, New Haven, Connecticut,
USA

Subject Area Volumes from the Second Edition

Climate & Oceans edited by Karl K. Turekian

Elements of Physical Oceanography edited by Steve A. Thorpe

Marine Biology edited by John H. Steele

Marine Chemistry & Geochemistry edited by Karl K. Turekian

Marine Ecological Processes edited by John H. Steele

Marine Geology & Geophysics edited by Karl K. Turekian

Marine Policy & Economics guest edited by Porter Hoagland, Marine Policy Center,
Woods Hole Oceanographic Institution, Woods Hole, Massachusetts

Measurement Techniques, Sensors & Platforms edited by Steve A. Thorpe

Ocean Currents edited by Steve A. Thorpe

ENCYLOPEDIA
OF
OCEAN SCIENCES: MARINE BIOLOGY

Editor

JOHN H. STEELE

BOSTON • HEIDELBERG • LONDON • NEW YORK • OXFORD
PARIS • SAN DIEGO • SAN FRANCISCO • SINGAPORE • SYDNEY • TOKYO
Academic Press is an imprint of Elsevier

ELSEVIER

ACADEMIC PRESS

Academic Press is an imprint of Elsevier
32 Jamestown Road, London NW1 7BY, UK
30 Corporate Drive, Suite 400, Burlington, MA 01803, USA
525 B Street, Suite 1900, San Diego, CA 92101-4495, USA

Notice
No responsibility is assumed by the publisher for any injury and/or damage to persons or property as a matter of products liability, negligence or otherwise, or from any use or operation of any methods, products, instructions or ideas contained in the material herein, Because of rapid advances in the medical sciences, in particular, independent verification of diagnoses and drug dosages should be made

British Library Cataloguing in Publication Data
A catalogue record for this book is available from the British Library

Library of Congress Control Number: 2009932544

ISBN: 978-0-08-096480-5

For information on all Elsevier publications
visit our website at www.elsevierdirect.com

Printed and bound by CPI Group (UK) Ltd, Croydon, CR0 4YY
Transferred to Digital Print 2011

CONTENTS

Marine Biology: Introduction   ix

PLANKTON & NEKTON

Plankton Overview   *M M Mullin*   3

Marine Plankton Communities   *G-A Paffenhöfer*   5

Plankton Viruses   *J Fuhrman, I Hewson*   13

Bacterioplankton   *H W Ducklow*   21

Protozoa, Radiolarians   *O R Anderson*   28

Protozoa, Planktonic Foraminifera   *R Schiebel, C Hemleben*   33

Copepods   *R Harris*   40

Gelatinous Zooplankton   *L P Madin, G R Harbison*   51

Krill   *E J Murphy*   62

Nekton   *W G Pearcy, R D Brodeur*   71

Cephalopods   *P Boyle*   78

Phytoplankton Size Structure   *E Marañón*   85

Primary Production Methods   *J J Cullen*   93

Primary Production Processes   *J A Raven*   100

Primary Production Distribution   *S Sathyendranath, T Platt*   105

Bioluminescence   *P J Herring, E A Widder*   111

BENTHOS

Benthic Organisms Overview   *P F Kingston*   123

Microphytobenthos   *G J C Underwood*   132

Phytobenthos   *M Wilkinson*   140

Benthic Foraminifera   *A J Gooday*   147

Meiobenthos   *B C Coull, G T Chandler*   159

Macrobenthos   *J D Gage*   165

Deep-sea Fauna   *P V R Snelgrove, J F Grassle*   176

Cold-Water Coral Reefs   *J M Roberts*   188

Benthic Boundary Layer Effects   *D J Wildish*   199

FISH BIOLOGY

Fish: General Review   *Q Bone*   209

Antarctic Fishes   *I Everson*   218

Coral Reef Fishes   *M A Hixon*   222

Deep-sea Fishes   *J D M Gordon*   227

Intertidal Fishes   *R N Gibson*   233

Mesopelagic Fishes   *A G V Salvanes, J B Kristoffersen*   239

Pelagic Fishes   *D H Cushing*   246

Salmonids   *D Mills*   252

Eels   *J D McCleave*   262

Fish Ecophysiology   *J Davenport*   272

Fish Feeding and Foraging   *P J B Hart*   279

Fish Larvae   *E D Houde*   286

Fish Locomotion   *J J Videler*   297

Fish Migration, Horizontal   *G P Arnold*   307

Fish Migration, Vertical   *J D Neilson, R I Perry*   316

Fish Predation and Mortality   *K M Bailey, J T Duffy-Anderson*   322

Fish Reproduction   *J H S Blaxter*   330

Fish Schooling   *T J Pitcher*   337

Fish Vision   *R H Douglas*   350

Fish: Demersal Fish (Life Histories, Behavior, Adaptations)   *O A Bergstad*   363

Fish: Hearing, Lateral Lines (Mechanisms, Role in Behavior, Adaptations to Life
    Underwater)   *A N Popper, D M Higgs*   372

MARINE MAMMALS

Marine Mammal Overview   *P L Tyack*   381

Baleen Whales   *J L Bannister*   391

Sperm Whales and Beaked Whales   *S K Hooker*   403

Dolphins and Porpoises   *R S Wells*   411

Seals   *I L Boyd*   424

Sirenians   *T J O'Shea, J A Powell*   431

Sea Otters   *J L Bodkin*   442

Sea Turtles   *F V Paladino, S J Morreale*   450

Marine Mammal Diving Physiology   *G L Kooyman*   458

Marine Mammal Evolution and Taxonomy   *J E Heyning*   465

Marine Mammal Migrations and Movement Patterns   *P J Corkeron, S M Van Parijs*   472

Marine Mammal Social Organization and Communication   *P L Tyack*   481

Marine Mammal Trophic Levels and Interactions   *A W Trites*   488

BIRDS

Seabirds: An Overview   *G L Hunt, Jr.*   497

Alcidae   *T Gaston*   503

Laridae, Sternidae and Rynchopidae   *J Burger, M Gochfeld*   510

Pelecaniformes   *D Siegel-Causey*   522

Phalaropes   *M Rubega*   531

Procellariiformes   *K C Hamer*   539

Sphenisciformes   *L S Davis*   546

Seabird Conservation   *J Burger*   555

Seabird Foraging Ecology   *L T Balance, D G Ainley, G L Hunt Jr.*   562

Seabird Migration   *L B Spear*   571

Seabird Population Dynamics   *G L Hunt*   582

APPENDIX

Appendix 9. Taxonomic Outline of Marine Organisms   *L P Madin*   589

INDEX

                                                                            601

MARINE BIOLOGY: INTRODUCTION

This volume is a selection of articles from the second, electronic, edition of the *Encyclopedia of Ocean Science*. It is one of nine volumes that focus on particular aspects of marine studies. Marine Biology not only covers a great variety of plant and animal species but refers to diverse aspects of their physical, chemical and human environment.

The volume is divided into the traditional sub-disciplines: Plankton, Benthos, Fish, Marine Mammals and Seabirds. Within each category, there are articles on the main taxonomic groups, but also articles dealing with important processes such as primary production of phytoplankton, fish locomotion, feeding and foraging, marine mammal diving physiology and seabird conservation.

Marine organisms have an intimate relation with their fluid environment. Ocean currents play a large role in determining their migrations and vertical mixing and advection control the nutrient supply that regulates their food production. Longer term changes in these physical processes cause major stresses on populations and communities. This close coupling of ocean physics and biology is a theme of many articles, especially those concerned with the impact of climatic changes on plankton, marine mammals and seabirds.

There are others stresses on marine communities. The general topic of marine pollution is dealt with in a separate volume. The impact of fisheries not only on commercial stocks of fish but also on the remainder of their ecosystems, is considered at length in a companion volume to this one dealing broadly with ecological processes.

Each section of this volume opens with an "Overview" article written by the Section Editor responsible for this theme within the Encyclopedia. These Section Editors were also involved in the selection of authors for the individual topics. The Editors of the Encyclopedia are in their debt for their work in ensuring the quality and coverage of these articles.

Given the breadth of topics under the rubric of Marine Biology and their inter-relation with other aspects of ocean science, this one volume must be considered as a summary or introduction. For this reason each article has, not only a further reading list, but also references to articles in the Encyclopedia or in other volumes in this series.

The articles in this volume could not have been produced without the considerable help of the members of the Editorial Advisory Board of the Encyclopedia's second edition, from which these articles were chosen. The board provided advice and suggestions about the content and authorship of particular subject areas covered in the Encyclopedia. In addition to thanking the authors of the articles in this volume, the Editors wish to thank the members of the Editorial Board for the time they gave to identify and encourage authors, to read and comment on (and sometimes to suggest improvements to) the written articles, and to make this venture possible.

John H. Steele
Editor

PLANKTON & NEKTON

PLANKTON OVERVIEW

M. M. Mullin, Scripps Institution of Oceanography, La Jolla, California, USA

The category of marine life known as plankton represents the first step in the food web of the ocean (and of large bodies of fresh water), and components of the plankton are food for many of the fish harvested by humans and for the baleen whales. The plankton play a major role in cycling of chemical elements in the ocean, and thereby also affect the chemical composition of sea water and air (through exchange of gases between the sea and the overlying atmosphere). In the parts of the ocean where planktonic life is abundant, the mineral remains of members of the plankton are major contributors to deep-sea sediments, both affecting the chemistry of the sediments and providing a micropaleontological record of great value in reconstructing the earth's history.

'Plankton' refers to 'drifting', and describes organisms living in the water column (rather than on the bottom – the benthos) and too small and/or weak to move long distances independently of the ocean's currents. However, the distinction between plankton and nekton (powerfully swimming animals) can be difficult to make, and is often based more on the traditional method of sampling than on the organisms themselves.

Although horizontal movement of plankton at kilometer scales is passive, the metazoan zooplankton nearly all perform vertical migrations on scales of 10s to 100s of meters. This depth range can take them from the near surface lighted waters where the phytoplankton grow, to deeper, darker and usually colder environments. These migrations are generally diurnal, going deeper during the day, or seasonal, moving to deeper waters during the winter months to return to the surface around the time that phytoplankton production starts. The former pattern can serve various purposes: escaping visual predators and scanning the watercolumn for food. (It should be noted that predators such as pelagic fish also migrate diurnally.) Seasonal descent to greater depths is a common feature for several copepod species and may conserve energy at a time when food is scarce in the upper layers. However, vertical migration has another role. Because of differences in current strength and direction between surface and deeper layers in the ocean, time spent in deeper water acts as a transport mechanism relative to the near surface layers. On a daily basis this process can take plankton into different food concentrations. Seasonally, this effective 'migration' can complete a spatial life cycle.

The plankton can be subdivided along functional lines and in terms of size. The size category, picoplankton (0.2–$2.0\,\mu m$), is approximately equivalent to the functional category, bacterioplankton; most phytoplankton (single-celled plants or colonies) and protozooplankton (single-celled animals) are nano- or microplankton (2.0–$20\,\mu m$ and 20–$200\,\mu m$, respectively). The metazoan zooplankton (animals, the 'insects of the sea') includes large medusae and siphonophores several meters in length. Size is more important in oceanic than in terrestrial ecosystems because most of the plants are small (the floating seaweed, *Sargassum*, being the notable exception), predators generally ingest their prey whole (there is no hard surface on which to rest prey while dismembering it), and the early life stages of many types of zooplankton are approximately the same size as the larger types of phytoplankton. Therefore, while the dependence on light for photosynthesis is characteristic of the phytoplankton, the concepts of 'herbivore' and 'carnivore' can be ambiguous when applied to zooplankton, since potential plant and animal prey overlap in size and can be equivalent sources of food. Though rabbits do not eat baby foxes on land, analogous ontogenetic role-switching is very common in the plankton.

Among the animals, holoplanktonic species are those that spend their entire life in the plankton, whereas many benthic invertebrates have meroplanktonic larvae that are temporarily part of the plankton. Larval fish are also a temporary part of the plankton, becoming part of the nekton as they grow. There are also terms or prefixes indicating special habitats, such as 'neuston' to describe zooplanktonic species whose distribution is restricted to within a few centimeters of the sea's surface, or 'abyssoplankton' to describe animals living only in the deepest waters of the ocean. Groups of such species form communities (see below).

Since the phytoplankton depend on sunlight for photosynthesis, this category of plankton occurs almost entirely from the surface to 50–$200\,m$ of the ocean – the euphotic depth (where light intensity is 0.1–1% of full surface sunlight). Nutrients such as nitrate and phosphate are incorporated into protoplasm in company with photosynthesis, and returned to dissolved form by excretion or remineralization of dead organic matter (particulate detritus). Since much of the latter process occurs after sinking of the detritus, uptake of nutrients and their regeneration are partially separated vertically. Where and when photosynthesis is proceeding actively and vertical mixing is not excessive, a near-surface layer of low nutrient concentrations is separated from a layer of abundant nutrients, some distance below the euphotic depth, by a nutricline (a layer in which nutrient concentrations increase rapidly with depth). Therefore, the spatial and temporal relations between the euphotic depth (dependent on light intensity at the surface and the turbidity of the water), the nutricline, and the pycnocline (a layer in which density increases rapidly with depth) are important determinants of the abundance and productivity of phytoplankton.

Zooplankton is typically more concentrated within the euphotic zone than in deeper waters, but because of sinking of detritus and diel vertical migration of some species into and out of the euphotic zone, organic matter is supplied and various types of zooplankton (and bacterioplankton and nekton) can be found at all depths in the ocean. An exception is anoxic zones such as the deep waters of the Black Sea, although certainly types of bacterioplankton that use molecules other than oxygen for their metabolism are in fact concentrated there.

Even though the distributions of planktonic species are dependent on currents, species are not uniformly distributed throughout the ocean. Species tend to be confined to particular large water masses, because of physiological constraints and inimical interactions with other species. Groups of species, from small invertebrates to active tuna, seem to 'recognize' the same boundaries in the oceans, in the sense that their patterns of distribution are similar. Such groups are called 'assemblages' (when emphasizing their statistical reality, occurring together more than expected by chance) or 'communities' (when emphasizing the functional relations between the members in food webs), though terms such as 'biocoenoses' can be found in older literature. Thus, one can identify 'central water mass,' 'subantarctic,' 'equatorial,' and 'boreal' assemblages associated with water masses defined by temperature and

salinity; 'neritic' (i.e. nearshore) versus 'oceanic' assemblages with respect to depth of water over which they occur, and 'neustonic' (i.e. air–sea interface), 'epipelagic,' 'mesopelagic,' 'bathypelagic,' and 'abyssopelagic' for assemblages distinguished by the depth at which they occur. Within many of these there may be seasonally distinguishable assemblages of organisms, especially those with life spans of less than one year.

Regions which are boundaries between assemblages are sometimes called ecotones or transition zones; they generally contain a mixture of species from both sides, and (as in the transition zone between subpolar and central water mass assemblages) may also have an assemblage of species that occur only in the transition region.

Despite the statistical association between assemblages and water masses or depth zones, it is far from clear that the factor that actually limits distribution is the temperature/salinity or depth that physically defines the water mass or zone. It is likely that a few important species have physiological limits confining them to a zone, and the other members of the assemblage are somehow linked to those species functionally, rather than being themselves physiologically constrained. Limits can be imposed on certain life stage, such as the epipelagic larvae of meso- or bathypelagic species, creating patterns that reflect the environment of the sensitive life stage rather than the adult. Conversely, meroplanktonic larvae, such as the phyllosome of spiny lobsters, can often be found far away from the shallow waters that are a suitable habitat for the adults.

See also

Bacterioplankton. Gelatinous Zooplankton. Protozoa, Planktonic Foraminifera.

Further Reading

Cushing DH (1995) *Population Production and Regulation in the Sea*. Cambridge: Cambridge University Press.

Longhurst A (1998) *Ecological Geography of the Sea*. New York: Academic Press.

Mullin MM (1993) *Webs and Scales*. Seattle: University of Washington Press.

MARINE PLANKTON COMMUNITIES

G.-A. Paffenhöfer, Skidaway Institute of
Oceanography, Savannah, GA, USA

Introduction

By definition, a community is an interacting population of various kinds of individuals (species) in a common location (*Webster's Collegiate Dictionary*, 1977).

The objective of this article is to provide general information on the composition and functioning of various marine plankton communities, which is accompanied by some characteristic details on their dynamicism.

General Features of a Plankton Community

The expression 'plankton community' implies that such a community is located in a water column. It has a range of components (groups of organisms) that can be organized according to their size. They range in size from tiny single-celled organisms such as bacteria (0.4–1-µm diameter) to large predators like scyphomedusae of more than 1 m in diameter. A common method which has been in use for decades is to group according to size, which here is attributed to the organism's largest dimension; thus the organisms range from picoplankton to macroplankton (**Figure 1**). It is, however, the smallest dimension of an organism which usually determines whether it is retained by a mesh, since in a flow, elongated particles align themselves with the flow.

A plankton community is operating/functioning continuously, that is, physical, chemical, and biological variables are always at work. Interactions among its components occur all the time. As one well-known fluid dynamicist stated, "The surface of the ocean can be flat calm but below that surface there is always motion of the water at various scales." Many of the particles/organisms are moving or being moved most of the time: Those without flagella or appendages can do so due to processes within or due to external forcing, for example, from water motion due to internal waves; and those with flagella/cilia or appendages or muscles move or create motion of the water in order to exist. Oriented motion is usually in the vertical which often results in distinct layers of certain organisms. However, physical variables also, such as light or density differences of water masses, can result in layering of planktonic organisms. Such layers which are often horizontally extended are usually referred to as patches.

As stated in the definition, the components of a plankton community interact. It is usually the case that a larger organism will ingest a smaller one or a part of it (**Figure 1**). However, there are exceptions. The driving force for a planktonic community originates from sun energy, that is, primary productivity's (1) direct and (2) indirect products: (1) autotrophs (phytoplankton cells) which can range from near 2 to more than 300-µm width/diameter, or chemotrophs; and (2) dissolved organic matter, most of which is released by phytoplankton cells and protozoa as metabolic end products, and being taken up by bacteria and mixo- and heterotroph protozoa (**Figure 1**). These two components mainly set the microbial loop (ML; (*see* Bacterioplankton and Protozoa, Planktonic Foraminifera)) in motion; that is, unicellular organisms of different sizes and behaviors (auto-, mixo-, and heterotrophs) depend on each other – usually, but not always, the smaller being ingested by the larger. Most of nutrients and energy are recirculated within this subcommunity of unicellular organisms in all marine regions of our planet (*see* Bacterioplankton, Phytoplankton Size Structure, and Protozoa, Planktonic Foraminifera for more details, especially the ML)These processes of the ML dominate the transfer of energy in all plankton communities largely because the processes (rates of ingestion, growth, reproduction) of unicellular heterotrophs almost always outpace those of phytoplankton, and also of metazooplankton taxa at most times.

The main question actually could be: "What is the composition of plankton communities, and how do they function?" **Figure 1** reveals sizes and relationships within a plankton community including the ML. It shows the so-called 'bottom-up' and 'top-down' effects as well as indirect effects like the above-mentioned labile dissolved organic matter (labile DOM), released by auto- and also by heterotrophs, which not only drives bacterial growth but can also be taken up or used by other protozoa. There can also be reversals, called two-way processes. At times a predator eating an adult metazoan will be affected by the same metazoan which is able to eat the predator's early juveniles (e.g., well-grown ctenophores capturing adult omnivorous copepods which have the ability to capture and ingest very young ctenophores).

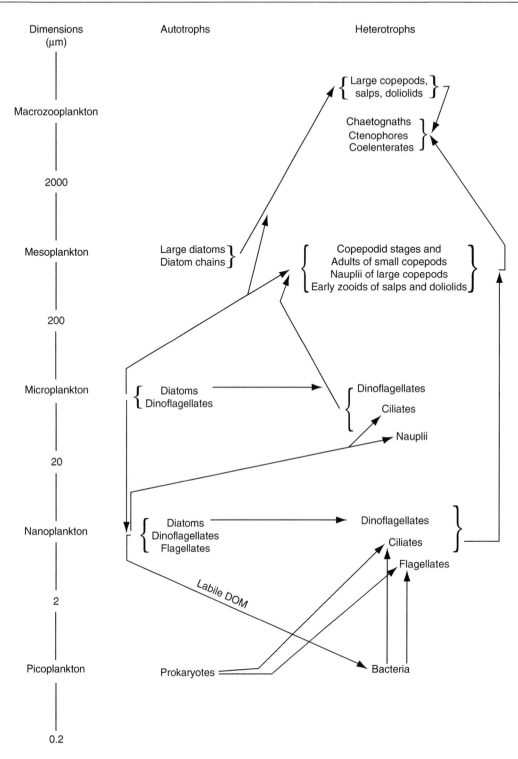

Figure 1 Interactions within a plankton community separated into size classes of auto- and heterotrophs, including the microbial loop; the arrows point to the respective grazer, or receiver of DOM; the figure is partly related to figure 9 from Landry MR and Kirchman DL (2002) Microbial community structure and variability in the tropical Pacific. *Deep-Sea Research II* 49: 2669–2693.

To comprehend the functioning of a plankton community requires a quantitative assessment of the abundances and activities of its components. First, almost all of our knowledge to date stems from *in situ* sampling, that is, making spot measurements of the abundance and distribution of organisms in the water column. The accurate determination of abundance and distribution requires using meshes or devices which quantitatively collect the respective organisms. Because of methodological difficulties and insufficient comprehension of organisms' sizes and activities, quantitative sampling/quantification of a community's main components has been often inadequate. The following serves as an example of this. Despite our knowledge that copepods consist of 11 juvenile stages aside of adults, the majority of studies of marine zooplankton hardly considered the juveniles' significance and this manifested itself in sampling with meshes which often collected merely the adults quantitatively. Second, much knowledge on rate processes comes from quantifying the respective organisms' activities under controlled conditions in the laboratory. Some *in situ* measurements (e.g., of temperature, salinity, chlorophyll concentrations, and acoustic recordings of zooplankton sizes) have been achieved 'continuously' over time, resulting in time series of increases and decreases of certain major community components. To date there are few, if any, direct *in situ* observations on the activity scales of the respective organisms, from bacteria to proto- and to metazooplankton, mainly because of methodological difficulties. In essence, our present understanding of processes within plankton communities is incomplete.

Specific Plankton Communities

We will provide several examples of plankton communities of our oceans. They will include information about the main variables affecting them, their main components, partly their functioning over time, including particular specifics characterizing each of those communities.

In this section, plankton communities are presented for three different types of marine environments: estuaries/inshore, continental shelves, and open ocean regions.

Estuaries

Estuaries and near-shore regions, being shallow, will rapidly take up and lose heat, that is, will be strongly affected by atmospheric changes in temperature, both short- and long-term, the latter showing in the seasonal extremes ranging from 2 to 32 °C in estuaries of North Carolina. Runoff of fresh water, providing continuous nutrient input for primary production, and tides contribute to rapid changes in salinity. This implies that resident planktonic taxa ought to be eurytherm as well as – therm. Only very few metazooplanktonic species are able to exist in such an environment (**Table 1**). In North Carolinian

estuaries, representative of other estuaries, they are the copepod species *Acartia tonsa*, *Oithona oculata*, and *Parvocalanus crassirostris*. In estuaries of Rhode Island, two species of the genus *Acartia* occur. During colder temperatures *Acartia hudsonica* produces dormant eggs as temperatures increase and then is replaced by *A. tonsa*, which produces dormant eggs once temperatures again decrease later in the year. Such estuaries are known for high primary productivity, which is accompanied by high abundances of heterotroph protozoa preying on phytoplankton. Such high abundances of unicellular organisms imply that food is hardly limiting the growth of the above-mentioned copepods which can graze on auto- as well as heterotrophs. However, such estuaries are often nursery grounds for juvenile fish like menhaden which prey heavily on late juveniles and adults of such copepods, especially *Acartia*, which is not only the largest of those three dominant copepod species but also moves the most, and thus can be seen most easily by those visual predators. This has resulted in diurnal migrations mostly of their adults, remaining at the seafloor during the day where they hardly eat, thus avoiding predation by such visual predators, and only entering the water column during dark hours. That then is their period of pronounced feeding. The other two species which are not heavily preyed upon by juvenile fish, however, can be affected by the co-occurring *Acartia*, because from early copepodid stages on this genus can be strongly carnivorous, readily preying on the nauplii of its own and of those other species.

Nevertheless, the usually continuous abundance of food organisms for all stages of the three copepod species results in high concentrations of nauplii which in North Carolinian estuaries can reach $100 l^{-1}$, as can their combined copepodid stages. The former is an underestimate, because sampling was done with a 75-µm mesh, which is passed through by most of those nauplii. By comparison, in an estuary on the west coast of Japan (Yellow Sea), dominated also by the genera *Acartia*, *Oithona*, and *Paracalanus* and sampling with 25-µm mesh, nauplius concentrations during summer surpassed $700 l^{-1}$, mostly from the genus *Oithona*. And copepodid stages plus adults repeatedly exceeded $100 l^{-1}$. Here sampling with such narrow mesh ensured that even the smallest copepods were collected quantitatively.

In essence, estuaries are known to attain among the highest concentrations of proto- and metazooplankton. The known copepod species occur during most of the year, and are observed year after year which implies persistence of those species beyond decades.

Table 1 Some characteristics of marine plankton communities

	Estuaries	Shelves	Open ocean gyres		Epipelagic subtropical
			Subarctic Pacific	Boreal Atlantic	Atlantic/Pacific
Physical variables	Wide range of temperature and salinity	Intermittent and seasonal atmospheric forcing	Steady salinity, seasonal temp. variability	Major seasonal variability of temperature	Steady temperature and salinity, continuous atmospheric forcing
Nutrient supply	Continuous	Episodic	Seasonal	Seasonal	Occasional
Phytoplankton abundance	High from spring to autumn	Intermittently high	Always low	Major spring bloom	Always low
Phytoplankton composition	Flagellates, diatoms	Flagellates, diatoms, dinoflagellates	Nanoflagellates	Spring: diatoms Other: mostly nanoplankton	Mostly prokaryotes, small nano- and dinoflagellates
Primary Productivity	High at most times	Intermittently high	Maximum in spring	Max. in spring and autumn	Always low
No. of metazoan species	≤ 5	~ 10–30	>10	>20	>100
Seasonal variability of metazoan abundance	High spring and summer, low winter	Highly variable	High	High	Low
Copepod Ranges	$N^a \sim 10$–$500\, l^{-1}$	<5–$50\, l^{-1}$			3–$10\, l^{-1}$
Abundance	$Cop^b \sim 5$–$100\, l^{-1}$	<3–$30\, l^{-1}$	Up to $1000\, m^{-3}$ Neocalanus	Up to $1000\, m^{-3}$ C. finmarchicus	300–$1000\, m^{-3}$
Dominant metazooplankton taxa	Acartia Oithona Parvocalanus	Oithona Paracalanus Temora Doliolida	Neocalanus Oithona Metridia	Calanus Oithona Oncaea	Oithona Clausocalanus Oncaea

[a] Nauplii.
[b] Copepodids and adult copepods.

Continental Shelves

By definition they extend to the 200-m isobath, and range from narrow (few kilometers) to wide (more than 100-km width). The latter are of interest because the former are affected almost continuously and entirely by the nearby open ocean. Shelves are affected by freshwater runoff and seasonally changing physical variables. Water masses on continental shelves are evaluated concerning their residence time, because atmospheric events sustained for more than 1 week can replace most of the water residing on a wide shelf with water offshore but less so from near shore. This implies that plankton communities on wide continental shelves, which are often near boundary currents, usually persist for limited periods of time, from weeks to months (**Table 1**). They include shelves like the Agulhas Bank, the Campeche Banks/Yucatan Shelf, the East China Sea Shelf, the East Australian Shelf, and the US southeastern continental shelf. There can be a continuous influx year-round of new water from adjacent boundary currents as seen for the Yucatan Peninsula and Cape Canaveral (Florida). The momentum of the boundary current (here the Yucatan Current and Florida Current) passing a protruding cape will partly displace water along downstream-positioned diverging isobaths while the majority will follow the current's general direction. This implies that upstream-produced plankton organisms can serve as seed populations toward developing a plankton community on such wide continental shelves.

Whereas estuarine plankton communities receive almost continuously nutrients for primary production from runoff and pronounced benthic-pelagic coupling, those on wide continental shelves infrequently receive new nutrients. Thus they are at most times a heterotroph community unless they obtain nutrients from the benthos due to storms, or receive episodically input of cool, nutrient-rich water from greater depths of the nearby boundary current as can be seen for the US SE shelf. Passing along the outer shelf at about weekly intervals are nutrient-rich cold-core Gulf Stream eddies which contain plankton organisms from the highly productive Gulf of Mexico. Surface winds, displacing shelf surface water offshore, lead to an advance of the deep cool water onto the shelf which can be flooded entirely by it. Pronounced irradiance and high-nutrient loads in such upwellings result in phytoplankton blooms which then serve as a food source for protozoo- and metazooplankton. Bacteria concentrations in such cool water masses increase within several days by 1 order of magnitude. Within 2–3 weeks most of the smaller phytoplankton (c. <20-μm width) has been greatly reduced, usually due to grazing by protozoa and relatively slow-growing assemblages of planktonic copepods of various genera such as *Temora*, *Oithona*, *Paracalanus*, *Eucalanus*, and *Oncaea*. However, quite frequently, the Florida Current which becomes the Gulf Stream carries small numbers of Thaliacea (Tunicata), which are known for intermittent and very fast asexual reproduction. Such salps and doliolids, due to their high reproductive and growth rate, can colonize large water masses, the latter increasing from ~5 to >500 zooids per cubic meter within 2 weeks, and thus form huge patches, covering several thousands of square kilometers, as the cool bottom water is displaced over much of the shelf. The increased abundance of salps (usually in the warmer and particle-poor surface waters) and doliolids (mainly in the deeper, cooler, particle-rich waters, also observed on the outer East China shelf) can control phytoplankton growth once they achieve bloom concentrations. The development of such large and dense patches is partly due to the lack of predators.

Although the mixing processes between the initially quite cool intruding bottom (13–20 °C) and the warm, upper mixed layer water (27–28 °C) are limited, interactions across the thermocline occur, thus creating a plankton community throughout the water column of previously resident and newly arriving components. The warm upper mixed layer often has an extraordinary abundance of early copepodid stages of the poecilostomatoid copepod *Oncaea*, thanks to their ontogenetical migration after having been released by the adult females which occur exclusively in the cold intruding water. Also, early stages of the copepod *Temora turbinata* are abundant in the warm upper mixed layer; while *T. turbinata*'s late juvenile stages prefer the cool layer because of the abundance of large, readily available phytoplankton cells. As in estuaries, the copepod genus *Oithona* flourishes on warm, temperate, and polar continental shelves throughout most of the euphotic zone.

Such wide subtropical shelves will usually be well mixed during the cooler seasons, and then harbor, due to lower temperatures, fewer metazooplankton species which are often those tolerant of wider or lower temperature ranges. Such wide shelves are usually found in subtropical regions, which explains the rapidity of the development of their plankton communities. They, however, are also found in cooler climates, like the wide and productive Argentinian/Brazilian continental shelf about which our knowledge is limited. Other large shelves, like the southern North Sea, have a limited exchange of water with the open ocean but at the same time considerable influx of runoff, plus nutrient supply from the benthos due to storm events, and thus can maintain identical plankton communities over months and seasons.

In essence, continental shelf plankton communities are usually relatively short-lived, which is largely due to their water's limited residence time.

Open Ocean

The open ocean, even when not including ocean margins (up to 1000-m water column), includes by far the largest regions of the marine environment. Its deep-water columns range from the polar seas to the Tropics. All these regions are under different atmospheric and seasonal regimes, which affect plankton communities. Most of these communities are seasonally driven and have evolved along the physical conditions characterizing each region. The focus here is on gyres as they represent specific ocean communities whose physical environment can be readily presented.

Gyres represent huge water masses extending horizontally over hundreds to even thousands of kilometers in which the water moves cyclonically or anticyclonically. They are encountered in subpolar, temperate, and subtropical regions. The best-studied ones are:

- subpolar: Alaskan Gyre;
- temperate: Norwegian Sea Gyre, Labrador–Irminger Sea Gyre;
- subtropical: North Pacific Central Gyre (NPCG), North Atlantic Subtropical Gyre (NASG).

The Alaskan Gyre is part of the subarctic Pacific (**Table 1**) and is characterized physically by a shallow halocline ($\sim$110-m depth) which prevents convective mixing during storms. Biologically it is characterized by a persistent low-standing stock of phytoplankton despite high nutrient abundance, and several species of large copepods which have evolved to persist via a life cycle as shown for *Neocalanus plumchrus*. By midsummer, fifth copepodids (C5) in the upper 100 m which have accumulated large amounts of lipids begin to descend to greater depths of 250 m and beyond undergoing diapause, and eventually molt to females which soon begin to spawn. Spawning females are found in abundance from August to January. Nauplii living off their lipid reserves and moving upward begin to reach surface waters by mid- to late winter as copepodid stage 1 (C1), and start feeding on the abundant small phytoplankton cells (probably passively by using their second maxillae, but mostly by feeding actively on heterotrophic protozoa which are the main consumers of the tiny phytoplankton cells). The developing copepodid stages accumulate lipids which in C5 can amount to as much as 50% of their body mass, which then serve as the energy source for metabolism of the females at depth, ovary development, and the nauplii's metabolism plus growth. While the genus *Neocalanus* over much of the year provides the highest amount of zooplankton biomass, the cyclopoid *Oithona* is the most abundant metazooplankter; other abundant metazooplankton taxa include *Euphausia pacifica*, and in the latter part of the year *Metridia pacifica* and *Calanus pacificus*.

In the temperate Atlantic (**Table 1**), the Norwegian Sea Gyre maintains a planktonic community which is characterized, like much of the temperate oceanic North Atlantic, by the following physical features. Pronounced winds during winter mix the water column to beyond 400-m depth, being followed by lesser winds and surface warming resulting in stratification and a spring bloom of mostly diatoms, and a weak autumn phytoplankton bloom. A major consumer of this phytoplankton bloom and characteristic of this environment is the copepod *Calanus finmarchicus*, occurring all over the cool North Atlantic. This species takes advantage of the pronounced spring bloom after emerging from diapause at > 400-m depth, by moulting to adult, and grazing of females at high clearance rates on the diatoms, right away starting to reproduce and releasing up to more than 2000 fertilized eggs during their lifetime. Its nauplii start to feed as nauplius stage 3 (N3), being able to ingest diatoms of similar size as the adult females, and can reach copepodid stage 5 (C5) within about 7 weeks in the Norwegian Sea, accumulating during that period large amounts of lipids (wax ester) which serve as the main energy source for the overwintering diapause period. Part of the success of *C. finmarchicus* is found in its ability of being omnivorous. C5s either descend to greater depths and begin an extended diapause period, or could moult to adult females, thus producing another generation which then initiates diapause at mostly C5. Its early to late copepodid stages constitute the main food for juvenile herring which accumulate the copepods' lipids for subsequent overwintering and reproduction. Of the other copepods, the genus *Oithona* together with the poecilostomatoid *Oncaea* and the calanoid *Pseudocalanus* were the most abundant.

Subtropical and tropical parts of the oceans cover more than 50% of our oceans. Of these, the NPCG, positioned between *c*. 10° and 45° N and moving anticyclonically, has been frequently studied. It includes a southern and northern component, the latter being affected by the Kuroshio and westerly winds, the former by the North Equatorial Current and the trade winds. Despite this, the NPCG has been considered as an ecosystem as well as a huge plankton community. The NASG, found between *c.* 15° and 40° N and moving anticyclonically, is of similar horizontal dimensions. There are close relations between subtropical and tropical communities; for example, the Atlantic south of Bermuda is considered close to tropical conditions. Vertical mixing in both gyres is limited. Here we focus on the epipelagic community which ranges from the surface to about 150-m depth, that is, the euphotic zone. The epipelagial is physically characterized by an upper mixed layer of *c.* 15–40 m of higher temperature, below which a thermocline with steadily decreasing temperatures extends to below 150-m depth. In these two gyres, the concentrations of phytoplankton hardly change throughout the year in the epipelagic (**Table 1**) and together with the heterotrophic protozoa provide a low and quite steady food concentration (**Table 1**) for higher trophic levels. Such very low particle abundances imply that almost all metazooplankton taxa depending on them are living on the edge, that is, are severely food-limited. Despite this fact, there are more than 100 copepod species registered in the epipelagial of each of the two gyres. How can that be? Almost all these copepod species are small and rather diverse in their behavior: the four most abundant genera have different strategies to obtain food particles: the intermittently moving *Oithona* is found in the entire epipelagial and depends on moving food particles (hydrodynamic signals); *Clausocalanus* is mainly found in the upper 50 m of the epipelagial and always moves at high speed, thus encountering numerous food particles, mainly via chemosensory; *Oncaea* copepodids and females occur in the lower part of the epipelagial and feed on aggregates; and the feeding-current producing *Calocalanus* perceives particles via chemosensory. This implies that any copepod species can persist in these gyres as long as it obtains sufficient food for growth and reproduction. This is possible because protozooplankton always controls the abundance of available food particles; thus, there is no competition for food among the metazooplankton. In addition, since total copepod abundance (quantitatively collected with a 63-μm mesh by three different teams) is steady and usually $< 1000 \, \text{m}^{-3}$ including copepodid stages (pronounced patchiness of metazooplankton has not yet been observed in these oligotrophic waters), the probability of encounter (only a minority of the zooplankton is carnivorous on metazooplankton) is very low, and therefore the probability of predation low within the metazooplankton. In summary, these steady conditions make it possible that in the epipelagial more than 100 copepod species can coexist, and are in a steady state throughout much of the year.

Conclusions

All epipelagic marine plankton communities are at most times directly or indirectly controlled or affected by the activity of the ML, that is, unicellular organisms. Most of the main metazooplankton species are adapted to the physical and biological conditions of the respective community, be it polar, subpolar, temperate, subtropical, or tropical. The only metazooplankton genus found in all communities mentioned above, and also all other studied marine plankton communities, is the copepod genus *Oithona*. This copepod has the ability to persist under adverse conditions, for example, as shown for the subarctic Pacific. This genus can withstand the physical as well as biological (predation) pressures of an estuary, the persistent very low food levels in the warm open ocean, and the varying conditions of the Antarctic Ocean. Large copepods like the genus *Neocalanus* in the subarctic Pacific, and *C. finmarchicus* in the temperate to subarctic North Atlantic are adapted with respective distinct annual cycles in their respective communities. Among the abundant components of most marine plankton communities from near shore to the open ocean are appendicularia (Tunicata) and the predatory chaetognaths.

Our present knowledge of the composition and functioning of marine planktonic communities derives from (1) oceanographic sampling and time series, optimally accompanied by the quantification of physical and chemical variables; and (2) laboratory/onboard experimental observations, including some time series which provide results on small-scale interactions (microns to meters; milliseconds to hours) among components of the community. Optimally, direct *in situ* observations on small scales in conjunction with respective modeling would provide insights in the true functioning of a plankton community which operates continuously on scales of milliseconds and larger.

Our future efforts are aimed at developing instrumentation to quantify *in situ* interactions of the various components of marine plankton communities. Together with traditional oceanographic methods we would go 'from small scales to the big picture', implying the necessity of understanding the functioning on the individual scale for a comprehensive understanding as to how communities operate.

See also

Bacterioplankton.   Copepods.   Gelatinous Zooplankton. Phytoplankton Size Structure. Plankton Overview. Protozoa, Planktonic Foraminifera.

Further Reading

Atkinson LP, Lee TN, Blanton JO, and Paffenhöfer G-A (1987) Summer upwelling on the southeastern continental shelf of the USA during 1981: Hydrographic observations. *Progress in Oceanography* 19: 231–266.

Fulton RS, III (1984) Distribution and community structure of estuarine copepods. *Estuaries* 7: 38–50.

Hayward TL and McGowan JA (1979) Pattern and structure in an oceanic zooplankton community. *American Zoologist* 19: 1045–1055.

Landry MR and Kirchman DL (2002) Microbial community structure and variability in the tropical Pacific. *Deep-Sea Research II* 49: 2669–2693.

Longhurst AR (1998) *Ecological Geography of the Sea*, 398pp. San Diego, CA: Academic Press.

Mackas DL and Tsuda A (1999) Mesozooplankton in the eastern and western subarctic Pacific: Community structure, seasonal life histories, and interannual variability. *Progress in Oceanography* 43: 335–363.

Marine Zooplankton Colloquium 1 (1998) Future marine zooplankton research – a perspective. *Marine Ecology Progress Series* 55: 197–206.

Menzel DW (1993) *Ocean Processes: US Southeast Continental Shelf*, 112pp. Washington, DC: US Department of Energy.

Miller CB (1993) Pelagic production processes in the subarctic Pacific. *Progress in Oceanography* 32: 1–15.

Miller CB (2004) *Biological Oceanography*, 402pp. Boston: Blackwell.

Paffenhöfer G-A and Mazzocchi MG (2003) Vertical distribution of subtropical epiplanktonic copepods. *Journal of Plankton Research* 25: 1139–1156.

Paffenhöfer G-A, Sherman BK, and Lee TN (1987) Summer upwelling on the southeastern continental shelf of the USA during 1981: Abundance, distribution and patch formation of zooplankton. *Progress in Oceanography* 19: 403–436.

Paffenhöfer G-A, Tzeng M, Hristov R, Smith CL, and Mazzocchi MG (2003) Abundance and distribution of nanoplankton in the epipelagic subtropical/tropical open Atlantic Ocean. *Journal of Plankton Research* 25: 1535–1549.

Smetacek V, DeBaar HJW, Bathmann UV, Lochte K, and Van Der Loeff MMR (1997) Ecology and biogeochemistry of the Antarctic Circumpolar Current during austral spring: A summary of Southern Ocean JGOFS cruise ANT X/6 of RV *Polarstern*. *Deep-Sea Research II* 44: 1–21 (and all articles in this issue).

Speirs DC, Gurney WSC, Heath MR, Horbelt W, Wood SN, and de Cuevas BA (2006) Ocean-scale modeling of the distribution, abundance, and seasonal dynamics of the copepod *Calanus finmarchicus*. *Marine Ecology Progress Series* 313: 173–192.

Tande KS and Miller CB (2000) Population dynamics of *Calanus* in the North Atlantic: Results from the Trans-Atlantic Study of *Calanus finmarchicus*. *ICES Journal of Marine Science* 57: 1527 (entire issue).

Webber MK and Roff JC (1995) Annual structure of the copepod community and its associated pelagic environment off Discovery Bay, Jamaica. *Marine Biology* 123: 467–479.

PLANKTON VIRUSES

J. Fuhrman, University of Southern California, Los Angeles, CA, USA
I. Hewson, University of California Santa Cruz, Santa Cruz, CA, USA

Introduction

Although they are the tiniest biological entities in the sea, typically 20–200 nm in diameter, viruses are integral components of marine planktonic systems. They are extremely abundant in the water column, typically 10^{10} per liter in the euphotic zone, and they play several roles in system function: (1) they are important agents in the mortality of prokaryotes and eukaryotes; (2) they act as catalysts of nutrient regeneration and recycling, through this mortality of host organisms; (3) because of their host specificity and density dependence, they tend to selectively attack the most abundant potential hosts, thus may 'kill the winner' of competition and thereby foster diversity; and (4) they may also act as agents in the exchange of genetic material between organisms, a critical factor in evolution and also in relation to the spread of human-engineered genes. Although these processes are only now becoming understood in any detail, there is little doubt that viruses are significant players in aquatic and marine plankton.

History

It has only been in the past 25 years that microorganisms like bacteria and small protists have been considered 'major players' in planktonic food webs. The initial critical discovery, during the mid-1970s, was of high bacterial abundance as learned by epifluorescence microscopy of stained cells, with counts typically $10^9 \, \mathrm{l}^{-1}$ in the plankton. These bacteria were thought to be heterotrophs (organisms that consume preformed organic carbon), because they apparently lacked photosynthetic pigments like chlorophyll (later it was learned that this was only partly right, as many in warm waters are in fact chlorophyll-containing prochlorophytes). With such high abundance, it became important to learn how fast they were dividing, in order to quantify their function in the food web. Growth rates were estimated primarily by the development and application of methods measuring bacterial DNA synthesis. The results of

these studies showed that bacterial doubling times in typical coastal waters are about 1 day. When this doubling time was applied to the high abundance, to calculate how much carbon the bacteria are taking up each day, it became apparent that bacteria are consuming a significant amount of dissolved organic matter, typically at a carbon uptake rate equivalent to about half the total primary production. However, the bacterial abundance remains relatively constant over the long term, and they are too small to sink out of the water column. Therefore, there must be mechanisms within the water to remove bacteria at rates similar to the bacterial production rate. In the initial analysis, most scientists thought that grazing by protists was the only significant mechanism keeping the bacterial abundance in check. This was because heterotrophic protists that can eat bacteria are extremely common, and laboratory experiments suggested they are able to control bacteria at near-natural-abundance levels. However, some results pointed to the possibility that protists are not the only things controlling bacteria. In the late 1980s, careful review of multiple studies showed that grazing by protists was often not enough to balance bacterial production, and this pointed to the existence of additional loss processes. About that same time, data began to accumulate that viruses may also be important as a mechanism of removing bacteria. The evidence is now fairly clear that this is the case, and it will be outlined below. This article briefly summarizes much of what is known about how viruses interact with marine microorganisms, including general properties, abundance, distribution, infection of bacteria, mortality rate comparisons with protists, biogeochemical effects, effects on species compositions, and roles in genetic transfer and evolution.

General Properties

Viruses are small particles, usually about 20–200 nm long, and consist of genetic material (DNA or RNA, single or double stranded) surrounded by a protein coat (some have lipid as well). They have no metabolism of their own and function only via the cellular machinery of a host organism. As far as is known, all cellular organisms appear to be susceptible to infection by some kind of virus. Culture studies show that a given type of virus usually has a restricted host range, most often a single species or genus, although some viruses infect only certain subspecies and <0.5% may infect more than one genus. Viruses have no motility of their own, and contact the host cell by diffusion. They attach to the host usually via some normally exposed cellular component, such as a transport protein or flagellum. There are three basic kinds of virus reproduction (**Figure 1**). In lytic infection, the virus attaches to a host cell and injects its nucleic acid. This nucleic acid (sometimes accompanied by proteins carried by the virus) causes the host to produce numerous progeny viruses, the cell then bursts, progeny are released, and the cycle begins again. In chronic infection, the progeny virus release is not lethal and the host cell releases the viruses by extrusion or budding over many generations. In lysogeny after injection, the viral genome becomes part of the genome of the host cell and reproduces as genetic material in the host cell line unless an 'induction' event causes a switch to lytic infection. Induction is typically caused by DNA damage, such as from ultraviolet (UV) light or chemical mutagens such as mitomycin C. Viruses may also be involved in killing cells by mechanisms that do not result in virus reproduction.

Observation of Marine Viruses

Viruses are so small that they are at or below the resolution limit of light microscopy (c. 0.1 μm). Therefore electron microscopy is the only way to observe any detail of viruses. Sample preparation requires concentrating the viruses from the water onto an electron microscopy grid (coated with a thin transparent organic film). Because viruses are denser than seawater, this can be done by ultracentrifugation, typically at forces of at least $100\,000 \times g$ for a few hours. It should be noted that under ordinary gravity, forces like drag and Brownian motion prevent viruses from sinking. To be observable the viruses must be made electron-dense, typically by staining with uranium salts. The viruses are recognized by their size, shape, and staining properties (usually electron-dense hexagons or ovals, sometimes with a tail), and counted. Typical counts are on the order of 10^{10} viruses per liter in surface waters, with abundance patterns similar to those of heterotrophic bacteria (see below). Recently, it has been found that viruses can also be stained with nucleic acid stains like SYBR Green I, and observed and counted by

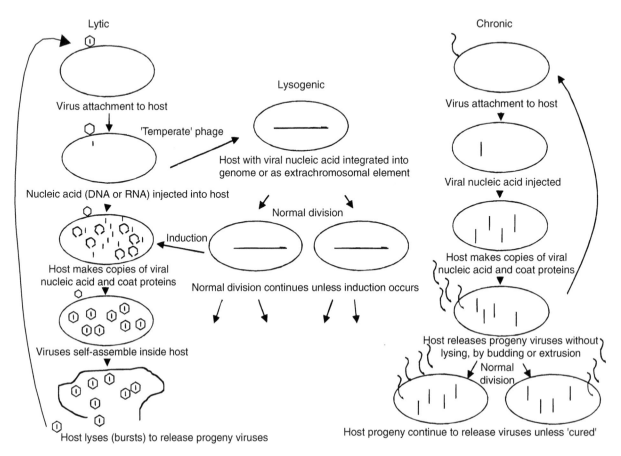

Figure 1 Virus life cycles. See text for explanation.

Figure 2 Epifluorescence micrograph of prokaryotes and viruses from 16 km offshore of Los Angeles, stained with SYBR Green I. The viruses are the very numerous tiny bright particles, and the bacteria are the rarer larger particles. Bacterial size is approximately 0.4–1 μm in diameter.

epifluorescence microscopy. This is faster, easier, and less expensive than transmission electron microscopy (TEM). Epifluorescence viewing of viruses is shown in **Figure 2**, a micrograph of SYBR Green I-stained bacteria and viruses, which dramatically illustrates the high relative virus abundance. Epifluorescence microscopy of viruses is possible even though the viruses are below the resolution limit of light because the stained viruses are a source of light and appear as bright spots against a dark background (like stars visible at night). Epifluorescence counts are similar to or even slightly higher than TEM counts from seawater.

What Kinds of Viruses Occur in Plankton?

Microscopic observation shows the total, recognizable, virus community, but what kinds of viruses make up this community, and what organisms are they infecting? Most of the total virus community is thought to be made up of bacteriophages (viruses that infect bacteria). This is because viruses lack metabolism and have no means of actively moving from host to host (they depend on random diffusion), so the most common viruses would be expected to infect the most common organism, and bacteria are by far the most abundant organisms in the plankton. Field studies show a strong correlation between viral and bacterial numbers, whereas the correlations between viruses and chlorophyll are weaker. This

suggests that most viruses are bacteriophages rather than those infecting phytoplankton or other eukaryotes. However, viruses infecting cyanobacteria (*Synechococcus*) are also quite common and sometimes particularly abundant, exceeding 10^8 per liter in some cases. Even though most of the viruses probably infect prokaryotes, viruses for eukaryotic plankton are also readily found. For example, those infecting the common eukaryotic picoplankter, *Micromonas pusilla*, are sometimes quite abundant, occasionally near 10^8 per liter in coastal waters. Overall, the data suggest that most viruses from seawater infect non-photosynthetic bacteria or archaea, but viruses infecting prokaryotic and eukaryotic phytoplankton also can make up a significant fraction of the total.

A great leap in our understanding of viral diversity has occurred with the application of molecular biological techniques. These involve the analysis of variability of DNA sequence of a single gene product (e.g., a capsid head protein, or an enzyme that makes DNA), or through a technique known as viriomics, whereby all genomes in a sample are cut up into small pieces, then all the pieces sequenced and reassembled. These studies have revealed that: (1) there is a lot of diversity of closely related viruses infecting the same or similar strains of microorganisms, a phenomenon known as 'microdiversity'; (2) RNA-containing viruses comprise a small percentage of the viral mix in oceanic waters, and are similar to viruses that infect insects and mollusks; and (3) that the diversity of viruses in a liter of seawater is astonishingly high, estimated at $>10^4$ different types for a sample from coastal California.

Virus Abundance

Total direct virus counts have been made in many planktonic environments – coastal, offshore, temperate, polar, tropical, and deep sea. Typical virus abundance is $1-5 \times 10^{10} \, l^{-1}$ in rich near-shore surface waters, decreasing to about $0.1-1 \times 10^{10} \, l^{-1}$ in the euphotic zone of offshore low-nutrient areas, and also decreasing with depth, by about a factor of 10. A typical deep offshore profile is shown in **Figure 3**. Seasonal changes are also common, with viruses following general changes in phytoplankton, bacteria, etc.

Virus:prokaryote ratios also provide an interesting comparison. In plankton, this ratio is typically 5–25, and commonly close to 10, even as abundance drops to low levels in the deep sea. Why this ratio stays in such a relatively narrow range is a mystery, but it does suggest a link and also tight regulatory mechanisms between prokaryotes and viruses.

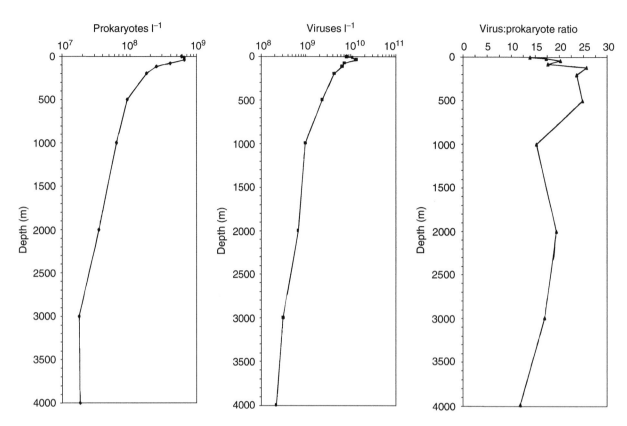

Figure 3 Depth profile of total prokaryote (bacteria–archaea) counts, total viral counts, and virus:prokaryote ratios from the Coral Sea (Apr. 1998), as determined by epifluorescence microscopy of SYBR Green-stained samples. Note the log scales of the counts.

Viral Activities

Viruses have no physical activity of their own, so 'viral activity' usually refers to lytic infection. However, before discussing such infection, lysogeny and chronic infection are considered briefly. Lysogeny, where the viral genome resides in the host's genome (**Figure 1**), is common. Lysogens (bacteria harboring integrated viral genomes) can easily be found and isolated from seawater, and lysogeny, which is linked to genetic transfer in a variety of bacteria, probably impacts microbial population dynamics and evolution. However, the induction rate appears to be low and seasonally variable under ordinary natural conditions, and lysogenic induction appears to be responsible for only a tiny fraction of total virus production in marine systems for most of the year. On the other hand, at this time we simply do not know if chronic infection is a significant process in natural systems. Release of filamentous (or other kinds of budding) viruses from native marine bacteria has not been noted in TEM studies of plankton, nor have significant numbers of free filamentous viruses (however, filamentous viruses have been observed in marine and freshwater sediments). But they could have been missed.

Regarding lytic infection, there are several studies with a variety of approaches that all generally conclude that viruses cause approximately 10–50% of total microbial mortality, depending on location, season, etc. These estimates are convincing, having been determined in several independent ways. These include: (1) TEM observation of assembled viruses within host cells, representing the last step before lysis; (2) measurement of viral decay rates; (3) measurement of viral DNA synthesis; (4) measurement of the disappearance rate of bacterial DNA in the absence of protists; (5) use of fluorescent virus tracers to measure viral production and removal rates simultaneously; and (6) direct observation of viral appearance in incubations where the viral abundance has been decreased several fold, yet host abundances remain undiluted.

Comparison to Mortality from Protists

Because the earlier thinking was that protists are the main cause of bacterial mortality in marine planktonic systems, it is useful to ask how the contribution of viruses to bacterial mortality compares to that of protists. Multiple correlation analysis of abundances of bacteria, viruses, and flagellates showed virus-induced mortality of bacteria could occasionally prevail over flagellate grazing, especially at high bacterial abundances. In more direct comparisons, measuring virus and protist rates by multiple independent approaches, the total mortality typically balances production, and viruses are found to be responsible for anything ranging from a negligible proportion to the majority of total mortality.

To sum up these studies, the consensus is that viruses are often responsible for a significant fraction of bacterial mortality in marine plankton, typically in the range 10–40%. Sometimes viruses may dominate bacterial mortality, and sometimes they may have little impact on it. It is unknown what controls this balance, but it probably includes variation in host abundance, because when hosts are less common, the viruses are more likely to be inactivated before diffusing to a suitable host, as well as the exact types of bacteria that are present and their palatability. Application of new molecular techniques, based on ribosomal RNA sequences, and variable regions within the host genomes have revealed that aquatic bacterial communities are typically dominated by a handful of bacterial taxa, while most taxa make up a tiny proportion of cell numbers. This would seem to support the notion that mostly dominant taxa are targets of viral attack.

Roles in Food Web and Geochemical Cycles

The paradigm of marine food webs has been revised a great deal in response to the initial discovery of high bacterial abundance and productivity. It is now well established that a large fraction of the total carbon and nutrient flux in marine systems passes through the heterotrophic bacteria via the dissolved organic matter. How do viruses fit into this picture? Three features of viruses are particularly relevant: (1) small size; (2) composition; and (3) mode of causing cell death, which is to release cell contents and progeny viruses to the surrounding seawater.

When a host cell lyses, the resultant viruses and cellular debris are made up of easily digested protein and nucleic acid, plus all other cellular components, in a nonsinking form that is practically defined as dissolved organic matter. This is composed of dissolved molecules (monomers, oligomers, and polymers), colloids, and cell fragments. This material is most probably utilized by bacteria as food. If it was a bacterium that was lysed in the first place, then uptake by other bacteria represents a partly closed loop, whereby bacterial biomass is consumed mostly by

other bacteria. Because of respiratory losses and inorganic nutrient regeneration connected with the use of dissolved organic substances, this loop has the net effect of oxidizing organic matter and regenerating inorganic nutrients (**Figure 4**). This bacterial–viral loop effectively 'steals' production from protists that would otherwise consume the bacteria, and segregates the biomass and activity into the dissolved and smallest particulate forms. The potentially large effect has been modeled mathematically, and such models show that significant mortality from viruses greatly increases bacterial community growth and respiration rates.

Segregation of matter in viruses, bacteria, and dissolved substances leads to better retention of nutrients in the euphotic zone in virus-infected systems because more material remains in these small nonsinking forms. In contrast, reduced viral activity leads to more material in larger organisms that either sink themselves or as detritus, transporting carbon and inorganic nutrients to depth. The impact can be particularly great for potentially limiting nutrients like N, P, and Fe, which are relatively concentrated in bacteria compared to eukaryotes. At least one study has demonstrated that in cyanobacterial cultures that are starved for trace metals, the Fe in viral lysate of another culture of equal density is taken up within an hour of supplement, which suggests a major role of viral lysis in the availability of limiting nutrients. Therefore, the activity of viruses has the possible effect of helping to support higher levels of biomass and productivity in the planktonic system as a whole.

There are other potential geochemical effects of viral infection and its resultant release of cell contents to the water, owing to the chemical and physical nature of the released materials and the location in the water column where the lysis occurs. For example, polymers released from lysed cells may facilitate aggregation and sinking of material from the euphotic zone. On the other hand, viral lysis of microorganisms within sinking aggregates may lead to the breakup of the particles, converting some sinking particulate matter into nonsinking dissolved material and colloids at whatever depth the lysis occurs. This contributes to the dissolution of sinking organic matter and its availability to free-living bacteria in the ocean's interior.

Viruses, particularly lysogens, have long been known to confer to hosts the ability to produce toxins – in fact cholera toxin is only produced by the cholera bacteria when lysogenized. Through study of viriomics in marine plankton, it is now known that viral genes may encode for several geochemically important enzymes including phosphorus uptake, and components of the photosynthetic apparatus. These appear to be used both under stable host–virus

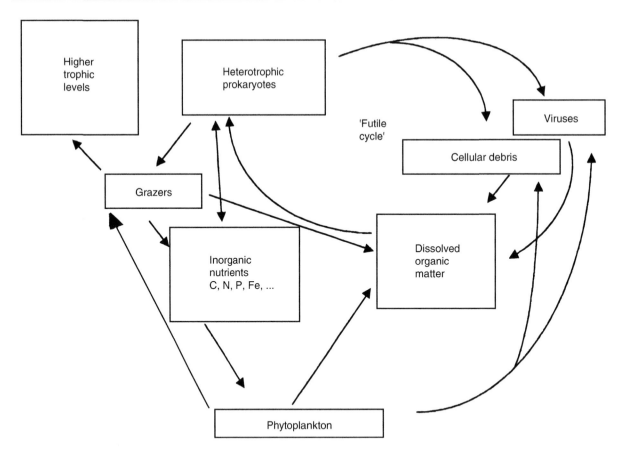

Figure 4 Prokaryote–viral loop within the microbial food web. Arrows represent transfer of matter.

conditions, as well as during host replication, when the enhanced substrate utilization capabilities inferred to hosts is used to produce new virus particles.

Effects on Host Species Compositions and Control of Blooms

Viruses mostly infect only one species or related species, and are also density dependent. Thus, the most common or dominant hosts in a mixed community are believed to be most susceptible to infection, and rare ones least so. Lytic viruses can increase only when the average time to diffuse from host to host is shorter than the average time that at least one member from each burst remains infectious. Therefore, when a species or strain becomes more abundant, it is more susceptible to infection. The end result is that viral infection works in opposition to competitive dominance. This may help to solve Hutchinson's 'paradox of plankton', which asks us how so many different kinds of phytoplankton coexist on only a few potentially limiting resources, when competition theory predicts one or a few competitive winners. Although there have been several possible explanations for this paradox, viral

activity may also help solving it, because as stated above, competitive dominants become particularly susceptible to infection whereas rare species are relatively protected. Extending this argument, one might conclude that viruses have the potential to control algal blooms, such as those consisting of coccolithophorids, and so-called 'red tides' of dinoflagellates. There is now evidence that at least under some circumstances this may be true. Declining blooms have been found to contain numerous infected cells.

Along similar lines, it is now commonly thought that viral infection can influence the species composition of diverse host communities even when they are responsible for only a small portion of the host mortality. This is again because of the near-species specificity of viruses in contrast to the relatively particular tastes of protists or metazoa as grazers. This conclusion is supported by mathematical models as well as limited experimental evidence.

Resistance

The development of host resistance to viral infection is a common occurrence in laboratory and medical situations. Such resistance, where hosts mutate to resist the viral attack, is well known from nonmarine experiments with highly simplified laboratory systems. However, the existence of an apparently high infection rate in plankton suggests that the rapid development resistance is not a dominant factor in the plankton. How can the difference between laboratory and field situations be explained?

Natural systems with many species and trophic levels have far more interactions than simple laboratory systems. One might expect that a species with a large fraction of mortality from one type of virus benefits from developing resistance. However, resistance is not always an overall advantage. It often leads to a competitive disadvantage from the loss of some important receptor, for example, involved in substrate uptake. Even resistance to viral attachment, without any receptor loss, if that were possible, would not necessarily be an advantage. For a bacterium in a low-nutrient environment whose growth may be limited by N, P, or organic carbon, unsuccessful infection by a virus (e.g., stopped intracellularly by a restriction enzyme, or with a genetic incompatibility) may be a useful nutritional benefit to the host organism, because the virus injection of DNA is a nutritious boost rich in C, N, and P. Even the viral protein coat, remaining outside the host cell, is probably digestible by bacterial proteases. From this point of view, one might even imagine bacteria using 'decoy' virus receptors to lure viral strains that cannot successfully infect them. With the proper virus and host distributions, the odds could be in favor of the bacteria, and if an infectious virus (i.e., with a protected restriction site) occasionally gets through, the cell line as a whole may still benefit from this strategy.

There are other reasons why resistance might not be an overall advantage. As described earlier, model results show that the heterotrophic bacteria as a group benefit substantially from viral infection, raising their production by taking carbon and energy away from larger organisms. Viruses also raise the overall system biomass and production by helping to keep nutrients in the lighted surface waters. However, these arguments would require invoking some sort of group selection theory to explain how individuals would benefit from not developing resistance (i.e., why not 'cheat' by developing resistance and letting all the other organisms give the group benefits of infection?). In any case, evidence suggests that even if resistance of native communities to viral infection may be common, it is not a dominant force, because there is continued ubiquitous existence of viruses at roughly 10 times greater abundance than bacteria and with turnover times on the order of a day (as discussed above). Basic mass balance calculations show that significant numbers of hosts must be infected and releasing viruses all the time. For example, with a typical lytic burst size of 50 and viral turnover time of 1 day, maintenance of a 10-fold excess of viruses over bacteria requires 20% of the bacteria to lyse daily. The lack of comprehensive resistance might be due to frequent development of new virulent strains, rapid dynamics or patchiness in species compositions, or to a stable coexistence of viruses and their hosts. All these are possible, and they are not mutually exclusive.

Lysogeny, which is common in marine bacterioplankton, also conveys resistance to superinfection (i.e., infection by the same, or similar virus). This may be an advantage so long as the prophage remains uninduced, in that viral genomes often also contain useful genes involved in membrane transport, photosynthesis, etc., that benefit the host. However, because the prophage is carried around in the host cell, this may also place an extra burden on the host cell machinery (i.e., it must also replicate the viral genome in addition to the host genome).

Genetic Transfer

Viruses can also play central roles in genetic transfer between microorganisms, through two processes. In an indirect mechanism, viruses mediate genetic transfer by causing the release of DNA from lysed host cells that may be taken up and used as genetic material by another microorganism. This latter process is called transformation. A more direct process is known as transduction, where viruses package some of the host's own DNA into the phage head and then inject it into another potential host. Transduction in aquatic environments has been shown to occur in a few experiments. Although transduction usually occurs within a restricted host range, recent data indicate that some marine bacteria and phages are capable of transfer across a wide host range. Although the extent of these mechanisms in natural systems is currently unknown, they could have important roles in population genetics, by homogenizing genes within a potential host population, and also on evolution at relatively long timescales. Gene transfer across species lines is an integral component of microbial evolution, as shown in the genomes of modern-day microbes that contain numerous genes that have obviously been transferred from other species. On shorter timescales, this process can be responsible for the dissemination of genes that may code for novel properties, whether introduced to native communities naturally or via genetic engineering.

Summary

It is now known that viruses can exert significant control of marine microbial systems. A major effect is on mortality of bacteria and phytoplankton, where viruses are thought to stimulate bacterial activity at the expense of larger organisms. This also stimulates the entire system via improved retention of nutrients in the euphotic zone. Other important roles include influence on species compositions and possibly also genetic transfer.

See also

Bacterioplankton. Primary Production Distribution. Primary Production Methods. Primary Production Processes.

Further Reading

Ackermann HW and Du Bow MS (1987) *Viruses of Prokaryotes, Vol. I: General Properties of Bacteriophages*. Boca Raton, FL: CRC Press.

Azam F, Fenchel T, Gray JG, Meyer-Reil LA, and Thingstad T (1983) The ecological role of water-column microbes in the sea. *Marine Ecology Progress Series* 10: 257–263.

Bratbak G, Thingstad F, and Heldal M (1994) Viruses and the microbial loop. *Microbial Ecology* 28: 209–221.

Breitbart M, Salamon P, Andresen B, *et al.* (2002) Genomic analysis of uncultured marine viral communities. *Proceedings of the National Academy of Sciences of the United States of America* 99: 14250–14255.

Culley A, Lang AS, and Suttle CA (2003) High diversity of unknown picorna-like viruses in the sea. *Nature* 424: 1054–1057.

Fuhrman JA (1992) Bacterioplankton roles in cycling of organic matter: The microbial food web. In: Falkowski PG and Woodhead AS (eds.) *Primary Productivity and Biogeochemical Cycles in the Sea*, pp. 361–383. New York: Plenum.

Fuhrman JA (1999) Marine viruses: Biogeochemical and ecological effects. *Nature* 399: 541–548.

Fuhrman JA (2000) Impact of viruses on bacterial processes. In: Kirchman DL (ed.) *Microbial Ecology of the Oceans*, pp. 327–350. New York: Wiley-Liss.

Fuhrman JA and Suttle CA (1993) Viruses in marine planktonic systems. *Oceanography* 6: 51–63.

Lindell D, Sullivan MB, Johnson ZI, Tolonen AC, Rohwer F, and Chisholm SW (2004) Transfer of photosynthesis genes to and from *Prochlorococcus* viruses. *Proceedings of the National Academy of Sciences of the United States of America* 101: 11013–11018.

Maranger R and Bird DF (1995) Viral abundance in aquatic systems – a comparison between marine and fresh waters. *Marine Ecology Progress Series* 121: 217–226.

Noble RT and Fuhrman JA (1998) Use of SYBR Green I for rapid epifluorescence counts of marine viruses and bacteria. *Aquatic Microbial Ecology* 14(2): 113–118.

Paul JH, Rose JB, and Jiang SC (1997) Coliphage and indigenous phage in Mamala Bay, Oahu, Hawaii. *Applied and Environmental Microbiology* 63(1): 133–138.

Proctor LM (1997) Advances in the study of marine viruses. *Microscopy Research and Technique* 37(2): 136–161.

Suttle CA (1994) The significance of viruses to mortality in aquatic microbial communities. *Microbial Ecology* 28: 237–243.

Thingstad TF, Heldal M, Bratbak G, and Dundas I (1993) Are viruses important partners in pelagic food webs? *Trends in Ecology and Evolution* 8(6): 209–213.

Wilhelm SW and Suttle CA (1999) Viruses and nutrient cycles in the sea – viruses play critical roles in the structure and function of aquatic food webs. *Bioscience* 49(10): 781–788.

Wommack KE and Colwell RR (2000) Virioplankton: Viruses in aquatic ecosystems. *Microbiology and Molecular Biology Reviews* 64(1): 69–114.

BACTERIOPLANKTON

H. W. Ducklow, The College of Willian and Mary, Gloucester Point, VA, USA

Introduction

Marine bacteria, unicellular prokaryotic plankton usually less than 0.5–1 μm in their longest dimension, are the smallest autonomous organisms in the sea – or perhaps in the biosphere. The nature of their roles in marine food webs and the difficulty of studying them both stem from their small size. A modern paradigm for bacterioplankton ecology was integrated into oceanography only following development of modern epifluorescence microscopy and the application of new radioisotopic tracer techniques in the late 1970s. It was not until a decade later, with the use of modern genomic techniques, that their identity and taxonomy began to be understood at all. Thus we are still in the process of constructing a realistic picture of marine bacterial ecology, consistent with knowledge of evolution, plankton dynamics, food web theory, and biogeochemistry. The lack of bacterioplankton compartments in most numerical models of plankton ecology testifies to out current level of ignorance. Nevertheless, much is now well known that was just beginning to be guessed in the 1980s.

Bacterioplankton are important in marine food webs and biogeochemical cycles because they are the principal agents of dissolved organic matter (DOM) utilization and oxidation in the sea. All organisms liberate DOM through a variety of physiological processes, and additional DOM is released when zooplankton fecal pellets and other forms of organic detritus dissolve and decay. By recovering the released DOM, which would otherwise accumulate, bacterioplankton initiate the microbial loop, a complicated suite of organisms and processes based on the flow of detrital-based energy through the food web. The flows of energy and materials through the microbial loop can rival or surpass those flows passing through traditional phytoplankton-grazer-based food chains. For further information on the topics summarized here, the reader may consult the Further Reading, especially the recent book edited by Kirchman.

Identity and Taxonomy

Most bacterial species cannot be cultivated in the laboratory and, until the development of culture-independent genomic methods, the identity of over 90% of bacterial cells enumerated under the microscope was unknown. Only those few cells capable of forming colonies on solid media (agar plates) could be identified by classical bacteriological techniques. However, since the application of molecular genomic methods to sea water samples in the mid-1980s, our understanding of marine bacterial systematics and evolution has undergone a profound revolution. In this approach, plankton samples including bacterioplankton cells are collected and lysed to yield a mixture of DNA strands reflecting the genetic composition of the original assemblage. Then individual genes on the DNA molecules can be cloned and amplified via the polymerase chain reaction (PCR) for further analysis. Theoretically, any gene complex can be cloned, and several major groups of genes have been studied to date – for example, genes controlling specific biogeochemical transformations like ammonium oxidation, nitrogen fixation, sulfate reduction, and even oxidation of xenobiotic pollutant molecules. The most useful and widely studied genes for elucidating evolutionary relationships among bacterioplankton have been the genes coding for small subunit ribosomal RNA (SSU rRNA), because they evolve relatively slowly and their characters have been conserved across all life forms during the course of evolution. By sequencing the base pairs making up individual SSU rRNA molecules, the similarity of different genes can be established with great sensitivity. To date, nearly 1000 individual microbial SSU rRNA genes have been cloned and sequenced, yielding an entirely new picture of the composition of marine communities.

The most important aspect of our understanding is that what we term 'bacterioplankton' really consists of two of the fundamental domains of life: the *Bacteria* and the *Archaea* (**Figure 1**). Domain *Archaea* is a group of microbial organisms with unique genetic, ultrastructural, and physiological characters that are about as different, genetically, from the *Bacteria* as either group is from higher life forms. Members of the *Archaea* may be typified by organisms from extreme habitats including anaerobic environments, hot springs, and salt lakes, but marine archaeal groups I and II are common in sea water. They make up about 10% of the microbial plankton in the

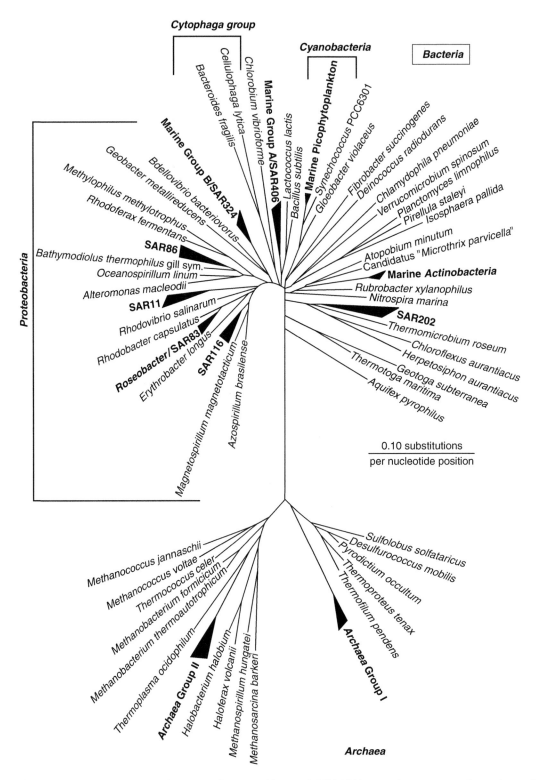

Figure 1 Dendrogram showing relationships among the most widespread SSU rRNA gene clusters among the marine prokaryotes (the 'bacterioplankton'). (Modified after Giovannoni SJ, in Kirchman (2000).)

surface waters of the oceans, and are relatively more numerous at greater depths, where they approach about half the total abundance. Since most of these organisms are known only from their RNA genes and have never been cultured, their physiology and roles in the plankton are almost entirely unknown.

Domain *Bacteria* contains all the familiar, cultur-able eubacterial groups and also a large number of unculturable, previously unknown groups. The main culturable groups include members of the *Proteobacteria*, marine oxygenic, phototrophic *Cyanobacteria*, and several other major groups including methylotrophs, planctomycetes, and the *Cytophaga–Flavobacterium–Bacteroides* group. But the most abundant genes recovered so far are not similar to those of the known culturable species. These include the most ubiquitous of all groups yet recovered, the SAR-11 cluster of the alpha *Proteobacteria*, which have been recovered from every bacterial clone library yet isolated. It appears to be the most widely distributed and successful of the *Bacteria*. The photosynthetic *Cyanobacteria*, including *Synecchococcus* spp. and the unicellular prochlorophytes, are functionally phytoplankton and they dominate the primary producer populations in the open sea, and at times in coastal and even estuarine regimes. They are treated elsewhere in this encyclopedia, so our discussion here is limited to heterotrophic forms of *Bacteria* and to the planktonic *Archaea*, although we cannot specify what many (or most) of them do. Genomic techniques are now being used to investigate bacterial and archaeal species succession during oceanographic events over various timescales, much as phytoplankton and higher organism successions have been observed for a century or more.

Nutrition and Physiology

Knowledge of the nutrition and physiology of naturally occurring bacterioplankton as a functional group in the sea is based partly on laboratory study of individual species in pure culture, but mostly on sea water culture experiments. Traditional laboratory investigations show that bacteria can only utilize small-molecular-weight compounds less than ~500 Daltons. Larger polymeric substances and particles must first be hydrolyzed by extracellular enzymes. In the sea water culture approach, samples with natural bacterioplankton assemblages are incubated for suitable periods (usually hours to a few days) while bacterial abundance is monitored, the utilization of various compounds with ^{14}C- or ^{3}H-labelled radiotracers is estimated, and the net production or loss of metabolites like oxygen, CO_2, and inorganic nutrients is measured. Such experiments, combined with size-fractionation using polycarbonate filters with precise and uniform pores of various diameter (0.2–10 μm), revealed that over 90% of added organic radiotracers are utilized by the smallest size fractions (<1 μm). Bacteria are overwhelmingly the sink for DOM in all habitats studied to date. Nutrient limitation of bacterial growth can

be identified by adding various compounds (e.g., ammonium, phosphate, or iron salts; monosaccharides and amino acids) singly or in combination to experimental treatments and comparing growth responses to controls. Using this approach, it has been learned that bacteria are effective competitors with phytoplankton for inorganic nutrients, including iron, which bacteria can mobilize by producing iron-binding organic complexes called siderophores. In general, bacterial growth in the sea, from estuaries to the central gyres, tends to be limited by organic matter. Sea water cultures most often respond to additions of sugars and amino acids, with the response sometimes enhanced if inorganic nutrients (including iron) are also added.

At larger scales, the ultimate dependence of bacteria on organic matter supply is indicated by significant correlations between bacterial standing stocks or production (see below) and primary production (PP) across habitats (**Figure 2**). At within-habitat scales and shorter timescales, significant relationships are less common, indicating time lags between organic matter production and its conversion by bacteria. Such uncoupling of organic matter production and consumption is also shown by transient accumulations of DOM in the upper ocean, where production processes tend to exceed utilization. It is not yet understood why DOM accumulates. Some fraction might be inherently refractory or rendered so by ultraviolet radiation or chemical condensation reactions in sea water. Deep ocean

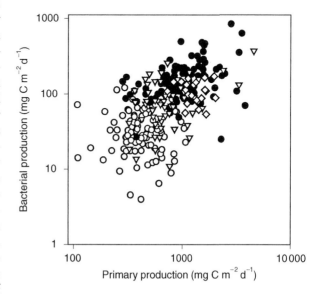

Figure 2 Bacterial production plotted against primary production for the euphotic zone in several major ocean regimes or provinces. The overall data set has a significant regression, but the individual regions do not. ○, Sargasso Sea; ●, Arabian Sea; ◇, equatorial Pacific; △, equatorial Pacific.

DOM has a turnover time of centuries to millennia, and seems to become labile (vulnerable to bacterial attack) when the ocean thermohaline circulation returns it to the illuminated surface layer. Alternatively, bacterial utilization of marine DOM, which generally has a high C:N ratio, might be limited by availability of inorganic nutrients. The latter hypothesis is supported by observations that DOM accumulation tends to be greater in the tropics and subtropics, where nitrate and phosphate are depleted in surface waters.

The efficiency with which bacteria convert organic matter (usually expressed in carbon units) into biomass can be estimated by comparing the apparent utilization of individual compounds or bulk DOM with increases in biomass or with respiration. Respiration is usually measured by oxygen utilization but precise new analytical techniques for measuring carbon dioxide make CO_2 production a preferable approach. Bacterial respiration (BR) is difficult to measure because water samples must first be passed through filters to remove other, larger respiring organisms, and because the resulting respiration rates are low, near the limits of detection of oxygen and CO_2 analyses. It is also not easy to estimate bacterial biomass precisely (see below). The conversion efficiency or bacterial growth efficiency (BGE) is the quotient of net bacterial production (BP) and the DOM utilization:

$$BGE = \frac{BP}{\Delta DOM} = \frac{BP}{BP + BR} \quad [1]$$

Bacteria have rather uniform biomass C:N composition ratios of 4–6. Intuitively, it seems reasonable to expect that they would utilize substrates with high C:N ratios at lower efficiency. Enrichment cultures initiated from natural bacterial assemblages grow in sea water culture in the laboratory on added substances with efficiencies of 30–90%. The BGE is inversely related to the C:N ratio of the organic substrate if just a single compound is being utilized, but when a mixture of compounds is present, as is probably always the case in the environment, there is no discernible relationship between the chemical composition of the materials being used and the BGE.

In the open ocean, BGE averages about 10–30%, a relatively low value that has important implications for our understanding and modeling of organic matter turnover and ocean metabolism. At larger scales, BGE appears to increase from $\sim 10\%$ to 50% along an offshore-to-onshore gradient of increasing primary productivity, probably reflecting greater organic matter availability. This pattern has been used to support an argument suggesting that in lake and oceanic systems with the lowest primary productivity, respiration exceeds production; that is, such oligotrophic systems might be net heterotrophic. This possibility has also been supported by results from careful light–dark bottle studies in which oxygen consumption exceeds production. This finding, however, is inconsistent with a large amount of geochemical evidence, for instance showing net oxygen production at the basin and seasonal to annual scale. Resolution of this debate probably rests on improved estimates of BGE.

Pure culture, sea water culture, and the latest genomic studies indicate fundamental metabolic and genetic differences among different bacterial populations, which can generally be grouped into two broad classes based on organic matter utilization. Native marine bacteria capable of utilizing DOM at concentrations below $100 \, \text{nmol} \, l^{-1}$, termed oligotrophs, cannot survive when DOM is greater than about $0.1–1 \, \text{mmol} \, l^{-1}$. Copiotrophic bacteria found in some habitats with higher ambient DOM levels thrive on concentrations far exceeding this threshold. Observations that copiotrophs shrink and have impressive survival capability under severe starvation conditions (thousands of days to, apparently, centuries) led some investigators to suggest that the dominant native marine bacteria are starving (nongrowing) copiotrophs in a survival mode, awaiting episodes of nutrient enrichment. A variable fraction of the total population usually does appear to be dormant, as indicated by autoradiography, vital staining, and RNA probes, but the timescales of the transition from active growth to dormancy and back again are not well defined. Maintenance of dormant cells in a population depends on strong predator preferences for actively growing cells and prey selection against the nongrowing cells. Most oligotrophs so far isolated in the laboratory under stringent low-DOM conditions appear to be unrelated to known bacterial groups.

Bacterial Biomass, Growth, and Production

The standing stock of bacteria is still most commonly assessed by epifluorescence microscopy, following staining of the cells with a fluorochrome dye. Flow cytometric determination is gradually taking over, and has several key advantages over microscopy: faster sample processing, improved precision, and discrimination of heterotrophic and phototrophic bacteria. There is a gradient in bacterial abundance proceeding from $\sim 10^{10}$ cells l^{-1} in estuaries to 10^9

cells l^{-1} in productive ocean regimes and 10^8 cells l^{-1} in the oligotrophic gyres (**Figure 3**). These horizontal gradients parallel gradients in primary production and organic matter fluxes, suggesting the overall importance of bottom-up controls on bacterial abundance. Chlorophyll a concentrations, indicative of phytoplankton biomass, vary somewhat more widely than bacterial abundance over basin to global scales, but within habitats, the variability of bacterial and phytoplankton biomass is about equal, reflecting the generally close coupling between the two groups and the similarity of removal processes (grazing, viral lysis, unspecified mortality) acting on them.

It is more difficult to estimate bacterial biomass, because we cannot measure the mass (e.g., as carbon) directly, and have to convert estimates of cell volumes to carbon instead. The best estimates now indicate $7–15 \times 10^{-15}$ g C cell^{-1} for oceanic cells and $15–25 \times 10^{-15}$ g C cell^{-1} for the slightly larger cells found in coastal and estuarine habitats. Thus the biomass gradient is steeper than the abundance gradient because the cells are larger inshore. **Table 1** shows data compiled from Chesapeake Bay and the Sargasso Sea off Bermuda, two well-studied sites that illustrate the contrasts in phytoplankton and bacterioplankton from a nutrient-rich estuary to the oligotrophic ocean gyres. Bacterial and phytoplankton biomass are much greater in the estuary, as

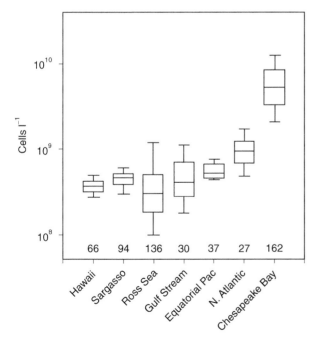

Figure 3 Bacterial abundance in the euphotic zone of several major ocean provinces. The box plots show the median, 10th, 25th, 75th and 90th centiles of the data. The number of samples is listed for each region. There is no statistical difference among the regions except for Chesapeake Bay.

expected. Interestingly, assuming a mean euphotic zone depth of 1 m in the Bay and 140 m off Bermuda, we find that the standing stocks of bacteria in these euphotic zones are ~ 10 and 50 mmol C m^{-2} in the estuaries and open sea, respectively. The oceanic euphotic zone is somewhat more enriched in bacteria than the more productive estuaries. Bacterial and phytoplankton stocks are nearly equal in the open sea, but phytoplankton exceeds bacterial biomass inshore. Carbon from primary producers appears to be more efficiently stored in bacteria in oceanic systems compared to estuarine ones.

Bacterial stocks in different environments can be assessed using the relationship

$$B_{max} = F/m \qquad [2]$$

where B_{max} is the carrying capacity in the absence of removal, F is the flux of utilizable organic matter to the bacteria, and m is their maintenance efficiency (the specific rate of utilization when all of F is used to meet cellular maintenance costs, with nothing left for growth). The problem is specifying values for F and m. The DOM flux can be evaluated by flow analysis and is about 20–50% of the net primary production (NPP) in most systems. Maintenance costs are poorly constrained and possibly very low if most cells are near a starvation state, but 0.01 d^{-1} is a reasonable value for actively growing cells. Thus for the oligotrophic gyres where the latest NPP estimates are about 200–400 mg C m^{-2} d^{-1}, we can calculate that B_{max} should be about $4–8 \times 10^9$ cells l^{-1}, an order of magnitude greater than observed. Removal processes must maintain bacterial stocks considerably below their maximum carrying capacity.

Bacteria convert preformed organic matter into biomass. This process is bacterial production, which can be expressed as the product of the biomass and the specific growth rate (μ)

$$BP = dB/dt = \mu B \qquad [3]$$

Like biomass, BP cannot be measured directly in mass units. Instead, metabolic processes closely coupled to growth are measured and BP is derived using conversion factors. The two most common methods follow DNA and protein synthesis using (^{3}H)thymidine and (^{3}H)leucine incorporation rates, respectively. The values for the conversion factors are poorly constrained and hard to measure, leading to uncertainty of at least a factor of two in the BP estimates. Few measurements were performed in the open sea before the 1990s. The Joint Global Ocean Flux Study (JGOFS) time-series station at Bermuda is perhaps the best-studied site in the ocean (**Table 1**).

Table 1 The biomass (*B*) and production rates (*P*) of bacterioplankton and phytoplankton at estuarine and open ocean locations[a]

Location	Biomass ($mmol\ m^{-3}$)		Production rate ($mmol\ m^{-3}\ d^{-1}$)		P/B (d^{-1})	
	Phytoplankton	Bacteria	Phytoplankton	Bacteria	Phytoplankton	Bacteria
Chesapeake Bay	5–400 (56)	1–80 (11)	20–47 (33)	0.1–50 (4)	0.07–1.9	0.01–2 (0.34)
Sargasso Sea	0.3–3.2 (1.0)	0.2–0.6 (0.4)	0.06–0.9 (0.3)	0.002–0.07 (0.02)	0.1–1 (0.3)	0.01–0.16 (0.06)

[a]The values are annual, euphotic zone averages derived from published reports. P/B is the specific turnover rate for the population. The data are presented as ranges with the mean of various estimates in parentheses. Ranges encompass observations and assumptions about conversion factors for deriving values from measurements (see text).

In the open sea, far removed from allochthonous inputs of organic matter, we can compare BP and PP directly, since all the organic matter ultimately derives from the PP. One difficulty is that BP itself is not constrained by PP, since if the recycling efficiency of DOM and the BGE are sufficiently high, BP can exceed PP. BP also commonly exceeds local PP in estuaries, where inputs of terrestrial organic matter are consumed by bacteria. Bacterial respiration, however, cannot exceed the organic matter supply and serves as an absolute constraint on estimates of BP. But as noted above, bacterial respiration is very hard to measure and there are many fewer reliable measurements than for BP itself. BR is usually estimated from the BGE. Rearranging eqn [1],

$$BR = \frac{(1 - BGE)BP}{BGE} \quad [4]$$

Most commonly, variations of eqn [1] have been used to estimate the total bacterial carbon utilization or demand (BCD BR + BP) from estimates or assumptions about BP and BGE. Earlier estimates and literature surveys suggested that BP was as high as 30% of PP. Combining this value with a BGE of 20% yields a BCD of 1.5 times the PP. This estimate in itself is possibly acceptable, if recycling of DOM is high, but then eqn [4] yields a BR of 1.2 times the total PP – an impossibility. More recent estimates of BP, typified by the Sargasso Sea data, suggest BP is about 10% of PP in the open sea. Applying this value and the mean BGE for the region (0.14), we find that BR consumes about 55% of the primary production in the Sargasso Sea, still a substantial figure. Similar calculations for other well-studied ocean areas suggest that zooplankton (including protozoans and microzooplankton) and bacteria consume nearly equal amounts of the total primary productivity. These estimates illustrate the biogeochemical importance of bacterioplankton in the ocean carbon cycle: although their growth efficiency is low,

bacteria process large amounts of DOM. DOM produced by a myriad of ecological and physiological processes must escape bacterial metabolism to enter long-term storage in the oceanic reservoir.

Role in Food Webs and Biogeochemical Cycles

The process of bacterivory (consumption of bacteria by bacteriovores) completes the microbial loop. Bacterioplankton cells are ingested by a great diversity of predators, but, because of the small size of the prey, most bacteriovores are small protozoans, typically $<5\ \mu m$ nanoflagellates and small ciliates. Bacterial cells only occupy about 10^{-7} of the volume of the upper ocean, indicating the difficulty of encountering these small prey. Larger flagellates, small ciliates, and some specialized larger predators can also ingest bacterial prey. The most important of the larger predators are gelatinous zooplankton like larvaceans, which use mucus nets to capture bacterial cells sieved from suspension. But most bacteriovores are also very small. Nanoflagellates can clear up to 10^5 body volumes per hour, thus making a living from harvesting small, rare bacterial prey, and generally dominating bacterivory in the sea. Protozoan bacterivory closely balances BP in less-productive oceanic regimes. Most crustacean zooplankton cannot efficiently harvest bacterioplankton unless the latter are attached to particles, effectively increasing their size. Bacterial prey enter marine food webs following ingestion by flagellates, and ingestion of the flagellates by other flagellates, ciliates, and copepods. This means that bacteria usually enter the higher trophic levels after several cycles of ingestion by consumers of increasing size, with attendant metabolic losses at each stage. The microbial loop and its characteristic long, inefficient food chains can be short-circuited by the gelatinous bacteriovores, which package bacterial cells into larger prey.

Compared to phytoplankton and to bacteriovores, bacteria are enriched relative to body carbon in nitrogen, phosphorus, protein, nucleic acids, and iron. Their excess nutritional content, coupled with the many trophic exchanges that bacterial biomass passes through as it moves in food webs, means that the microbial loop is primarily a vehicle for nutrient regeneration in the sea rather than an important source of nutrition for the upper trophic levels. The main function of bacteria in the microbial loop is to recover 'lost' DOM, enrich it with macro- and micronutrients, and make it available for regeneration and resupply to primary producers. Lower bacterial production estimates (see above) would also tend to decrease the importance of bacteria as a subsidy for higher consumers.

In estuaries and other shallow near-shore habitats, BP is not as closely balanced by planktonic bacteriovores as in ocean systems. In these productive habitats, bacteria are larger and more often associated with particles, so they are vulnerable to a wider range of grazers. Bacteria can also be consumed by mussels, clams, and other benthic suspension feeders. External subsidies of organic matter mean that BP is higher inshore, so bacteria are a more important food source in coastal and estuarine food webs than in oceanic waters. In these productive systems where bacterial abundance is greater, more of the bacterial stock is also attacked and lysed by viruses, resulting in release of DOM and nutrients instead of entry into food webs. The relative importance of viruses and bacteriovores in removing bacteria is not yet well known, but has important implications for food web structure.

Bacteria are the major engines of biogeochemical cycling on the planet, and serve to catalyze major transformations of nitrogen and sulfur as well as of carbon. They participate in the carbon cycle in several ways. Their principal role is to serve as a sink for DOM, and thus regulate the export of DOM from the productive layer. Bacteria also have intensive hydrolytic capability and participate in decomposition and mineralization of particles and aggregates. Bacteria rapidly colonize fresh particulate matter in the sea, and elaborate polymeric material that helps to cement particles together, so they both reduce particle mass by enzymatic hydrolysis and promote particle formation by fostering aggregation. The balance of bacterial activity for forming particles and accelerating particle sedimentation or, in contrast, decomposing particles and reducing it, is not clear. Larvaceans and other giant, specialized bacteriovores, centimeters to meters in size, can repackage tiny bacterial cells into large, rapidly sinking fecal aggregates, thus feeding the ocean's smallest organisms into the biological carbon pump.

Conclusions

Knowledge of the dynamics of bacterioplankton, their identity, roles in food webs and biogeochemical cycles is now becoming better known and integrated in a general theory of plankton dynamics, but these aspects are not yet common in plankton ecosystem models. The differential importance of bacteria in plankton food webs in coastal and oceanic systems might serve as a good test of our understanding in models. The dynamics of DOM are only crudely parametrized in most models, and explicit formulation of bacterial DOM utilization may help in better characterizing DOM accumulation and export. Other interesting problems such as the effects of size-selective predation, bacterial community structure, and species succession are just beginning to be explored. Exploration of marine bacterial communities together with molecular probes and numerical models should lead to a new revolution in plankton ecology.

See also

Copepods. Marine Mammal Trophic Levels and Interactions. Primary Production Methods. Primary Production Processes. Protozoa, Planktonic Foraminifera. Protozoa, Radiolarians.

Further Reading

Azam F (1998) Microbial control of oceanic carbon flux: the plot thickens. *Science* 280: 694–696.

Carlson C, Ducklow HW, and Sleeter TD (1996) Stocks and dynamics of bacterioplankton in the northwestern Sargasso Sea. *Deep-Sea Research II* 43: 491–516.

del Giorgio P and Cole JJ (1998) Bacterial growth efficiency in natural aquatic systems. *Annual Review of Ecological Systems* 29: 503–541.

Ducklow HW and Carlson CA (1992) Oceanic bacterial productivity. *Advances in Microbial Ecology* 12: 113–181.

Kemp PF, Sherr B, Sherr E, and Cole JJ (1993) *Handbook of Methods in Aquatic Microbial Ecology.* Boca Raton, FL: Lewis Publishers.

Kirchman DL (2000) *Microbial Ecology of the Oceans.* New York: Wiley.

Pomeroy LR (1974) The ocean's food web, a changing paradigm. *BioScience* 24: 499–504.

PROTOZOA, RADIOLARIANS

O. R. Anderson, Columbia University, Palisades, NY, USA

Introduction

Radiolarians are exclusively open ocean, silica-secreting, zooplankton. They occur abundantly in major oceanic sites worldwide. However, some species are limited to certain regions and serve as indicators of water mass properties such as temperature, salinity, and total biological productivity. Abundances of total radiolarian species vary across geographic regions. For example, maximum densities reach 10 000 per m^3 in some regions such as the subtropical Pacific. By contrast, densities range about 3–5 per m^3 in the Sargasso Sea. Radiolarians are classified among the Protista, a large and eclectic group of eukaryotic microbiota including the algae and protozoa. Algae are photosynthetic, single-celled protists, while the protozoa obtain food by feeding on other organisms or absorbing dissolved organic matter from their environment. Radiolarians are single-celled or colonial protozoa. The single-celled species vary in size from $<100\,\mu\text{m}$ to very large species with diameters of 1–2 mm. The larger species are taxonomically less numerous and include mainly gelatinous species found commonly in surface waters. The smaller species typically secrete siliceous skeletons of remarkably complex design (**Figure 1**). The skeletal morphology is species-specific and used in taxonomic identification. Larger, noncolonial species are either skeletonless, being enclosed only by a gelatinous coat, or produce scattered siliceous spicules within the peripheral cytoplasm and surrounding gelatinous layer. Colonial species contain numerous radiolarian cells interconnected by a network of cytoplasmic strands and enclosed within a clear, gelatinous envelope secreted by the radiolarian. The colonies vary in size from several centimeters to nearly a meter in length. The shape of the colonies is highly variable among species. Some are spherical, others ellipsoidal, and some are elongate ribbon-shaped or cylindrical forms. These larger species of radiolarians are arguably, the most diverse and largest of all known protozoa. Many of the surface-dwelling species contain algal symbionts in the peripheral cytoplasm that surrounds the central cell body. The algal symbionts provide some nutrition to the radiolarian host by secretion of photosynthetically produced organic products. The food resources

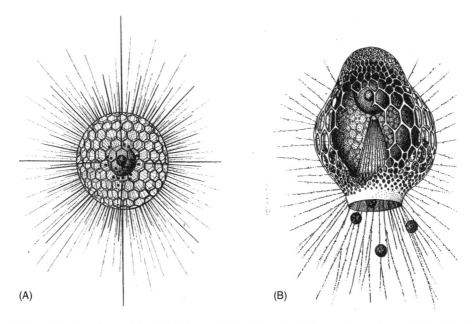

(A)   (B)

Figure 1 Morphology of polycystine radiolaria. (A) Spumellarian with spherical central capsule and halo of radiating axopodia emerging from the fusules in the capsular wall and surrounded by concentric, latticed, siliceous shells. (B) Nassellarian showing the ovate central capsule with conical array of microtubules that extend into the basally located fusules and external axopodia protruding from the opening of the helmet-shaped shell. Reproduced with permission from Grell K (1973) *Protozoology*. Berlin: Springer-Verlag.

are absorbed by the radiolarian and, combined with food gathered from the environment, are used to support metabolism and growth. Radiolarians that dwell at great depths in the water column where light is limited or absent typically lack algal symbionts. The siliceous skeletons of radiolarians settle into the ocean sediments where they form a stable and substantial fossil record. These microfossils are an important source of data in biostratigraphic and paleoclimatic studies. Variations in the number and kind of radiolarian species (based on skeletal form) in relation to depth in the sediment provide information about climatic and environmental conditions in the overlying water mass at the time the radiolarian skeletons were deposited at that geographic location. The radiolarians are second only to diatoms as a major source of biogenic opal (silicate) deposited in the ocean sediments.

Cellular Morphology

The radiolarian cell body contains a dense mass of central cytoplasm known as the central capsule (**Figure 2**). Among the organelles included in the central capsule are the nucleus, or nuclei in species with more than one nucleus, most of the food reserves, major respiratory organelles, i.e., mitochondria, Golgi bodies for intracellular secretion, protein-synthesizing organelles, and vacuoles. The central capsule is surrounded by a nonliving capsular wall secreted by the radiolarian cytoplasm. The thickness of the capsular wall varies among species. It may be thin or in some species very reduced, consisting of only a sparse deposit of organic matter contained within the surrounding cytoplasmic envelope. In others, the wall is quite thick and opalescent with a pearl-like appearance. The capsular wall contains numerous pores through which cytoplasmic strands (fusules) connect to the extracapsular cytoplasm. The extracapsular cytoplasm usually forms a network of cytoplasmic strands attached to stiffened strands of cytoplasm known as axopodia that extend outward from the fusules in the capsular wall. The central capsular wall and axopodia are major defining taxonomic attributes of radiolarians. A frothy or gelatinous coat typically surrounds the central capsule and supports the extracapsular cytoplasm. Algal symbionts, when present, are enclosed within perialgal vacuoles produced by the extracapsulum. In most species, the algal symbionts are exclusively located in the extracapsulum. Thus far, symbionts have been observed within the central capsular cytoplasm in only a few species. Food particles, including small algae and

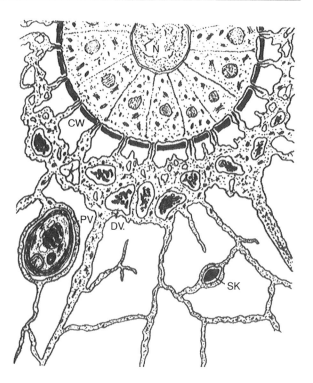

Figure 2 Cytoplasmic organization of a spumellarian radiolarian showing the central capsule with nucleus (N), capsular wall (CW) and peripheral extracapsulum containing digestive vacuoles (DV) and algal symbionts in perialgal vacuoles (PV). The skeletal matter (SK) is enclosed within the cytokalymma, an extension of the cytoplasm, that acts as a living mold to dictate the shape of the siliceous skeleton deposited within it. Reproduced with permission from Anderson OR (1983) *Radiolaria*. New York: Springer-Verlag.

protozoa or larger invertebrates such as copepods, larvacea, and crustacean larvae, are captured by the sticky rhizopodia of the extracapsulum. The cytoplasm moves by cytoplasmic streaming to coat and enclose the captured prey. Eventually, the prey is engulfed by the extracapsular cytoplasm and digested in digestive vacuoles (lysosomes). These typically accumulate in the extracapsulum near the capsular wall. Large prey such as copepods are invaded by flowing strands of cytoplasm and the more nutritious soft parts such as muscle and organ tissues are broken apart, engulfed within the flowing cytoplasm and carried back into the extracapsulum where digestion takes place. The siliceous skeleton, when present, is deposited within cytoplasmic spaces formed by extensions of the rhizopodia. This elaborate framework of skeletal-depositing cytoplasm is known as the cytokalymma. Thus, the form of the skeleton is dictated by the dynamic streaming and molding action of the cytokalymma during the silica deposition process. Consequently, the very elaborate and species-specific form of the skeleton is determined by the dynamic activity of the radiolarian and is not simply a consequence of passive physical chemical processes taking place at interfaces among the frothy components of the cytoplasm as was previously proposed by some researchers.

Taxonomy

Radiolarians are included in some modern classification schemes in the kingdom Protista. However, the category of radiolaria as such is considered an artificial grouping. Instead of the group 'Radiolaria', two major subgroups previously included in 'Radiolaria' are placed in the kingdom Protista. These are the Polycystina and the Phaeodaria. Polycystina are radiolarians that contain a central capsule with pores that are rather uniform in shape and either uniformly distributed across the surface of capsular wall, or grouped at one location. The Phaeodaria have capsular walls with two distinctive types of openings. One is much larger and is known as the astropyle with an elaborately organized mass of cytoplasm extending into the extracapsulum. The other type is composed of smaller pores known as parapylae with thin strands of emergent cytoplasm. Some Phaeodaria also have skeletons that are enriched in organic matter compared with the skeletons of the Polycystina. Among the Polycystina, there are two major taxonomic groups, the Spumellaria and Nassellaria, assigned as orders in some taxonomic schemes. Spumellaria have central capsules that are usually spherical or nearly so at some stage of development and have pores distributed uniformly over the entire surface of the capsular wall. All known colonial species are members of the Spumellaria. Although expert opinion varies, there are two families and about 10 genera of colonial radiolarians. There are seven widely recognized families of solitary Spumellaria with scores of genera. Nassellaria have central capsules that are more ovate or elongated and the fusules are located only at one pole of the elongated capsular wall. This pore field is called a porochora and the fusules tend to be robust with axopodia that emerge through outward-directed collar-like thickenings surrounding the pore rim. Moreover, the skeleton of the Nassellaria, when present, tends to be elongated and forms a helmet-shaped structure, often with an internal set of rods forming a tripod to which the external skeleton is attached. Current systematics include seven major families with numerous genera. Spumellarian skeletons are typically more spherical, or based on a form that is not derived from a basic tripodal or helmet-like architectural plan. The shells of the Phaeodaria are varied in shape. Some species have ornately decorated open lattices resembling geodesic structures composed of interconnected, hollow tubes of silica. Other species have thickened skeletons resembling small clam shells with closely spaced pores on the surface. There are 17 major families with scores of genera. Since many species of radiolarians were first identified from sediments based solely on their mineralized skeletons, much of the key taxonomic characteristics include skeletal morphology. Increasingly, evidence of cytoplasmic fine structure obtained by electron microscopy and molecular genetic analyses is being used to augment skeletal morphology in making species discriminations and constructing more natural evolutionary relationships. It is estimated that there are several hundred valid living species of radiolarians.

Biomineralization

Biomineralization is a biological process of secreting mineral matter as a skeleton or other hardened product. The skeleton of radiolaria is composed of hydrated opal, an oxidized compound of silicon (nominally $SiO_2 \cdot nH_2O$) highly polymerized to form a space-filling, glassy mass incorporating a variable number (n) of water molecules within the molecular structure of the solid. Electron microscopic evidence indicates that some organic matter is incorporated in the skeleton during early stages of deposition, but on the whole, the skeleton is composed mostly of pure silica. Electron microscopic, X-ray dispersive analysis shows that a small amount of divalent cations such as Ca^{2+} may be incorporated in the final veneer deposited on the surface to enhance the hardness of the skeleton. During deposition of the skeleton, the cytoplasm forms the living cytokalymma, i.e., the cytoplasmic silica-depositing mold, by extension of the surface of the rhizopodia. The cytokalymma enlarges as silica is deposited within it, gradually assuming a final form that dictates the morphology of the internally secreted skeleton. Small vesicles are observed streaming outward from the cell body into the cytoplasm of the cytokalymma and these may bring silica to be deposited within the skeletal spaces inside the cytokalymma. The cytoplasmic membrane surrounding the developing skeleton appears to act as a silicalemma or active membrane that deposits the molecular silica into the skeletal space. During deposition, the dynamic molding process is clearly evident as the living cytoplasm continuously undergoes transformations in form, gradually approximating the ultimate geometry of the species-specific skeleton being deposited by the radiolarian cell. In general, species with multiple, concentric, lattice shells surrounding the central capsule appear to lay down the lattices successively, progressing outward from the innermost shell.

The process of skeletal construction has been documented in fair detail for a few species, most notably among the colonial radiolaria. Two forms of growth have been identified. Bar growth is a process of depositing silica as rodlets within a thin tubular network of cytoplasm formed by the cytokalymma. The rodlets become connected during silicogenesis and further augmented with silica to form a porous lattice with typically large polygonal pores. The pores, once formed, may be further subdivided into smaller pores by additional bar growth that spans the opening of the pore. Rim growth occurs by deposition of silica as curved plates that are differentially deposited at places to form rounded pores. At maturity, these are typically spherical skeletons with rather regular, rounded pores scattered across the surface. For both types of skeletons, in some species, the ratio of the bar width between the pores to the pore diameter is a taxonomic diagnostic feature.

The rate of silica biomineralization in some species has been determined by daily observation of growth of individuals in laboratory culture using light microscopy. The amount of silica in the skeleton of a living radiolarian is mathematically related to the size of the skeleton. For example, in the spumellarian species *Spongaster tetras* with a rectangular, spongiose skeleton, the amount of silica (W) in micrograms (μg) as related to the length of the major diagonal axis of the quadrangular shell (L) in micrometers (μm) is approximated as follows:

$$W = \left(3.338 \times 10^{-6}\right) \cdot L^{2.205} \qquad [1]$$

The average daily growth in cultures of an *S. tetras* is 3 μm with an average daily gain in weight of *c.* 8 ng. The total weight gain for one individual radiolarian during maturation is about 0.1 μg. Silica deposition during maturation appears to be sporadic and irregular, varying from one individual to another, with periods of rapid deposition followed by plateaus in growth. The amount of skeletal opal produced by *S. tetras* alone in the Caribbean Sea, for example, is *c.* 42 μg per m^3 of sea water, with a range of 8–61 μg per m^3. Peak production occurred in mid-summer (June to July). The rate of total radiolarian-produced biogenic opal settling into the ocean sediments at varying oceanic locations has been estimated in the range of 1–10 mg per m^2 per day.

Reproduction

Protozoa reproduce by either asexual or sexual reproduction. Asexual reproduction occurs by cell division during mitosis to produce two or more genetically identical offspring. Sexual reproduction occurs by the release of haploid gametes (e.g., sperm and egg cells) that fuse to produce a zygote with genetic characteristics contributed by both of the parent organisms. Thus, sexual reproduction permits new combinations of genetic material and the offspring are usually genetically different from the parents. There is evidence that some colonial radiolaria have asexual reproduction. The central capsules within the colony have been observed to divide by fission. This increases the number of central capsules and allows the colony to grow in size. The colony may also break into parts, thus increasing the total numbers of colonies at a given location. In most species of radiolaria, reproduction occurs by release of numerous flagellated swarmer cells that are believed to be gametes. The nucleus of the parent radiolarian undergoes multiple division and the entire mass of the parent cell is converted into uninucleated flagellated swarmers. These are released nearly simultaneously in a burst of activity, and presumably after dispersal fuse to form a zygote. The details of gamete fusion and the early ontogenetic development of radiolaria are poorly understood and require additional investigation. Ontogenetic development of individuals from very early stages to maturity has been documented in laboratory cultures and the stages of skeletal deposition are well understood for several species, as explained above in the section on biomineralization.

Physiological Ecology and Zoogeography

The physiological ecology of radiolaria has been studied by collecting samples of radiolaria and other biota at varying geographical locations in the world oceans to determine what abiotic and biotic factors are correlated with and predict their abundances, and by experimental studies of the physical and biological factors that promote reproduction, growth, and survival of different species under carefully controlled laboratory conditions. Temperature appears to be a major variable in determining abundances of some species of radiolaria. For example, high latitude species that occur abundantly at the North or South Poles are also found at increasing depths in the oceans toward the equator. Since the water temperature in general decreases with depth, these organisms populate broad depth regions within the water column that match their physiological requirements. Species that occur in subtropical locations, where the water is intermediate in temperature based on a global range, are found at the equator at intermediate water depths that are cooler than the warm surface water. Some species are characteristically most abundant in only warm, highly productive water masses. For example, some species of colonial radiolaria occur typically in surface water near the equator in the Atlantic Ocean, while others are most abundant at higher latitudes in the Sargasso Sea where usually the water is also less productive. Upwelling regions where deep, nutrient-enriched sea water is brought to the surface are typically highly productive regions for radiolaria, as occurs for example along the Arabian, Chilean, and California coast lines. Shallow-water dwelling species have been categorized into seven zoogeographic zones based on water mass properties: (1) SubArctic at high northern latitudes; (2) transition region as occurs in the North Pacific drift waters; (3) north central region, typical of waters within the large anticyclonic circulation of the North Pacific; (4) equatorial region in locations occupied by the North and South Equatorial Current systems; (5) south central water mass, as in the South Pacific anticyclonic circulation pattern; (6) subAntarctic, a water regime bounded on the north by the Subtropical Convergence and on the south by the Polar Convergence; and (7) Antarctic, bounded by the Polar Convergence on the north and the Antarctic Continent on the south.

The growth requirements of some species have been studied extensively in laboratory cultures. For example, the following three surface- to near-surface-dwelling species exhibit a range of optimal growth conditions. *Didymocyrtis tetrathalamus*, with a somewhat hour-glass-shaped skeleton (150 μm), prefers cooler water (21–27°C) and salinities in the range of 30–35 ppm. *Dictyocoryne truncatum*, a spongiose triangular-shaped species (300 μm), is more intermediate in habitat requirements with optimal temperature of 28°C and salinity of 35 ppm. *Spongaster tetras*, a quadrangular, spongiose species (300 μm), prefers warmer, more saline water (c. 28°C and 35–40 ppm). The temperature tolerance ranges (in °C) for the three species also show a similar pattern of increasing preference for warmer water, i.e., 10–34, 15–28, and 21–31, respectively.

The prey consumed by radiolarians varies substantially among species, but many of the polycystine species appear to be omnivorous, consuming both phytoplankton and zooplankton prey. The smaller species consume microplankton and bacteria. Larger species are capable of capturing copepods and small invertebrates. Phaeodaria, especially those species dwelling at great depths in the water column, appear to consume detrital matter in addition to preying on plankton in the water column. The broad range of prey accepted by many of the radiolarians studied thus far suggests that they are opportunistic feeders and are capable of adapting to a broad range of trophic conditions.

The role of algal symbionts, when present, has been debated for some time – beginning with their discovery in the mid-nineteenth century. At first, it was supposed that the green symbionts may largely provide oxygen to the host. However, most radiolaria dwell in fairly well-oxygenated habitats and it is unlikely that photosynthetically derived oxygen is necessary. The other competing hypothesis was that the symbionts provide organic nourishment to the host. Modern physiological studies have confirmed that the algal symbionts provide photosynthetically produced nutrition for the host. Biochemical analyses combined with ^{14}C isotopic tracer studies have shown that stores of lipids (fats) and carbohydrates in the host cytoplasm contain carbon derived from algal photosynthetic activity. Well-illuminated, laboratory cultures of symbiont-bearing radiolaria survive for weeks without addition of prey organisms. Some of the algal symbionts are digested as food and can be replaced by asexual reproduction of the algae, but it appears that much of the nutrition of the host comes from organic nutrients secreted into the host cytoplasm by the algal symbionts. This readily available, 'internal' supply of autotrophic nutrition makes symbiont-bearing radiolaria much less dependent on external food sources and may account in part for their widespread geographic distribution, including some oligotrophic water masses such as the Sargasso Sea.

Further Reading

Anderson OR (1983) *Radiolaria*. New York: Springer-Verlag.

Anderson OR (1983) The radiolarian symbiosis. In: Goff LJ (ed.) *Algal Symbiosis: A Continuum of Interaction Strategies*, pp. 69–89. Cambridge: Cambridge University Press.

Anderson OR (1996) The physiological ecology of planktonic sarcodines with applications to paleoecology: Patterns in space and time. *Journal of Eukaryotic Microbiology* 43: 261–274.

Cachon J, Cachon M, and Estep KW (1990) Phylum Anctinopoda Classes Polycystina (= Radiolaria) and Phaeodaria. In: Margulis L, Corliss JO, Melkonian M, and Chapman DJ (eds.) *Handbook of Protoctista*, pp. 334–379. Boston: Jones and Barlett.

Casey RE (1971) Radiolarians as indicators of past and present water masses. In: Funnell BM and Riedel WR (eds.) *The Micropaleontology of Oceans*, pp. 331–349. Cambridge: Cambridge University Press.

Steineck PL and Casey RE (1990) Ecology and paleobiology of foraminifera and radiolaria. In: Capriulo GM (ed.) *Ecology of Marine Protozoa*, pp. 46–138. New York: Oxford University Press.

PROTOZOA, PLANKTONIC FORAMINIFERA

R. Schiebel and C. Hemleben, Tübingen University, Tübingen, Germany

Introduction

Planktonic foraminifers are single celled organisms (protozoans) sheltered by a test (shell) made of calcite, with an average test diameter of 0.25 mm. They live in surface waters of all modern open oceans and deep marginal seas, e.g., Mediterranean, Caribbean Sea, Red Sea, and Japan Sea, and are almost absent from shelf areas including the North Sea and other shallow marginal seas. Planktonic foraminifers constitute a minor portion of the total zooplankton, but are the main producers of marine calcareous particles deposited on the ocean floor and form the so-called 'Globigerina ooze.'

Planktonic foraminifers (Greek: foramen = opening, ferre = carry) first appeared in the middle Jurassic, about 170 million years ago (Ma), and spread since the mid-Cretaceous over all world oceans. Times of main appearance of new species in the Aptian (120 Ma), the Turonian (90 Ma), the Paleocene (55 Ma), and the Miocene (20 Ma), alternate with phases of main extinction in the Cenomanian (95 Ma), at the Cretaceous/Tertiary boundary (60 Ma), and in the Upper Eocene (40 Ma). Modern planktonic foraminifers have evolved since the early Tertiary, when first spinose species occurred directly after the Cretaceous/Tertiary boundary. Approximately 450 fossil and 50 Recent species are known, not including species based on molecular biology investigations. The appearance and radiation of new species seem to correlate with the development of new realms and niches, linked to plate tectonics and paleoceanographic changes. The geographical distribution and main events in planktonic foraminiferal evolution are associated in general with water mass properties, e.g., availability of food or temperature. The reproductive strategies depend highly on their life habitat in the photic zone or slightly below. The life span of planktonic foraminifers varies between 14 days and a year, mostly linked to the lunar cycle. Most living species bear symbionts requiring a habitat in the upper to middle photic zone. Their feeding habit depends on the spinosity (spinose versus nonspinose species) in respect to the size and class of prey. Predators that are specialized on planktonic foraminifers are not known.

History

With the technological improvement of microscopes d'Orbigny in 1826 was able to describe the first planktonic foraminiferal species, *Globigerina bulloides*, from beach sands, and classified it as a cephalopod. In 1867 Owen described the planktonic life habit of these organisms. Following the Challenger Expedition (1872–1876) the surface-dwelling habitat of planktonic foraminifers was recognized. Rhumbler first described the biology of foraminifers in 1911. In the first half of the twentieth century, foraminifers were widely used for stratigraphic purposes, and many descriptions were published, mainly by Josef A. Cushman and co-workers. Studies on the geographic distribution of individual foraminiferal species are based on samples from the living plankton since the work of Schott in 1935. Planktonic foraminifers have been used since the beginning of the twentieth century to date marine sediments drilled by oil companies, and later on through the Deep-Sea Drilling and Ocean Drilling Programs. In addition, extensive studies on distribution, ecology of live and fossil faunas were carried out to understand the changing marine environment. The ecological significance has been applied in paleoecological and paleoceanographic settings and yielded subtle information on ancient oceans and the Earth's climate. Recent investigation still focuses on evolution and population dynamics. Modern techniques, e.g., polymerase chain reaction (PCR), are being used to reveal the genetic code, and their relation to morphological classification tests needs to be checked.

Methods

Planktonic foraminifers are sampled from the water column by plankton nets of various design, with a mesh size of 0.063–0.2 mm, by employing plankton recorders, water samplers, pumping systems, or collection by SCUBA divers. To study faunas from sediment samples or consolidated rock, the surrounding sediment has to be disaggregated by hydrogen peroxide, tensids, acetic acid (pure), or physical methods, and washed over a sieve (0.03–0.063 mm). Shells may be studied under a binocular microscope, or with a scanning electron microscope for more detail. Transmission electron microscopy is suited to the study of cytoplasm at high resolution. Some species have already been cultivated under laboratory conditions. For faunistic analysis live and dead specimens are distinguished by their content of cytoplasm. For statistical significance on average 300 individuals have to be classified and counted.

According to the distribution and ecology of modern planktonic foraminifers, and due to the fact that their calcitic tests contribute substantially to the microfossil faunal record of marine sediments, planktonic foraminifers are used in reconstructing the climatic, ecological, and geological history of the Earth. Physical factors that determine the modern faunal composition are related to the fossil assemblages by multiple regression statistical techniques (transfer functions) to yield a confident estimate on ancient environmental parameters.

Stable isotopic ($^{18/16}$O and $^{13/12}$C) and trace element ratios of the calcareous (calcite) shell display mostly the composition of the ambient water. These so-called proxies of the physical, chemical, and biological state of modern and ancient oceans are used to reconstruct productivity, temperature, and salinity of paleo-water masses, and to determine the relative age of marine sediments. Laboratory experiments and field calibration are carried out for synoptical evaluation of physical and chemical controls over the geochemical composition of foraminiferal calcite. The metabolic fractionation of isotopes (vital effect) that are included in the foraminiferal shell, varies between species, and depends on water temperature and carbonate (CO_3^{2-}) concentration. The radioactive ^{14}C isotope gives an absolute age of the shell, limited to the last approximately 40 000 years.

Molecular biology methods have recently been used to investigate foraminiferal rRNA genes (rDNA) after DNA extraction, amplification by PCR and normally automated sequencing.

Cellular Structure

Planktonic foraminifers have a single cell that builds calcareous shells and forms chambered tests. Chamber formation, resulting from deposition of calcite, takes place within a cytoplasmic envelope produced by rhizopodia that also secrete a primary organic membrane. A calcitic bilamellar wall is formed at the primary organic membrane. The only exception is the monolamellar genus *Hastigerina* (**Figure 1**). The proximal side of the POM consists of two to three calcite layers whereas the distal (outer) layer reveals as many layers as there are chambers. Layered pustules are built within the outer layers of the wall, concurrent with successive stages of calcite

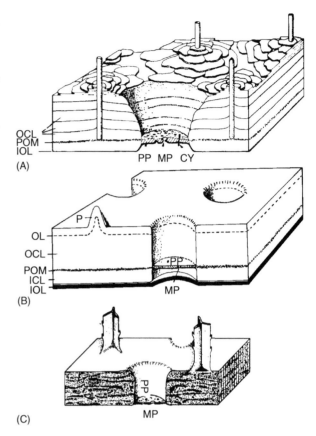

Figure 1 Schematic diagrams of wall structures and pores in bilamellar spinose (A), bilamellar nonspinose (B), and monolamellar (C) planktonic foraminifera (courtesy Cushman Foundation). CY, foraminiferal cytoplasm; ICL, inner calcite layer; IOL, inner organic lining; MP, micropore; OCL, outer calcite layer; OL, outer organic layer; POM, primary organic membrane; P, pustule; PP, pore plate.

lamination. Spines are not layered and are lodged as plugs within the wall.

Intrashell cytoplasm is differentiated from a reticulate or rhizopodial type on the outer shell. Planktonic foraminiferal cell organelles, e.g., nucleus, mitochondria, peroxisomes, Golgi complex, endoplasmic reticulum, annulate lamellae, vacuolar system (**Figure 2**), are typical of those observed in other eukaryotic cells. A fibrillar system seems to be unique among known protozoa, and is suspected to be a floating device or calcifying organelle. Food in the form of lipids and starch is stored in special vacuoles.

Chambers are connected by openings (foramen) between them and have sealed pores in the chamber wall which faces the external environment. Gas exchange between cell and the ambient sea water takes place through these pores; the aperture(s) serves for cytoplasmic contact with the surrounding water, mainly to exchange food particles and waste products. Different types of spines, pores, wall structures, and test morphology may adapt the species to certain

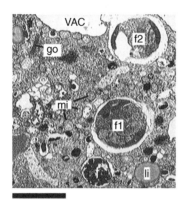
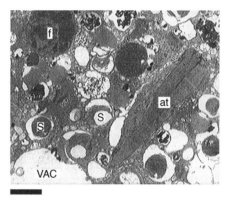

Figure 2 Transmission electron microscopical sections of *Globigerinella siphonifera* cytoplasm including cell organelles. at, animal tissue; f, food vacuole; f1, food vacuole including fresh green algae; f2, food vacuole including partly digested green algae; go, Golgi complex; li, lipid droplet; mi, mitochondria; s, symbiont; vac, empty vacuole. Scalebar = 3 μm.

environments and are of taxonomic significance. Spines allow anastomosing cytoplasm to stretch far out of the test, to form rhizopodial nets for capturing prey, and to carry prey and symbionts as on a conveyor belt to support the cell.

Reproduction and Ontogeny

Planktonic foraminifers probably display only sexual reproduction without the diploid generation. Shallow-dwelling species have been shown to reproduce once per month (*Globigerina bulloides*), or within two weeks (*Globigerinoides ruber*), triggered by the synodic lunar cycle. During reproduction adults release gametes (several hundred thousand) to form offspring with a first calcified chamber (proloculus), which is 8–34 μm in size. The prolocular ontogenetic stage consists of a first (protoconch) and second chamber (deuteroconch). First juvenile chambers are formed on a subdaily rate. During ontogeny the rate of chamber formation gradually decreases. The neanic stage is marked by substantial changes in morphology, and occasional changes in selection of diet and depth habitat, which might explain the relative enrichment of $\delta^{13}C$ with increasing test size. Maturity is reached when the adult stage is reached and tests consist of 10–20 chambers, with a size of 0.1–2 mm (about 0.25 mm on average). The terminal stage is related to reproduction and marked by chamber alterations such as shedding of spines and partial wall thickening. The empty adult test sinks towards the seafloor forming the 'Globigerina Ooze.'

Symbionts, Commensals, and Parasites

Species that bear symbionts (mostly spinose species) are bound to light and live in the euphotic zone of the ocean. Some species without symbionts live in the deep ocean, and only ascend to the sea surface once a year to reproduce (e.g., *Globorotalia truncatulinoides*). Symbionts associated with spinose (**Figure 3**) and occasionally with nonspinose species are dinoflagellates and chrysophycophytes, which may contribute photosynthetic compounds to the host and provide energy to drive the calcification process. This is especially important for the use of $\delta^{13}C_{SHELL}$ in paleo-reconstructions because the $\delta^{13}C$, among other parameters, is determined by the symbiont activity, which is directly correlated to the light level in the water column. Commensalistic dinophytes acquire nutrients as metabolic by-products from the host. Parasites, such as sporozoans, may dart in and feed on foraminiferal cytoplasm.

Molecular Biology

Most recently, methods widely used in molecular biology have also been applied to planktonic foraminifers. A general question is based on the bipolar distribution of the group, and the genetic relationships of species in both hemispheres, especially those clearly distributed in both arctic and antarctic cold waters. Species diversity, based on molecular genetics, is greater than would be expected by applying the traditional concept of morphotaxa. This points towards a larger variety of genetically defined populations, which may permit these to be classified as cryptic sibling species. In addition, the most recent results indicate a polyphyletic origin linked to benthic foraminifers. The molecular methods being used explore the small and large subunits of ribosomal DNA (SSU and LSU rDNA) sequence variability. The results achieved by these molecular methods are manyfold and as yet cannot be explained consistently. However, the large potential of this new field

Figure 3 Spinose planktonic foraminifer *Orbulina universa*. Spines allow cytoplasm to stretch far out of the test, to form a rhizopodial net. Symbionts are carried out by cytoplasm streaming during the light period, and are withdrawn into the test during darkness. Diameter of the test is about 0.5 mm (without spines).

will certainly aid in unraveling the distribution pattern, ecological adaption, speciation, and phylogeny of planktonic foraminifers.

Trophic Demands

Planktonic foraminifers are basically omnivorous. Spinose species prefer a wide variety of animal prey, including larger metazoans such as copepods, pteropods, and ostracods. Cannibalism has also been reported and bacteria are suspected to form part of the diet. Nonspinose species are largely herbivorous. However, in addition to diatoms, which seem to be the major diet, dinoflagellates, thecate algae, and eukaryotic algae, and also muscle tissue and other animal tissue has been found in food vacuoles. The position of planktonic foraminifers in the marine food web is, therefore, different compared to other protozoans, and occasionally ranges above the basic level of heterotrophic consumers. Predators specialized on planktonic foraminifers are not known, but

tests have been found in pteropods, salps, shrimp, and other metazooplankton. Species are spatially and temporally distributed according to diet and temperature and are sensitive to environmental impacts.

Ecology and Distribution

Different faunal groups are characteristic of various oceanic realms. Species are bound to their typical depth habitat in the water column, permitting separation of potentially competing species, and faunal composition changes on a temporal and spatial scale. The vertical separation of the habitats of different species is more evident in warmer than in colder waters; physical and biotic conditions between the sea surface and bathyal depth vary more in subtropical and tropical regions than at high latitudes. Faunal provinces roughly follow a latitudinal pattern, displaying the water temperature and salinity. However, on a finer scale the amount and quality of light, turbidity of the ambient water, trophic state, and distribution of predators play an important role. Only two Recent species (*Neogloboquadrina pachyderma* and *Turborotalita quinqueloba*) are frequent in polar regions. In general, assemblages of planktonic foraminifers occur in five major faunal provinces: (1) polar, (2) subpolar, (3) transition, (4) subtropical, and (5) tropical. Faunal mixing occurs due to hydrodynamic features (e.g., upwelling or current systems) and additional provinces are (6) upwelling, (7) subtropic/tropic, and (8) transitional/subpolar (**Figure 4**). Special environments like the upwelling of nutrient-rich water masses are characterized by high numbers of *Globigerina bulloides*. Typical faunas exist along the margins of the subtropical gyres and at hydrographic frontal systems. The highest diversity is recorded from temperate to subtropical waters (**Figure 5**). A seasonal distribution pattern of planktonic foraminifers is most pronounced in high and mid-latitudes, displaying the phytoplankton succession and associated food chain. Due to meso-scale and local features and a certain reproduction pattern, the distribution of planktonic foraminifers is patchy on various temporal and spatial scales.

The highest numbers (> 1000 specimens per m^3) of adult tests (> 0.1 mm) are recorded in areas and during times of highest primary production, which are upwelling areas and seasonal blooms in the temperate and polar oceans (**Figure 6**). High numbers of individuals correlate with maximum amounts of chlorophyll in the upper ocean and to the deep chlorophyll maximum at the base of the surface mixed layer of the ocean. In the mesotrophic to oligotrophic ocean 1–50 specimens per m^3 occur, and

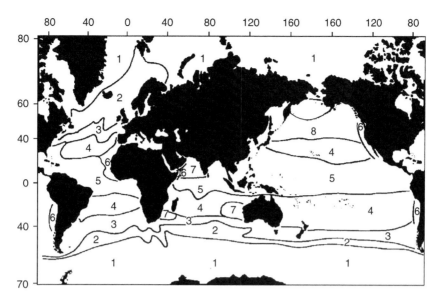

Figure 4 Foraminiferal provinces. 1, polar; 2, subpolar; 3, transitional; 4, subtropic; 5, tropic; 6, upwelling; 7, subtropic/tropic; 8, transitional/subpolar (after Hemleben *et al.*, 1989).

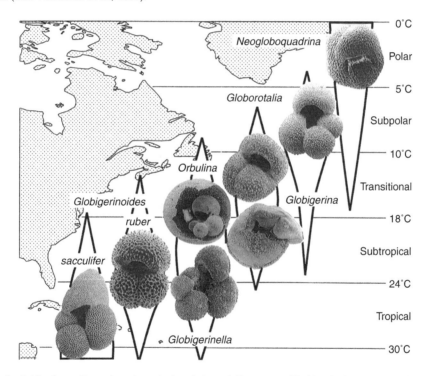

Figure 5 Schematic distribution pattern of modern planktonic foraminifers exemplified for the North Atlantic (according to Hemleben *et al.*, 1989). Species from the lower left to the upper right are *Globigerinoides sacculifer*, *Globigerinoides ruber*, *Globigerinella siphonifera*, *Orbulina universa* (spherical stage cracked open to show the interior preadult test), *Globorotalia truncatulinoides*, *Globorotalia inflata*, *Globigerina bulloides*, and *Neogloboquadrina pachyderma*.

from blue waters (e.g., eastern Mediterranean) less than one specimen per m³ is reported.

Sedimentation

Global calcite production of planktonic foraminifers amounts to about two Gigatons per year, from which only 1–2% reaches the deep sea floor. Planktonic foraminiferal shells dissolve while settling through the water column. Preservation of tests depends on the biogeochemistry of the ambient water and on the resting time of tests in the water column. Due to the low sinking velocity and long time of exposition, small and thin-walled tests (about 100 m per day) are

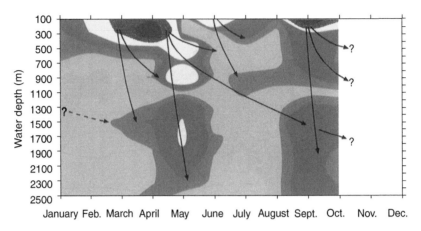

Figure 6 The calcite flux of planktonic foraminiferal shells from the upper water column to the deep sea is determined by population dynamics and settling velocity of empty tests. In the eastern North Atlantic maximum standing stocks of more than 400 specimens per m^3 occur in the upper 300 m during the phytoplankton spring bloom (March through May). A minor maximum in abundance during fall (September to October) is due to redistribution of chlorophyll and entrainment of nutrients and resulting phytoplankton growth in surface waters. During summer and winter the number of specimens may not exceed 10–50 per m^3. Most species live in surface waters. Only a few species live at a depth of 100–500 m. Abyssal species are rare in the transitional North Atlantic and more frequent in the subtropical realm. As a result of population dynamics and differential settling velocity of tests (100–1500 m day^{-1}), the test flux occurs in pulses (arrows), being highest during spring and fall exceeding 60 mg m^{-2} d^{-1} (red; orange $= 30$–60; yellow $= 10$–30; dark green $= 3$–10; light green $= 1$–3; blue $= <1$ mg m^{-2} d^{-1}). Remineralization of tests is highest between 200 and 700 m depth. Below 700 m major $CaCO_3$ flux is restricted to mass sinking events during high-productivity periods. November and December have so far not been sampled.

preferentially removed from the settling assemblage, and mainly large and fast sinking tests (up to 1500 m per day) are deposited at the seafloor. Mass sinking of aggregates (marine snow) during seasons of enhanced biological productivity includes planktonic foraminiferal test, which balances the fossil faunal record towards species assemblages that reflect high productivity, e.g., seasonal upwelling and spring blooms (**Figure 6**). A substantial amount of planktonic foraminiferal shells is remineralized far above the calcite lysocline, between 200 and 700 m water depth. Below the calcite compensation depth virtually no calcareous particles are preserved.

Application

As major contributors to the vertical $CaCO_3$ flux, planktonic foraminiferal shells cause a substantial portion of $CaCO_3$ burial in deep-sea sediments. As a component of the marine carbon turnover and vertical flux, planktonic foraminifers are of major interest for paleoclimatologists, because their tests carry fossil information on climates since the mid-Cretaceous. Their faunal composition, details of the test morphology, and their stable isotope and element ratios, provide detailed information on paleotemperature (δ^{18}O, Mg/Ca, Sr/Ca, δ^{44}Ca), paleoproductivity (δ^{13}C), paleo-pH (δ^{11}B), nitrate (NO_3^-) concentration of seawater (δ^{15}N), and paleo-CO_2 levels by estimating the vertical flux and burial

rates of $CaCO_3$ of planktonic foraminiferal and other marine calcite-sequestrating organisms (mainly coccolithophorids and pteropods). Their role in the marine and global carbon budget which still needs to be quantified, provides great potential information on marine biogeochemistry.

See also

Benthic Foraminifera. Protozoa, Planktonic Foraminifera.

Further Reading

Bé AHW (1977) An ecological, zoogeographic and taxonomic review of Recent planktonic Foraminifera. In: Ramsay ATS (ed.) *Oceanic Micropaleontology*, vol. 1, pp. 1–100. London, New York, San Francisco: Academic Press.

Bolli HM, Saunders JB, and Perch-Nielsen K (1985) *Plankton Stratigraphy*. Cambridge: Cambridge University Press.

Darling K, Wade CM, and Stewart IA (2000) Molecular evidence for genetic mixing of Arctic and Antarctic subpolar populations of planktonic foraminifers. *Nature* 405: 43–47.

Fischer G and Wefer G (1999) *Use of Proxies in Paleoceanography*. Berlin: Springer-Verlag.

Hemleben Ch, Spindler M, and Anderson OR (1989) *Modern Planktonic Foraminifera*. New York: Springer-Verlag.

Kennet JP and Shrinivasan MS (1983) *Neogene Planktonic Foraminifera. A Phylogenetic Atlas.* Stroudsburg, PA: Hutchinson Ross.

Loeblich AR Jr and Tappan H (1987) *Foraminiferal Genera and Their Classification.* New York: Van Nostrand Reinhold.

Olsson RK, Hemleben Ch, Berggren WA, and Huber B (eds.) (1999) *Atlas of Paleocene Planktonic Foraminifera.* *Smithsonian Contributions to Paleobiology 85.* Washington, DC: Smithsonian Institution Press.

Spero HJ, Lea DW, and Young JR (1996) Experimental determination of stable isotope variability in Globigerina bulloides: implications for paleoceanographic reconstructions. *Marine Micropaleontology* 28: 231–246.

COPEPODS

R. Harris, Plymouth Marine Laboratory, Plymouth, UK

Introduction

Copepods are microscopic members of the phylum Crustacea, the taxonomic group that includes crabs, shrimps and lobsters and is the only large class of arthropods that is primarily aquatic. The name copepod comes from the Greek words *kope* (an oar) and *podos* (foot), the majority of members of the group having five pairs of flat paddle like swimming legs. About 10 000 species are currently known, and their numerical dominance as members of the marine plankton means that they are probably the most numerous metazoan – multicellular – animals on earth. In addition to forming a major component of marine plankton communities, copepods are also found in sea-bottom sediments, as well as associated with many marine plants and animals. They play a pivotal role in marine ecosystems by controlling phytoplankton production through grazing, and by providing a major food source for larval and juvenile fish. This article will place particular emphasis on the dominant group of planktonic copepods, known as the Calanoida (**Figure 1**), playing a central role in these processes inthe world's oceans.

Taxonomy

There are 10 taxonomic orders of copepods, of which 9 have marine representatives. Of these the most important marine orders are the Calanoida, Cyclopoida, and Harpacticoida. Calanoid copepods are primarily pelagic, 75% of the known species are marine, and some are benthopelagic or commensal. The group includes the species *Calanus finmarchicus* (Gunnerus), a dominant component of North Atlantic boreal ecosystems, first named nearly 250 years ago as *Monoculus finmarchicus* by Johan Ernst Gunnerus, Bishop of Trondheim in Norway (**Figure 2**). The Cyclopoida include pelagic commensal and parasitic species (**Figure 3**). Harpacticoid copepods are predominantly marine, with only 10% of species being freshwater. Most are benthic, with a few pelagic and commensal representatives, they represent the most abundant component of the meiofauna after nematode worms. The Platycopoida and Misophrioida are primarily benthopelagic

groups, the latter having two pelagic species. The Poecilostomatoida and Siphonostomatoida are commensal or parasitic groups. Finally, the Monstrilloidaare exclusively marine, with parasitic juveniles, but a pelagic adult stage.

Morphology

Most copepods are small, requiring study with a microscope. Small planktonic cyclopoids may be only 0.2 mm long and similarly harpacticoid copepods found in the interstitial space of sandy sediments are among the smallest Metazoa. In contrast, some large deep-sea calanoids, such as *Valdiviella*, may exceed 20 mm in length. *Calanus finmarchicus* is often said to be about the size of a grain of rice (**Figure 2**). Parasitic forms are generally larger than the free living copepods. For example, species of the genus *Penella*, which is parasitic on fish and whales, may be over 30 cm in length.

The body of a free-living copepod (**Figure 4**) is normally cylindrical, and is distinctly segmented. The head, which is the site of the median naupliar eye, is either rounded or may bear a pointed rostrum. The presence of at least two pairs of swimming legs is characteristic of nearly all copepods at some stage in their life cycle. Similarly, antennules with up to 27 (**Figure 5**) segments are general in the order, though segmentation may often be reduced. The body is divided into the prosome, which may be further subdivided into the cephalosome and metasome, and the urosome. The feeding appendages are on the cephalosome; in the calanoids these comprise, from anterior to posterior, the antennule, antenna, mandible, maxillule, maxilla, and maxilliped (**Figure 6**). The swimming legs are attached to the metasome in adult calanoids, one pair for each of the five segments of the metasome. The urosome contains the genital and anal segments, and ends with the furca orcaudal rami, a series of spines or fine hairs. Most copepods are pale and transparent, though some species, particularly those living at the sea surface or in the deep sea, may be pigmented blue, red, orange, or black.

The early developmental stages in the life history are the nauplii (**Figure 7**), with reduced numbers of appendages. In calanoids there are six naupliar stages, NI to NVI. Copepods, like all other Crustacea, molt by shedding their exoskeleton as they grow. Hence there is amolt between each naupliar stage. Molting from NVI involves a radical change

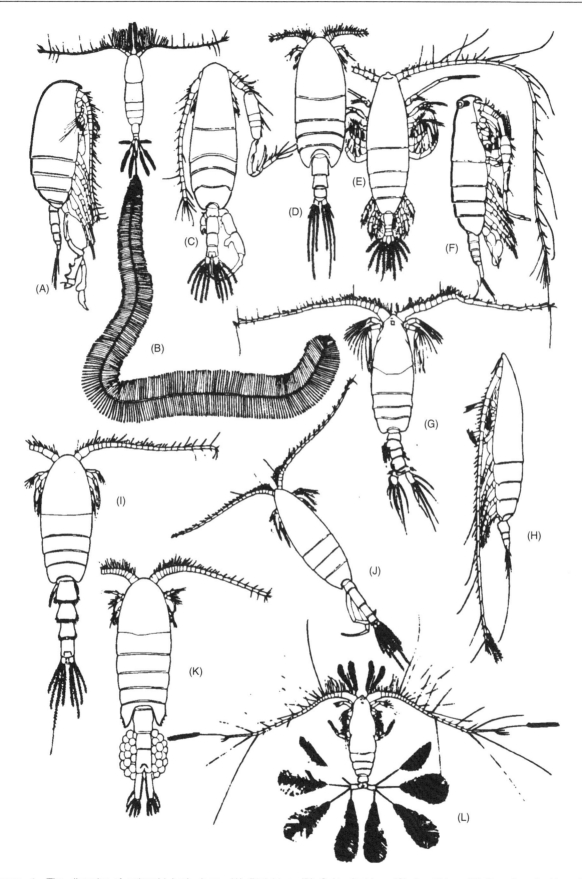

Figure 1 The diversity of calanoid body form. (A) Diaixidae; (B) Calocalanidae; (C) Acartiidae; (D) Pseudocyclopidae; (E) Augaptilidae; (F) Pontellidae; (G) Metridinidae; (H) Eucalanidae; (I) Stephidae; (J) Euchaetidae; (K) Temoridae; (L) Calocalanidae. (Permission from Huys and Boxshall, 1991.)

Figure 2 Gunnerus' sketches of *Calanus*. The smallest shows the natural size. (Permission from Marshall and Orr, 1955.)

inmorphology (metamorphosis) to the first copepodite stage. The copepodites, of which there are normally six stages (CI–CVI) are like small adults, and gradually develop adult characteristics during successive molts. The adult stage is CVI, and no further molts occur.

Distribution and Habitats

As has already been noted, copepods are probably the most numerous multicellular organisms on earth. They are found throughout the marine and estuarine environments of the world's oceans. Species inhabiting coastal and brackish waters have wider tolerances of environmental variables than, for example, deep sea copepods, which are specifically adapted to the special conditions of this environment. Generally, copepods are more abundant in coastal and productive upwelling environments than in the oligotrophic open ocean. Over the deep ocean,

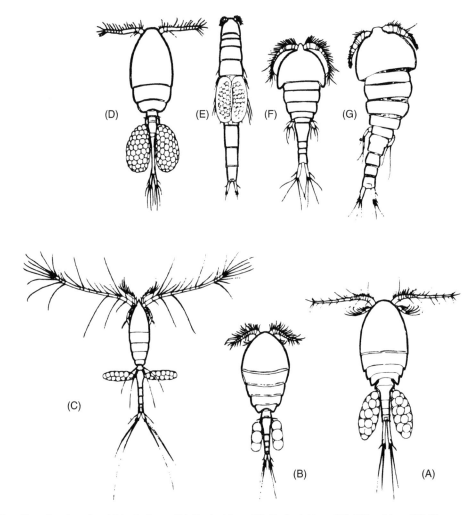

Figure 3 The diversity of cyclopoid body form. (A) Cyclopidae; (B) Cyclopinidae; (C) Oithonidae; (D) Thespessiopsyllidae; (E) Asidicolidae; (F) Archinotodelphyidae; (G) Mantridae. (Permission from Huys and Boxshall, 1991.)

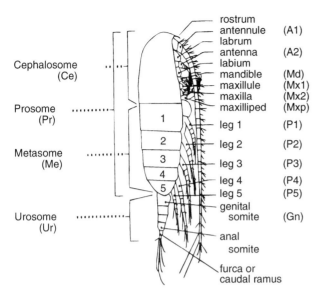

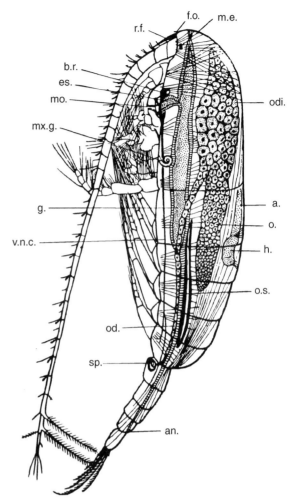

Figure 4 Diagrammatic illustration of the external morphology and appendages of a female calanoid copepod. The metasome has five clearly defined segments, numbered 1–5; this species has five pairs of swimming legs and so these five metasome segments are synonymous with pedigerous segments 1–5. Legs 1–5 are the swimming legs. (Permission from Mauchline, 1998.)

where the water column may extend to 8000 m, the abundance of copepods is highest in the surface layers, and then decreases almost exponentially. The number of species occurring in a particular environment varies. In some, for example brackishtide-pools, a single species may dominate. In contrast, assemblages in the open ocean normally exceed 100 species (**Figure 8**).

In addition to those that dominate the plankton, copepods also live in marine sediments, forming a major component of the meiofauna. They are found in all sediments from muds to coarse sands, and from the intertidal zone to the deep ocean. Harpactocoids are the dominant copepod component of the meiofauna. This group is also abundant on intertidal and subtidal macroalgae.

Apart from free-living planktonic and benthic forms, almost half of the described species live in association with other marine animals. Copepods parasitize almost every phylum of marine animals, many as ectoparasites living on the external body surface, though others have exploited, for example, the internal surfaces of the gills of fish. In the majority of cases it is the adult copepods that are parasites, but the Monstrilloida are an exception, as the naupliar stages are internal parasites of polychaete worms and gastropod mollusks. The adults live in the plankton, but do not feed.

Specialized habitats include marine caves that are home to a number of platycopoid and calanoid species, all living in association with the bottom

Figure 5 Diagram of the internal anatomy of a female *Calanus* from the side. a., aorta; an., anus; br., brain; f.o., frontal organ; g., gut; h., heart; m.e., median eye; mo., mouth; mx.g., maxillary gland; o., ovary; o.di., oviducal diverticula; od., oviduct; es., esophagous; o.s., oil sac; r.f., rostral filament; sp., spermathecal sac; v.n.c. ventral nerve cord. (Permission from Marshall and Orr, 1955.)

sediments. Other such hyperbenthic copepods, living close to the sediment surface, are also found throughout shallow and deep seas. Deep-sea hydrothermal vents also have an associated copepod fauna, which is only now being described.

Other interfaces in the marine environment that provide specialized habitats for copepods are the under ice environment in Polar regions, and the sea surface. The under-ice habitat supports a rich growth of microalgae, and in turn this food source is exploited by a large number of copepods. The sea surface habitat is that of the neuston, the group of animals and plants living in the extreme surface film. The calanoids of the family Pontellidae, such as members of the genera *Pontella* and *Anomalocera* are the commonest neustonic copepods. Many have

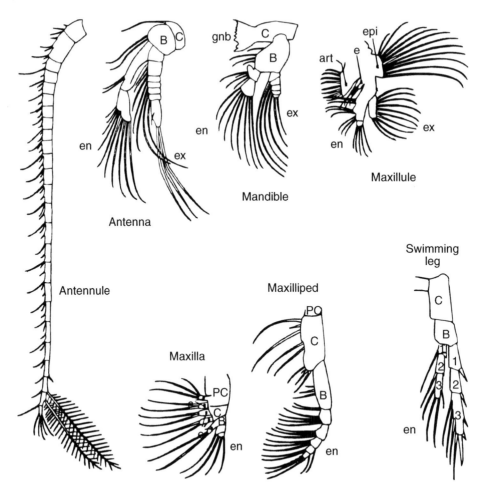

Figure 6 Diagrammatic representations of the appendages of a calanoid copepod. The swimming legs usually have developed endopods and exopods with upto three segments, numbered 1–3 here. art, arthrite; B, basis; C, coxa;e, endite; en, endopod; epi, epipodite; ex, exopod; gnb, gnathobase; PC, praecoxa. (Permission from Mauchline, 1998.)

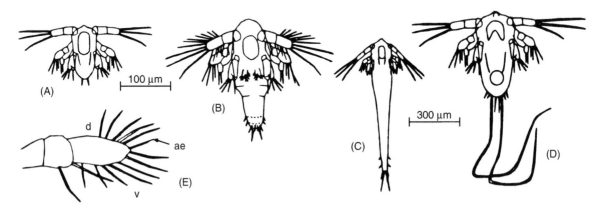

Figure 7 Nauplii of calanoid copepods. (A) *Clausocalanus furcatus*, stage I (NI); (B) *Paracalanus aculeatus*, NV; (C) *Rhincalanus cornutus*, NIV; (D) *Euchaeta marina*, NVI; (E) antennule showing dorsal (d) and ventral (v) setae and terminal aesthetasc (ae). (Permission from Mauchline, 1998.)

strong blue pigmentation, which may be associated with protection against surface ultraviolet radiation, and also attachment structures on the back of the head by which the copepod suspends attached to the surface film. A few species can move with such vigor that they can hurl themselves out of the water, and a shoal of these creatures can appear like a rain shower on the surface of the sea.

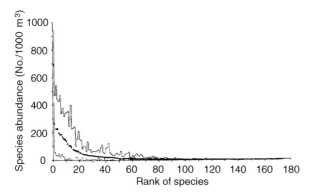

Figure 8 Abundances of copepod species in the open Pacific Ocean showing the species order. Data points are the overall means. Gray bars show the range of seven individual cruise mean abundances per species. (Permission from McGowan JA and Walker PW (1985) Dominance and diversity maintenance in an oceanic ecosystem. *Ecological Monographs* 55: 113–118.)

Feeding

The majority of planktonic copepods were originally thought to be exclusively herbivorous, filtering phytoplankton fromsea water with the fine hairs of the mouth parts. In contrast, carnivorous copepods have more robust spines on the mouth parts. More recently it has been appreciated that many copepods are omnivores, feeding on a wide range of naturally occurring particulate material, phytoplankton, small planktonic animals of the microzooplankton, and detritus.

The feeding appendages are the antennules,antennae, mandibles, maxillules and maxillae (**Figure 5**). These are often considerably reduced in adult males, which may not feed. The mouth parts of ectoparasites are adapted for piercing or sucking. Internal parasites have often lost their mouth parts, and food is absorbed directly from the host. Among planktonic calanoids there are three general mouth part patterns, related to feeding ecology: the true filter-feeders, the omnivores, and the true carnivores. The antennules are particularly involved in carnivorous feeding, having sensory organs that function in prey detection.

Spacing between the hairs (the setae) of the maxillae has been considered to indicate the size of particles that can be filtered by a copepod (**Figure 9**). However, the model of copepod filter-feeding as a mechanical process depending on the morphological characteristics of the maxillae is no longer accepted. Direct studies of feeding behavior using high-speed microcinematography and video observations have shown that feeding behavior is complex, taking account of the viscous, low-Reynolds-number, environment that these small organisms inhabit. Feeding behavior, and adaptations of the appendages,

Figure 9 Left maxilla of *Calanus helgolandicus* female from the right. A, B and C represent the sizes of three algal cells: (A) *Nannochloris oculata*; (B) *Syracosphaera elongata*; (C) *Chaetoceros decipiens*. (Permission from Marshall SM and Orr AP (1956) On the biology, of *Calanus finmarchicus* IX. Feeding and digestion in the young stages. *Journal of the Marine Biological Association of the United Kingdom* 35: 587–603.)

enable copepods to exploit particles such as detritus and phytoplankton, a few micrometers in size, while at the other extreme they can feed on other members of the zooplankton such as other copepods, chaetognaths, and fish larvae. Particles may be rejected during the feeding process, resulting in food selectivity.

The feeding rate of planktonic copepods is dependent on type and size of food particle (**Figure 10**), as well as environmental factors such as temperature, light, and turbulence. The latter, in particular, can affect therate of encounter between predator and prey.

Growth and Development

Copepods grow by molting, as do all other Crustacea. Normally the nauplius stage NI hatches from the egg; naupliar growth involves five molts to the sixth nauplius (NVI), and then after metamorphosis to copepodite stage one (CI) there are a further five molts until the adult, CVI stage, is reached. In a few groups the egg hatches directly into one of the later naupliar stages, for example NII.

The development rate of copepod eggs is dependent on temperature within any one species. The relationship between development time D (days) and

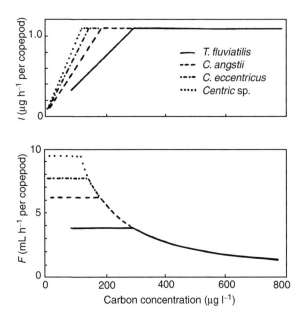

Figure 10 The effect of size (species) and concentration (as carbon) of food particles on ingestion rate, *I*, and volume swept clear, *F*, of adult females of *Calanus*. (Permission from Frost BW (1972) Effects of size and concentration of food particles on the feeding behavior of the marine planktonic copepod *Calanus pacificus*. *Limnology and Oceanography* 17: 805–815.)

temperature T (°C) is generally described by the empirical equation [1], in which α and b are fitted constants.

$$D = a(T - \alpha)^b \qquad [1]$$

Egg development times of egg sac-carrying groups are longer than those of free spawners.

A number of models of development have been applied to the naupliar and copepodite stages of copepods. Equiproportional development considers that the duration of each developmental stage is proportional to the egg development time, determined by the equation above, at the same temperature. The isochronal model of development describes those species for which all stages have almost the same duration, and development proceeds linearly with time. In sigmoidal development, the development rate of the early nauplier stages is significantly slower, and the later copepodite stages also have a longer relative development duration.

Growth rates of copepods are temperature-dependent, and are most usefully expressed as the weight-specific growth rate (per day, d^{-1}), which is the increase in body weight per day as a proportion of the body weight of the developmental stage being considered (**Figure 11**). An adequate food supply, both quantitative and qualitative, is clearly necessary for proper development and growth. Ultimate body

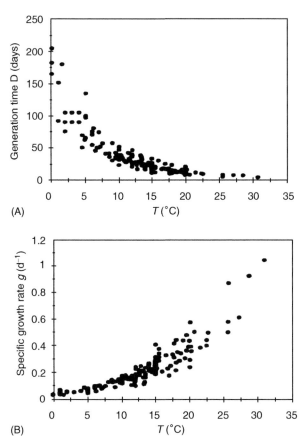

Figure 11 (A) The generation D time (days) of different species of copepods related to environmental temperature T (°C). The relationship is described by the equation $D = 128.8 \ e^{-0.120T}$ (B) The specific growth rate g (d^{-1}) of species of copepods calculated from the weight of egg and the adult and the generation time of each species and related to environmental temperature T (°C). The equation for the relationship is $g = 0.0445e^{0.111T}$. (Permission from Huntley ME and Lopez MDG (1992) Temperature-dependent production of marine copepods: a global synthesis. *American Naturalist* 140: 201–242.)

size, either as length or weight, is dependent on both temperature and food conditions. It has been suggested that for small planktonic copepods growth is optimized, and food is utilized more efficiently, at higher temperatures, whereas larger forms optimize growth and food utilization at lower temperature. This may explain some aspects of the geographical and vertical distribution patterns of copepods.

Metabolism

Copepods have a variety of digestive enzymes in the gut, and there are both diel and seasonal changes in enzyme activity. In particular, overwintering animals in diapause may have considerably reduced digestive enzyme activity. The proportion of the ingested food that is assimilated, and therefore available for subsequent metabolism, ranges from 60% to 90% in herbivores, the remaining 10% to 40% being released as fecal pellets. The soluble excretory products of metabolism are generally excreted as ammonia or urea and dissolved phosphorus compounds, and this process is important in nutrient regeneration cycles supporting phytoplankton growth in marine ecosystems. There are no gills in free-living copepods, and respiratory exchange is supported by direct uptake of dissolved oxygen from sea water. Apart from in the calanoids and some parasitic species there is no heart or circulatory system. Both respiration and excretion are closely coupled to feeding activity and often exhibit diel cycles.

Reproduction

Mating behavior follows a generally similar sequence in all copepods. Initially the male is attracted to the female, often by chemical attractants, pheromones (**Figure 12**), Then the male captures the female, adjusts to the mating position, and finally transfers

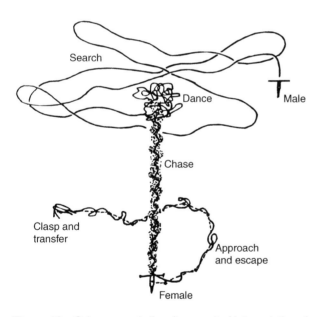

Figure 12 *Calanus marshallae*. A conceptual interpretation of mate-attraction–mate-search behavior. The sequence of events is (1) a female generates a vertical pheromone trail; (2) a male is alerted by pheromone to females in the general vicinity and swims in smooth loops of mostly horizontal orientation; (3) on crossing a pheromone trail, the male performs a dance (or sometimes does not); (4) the male chases down the pheromone trail to the female; (5) the female jumps away repeatedly with the male pursuing, sometimes bumping her; and (6) a mating clasp is established and a spermatophore is transferred from the male to the female. (Permission from Tsuda A and Miller CB (1998) Mate-finding behavior in *Calanus marshallae*. *Philosophical Transactions of the Royal Society of London*, series B 353: 713–720.)

and attaches a package of sperm, the spermatophore, to the female. Most species of planktonic calanoids lay their eggs directly into the water. However, harpacticoids and cyclopoids usually carry the eggs in a single or paired egg sacs and a number of calanoid genera, for example *Euchaeta*, *Eurytemora*, and *Pseudocalanus* also carry egg sacs. Individual eggs are usually spherical, ranging in size from 0.2 to over 0.6 mm, the eggs within egg sacs often being relatively larger than those that are freely spawned. Females of some freely spawning species, when well fed, produce over 100 eggs in a day. The daily rates of egg production expressed as a proportion of the female body weight are around 0.17 for copepods carrying their eggs, and 0.2 for free spawners. The lifetime fecundity of egg sac-bearers is lower than that of the free spawners; the latter have been observed in laboratory studies to produce over 2000 eggs in a female's lifetime.

A number of groups, including calanoids of the families Acartiidae, Centropagidae, Temoridae, and Pontellidae, produce resting eggs, often distinguishable from the normal eggs by having a thicker outer coating. These diapause eggs sink to the seabed and may become buried in bottom sediments until conditions are appropriate for hatching. It has been estimated that diapause eggs may remain viable in sediment, capable of hatching, for up to 40 years.

Behavior

Perhaps the most striking aspect of the behavior of planktonic copepods is that of diel vertical migration. This behavior, characteristic of most planktonic organisms, involves the population remaining at depth during the daytime. As night falls, the copepods actively migrate upward to spend some hours in the surface during the hours of darkness, before descending at dawn to the original daytime depth (**Figure 13**). Although this is a general phenomenon, there are many variations, depending both on species, and the influence of environmental factors.

Light is the dominant environmental factor controlling diel vertical migration, with populations following diel changes in light intensity, isolumes (layers of constant light intensity). Predator avoidance related to light is thought to be one of the major adaptive advantages of diel vertical migration. By migrating to deeper darker layers by day, copepods minimize mortality from visual predators, in particular fish. Predator avoidance has to be balanced against the need to feed and, as the phytoplankton is concentrated in the surface layers, migration to the surface by night is generally associated with active

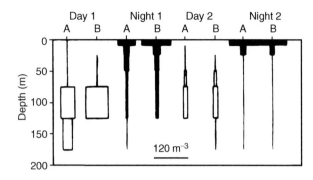

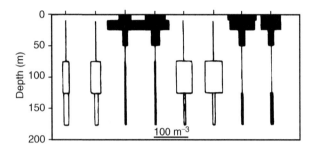

Figure 13 Vertical distribution of adult females of *Calanus pacificus* (A) and *Metridia lucens* (B), 5 and 6 August 1986. (Permission from Dagg MJ, Frost BW and Walser WE Jr (1989) Copepod diel migration, feeding, and the vertical flux of pheopigments. *Limnology and Oceanography* 34: 1062–1071.)

feeding, diel cycles of digestive enzyme activity, and diel feeding rhythms. Where invertebrate predators detecting prey nonvisually are dominant, the phasing of migration may be reversed.

Vertical migratory behavior involves active swimming. Most copepods swim by rapid beating of the appendages, the antennae, mandibular palps, the maxillules, and the maxillae. In some species of planktonic calanoids, such as the genera *Metridia*, *Centropages*, and *Temora*, these movements result in a smooth continuous swimming behavior. In others, periods of active swimming are interspersed with inactivity when the animal sinks. This hop-and-sink behavior is characteristic of *Calanus finmarchicus*. Rapid jumping, often as apredator-avoidance behavior, involves strong strokes of the antennules and the swimming legs. This results in very rapid jumps, which propel the copepod several body lengths from the source of stimulus. The benthic harpacticoids and some cyclopoids crawl over or burrow through sediment. The thoracic limbs are used in crawling, and this is accompanied in harpacticoids, by sideways undulations of the body.

Nauplii use three pairs of appendages in swimming: the antenules, the antennae, and the mandibles. Three swimming behaviors have been recognized; a slow gliding motion propelled by the antennae and mandibles; a rapid darting behavior driven by all three pairs of appendages beating together; a cruise and pause behavior.

Swimming and feeding behaviors are interdependent in herbivores, omnivores, and carnivores. Swimming speeds of planktonic calanoids generally range from 1 to 20 mm s^{-1} which is equivalent to 1–5 body lengths per second. Estimates based on field studies of oceanic diel vertical migrators, such as *Pleuromamma*, range from 10 to 50 mm s^{-1}, representing the ability of such copepods to migrate at rates in excess of 100 m h^{-1}.

The predominant sensory mechanisms are mechanoreception and chemoreception, and receptor structures are found on the antennules. The antennules, particularly of males, are covered with sensory structures, aesthetascs, which are important in detecting water movement, food, predators, and potential mates. Detection of mechanical stimuli appears only to operate over short distances, often less than one body length. Chemoreception probably operates over longer distances and is involved in mate detection and response to food concentrations and to predators.

Many planktonic copepods are bioluminescent. The families Megacalanidae, Lucicutiidae, Heterohabdidae, Augaptilidae, and Metridinidae have luminescent glands that produce luminous glandular secretions. The number of light organs varies from 10 to 70, and they may be distributed widely over the body surface. The function of copepod bioluminescence is not certain. It may deter predators in the dark water column of the deep sea, and may act as a warning signal between individuals of the same species.

Life Histories

Copepods inhabit a wide range of environments, from the tropics to the Polar regions, and from the intertidal zone to the deep ocean, and their life histories accordingly vary considerably. In the tropics and subtropics there is no seasonality in breeding, most species breeding continuously. An exception to this pattern occurs in the upwelling system off the Gulf of Guinea, and in the Benguela Current off south-west Africa, where the dominant calanoid, *Calanoides carinatus* enters a diapause resting stage, at copepodite stage CV, at the end of the cold season and sinks to colder water until the next season. At high latitudes, diapause and overwintering strategies are the dominant responses to the highly seasonal environment (**Figure 14**). Breeding periods are restricted, often only one generation is produced each year, and growth and development rates are slowed. The most common diapause stage is copepodite CV.

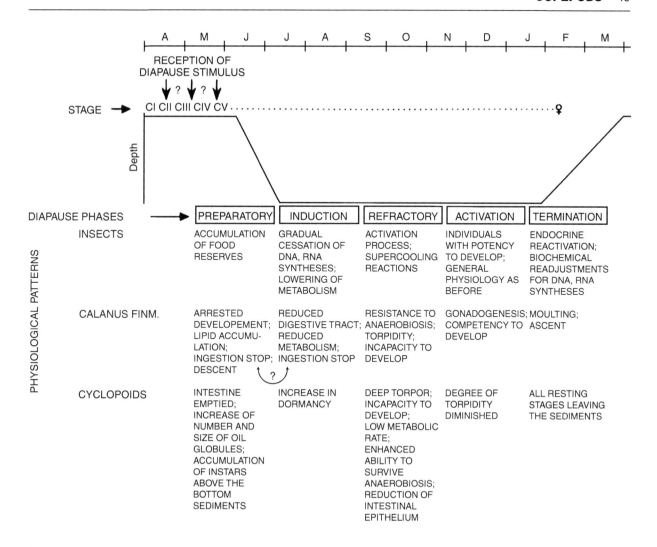

Figure 14 Generalized pattern of seasonal ontogenetic migration and physiological changes during overwintering of *Calanus finmarchicus* in relation to diapause phases in insects and comparison with insect and cyclopoid diapause. (Permission from Hirche H-J (1996) Diapause in the marine copepod, *Calanus finmarchicus* – a review. *Ophelia* 44: 129–143.)

Animals in this state overwinter in deep water with delayed development and reduced respiration and excretion rates, usually do not feed, and often show reduced digestive enzyme activity and changes in the digestive epithelium of the gut. Metabolism is sustained by the extensive lipid reserves, which in high-latitude copepods may exceed 75% of the total body weight, and these reserves give the body a brilliant red coloration in some species. Lipid stores can fuel egg laying in the spring before the spring phytoplankton bloom in species such as *Cala nus glacialis* and *Calanus hyperboreus*, ensuring that the resultant nauplii are able to exploit the spring pulse of phytoplankton production.

In the bathypelagic environment of the deepsea below 500 m, copepods do not undertake diel vertical migration, and there is reduced seasonality with depth. The life histories of deep sea copepods are relatively little known, but the majority probably breed continuously throughout the year, with slow development rates and long generation times, reflecting the low-temperature environment.

Copepods as Pests

Some copepods are economically important pests. An example is the salmon louse *Lepeophtheirus salmonis* (Krøyer), which may have significant impact on the economics of salmon aquaculture. The copepods breed rapidly in the high fish densities of the salmon cages and may kill the fish either directly or by causing skin damage that in turn makes the fish susceptible to disease. Gill-parasitic copepods such as *Lernaeocera* may have significant effects on commercial fish species, and the shell fish parasites such as *Myticola intestinalis* may also have economic effects.

Biogeochemical Role

The production of fecal pellets by copepodsis an important source of sedimented material for benthic organisms and plays a significant role in nutrient cycling and in vertical flux of biogenic elements to the deep ocean. Fecal pellet production rates of actively feeding copepods may exceed ten pellets per hour. Such rates, combined with the abundance of copepods in some ecosystems, mean that a significant component of the small particulate food captured is transformed into much larger packages represented by the fecal pellets. These may have sinking rates greater than $100\,\mathrm{md}^{-1}$, the rate being dependent both on size and composition, derived from the diet, of the pellets. Many of these rapidly sinking pellets may exit the surface layers of the ocean, and either reach the sea floor of the continental shelves or enter the bathypelagic zone. This pellet flux is sogreat that sinking pellets may form a significant part of the diet of other members of the plankton, including copepods, and of benthic organisms.

Role in the Ecosystem

Planktonic copepods, through their grazing activity, are one of the major controls on the growth of phytoplankton, and quantitative understanding of grazing processes is central to modeling marine ecosystem dynamics. Similarly, copepods play a pivotal role in nutrient cycles, by excreting dissolved nitrogen and phosphorus compounds, which are then utilized by phytoplankton to support growth, and hence primary production.

Pelagic cyclopoid and calanoid copepods form the first link in the marine food chain that leads from the single-celled plants of the phytoplankton to the fishes and marine mammals that form the exploitable living resources of the world's oceans. Nauplii through to adult stages of copepods are the typical food of nearly all larvae of commercially exploited marine fish. Some adult fish, such as herring, continue to feed on them. Similarly, the harpacticoid copepods of the meiofauna are a food source for bottom-feeding flatfish.

Copepods have been subjected to limited commercial exploitation. Although they are extremely abundant, their small size makes direct harvesting impractical. Limited fisheries for *Calanus* species, in areas of high coastal abundance, have provided dietary supplements for salmon aquaculture and for pet food.

See also

Fish Migration, Vertical. Fish Feeding and Foraging. Fish Larvae. Gelatinous Zooplankton. Meiobenthos. Plankton Overview.

Further Reading

Boxshall GA and Schminke HK (eds.) (1988) *Biology of Copepods*. Dordrecht: Kluwer.

Corner EDS and O'Hara SCM (eds.) (1986) *The Biological Chemistry of Marine Copepods*. Oxford: Clarendon Press.

Ferrari FD and Bradley BP (1994) *Ecology and Morphology of Copepods*. Dordrecht: Kluwer.

Gotto RV (1979) The association of copepods with marine invertebrates. *Advances in Marine Biology* 16: 1–109.

Hardy A (1956) *The Open Sea: Its Natural History, Part 1: The World of Plankton*. London: Collins.

Harris RP (ed.) (1995) Zooplankton Production. *ICES Journal of Marine Science* 52: 261–773.

Harris RP, Wiebe PH, Lenz J, Skjoldal HR, and Huntley M (eds.) (2000) *ICES Zooplankton Methodology Manual*. London: Academic Press.

Huys R and Boxshall GA (1991) *Copepod Evolution*. London: The Ray Society.

Kerfoot CW (1980) *Evolution and Ecology of Zooplankton Communities*. Hanover, NH: University Press of New England.

Marshall SM (1973) Respiration and feeding in marine copepods. *Advances in Marine Biology* 11: 57–120.

Marshall SM and Orr AP (1955) *The Biology of a Marine Copepod, Calanus finmarchicus (Gunnerus)*. London: Oliver and Boyd.

Mauchline J (1998) The biology of calanoid copepods. In: Blaxter JHS, Southward AJ, and Tyler PA (eds.) *Advances in Marine Biology*, 33, 1–710.

Raymont JEG (1983) *Plankton and Productivity in the Oceans*, 2nd edn; vol. 2, *Zooplankton*. Oxford: Pergamon Press.

GELATINOUS ZOOPLANKTON

L. P. Madin and G. R. Harbison, Woods Hole
Oceanographic Institution, Woods Hole, MA, USA

Introduction

Gelatinous zooplankton comprise a diverse group of organisms with jellylike tissues that contain a high percentage of water. They have representatives from practically all the major, and many of the minor phyla, ranging from protists to chordates. The fact that so many unrelated groups of animals have independently evolved similar body plans suggests that gelatinous organisms reflect the nature of the open ocean environment better than any other group. Whether as predators or grazing herbivores, they seem particularly well adapted to life in the oligotrophic regions of the world oceans, where their diversity and abundance relative to crustacean zooplankton is often greatest.

The gelatinous body plan has evolved in a world where physical parameters are relatively constant but food resources are sparse or unpredictable. Gelatinous zooplankton exhibit many common adaptations to this habitat.

- Transparent tissues provide concealment in the upper layers of the ocean, an environment without physical cover. Transparency is less common below the photic zone.
- The high water content of gelatinous tissues gives the organisms a density very close to that of sea water. The resulting neutral buoyancy decreases the energy required to maintain depth, but may actually require more energy overall, because of drag.
- The environment lacks physical barriers, strong turbulence, and current shears, so that gelatinous bodies do not need great structural strength. However, fragility makes many species difficult to sample or handle, and excludes most from more energetic coastal environments.
- High water content and noncellular jelly permit rapid growth and large body sizes, which can act as, or produce, large surfaces for the collection of food.
- Relatively large size makes gelatinous animals too big to be attacked by some predators, while their high water content reduces the food value of their tissues, which may also deter predation. Large size also permits commensal crustaceans to live on or in the body.

Thus, as we look at the diversity of gelatinous zooplankton, we should keep in mind the forces that have led to their remarkable convergence. It is impossible to deal in a short article with the entire range of phyla that have gelatinous representatives, so some of the major groups will be highlighted.

Taxonomic Groups

Radiolaria

Species of polycystine radiolarians form large gelatinous colonies up to several meters in length. Thousands of individual protists are embedded in a common gelatinous matrix from which their pseudopodia extend into the water. The combined efforts of individuals in the colony enable relatively large plankton (such as copepods) to be captured and ingested. In addition to the protistan members of the colony, the matrix also contains symbiotic dinoflagellates (zooxanthellae) that grow on the metabolites of the radiolarians. In turn, the radiolarians digest the zooxanthellae, so that these colonies are planktonic homologues of coral reefs.

Medusae

The phylum Cnidaria has many gelatinous representatives, comprising various groups of medusae and the strictly oceanic siphonophores (see below). What are commonly called jellyfish are medusae belonging to three Classes of the Cnidaria — the Hydrozoa, the Scyphozoa, and the Cubozoa. Since the morphology and life history of all three groups is broadly similar, it is practical to treat them together here. There are perhaps 1000 species of hydro- and scyphomedusae, probably with more to be discovered, especially in deep or polar waters. All are carnivorous, capturing prey with specialized stinging cells, called nematocysts. A wide variety of prey is eaten by different medusae, ranging from larval forms and small crustaceans to other gelatinous animals and large fish. Many epipelagic medusae also harbor zooxanthellae, and presumably they share their resources in the same way as the polycystine radiolarians. Many of these medusae are part of a life history that alternates between a sessile, benthic, asexually reproducing polyp and a sexually reproducing and dispersing planktonic medusa. However, many oceanic species have lost the polyp stage and evolved instead a variety of sexual and asexual reproductive mechanisms that do not require a benthic habitat. There are several classification schemes for Cnidarians; the group names given here are common usage, but these vary in different taxonomies.

Anthomedusae This order of hydromedusae includes small species ranging in size from less than 1 mm to several centimeters. The umbrella is usually shaped like a tall bell, and gonads are almost always found on the sides of the central stomach. There are four radial canals connecting the stomach to a marginal ring canal. Tentacles occur in varying numbers around the umbrella margin and sometimes around the mouth. Anthomedusae alternate with polyp forms, but some also bud medusae directly (**Figure 1A**).

Leptomedusae These medusae are generally flatter than a hemisphere. They usually have four radial canals, but sometimes eight or more, or canals that are branched. Gonads are located on the radial canals, and there may be various sense organs on the margin. The stomach is sometimes flat, and sometimes mounted on a peduncle that can be quite long. There are tentacles around the margin but not the mouth. Leptomedusae also alternate with hydroids, but some species produce new medusae by budding or fission (**Figure 1B**).

Limnomedusae Both high and low umbrella shapes are found in this order. There are usually four radial canals, sometimes branched. Gonads are either on the stomach or on the radial canals, and there is alternation of generations. Species of limnomedusae live in brackish, fresh (one species), or marine environments.

Trachymedusae These medusae in the order Trachylina do not alternate generations but develop young medusae directly from planula larvae, or by asexual budding. The umbrella is often high, with stiff mesoglea and well-developed muscle fibers. Most have eight unbranched radial canals and gonads located on them. Many trachymedusae live in deep water and are heavily pigmented (**Figure 1D**).

Narcomedusae Also in the Trachylina, narcomedusae have direct development from planulae, with a larval stage that is often parasitic on other medusae. There are no radial canals, but the flat central stomach is very wide and, in some genera, extends into radial stomach pouches. Tentacles are solid and stiff, and often extend aborally. Narcomedusae are common in epipelagic and mesopelagic environments; some are strong vertical migrators (**Figure 1C**).

Coronatae This order of scyphomedusae includes mainly deep-water species. The umbrella is divided into a high central part and a thinner marginal part by a coronal groove. The margin of the bell is divided into lappets; sense organs and solid tentacles arise from the cleft between lappets. The mouth has simple lips and the gastrovascular cavity is often deeply pigmented (**Figure 1G**).

Semaeostomae The familiar large jellyfish are mainly in this order of the Scyphozoa. The umbrella margin is divided into lappets, and bears sense organs and hollow tentacles. There is no coronal groove around the umbrella. The mouth opening is surrounded by four long oral arms, often frilled. Gonads are in folds of the subumbrella (**Figure 1E**).

Rhizostomae Medusae in this order of the Scyphozoa are mainly coastal species and can attain large size. They lack tentacles for prey capture, and instead ingest small particles carried into numerous small mouth openings by water currents. Some species in tropical waters host intracellular symbiotic algae.

Cubomedusae Medusae in the class Cubozoa also alternate with a benthic polyp form, although details of their life cycles are poorly known. Cubomedusae can be quite large, and have the most virulently toxic nematocysts of any Cnidarians. Some species are responsible for human fatalities. Cubomedusae are also unusual in possessing complex, image-forming eyes, which are not as well developed in other medusae (**Figure 1F**).

Siphonophores

The Order Siphonophora comprises a large and diverse group of predatory Cnidarians in the Class Hydrozoa. Their complex life cycles and colonial morphology are very different from the relatively simple hydromedusae and it is practical to consider the siphonophores as a separate group. The colonial, or polygastric, phase of the life cycle is the largest and most familiar. In this stage, siphonophores consist of an assemblage of medusoid and polypoid zooids, which are budded asexually from a founding larval polyp. The colony may include a gas float, nectophores or swimming bells, and a series of stem

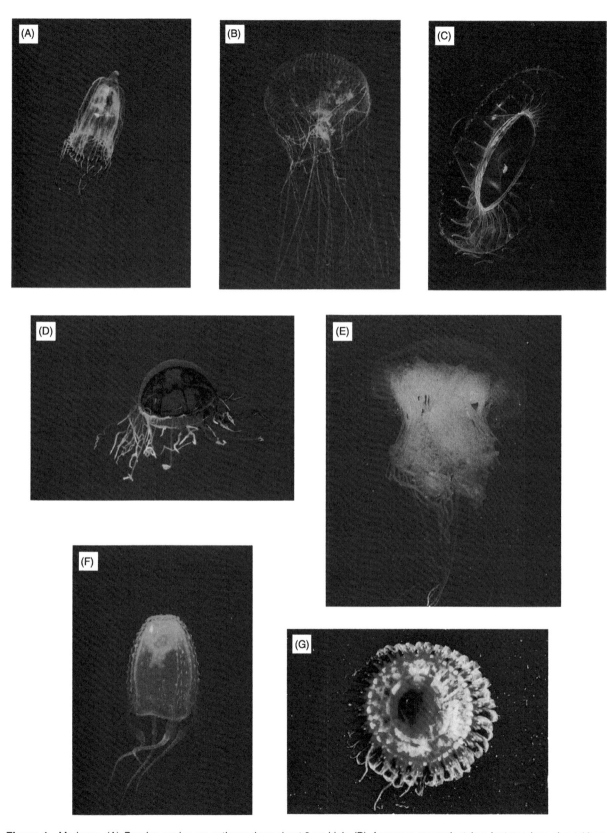

Figure 1 Medusae. (A) *Pandea conica*, an anthomedusa about 2 cm high. (B) *Aequorea macrodactyla*, a leptomedusa about 10 cm in diameter. (C) *Cunina globosa*, a narcomedusa about 5 cm in diameter. (D) *Benthocodon hyalinus*, a trachymedusa about 3 cm in diameter. (E) *Cyanea capillata*, a semaeostome scyphomedusa which can attain 1 m in diameter. (F) *Carybdea alata*, a cubomedusa up to 15 cm high. (G) *Atolla wyvillei*, a coronate scyphomedusa up to 25 cm diameter. (All photographs by L. P. Madin.)

groups made up of feeding polyps and tentacles. In some siphonophores the stem groups break off as dispersal and sexually reproductive stages called eudoxids. The colony can be thought of as an overgrown, polymorphic juvenile stage that eventually bears the sexually reproductive adults. These are small medusoid zooids called gonophores, which produce gametes. Siphonophores range in size from a few millimeters to over 30 m in length, and occur throughout the water column. All are predators on other small zooplankton, and many genera are known to be luminescent.

The colonies are fragile, and usually break up into their various units when collected in plankton nets. For this reason, much of the taxonomy is based on the morphology of the pieces, principally nectophores, and the appearance of the intact colonial stage is not always known. The Order Siphonophora is divided into three suborders and 15 families.

Cystonectae This suborder includes siphonophores that possess a float but no swimming bells, so they are at the mercy of ocean currents. The Portuguese man-of-war is the most familiar example. It has a float so large that the animal rests on the surface, but most cystonect species have smaller floats and are wholly submerged. Cystonects have virulent nematocysts and capture large, soft-bodied prey such as fish and squids (**Figure 2A**).

Physonectae These siphonophores have more complex colonies, comprising a small apical float, numerous swimming bells that form a nectosome, and a stem containing several groups of gastrozooids, tentacles, bracts, etc. The stem typically contracts when the animal is swimming, and then relaxes so that the stem and tentacles extend to maximum length for fishing. This group is a major contributor to the deep scattering layer in many regions of the ocean. The largest siphonophores (the Apolemiidae, over 30 m long) are found in this group. Physonects prey mainly on small zooplankton, and many species are strong swimmers and vertical migrators. (**Figure 2C**).

Calycophorae In this group, which contains the largest number of species, the float is absent and the nectophores are reduced, usually to two. A sequence of stem groups is budded and breaks free as eudoxids. Calycophorans are the most diverse, widely distributed, and abundant siphonophores. They catch small zooplankton and, when feeding, their tentacles form complex three-dimensional structures in the water, reminiscent of spider webs (**Figure 2B**).

Ctenophores

Ctenophores are exclusively marine gelatinous animals all but a few of which are holoplanktonic. Although they superficially resemble the Cnidaria, morphological and molecular studies indicate that Cnidarians and ctenophores are not closely related. Ctenophores are predators that use tentacles equipped with 'glue' cells or colloblasts to capture prey. The name 'ctenophore' is Greek for 'comb bearer,' referring to the comb-like plates of fused cilia that are used for propulsion. All ctenophores initially have eight meridional rows of comb plates, although in some groups these are lost or reduced during development. The vast majority of ctenophore species fall into six orders.

Cydippida This group contains many species with paired tentacles that exit the body through tentacle sheaths. Species in the family Pleurobrachiidae catch prey ranging from small crustaceans to fish, while members of the Lampeidae feed mainly on large gelatinous animals like salps. The members of one species of cydippid, *Haeckelia rubra*, eat medusae, and retain the nematocysts of their prey ('kleptocnidae') for defensive use in their own tentacles. Before this behavior was known, these nematocysts were considered strong evidence for a close relationship between Cnidarians and ctenophores (**Figure 2F**).

Platyctenida This group is primarily benthic and is distributed widely from the Arctic to the Antarctic. Members of the family Ctenoplanidae have comb rows as adults and are found in the plankton in the Indo-Pacific; all other species in the order lose their comb rows as adults, and live primarily as creeping benthic organisms. Platyctenes have functional tentacles that capture prey.

Thalassocalycida This order contains a single species, *Thalassocalyce inconstans*, which lives in the midwater zone. It superficially resembles a medusa in overall shape, but can easily be distinguished by its eight comb rows and paired tentacles.

Lobata Members of this order all have oral lobes and auricles, specialized structures that are used in feeding. Most lobates move through the water with their oral lobes widely spread to form a sort of basket. Small prey, such as crustaceans, are trapped on the mucus-covered oral lobes and tentilla, which stream over the body or extend onto the oral lobes. Ctenophores in the family Ocyropsidae lack

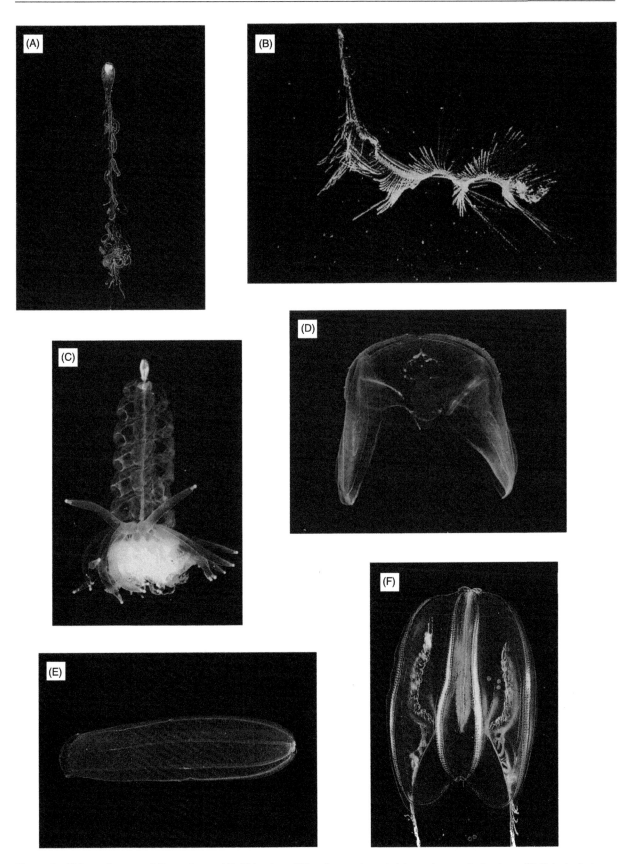

Figure 2 Siphonophores and Ctenophores. (A) *Rhizophysa filiformis*, a cystonect siphonophore up to 2 m long. (B) *Sulculeolaria* sp. a calycophoran siphonophore up to 1 m long. (C) *Physophora hydrostatica*, a physonect siphonophore about 10 cm high. (D) *Ocyropsis maculata*, a lobate ctenophore about 8 cm in diameter. (E) *Beroe cucumis*, a beroid ctenophore up to 25 cm long. (F) *Mertensia ovum*, a cydippid ctenophore about 4 cm high. (Photographs A, C, D, F by G. R. Harbison; B, E by L. P. Madin.)

tentacles, and capture prey by enclosing them in their muscular oral lobes (**Figure 2D**).

Cestida These ctenophores are shaped like long, flat belts. They appear to be related to the Lobata, but lack oral lobes and auricles. There are only two genera (*Cestum* and *Velamen*) in one family (Cestidae). The comb rows extend along the aboral edge of the ribbonlike body, propelling the animal with the oral edge forward. Small prey are captured by the fine branches of the tentacles that lie over the flat sides of the body. Cestids are characteristic of oceanic, epipelagic environments.

Beroida Beroids lack tentacles altogether. Their large stomodaeum occupies most of the space in the body. All beroids are predators on other ctenophores, and occasionally salps. They capture prey by engulfing them, and can bite off pieces of the prey with specialized macrocilia located immediately behind the mouth (**Figure 2E**).

Heteropods

The Phylum Mollusca contains many gelatinous representatives, and the gelatinous body plan has apparently arisen independently in several groups. The Heteropoda is a superfamily of prosobranch gastropods that includes the families Atlantidae, Carinariidae, and Pterotracheidae. Heteropods are visual predators with well-developed eyes and a long proboscis containing a radula. Atlantid heteropods have thin, flattened shells into which they can completely withdraw their bodies. They feed on small crustaceans and other mollusks. The family Carinariidae includes eight species in three genera, *Carinaria*, *Pterosoma*, and *Cardiapoda*. These heteropods have a greatly reduced shell, enclosing only a small fraction of the body. Carinariids feed primarily on other gelatinous organisms, such as salps, doliolids and chaetognaths. In the family Pterotracheidae, with two genera — *Pterotrachea* (four species) and *Firoloida* (one species) — the shell is completely absent (**Figure 3D**).

Pteropods

This molluskan group comprises two orders in the gastropod subclass Opisthobranchia. The foot in pteropods is modified into two wingshaped paddles responsible for swimming; their fluttering gives rise to the common name for pteropods, sea butterflies.

Thecosomata This group contains the shelled pteropods, some of which (Euthocosomata) have calcareous shells and are not truly gelatinous, and others of which (Pseudothecosomata) have gelatinous shells and tissues. There are over 30 species of euthecosome pteropods, in two families, the Limacinidae and the Cavoliniidae. Thecosome pteropods feed by collecting particulate food on the surface of a mucous web or bubble, produced by mucus glands on the wings, and held above the neutrally buoyant and motionless animal. The mucus is periodically retrieved and ingested along with adhering particles, then replaced by a newly secreted web. Some cavoliniids have brightly colored mantle appendages that may aid in maintaining neutral buoyancy or serve as warning devices to predators. When disturbed, animals lose their neutral buoyancy and rapidly sink (**Figure 3A**).

The Pseudothecosomata includes three families, the Peraclididae (one genus), the Cymbuliidae (three genera), and the Desmopteridae (one genus). Pseudothecosomes are larger than euthecosomes, and their mucous webs are correspondingly larger, reaching over a meter across in *Gleba cordata* (**Figure 3C**).

Gymnosomata Members of this order are poorly known, largely because they have no shells and contract into shapeless masses when preserved. Most species live in the deep sea, and only a few of the approximately 50 species have been observed alive. Gymnosomes appear to be highly specialized predators on particular species of thecosome pteropods. The order is divided into two suborders, the Gymnosomata and the Gymnoptera. The four families of the Gymnosomata include the Pneumodermatidae (seven genera and 22 species) with sucker-bearing arms similar to cephalopod tentacles; the Notobranchaeidae (one genus and eight species), with suckerless feeding arms called buccal cones; the Clionidae (eight genera and 16 species); and the Cliopsidae (two genera and three species) (**Figure 3B**).

There are two families in the Gymnoptera, the Hydromylidae (one genus, one species) and the Laginiopsidae (one genus, one species). These groups are very different from each other and from other gymnosome pteropods, and some may not actually belong in the Order Gymnosomata.

Cephalopods

Although many cephalopods are active, muscular swimmers, there are several gelatinous and/or transparent genera. The family Cranchiidae is composed entirely of gelatinous species, including the genera *Taonius*, *Megalocranchia*, and *Teuthowenia*. These relatively large, slow-moving squids probably

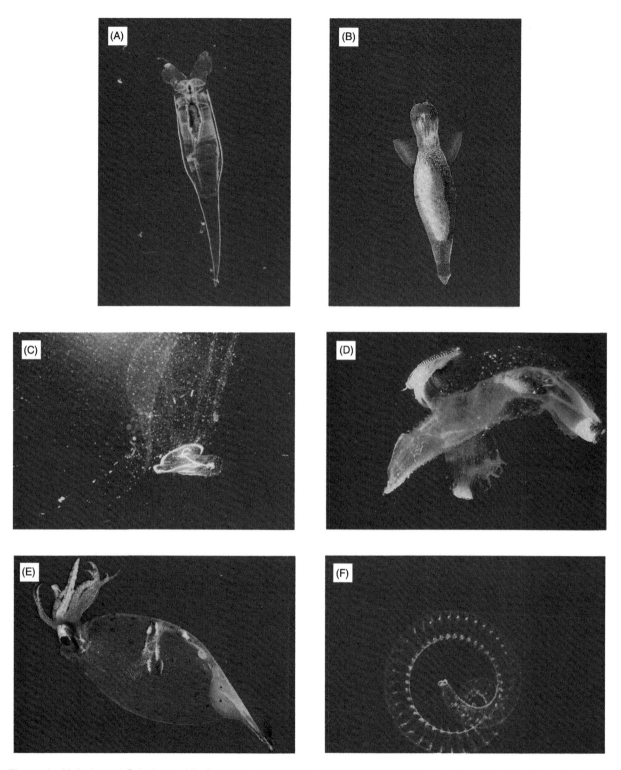

Figure 3 Mollusks and Polychaete. (A) *Cuvierina columnella*, a euthecosome pteropod about 1 cm high. (B) *Clione limacina*, a gymnosome pteropod about 2 cm high. (C) *Corolla spectabilis*, a pseudothecosome pteropod about 10 cm in diameter, with mucous web in background. (D) *Carinaria* sp. a heteropod about 10 cm long. (E) *Teuthowenia megalops*, a cranchiid cephalopod about 10 cm long. (F) Alciopid polychaete worm, up to 1 m long. (Photographs A, B, C, E, F by G. R. Harbison; D by L. P. Madin.)

capture prey through stealth rather than active pursuit. Vitreledonelliid octopods are also gelatinous (**Figure 3E**).

Polychaete Worms

Two major groups of planktonic polychaetes are gelatinous, the Alciopidae and the Tomopteridae. Both are in the order Phyllodocida, although they are probably not closely related. Alciopids are characterized by well-developed eyes with lenses. Many have ink glands along the sides of their bodies, which may function analogously to the ink glands of cephalopods. Their habits are poorly known, but they may feed on gelatinous prey. Alciopids may attain lengths of nearly a meter. Tomopterids do not have well-developed eyes, but probably use chemoreception to locate prey. Some release luminous secretions from glands along their body when disturbed. Deep-sea tomopterids may be 25 cm long, but most shallow species are much smaller. (**Figure 3F**).

Crustaceans

Although arthropods cannot really be considered gelatinous because of their exoskeletons, there are some examples of very transparent bodies, presumably also an adaptation for concealment. The most notable examples are species of the hyperiid amphipods *Cystisoma* and *Phronima*. Species of *Cystisoma* are large and transparent, and the enormous retinas of the compound eye are lightly tinted. Although the retinas of species of *Phronima* are darkly pigmented, the rest of the body is transparent. It is likely that the transparency of these species is a form of protective coloration, since they live on transparent gelatinous hosts.

Holothurians

Although the majority of holuthurian species are rather sedentary benthic deposit feeders, there are several deep-sea genera of swimming or drifting holothurians with gelatinous bodies. Species in the genera *Peniagone* and *Enypniastes* feed on bottom deposits, but can swim up into the water column. The genus *Pelagothuria* appears to be wholly pelagic, with a morphology that suggests it collects and feeds on sinking particulate matter. Few pelagic holothurians have been observed alive and little is known of their life history or behavior.

Pelagic Tunicates

The subphylum Urochordata includes two classes of pelagic tunicates, the Thaliacea and the Larvacea or Appendicularia. Thaliaceans (including the orders Pyrosomida, Doliolida, and Salpida) are relatively large animals with more or less barrel-shaped bodies. They pump a current of water through their bodies and strain phytoplankton and other small particulates from it with a filter made of mucous strands. The same current provides jet propulsion. Thaliaceans have complex life cycles with alternating generations and multiple zooid types. The class Larvacea comprises a single order of small organisms that filter food particles using an external mucous structure called a house. Both Thaliaceans and Larvaceans are widely distributed in the oceans, and are sometimes extremely abundant.

Pyrosomida Pyrosomes form colonies made up of numerous small ascidian-like zooids embedded in a stiff gelatinous matrix or tunic. The colony is tubular, with a single terminal opening. Water is pumped by ciliary action through each zooid, and suspended food particles are retained on the branchial basket within the body. The excurrent water from each zooid passes into the lumen of the colony, forming a single exhalent current that provides jet propulsion. Most pyrosome colonies are a few centimeters to a meter in length, but colonies of at least one species can attain lengths of 20 m.

Doliolida This order of the Thaliacea comprises six genera and 23 species of small (2–10 mm), barrel-shaped animals with circumferential muscle bands. The filter feeding mechanism is similar to that of pyrosomes, with currents generated by ciliary beating passing through a mucous net supported on the branchial basket. The life cycle involves five asexual stages and one sexual stage, several of which occur together as parts of large colonies of thousands of zooids. These colonies may attain lengths over 1 m, but are fragile and are rarely collected intact. In most genera of doliolids, the life cycle begins with a sexually produced larva, which becomes the oozooid stage. This stage feeds initially, but then begins budding off the trophozooid and phorozooid stages, thus forming the colony. During this process the oozooid loses its branchial basket and gut, and transforms into the 'old nurse' stage, whose function is to swim by jet propulsion and pull the attached colony along behind it. Contractions of the body muscles produce short exhalent pulses that move the colony rapidly. The trophozooids in the colony filter-feed to support themselves and the nurse. The phorozooids grow attached to the colony, but then break free to lead independent lives and produce asexually a small group of gonozooids. These eventually break free from the phorozooid, and become the sexually reproducing stages (hermaphrodites?) that produce the larvae and begin the whole cycle again (**Figure 4B**).

Salpida This order (with 12 genera and about 40 species) is of larger filter-feeding animals, also with circumferential muscle bands. The salps alternate between two forms, an asexually budding solitary (oozooid) stage and a sexually reproducing aggregate (blastozooid) stage. The aggregate salps usually remain connected together in chains or whorls of various types. Swimming is by jet propulsion, produced by a pulsed water current generated by rhythmic contraction of body muscles.

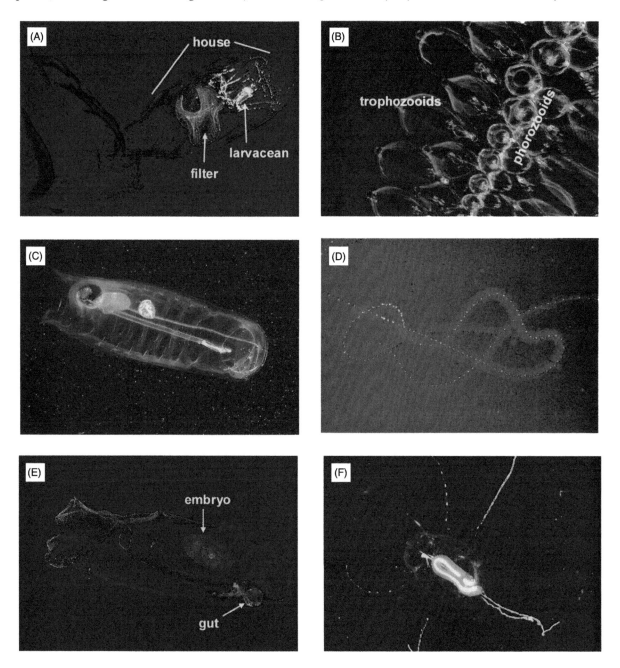

Figure 4 Pelagic Tunicates. (A) *Megalocercus huxleyi*, a larvacean of about 5 mm body length, house length about 4 cm. (B) *Dolioletta gegenbauri*, portion of a colony showing gastrozooids and phorozooids, individuals 2–5 mm long, colonies up to 1 m. (C) *Salpa maxima*, solitary generation salp, up to 25 cm long. (D) *Salpa maxima*, chain of aggregate generation salps; orange dots are guts of salps; individuals are to 15 cm, chains up to 10 m long. (E) *Pegea socia*, aggregate generation salp with attached embryo of solitary generation; aggregate 7 cm, embryo about 1 cm. (F) *Traustedtia multitentaculata*, solitary generation salp with appendages of uncertain function, about 3 cm long. (All photographs by L. P. Madin.)

Food particles are strained from the water passing through the body cavity by a mucous filter, which is continuously secreted and ingested. The individual animals range in size from 5 to over 100 mm, and chains can be several meters long (**Figures 4C–F**).

Larvacea This class (also called Appendicularia) is divided into three families (with 15 genera and 70 species) of small (1–10 mm) animals consisting of a trunk and long, flat tail. Larveaceans are also filter feeders on small particulates but are unique among tunicates in the use of an external concentrating and filtering structure called the house. The house surrounds the animal, and contains a complex set of channels and filters made of mucous fibers and sheets. Water is pumped into the house by the oscillation of the larvacean's tail; the exhalent stream provides slow jet propulsion in some species. Particles are sieved from the flow as it passes through the internal filter; they accumulate and are aspirated at intervals into the pharynx of the larvacean via a mucous tube. The complex house is formed as a mucous secretion on the body of the larvacean, produced by specialized secretory cells. It is inflated with sea water, pumped into it by action of the tail, until it attains its full size, with all the internal structures expanding in proportion. Houses eventually become clogged with particulates and fecal pellets, and are then jettisoned. The larvacean expands a new house (there may be several house rudiments on its body, awaiting expansion) and resumes filter feeding. The abandoned houses can be an important source of marine snow and serve as food for various planktonic scavengers (**Figure 4A**).

Ecology of Gelatinous Zooplankton

Gelatinous zooplankton are found in all of the oceans of the world, from the tropics to polar regions. They also occur at all depths, and many of the largest and most delicate species have been collected in recent years from the mesopelagic and bathypelagic parts of the ocean. The absence of turbulence in the deep sea probably allows these species to attain such large sizes, but there are also robust species that thrive in surface and coastal environments. Examples include the Portuguese man-of-war (*Physalia physalis*), which lives at the air–water interface and can ride out hurricanes, and the ctenophore *Mnemiopsis leidyi*, which lives in estuaries with strong tidal currents and turbulence.

In general, gelatinous organisms have been rather neglected by zooplankton ecologists, primarily because their delicacy makes them difficult to sample and study. Most are damaged or destroyed in conventional plankton nets, and many deep-water siphonophores and ctenophores are too delicate to be captured intact even with the most gentle of techniques. Much recent progress in understanding their biology has been based on *in situ* methods of study using SCUBA diving, submersibles, or remote vehicles. These methods permit observation of undisturbed behavior and collection of intact living specimens. Advances in culture techniques and laboratory measurements have improved our understanding of energetics, reproduction, and life history of some species, but most remain only partially understood.

Gelatinous animals occupy every trophic niche, ranging from primary producers (symbiotic colonial radiolaria) to grazers (pteropods and pelagic tunicates) and predators (medusae, siphonophores, and ctenophores). In all these niches, the gelatinous body plan confers advantages of size and low metabolic costs. In addition to attaining large sizes with relatively little food input, gelatinous organisms such as medusa and ctenophores are able to 'de-grow' when deprived of food. Metabolic rates remain unchanged, and the animal simply shrinks until higher food levels allow it to resume growth. This energetic flexibility is probably important to the success of gelatinous species in the oligotrophic open ocean and deep sea. Many species of medusae and siphonophores, for example, appear able to survive at low population densities spread over very large areas.

In other cases the efficiency of their feeding, growth, and reproduction allows gelatinous species to outcompete other types of zooplankton and form dense populations over large areas, which can have considerable impact on ecosystems. A dramatic recent example was the accidental introduction of the ctenophore *Mnemiopsis leidyi* into the Black Sea from the eastern seaboard of the Americas. In the late 1980s, this ctenophore reproduced in prodigious quantities, and the resulting predation on zooplankton and larval fishes led to the collapse of pelagic fisheries in the Black Sea. These fisheries have to some extent recovered, but seasonal blooms of this ctenophore continue to occur in the Black Sea, just as they do on the eastern shores of the Americas. Reports of *Mnemiopsis* in the Caspian Sea suggest that the pattern may be repeated.

Many other gelatinous species form dramatic seasonal blooms, such as the medusa *Chrysaora quinquecirrha* in the Chesapeake Bay, the salp *Thalia democratica* off Florida, Georgia, and Australia, and the medusa *Pelagia noctiluca* in the Mediterranean. In the Southern Ocean, immense populations of the salp *Salpa thompsoni* alternate with those of the

Antarctic krill *Euphausia superba*. The formation of large aggregations through rapid reproduction appears to be a common strategy for taking advantage of favorable conditions. Dense populations are sometimes further concentrated by wind or current action, or are transported close to the coast from their normal habitats farther offshore. The combination of rapid growth and advection can cause the sudden appearance of swarms of medusae, ctenophores, or salps in coastal waters. Although these blooms may sometimes have serious or even catastrophic effects on other organisms, including fisheries or human activities, they are a natural part of the life histories of the species, and not events for which remedial action is needed, or even possible. Gelatinous zooplankton are normal components of virtually all planktonic ecosystems. They are among the most common and typical animals in the oceans, whose biology and ecological roles are now becoming better understood.

See also

Plankton Overview.

Further Reading

Bone Q (1998) *The Biology of Pelagic Tunicates*. Oxford: Oxford University Press.

Harbison GR and Madin MP (1982) The Ctenophora. In: Parker SB (ed.) *Synopsis and Classification of Living Organisms*. New York: McGraw Hill.

Lalli CM and Gilmer RW (1989) *Pelagic Snails*. Stanford, CA: Stanford University Press.

Mackie GO, Pugh PR, and Purcell JE (1987) Siphonophore biology. *Advances in Marine Biology* 24: 98–263.

Needler-Arai M (1996) *A functional biology of Scyphozoa*. London: Chapman and Hall.

Wrobel D and Mills CE (1998) *Pacific Coast Pelagic Invertebrates: A Guide to the Common Gelatinous Animals*. Monterey: Sea Challengers and Monterey Bay Aquarium.

KRILL

E. J. Murphy, British Antarctic Survey, Marine Life
Sciences Division, Cambridge, UK

Introduction

Krill play a major role in the transfer of energy
in marine food webs, being important consumers
of phytoplankton and other zooplankton, and prey
of many higher trophic level predators that are
often commercially important. The importance of
krill in the diet of marine predators is reflected
in their name; 'krill' comes from the Norwegian
whaler's description of the larger food of the great
whales. Krill form an order within the Crustacea,
the Euphausiacea, which comprises over 80 species
in 10 genera. Detailed keys are available to identify
individual species that have broadly similar body
pattern (**Figure 1**). The euphausiids occur in a wide
range of habitats – coastal, oceanic, and deep-
oceanregions – and their distributions also extend
into the ice-covered regions of the Arctic and the
Antarctic. Krill are generally more abundant in
higher latitudes and can occur in such large numbers
near the surface that they discolor the water.

The phylogenetic relationships of many of the
euphausiids are unknown, but for some of the
key oceanic species their evolutionary development
appears to have been associated with the formation
of the major circulation patterns of the world's
oceans. This link to large-scale ocean circulation
patterns is also reflected in the population distributions and life histories of the euphausiids. Many of
the oceanic krill species occur over broad regions
in which the centers of the populations tend to be
associated with restricted features of the ocean circulation. However, the patterns of flow often result
in transport of krill out of their main breeding
regions to areas where they do not breed successfully.
This also appears to be crucial to their role in many
food webs, providing energy input into regions
remote from their own main areas of production.
The observation that krill are often transported into
regions where they do not reproduce also highlights
the colonization potential of the group should any
changes occur in patterns of ocean circulation.

There are several features that mark the euphausiids as unusual plankton. A number of species are
relatively large with a long life span compared with
other zooplankton. The largest of the krill grow to
over 60 mm and can live for more than 5 years.
Another key feature is that in a number of the species
the individuals form dense aggregations known ass-
warms. In some of the larger euphausiids these
swarms might more appropriately be thought of as
schools, similar to those formed by small fish, where
members of the aggregation are aligned and show
coherent patterns of behavior.

In the Antarctic the term 'krill' is often used to
denote a single species: the Antarctic krill, *Euphausia
superba* Dana (**Figure 1A**). This is, as its name suggests, the 'superb' krill that is large in size, occurs in
vast numbers in the Southern Ocean, and is central
to the Antarctic food web. It is the food of not only
the now greatly depleted populations of whales
but also many of the seals, penguins and other sea
birds, and of fish and squid. It is the most studied
species and much of the available information on
euphausiids in general is based on knowledge of the
Antarctic krill, so it is important to remember that
this is something of an extreme representative of the
group.

A number of the euphausiids have been exploited
in fisheries. As krill are typically a low trophic level
species there has been recognition of the potential
impact this could have on the higher trophic levels of
marine food webs. The pivotal role of krill in marine
food webs has meant that, particularly in the Antarctic, an ecosystem approach to the management of
krill fisheries is being developed that has relevance to
the sustainable management of marine ecosystems
globally.

Species Separation and Geographical Distributions

Euphausiids are found throughout the oceans of the
world, but their distributions highlight marked differences in habitat and life history amongst apparently similar species. There is a continuing debate
about the exact number of species of euphausiids and
the degree of separation of subgroups. There are
indications from evolutionary studies of mitochondrial DNA that vicariant speciation (separation by
formation of a natural barrier) has been important in
the development of euphausiid species in the Antarctic. The generation of the Antarctic Polar Front
about 25–22 Ma probably led to the separation of

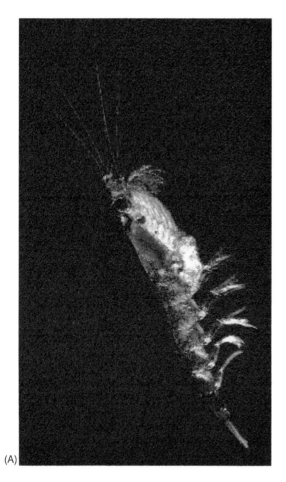

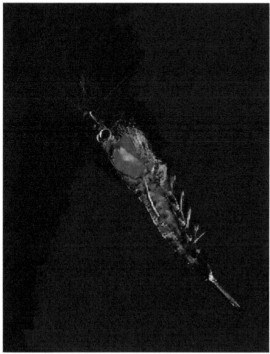

Figure 1 Two krill species: (A) the Antarctic krill, *Euphausia superba* and (B) a North Atlantic krill, *Meganyctiphanes norvegica.*

the 'Antarctic clade' (*E. superba* and *E. crystallorophias*) from the sister clade of *E. vallentini* and *E. frigida* dated at about 20 Ma.

Although some euphausiid species occur in coastal and bathypelagic regions (1000–2500 m), most are found in oceanic epipelagic (0–200 m) and mesopelagic regions (200–1000 m). Although broad distributions have been described for many of the euphausiids, because these animals frequently occur in only relatively low numbers their local distributions are often not well defined. Generally, there is a trend of increasing abundance of krill at higher latitudes. However, there are variations in this pattern, with strong links between the ocean current systems and the regional distribution of krill species.

A feature of the euphausiids is that in Southern Ocean and Southern Hemisphere regions many of the key species occur across the full longitudinal range (**Figure 2**). In the Southern Ocean the ocean circulation is circumpolar, so the same basic pattern of species distribution is found throughout the connected ocean. The key species in the mainly ice-covered regions is *E. crystallorophias* which inhabits the Antarctic continental shelf, although on occasion it has been found transported northward by the major current flows. Further to the north in the seasonally ice-covered areas of the main flow regions of the Antarctic Circumpolar Current are *E. superba* and *E. frigida*, with *Thysanossa vicina* and *T. macrura* extending northwards to the Antarctic Polar Front. All of these species have heterogeneous distributions in the region. For *E. superba* there appear to be centers of population in which they can spawn and reproduce successfully, separated by and possibly connected through, regions that are not favorable to breeding but in which krill are found (**Figure 3**). *E. triacantha* overlaps the northern limit of *E. superba* in the south and in the north it overlaps the southern limit of the range of *E. vallentini* extending north to south of 40°S. Further north still are less abundant species such as *E. longirostris* and *E. lucens* that extend north of 40°S in areas encompassed by the eastward flows in the southern regions of the main ocean gyres of the Pacific, Atlantic, and Indian Oceans. To the north of this *E. similis* occurs in all three ocean basin regions, extending from about 50–60°S to 30°S, but the species is also present further north in the north-west region of the Indian Ocean to the north of Madagascar. Across the subtropical and tropical regions there is a wide range of species. One that occurs in all the ocean basins is *E. tenera*, where it has a wide distribution but is not very abundant in any region.

There are a number of other species found in both the Atlantic and Pacific Oceans. Some species,

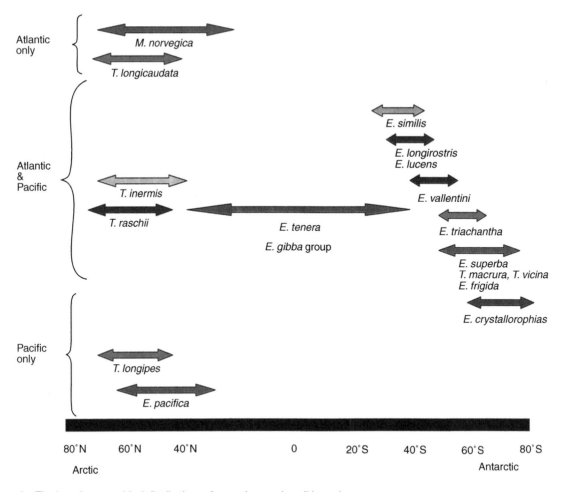

Figure 2 The broad geographical distributions of some key euphausiid species.

particularly in the central North Pacific and North Atlantic, are abundant but only found in one of the ocean basins. Key species that show this pattern are *E. pacifica* and *T. longipes/T. inspinata* that occur only in the northern North Pacific, while in the North Atlantic *Meganyctiphanes norvegica* and *T. longicaudata* are dominant (**Figure 2**).

In the northern North Atlantic and North Pacific there are species that occur in both oceans and through into the Arctic regions in the far north. In particular there are two important species, *T. raschii* and *T. inermis*, with distributions extending from about 45°N to about 80°N, although breeding is largely restricted to areas south of 70°N.

As well as geographical differences there are also marked differences in vertical distribution and many of the species show some form of vertical migration. For example, *E. pacifica* occurs mainly above 300 m during the day, moving nearer the surface (<150m) at night, while *M. norvegica* occurs between 100 and 500 m during the day and vertically migrates to shallower depths at night, and in the south

E. superba occurs mainly above about 250 m and migrates nearer the surface at night.

Growth, Development, Physiology

Krill species show a range of development strategies that vary between species, and also with the environmental conditions to which they are exposed. Studies of krill population dynamics and development are made difficult because of problems in determining the age of a number of the species. Traditional techniques involving the analysis of the population age structure are still relied upon and a range of mathematical techniques have been employed to distinguish different cohorts in length–frequency size distributions. These are not always definitive and a range of other techniques has been explored such as using age pigment analyses, multiple-morphometric analyses, analyses of structures in the eye, and laboratory maintenance of live specimens. None of these techniques has so far

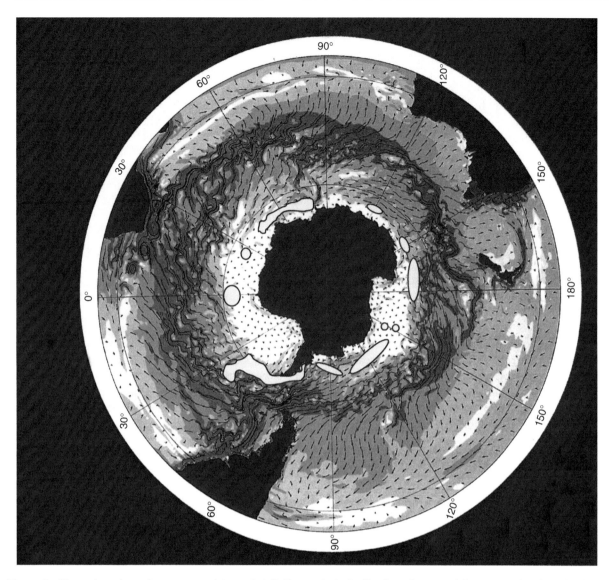

Figure 3 The main regions of occurrence of Antarctic krill *E. superba* in the Southern Ocean and the pattern of surface circulation fromthe FRAM model (FRAM Group).

provided a good and practical solution to determining the age of krill. However, there is general agreement about the broad characteristics of growthand development of many of the key species.

In the Southern Ocean, early studies of *E. superba* indicated a 2–3 year life cycle based mainly on samplesfrom open-ocean regions. However, further detailed analyses of the size-structure and development of *E. superba* populations have led to a revision in the life-span up to >5 years with suggestions that in some areas 5–7 year classes can be identified. Laboratory experiments have maintained krill obtained from the sea for >6 years, indicating that a total age of 7–8 years is probably possible in the wild. The development and growth of the krill will depend on the conditions to which they are exposed. *E. superba* can reach a size of >60 mm with

indications that growth may be very plastic, as in other euphausiid species, varying with the environmental conditions. Thus, krill in more northern and warmer regions may grow more rapidly and develop earlier than krill further south. In these more northern regions, such as around the Island of South Georgia which lies at about 54°S, near the Antarctic Polar Front, the krill do not reproduce successfully, with few indications of any viable larvae being found in thearea. The *E. superba* population in these regions is probably maintained by advection inputs from further south in the Southern Scotia Sea, Weddell Searegion, and from around the Antarctic Peninsula.

Such a plastic range of development is illustrated clearly in the northern species *T. inermis* and *T. raschii*. These species have a maximum age of

1 year at the south of their distribution (about 45°N), where as further north they survive to over 2 years old, spawning in each year, although females may not mature until their second year. Continuing northward, the maximum age increases and maturation is delayed further with spawning delayed to year 3, and a maximum age of 3 years. In the high Arctic waters the krill still mature but do not spawn and it is the water circulation bringing krill from further south that maintains the species in these areas. Across this range T. inermis grows to over 20 mm, but the rates involved vary depending on the conditions, with slower growth and development occurring further north in their range.

M. norvegica is one of the most abundant North Atlantic krill and individuals can reach a maximum size of over 45 mm in some regions. This species shows less age variation across its range than the more northern Thysanossa species, but the variation is still significant. In the south of its range individuals live up to about 1 year and spawn only once, whereas further north they reach over 2 years of age, spawning more than once. Like T. inermis, this species does not spawn in the extreme northern part of its range, so advection in the current systems is again important in maintaining the distribution.

In the Pacific E. pacifica also shows this plastic character of changing maximum age with environmental variation. At the southern limit of its range individuals have a very short life span of only 6–8 months, whereas further north the maximum age is extended to about 15–21 months. In the most northern parts of the range the krill survive to over 2 years old and probably spawn twice. The maximum size across the range is about 20–22 mm, but growth is slower in the regions further north.

As well as these general changes in development and life span, there are also sex-related differences. For example, in the northern Thysanossa species the males mature at just over 1 year old while the females mature mainly at over 2 years of age. In E. pacifica both sexes mature and spawn at 1 year old, but females may continue to survive and spawn at over 2 years old. In E. superba the situation can be different, with females spawning and maturing earlier at 2 years old, while males may not mature until over 3 years old.

The euphausiids have the potential for rapid growth and development under suitable conditions, moulting as they increase in size. So, for example, E. superba has an energy input of perhaps 20% of body carbon per day or greater, sustained by a high and effective rate of filtration. This level of energy input can result in growth rates of $>0.1 \, \text{mm} \, \text{d}^{-1}$,

particularly for the younger age groups. Krill have the capacity for a large and sustained reproductive output under good conditions, with continuous or multiple spawning occurring through the season in some species.

A key question for many of the species is how they survive during winter when food appears scarce, particularly in the extremely seasonal environments of the polar oceans. Studies of polar species show that the krill utilize stored lipids as a major energy source, but the dynamics of storage and utilization vary greatly between species. The lipids are accumulated primarily for winter survival or reproduction, but they may also provide a small degree of buoyancy that may help reduce the costs of swimming.

In the Southern Ocean, the diet of E. superba varies with age. Phytoplankton sources are important for the early stages while older groups utilize more animal-based food sources or detritus. Lipids are utilized in winter, but a strong seasonal bloom of production is necessary for reproduction. For E. superba the suggestion is that winter survival is dependent not only on reduced metabolic rate, a potential reduction in size, and use of lipids, but also on the use of alternative food sources. Antarctic krill have been observed to get smaller during poor feeding conditions in the laboratory, but it is unclear how much this occurs in the ocean. Larval E. superba are dependent on sea ice as a habitat and the observation of krill grazing algae associated with the ice indicates that they can utilize this as an alternative food source. It remains unclear how important sea ice algae are for maintaining adult E. superba and this is likely to be a variable contribution to the diet depending on opportunity of access to the right feeding conditions. E. superba are also known to graze other components of the plankton, including copepods, so that a range of possible feeding strategies is likely to be open to them, depending on opportunity.

E. crystallorophias occupies the area further south in the Antarctic where the spawning appears to occur before the main bloom, suggesting that lipid stores are used for survival and for reproduction. T. macrura has a similar distribution to E. superba but spawns earlier so it again is dependent on lipid stores for reproduction, but may also utilize other food sources to get through the winter. In northern regions T. inermis converts phytoplankton rapidly into lipids to cope with the seasonal environment, but also utilizes other available organic material such as detritus. M. norvegica also builds up high lipid stores but is more carnivorous, utilizing lipid-rich copepods.

Overall, there appears to be a pattern from high to lower latitudes in the strategies of feeding and energy storage. Truly polar species such as *E. crystallorophias* and *T. inermis* rely totally on the seasonal phytoplankton bloom and lipid stores, whereas species such as *E. superba*, *T. raschii* and *T. macrura* survive winter utilizing alternative food sources and require the bloom to reproduce. Further away from the polar regions, species such as *T. longicaudata* and *M. norvegica* are more carnivorous, utilizing copepods as their main food source.

Spatial Distribution

At large scales there are heterogeneities in the distributions of euphausiids that extend for tens or hundreds of kilometers. Within these broad aggregations there are also more dense regions where krill form patches, swarms, or schools (**Figure 4**) forming a distribution generated by a very dynamic system, with aggregation and dispersal over a wide range of scales. These patches can be very dense and compact and it has been suggested that it is likely that on occasion all species of euphausiids aggregate to some extent. This ability to form such dense aggregations is certainly found in a number of species, particularly *E. superba*, but also *M. norvegica*, *Nyctyphanes australis*, *E. pacifica* and *E. lucens*.

The generation of such patchy distributions is the result of interactions between biological and physical processes over a range of scales. Over small scales and in very dense aggregations, behavior probably dominates. Swimming speeds can be high – ~ 20 cm s^{-1} in short bursts in Antarctic krill – so individuals have a marked ability to undertake directed movement at least over relatively short spatial scales. The formation of these smaller aggregations, along with diurnal vertical migration, is considered to be mainly a predator avoidance effect. However, it will also lead to changes in the dynamics of the interaction of krill with their food and may generate complex outcomes in the dynamics of planktonic systems.

At larger scales physical processes probably dominate so that aggregations are dependent on physical concentration mechanisms in areas of shelf-breaks, around islands, in ice edge regions or associated with eddies. This larger-scale aggregation may be a precondition for behavioral effects to dominate at smaller scales. Krill within aggregations appear to share more similar characteristics in size and maturity compared with those in other aggregations in the same area. The densities of krill within these aggregations can be well in excess of 10 000 m^{-3};

0.5 km

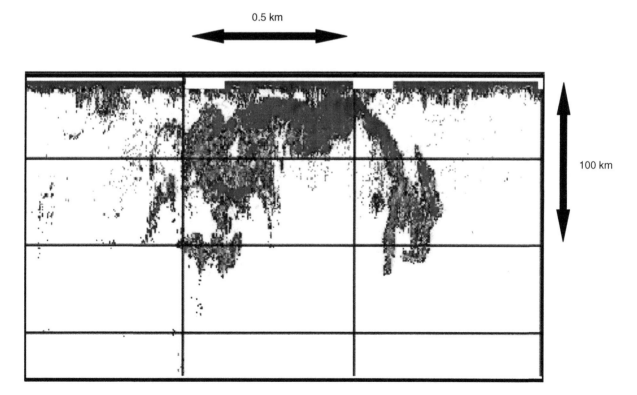

100 km

Figure 4 A hydroacoustic trace of an aggregation of Antarctic krill, *E. superba*.

over 50 000 m^{-3} has been estimated for *E. superba* and *E. pacifica* and over 500 000 m^{-3} recorded for *N. australis*, *E. lucens* and *M. norvegica*.

Some of the aggregations are extremely large and can account for a considerable proportion of the total biomass in an area. So for example, in one survey of Antarctic krill >10% of the regional biomass was recorded in just one aggregation that extended about 1 km horizontally. This large aggregation was observed in the vicinity of a large number of whales, suggesting that some of the very large, dense aggregations may be the result of intense predator–prey interaction and emphasizes the dynamic nature of the spatial distribution of krill. This makes the design of krill distribution surveys using nets or hydroacoustic techniques challenging and the survey data require careful interpretation and analysis.

A number of the species undertake diurnal vertical migration, rising to nearer the surface and dispersing at night. This has been shown clearly in *M. norvegica*, whereas in *E. superba* vertical migration appears to be highly variable and may depend on local physical conditions, surface predator affects, and predation effects from below, particularly in areas of the shelf. In addition to the importance of predation effects the behavioral tracking of particular isolumes has been suggested as a mechanism involved in diurnal vertical migration and some species appear to show an endogenous rhythm.

Seasonal changes in the pattern of aggregation and vertical migration have also been noted in some species. In one area it has been observed that during spring aggregations of *E. superba* are of the order of 0.7–2 km in length, whereas they are smaller and moredense in summer, and larger and less dense in autumn and winter. In the same study many of the swarms occurred in the upper 70 m during the summer, while in winter many were below 100 m deep. Other studies have found no such vertical change in depth distribution during the year, although diurnal vertical migration did change, being marked only during the spring and autumn.

Role in the Food Web

Krill as Consumers

Krill show a range of feeding strategies from complete herbivory to total carnivory, with a full range of capabilities and flexible feeding strategies in between. In the Antarctic they can have a major impact on the large diatoms that form the major components of the intense blooms associated with the summer retreat of the sea ice. Antarctic krill are also known to consume copepods and negative correlations have been shown between the krill occurrence and the density of copepods and the phytoplankton concentration. This indicates that euphausiids are important in the plankton dynamics of these regions. They generate large fecal pellets which have high rates of sinking, suggesting that they can be important in the export of carbon from the surface layers. Rapid grazing of diatom blooms in some areas can therefore lead to a rapid flux of material to deeper ocean regions. The highly aggregated nature of the distributions is also likely to be important in determining the plankton dynamics, not just in terms of producing an interactive mosaic of production and consumption, but also in terms of the nutrient regime. Large krill swarms will generate high concentrations of ammonia that may favor the production of particular size groups of phytoplankton, leading to complex interactions in the plankton. Their role as consumers continues to be studied, but across the order they clearly have the capacity to feed on a wide range of food sources including diatoms, coccolithophores, dinoflagellates, chaetognaths, copepods, and other crustaceans; cannibalism has also been shown in some species. On the basis of observed variations in feeding strategies, it has been suggested that most species of euphausiids can adapt their feeding to utilize what is available, modifying their feeding strategies depending on the food they encounter.

Krill as Prey

Krill are prey of many higher trophic level predators and as such play a key role throughout the oceans by transferring energy up the food chain. The baleen whales are the most well known predator, eating krill throughout their range, and despite the massive depletion of their populations due to harvesting they are still important krill predators. So for example, dense aggregations of *M. norvegica* and *T. raschii* in the Gulf of St Lawrence are associated with high abundances of fish (capelin) and a range of whale species including minkes, fin, blues, humpbacks, sperm, and beluga.

Seals are also major predators of euphausiids in many areas. In Arctic waters, for example, harp seals consume *M. norvegica* and *T. inermis*, while in the Southern Ocean, crabeater seals consume *E. crystallorophias*. Further north, around some of the sub Antarctic islands such as South Georgia, the previously exploited fur seal populations that are now very large consume a considerable quantity of *E. superba*.

Euphausiids also comprise a key component of the diet of a wide range of fish species, many of which are, or were, exploited. In the North Atlantic these include herring, cod, haddock, whiting, and mackerel. *M. norvegica* is probably the key species consumed, but others such as *T. raschii, T. inermis* and *T. longicaudata* are also important. In the North Pacific and adjacent regions *E. pacifica* is eaten by most commercial fish species, including Pacific cod, walleye pollack, chub mackerel, and sand lance. Other krill taken include *T. raschii* and *T. inermis*. The importance of euphausiids in the diet of many commercially exploited fish species is seen throughout the world. So for example around Australia *N. australis* is eaten by bluefin tuna and striped tuna, while in the Antarctic *E. superba* is eaten by the Mackerel icefish.

Seabirds are also important predators of euphausiids throughout the world. In the North Atlantic a wide range of bird species consume *M. norvegica* and *T. inermis* including gulls, puffins, kittiwakes, and fulmars. The importance of sea-birds as predators of krill is highlighted in the Southern Ocean where *E. superba* is a key item in the diet and consumed in vast numbers by penguins (gentoos, macaronis, and Adelies) and by flying sea-birds including albatrosses such as grey-headed and black-browed albatross.

This broad range view of euphausiids as prey emphasizes the important role that krill play in transferring energy to higher trophic levels in marine food webs worldwide. One of the key reasons for the importance of euphausiid species in food webs is the heterogeneity of krill distribution on a range of spatial and temporal scales. Different predators exploit the aggregation pattern with different foraging strategies, so exploiting different scales of pattern in the prey field. The pattern generated by the biological or biological–physical interactions will thus determine which predators can exploit the prey and hence the structure of the food web.

Krill Fisheries

There are extensive fisheries for *E. superba* in the Southern Ocean, while in the Pacific off Japan there are important fisheries for *E. pacifica*. There is a more limited *E. pacifica* fishing off western Canada and there have also been intermittent fisheries for other species. The Southern Ocean fishery for *E. superba* is the largest and started at beginning of the 1970s, peaking in the early 1980s at 0.5 million tonnes. The fishery has since declined with changes in its economic basis. Catches over recent years have been <100 000 tonnes. The *E. superba* fishery in the Scotia Sea region is linked to seasonal seaice changes. During winter the fishery operates in the north

around South Georgia, it moves further south in the spring with the ice, to the area near the South Orkney Islands, and then during summer the fishery exploits krill around the Antarctic Peninsula.

The management regime for *E. superba* takes account of krill recruitment variability, growth, and mortality to examine effects of various harvesting levels. Decision rules are included tomaintain stocks at a level that takes into account the dependent predators. At the current time, catch levels are much lower than the allowable catch (<10%) and future expansion of the fishery depends on the development of new products utilizing krill. An ecosystem approach is being developed for managing Southern Ocean fisheries and extensive predatormonitoring programs are operating. The challenge here is to develop management decision rules that consider not just the target species, the krill, but also incorporate ecological information from a number of levels in the food web,taking into account dependent species as well as environmental links. This ecosystem rather than species-based approach is one that will be increasingly relevant elsewhere.

Krill Variability

There is considerable evidence of the importance of variation in the physical environment and circulation systems of the oceans in determining the distribution and abundance of krill. For example, links have been noted between variations in the oceanographic regimes associated with El Niño events and the recruitment of *E. pacifica* in the North Pacific, while water temperature variations have been linked to 2–3 year variations in *T. inermis* populations. Biological processes associated with the environmental variation are also important in generating the variation observed in euphausiid populations. For example, there are marked interannual variations in the abundance of *E. superba* in the Southern Ocean, where recruitment strength has been linked to variations in the extent and concentration of sea ice. The current view is that increased sea ice cover and extent lead to favorable conditions for spawning and larval survival. The sea ice is thought to provide better over winter conditions for the krill. Salps compete with krill for phytoplankton – in poor sea ice years salp numbers are increased and krill recruitment is reduced. Further north in their range, *E. superba* abundance is dependent on the transport of krill in the ocean currents as well as fluctuations in the strength of particular cohorts.

Given the importance of euphausiids in marine food webs throughout the world's oceans, they are potentially important indicator species for detecting and understanding climate change effects. Changes in ocean circulation or environmental regimes will be reflected in changes in growth, development, recruitment success, and distribution. These effects may be most notable at the extremes of their distribution where any change in the pattern of variation will result in major changes in food web structure. Given their significance as prey to many commercially exploited species, this may also have a major impact on harvesting activities. A greater understanding of the large-scale biology of the euphausiids and the factors generating the observed variability is crucial. Obtaining good long-term and large-scale biological and physical data will be fundamental to this process.

See also

Baleen Whales. Copepods. Phalaropes. Plankton Overview. Seals. Sperm Whales and Beaked Whales.

Further Reading

Constable AJ, de la Mare W, Agnew DJ, Everson I, and Miller D (2000) Managing fisheries to conserve the Antarctic marine ecosystem: practical implementation of the Convention on the Conservation of the Antarctic Marine Living Resources (CCAMLR). *ICES Journal of Marine Science* 57: 778–791.

Everson I (ed.) (2000) *Krill: Biology, Ecology and Fisheries*. Oxford: Blackwell Science.

Everson I (2000) Introducing krill. In: Everson I (ed.) *Krill: Biology, Ecology and Fisheries*. Oxford: Blackwell Science.

Falk-Petersen S, Hagen W, Kattner G, Clarke A, and Sargent J (2000) Lipids, trophic relationship, and biodiversity in Arctic and Antarctic krill. *Canadian Journal of Fisheries and Aquatic Sciences* 57: 178–191.

Mauchline JR (1980) The biology of the Euphausids. *Advances in Marine Biology* 18: 371–677.

Mauchline JR and Fisher LR (1969) The biology of the Euphausids. *Advances in Marine Biology* 7: 1–454.

Miller D and Hampton I (1989) *Biology and Ecology of the Antarctic Krill*. BIOMASS Scientific Series, 9. Cambridge: SCAR & SCOR.

Murphy EJ, Watkins JL, Reid K, *et al.* (1998) Interannual variability of the South Georgia marine ecosystem: physical and biological sources of variation. *Fisheries Oceanography* 7: 381–390.

Siegel V and Nichol S (2000) Population parameters. In: Everson I (ed.) *Krill: Biology, Ecology and Fisheries*. Oxford: Blackwell Science.

NEKTON

W. G. Pearcy, Oregon State University, Corvallis, OR, USA

R. D. Brodeur, Northwest Fisheries Science Center, Newport, OR, USA

Introduction

Marine nekton are the swimmers in the sea, as opposed to plankton that drift with currents. Although the main criterion is swimming ability, most definitions of nekton, as with plankton, are not precise. They are based on swimming speeds, lengths, Reynolds number, shape to reduce drag, or simply those animals too swift or agile to be captured in plankton nets.

Nekton include fishes, cephalopods, seabirds, marine mammals, reptiles, and crustaceans. They inhabit all the ecological zones of the ocean, from the epipelagic near-surface waters to the deep-sea abyss, including both pelagic and epibenthic animals. Methods of assessing the distribution and abundances of nekton include nets, baited hooks or traps, acoustics, visual systems, lasers, egg/larval surveys, tagging/recapture, and food habits of their predators.

Because this edition includes articles on the biology of fishes, fisheries, birds, and marine mammals, the emphasis here will be on the smaller nekton, or micronekton (usually 2–20 cm in length). Although most micronekton are not harvested, krill (euphausiids) and lanternfishes are sometimes fished commercially. Some problems for nektonic animals in the sea include overcoming the forces of drag while moving through a viscous medium, counteracting sinking through buoyancy adaptations, feeding, and avoiding being eaten. The reader is referred to articles on biology of fishes that address some of these subjects.

The major taxonomic groups of micronekton are fishes, cephalopods, and large crustaceans. They are important components of marine ecosystems and vital links between zooplankton and higher trophic levels. Many are carnivorous; some are herbivores. As a group they are important to us as food, either directly through fisheries or indirectly as food for commercially important species. Fishes, squids, and euphausiids are important prey for seabirds, larger fishes, and marine mammals. Oceanic and coastal micronekton are important, but poorly understood, intermediate links in the food webs from zooplankton to apex predators for many marine ecosystems including the tropical Pacific, the Gulf of Alaska, and the Barents Sea (**Figure 1**). Euphausiids are especially important as prey for many species of fishes, seabirds, and marine mammals. In the Antarctic, krill (mainly *Euphausia superba*; **Figure 2**) are a major food for whales, penguins, and some pinnipeds. Although most euphausiids are herbivorous, most of the micronektonic fishes and squids are carnivorous, preying on euphausiids, copepods, amphipods, and other zooplankton.

Distributions

Many studies have shown that the distributions of mesopelagic micronekton, like zooplankton, generally coincide with specific water masses in the ocean, often forming faunal assemblages. However, some species are found in all oceans while others have distributions restricted to waters of the continental shelf or slope. Species richness of the micronektonic fauna usually increases from high to low latitudes.

The biology and morphology of micronekton differ among the different pelagic realms: epipelagic (0–200 m), mesopelagic (200–1000 m), bathy/abyssopelagic (> 1000 m), and epibenthic. Epipelagic micronekton include the juvenile phases of many species, as well as the mature stages of small fishes such as Engraulidae (anchovies), Clupeidae (herrings, sardines), and Osmeridae (smelts) (see **Figure 1**). These epipelagic species are most common over productive regions of the continental shelves and coastal upwelling regions where they are often subjected to commercial fisheries.

Midwater Micronekton

Many families of fishes, squids, and crustaceans are considered micronekton (**Figure 3**). Myctophidae (lanternfishes) and Gonostomatidae (bristlemouths) are common families of mesopelagic fishes. It has been estimated that the worldwide biomass of mesopelagic fishes alone is over 900 million metric tons. Many of these animals have distributions in oceanic (waters beyond the continental shelves) in all the world's oceans. Generally, mesopelagic micronekton are most abundant over the continental slopes and their numbers decrease both inshore and offshore. Diel migrations also provide an abundance of food resource for shallow-water or coastal fishes.

Tropical Pacific

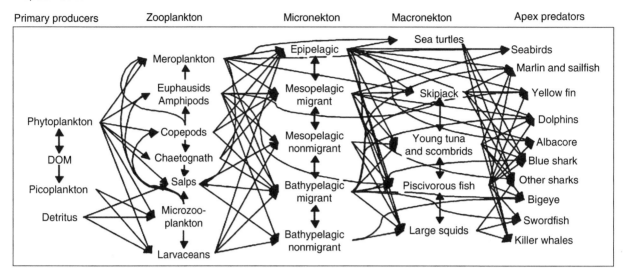

Gulf of Alaska

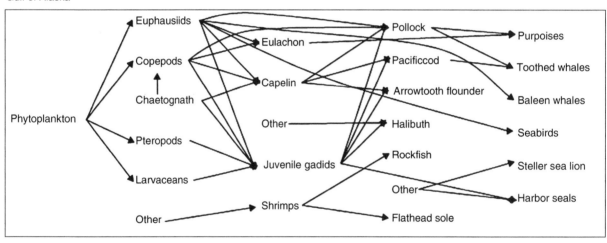

Barents Sea

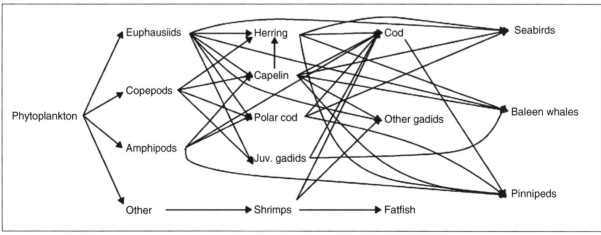

Figure 1 Simplified food web for tropical Pacific, Gulf of Alaska, and Barents Sea showing the intermediate position of micronekton in the food webs. Adapted from Ciannelli L, Hjermann DO, Lehodey P, Ottersen G, Duffy-Anderson JT, and Stenseth NC (2005) Climate forcing, food web structure, and community dynamics in pelagic marine ecosystems. In: Belgrano A, Scharler UM, Dunne J, and Ulanowicz RE (eds.) *Aquatic Food Webs, an Ecosystem Approach*, pp. 143–169. Oxford, UK: Oxford University Press.

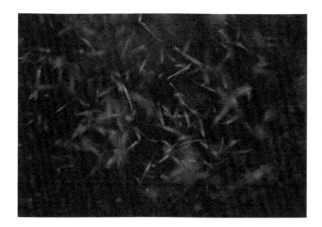

Figure 2 Euphausiid swarm. Photo courtesy of Bruce Robison, Monterey Bay Aquarium Research Institute (MBARI).

Off Oregon, rockfishes (*Sebastes* spp.) often prey on euphausiids and sergestid shrimp and myctophid fishes that are carried by currents inshore and then trapped against the shallow seafloor or offshore banks or seamounts during daytime. Off Hawaii, the animals forming layers are impinged along the sides or on the tops of seamounts where they are easy prey (**Figure 4**).

In Hawaiian coastal waters, a mesopelagic boundary layer community consists of a distinct resident community of micronekton distributed along a narrow band where the upper slope meets the oceanic realm. Based on acoustical measurements from moorings, it exhibits both diel vertical and horizontal migrations. Remarkable inshore migrations at night extend within 1 km from shore. Again, this provides a trophic link between neritic and oceanic systems.

Diel Vertical Migrations

Many species of epipelagic and mesopelagic micronekton undertake diel, nocturnal, or daily vertical migrations, swimming toward the surface at night and descending into deeper water during periods of daylight. This common and unique behavior of mesopelagic micronekton is shown in **Figure 4**, where a distinct sound-scattering layer migrates toward the surface during twilight, while other layers do not migrate. Interestingly, not all individuals migrate even within the same species. Some are basically nonmigratory. Some studies have observed diel migrations of sound scatterers between the depths of 500 and 1000 m. However, there is little evidence for migrations of animals living below about 1000 m in the open ocean at bathy/abyssopelagic depths.

These diel migrations were first observed in the open ocean by physicists who were perplexed by sound-scattering layers at mid-depths during the day that looked like false bottoms, and often moved toward the surface at night. They called them 'deep-scattering layers'. Different animals reflect sound depending on the frequency of sound used and the sound velocity and density contrast of the animals. We know that the animals that reflect the sound (10–50 kHz) in these layers are usually fishes or siphonophores, often with gas-filled swimbladders or floats that effectively resonate sound. In the open ocean, migrations are generally from mesopelagic depths (200–1000 m) into epipelagic waters, or even to the surface, at night. Because of the huge biomass involved, these migrations of micronekton account for the largest daily movement of biomass on Earth (except for the masses of *Homo sapiens* that commute daily!).

Sunlight is the primary sensory cue that initiates or guides diel vertical migrations. This conclusion is supported by the conformity of the vertical movements of scattering layers and specific isolumes or light intensities, as well as responses to artificial light or solar eclipses. However, endogenous rhythms or changes of light intensity have also been suggested as triggers to initiate migrations.

Adaptive significance of diel vertical migrations Why migrate vertically? The main theories concern the predator following prey and the latter avoiding predation. Food is available in greater quantity in shallower waters that are rich with zooplankton but these sunlit waters are dangerous during the daytime because of large visual predators. Hence this behavior is thought to be an energetic trade-off between the risk of predation and feeding. Vulnerability to large visual predators is reduced at night in near-surface water and in deep, dimly lit water during the daytime. Other theories involve an energy bonus for animals feeding in warmer near-surface waters and digesting and assimilating food in deeper cooler waters. Mid-water fishes often migrate vertically in the Antarctic even when the water column has near-uniform temperature and their major prey, *E. superba*, remains near the surface and does not migrate. This suggests that avoidance of visual predators is a major selection factor for diel vertical migrations in these waters.

These ubiquitous vertical migrations in the open ocean contribute to the vertical transport of energy and elements. Carbon in zooplankton, as well as anthropogenic materials such as pesticides or radio-isotopes, are ingested by micronekton in near-surface waters and are defecated in deeper water. This rapid transport of materials has been called a 'biological pump'.

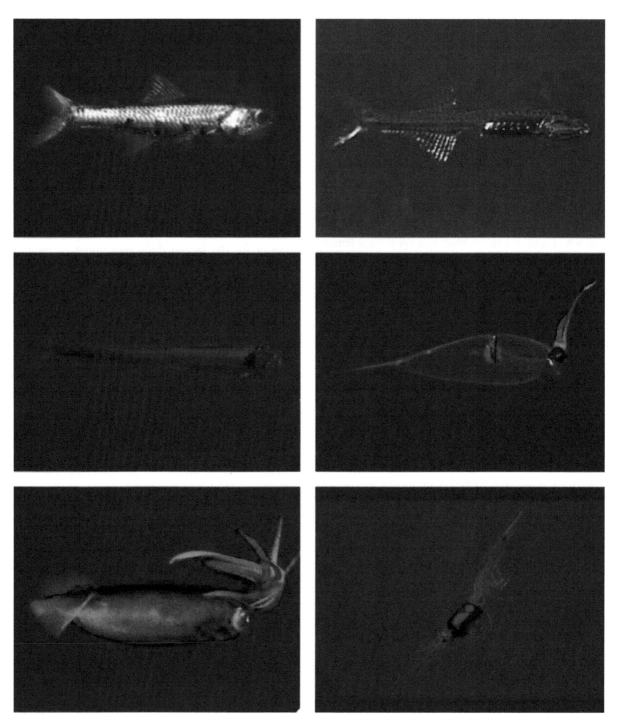

Figure 3 Photos of *in situ* mesopelagic fishes, squids, and shrimp taken from a remotely operated vehicle (ROV) Tiburon. All images © MBARI. From top, left to right: lanternfish *Stenobrachius leucopsarus*, bristlemouth *Cyclothone signata*, viperfish *Chauliodus macouni*, squids *Galiteuthis*, Gonatidae, and shrimp *Sergestes similis*. All images © MBARI.

Bioluminescence

Many species of oceanic micronekton, especially in the meso/bathypelagic regions, are bioluminescent. These include most families of euphausiids, many pelagic mysids, shrimps, fishes, and squids. Bioluminescence is produced by simple or complex light organs, photophores, or luminous secretions, either by symbiotic bacteria or intrinsic organs under nervous control. The color of most bioluminescence in mesopelagic animals is blue-green, *c.* 470–490 nm. Notable exceptions are some stomiatoid fishes that emit a red light, a frequency that is invisible to most

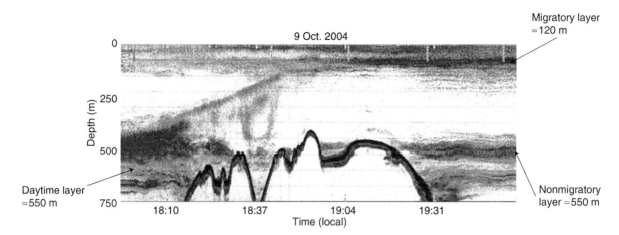

Figure 4   A 38-kHz echogram off Hawaii showing the complexity of sound-scattering layers, with a layer migrating toward the surface at twilight and nonmigratory layers near the surface and along the flanks of the seamount.

deep-water nekton – it functions as an underwater snooperscope.

Many of the mesopelagic fishes, squids, and crustaceans have photophores arrayed along the ventral portions of their bodies (see fishes in **Figure 3**). This is thought to provide a ventral 'countershading' or counter-illumination that reduces the silhouette of their bodies from predators that lurk below. This theory is supported by experiments that have shown that the intensity of light emitted by the photophores closely matches downwelling light from above. Other bioluminescent organs are thought to be lures to attract prey or to promote schooling or mate recognition.

Bioluminescence occurs in the ocean, even down to the deepest depths. If a light meter is lowered to measure light intensity, downwelling light from the sun decreases exponentially, but then light intensity may increase with depth at mid-depths from flashes of bioluminescent organisms. Because bioluminescence is the predominant source of light below 500–1000 m, fishes and crustaceans in the deep ocean usually have either jet-black (fishes) or crimson (crustaceans) pigments, presumably to absorb bioluminescent flashes. Here silvery or reflective bodies, as found in most epipelagic fishes, would reflect these flashes and stand out as mirrors that would attract predators.

Life History and Biology

In the large volume that comprises the mid-water domain of the world's oceans, finding and recognizing mates to reproduce with can be difficult, especially when overall abundances are low. Many fishes depend on random encounters of males and females, perhaps recognizing their own kind by their

photophore patterns. Some male anglerfishes have evolved a bizarre behavior where they actually attach as a dwarf form to a female once they locate them. They then draw nutrition from the female and continue in this parasitic relationship for the rest of their lives.

Reproductive patterns are not well known in many mid-water nekton. Myctophids are known to spawn year-round in the tropics but spawn during a single season at higher latitudes. Many myctophids undertake extended seasonal north–south migrations between feeding and spawning grounds, as seen in epipelagic fishes. Most larvae and early juveniles occur in the epipelagic zone, are often nonmigratory, and are found progressively deeper as they grow older, metamorphosing into the adult stage in the mesopelagic zone. They usually mature after 1–3 years but their typical life spans are not generally known. With the exception of Antarctic krill, which can live to be about 7 years old, most nektonic crustaceans and squid probably live only a few years.

Food Habits

Feeding modes in midwater nekton can range from filter-feeding to highly carnivorous. Most deep-water animals are carnivorous (note the viperfish in **Figure 5**). Numerous morphological or behavioral adaptations have evolved to overcome the paucity of food in the deep ocean. Bioluminescence is used to both locate prey, as discussed previously, and attract prey. Anglerfishes often use lighted 'lures' suspended over the mouth by fin rays modified to work as 'fishing poles' to attract gullible prey. Most species have upward-facing eyes and mouths to locate overhead prey that may be backlit against the

Figure 5   Head-on view of the viperfish, *Chauliodus macouni*. Image © 2006 MBARI.

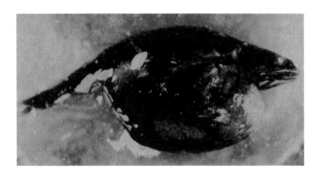

Figure 6   Whiptail gulper, *Saccopharynx*. Note large mouth and extended stomach.

downwelling light. In many cases, the jaws are un-hinged or extremely large relative to the size of the organism to swallow large prey. In an extreme ex-ample, the whiptail gulper fish (**Figure 6**) possesses a mouth and stomach that are highly extensible to consume prey bigger than themselves similar to snakes, and then gradually digest the prey item until the next meal comes along, perhaps months later.

Given the large number of piscivorous predators in these regions, it is not surprising that many micro-nekton have developed adaptations for predator avoidance. Their bioluminescent countershading and light absorptive pigmentation reduce their visibility to most predators. Bioluminescent flashes may also confuse predators. Some animals align themselves vertically in the water column to minimize their

cross-sectional area visible to predators from below. The squid *Galiteuthis* (see **Figure 3**), which has a transparent body, often holds its opaque arms and tentacles directly over the large downward-oriented light organs on its head, presumably to minimize the silhouette of its arms and tentacles by counter-illumination. Other elongate fishes curl up into a ball to mimic other unpalatable species or appear larger than they actually are to a predator. Squids have developed a different strategy in that they can release a black or luminous ink that may temporarily con-fuse their predators or act as 'decoys', allowing their getaway from predators.

See also

Cephalopods.

Further Reading

Benoit-Bird KJ and Au WWL (2006) Extreme diel horizontal migrations by a tropical nearshore resident micronekton community. *Marine Ecology Progress Series* 319: 1–14.

Brodeur RD and Yamamura O (2005) Micronekton of the North Pacific. *PICES Scientific Report* 30: 115pp.

Ciannelli L, Hjermann DO, Lehodey P, Ottersen G, Duffy-Anderson JT, and Stenseth NC (2005) Climate forcing, food web structure, and community dynamics in pelagic marine ecosystems. In: Belgrano A, Scharler UM, Dunne J and Ulanowicz RE (eds.) *Aquatic Food Webs, an Ecosystem Approach*, pp. 143–169. Oxford, UK: Oxford University Press.

Marshall NB (1971) *Explorations in the Life of Fishes.* Cambridge, MA: Harvard University Press.

Nixon M and Messenger JB (eds.) (1977) *Symposia of the Zoological Society of London, No. 38: The Biology of Cephalopods.* New York: Academic Press.

Robison BH (2003) What drives the diel vertical migrations of Antarctic midwater fishes? *Journal of the Marine Biological Association of UK* 83: 639–642.

CEPHALOPODS

P. Boyle, University of Aberdeen, Aberdeen, UK

Introduction

The Cephalopoda is the class of the Mollusca comprising the octopuses, cuttlefish, squid, and their allies. Exclusively marine and present in all of the world's oceans and seas, their lineage can be traced from the Ordovician to the present due to fossilization of their large, heavy, chambered shells. The Pearly Nautilus (*Nautilus* spp.) of the Indo-Pacific region is the only surviving relative of this ancient ancestry (10–12 000 extinct species) Modern living cephalopods (subclass Coleoidea), having reduced or lost the ancestral shell, are represented by only about 650–700 species. These are characteristically large, active, soft-bodied predators, with complex behavioral and physiological capabilities. Occupying a wide range of benthic and pelagic habitats they are abundant in productive shelf regions, where genera such as *Octopus* and the common cuttlefish *Sepia* are each credited with over 100 species. The greatest diversity of form and biomass of cephalopods is oceanic and mesopelagic in distribution, but the biology of these offshore species is little understood and generalizations are based mostly on coastal forms. Now, and throughout their evolutionary history, representatives of the cephalopods reach the largest of all invertebrate body sizes.

With the exception of *Nautilus* (and some of the deep-sea forms), cephalopods generally share common life cycle features. The large eggs hatch directly to free-swimming juvenile forms resembling the adult (paralarvae). Growth is very rapid (exponential and logarithmic phases) and adult size is reached in about a year (6–24 months). The sexes are separate and there are complex arrangements for mating and fertilization. After spawning of the fertilized egg masses, either attached to the bottom or freely into the water column, most individuals of both sexes die within a short period of time afterwards. Although there are some variations between species in the timing of breeding and the duration of spawning, uniseasonal breeding appears to be more or less universal. The consequences of this life cycle at the population level are that there is little overlap of generations, the species biomass present at any time tends to build and crash seasonally. Distribution and abundance of the shelf species at least is thus highly dependent on inter-annual conditions for recruitment and growth.

Cephalopods have a very significant role in the trophic relations of marine ecosystems. Universally predatory, they consumea wide variety of fish, crustacea and other invertebrates. Cephalopods themselves are also preyed upon by many other large marine organisms such as fish, marine mammals of all sorts, and many oceanic birds. Conservative estimates of consumption of cephalopods by these predators considerably exceed 100 million tonnes annually. Human fisheries for cephalopods have increased steadily and are reaching about 3 million tonnes annually. Critical assessment of the role of cephalopods in the world's oceans is compromised by the relative lack of information on the oceanic and deep-water forms and the consequent extrapolation of knowledge from the better-known coastal species.

Diagnosis of the Cephalopoda (Table 1)

As a class of the Mollusca cephalopods share fundamental features of their body layout and development with the other classes (gastropods, bivalves, chitons, etc.) including absolutely characteristic molluskan features such as the radula (feeding organ). Other typically molluskan features such as the calcareous shell are reduced or absent (it remains as the 'cuttlebone' in Sepia and the gladius or 'pen' in squid) and there are no specialized larval forms (no molluscan trochophore or veliger). Cephalopods have also developed quite unique systems of locomotion and mobility (jet propulsion, suckers), brain development, color change (chromatophores), and light production (photophores).

Although the molluskan relationships of cephalopods are without doubt, the scale and dynamics of their extant populations are better understood in terms of comparison with the teleost fishes – co-evolution and competition for the most productive marine environments.

Biology

Buoyancy and Jet Propulsion

The reduction and loss of a calcareous shell in the modern cephalopods (Coleoidea) has allowed the evolution of their highly mobile, active lifestyles,

Table 1 Diagnosis of the Cephalopoda: Classification is not entirely consistent between different authors and only the principal categories are given here, some common genera are listed and the common names of classification categories are shown in bold. (Abbreviated with permission from *Boyle, 1983* (*Cephalopod Life Cycles*, Vol 1, 1–8, Academic Press, London) with additional information on the numbers of living species from *Nesis, 1987*.)

Class **CEPHALOPODA**
All marine; bilaterally symmetrical; primitively with a chambered external shell. Radula enclosed within chitinous mandibles ('beaks'); ring of prehensile appendages around mouth; one or two pairs of gills; water circulation in mantle cavity expelled through ventral 'funnel' tube for jet propulsion. Centralized nervous system; highly organized sense organs; complex behavior. Sexes separate; sperm transferred in complex spermatophores; eggs large and yolky; direct development (no veliger).

    Subclass **Nautiloidea**
        Cephalopods with straight or coiled external shells; appearing in Cambrian period; numerous species populous throughout warm seas; all extinct except for one family (Nautilidae) now limited to Indo-Pacific Oceans. Body occupies terminal chamber of shell and by retraction displaces water from mantle cavity for jet propulsion; inner chambers form buoyancy organ by adjusting contained fluid/gas spaces; numerous unsuckered appendages. **Nautiluses**. 1 family, 1 genus, 3–6 species.

    Subclass **Ammonoidea**
        Chambered external shells; usually coiled and with complex septa and sutures separating chambers. Very numerous from the Devonian to the Cretaceous periods but now all extinct. **Ammonites**.

    Subclass **Coleoidea**
        Modern forms. Devonian period to present. Shell internal and reduced. Muscular fins (absent in incirrate octopods) and muscular mantle forming a sac enclosing the viscera; large mantle cavity; ink sac typically present; skin containing pigment organs (*chromatophores*) variably expanded by neuromuscular control.

        Order **BELEMNOIDEA**
            Internal shell, straight and with a solid posterior portion; commonly fossilized; all extinct. **Belemnites**.

        Order **SEPIODEA**
            Calcareous chambered shell present internally and functioning as a buoyancy organ in some genera (*Spirula, Sepia*); shell greatly reduced to a purely organic pen in others (*Sepiola, Euprymna, Sepietta, Idiosepius*). Eight suckered arms plus two long tentacles with suckered club; suckers pedunculate with horny rims. **Spirula, Cuttlefish** and **Sepiolids**. 5 families, 20 genera, 150–180 species.

        Order **TEUTHOIDEA**
            Shell reduced to chitinous 'pen' (gladius) lying dorsally. Elongate body usually finned. Eight suckered arms plus two long tentacles with suckered club, suckers pedunculate with horny rims, some with hooks. **Squid**.

            Suborder **Myopsida**
                Eyes with transparent corneal covering. Eggs spawned in masses attached to seabed. Typical of the continental shelf and including many abundant and valuable fished genera of the family Loliginidae (e.g., *Loligo, Sepioteuthis, Loliolus, Alloteuthis*). 2 families, 8 genera, 42–51 species.

            Suborder **Oegopsida**
                Eyes without corneal covering, a large assemblage of many families. Oceanic and midwater, seldom over the shelf or near coasts. Eggs apparently spawned in midwater gelatinous masses. Includes many genera significant in the diets of top predators (e.g., *Gonatus, Histioteuthis*) and the giant squids *Architeuthis* spp.; includes the family Ommastrephidae with most of the genera significant to shelf-break and oceanic fisheries (e.g., *Illex, Todarodes, Todaropsis, Martialia, Dosidicus, Ommastrephes, Nototodarus, Sthenoteuthis*). 23 families, 77–81 genera, 200–230 species.

        Order **OCTOPODA**
            Internal shell drastically reduced and split into two lateral rods or absent. Eight arms only with nonpedunculate suckers. Globular body with or without fins. **Octopuses**.

            Suborder **Cirrata**
                Deep-water benthic or benthopelagic animals, gelatinous, jelly-like tissues. Locomotion by a pair of paddle-shaped fins, arms with reduced suckers and bearing lateral cirri (e.g., *Opisthoteuthis, Cirroteuthis, Stauroteuthis, Grimpoteuthis*). 2–3 families, 7–8 genera, 28–33 species.

            Suborder **Incirrata**
                A large group with several pelagic families (e.g., *Argonauta, Tremoctopus*). The common octopods of coastal waters all belong to one benthic family (Octopodidae, e.g., *Octopus, Benthoctopus, Eledone, Pareledone*). 9 families, 33–35 genera, 165–180 species.

        Order **VAMPYROMORPHA**
            The **vampire squid**, a subtropical, bathypelagic species. Eight long arms united by a swimming web, two small tendril-like arms in dorso-lateral position. 1 family, 1 genus, 1 species.

quite distinct from the other molluscan classes. Cuttle-fish (*Sepia*) and *Spirula* retain an internal remnant of the shell which functions as a buoyancy organ. The distribution of gas and fluid space within the chambers is controlled osmotically and allows neutral buoyancy to be achieved. The physiological mechanism in these Sepiodea appears to be similar to that used by *Nautilus* for controlled vertical movements through 1000 m of the water column and is thought to be the ancestral buoyancy mechanism common to the extinct nautiloids and ammonoids. The squids and octopuses have lost this mechanism entirely, most of them are negatively buoyant, but mesopelagic forms commonly reduce their density by chemical means such as retention of ammonium ions and loss of protein.

Active locomotion of the pelagic species is by jet propulsion – regular spasmodic forcing of water from the muscular mantle through the ventral funnel which can be directed forwards, backwards or side-to-side to allow great maneuverability. Paired fins contribute to directional control of swimming and their undulations are used for 'hovering' and slow swimming. The common coastal octopuses (Incirrata) are mainly benthic, using the suckered arms for relaxed scrambling over the bottom, and jet propulsion only for rapid attacking or escape movements.

Brain and Senses

The nervous system of cephalopods is still arranged in the basic molluscan layout as ganglionic masses grouped around the esophagus. It is centralized and developed to a much greater degree than that of other Mollusca. Coupled with large and complex sense organs, especially the eyes and statocysts (gravity and movement senses), the central nervous system supports an extensive and flexible repertoire of behavior unequalled by other invertebrate taxa. Especially in *Octopus*, the capability of the animal to discriminate between environmental cues and to make appropriate behavioral responses has been intensively studied and the detailed neuroanatomy has shown how the motor, sensory, and integrative functions of the brain are spatially located in its many subdivisions. Learning the significance of environmental cues and adapting its behavior accordingly is highly developed in *Octopus*.

Color and Pattern

The most remarkable, and immediately visible manifestation of the behavior and responses of cephalopods, is their ability to control and change the colors, pattern, and texture of body surface. The skin of cephalopods is a delicate epithelial surface beneath which are layers of connective tissue, active colored cells (chromatophores), passive reflecting bodies (iridophores, leucophores), and a complex system of muscle fibers for moving the skin over the underlying somatic muscle surface. This capability for altering the appearance of the animal is present throughout the Cephalopoda but is expressed to the greatest degree among the coastal octopuses, cuttlefish, and loliginid squid.

The unique functional components of color change in the cephalopod skin are the chromatophores. Each one consists of a single cell within which is an elastic sac of pigment (yellow-red-brown-black). Inserted onto the pigment sac is a series of muscle fibers (25–30) radiating out into the surrounding connective tissue. In the relaxed state, the pigment sac is passively retracted to a microscopic dark point, the muscle fibers are extended, and the skin surface appears white due to reflection from the underlying somatic muscle. In the active state, the chromatophore muscle fibers contract, extending the pigment sac and spreading the area covered by its contained pigment. The pigment now screens the underlying muscle, the incident light is selectively absorbed by the pigment and the reflected wavelengths give color to the surface.

The chromatophore muscles responsible for these pigment movements are innervated by fine nerve fibers ramifying throughout the skin. Contraction of the individual muscles may take only 200–300 ms, and the animal may change its complete appearance in a few seconds. In addition to these active chromatophores there may be several classes of passively 'reflecting cells' responsible for colors in the blue-green range (iridophores) or white by scattering of all wavelengths (leucophores). The arrangement of these layers, overlaying each other throughout the depth of the skin, allows almost infinite combinations of effects. Since the chromatophores are innervated directly from the brain, their activity can be controlled to express a great variety of pattern and contrast.

The use of color, contrast, and textural change in the intra- and inter-specific behavior patterns of cephalopods has been described for many species. These capabilities are mostly involved in crypsis (camouflage), mating activities, prey and predator responses, and are generally assumed to be of great survival significance and selective value. Surprisingly, there is no evidence that cephalopods themselves can discriminate colors, they respond mostly to contrast and pattern information. The scientific literature on cephalopod behavior is dominated by their visual capabilities. It is certain that they have also developed senses for tactile, vibration, and chemical stimuli, but little is known about the significance of these senses to behavior.

Escape and Luminescence

Squid have evolved a rapid 'escape response' behavior which is mediated by three sets of nerve cells with exceptionally large fibers <1 mm in diameter ('giant fibers') and specialized connections (synapses). This 'giant fiber system' distributes the motor commands from the brain simultaneously to all parts of the mantle musculature, and synchronously to each side, ensuring the maximum power of mantle contraction and speed of escape.

In common with fish and other invertebrate life of the deep sea, most of the mesopelagic squid show various forms of luminescent display. Most commonly present as a pattern of light-emitting organs distributed on the surface, symbiotic luminescent bacteria are also present in some of the internal organs or may be released into water as a luminescent cloud (*see* Bioluminescence). The functions of these systems are not fully understood, but presumably they are involved in counter-shading of the animal against surface illumination, sexual signalling, or predator–prey encounters.

The Life Cycle

Feeding and Growth

The life cycles of coastal cephalopods share many features. All are predators, feeding on a wide range of species especially crustacea and fish, many are also cannibalistic on smaller members of their own species. They ingest food at high rates, ranging between 1.5 and 15% body weight per day in different species and at a range of temperatures. Individual growth rates have been estimated to range from 1% to over 10% body weight per day, with estimates for gross growth efficiency (growth increment as a percentage of food ingested) between 10 and 70% per day for animals in captivity. Feeding and growth rates decline at large body sizes and at lower temperatures, and gross growth efficiency is generally lower in active squid species than the more sedentary octopuses and cuttlefish.

Feeding generally entails visual orientation and forward strike at the prey, gripping and pulling it in towards the mouth with the tentacles and arms. Squid and cuttlefish bite immediately into the tissues with the powerful chitinous mandibles (beaks), ingesting the most accessible parts and often releasing the dead remains partially eaten. Octopuses, in contrast, have evolved elaborate methods of prey handling – particularly effective on crustacea – involving external toxins and enzymes, before cleanly extricating the flesh from the carapace. After capture, a minute penetration of the carapace is made (<1 mm long) and a cocktail of compounds, including protease and chitinase enzymes together with paralyzing toxins, is injected. As well as subduing the prey, the enzymes have the effect of releasing the attachments of the crustacean tissues, allowing them to be selectively eaten.

Reproduction

All cephalopods are diecious, the sexes are separate and no hermaphroditism or sex change is described.

Among the better-studied coastal species, sexual maturation occurs rapidly, often at wide range of body sizes, and is usually associated with slowing or cessation of growth. In females, maturation is primarily a process of egg growth due to the accumulation of large amounts of lipoprotein yolk. Males mature spermatozoa, package them into complex spermatophores, and store them in the spermatophoric (Needham's) sac. Individual matings occur in which the male transfers the spermatophores directly to a receptive female. In many species there may be complex reproductive behavior, allowing the possibility of mate selection by either sex, In addition, females may mate with several males and usually have the capacity for storage of the transferred sperm and the possibility of multiple paternity of offspring.

Spawning and Death

Coastal squid (family Loliginidae), octopuses (family Octopodidae), and all of the cuttlefish (order Sepiidae) encapsulate their eggs, often in tough secreted coatings, and attach them to the bottom or other hard surfaces in clusters or strings. Some octopuses (e.g., *Octopus vulgaris*), subsequently stay with the egg mass, protecting it from epigrowths and defending it against predators. Most of the oceanic squid families (suborder Oegopsida) apparently spawn their eggs in fragile mid-water masses, but very little is known about the details of their mating and spawning habits.

Compared with other molluscs, cephalopod eggs are large (1–25 mm long) and yolky. Fecundity is estimated to be as low as 10–25 eggs/female for some sepiolids and octopus; 10 000–100 000 for most coastal squid and octopus; and 100 000 to over 1 million for oceanic squid (e.g., family Ommastrephidae). The eggs hatch directly, without any specialized larval forms, to an active swimming miniature of the adult form. Because the hatchling may have different habits and occupy a different ecological zone from the adult, they are usually referred to as 'paralarvae'.

Reproduction in many cephalopods occurs with a seasonal peak which is often rather inconsistent in duration and timing. In most populations some breeding individuals may be found throughout the year. Evidence from captive individuals and field populations consistently shows that modern cephalopods (with the exception of *Nautilus*) have only one breeding season. Taking account of some variations, such as 'batch spawning' (the release of the reproductive output in several episodes over a short period of time), and the apparently 'continuous' release of single eggs by deep-water octopods (suborder Cirrata); there is no indication that, after spawning, there is regeneration of the 'spent' gonads for a subsequent breeding season.

Instead, there is every indication that spawning in both sexes marks the end of their lives and that death soon follows (called semelparous reproduction).

Ecology

Population Biology

The cephalopod life cycle paradigm of fast growth, single breeding, and short life has profound consequences for the population biology of any species.

In its most extreme expression, if breeding is more or less synchronous and strongly seasonal throughout the population, it means that there will be only one adult size mode; biomass will strongly build and crash; and there will be little overlap of generations to buffer the influence on recruitment of environmental variables. In fact, tendencies to asynchronous breeding and lack of seasonality coupled with plasticity of feeding and growth characters, operate to spread the risks to the population of this life cycle and are shown schematically in **Figure 1**.

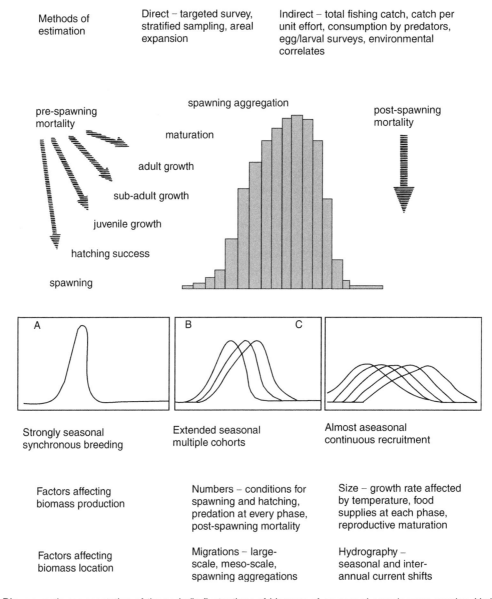

Figure 1 Diagrammatic representation of the periodic fluctuations of biomass of an annual semelparous species. Variations in the pattern of breeding and recruitment are suggested in (A), (B) and (C). Methods of estimation and the factors affecting biomass production are summarized. (Reproduced with permission from Boyle PR and Boletzky SV (1996) Cephalopod populations: definition and dynamics. *Philosophical Transactions of the Royal Society of London: Series B* 351: 985–1002.)

Trophic Relations

Numerous studies have shown that cephalopods are of great significance in the diets of many marine top predators. This evidence arises from the presence of the indigestible mandibles or 'beaks' in the gut contents of predators such as large fish, many seals, whales, and oceanic birds. As many as 30 000 have been found in the stomach of a sperm whale. The shape, size, and features of the upper beaks can be used to identify the cephalopod family or species, and estimate the size of the individual from which it came. Despite inherent errors in the procedure due to the unknown residence time of beaks in the gut, these beaks can be used to estimate the species composition and relative biomass of different cephalopods in the predator diet.

No cumulative estimate of consumption by these large vertebrate predators is available, but taking into account the estimated predator population size and consumption rates it has been variously estimated that sperm whales alone could consume $213–320 \times 10^6$ t of cephalopods from open ocean areas.

Fisheries

Human fisheries for cephalopods have been recorded at least since classical times. Coast dwellers throughout the world still catch octopuses with traditional traps of pots or baskets, while cuttlefish and squid are taken on simple hand lures. The quantities taken by these hand capture fisheries are usually unrecorded.

Cephalopods are also valuable fishery products traded on a global market. Using large bottom trawls specially tuned for cephalopods, oceanic drift nets (now banned in many areas), and highly efficient mechanized jigging vessels, the annual commercial harvest has risen steadily from about 1 million tonnes in 1970 to around 3 million tonnes by the mid-1990s. This was achieved largely by the extension of fishing to previously unexploited areas (North Pacific, South Atlantic) and also from increased catches in areas where the teleost fishes have been heavily exploited (Saharan Bank).

Role in the Oceans

Combining data from commercial harvest fisheries with crude estimates of total consumption by predators, the global biomass (standing stock of adults and sub-adults) of cephalopods in the oceans has been variously estimated to lie between 193 and 375×10^6 t. These figures have been derived by tentatively accumulating the estimate ranges for mesopelagic squid ($150–300 \times 10^6$ t); oceanic epipelagic squid ($30–50 \times 10^6$ t); slope/shelf-edge squid ($8–15 \times 10^6$ t); and shelf sepioids and octopuses ($5–10 \times 10^6$ t).

Whether or not these apparently massive estimates of biomass are realistic, their compatibility with other global estimates of marine productivity and consistency with the productive potential of cephalopods is uncertain. Overestimation of cephalopod frequency in predator diets could arise because cephalopod remains persist longer than those of other prey. Another source of uncertainty is the scaling up of limited predator and fishery data to ocean basin scales and the low carbon content (watery tissues) of many mesopelagic cephalopods. Some studies suggest the cephalopod biomass in the open sea (nektonic) to be about half that of fish and the mismatch between direct sampling with nets and indirect sampling from higher predators in this environment is well known.

Conclusions

Cephalopods are undoubtedly one of the most charismatic groups of marine animals. Sharing a basic body with the other molluscs they have evolved very distinctive biological characters, advanced behavior, and life cycle patterns. Cephalopods comprise a major sector of marine biomass, having central significance to higher tropic levels and global fisheries. Little is understood about the biology of the oceanic and mesopelagic species and, consequently major uncertainties remain surrounding the quantitative role of cephalopods in the world's oceans.

See also

Bioluminescence.

Further Reading

Boucaud-Camou E (ed.) (1991) *La Seiche*. Caen: Université de Caen.

Boyle PR (ed.) (1983) *Cephalopod Life Cycles*, vol. 1. *Species Accounts*. London: Academic Press.

Boyle PR (1986) Neural control of cephalopod behaviour. In: Willows AOD (ed.) *The Mollusca*, vol. 9(2), pp. 1–99. New York: Academic Press.

Boyle PR (ed.) (1987) *Cephalopod Life Cycles*, vol. 2. *Comparative Reviews*. London: Academic Press.

Clarke MR (ed.) (1996) The role of cephalopods in the world's oceans. *Philosophical Transactions of the Royal Society of London, Series B* 351: 977–1112.

Hanlon R and Messenger J (1996) *Cephalopod Behaviour*. Cambridge: Cambridge University Press.

Nesis KN (1987) *Cephalopods of the World: Squids, Cuttlefishes, Octopuses and Allies*. TFH Publications (English translation of the original Russian editon, 1982 VAAP Copyright Agency of the USSR for Light and Food Industry Publishing House, Moscow).

Nixon M and Messenger JB (eds.) *The Biology of Cephalopods*. Symposium of the Zoological Society, London, no. 38. London: Academic Press.

Rodhouse PG, Dawe EG, and O'Dor RK (eds.) (1998) *Squid Recruitment Dynamics. The Genus* Illex *as a Model, the Commercial* Illex *species and Influences on Variability*. FAO Fisheries Technical Paper, no. 376.

Roper CFE, Sweeney MJ, and Nauen CE (1984) *FAO Species Catalogue*, vol. 3. *Cephalopods of the World. An Annotated and Illustrated Catalogue of Species of Interest to Fisheries*. FAO Fisheries Synopsis.

Saunders WB and Landman NH (eds.) (1987) *NAUTILUS: The Biology and Paleobiology of a Living Fossil*. New York: Plenum Publishing.

Wells MJ (1978) *Octopus: Physiology and Behaviour of an Advanced Invertebrate*. London: Chapman and Hall.

PHYTOPLANKTON SIZE STRUCTURE

E. Marañón, University of Vigo, Vigo, Spain

Introduction

Phytoplankton are unicellular organisms that drift with the currents, carry out oxygenic photosynthesis, and live in the upper illuminated waters of all aquatic ecosystems. There are approximately 25 000 known species of phytoplankton, including eubacterial and eukaryotic species belonging to eight phyla. This phylogenetically diverse group of organisms constitutes the base of the food chain in most marine ecosystems and, since their origin more than 2.8×10^9 years ago, have exerted a profound influence on the biogeochemistry of Earth. Currently, phytoplankton are responsible for the photosynthetic fixation of around 50×10^{15} g C annually, which represents almost half of global net primary production on Earth. Some 20% of phytoplankton net primary production is exported toward the ocean's interior, either in the form of sinking particles or as dissolved material. The mineralization of this organic matter gives way to an increase with depth in the concentration of dissolved inorganic carbon. The net effect of this phytoplankton-fueled, biological pump is the transport of CO_2 from the atmosphere to the deep ocean, where it is sequestered over the timescales of deep-ocean circulation (10^2–10^3 years). A small fraction ($<1\%$) of the organic matter transported toward the deep ocean escapes mineralization and is buried in the ocean sediments, where it is retained over timescales of $>10^6$ years. It has been calculated that thanks to the biological pump the atmospheric concentration of CO_2 is maintained 300–400 ppm below the levels that would occur in the absence of marine primary production. Thus, phytoplankton play a role in the regulation of the atmospheric content of CO_2 and therefore affect climate variability.

The cell size of phytoplankton ranges widely over at least 9 orders of magnitude, from a cell volume around $0.1 \, \mu m^3$ for the smallest cyanobacteria to more than $10^8 \, \mu m^3$ for the largest diatoms. Cell size affects many aspects of phytoplankton physiology and ecology over several levels of organization, including individuals, populations, and communities. Phytoplankton assemblages are locally diverse: typically, several hundreds of species can be found in just a liter of seawater. Therefore, a full determination of the biological properties of all species living in a given water body is not possible. The study of cell size represents an integrative approach to describe the structure and function of the phytoplankton community and to understand its role within the pelagic ecosystem and the marine biogeochemical cycles. This article starts with a review of the relationship between cell size and phytoplankton metabolism and growth. It follows with a description of the general patterns of variability in the size structure of phytoplankton in the ocean. Next, the different mechanisms involved in bringing about these patterns are examined. The article ends with a consideration of the ecological and biogeochemical implications of phytoplankton size structure.

Phytoplankton Cell Size, Metabolism, and Growth

Cell Size and Resource Acquisition

Like any other photoautotrophic organisms, phytoplankton must take up inorganic nutrients and absorb light in order to synthesize new organic matter. Both nutrient uptake and light absorption are heavily dependent on cell size. The supply of nutrients to the cell may become diffusion-limited when nutrient concentrations are low, if the rate of nutrient uptake exceeds the rate of molecular diffusion and a nutrient-depleted area develops around the cell. Assuming a spherical cell shape and applying Fick's first law of diffusion, the uptake rate (U, $mol \, s^{-1}$) can be expressed as

$$U = 4\pi r D \Delta C \qquad [1]$$

where r is the cell radius (μm), D is the diffusion coefficient ($\mu m^2 \, s^{-1}$), and ΔC ($mol \, \mu m^3$) is the nutrient concentration gradient between the cell's surface and the surrounding medium. The specific uptake rate (uptake per unit of cell volume) will then be

$$U/V = 4\pi r D \Delta C (4/3 \times \pi r^3)^{-1} = 3D\Delta C r^{-2} \quad [2]$$

Equation [2] indicates that specific uptake rate decreases with the square of cell radius. However, specific (i.e., normalized to mass or volume) metabolic rates in phytoplankton, and therefore specific resource requirements, decrease with cell size much more slowly. Typically, specific metabolic rates are proportional to cell mass or volume elevated to a power between $-1/3$ and 0 (see below), or to

r elevated to a power between -1 and 0. Therefore, large cell size is a major handicap when nutrient concentrations are low. The nutrient concentration below which phytoplankton growth starts to be nutrient-limited can be calculated, for a range of cell sizes and growth rates, as follows. Nutrient limitation occurs when $U < \mu \times Q$, where μ is the specific growth rate (s^{-1}) and Q is the cell's quota for the particular nutrient $(mol\,cell^{-1})$. Q can be computed by applying an empirical relationship between cellular nitrogen content and V $(pgN = 0.017\,2\,V^{1.023})$ and D is assumed to be $1.5\,cm^2\,s^{-1}$. The threshold for nutrient limitation increases exponentially with cell radius (**Figure 1**). In this particular example, if the ambient nitrogen concentration is $10\,nM$, a cell of $r = 1\,\mu m$ could grow at growth rates well above $1\,d^{-1}$ without suffering diffusion limitation, whereas a cell of $r = 6\,\mu m$ would already experience diffusion limitation at $\mu = 0.1\,d^{-1}$. Processes such as turbulence, sinking, and swimming, which enhance the advective transport of nutrients toward the cell surface, partially compensate for the negative impact of large cell size on diffusive nutrient fluxes. The effect, however, is small and does not alter the fact that larger cells are at a disadvantage over smaller cells for nutrient uptake.

Light absorption in phytoplankton is a function of cell size and the composition and concentration of pigments. The amount of light absorbed per unit pigment decreases with cell size and with intracellular pigment concentration, because the degree of self-shading among the different pigment molecules (the so-called package effect) increases. The optical absorption cross-section of a phytoplankton cell is given by

$$a^* = (3/2)(a_s^* Q \rho) \qquad [3]$$

where

$$Q = 1 + 2(e^{-\rho}/\rho) + 2(e^{-\rho} - 1)/(\rho^2) \qquad [4]$$

and

$$\rho = a_s^* c_i d \qquad [5]$$

Units of a^* and a_s^* are $m^2\,(mg\,chl\,a)^{-1}$, Q and ρ are coefficients without dimensions, c_i is the intracellular chlorophyll a concentration $(mg\,chl\,a\,m^{-3})$, and d is cell diameter (m).

The package effect can be quantified as the ratio between the actual absorption inside the cell (a^*) and the maximum absorption possible, determined for the pigment in solution (a_s^*). The higher the package effect, the lower the value of a^*/a_s^*. The data shown in **Figure 2**, calculated assuming $a_s^* = 0.04\,m^2$ $(mg\,chl\,a)^{-1}$ and a range of c_i values between 10^6 and $10^8\,mg\,chl\,a\,m^{-3}$, indicate that the package effect increases with cell size, which means that, other things being equal, larger phytoplankton are expected to be less efficient than smaller cells at absorbing light. This effect is much stronger under low

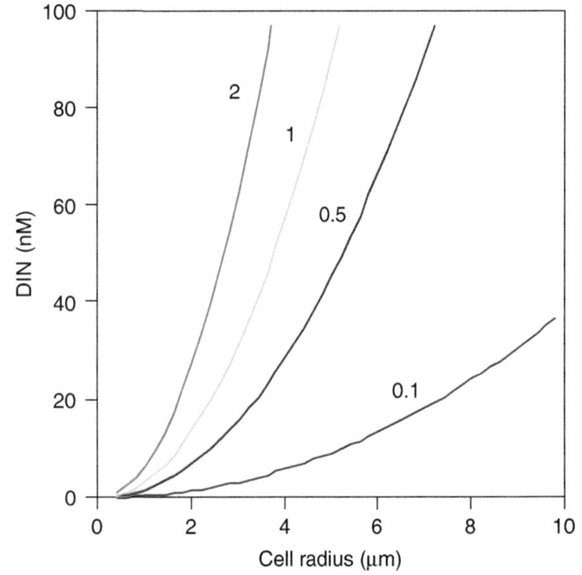

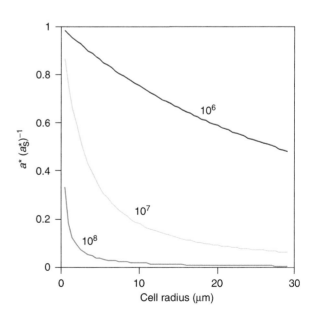

Figure 1 Relationship between cell radius and the concentration of dissolved inorganic nitrogen (DIN) below which phytoplankton growing at rates of 0.1, 0.5, 1, and $2\,d^{-1}$ begin to suffer diffusion limitation of growth.

Figure 2 Relationship between cell radius and the package effect for intracellular chlorophyll a concentrations of 10^6, 10^7, and $10^8\,mg\,m^{-3}$.

light conditions, when c_i tends to be larger as a result of photoacclimation.

The general allometric theory predicts a reduction in metabolic rates (R) with body size (W, in units of biomass or volume). From unicells to large mammals, individual metabolic rates (R) scale as:

$$R = aW^b \qquad [6]$$

which is equivalent to

$$\log R = \log a + b \log W \qquad [7]$$

where b, the size-scaling exponent, usually takes a value of 3/4 and a, the intercept of the log–log relationship, is a taxon-related constant. When biomass-specific metabolism or growth rates are considered, b takes a value near $-1/4$. Equations [6] and [7] mean that a 10-fold increase in cell size is associated with only a 7.5-fold increase in individual metabolic rate, and therefore larger organisms have a slower metabolism. In the case of phytoplankton, one would expect that the geometrical constraints on resource acquisition and use would result in a reduction in specific metabolism with cell size. However, several experimental studies with laboratory cultures and natural phytoplankton assemblages suggest that the relationship between photosynthesis and cell size cannot be predicted by a single scaling model and that phytoplankton metabolism often departs from the 3/4-power rule. For instance, the size scaling of photosynthesis in cultured diatoms has been shown to depend on light availability (**Figure 3**). Light limitation leads not only to low photosynthetic rates (irrespective of cell size) but also to a reduction in the size scaling exponent b, indicating a faster decrease in specific photosynthesis with cell size as a result of a stronger package effect in larger cells. In addition, changes in taxonomic affiliation of phytoplankton species along the size spectrum may interfere with the effects of cell size *per se*. If we consider natural phytoplankton assemblages, which include a mixture of species from diverse taxonomic groups, the scaling between phytoplankton photosynthesis is approximately isometric (**Figure 3**). This means that, under natural conditions at sea, the expected slowdown of metabolism as cell size increases does not seem to occur. This pattern results from the fact that larger species possess strategies that allow them to sustain higher metabolic rates than expected for their size (see below). Thus, both resource availability and taxonomic variation along the size spectrum help explain why the scaling of phytoplankton growth rates and cell size is quite variable. However, before turning our attention to the size scaling of

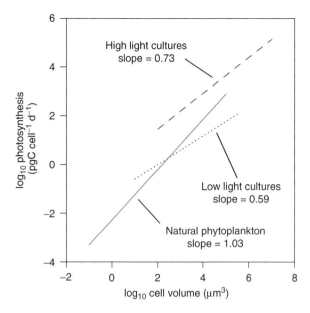

Figure 3 Scaling relationship between photosynthesis per cell and cell size in diatom cultures growing under high-light conditions, diatom cultures growing under light-limited conditions, and natural phytoplankton assemblages in the ocean.

phytoplankton growth rates, we need to look into the relationship between cell size and loss rates, because the net growth rate of any population ultimately depends on the balance between production and loss processes.

Cell Size and Loss Processes

Respiration and exudation are the main metabolic loss processes for phytoplankton. In most organisms, individual respiration rates increase as the 3/4-power of body size (e.g., $b = 3/4$). However, several studies of the size scaling of phytoplankton respiration in algal cultures show that, although taxon-related differences exist, b tends to be significantly higher than 3/4, indicating that respiration increases with cell size more steeply than predicted by the general allometric theory. This effect has been attributed to the fact that smaller algae seem to have lower respiration rates than expected for their size, perhaps as an energy-saving strategy in organisms with a comparatively small capacity for accumulation of reserves. As far as exudation is concerned, theory predicts that smaller cells, on account of their higher surface-to-volume ratios, should suffer a relatively greater loss of cellular compounds through the cell membrane. However, both laboratory work with cultures and experimental studies at sea have failed to provide conclusive evidence as to the existence of higher relative rates of exudation in smaller cells as compared to larger cells.

One of the most unquestionable effects of large cell size for phytoplankton is that it increases sinking velocity, which, for nonmotile cells, implies a reduction of their residence time in the euphotic layer. According to Stokes' law, the sinking velocity of a spherical particle increases in proportion to the square of its radius. Assuming an excess cell density over medium density of $50\,g\,l^{-1}$, a picophytoplankton cell of $r = 0.5\,\mu m$ will have a sinking velocity of only $2-3\,mm\,day^{-1}$, compared with $20-30\,m\,day^{-1}$ for a microphytoplankton cell of $r = 50\,\mu m$. Although many large phytoplankton species have acquired different strategies to cope with sinking (such as motility, buoyancy control, departure from spherical shape, etc.), smaller species are clearly at an advantage over their larger counterparts to remain within the euphotic zone. This advantage is particularly relevant in strongly stratified water columns, where the absence of upward water motion makes it unlikely for cells sinking below the pycnocline to return to the upper, well-illuminated waters.

While small cell size is superior in terms of resource acquisition and avoidance of sedimentation, large cell size can provide a major competitive advantage because it offers a refuge from predation. The main reason is that the generation time of predators increases with body size more rapidly than the generation time of phytoplankton. Small phytoplankton are typically consumed by unicellular protist herbivores, which have generation times similar to those of phytoplankton (in the order of hours to days). In contrast, larger phytoplankton are mostly grazed by metazoan herbivores, such as copepods and euphasiaceans, which have much longer generation times (in the order of weeks to months). When nutrients are injected into the euphotic layer, smaller phytoplankton are efficiently controlled by their predators and their abundance seldom increases substantially. On the contrary, larger phytoplankton, thanks to the time lag between their growth response and the numerical response of their predators, are able to form blooms and carry on growing until nutrients are exhausted. As discussed below, this trophic mechanism plays a role in determining the size structure of phytoplankton communities in contrasting marine environments.

Cell Size and Growth Rates

Experiments with cultured and natural phytoplankton species have yielded quite variable values for the scaling exponent in the equation relating growth rate (units of $time^{-1}$) to cell size. The most frequently reported values for b range between -0.1 and -0.3. It has been noted that the size scaling of

phytoplankton growth rates is relatively weak (that is, b tends to take a less negative value than $-1/4$). Furthermore, although the slope of the growth versus size relationship may be similar in different taxonomic groups, the intercept frequently is not: for instance, diatoms consistently have higher growth rates than other species of the same cell size. Another feature in the size scaling of phytoplankton growth is that very small cells (less than $5\,\mu m$ in diameter) depart from the inverse relationship between cell size and growth rate, showing slower rates than expected for their size. This pattern probably results from the influence of nonscalable components (such as the genome and the membranes), which progressively take up more space as cell size decreases, thus leaving less cell volume available for rate-limiting catalysts and the accumulation of reserves.

Although experimental determinations in natural conditions are still scarce, the available data suggest that phytoplankton populations in nature tend to exhibit less negative size scaling exponents in the power relation between growth rate and cell size. In fact, there is evidence to suggest that this size-scaling exponent may even become positive under favorable conditions for growth, such as high nutrient and light availability. This means that, when resources are plentiful, larger phytoplankton may grow faster than their smaller relatives. Several strategies allow larger species, and diatoms in particular, to achieve high growth rates (e.g., $> 1\,day^{-1}$) in nature in spite of the geometrical constraints imposed by their size. These strategies include the increase in the effective surface-to-volume ratio, due to changes in cell shape and the presence of the vacuole; the accumulation of nonlimiting substrates to increase cell size and optimize nutrient uptake; and the ability to sustain high specific uptake rates and store large amounts of reserves under conditions of discontinuous nutrient supply.

Patterns of Phytoplankton Size Structure in the Ocean

Size-Fractionated Chlorophyll a

Given that chlorophyll a (chl a) serves as a proxy for phytoplankton biomass, a common approach to study the relative importance of phytoplankton with different cell sizes is to measure the amount of chl a in size classes, usually characterized in terms of equivalent spherical diameter (ESD) of particles. The most frequently considered classes are the picophytoplankton (cells smaller than $2\,\mu m$ in ESD), the nanophytoplankton (cells with an ESD between 2 and $20\,\mu m$), and the microphytoplankton (cells with an ESD larger than $20\,\mu m$). When we plot together hundreds of

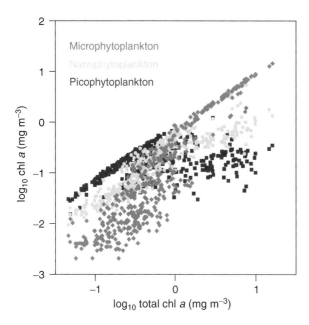

Figure 4 Chlorophyll *a* concentration in picophytoplankton, nanophytoplankton, and microphytoplankton vs. total chlorophyll *a* concentration in samples obtained throughout the euphotic layer in coastal and oceanic waters of widely varying productivity.

measurements of size-fractionated chl *a* concentration, obtained throughout the euphotic layer in coastal and oceanic waters of widely varying productivity, several consistent patterns emerge (**Figure 4**). In relatively poor waters, where total chl *a* concentrations are below 0.8–1 mg m^{-3}, picophytoplankton account for up to 80% of total chl *a*, while microphytoplankton typically contributes less than 10%. As total chl *a* increases, the concentration of picophytoplankton chl *a* reaches a plateau at around 0.5 mg m^{-3} and then decreases in very rich waters. Similarly, nanophytoplankton chl *a* rarely increases beyond 1 mg m^{-3}. By contrast, microphytoplankton chl *a* continues to increase, so that at total chl *a* levels above 2 mg m^{-3} this size class accounts for more than 80% of total chl *a*, while picophytoplankton contribute less than 10%. Compared to the other size fractions, nanophytoplankton show smaller variability in their relative contribution to total chl *a*, which normally falls within the range 20–30%. The patterns shown in **Figure 4** reflect both temporal and spatial variability in total phytoplankton biomass and the relative importance of each size class. Thus, the oligotrophic waters of the subtropical gyres are typically dominated by picophytoplankton, whereas in upwelling areas and coastal, well-mixed waters microphytoplankton usually account for most of the photosynthetic biomass. Similarly, in ecosystems that experience marked seasonal variability, microphytoplankton dominate the episodes of intense algal growth and biomass, such as the spring bloom. Small nano- and picophytoplankton

become more important during periods of prolonged stratification and low nutrient availability, as well as during conditions of intense vertical mixing that lead to light limitation of growth.

Size–Abundance Spectra

Although the partition into discrete classes is useful to describe broadly the size structure of the community, the different phytoplankton species are in fact characterized by a continuum of cell sizes that is best represented with a size–abundance spectrum. In this approach, the abundance (N) and cell size (W) of all species present in a sample are determined, using flow cytometry for picophytoplankton and small nanoplankton and optical microscopy for large nanoplankton and microphytoplankton. Size–abundance spectra are constructed by distributing the abundance data along an octave scale of cell volume. The abundance of all cells within each size interval is summed and the resulting abundance is plotted on a log-log scale against the nominal size of the interval. When these spectra are constructed at the local scale, it is common that the relationship between abundance and cell size shows irregularities, for example, departures from linearity. This is the case of large blooms, when one or a few species make up a major fraction of total phytoplankton abundance, which translates into a bump in the size–abundance spectrum. However, when numerous observations, collected over longer spatial and temporal scales, are put together, good linear relationships are usually obtained. In these cases, the slope of the linear relationship between log N and log W (the size-scaling exponent in the power relationship between N and W) is a general descriptor of the relative importance of small versus large cells in the ecosystem. **Figure 5** illustrates the differences in the phytoplankton size–abundance spectrum between two contrasting ecosystems such as the oligotrophic subtropical gyres of the Atlantic Ocean ($N \propto W^{-1.25}$) and the productive waters of the coastal upwelling region in the NW Iberian peninsula ($N \propto W^{-0.90}$). Typically, the slopes of the size–abundance spectrum in oligotrophic, open ocean waters are in the range -1.1 to -1.4, while values between -0.6 and -0.9 are measured in coastal productive environments. The size spectrum extends further to the left in oligotrophic ecosystems, reflecting the presence of the prochlorophyte *Prochlorococcus*. At 0.1–0.2 μm^3 in volume, this species is the smallest and most abundant photoautotrophic organism on Earth, and dominates picophytoplankton biomass in the oligotrophic open ocean. When combined with size-scaling relationships for biomass and metabolism, size–abundance spectra allow us to determine the variability in the flow of

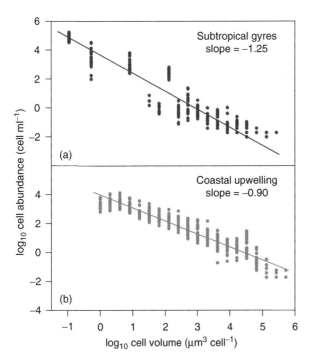

Figure 5 Size–abundance spectra in natural phytoplankton assemblages from (a) the Atlantic subtropical gyres and (b) coastal upwelling waters off NW Iberian peninsula. Spectra were constructed by combining measurements collected throughout the euphotic layer during different sampling surveys. Reprinted with permission from Marañón E, Cermeño P, Rodríguez J, Zubkov MV, and Harris RP (2007) Scaling of phytoplankton photosynthesis and cellsize in the ocean. *Limnology and oceanography* 54(5): 2194. Copyright (2008) by the Americal Society of Limnology and Ocenography, Inc.

materials and energy along the size spectrum. For instance, if phytoplankton photosynthesis per cell (P) in the ocean scales isometrically with cell size ($P \propto W^1$), the above-mentioned scaling exponents for phytoplankton abundance (between -0.6 and -0.9 for eutrophic waters and between -1.1 and -1.4 for oligotrophic ones) imply that total photosynthesis per unit volume ($N \times P$) must increase with cell size in productive ecosystems, while it decreases in unproductive ones.

Factors Controlling Phytoplankton Size Structure

Given their superior ability to avoid diffusion limitation of nutrient uptake, it is no surprise that picophytoplankton dominate in the oligotrophic waters of the open ocean. However, the competitive advantage of being small, although reduced, is not eliminated under resource sufficient conditions. One can therefore ask why is it that picophytoplankton do not dominate also in nutrient-rich environments. This is tantamount to asking what mechanisms are

responsible for the increased importance of larger cells under conditions of high nutrient availability, such as those found within upwelling regions, frontal regions, and cyclonic eddies. Hydrodynamic processes have been suggested to play a role, since upwelling water motion causes a retention of larger cells, counteracting their tendency to sink out of the euphotic layer. However, upwelling of subsurface waters also contributes to the injection of nutrients into surface waters and therefore the physical transport mechanism cannot be easily distinguished from a direct nutrient effect. These two processes, however, have been separated experimentally during iron addition experiments in high-nutrient, low-chlorophyll regions. Invariably, iron addition brings about an increase in phytoplankton biomass and a marked shift toward a dominance by larger species, usually chain-forming diatoms. Since hydrodynamics remain unaltered during these experiments, these observations highlight the role of the nutrient field in determining phytoplankton size structure. Two main, nonexclusive processes contribute to the selective growth and accumulation of larger cells in high-light and nutrient-rich environments. First, larger phytoplankton are capable of sustaining higher biomass-specific metabolic rates and growth rates than smaller cells when resources are abundant. Second, larger phytoplankton are less efficiently controlled by grazing, as a result of the difference between their generation time and that of their predators.

Ecological and Biogeochemical Implications of Phytoplankton Size Structure

Phytoplankton size structure affects significantly the trophic organization of the planktonic ecosystem and, therefore, the efficiency of the biological pump in transporting atmospheric CO_2 toward deep waters (**Table 1**). In communities dominated by picophytoplankton, where resource limitation leads to low phytoplankton biomass and production, the dominant trophic pathway is the microbial food web. Given that the growth rates of picophytoplankton and their protist microbial grazers (dinoflagellates, ciliates, and heterotrophic nanoflagellates) are similar, trophic coupling between production and grazing is tight, most of phytoplankton daily primary production is consumed within the microbial community, and the standing stock of photosynthetic biomass is relatively constant. Phytoplankton exudation and microzooplankton excretion contribute to an important production of dissolved organic matter, which fuels bacterial production. In turn,

Table 1 General ecological and biogeochemical properties of plankton communities in which phytoplankton are dominated by small vs. large cells

Phytoplankton dominated by	Small cells	Large cells
Total phytoplankton biomass	Low	High
Total primary production	Low	High
Dominant trophic pathway	Microbial food web	Classic food chain
Main loss process for phytoplankton	Grazing by protists	Sedimentation and grazing by metazoans
Photosynthesis-to-respiration ratio	~1	>1
f-ratio and e-ratio	5–15%	>40%
Main fate of primary production	Recycling within the euphotic layer	Export toward deep waters

bacteria are efficiently controlled by protist microbial grazers. The resulting, complex food web is characterized by intense recycling of matter and low efficiency in the transfer of primary production toward larger organisms such as mesozooplankton or fish. Photosynthetic production of organic matter is balanced by the respiratory losses with the microbial community. In addition, the small size of microbial plankton implies that losses through sedimentation are unimportant. As a result, little newly produced organic matter escapes the euphotic later. In contrast, plankton communities dominated by large phytoplankton, such as chain-forming diatoms, are characterized by enhanced sinking rates and simpler trophic pathways, where phytoplankton are grazed directly by mesozooplankton (the so-called classic food chain). Phytoplankton photosynthesis exceeds community respiration, leaving an excess of organic matter available for export. Thus, a major fraction of phytoplankton production is eventually transported toward deep waters, either directly through sinking of ungrazed cells, or indirectly through sedimentation of packaged materials such as aggregates and zooplankton fecal pellets. It must be noted that the microbial trophic pathway is always present in all planktonic communities, but its relative importance decreases in productive waters because of the addition of the classic food chain.

The ecological properties outlined above for phytoplankton assemblages dominated by small versus large cells dictate the biogeochemical functioning of the biological pump in contrasting marine environments. In stable, oligotrophic ecosystems, where small photoautotrophs dominate, primary production sustained by nutrients coming from outside the euphotic layer (new production) is small, and so is the ratio between new production and total production (the f-ratio), as well as the ratio between exported production and total production (the e-ratio). Typical values of the f- and e-ratios in these systems are in the range 5–15%. Given that, in the long term, only new production has the potential to contribute to the transport of biogenic carbon toward the deep ocean, the biological pump in these systems has a low efficiency and the net effect of the biota on the ocean–atmosphere CO_2 exchange is small. By contrast, phytoplankton assemblages dominated by larger cells are typical of dynamic environments that are subject to perturbations leading to enhanced resource supply. In these systems, production and consumption of organic matter are decoupled, new and export production are relatively high (e- and f-ratios >40%), and the biological pump effectively transports biogenic carbon toward the ocean's interior, thus contributing to CO_2 sequestration.

Nomenclature

a^*	absorption coefficient of pigments *in vivo*
a_s^*	absorption coefficient of pigments in solution
c_i	intracellular chlorophyll *a* concentration
d	cell diameter
D	nutrient diffusion coefficient
N	cell abundance
P	photosynthesis per cell
Q	cellular nutrient quota
r	cell radius
R	metabolic rate
U	nutrient uptake rate
V	cell volume
W	cell size (volume or weight)
μ	growth rate

See also

Primary Production Distribution. Primary Production Methods. Primary Production Processes.

Further Reading

Agawin NSR, Duarte CM, and Agustí S (2000) Nutrient and temperature control of the contribution of picoplankton to phytoplankton biomass and production. *Limnology and Oceanography* 45: 591–600.

Banse K (1976) Rates of growth, respiration and photosynthesis of unicellular algae as related to cell size – a review. *Journal of Phycology* 12: 135–140.

Blasco D, Packard TT, and Garfield PC (1982) Size dependence of growth rate, respiratory electron transport system activity, and chemical composition in marine diatoms in the laboratory. *Journal of Phycology* 18: 58–63.

Brown JH, Gillooly JF, Allen AP, Savage VM, and West GB (2004) Toward a metabolic theory of ecology. *Ecology* 85: 1771–1789.

Cermeño P, Marañón E, Rodríguez J, and Fernández E (2005) Large-sized phytoplankton sustain higher carbon-specific photosynthesis than smaller cells in a coastal eutrophic ecosystem. *Marine Ecology Progress Series* 297: 51–60.

Chisholm SW (1992) Phytoplankton size. In: Falkowski PG and Woodhead AD (eds.) *Primary Productivity and Biogeochemical Cycles in the Sea*, pp. 213–237. New York: Plenum.

Falkowski PG, Laws EA, Barber RT, and Murray JW (2003) Phytoplankton and their role in primary, new, and export production. In: Fasham MJR (ed.) *Ocean Biogeochemistry – The Role of the Ocean Carbon Cycle on Global Change*, pp. 99–122. Berlin: Springer.

Finkel ZV (2001) Light absorption and size scaling of light-limited metabolism in marine diatoms. *Limnology and Oceanography* 46: 86–94.

Kiørboe T (1993) Turbulence, phytoplankton cell size, and the structure of pelagic food webs. *Advances in Marine Biology* 29: 1–72.

Legendre L and Rassoulzadegan F (1996) Food-web mediated export of biogenic carbon in oceans. *Marine Ecology Progress Series* 145: 179–193.

Li WKW (2002) Macroecological patterns of phytoplankton in the north western North Atlantic Ocean. *Nature* 419: 154–157.

Marañón E, Holligan PM, Barciela R, *et al.* (2001) Patterns of phytoplankton size-structure and productivity in contrasting open ocean environments. *Marine Ecology Progress Series* 216: 43–56.

Marañó E, Cermeño P, Rodríguez J, Zubkov MV, and Harris RP (2007) Scaling of phytoplankton photosynthesis and cellsize in the ocean. *Limnology and oceanography* 54(5): 2194.

Platt T, Lewis M, and Geider R (1984) Thermodynamics of the pelagic ecosystem: Elemental closure conditions for biological production in the open ocean. In: Fasham MJR (ed.) *Flows of Energy and Material in Marine Ecosystems*, pp. 49–84. New York: Plenum.

Raven JA (1998) Small is beautiful: The picophytoplankton. *Functional Ecology* 12: 503–513.

Rodríguez J, Tintoré J, Allen JT, *et al.* (2001) Mesoscale vertical motion and the size structure of phytoplankton in the ocean. *Nature* 410: 360–363.

Sheldon RW, Prakash A, and Sutcliffe WH (1972) The size distribution of particles in the ocean. *Limnology and Oceanography* 17: 327–341.

Tang EPY (1995) The allometry of algal growth rates. *Journal of Plankton Research* 17: 1325–1335.

PRIMARY PRODUCTION METHODS

J. J. Cullen, Department of Oceanography, Halifax, Canada

Introduction

Primary production is the synthesis of organic material from inorganic compounds, such as CO_2 and water. The synthesis of organic carbon from CO_2 is commonly called carbon fixation: CO_2 is fixed by both photosynthesis and chemosynthesis. By far, photosynthesis by phytoplankton accounts for most marine primary production. Carbon fixation by macroalgae, microphytobenthos, chemosynthetic microbes, and symbiotic associations can be locally important.

Only the measurement of marine planktonic primary production will be discussed here. These measurements have been made for many decades using a variety of approaches. It has long been recognized that different methods yield different results, yet it is equally clear that the variability of primary productivity, with depth, time of day, season, and region, has been well described by most measurement programs. However, details of these patterns can depend on methodology, so it is important to appreciate the uncertainties and built-in biases associated with different methods for measuring primary production.

Definitions

Primary production is centrally important to ecological processes and biogeochemical cycling in marine systems. It is thus surprising, if not disconcerting, that (as discussed by Williams in 1993), there is no consensus on a definition of planktonic primary productivity, or its major components, net and gross primary production. One major reason for the problem is that descriptions of ecosystems require clear conceptual definitions for processes (e.g., net daily production of organic material by phytoplankton), whereas the interpretation of measurements requires precise operational definitions, for example, net accumulation of radiolabeled CO_2 in particulate matter during a 24 h incubation. Conceptual and operational definitions can be reconciled for particular approaches, but no one set of definitions is sufficiently general, yet detailed, to serve as a framework both for measuring planktonic primary

production with a broad variety of methods and for interpreting the measurements in a range of scientific contexts. It is nonetheless useful to define three components of primary production that can be estimated from measurements in closed systems:

- **Gross primary production** (P_g) is the rate of photosynthesis, not reduced for losses to excretion or to respiration in its various forms
- **Net primary production** (P_n) is gross primary production less losses to respiration by phytoplankton
- **Net community production** (P_{nc}) is net primary production less losses to respiration by heterotrophic microorganisms and metazoans.

Other components of primary production, such as new production, regenerated production, and export production, must be characterized to describe food-web dynamics and biogeochemical cycling. As pointed out by Platt and Sathyendranath in 1993, in any such analysis, great care must be taken to reconcile the temporal and spatial scales of both the measurements and the processes they describe.

Marine primary production is commonly expressed as grams or moles of carbon fixed per unit volume, or pet unit area, of sea water per unit time. The timescale of interest is generally 1 day or 1 year. Rates are characterized for the euphotic zone, commonly defined as extending to the depth of 1% of the surface level of photosynthetically active radiation (PAR: 400–700 nm). This convenient definition of euphotic depth (sometimes simplified further to three times the depth at which a Secchi disk disappears) is a crude and often inaccurate approximation of where gross primary production over 24 h matches losses to respiration and excretion by phytoplankton. Regardless, rates of photosynthesis are generally insignificant below the depth of 0.1% surface PAR.

Photosynthesis and Growth of Phytoplankton

Primary production is generally measured by quantifying light-dependent synthesis of organic carbon from CO_2 or evolution of O_2 consistent with the simplified description of photosynthesis as the reaction:

$$CO_2 + 2H_2O \overset{\sim 8h\nu}{\rightarrow} (CH_2O) + H_2O + O_2 \quad [1]$$

Absorbed photons are signified by $h\nu$ and the carbohydrates generated by photosynthesis are represented as CH_2O. Carbon dioxide in sea water is found in several chemical forms which exchange quickly enough to be considered in aggregate as total CO_2 (TCO$_2$). In principle, photosynthesis can be quantified by measuring any of three light-dependent processes: (1) the increase in organic carbon; (2) the decrease of TCO$_2$; or (3) the increase of O_2. However, growth of phytoplankton is not so simple: since phytoplankton are composed of proteins, lipids, nucleic acids, and other compounds besides carbohydrate, both photosynthesis and the assimilation of nutrients are required. Consequently, many chemical transformations are associated with primary production, and eqn [1] does not accurately describe the process of light-dependent growth.

It is therefore useful to describe the growth of phytoplankton (i.e., net primary production) with a more general reaction that describes how transformations of carbon and oxygen depend on the source of nutrients (particularly nitrogen) and on the chemical composition of phytoplankton. For growth on nitrate:

$$1.0NO_3^- + 5.7CO_2 + 5.4H_2O$$
$$\rightarrow (C_{5.7}H_{9.8}O_{2.3}N) + 8.25O_2 + 1.0OH^- \qquad [2]$$

The idealized organic product, $C_{5.7}H_{9.8}O_{2.3}N$, represents the elemental composition of phytoplankton. Ammonium is more reduced than nitrate, so less water is required to satisfy the demand for reductant:

$$1.0NH_4^+ + 5.7CO_2 + 3.4H_2O$$
$$\rightarrow (C_{5.7}H_{9.8}O_{2.3}N) + 6.25O_2 + 1.0H^+ \qquad [3]$$

The photosynthetic quotient (PQ; mol mol^{-1}) is the ratio of O_2 evolved to inorganic C assimilated. It must be specified to convert increases of oxygen to the synthesis of organic carbon. For growth on nitrate as described by eqn [2], PQ is 1.45 mol mol^{-1}; with ammonium as the source of N, PQ is 1.10. The photosynthetic quotient also reflects the end products of photosynthesis, the mixture of which varies according to environmental conditions and the species composition of phytoplankton. For example, if the synthesis of carbohydrate is favored, as can occur in high light or low nutrient conditions, PQ is lower because the reaction described in eqn [1] becomes more important. Uncertainty in PQ is often ignored. This can be justified when the synthesis of organic carbon is measured directly, but large errors can be introduced when attempts are made to infer carbon fixation from the dynamics of oxygen. Excretion of organic material would have a small influence on PQ and is not considered here.

Approaches

Primary production can be estimated from chlorophyll (from satellite color or *in situ* fluorescence) if carbon uptake per unit of chlorophyll is known. Therefore, 'global' estimates of primary production depend on direct measurements by incubation. The technical objectives are to obtain a representative sample of sea water, contain it so that no significant exchange of materials occurs, and to measure light-dependent changes in carbon or oxygen during incubations that simulate the natural environment. Methods vary widely, and each approach involves compromises between needs for logistical convenience, precision, and the simulation of natural conditions. Each program of measurement involves many decisions, each of which has consequences for the resulting measurements. Several options are listed in **Tables 1** and **2** and discussed below.

Light-dependent Change in Dissolved Oxygen

The light-dark oxygen method is a standard approach for measuring photosynthesis in aquatic systems, and it was the principal method for measuring marine primary production until it was supplanted by the ^{14}C method, which is describged below. Accumulation of oxygen in a clear container (light bottle) represents net production by the enclosed community, and the consumption of oxygen in a dark bottle is a measure of respiration. Gross primary production is estimated by subtracting the dark bottle result from that for the light bottle. It is thus assumed that respiration in the light equals that in the dark. As documented by Geider and Osborne in their 1992 monograph, this assumption does not generally hold, so errors in estimation of the respiratory component of P_g must be tolerated unless isotopically labelled oxygen is used (see below).

Methods based on the direct measurement of oxygen are less sensitive than techniques using the isotopic tracer ^{14}C. However, careful implementation of procedures using automated titration or pulsed oxygen electrodes can yield useful and reliable data, even from oligotrophic waters of the open ocean. Interpretation of results is complicated by containment effects common to all methods for direct measurement of primary production (see below). Also, a value for photosynthetic quotient must be assumed in order to infer carbon fixation from oxygen production. Abiotic consumption of oxygen through photochemical reactions with dissolved organic matter can also contribute to the measurement, primarily near the surface, where the effective ultraviolet wavelengths penetrate.

Table 1 Measurements that can be related to primary production

Measurement	Advantages	Disadvantages	Comments
Change in TCO_2	Direct measure of net inorganic C fixation	Relatively insensitive: small change relative to large background	Not generally practical for open-ocean work
Change in oxygen concentration (high precision titration)	Direct measures of O_2 dynamics can yield estimates of net and gross production	Small change relative to large background. Interpretation of light-dark incubations is not simple	Very useful if applied with great care. Requires knowledge of PQ to convert to C-fixation
Incorporation of ^{14}C-bicarbonate into organic material (radioactive isotope)	Very sensitive and relatively easy. Small volumes can be used and many samples can be processed	Tracer dynamics complicate interpretations. Radioactive – requires special precautions and permission	The most commonly used method in oceanography
Incorporation of ^{13}C-bicarbonate into organic material (stable isotope)	No problems with radioactivity	Less sensitive and more work than ^{14}C method. Larger volumes required	A common choice when ^{14}C method is impractical
Measurement of $^{18}O_2$ production from $H_2^{18}O$	Measures photosynthesis without interference from respiration	Requires special equipment	A powerful research tool, not generally used for routine measurements

Table 2 Approaches for incubating samples for the measurement of primary production

Incubation system	Advantages	Disadvantages	Comments
Incubation in situ	Best simulation of the natural field of light and temperature	Limits mobility of the ship. Vertical mixing is not simulated. Artifacts possible if deployed or recovered in the light	Not perfect, but a good standard method if a station can be occupied all day
Simulated in situ	Many stations can be surveyed. Easy to conduct time-courses and experimental manipulations	Special measures must be taken to stimulate spectral irradiance and temperatureo. Vertical mixing not simulated	Commonly used when many stations must be sampled. Significant errors possible if incubated samples are exposed to unnatural irradiance and temperature
Photosynthesis versus irradiance (P versus E) incubator (^{14}C)	Data can be used to model photosynthesis in the water column. With care, vertical mixing can be addressed	Extra expenses and precautions are required. Spectral irradiance is not matched to nature. Results depend on timescale of measurement. Analysis can be tricky	A powerful approach when applied with caution

Light-dependent Change in Dissolved Inorganic Carbon

Changes in TCO_2 during incubations of sea water can be measured by several methods. Uncertainties related to biological effects on pH-alkalinity-TCO_2 relationships are avoided through the use of coulometric titration or infrared gas analysis after acidification. Measurement of gross primary production and net production of the enclosed community is like that for the light-dark oxygen method, but there is no need to assume a photosynthetic quotient. However, precision of the analyses is not quite as good as

for bulk oxygen methods. Extra procedures, such as filtration, would be required to assess precipitation of calcium carbonate (e.g., by coccolithophores) and photochemical production of CO_2. These processes cause changes in TCO_2 that are not due to primary production. The TCO_2 method is not used routinely for measurement of primary production in the ocean.

The ^{14}C Method

Marine primary production is most commonly measured by the ^{14}C method, which was introduced by Steemann Nielsen in 1952. Samples are collected and the dissolved inorganic carbon pool is labeled with a known amount of radioactive ^{14}C-bicarbonate. After incubation in clear containers, carbon fixation is quantified by liquid scintillation counting to detect the appearance of ^{14}C in organic form. Generally, organic carbon is collected as particles on a filter. Both dissolved and particulate organic carbon can be quantified by analyzing whole water after acidification to purge the inorganic carbon. It is prudent to correct measurements for the amount of label incorporated during incubations in the dark. The ^{14}C method can be very sensitive, and good precision can be obtained through replication and adequate time for scintillation counting. The method has drawbacks, however. Use of radioisotopes requires special procedures for handling and disposal that can greatly complicate or preclude some field operations. Also, because ^{14}C is added as dissolved inorganic carbon and gradually enters pools of particulate and dissolved matter, the dynamics of the labeled carbon cannot accurately represent all relevant transformations between organic and inorganic carbon pools. For example, respiration cannot be quantified directly. The interpretation of ^{14}C uptake (discussed below) is thus anything but straightforward.

The ^{13}C Method

The ^{13}C method is similar to the ^{14}C method in that a carbon tracer is used. Bicarbonate enriched with the stable isotope ^{13}C is added to sea water and the incorporation of CO_2 into particulate matter is followed by measuring changes in the $^{13}C{:}^{12}C$ ratio of particles relative to that in the TCO_2 pool. Isotope ratios are measured by mass spectrometry or emission spectrometry. Problems associated with radioisotopes are avoided, but the method can be more cumbersome than the ^{14}C method (e.g., larger volumes are generally needed) and some sensitivity is lost.

The ^{18}O Method

Gross photosynthesis can be measured as the production of ^{18}O-labeled O_2 from water labeled with this heavy isotope of oxygen (see eqn [1]). Detection is carried out by mass spectrometry. Net primary production of the enclosed community is measured as the increase of oxygen in the light bottle, and respiration is estimated by difference. In principle, the difference between gross production measured with ^{18}O and gross production from light-dark oxygen changes is due to light-dependent changes in respiration and photochemical consumption of oxygen. Respiration can also be measured directly by tracking the production of $H_2^{18}O$ from $^{18}O_2$.

The ^{18}O method is sufficiently sensitive to yield useful results even in oligotrophic waters. It is not commonly used, but when the measurements have been made and compared to other measures of productivity, important insights have been developed.

Methodological Considerations

Many choices are involved in the measurement of primary production. Most influence the results, some more predictably than others. A brief review of methodological choices, with an emphasis on the ^{14}C method, reveals that the measurement of primary production is not an exact science.

Sampling

Every effort should be made to avoid contamination of samples obtained for the measurement of primary production. Concerns about toxic trace elements are especially important in oceanic waters. Trace metal-clean procedures, including the use of specially cleaned GO-FLO sampling bottles suspended from Kevlar^{TM} line, prevent the toxic contamination associated with other samplers, particularly those with neoprene closure mechanisms. Frequently, facilitates for trace metal-clean sampling are unavailable. Through careful choice of materials and procedures, it is possible to minimize toxic contamination, but enrichment with trace nutrients such as iron is probably unavoidable. Such enrichment could stimulate the photosynthesis of phytoplankton, but only after several hours or longer.

Exposure of samples to turbulence during sampling can damage the phytoplankton and other microbes, altering measured rates. Also, significant inhibition of photosynthesis can occur when deep samples acclimated to low irradiance are exposed to bright light, even for brief periods, during sampling.

Method of Incubation

Samples of seawater can be incubated *in situ*, under simulated *in situ* (SIS) conditions, or in incubators illuminated by lamps. Each method has advantages and disadvantages (**Table 2**).

Incubation *in situ* ensures the best possible simulation of natural conditions at the depths of sampling. Ideally, samples are collected, prepared, and deployed before dawn in a drifting array. Samples are retrieved and processed after dusk or before the next sunrise. If deployment or retrieval occur during daylight, deep samples can be exposed to unnaturally high irradiance during transit, which can lead to artifactually high photosynthesis and perhaps to counteracting inhibitory damage. Incubation of samples *in situ* limits the number of stations that can be visited during a survey, because the ship must stay near the station in order to retrieve the samples. Specialized systems both capture and inoculate samples *in situ*, thereby avoiding some logistical problems.

Ship operations can be much more flexible if primary productivity is measured using SIS incubations. Water can be collected at any time of day and incubated for 24 h on deck in transparent incubators to measure daily rates. The incubators, or bottles in the incubators, are commonly screened with neutral density filters (mesh or perforated metal screen) to reproduce fixed percentages of PAR at the surface. Light penetration at the station must be estimated to choose the sampling depths corresponding to these light levels. Cooling comes from surface sea water. This system has many advantages, including improved security of samples compared with *in situ* deployment, convenient access to incubations for time-course measurements, and freedom of ship movement after sampling. Because the spectrally neutral attenuation of sunlight by screens does not mimic the ocean, significant errors can be introduced for samples from the lower photic zone where the percentage of surface PAR imposed by a screen will not match the percentage of photosynthetically utilizable radiation (PUR, spectrally weighted for photosynthetic absorption) at the sampling depth. Incubators can be fitted with colored filters to simulate subsurface irradiance for particular water types. Also, chillers can be used to match subsurface temperatures, avoiding artifactual warming of deep samples.

Artificial incubators are used to measure photosynthesis as a function of irradiance (*P* versus *E*). Illumination is produced by lamps, and a variety of methods are used to provide a range of light levels to as many as 24 or more subsamples. Temperature is controlled by a water bath. The duration of incubation generally ranges from about 20 min to several hours, and results are fitted statistically to a *P* versus *E* curve. If *P* versus *E* is determined for samples at two or more depths (to account for physiological differences), results can be used to describe photosynthesis in the water column as a function of irradiance. Such a calculation requires measurement of light penetration in the water and consideration of spectral differences between the incubator and natural waters. Because many samples, usually of small volume, must be processed quickly, only the ^{14}C method is appropriate for most *P* versus *E* measurements in the ocean.

Containers

Ideally, containers for the measurement of primary production should be transparent to ultraviolet and visible solar radiation, completely clean, and inert (**Table 3**). Years ago, soft glass bottles were used. Now it is recognized that they can contaminate samples with trace elements and exclude naturally occurring ultraviolet radiation. Glass scintillation vials are still used for some *P* versus *E* measurements of short duration; checks for effects of contaminants are warranted. Compared with soft glass, laboratory-grade borosilicate glass bottles (e.g., Pyrex^{TM}) have better optical properties, excluding only UV-B (320–400 nm) radiation. Also, they contaminate less. Laboratory-grade glass bottles are commonly used for oxygen measurements. Polycarbonate bottles are favored in many studies because they are relatively inexpensive, unbreakable, and can be cleaned meticulously. Polycarbonate absorbs UV-B and some UV-A 320–400 nm radiation, so near-surface inhibition of photosynthesis can be underestimated. The error can be significant very close to the surface, but not when the entire water column is considered. Teflon^{TM} bottles, more expensive than polycarbonate, are noncontaminating and they transmit both visible and UV (280–4000 nm) radiation. When the primary emphasis is an assessing effects of UV radiation, incubations are conducted in polyethylene bags or in bottles made of quartz or Teflon^{TM}.

The size of the container is an important consideration. Small containers (≤ 50 ml) are needed when many samples must be processed (e.g. for *P* versus *E*) or when not much water is available. However, small samples cannot represent the planktonic assemblage accurately when large, rare organisms or colonies are in the water. Smaller containers have greater surface-to-volume ratios, and thus small samples have greater susceptibility to contamination. If it is practical, larger samples should be used for the measurement of primary production. The problems with large samples are mostly logistical. More water, time, and materials are needed, more radioactive waste is generated, and some measurements can be compromised if handling times are too long.

Duration of Incubation

Conditions in containers differ from those in open water, and the physiological and chemical differences between samples and nature increase as the incubations proceed. Unnatural changes during incubation include: extra accumulation of phytoplankton due to exclusion of grazers; enhanced inhibition of photosynthesis in samples collected from mixed

Table 3 Containers for incubations

Container	Advantages	Disadvantages	Comments
Polycarbonate bottle	Good for minimizing trace element contamination Nearly unbreakable Affordable	Excludes UV radiation Compressible, leading to gas dissolution and filtration problems for deep samples	Many advantages for routine and specialized measurements at sea
Laboratory grade borosilicate glass (e.g., Pyrex™)	More transparent to UV Incompressible	More trace element contamination Breakable	A reasonable choice if compromises are evaluated
Borosilicate glass scintillation vials	Inexpensive Practical choice for P versus E	Contaminate samples with trace elements and Si Exclude UV radiation	Can be used with caution for short-term P versus E measurements
Polyethylene bag	Inexpensive Compact UV-transparent	More difficult to handle Requires caution with respect to contamination	Used for special projects, e.g., effect of UV
Quartz, Teflon™	UV-transparent Teflon™ does not contaminate	Relatively expensive	Used for work assessing effects of UV
Small volume (1–25 ml)	Good for P versus E Samples can be processed by acidification (no filtration)	Cannot sample large, rare phytoplankton evenly Containment effects more likely	Used for P versus E with many replicates
Large volume (1–20 l)	Some containment artifacts are minimized Potential for time-course measurements	More work Longer filtration times with possible artifacts	Required for some types of analysis, e.g., [13]C

layers and incubated at near-surface irradiance; stimulation of growth due to contamination with a limiting trace nutrient such as iron; and poisoning of phytoplankton with a contaminant, such as copper. When photosynthesis is measured with a tracer, the distribution of the tracer among pools changes with time, depending on the rates of photosynthesis, respiration, and grazing. All of these effects, except possibly toxicity, are minimized by restricting the time of incubation, so a succession of short incubations, or P versus E measurements, can in principle yield more accurate data than a day-long incubation. This requires much effort, however, and extrapolation of results to daily productivity is still uncertain. The routine use of dawn-to-dusk or 24 h incubations may be subject to artifacts of containment, but it has the advantage of being much easier to standardize.

Filtration or Acidification

Generally, an incubation with ^{14}C or ^{13}C is terminated by filtration. Labeled particles are collected on a filter for subsequent analysis. Residual dissolved inorganic carbon can be removed by careful rinsing with filtered sea water; exposure of the filter to acid purges both dissolved inorganic carbon and precipitated carbonate. The choice of filter can influence the result. Whatman GF/F glass-fiber filters, with nominal pore size 0.7 μm, are commonly used and widely (although not universally) considered to capture all sizes of phytoplankton quantitatively. Perforated filters with uniform pore sizes ranging from 0.2 to 5 μm or more can be used for size-fractionation. Particles larger than the pores can squeeze through, especially when vacuum is applied. The filters are also subject to clogging, leading to retention of small particles.

Labeled dissolved organic carbon, including excreted photosynthate and cell contents released through 'sloppy feeding' of grazers, is not collected on filters. These losses are generally several percent of total or less, but under some conditions, excretion can be much more. When ^{14}C samples are processed with a more cumbersome acidification and bubbling technique, both dissolved and particulate organic carbon is measured.

Interpretation of Carbon Uptake

Because the labeled carbon is initially only in the inorganic pool, short incubations with [14]C (≤ 1 h) characterize something close to gross production. As incubations proceed, cellular pools of organic carbon are labeled, and some ^{14}C is respired. Also, some excreted ^{14}C organic carbon is assimilated by heterotrophic microbes, and some of the phytoplankton are consumed by grazers. So, with time, the measurement comes closer to an estimate of the net primary production of the enclosed community (**Table 4**).

Table 4 Incubation times for the measurement of primary production

Incubation time	Advantages	Disadvantages	Comments
Short (≤ 1 h)	Little time for unnatural physiological changes	Usually requires artificial illumination Uncertain extrapolation to daily rates in nature	Closer to P_g
1–6 h	Convenient Appropriate for some process studies	Uncertain extrapolation to daily rates in nature	Used for P versus E, especially with larger samples
Dawn–dusk	Good for standardization of methodology	Limits the number of stations that can be sampled Containment effects Vertical mixing is not simulated, leading to artifacts	A good choice for standard method using *in situ* incubation Closer to P_{nc} near the surface; close to P_g deep in the photic zone
24 h	Good for standardization of methodology	Results may vary depending on start time. Longer time for containment effects to act	A good standard for SIS incubations. Close to P_{nc} near the surface; closer to P_g deep in the photic zone

However, many factors, including the ratio of photosynthesis to respiration, influence the degree to which ^{14}C uptake resembles gross versus net production. Consequently, critical interpretation of ^{14}C primary production measurements requires reference to models of carbon flow in the system.

Conclusions

Primary production is not like temperature, salinity or the concentration of nitrate, which can in principle be measured exactly. It is a biological process that cannot proceed unaltered when phytoplankton are removed from their natural surroundings. Artifacts are unavoidable, but many insults to the sampled plankton can be minimized through the exercise of caution and skill. Still, the observed rates will be influenced by the methods chosen for making the measurements. Interpretation is also uncertain: the ^{14}C method is the standard operational technique for measuring marine primary production, yet there are no generally applicable rules for relating ^{14}C measurements to either gross or net primary production.

Fortunately, uncertainties in the measurements and their interpretation, although significant, are not large enough to mask important patterns of primary productivity in nature. Years of data on marine primary production have yielded information that has been centrally important to our understanding of marine ecology and biogeochemical cycling. Clearly, measurements of marine primary production are useful and important for understanding the ocean. It is nonetheless prudent to recognize that the measurements themselves require circumspect interpretation.

See also

Primary Production Distribution. Primary Production Processes.

Further Reading

Geider RJ and Osborne BA (1992) *Algal Photosynthesis: The Measurement of Algal Gas Exchange*. New York: Chapman and Hall.

Morris I (1981) Photosynthetic products, physiological state, and phytoplankton growth. In: Platt T (ed.) Physiological Bases of Phytoplankton Ecology. *Canadian Bulletin of Fisheries and Aquatic Science* 210: 83–102.

Peterson BJ (1980) Aquatic primary productivity and the ^{14}C–CO$_2$ method: a history of the productivity problem. *Annual Review of Ecology and Systematics* 11: 359–385.

Platt T and Sathyendranath S (1993) Fundamental issues in measurement of primary production. *ICES Marine Science Symposium* 197: 3–8.

Sakshaug E, Bricaud A, Dandonneau Y, *et al.* (1997) Parameters of photosynthesis: definitions, theory and interpretation of results. *Journal of Plankton Research* 19: 1637–1670.

Steemann Nielsen E (1963) Productivity, definition and measurement. In: Hill MW (ed.) *The Sea*, vol. 1, pp. 129–164. New York: John Wiley.

Williams PJL (1993a) Chemical and tracer methods of measuring plankton production. *ICES Marine Science Symposium* 197: 20–36.

Williams PJL (1993b) On the definition of plankton production terms. *ICES Marine Science Symposium* 197: 9–19.

PRIMARY PRODUCTION PROCESSES

J. A. Raven, Biological Sciences, University of Dundee, Dundee, UK

Introduction

This article summarizes the information available on the magnitude of and the spatial and temporal variations in, marine plankton primary productivity. The causes of these variations are discussed in terms of the biological processes involved, the organisms which bring them about, and the relationships to oceanic physics and chemistry. The discussion begins with a definition of primary production.

Primary producers are organisms that rely on external energy sources such as light energy (photolithotrophs) or inorganic chemical reactions (chemolithotrophs). These organisms are further characterized by obtaining their elemental requirements from inorganic sources, e.g. carbon from inorganic carbon such as carbon dioxide and bicarbonate, nitrogen from nitrate and ammonium (and, for some, dinitrogen), and phosphate from inorganic phosphate. These organisms form the basis of food webs, supporting all organisms at higher trophic levels. While chemolithotrophy may well have had a vital role in the origin and early evolution of life, the role of chemolithotrophs in the present ocean is minor in energy and carbon terms (**Table 1**), but is very important in biogeochemical element cycling, for example in the conversion of ammonium to nitrate.

Quantitatively a much more important process in primary productivity on a global scale is photolithotrophy (**Table 1**). Essentially all photolithotrophs which contribute to net inorganic carbon removal from the atmosphere or the surface ocean are O_2-evolvers, using water as electron donor for carbon dioxide reduction, according to eqn [1]:

$$CO_2 + 2H_2^*O + 8 \text{ photons} \to (CH_2O) + H_2O + {}^*O_2 \qquad [1]$$

The contribution of O_2-evolving photolithotrophy from terrestrial environments is greater than that in the oceans, despite the sea occupying more than two-thirds of the surface of the planet (**Table 1**). Sunlight is attenuated by sea water to an extent which limits primary productivity to, at most, the top 300 m of the ocean. Since only a few percent of the ocean floor is within 300 m of the surface, the role of benthic primary producers (i.e. those attached to the ocean floor) is small in terms of the total marine primary production (**Table 1**). Despite the relatively small area of benthic habitat for photolithotrophs in the ocean as a percentage of the total sea area (2%), benthic primary productivity producers account for almost 10% of marine primary productivity (**Table 1**).

Until very recently it has been assumed that the photosynthetic primary producers in the marine phytoplankton are the O_2-evolvers with two photochemical reactions involved in moving each electron from water to carbon dioxide, although molecular genetic data from around 1990 indicated the presence of erythrobacteria in surface ocean waters. Recent work has shown that both rhodopsin-based and bacteriochlorophyll-based phototrophy is widespread in the surface ocean. This phototrophy does not involve O_2 evolution and, while it does not necessarily involve net carbon dioxide fixation, it may impact on surface ocean carbon dioxide dynamics. Thus, growth of prokaryotes using dissolved organic carbon can occur with less carbon dioxide produced per unit dissolved organic carbon incorporated into organic carbon by the use of energy from photons to replace energy that would otherwise be transformed by oxidation of dissolved organic carbon. It is probable that these phototrophs which do not evolve O_2 contribute <1% to gross carbon dioxide fixation by the surface ocean.

This article investigates the reasons for this constrained planktonic primary production in the ocean in terms of the marine pelagic habitat and the diversity of the organisms involved in terms of their phylogeny and life-form.

The Habitat

The surface ocean absorbs solar radiation via the properties of sea water, as well as of any dissolved organic material and of particles. A very small fraction (≤1% in most areas) of the 400–700 nm component is converted to energy in organic matter in photosynthesis, while the rest is converted to thermal energy. In the absence of wind shear and ocean currents, themselves ultimately caused by solar energy input, the thermal expansion of the surface water would cause permanent stratification, except near the poles in winter.

Table 1 Net primary productivity of habitats, and area of the habitats, on a world basis[a]

Habitat	Total area (m^2)	Organisms	Global net primary productivity (10^{15}g C year^{-1})
Marine phytoplankton	370×10^{12}	Cyanobacteria and microalgae	46
Marine planktonic chemolithotrophs converting to and to NH_4^+ to NO_2^- and NO_2^- to NO_3^-	370×10^{12}	Bacteria	≤ 0.19
Marine benthic[b,c]	6.8×10^{12}	Cyanobacteria and microalgae	0.34
		Macroalgae	3.4
Marine benthic[b]	0.35×10^{12}	Angiosperms (salt marshes plus beds of seagrasses)	0.35
Inland waters	2×10^{12}	Phytoplankton (cyanobacteria and microalgae), benthic algae and higher plants	0.58
Terrestrial[d]	150×10^{12}	Mainly higher plants	54

[a] From Raven (1991, 1996) and Falkowski et al. (2000). All values are for photolithotrophs unless otherwise indicated.
[b] Area of the habitat the marine benthic cyanobacteria, algae, and marine benthic angiosperm is in series with that of the overlying phytoplankton habitat. The benthic habitat area is included in the habitat area for marine phytoplankton.
[c] The marine benthic cyanobacterial and algal category includes cyanobacteria and algae symbionts with protistans and invertebrates.
[d] As well as higher plants the terrestrial productivity involves cyanobacteria and microalgae, both lichenized and free-living, although there seem to be no estimates of the magnitude of nonhigher plant productivity.

Such an ocean is approximated by most parts of the tropical ocean, where ocean currents and wind are inadequate to cause breakdown of thermal stratification; the upper mixed layer shows very little seasonal variation. By contrast, at higher latitudes the varying solar energy inputs throughout the year, combined with wind shear and ocean current influences, lead to stratification with a relatively shallow upper mixed layer in the (local) summer and a much deeper one in the (local) winter, usually giving a winter mixing depth so great that net primary production is not possible as a result of the inadequate mean photon flux density (light-energy) incident on the cells.

A very important impact of stratification is the isolation of the upper mixed layer, where inorganic nutrients are taken up by phytoplankton, from the lower, dark, ocean where nutrients are regenerated by heterotrophy. The movement of organic particles from the upper to lower zones is gravitational. While there is significant recycling of inorganic nutrients in the upper mixed layer via primary productivity and activities of other parts of the food web, ultimately there is loss of particles containing nutrient elements across the thermocline. Seasonal variations in mixing depth, and upwellings, are the main processes bringing nutrient solutes back to the euphotic zone.

Global biogeochemical cycling considerations suggest that the nutrient element that limits the extent of global primary production each year is, over long time periods, phosphorus. This element has a shorter residence time than the other nutrients (such as iron) which are supplied solely from terrestrial sources. Nitrogen, by contrast, is present in the atmosphere and dissolved in the ocean as dinitrogen in such large quantities that any limitation of marine phytoplankton primary productivity by the availability of such universally available nitrogen formed as ammonium and nitrate could be offset by diazotrophy, i.e. biologically dinitrogen fixation, which can only be brought about by certain Archea and Bacteria. In the ocean the phytoplanktonic cyanobacteria are the predominant diazotrophs, as the free-living *Trichodesmium* and as symbionts such as *Richia* in such diatoms as *Hemiaulis* and *Rhizosolenia*. Diazotrophy needs energy (ultimately from solar radiation) and trace elements such as iron (always), molybdenum (usually), and vanadium (sometimes). These trace elements all have longer oceanic residence times than phosphorus, and so are less likely to limit primary productivity than is phosphorus over geologically significant time intervals. However, the balance of evidence for the present ocean suggests that nitrogen is a limiting resource for the rate and extent of primary productivity over much of the world ocean, while iron seems to be the limiting nutrient in the 'high nutrient (nitrogen, phosphorus), low chlorophyll' areas of the ocean. Even where nitrogen does appear to be limiting, this could be a result of restricted iron supply which restricts the assimilation of combined nitrogen, and especially of nitrate.

Processes at the Cell Level

The photosynthetic primary producers in the marine plankton show great variability in taxonomy, and in size and shape. The taxonomic differences reflect phylogenetic differences, including the prokaryotic bacteria (cyanobacteria, embracing the chlorophyll b-containing chloroxybacteria) and a variety of phyla (divisions) of Eukaryotes. The Eukaryotes include members of the Chlorophyta (green algae), Cryptophyta (cryptophytes), Dinophyta (dinoflagellates), Haptophyta (*Phaeocystis* and coccolithophorids), and Heterokontophyta (of which the diatoms or Bacillariophyceae are the most common marine representatives).

The phylogenetic differences determine pigmentation, with the ubiquitous chlorophyll a accompanied by phycobilins in cyanobacteria *sensu stricto*, chlorophyll b in Chloroxybacteria and green algae, chlorophyll(s) c together with significant quantities of light-harvesting carotenoids in dinoflagellates, haptophytes, and diatoms, and chlorophyll c with phycobilins in cryptophytes. These differences in pigmentation alter the capacity for a given total quantity of pigment per unit volume of cells for photon absorption in a given light field, noting that the deeper a cell lives in open-ocean water, the less longer wavelength (red-orange-yellow) light is available relative to blue-green light. This effect of different light-harvesting pigments on light absorption capacity is greatest in very small cells as a result of the package effect.

Another phylogenetic difference among phytoplankton organisms is a dependence on Si (in diatoms) and on large quantities of Ca (in coccolithophorids) in those algae which have mineralized skeletons. Furthermore, some vegetative cells move relative to their immediate aqueous environment using flagella (almost all planktonic dinoflagellates, some green algae). Movement relative to the surrounding water occurs in any organism which is denser than the surrounding water sinking (e.g. by many mineralized cells) or less dense than the surrounding water (buoyancy engendered by cyanobacterial gas vacuoles or the ionic content of vacuoles in large vacuolate cells). The variation in cell size among marine phytoplankton organisms is also partly related to taxonomy. The smallest marine phytoplankton cells are prokaryotic, with cells of *Prochlorococcus* (cyanobacteria *sensu lato*) as small as 0.5 μm diameter, while cells of the largest diatoms (*Ethmosdiscus* spp.) and the green *Halosphaera* are at least 1 mm in diameter, i.e. a range of volume from $6.25 \times 10^{-20} \, m^3$ to more than $4.19 \times 10^{-9} \, m^3$. This means a volume of the largest cells which is almost 10^{11} that of the smallest; allowing for the vacuolation of the large cells gives a ratio of almost 10^{10} for cytoplasmic volume. The size range for phytoplankton organisms is expanded by considering colonial organisms (e.g. the cyanobacterium *Trichodesmium*, and the haptophyte *Phaeocystis*) to a range of cytoplasmic volumes up to almost 10^{12}. At the level of the cell size cyanobacteria have a limited volume range, while in organisms size the range is at least 10^{12}; for haptophytes it is at least 10^{11}. Cell (or organism) size is, on physicochemical principles, very important for the effectiveness of light absorption per unit pigment, nutrient uptake as a result of surface area per unit volume, and of diffusion boundary layer thickness, and rate of vertical movement relative to the surrounding water for a given difference in density between the organisms and their environment. These physicochemical predictions are, to some extent, modified by the organisms by, for example, changed pigment per unit volume and light scattering, and modulation of density.

Determinants of Primary Productivity

Despite these variations in phylogenetic origin and in the size of the organisms, that can be related to seasonal and spatial variations over the world ocean, it is not easy to find consistent spatial and temporal variations in the 'major element' ratios (C : N : P, 106 : 16 : 1 by atoms) or Redfield ratio in space or time. This means that we should not look to differences in the phylogeny or size of phytoplankton organisms to account for differences in the requirement for major nutrients (C, N, P) in supporting primary productivity. What is less clear is the possible variations in trace element (Fe, Mn, Zn, Mo, Cu, etc.) requirements in relation to the properties of different bodies of water in the world ocean. The trace metals are essential catalysts of primary productivity through their roles in photosynthesis, respiration, nitrogen assimilation, and protection against damaging active oxygen species. Geochemical evidence for limitation by some factor other than nitrogen and phosphorus is indicated for 'high nutrient' (i.e. available nitrogen and phosphorus) 'low chlorophyll' (i.e. photosynthetic biomass and hence productivity), or HNLC, regions of the ocean (north-eastern sub-Antarctic Pacific; eastern tropical Pacific; Southern Ocean).

Before seeking other geochemical limitations on primary production to explain why these apparently available sources of nitrogen and phosphorus have not been used in primary productivity, we need to consider geophysical or 'bottom up' (mixing depth, surface photon flux density) and ecological or 'top down' factors (involvement of grazers or pathogens). While these nongeochemical 'bottom up' (control of production of biomass) and 'top down' (removal of the product of primary production) constraints on the use of nitrate and phosphate are, in principle, causes of this HNLC phenomenon, *in situ* Fe enrichments show that addition of this trace element causes drawdown of nitrate and phosphate, increases in chlorophyll and primary productivity, and increased abundance, and contribution to primary productivity of large diatoms.

These IRONEX and SOIREE experiments strongly support the notion that iron limits primary productivity in HNLC regions, as well as suggesting that Fe enrichment can impact differentially on primary producers as a function of their taxonomy and cell size. The increased importance of large diatoms as a result of Fe enrichment can be a result of the diffusion boundary layer thicknesses and surface area per unit volume, rather than of the biochemical demand for Fe to catalyze a given rate of metabolism per unit cell volume. While data are not abundant, theoretical considerations suggest that cyanobacteria should, other things being equal, have higher requirements for Fe for growth than do diatoms, haptophytes, or green algae. This prediction contrasts with observations (and production) for major nutrients such as organic C, N, and P, where cell quotas are much less variable phylogenetically than are those for micronutrients. There is, of course, much less elasticity possible for the C content of cells than for other, less abundant, nutrient elements, with the same applying to a lesser extent to N and P. It is clear that the cost of N, P, or Fe in fixing carbon dioxide is higher for growth at low (limiting) as opposed to high (saturating) photon flux densities.

To broaden the issues of limitations on primary productivity, the ultimate limitation on primary productivity in the ocean is presumably the 'geochemical' limiting element P, i.e. the nutrient element with the shortest residence time in the ocean. It has been plausibly argued that, in the short term, nitrogen has become a limiting nutrient indirectly by the short-term (geologically speaking) Fe limitation. Thus, 'new' production, depending on nitrate upwelled or eddy-diffused from the deep ocean, has a greater Fe requirement than the NH_4^+ (or organic N) assimilation in 'recycled' production in which primary production is chemically fuelled by N, P, and Fe generated by zooplankton, and more importantly, by Fe limitation (at least in the geological short-term) is seen as restricting diazotrophy. In the context of the balance of diazotrophy plus atmospheric and riverine inputs of combined N, and denitrification and sedimentation loss of combined N, Fe limitation can reduce the combined N availability relative to that of P to below the 16:1 atomic ration of the Redfield ratio. This, then, restricts the N:P ration in

upwelled sea water. Even more immediate Fe limitation is seen in the HNLC ocean, as discussed above.

Conclusions

Marine primary production accounts for almost half of the global primary production, and is carried out by a much greater phylogenetic range of organisms than is the case for terrestrial primary production. As on land, almost all marine primary production involves O_2-evolving photolithotrophs. Marine phytoplankton has a volume of cells of 6.10^{-20}–4.10^{-9} m^3. While the primary production in the oceans is, on geological grounds, ultimately limited by P, proximal (shorter-term) limitation involves N or Fe.

Glossary

HNLC high nutrient, low chlorophyll. Areas of the ocean in which combined nitrogen and phosphate are present at concentrations which might be expected to give higher rates of primary production and levels of biomass, than are observed unless some 'top down' or 'bottom up' limitation is involved.

IRONEX iron enrichment experiment. Two releases of $FeSO_4$, with SF_6 as a tracer, south of the Galapagos in the Eastern Equatorial Pacific HNLC area.

Photon flux density. Units are mol photon m^{-2} s^{-1}. Means of expressing incident irradiance in terms of photons, i.e. the aspect of the particle/wave duality of electromagnetic radiation which is appropriate for consideration of photochemical reactions such as photosyntheses. For O_2-evolving photosynthetic organisms the appropriate wavelength range is 400–700 nm.

SOIREE southern ocean iron release experiment. A Southern Ocean analogue of IRONEX, performed between Australasia and Antarctica.

See also

Primary Production Distribution. Primary Production Methods.

Further Reading

Falkowski PG and Raven JA (1997) *Aquatic Photosynthesis*. Malden: Blackwell Science.

Falkowski PG, *et al.* (2000) The global carbon cycle: a test of our knowledge of Earth as a system. *Science* 290: 291–296.

Fuhrman JA (1999) Marine viruses and their biogeochemical and ecological effects. *Nature* 399: 541–548.

Martin JH (1991) Iron, Liebig's Law and the greenhouse. *Oceanography* 4: 52–55.

Platt T and Li WKW (eds.) (1986) Photosynthetic Picoplankton. *Canadian Bulletin of Fisheries and Aquatic Sciences 214.*

Raven JA (1991) Physiology of inorganic C acquisition and implications for resource use efficiency by marine phytoplankton: relation to increased CO_2 and temperature. *Plant Cell Environment* 14: 779–794.

Raven JA (1996) The role of autotrophs in global CO_2 cycling. In: Lidstrom ME and Tabita FR (eds.) *Microbial Growth on C_1 Compounds*, pp. 351–358. Dordrecht: Kluwer Academic Publishers.

Raven JA (1998) Small is beautiful: the picophytoplankton. *Funct. Ecol* 12: 503–513.

Redfield AC (1958) The biological control of chemical factors in the environment. *American Scientist* 46: 205–221.

Stokes T (2000) The enlightened secrets across the ocean. *Trends in Plant Science* 5: 461.

Van den Hoek C, Mann DG, and Jahns HM (1995) *Algae. An Introduction to Phycology.* Cambridge: Cambridge University Press.

PRIMARY PRODUCTION DISTRIBUTION

S. Sathyendranath and T. Platt,
Dalhousie University, Nova Scotia, Canada

Introduction

The study of the distribution of primary production in the world oceans preoccupied biological oceanographers for most of the twentieth century. Understanding the distribution remains a problem for which there is no facile answer; it is a many-faceted problem to which the approaches differ depending on the temporal and spatial scales at which understanding is sought. The issues relate to limitations of measurement techniques, interpretation of measurements that are available, and paucity of data.

In this context, the first point to bear in mind is that every method for the measurement of primary production has an intrinsic timescale associated with it (**Table 1**). Space and timescales are inextricably linked in oceanography, such that small timescales are always associated with small length scales, and large timescales with large length scales. Therefore, in studying the distribution of primary production in the ocean, one has to be careful that measurements are made at time and space scales appropriate for the problem at hand. Another important point is that all the available techniques do not measure the same quantity the measured quantity may be gross primary production, net primary production, or net community production (if carbon-based methods are used), or total production, new production, or regenerated production (if nitrogen-based methods are used) (**Table 1**). A careful analysis quickly reveals that no methods are currently available for estimating the same component of primary production at all scales of interest in oceanography. It follows that, in comparing results (or in combining methods), one has to make sure that like things are being compared (or combined). Failure to do this can lead to perplexing results.

Table 1 Various methods that can be used to estimate primary production in the ocean. The nominal timescales applicable to the results are also given. The components of primary production P_g (gross primary production), P_n (net primary production), and P_c (net community production) are based on rates of carbon uptake; P_T (total primary production), P_r (regenerated production) and P_{new} (new production) refer to nitrogen uptake rates. Sedimentation rate refers to the flux of organic particles from the photic zone

Method	Nominal component of production	Nominal timescale
In vitro		
^{14}C assimilation	$P_T (\equiv P_n)$	Hours to 1 day (duration of incubation)
O_2 evolution	P_T	Hours to 1 day (duration of incubation)
$^{15}NO_3$ assimilation	P_{new}	Hours to 1 day (duration of incubation)
$^{15}NH_4$ assimilation	P_r	Hours to 1 day (duration of incubation)
$^{18}O_2$ evolution	$P_{new} (\equiv P_c)$	Hours to 1 day (duration of incubation)
Bulk property		
NO_3 flux to photic zone	P_{new}	Hours to days
O_2 utilization rate (OUR) below photic zone	P_{new}	Seasonal to annual
Net O_2 accumulation in photic zone	P_{new}	Seasonal to annual
$^{238}U/^{234}Th$	P_{new}	1–300 days
$^3H/^3He$	P_{new}	Seasonal and longer
Optical		
Double-flash fluorescence	P_T	<1 s
Passive fluorescence	P_T	<1 s
Remote sensing	$P_{T,new}$	Days to weighted annual
Upper and lower limits		
Sedimentation rate below photic zone	$P_{new} (\equiv P_c)$: (lower limit)	Days to months (duration of trap deployment)
Optimal energy conversion of photons absorbed	P_T (upper limit)	Instantaneous to annual
Depletion of winter accumulation of NO_3	P_{new} (lower limit)	Seasonal

Precise measurements of primary production are time consuming, and paucity of data has been a constant difficulty. This necessitates the use of extrapolation schemes to produce large-scale distributions from a small number of observations. When undertaking such extrapolations, it would be desirable to test the extrapolation schemes by comparing the estimated distributions at large scales against some other independent estimates. However, the lack of techniques for measuring the same component of primary production at different time scales confounds efforts to make independent validations of extrapolation protocols.

A profitable approach to dealing with these matters may be to begin by examining the factors that influence primary production. One may anticipate that the distribution of primary production would be influenced by the variability in the forcing fields.

Factors that Influence Variations in Primary Production

Primary production is the rate of carbon fixation by photosynthesis. The primary forcing variable is therefore light. The light field varies with depth in the ocean, with time of day, and with time of year. Correspondingly, primary production in the ocean exhibits a strong depth dependence and a strong time dependence. The temporal variations occur on several scales, ranging from seconds (response to clouds and vertical mixing) to diurnal, seasonal, and annual. Adaptations of phytoplankton populations to various light regimes also influence primary production. The adaptations may involve, for example, change in the concentration of chlorophyll-*a* per cell, change in the number of chlorophyll-*a* molecules per photosynthetic unit, or changes in the concentrations of auxiliary pigments, which may play either a photoprotective or a photosynthetic role. Understanding photo-adaptation requires that we know the light history of the cells, in addition to the current light levels.

Light acts on the state variable, the biomass of phytoplankton. It has been a common practice, in this field, to treat the concentration of the main phytoplankton pigment, chlorophyll-*a*, as an index of phytoplankton biomass, because it (or a variant called divinyl chlorophyll-*a*) is present in all types of phytoplankton, because of the fundamental role it plays in the photosynthetic process, and because it is easy to measure. The tendency would be for primary production to increase with chlorophyll-*a* concentration, though the rate of increase would vary depending on other factors.

A third factor that determines the primary production in the ocean is the availability of essential nutrients such as nitrogen. In a stratified, oceanic water column, the upper illuminated layer is typically low in nutrients, with the deeper layers acting as a reservoir of nutrients. Mixing events bring these nutrients to the surface layer, enhancing primary production. In temperate and high latitudes, deep mixing events in winter, and subsequent stratification as the surface warming trend begins, lead to the well-known phenomenon of the spring bloom and more generally to a pronounced seasonal cycle in primary production. On the seasonal cycle are superimposed the effects of sporadic mixing events in response to passing storms. The short-term increases in primary production associated with these sporadic events are often missed by sampling schemes designed to record the seasonal cycle.

Temperature is another factor that influences primary production. It is believed that temperature controls the enzyme-mediated dark-reaction rates of photosynthesis. Thus, from laboratory experiments, it has been shown that photosynthetic rates in phytoplankton increase with temperature up to an optimal temperature, after which they decrease. However, the details of this response may differ with species. In nature, the tendency of primary production rates to increase with temperature is confounded by another effect: upwelling waters with high nutrients tend to have a low temperature, and the increase in primary production in response to the nutrient supply may in fact supersede the counteracting effect of temperature.

The recent years have seen an increasing appreciation of the role of micronutrients such as iron as limiting resources for primary production. The idea that iron limits production in certain marine environments was aired in the early twentieth century. However, it was in the 1980s that this idea gained renewed momentum, with the pioneering work by John Martin. It is now believed that iron limits primary production in large tracts of the ocean (the Southern Ocean, the Equatorial Pacific, the subarctic Pacific), leading to regimes known as the high-nutrient, low-chlorophyll regimes; these are environments where the nitrogen in the upper mixed layer is apparently never used up, and the phytoplankton biomass remains low throughout the year. Other contributing factors for the presence of high-nutrient, low-chlorophyll regimes include top-down control of phytoplankton biomass (and hence productivity) by zooplankton grazing, and the supply of nutrients by physical processes that exceeds the demands of biological production. It has been postulated that the variations in the distribution of primary production in response to changes in the availability of iron over geological timescales, and the consequent changes in the draw-down of carbon dioxide from the atmosphere into the ocean, may be implicated in climate change.

Species composition and species succession also influence rates of primary production. For example, the size distribution of the cells and the pigment composition of the cells can both change with species composition, thus influencing nutrient uptake and light absorption for photosynthesis.

In view of the large number of factors that influence the distribution of primary production at so many temporal and spatial scales, it is desirable to apply mathematical modeling techniques to organize and formalize the study of the distribution of primary production in the world oceans. Light-dependent models of primary production are particularly useful, not only because the models are based on the first principles of plant physiology, but also because of their direct applicability to remote sensing of primary production. According to such models, primary production P at a given location (x, y, z) and a given time (t) can be formalized as:

$$P(x, y, z, t) = B(x, y, z, t)f(I(x, y, z, t)) \quad [1]$$

where B is the phytoplankton biomass indexed as chlorophyll-a concentration, and the function f describes the biomass-specific, photosynthetic response of phytoplankton to available light I. The response function f is known to have three phases: a light-limited phase at low-light levels when production increases linearly as a function of available light; a saturation phase at high-light levels, when production becomes independent of light; and a photo-inhibition phase at extremely high light levels, when production is actually reduced by increasing light. At the most, three parameters (or only two, if the photoinhibitory phase is negligible, which is often the case) are required to describe such a response. Often, these are taken to be the initial slope ^{B} (typical units: mg C (mg chl$a$)$^{-1}$ h^{-1} (W m^{-2})$^{-1}$); the saturation parameter P_m^B (typical units: mg C (mg chl$a$)$^{-1}$ h^{-1}); and the photoinhibition parameter β^B (typical units: mg C (mg chl$a$)$^{-1}$ h^{-1} (W m^{-2})$^{-1}$).

In such models, light and biomass are taken, justifiably, to be the principal agents responsible for variations in primary production. The effects of other factors (temperature, nutrients, micronutrients, species, light history and photoadaptation) may be accounted for indirectly, through their influence on the parameters of the response function (the photosynthesis–irradiance curve). The success of such models depends on how well we are able to describe, mathematically, the fields of biomass and light, as well as the parameters of the photosynthesis–irradiance curve. Such models could be taken to a higher level of sophistication if the photosynthesis parameters were in turn expressed as functions of the various factors that influence them. General relationships valid in the natural environment, that would account for the influences of all known factors on photosynthesis–irradiance parameters, have eluded scientists so far. Therefore, assignment of parameters has to rely heavily on direct observations.

With this background, we can look at what we know of the distribution of primary production in the world ocean.

Vertical Distribution

It is only the upper, illuminated part of the water column that contributes to primary production. A useful rule of thumb is that practically all the primary production in the water column occurs within the euphotic zone, commonly defined as the zone that extends from the surface to the photic depth at which light is reduced to 1% of its surface value (though there are some arguments for using 0.1% light level as a more rigorous boundary on the euphotic zone). However, it is important to realize that photosynthesis is a quantum process, which depends on the absolute magnitude of light available, rather than on relative quantities as indicated by the photic depth. Thus, regardless of the definition of photic depth that may be used, it will only tell us that the production below that particular depth horizon will be small compared with that above; it tells us nothing about the absolute magnitude of production in the water column. The 1% light level may occur at depths exceeding 100 m in oligotrophic open-ocean waters, whereas it may be less than 10 m in eutrophic or turbid waters. It is noteworthy that phytoplankton themselves are a major factor responsible for modifying the optical properties of sea water, and hence the rate of penetration of solar radiation into the ocean.

Another useful depth horizon that is relevant in the study of primary production is the critical depth. If production and loss terms of phytoplankton (grazing, sinking, decay) from the surface to some finite depth are integrated, then the integrated production and loss terms become equal to each other at some depth of integration, which is known as the critical depth. The concept of critical depth was formalized by Sverdrup in 1956. If the mixed-layer depth is shallower than the critical depth, then production in the layer will exceed losses, which is favorable for the accumulation of biomass in the layer, which would, in turn, further enhance mixed-layer production. If the mixed layer is deeper than the critical depth, the conditions would be unfavorable for the formation of blooms. Thus, the vertical distribution of production relative to the mixed-layer depth plays an important role in determining the potential for enhanced production.

The concept of critical depth is built on the fact that, within the mixed layer, production is typically a decreasing function of depth (because the available light decreases exponentially with depth), whereas the biomass and the loss terms are uniformly distributed within the mixed layer. However, it would be erroneous to suppose that the maximum primary production always occurs at the surface of the ocean. In fact, the maximum primary production may well occur at some subsurface depth, where the nutrient availability and light levels are optimal. In a stratified water column, several factors (species composition, photoadaptation, nutrient supply, and light levels) conspire to produce a deep-chlorophyll maximum. Depending on the physiological status of the phytoplankton in the deep-chlorophyll maximum, there may be a subsurface maximum in primary production, which may occur at the same depth as the chlorophyll maximum, or at a shallower depth. If the light levels at the surface are sufficiently high to induce photoinhibition, then also the maximum in production would occur at some subsurface level.

Horizontal Distribution

Primary production varies markedly with region and with season. Upwelling regions (e.g., waters off north-west Africa, the north-west Arabian Sea off Somalia, the equatorial divergence zone, the northeast Pacific off California and Oregon, the south-east Pacific off Peru, and south-west Africa) are typically more productive than the central gyres of the major ocean basins, because of the high levels of nutrients that are brought to the surface by upwelling. In general, coastal regions are more productive than open-ocean waters, also due to the availability of nutrients. Temperate and high latitudes show a pronounced seasonal maximum in spring (a consequence of the high supply of nutrients to the mixed layer during deep mixing events in winter, followed by a shallowing of the mixed layer and an increase in incoming solar radiation as the seasons progress). In summer, the depletion of nutrients in the mixed layer leads to reduced production and biomass, and to the migration of the chlorophyll maximum to below the

mixed layer. This is often followed in fall by a secondary peak in production, as a result of increased nutrient supply as the mixed layers begin to deepen, and also perhaps a decrease in the grazing pressure. A notable exception to these seasonal cycles is the high-nutrient, low-chlorophyll regions mentioned earlier. Seasonal cycles are far less pronounced in tropical waters, unless they are influenced by seasonal upwelling, as is the case in the Arabian Sea. A summary of what we know of the distribution of primary production in the world's oceans, and of the physical and biological factors that influence it is available in the literature.

Given the highly dynamic nature of the distribution of primary production in the oceanic environment and the extreme paucity of measurements, it was only in the 1970s that compilations of measurements were first used to evaluate the distribution of marine primary production at the global scale. These studies had to combine observations from many years to maximize the number of measurements, so that it was impossible to examine interannual variability, or trends, in primary production. This inherent limitation to the study of the distribution of primary production at large scales was only addressed, at least partially, with the advent of remote-sensing techniques, which are examined briefly below.

The Use of Remote Sensing

The last two decades of the twentieth century saw the development of remote sensing to study the distribution of phytoplankton in the ocean. This technology uses subtle variations in the color of the oceans, as monitored by a sensor in space, to quantify variations in the concentrations of chlorophyll-a in the surface layers of the ocean. Since polar-orbiting satellites can provide global coverage at high spatial resolution (1 km or better), it became possible for the first time to see (cloud cover permitting) in great wealth of detail the variations in phytoplankton distribution at synoptic scales. The next logical step in the exploitation of ocean-color data was taken a few years later, when these fields of biomass were converted into fields of primary production.

The procedure that has met with the most success builds on models of photosynthesis as a function of available light. The light available at the sea surface can also be estimated using remote-sensing methods: geostationary satellites monitor cloud type and cloud cover, which are used in combination with atmospheric light-transmission models to estimate light available at the sea surface. Optical properties of the water, derived from ocean-color data are then used to compute light available at depth in the ocean. Thus, information on the variations in the forcing field (light), and the state variable on which light operates (phytoplankton biomass) in the process of photosynthesis, are directly available through remote sensing. If these satellite data are supplemented with best estimates of photosynthesis–irradiance parameters based on *in situ* observations, then all the major elements are in place for computations of primary production by remote sensing. These computations can be further refined by incorporating information on vertical structure in biomass, also derived from *in situ* observations. In regions where deep chlorophyll maxima are a consistent feature, ignoring their presence can lead to some systematic errors in the results.

The remote-sensing approach to computation of primary production has several advantages. It makes use of the vast database on light and biomass available through remote sensing, and in addition, makes use of all the available information on plant physiology obtained through *in situ* observations to define the parameters of the photosynthesis–irradiance curve. It has the potential to monitor the distributions of primary production at very large spatial scales, and over several timescales, ranging from the daily to interannual. It emerges as the method of choice for studying the distribution of primary production at large scales, especially when it is considered that remote sensing provides information on chlorophyll-*a* concentration, which can vary over four decades of magnitude, and on the highly variable light field at the surface of the ocean. In this method, the more sparse *in situ* observations are made use of only to define the parameters of the photosynthesis–irradiance curve and the parameters that are used to describe the vertical structure in biomass, which have a much lower range of variability compared with chlorophyll-*a* concentration.

The method is not without its problems, however. The major issues relate to differences in the time and space scales at which satellite and *in situ* measurements are made. This necessitates the development of methods for seamless integration of the two types of data streams. Ideally, the methods would bring to bear variations in the oceanographic environment

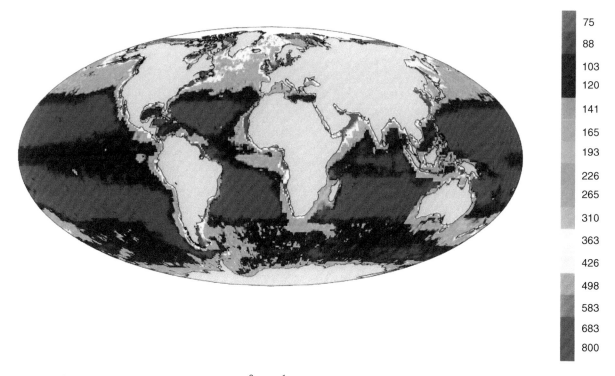

Figure 1 Distribution of primary production (g C m^{-2} year^{-1}) in the world oceans, as estimated using remotely sensed ocean-colour data (Longhurst *et al.*, 1995). Areas of high production are seen in northern high latitudes; these reflect the impact of spring blooms on the distribution of production integrated over an annual cycle. Areas of upwelling (e.g., equatorial upwelling, the Benguela upwelling, the Somali upwelling, the California upwelling and the Peru upwelling) also emerge as areas of high production. The coastal shelf regions are also locations of high production. However, it is recognized that the remote-sensing method used in this calculation could overestimate the biomass, and hence the production in some coastal areas. Recent improvements in ocean-color technology offer the potential to reduce this type of error in the computation.

and the physiological responses of phytoplankton to these variations. This is an active area of research. The first computation of global oceanic primary production using the remote-sensing approach appeared in the literature in 1995 (**Figure 1**). Other, similar computations have since appeared in the literature. It is a method that will continue to improve, with improvements in satellite technology as well as in the techniques for extrapolation of local biological measurements to large scales.

See also

Primary Production Methods. Primary Production Processes.

Further Reading

Chisholm SW and Morel FMM (eds.) (1991) *What Controls Phytoplankton Production in Nutrient-Rich Areas of the Open Sea*. Lawrence KS: American Society of Limnology and Oceanography.

Falkowski PG and Woodhead AD (eds.) (1992) *Primary Productivity and Biogeochemical Cycles in the Sea*. New York: Plenum Press.

Geider RJ and Osborne BA (1992) *Algal Photosynthesis*. New York: Chapman Hall.

Li WKW and Maestrini SY (eds.) (1993) *Measurement of Primary Production from the Molecular to the Global Scale, ICES Marine Science Symposia, vol. 197*. Copenhagen: International Council for the Exploration of the Sea.

Longhurst A (1998) *Ecological Geography of the Sea*. San Diego: Academic Press.

Longhurst A, Sathyendranath S, Platt T, and Caverhill C (1995) An estimate of global primary production in the ocean from satellite radiometer data. *Journal of Plankton Research* 17: 1245–1271.

Mann KH and Lazier JRN (1991) *Dynamics of Marine Ecosystems. Biological–Physical Interactions in the Oceans*. Cambridge, USA: Blackwell Science.

Platt T and Sathyendranath S (1993) Estimators of primary production for interpretation of remotely sensed data on ocean color. *Journal of Geophysical Research* 98: 14561–14576.

Platt T, Harrison WG, Lewis MR, *et al.* (1989) Biological production of the oceans: the case for a consensus. *Marine Ecolog Progress Series* 52: 77–88.

BIOLUMINESCENCE

P. J. Herring, Southampton Oceanography Centre, Southampton, UK
E. A. Widder, Harbor Branch Oceanographic Institution, Fort Pierce, FL, USA

Introduction

Bioluminescence is the capacity of living organisms to emit visible light. In doing so they utilize a variety of chemiluminescent reaction systems. It has historically been confused with phosphorescence and the latter term is still frequently (and erroneously) used to describe marine bioluminescence. Some terrestrial species (e.g., fireflies) have the same ability, but this adaptation has been most extensively developed in the oceans. Bioluminescent species occur in only five terrestrial phyla, and only in one of these (Arthropoda, which includes the insects) are there many examples. In contrast, bioluminescence occurs in 14 marine phyla, many of which include numerous luminescent species (**Table 1**). All oceanic habitats, shallow and deep, pelagic and benthic, include bioluminescent species, but the phenomenon is commonest in the upper 1000 m of the pelagic environment.

Biochemistry

Bioluminescence involves the oxidation of a substrate (luciferin) in the presence of an enzyme (luciferase). The distinctive feature of the reaction is that most of the energy generated is emitted as light rather than as heat. There are many different, and unrelated, kinds of luciferin, and biochemical and taxonomic criteria indicate that bioluminescence has been independently evolved many times. Marine animals are unusual, however, in that many species in at least seven phyla use the same luciferin. This compound is known as coelenterazine because it was first identified in jellyfish (coelenterates) and its molecular structure is derived from a ring of three amino acids (two tyrosines, and a phenylalanine). Nevertheless, many other marine organisms use different luciferins. In some animals (e.g., jellyfish) the luciferin/luciferase system can be extracted in the form of a stable 'photoprotein' that will emit light when treated with calcium.

Microorganisms

Bioluminescent organisms are found in all of the oceans of the world and at all depths. The prevalence of the phenomenon has long been known to seafarers, as the light seen at night in the wake or bow wave of their vessels. Three kinds of single-celled marine organisms include species that produce light, namely bacteria, dinoflagellates, and radiolarians, all with different luciferins. Individual luminous bacteria do not luminesce unless there are a lot of them together – colonies therefore become bright. This is because luciferase production is switched on only by the accumulation in the environment of a critical concentration of a chemical released by the bacteria (an autoinducer). Luminous bacteria are to be found free in the ocean but are more commonly encountered as glowing colonies on either marine snow or fecal pellets, or, as luminous symbionts, in the light organs of some fish and squid (see below).

There are many species of luminous dinoflagellates and they are the usual cause of sea surface luminescence, visible in the bow wave or wake of a boat or the turbulence caused by a swimmer, whether man, fish, or dolphin. They can accumulate in dense 'blooms,' some dense enough to be recognized as red tides, and individual dinoflagellates flash when subject to sufficient shear force (e.g., in turbulence). Because they live close to the surface, their light would be invisible by day. In fact most species have a circadian rhythm that conserves the luminescence by turning it off during the day. These organisms, and probably the radiolarians too, defend themselves against planktonic predators by their flashing, which has the added 'burglar alarm' benefit of alerting larger predators to the presence of the original grazer.

Plankton

Other common planktonic luminous organisms are copepod and ostracod crustaceans, cnidarians (jellyfish and siphonophores) and comb jellies. Copepods are in effect the insects of the sea and are the commonest planktonic animals. Many species are luminous. Most of them do not flash but have glands on their limbs or bodies from which they squirt gobbets of luminous secretion into the water as a defensive distraction. Ostracods, though less abundant, also produce luminous droplets from groups of gland cells. Usually this is a defense, but the males of some shallow-water species of *Vargula* swim up off

Table 1 Representative examples of bioluminescent marine organisms

Organism	Typical genera	Type of luminescence
Bacteria	*Photobacterium*	Glow
Dinoflagellates	*Ceratium, Lingulodinium (Gonyaulax), Noctiluca, Pyrocystis*	Flashes
Radiolarians	*Collozoum, Collosphaera, Thalassicolla*	Flashes or glows
Cnidarians		
Medusae	*Aequorea, Solmissus, Atolla, Periphylla, Pelagia, Halicreas*	Flashes, scintillating secretions, multiple waves of light
Siphonophores	*Hippopodius, Vogtia, Agalma, Praya, Nanomia, Halistemma*	Flashes and glows, multiple waves of light
Sea pens	*Renilla, Stylatula, Pennatula*	Flashes, multiple waves of light
Polyps	*Obelia, Campanularia*	Flashes, waves of light
Ctenophores	*Beroe, Cestum, Euplokamis, Kiyohimea*	Flashes, waves of light, luminous secretions
Molluscs		
Nudibranchs	*Phyllirrhoe*	Flashes
Pulmonates	*Planaxis*	Flashes, glows
Bivalves	*Pholas*	Secretion
Squid	*Sepiola[a], Heteroteuthis, Abralia, Cranchia, Chiroteuthis*	Flashes, glow, secretions
Octopods	*Japetella, Stauroteuthis*	Glows
Polychaete worms	*Tomopteris, Chaetopterus, Polynoe, Polycirrus, Odontosyllis*	Glows, flashes, waves of light, secretions
Pycnogonids (sea spiders)	*Collossendeis*	Glows
Crustaceans		
Copepods	*Pleuromamma, Metridia, Euaugaptilus, Lucicutia, Oncaea*	Secretions, flashes
Ostracods	*Vargula, Conchoecia*	Flashes, secretions
Amphipods	*Scina, Cyphocaris*	Flashes, secretions
Mysids	*Gnathophausia*	Secretions
Euphausiids	*Euphausia*	Glows, flashes
Decapod shrimp	*Acanthephyra, Heterocarpus, Thalassocaris, Sergestes, Hymenopenaeus*	Secretions, glows
Echinoderms		
Brittle stars	*Ophiacantha, Amphiura, Ophiomusium*	Flashes, waves of light, glows
Starfish	*Plutonaster, Benthopecten, Brisinga*	Glows
Crinoids (sea lilies)	*Thalassometra, Thaumatocrinus*	Glows
Holothurians (sea cucumbers)	*Paroriza, Laetmogone, Kolga, Enypniastes, Pannychia*	Glows, waves of light
Tunicates		
Larvaceans	*Oikopleura, Megalocercus*	Flashes
Thaliaceans (sea squirts)	*Pyrosoma[a], Clavelina*	Glows, slow flashes
Fishes		
Sharks	*Isistius, Euprotomicrus*	Glows
Eels	*Saccopharynx, Lumicongera &*	Glows?
Other fishes: Bathylagids	*Opisthoproctus[a], Winteria[a]*	Glows
Gonostomatids	*Cyclothone, Gonostoma, Vinciguerria*	Glows
Sternoptychids (hatchet fishes)	*Argyropelecus, Sternoptyx*	Glows
Stomiiforms (dragon fish, loose-jaws)	*Astronesthes, Melanostomias, Pachystomias Malacosteus, Chauliodus, Stomias, Idiacanthus*	Flashes, glows
Myctophids (lantern fishes)	*Electrona, Myctophum, Diaphus, Lampanyctus*	Flashes, glows
Ceratioids (angler fishes)	*Ceratias[a], Oneirodes[a], Himantolophus[a], Linophryne[a]*	Glows, flashes
Morids (deep sea cods)	*Physiculus[a]*	Glows
Macrourids (rattails)	*Coelorhynchus[a], Macrourus[a], Nezumia[a]*	Glows?
Anomalopids (flashlight fishes)	*Anomalops[a], Photoblepharon[a]*	Flashes, glows
Monocentrids (pinecone fishes)	*Cleidopus[a], Monocentris[a]*	Glows, flashes
Apogonids	*Apogon[a], Siphamia[a], Howella[a]*	Glows?
Leiognathids (pony fishes)	*Gazza[a], Leiognathus[a]*	Glows, flashes

[a]Symbiotic luminous bacteria.

the bottom to signal to the females. They encode a luminous message in the combination of the frequency of their light puffs, their swimming trajectory, and the timing of their displays. The displays are equivalent to complex smoke signals, or sky-writing, using light. Occasionally both copepods and ostracods may swarm in such numbers that their secretions light up the wave crests or the entire ocean surface. The luciferin of *Vargula* (previously named *Cypridina*) was the first to be identified and is a tri-peptide similar to coelenterazine, but made up of three different amino acids. Certain other ostracods use coelenterazine instead.

Copepods and ostracods, like bacteria, dino-flagellates, and most other marine organisms, produce blue or blue-green luminescence (**Table 1**). These wavelengths penetrate oceanic water best, so they are visible at the greatest range. Many cnidarians and comb jellies also produce blue light, but in a few the luminescence is a vivid green. These animals have incorporated a green fluorescent protein into the luminous cells, or photocytes. The energy from the luciferin–luciferase reaction is transferred to the fluor and is therefore made visible as green light. Some species of jellyfish, siphonophores, and comb jellies can not only flash but also pour out a luminous secretion. The secretion may include scintillating particles, which flash independently in the water. In other species of cnidarians the light-emitting cells (photocytes) are situated all over the surface of the body and a stimulus can set off one or more waves of light that may circle over the surface for several seconds. None of these animals has image-forming eyes, so their bioluminescent displays must be aimed at other animals, probably as a defense against predators or simply to protect their very fragile tissues from accidental damage by a blundering contact.

There are many luminous worms, though most of them spend their time on the sea floor. Syllid worms (fireworms) come to the surface in shallow waters for a luminous mating display, whose timing is linked to

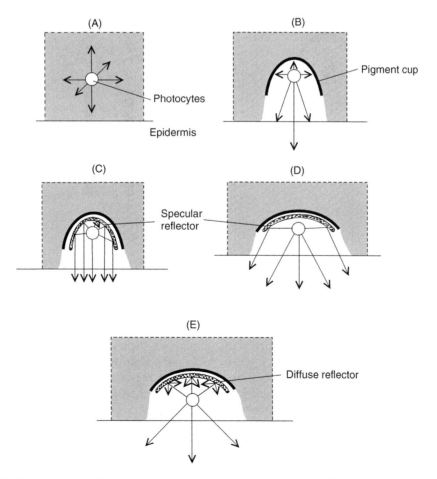

Figure 1 The effects of pigment and reflectors on light emission from photophores: (A) point source emission of a group of photocytes or bacteria is isotropic; (B) pigment cup restricts the solid angle of emission, but absorbs some of the light; (C)–(E) reflectors of different geometries provide a more efficient emission, whether they are specular (C, D) or diffuse (E). Arrows indicate possible ray paths. (From Herring (1985) with permission.)

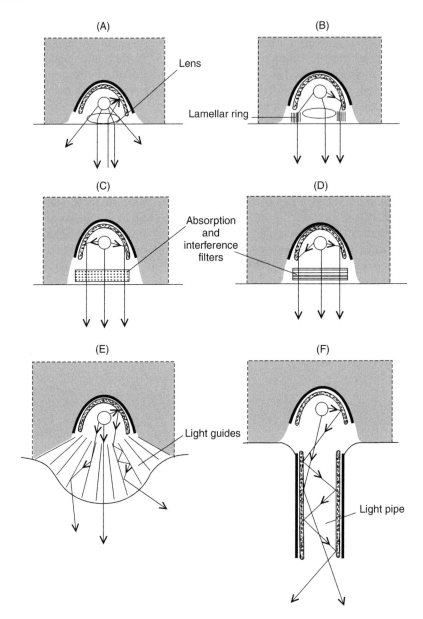

Figure 2 Effects of accessory optical structures in photophores: (A) lens alone; (B) lens and lamellar ring (e.g., euphausiid shrimp); (C) pigment filter; (D) interference filter; (E) light guide diffuser (e.g., some squid); (F) light pipe (e.g., some anglerfishes). (From Herring (1985) with permission.)

the phase of the moon. They have a greenish light, while the pelagic worm *Tomopteris* is very unusual in producing yellow light (**Table 1**). Scale worms when attacked can shed their scales, which then flash independently. A similar tactic is used by luminous brittlestars; when grasped they shed their arm tips, leaving them to flash and writhe in the predator's grip, like the lizard that sheds its tail. Many other echinoderms (relatives of brittlestars) are bioluminescent, including sea cucumbers, sea stars and sea lilies. Most of these live on the deep-sea floor and, like the jellies, lack image-forming eyes. Other bottom-living luminous animals include species of sea-

spiders, acorn worms, snails and clams, as well as cnidarians such as sea pens and gorgonians.

In the plankton and the nekton (those animals that can swim reasonably well) are many other luminous animals, including arrow worms and *Pyrosoma*. The latter forms a cylindrical colony of sea-squirt-like individuals, each of which has two patches of luminous cells. The cells contain bacteria-like organelles, which are uniquely intracellular. The colonies will respond to illumination by producing a slow glow of several seconds duration, and are often seen at night from the decks of ships. Only among the crustaceans, fish, and squid are the photocytes frequently associated with accessory optical structures, including reflectors, lenses, collimators, light guides, and filters (**Figure 1** and **Figure 2**). The result is a complex light organ or photophore.

Photophores have not been developed in luminous amphipods nor in the mysid *Gnathophausia*, but those in euphausiid and many decapod shrimps are very elaborate structures. In these animals the photophores are located on the underside of the body and eyestalks and provide a ventral illumination. Predators from below would normally see the shrimp as a silhouette against the dim downwelling daylight but, by emitting light of the same color and intensity as the daylight, the shrimp matches the background, a tactic known as counterillumination camouflage. If the shrimp were to change its orientation in the water, tilting up or down, its luminous output would no longer match the background. All euphausiids and some decapods get over this problem by rotating the photophores in the plane of pitch so that they remain directed vertically downwards and maintain the camouflage.

Many deep-sea decapod shrimps (and the mysid *Gnathophausia*) will squirt an intense cloud of luminescence into the water if they are startled and then disappear into the surrounding darkness. Some of the species living in the upper 1000 m have both squirted luminescence and ventral photophores. The color of light from the two sources is slightly different; the photophores necessarily match the spectral content of daylight, but the squirts are rather bluer and of broader bandwidth.

Squid and Octopods

At least one squid (*Heteroteuthis*) also produces a squirt of luminescence. It is not luminous ink but material from a special luminous gland. This squid can also produce a steady glow from within the gland. The complexity of photophores in different squid is quite remarkable; a single individual may have several different types on different parts of the body. Many of them are for counterillumination camouflage, being typically located beneath the eye, and sometimes under the liver, two opaque structures that need to be camouflaged. The photophores are able to match the intensity of downwelling light over a considerable range. Other squid have photophores in or on the arms and/or tentacles, sometimes with specialized photophores right at the tips. As they become mature, the females of some squid develop large photophores at the tips of certain arms, presumably as a signal for the males. Females of some pelagic octopods develop an analogous sexual photophore, in the form of a luminous ring round the mouth, as they become ripe, and lose it again when they have spawned. Deep-water octopods may have lights on the arms instead of suckers. Some shallow squids culture luminous bacteria (*Photobacterium fischeri*) in large paired ventral photophores. Bacteria from the female are shed into the water around the egg masses and reinfect the newly hatched larvae, which have special structures for acquiring the symbionts from the water.

Fishes

The variety of photophores in squid is exceeded only by those in fishes. Several groups of fish use luminous bacterial symbionts as their source of light. Shallow-water species (e.g., ponyfish and pinecone fish) utilize bacteria (*Photobacterium leiognathi* and *P. fischeri*, respectively) that grow best at warm temperatures. Deep-sea fishes (e.g., rattails and spookfish) have a different symbiont (*P. phosphoreum*) that does better in colder water. All these fishes have photophores

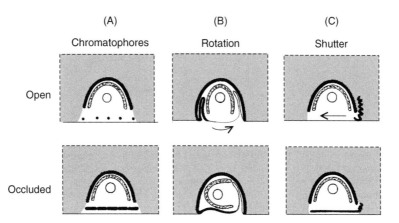

Figure 3 Three means whereby a photophore can be occluded: (A) chromatophores; (B) rotation; (C) shutter. (From Herring (1985) with permission.)

that open into the gut; their symbionts are extra-cellular and can be grown in laboratory cultures. It is assumed that the symbionts are somehow selected from the normal gut flora. Two particular families of fishes, the shallow-water flashlight fishes and deep-sea anglerfishes, have photophores that do not open to the gut, though, like all the bacterial light organs of squid and other fishes, they do open to the sea water via pores. The bacteria of these two groups of fishes are also extracellular but cannot yet be cultured. They do not belong to any known species, though they are closely related to the other symbionts. It is not known how they are reacquired in each generation. Bacteria glow continually, so these photophores have to be occluded to turn the light off (**Figure 3**).

Most fish do not use bacteria but use their own luciferin/luciferase system. There are a few exceptions, which cannot make the luciferin but have to have it in their diet, like a vitamin. The best-known is the midshipman fish *Porichthys*, which has numerous, complex, ventral photophores. It uses *Vargula* luciferin, and if deprived of dietary Vargula it does not luminesce. The luminescence returns if it is fed either whole *Vargula* or the pure luciferin. Populations of *Porichthys* that have no *Vargula* in their region are nonluminescent, even though they have photophores. The mysid *Gnathophausia* seems to have a similar dietary requirement, in this case for the luciferin coelenterazine.

Other fishes probably synthesize their own luciferin. Their photophores can be extremely elaborate and a single fish may have thousands of tiny simple photophores, as well as a much smaller number of large complex ones. Most of those fishes in the upper 1500 m have counterillumination camouflage

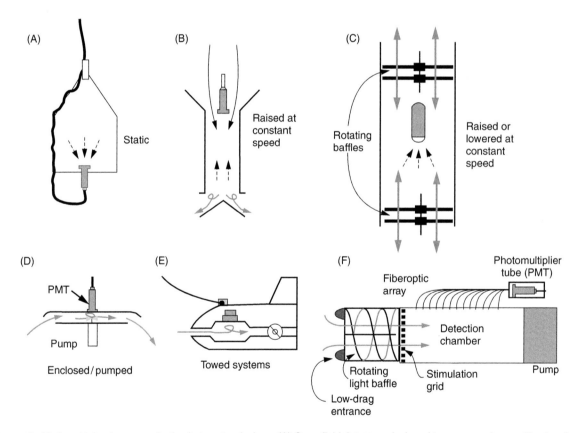

Figure 4 Various bioluminescence bathyphotometer designs. (A) Open field detectors designed to measure downwelling irradiance also measure bioluminescence stimulated by motion of the detector system. (B) An early sounding bathyphotometer that was raised at constant speed. Water was entrained by the upper funnel and bioluminescence was primarily triggered by turbulent flow at the exit baffle. (C) A refinement of the device in (B), equipped with entry and exit baffles that also provide excitation as water is entrained by raising or lowering. (D) Generic sketch of a low-volume enclosed and pumped bathyphotometer in which excitation is provided by pump impeller. Detector chamber volume about is 50 ml with indeterminate flow path and maximum flow rate of 1 liter s^{-1}. This device could be used in either a moored or profiling configuration. (E) Generic towed system with excitation provided by entry baffle and flow provided either by forward motion or pump downstream from detector chamber. (F) More recent design of a high-flow-rate (up to 44 s^{-1}), large inlet bathyphotometer (12 cm ID) with a large volume detection chamber (>11 litres) and hydrodynamically defined excitation using a grid at the inlet. (Adapted with permission from Case JF, Widder EA, Bernstein SA *et al.* (1993) Assessment of marine bioluminescence. *Naval Research Reviews* 45: 31–41.)

photophores along the ventral surface of the body; the shallower species (e.g., hatchetfishes) cover the whole ventral surface with large photophores; the deeper ones (dragon fishes) have fewer, smaller, ventral photophores. In the large family of lantern-fishes shallow-living and deep-living species have equivalent differences in the size and number of their ventral photophores. Many stomiiform fishes have a large postorbital photophore, behind or under each eye, very similar in position to the bacterial photo-phore of flashlight fishes. Both kinds of fish probably use them to illuminate prey in the surrounding water, and both can hide the white reflective surface of the photophore by rotating it or drawing a fold of black skin over its aperture. Stomiiform males usually have much larger postorbital photophores than females. Male and female lanternfishes have special sexually dimorphic photophores on the tail or head in add-ition to the ventral camouflage ones. Male angler-fishes have no photophores; the female's bacterial ones can be very complex, with light pipes trans-mitting the light from the bacterial core to quite distant apertures. The lights are presumed to act as lures, perhaps both for prey and for males. Many stomiiform fishes also have long and complex lumi-nous barbels, whose function is also assumed to be that of a lure, perhaps mimicking particular kinds of luminous plankton.

Almost all of these animals produce blue lumi-nescence, but there are a very few remarkable deep-sea fish that produce both blue and red light (*Malacosteus*, *Pachystomias*, *Aristostomias*). They have the usual complement of body photophores, including a blue-emitting postorbital photophore, but they also have a suborbital red-emitting one. The red-emitting photo-phores contain large amounts of red fluorescent material and it is presumed that this acts as a fluor, rather like the green fluorescent protein of some jel-lyfish. The red light will be invisible to most other animals in the deep sea, which have only blue-sensitive visual pigment, but these fishes also have a red-sensitive visual pigment. They have in effect a private wavelength, either for communication or, like a sniperscope, for illuminating prey.

Measurements of Bioluminescence

Some of these organisms are the main contributors to the 'stimulable bioluminescent potential' of the water, i.e., the maximum amount of light that can be produced by turbulence in the water. Stimulated bioluminescence is most obvious in the wakes and bow waves of ships, but measurements of its vertical and horizontal distribution can give a quick indi-cation of the planktonic biomass as well as an

indication of the signal a fish shoal or a submarine might produce as it travels through the waters. Oceanographic measurements of bioluminescence were first made in the 1950s when sensitive light meters, lowered into the depths to measure the penetration of sunlight, recorded flashes of lumi-nescence. Later, when it became apparent that it was actually the movement of the light meter that was stimulating the bioluminescence, detector systems known as bathyphotometers were developed. These instruments have taken a variety of forms, with the most common design elements being a light detector viewing a light-tight chamber through which water is drawn either by movement of the bathyphotometer or by a pump (**Figure 4**). Light is stimulated as the bioluminescent organisms in the water experience turbulence, which is generated as the water passes through one or more constrictions or is stirred with a pump impeller. Units of measurements depend on the method of calibration and the residence time of the luminescent organism in the chamber. When resi-dence times are short compared to the duration of the flash, the amount of light measured is a function of the detection chamber volume, so the light meas-ured by the light detector (in photons s^{-1} or watts) is divided by the chamber volume and reported as photons s^{-1} per unit volume or watts per unit vol-ume. On the other hand, when the residence time is long enough for an entire flash to be measured, the light measured is a function of the volumetric flow rate (volume s^{-1}) through the chamber rather than the chamber volume and the light measured must be divided by flow and reported as photons per unit volume.

Bathyphotometers come in a variety of configur-ations, including profiling systems, towed systems, and moored systems. The 'stimulable biolumin-escence potential' measured with a given bath-yphotometer will depend on the organisms it samples. Low-flow-rate systems with small inlets will preferentially sample slow swimmers such as dino-flagellates, while higher flow rates and larger inlets will also sample zooplankton such as copepods and ostracods. Bathyphotometer measurements of stimulated bioluminescence have been made in most of the major oceans of the world. These measure-ments have generally been made in the upper 100 m of the water column at night. There is considerable seasonal variability in the amount of light measured, with average values ranging from approximately 10^9 to 10^{11} photons l^{-1}. There is also a pronounced diel rhythm of stimulable bioluminescence, with the photon flux measured in surface waters being greatly reduced or absent during the day. This is a con-sequence of the circadian rhythm of stimulable bioluminescence found in many dinoflagellates, as well as of diel vertical migration, which results in many luminescent species of plankton and nekton moving into surface waters only at night.

In most cases where the organisms responsible for the stimulable bioluminescence potential have been sampled, they have been found to be primarily dinoflagellates, copepods, and ostracods. Euphausiids too may be significant sources of bioluminescence in the water column but will only be sampled by very high-flow-rate systems. Gelatinous zooplankton, such as siphonophores and ctenophores, represent another potentially significant source of bioluminescence but are often overlooked because they are destroyed by the nets and pumps that oceanographers generally depend on for sampling the water column. All these organisms represent significant secondary producers and measurement of their bioluminescence provides a rapid means of assessing their distribution patterns, in the same way that fluorescence measurements have provided valuable information on the fine-scale distribution patterns of primary producers. As with fluorescence measurements, the primary method used to determine which organisms are responsible for the light emissions has been to collect samples from regions of interest with nets or pumps.

More recently there has also been some progress in developing computer image recognition programs that can identify luminescent organisms by their unique bioluminescent 'signatures.' Potential identifying properties of the light emissions include ntensity, kinetics, spatial pattern, and spectral distribution. Flash intensities are highly variable; while a single bacterium may emit only 10^4 photons s^{-1} a single dinoflagellate can emit more than 10^{11} photons s^{-1} at the peak of a flash (approximately 0.1 mW). Some of the brightest sources of luminescence are found among the jellies; some comb jellies, for example, have been found to emit more than 10^{12} photons s^{-1}. Flash durations are also highly variable and can be tens of milliseconds (e.g., the flash from the 'stern chaser' light organs on the tail of a lantern fish) to many seconds (e.g., in many jellyfish). The vast majority of planktonic organisms such as dinoflagellates, copepods, and ostracods, have flash durations of between 0.1 and 1 s. The number of flashes that a single organism can produce depends on the amount of luminescent material that is stored and the manner and rate of excitation. While some organisms produce only a flash or two in response to prolonged stimulation, others may respond with tens to hundreds of flashes until their luminescent chemical stores are exhausted and/or their excitation pathways are fatigued. Full recovery

of luminescent capacity can occur in a matter of hours to days depending on the availability of substrates for resynthesis of the luminescent chemicals. Spatial patterns of bioluminescence vary from essentially point sources for the smaller plankton to highly identifiable outlines and/or species-specific photophore patterns for many of the nekton. As indicated earlier, most marine bioluminescence is blue; however, there are often subtle differences in spectral distributions that could aid in identifications.

Bioluminescent Phenomena

Sometimes the bioluminescent plankton are responsible for dramatic surface phenomena. Luminescent wave crests have already been noted, but occasionally the sea may appear to be glowing uniformly. This 'milky sea' phenomenon has been described as like 'sailing through a field of snow' and is particularly common in the north-west Indian Ocean at the time of the south-west monsoon. It is probably the result of luminous bacteria growing on an oily surface scum. Other luminous phenomena include erupting balls of light exploding at the surface (probably fish schools coming up through dense luminous plankton and scattering at the surface) and, most dramatic of all, 'phosphorescent wheels.' These appear first as parallel bands of light racing across the sea surface and then change to become vast rotating wheels whose spokes may appear to extend to the horizon and which travel past the vessel at 50–100 km h^{-1}! They occur only in less than 200 m of water and are most frequent in the Arabian Gulf. Explanations invoke stimulation of the surface bioluminescent plankton either by the ships engines or by seismic activity in the region. Neither alternative is wholly convincing.

Applications of Bioluminescence

Bioluminescence plays a major role in the ecology of the ocean at all depths. Its quantification and distribution can provide oceanographers with a rapid biological marker for the proximity of physical features such as fronts and eddies, as well as an indication of the presence of particular species in the zooplankton and nekton communities. Aerial surveys with intensified videocameras have been used to find near-surface shoals of commercial fishes in several parts of the world, and in time of war (hot or cold) can monitor the night-time movements of surface vessels, torpedoes and submarines. More profitably, the use of bioluminescence has extended well beyond the oceans and into less obvious fields such as biomedical assays, pollution monitoring, and neuromuscular and developmental physiology. Bio-luminescent systems extracted from marine organisms are now used widely as intracellular markers whose light emission signals a particular biochemical event or the presence of potentially damaging radicals such as active oxygen. Photoproteins extracted from jellyfish have provided much of the information on the role of intracellular calcium. The green fluorescent protein, also from jellyfish, is widely used as an intracellular marker. These systems have been cloned and manipulated genetically to extend their biomedical usefulness. The genes controlling the bioluminescence of marine bacteria have also been identified and cloned. They and the jellyfish genes can be inserted into other organisms as 'reporter' genes. These 'report' on the activation of other genes, to which they are attached, by causing light emission that can easily be monitored. Changes in the light emission of cultures of bioluminescent marine bacteria or dinoflagellates are also used to monitor a wide range of toxic pollutants. The bioluminescence that plays such an important part in the ecology of the oceans now has a plethora of other uses in the terrestrial world.

See also

Cephalopods. Copepods. Deep-sea Fishes. Fish Migration, Vertical. Fish Larvae. Gelatinous Zooplankton. Krill. Mesopelagic Fishes. Plankton Viruses. Protozoa, Planktonic Foraminifera. Protozoa, Radiolarians.

Further Reading

Buskey EJ (1992) Epipelagic planktonic bioluminescence in the marginal ice zone of the Greenland Sea. *Marine Biology* 113: 689–698.

Harvey EN (1952) *Bioluminescence*. New York: Academic Press.

Hastings JW and Morin JG (1991) Bioluminescence. In: Prosser CL (ed.) *Neural and Integrative Animal Physiology*, pp. 131–170. New York: Wiley-Liss.

Herring PJ (1977) Bioluminescence in marine organisms. *Nature, London* 267: 788–793.

Herring PJ (ed.) (1978) *Bioluminescence in action*. London: Academic Press.

Herring PJ (1985) How to survive in the dark: bioluminescence in the deep sea. In: Laverack MS (ed.) *Physiological Adaptations of Marine Animals*, pp. 323–350. Cambridge: The Company of Biologists.

Lapota D, Geiger ML, Stiffey AV, Rosenberger DE, and Young DK (1989) Correlations of planktonic bioluminescence with other oceanographic parameters from a Norwegian fjord. *Marine Ecology Progress Series* 55: 217–227.

Widder EA (1999) Bioluminescence. In: Archer SN (ed.) *Adaptive Mechanisms in the Ecology of Vision*, pp. 555–581. Leiden: Kluwer Academic Publishers.

BENTHOS

BENTHIC ORGANISMS OVERVIEW

P. F. Kingston, Heriot-Watt University, Edinburgh, UK

Introduction

The term benthos is derived from the Greek word βαος (vathos, meaning depth) and refers to those organisms that live on the seabed and the bottom of rivers and lakes. In the oceans the benthos extends from the deepest oceanic trench to the intertidal spray zone. It includes those organisms that live in and on sediments, those that inhabit rocky substrata and those that make up the biodiversity of coral reefs.

The benthic environment, sometimes referred to as the benthal, may be divided up into various well defined zones that seem to be distinguished by depth (**Figure 1**).

Physical Conditions Affecting the Benthos

In most parts of the world the water level of the upper region of the benthal fluctuates, so that the animals and plants are subjected to the influence of the water only at certain times. At the highest level, only spray is involved; in the remainder of the region, the covering of water fluctuates as a result of tides and wind and other atmospheric factors. Below the level of extreme low water, the seafloor is permanently covered in water.

Water pressure increases rapidly with depth. Pressure is directly related to depth and increases by one atmospheric (101 kPa) per 10 m. Thus at 100 m, water pressure would be 11 atmospheres and at 10 000 m, 1001 atmospheres. Water is essentially incompressible, so that the size and shape of organisms are not affected by depth, providing the species does not possess a gas space and move between depth zones.

Light is essential to plants and it is its rapid attenuation with increasing water depth that limits the distribution of benthic flora to the coastal margins. Daylight is reduced to 1% of its surface values between 10 m and 30 m in coastal waters; in addition, the spectral quality of the light changes with depth, with the longer wavelengths being absorbed first. This influences the depth zonation of plant species, which is partially based on photosynthetic pigment type.

Surface water temperatures are highest in the tropics, becoming gradually cooler toward the higher latitudes. Diurnal changes in temperature are confined to the uppermost few meters and are usually quite small (3°C); however, water temperature falls with increasing depth to between 0.5°C and 2°C in the abyssal zone.

The average salinity of the oceans is 34.7 ppt. Salinity values in the deep ocean remain very near to this value. However, surface water salinity may vary considerably. In some enclosed areas, such as the Baltic Sea, salinity may be as low as 14 ppt, while in

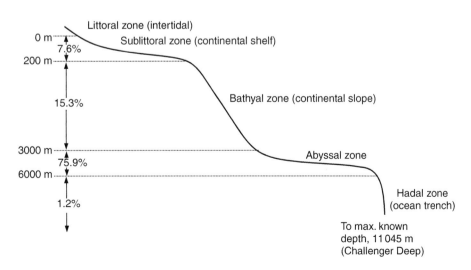

Figure 1 Classification of benthic environment with percentage representation of each depth zone.

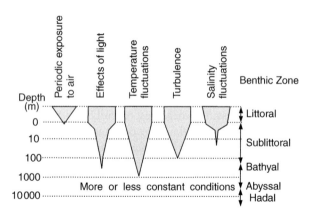

Figure 2 Diagram summarizing the influence of physical factors on the benthos. Relative influence of each factor indicated by width of shaded area.

areas of high evaporation, such as the Arabian Gulf, salinities in excess of 50 ppt may be reached. Salinity profiles may become quite complex in estuarine conditions or in coastal regions where there is a high fresh-water run off from the land (e.g., fiordic conditions) (**Figure 2**).

The composition of the benthos is profoundly affected by the nature of the substratum. Hard substrata tend to be dominated by surface dwelling forms, providing a base for the attachment of sessile animals and plants and a large variety of micro-habitats for organisms of cryptic habits. In contrast, sedimentary substrata are dominated by burrowing organisms and, apart from the intertidal zone and the shallowest waters, they are devoid of plants.

Hard substrata are most common in coastal waters where there are strong tidal currents and

surface turbulence. Farther offshore, and in the deep sea, the seabed is dominated by sediments, hard substrata occurring only on the slopes of seamounts, oceanic trenches, and other irregular features where the gradient of the seabed is too steep to permit significant accumulation of sedimentary material. Thus, viewed in its entirety, the seabed can be considered predominantly a level-bottom sedimentary environment (**Figure 3**).

Classification of the Benthos

It was the Danish marine scientist, Petersen working in the early part of the twentieth century, who first defined the two principal groups of benthic animals:

- the epifauna, comprising all animals living on or associated with the surface of the substratum, including sediments, rocks, shells, reefs and vegetation;
- the infauna comprising all animals living within the substratum, either moving freely through it or living in burrows and tubes or between the grains of sediments.

According to the great benthic ecologist Gunnar Thorson, the epifauna occupies less than 10% of the total area of the seabed, reaching its maximum abundance in the shallow waters and intertidal zones of tropical regions. However, the infauna, which Thorson believed occupies more than half the surface area of the planet, is most fully developed sublittorally. Nevertheless, the number of epifaunal species is far greater than the number of infaunal species. This is because the level-bottom habitat

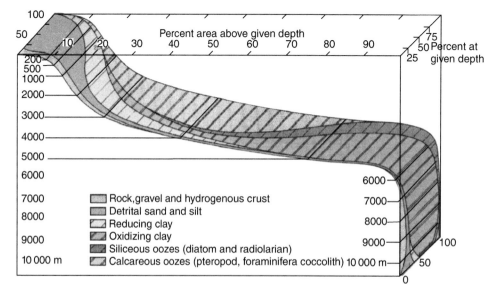

Figure 3 Distribution of sediment types in the ocean. (After Brunn, in Hedgepeth (1957).)

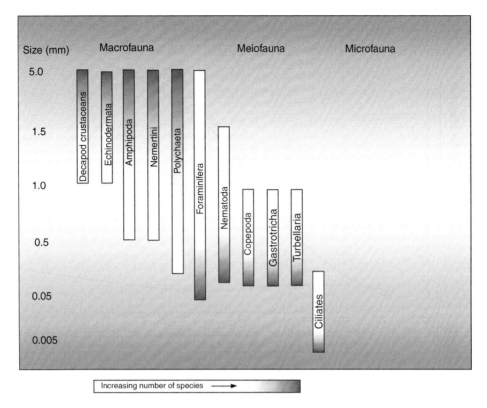

Figure 4 Principal taxa associated with each of the major categories of infauna.

provides a more uniform environment than hard substrata, with fewer potential habitat types to support a wide diversity of species. The epifauna of polar regions are not generally as well developed as those in the lower latitudes because of the effects of low temperatures in shallow waters and the effects of ice and meltwater in the intertidal region. The infauna largely escape these effects, exhibiting less latitudinal variation in number of species (**Figure 4**).

The infauna is further classified on the basis of size, the size categories broadly agreeing with major taxa that characterize the groups (**Figure 5**):

- macrofauna – animals that are retained on a 0.5 mm aperture sieve;
- meiofauna – animals that pass a 0.5 mm sieve but are retained on 0.06 mm sieve;
- microfauna – animals that pass a 0.06 mm sieve.

Petersen was also the first marine biologist to quantitatively sample soft-bottom habitats. After examining hundreds of samples from Danish coastal waters, he was struck by the fact that extensive areas of seabed were occupied by recurrent groups of species. These assemblages differed from area to area, in each case only a few species making up the bulk of the individuals and biomass. This contrasted with nonquantitative epifaunal dredge samples taken over the same range of areas, for which faunal lists

might be almost identical. Petersen proposed the infaunal assemblages that he had distinguished as communities and named them on the basis of the most visually dominant animals (**Table 1**).

Following the publication of Petersen's work in 1911–12, other marine biologists began to investigate benthic infauna quantitatively, and it began to emerge that there existed parallel bottom communities in which similar habitats around the world supported similar communities to those found by Petersen. These communities, although composed of different species, were closely similar, both ecologically and taxonomically, the characteristic species belonging to at least the same genus or a nearby taxon (**Table 2**).

Although the concept of parallel bottom communities appeared to hold good for temperate waters, in tropical regions such communities are not clearly definable, because of the presence of very large numbers of species and the small likelihood of any particular species or group of species dominating.

Not all benthic ecologists believe in Petersen's communities as functional biological units, since it is clear that abiotic factors such as sediment type must play a central role in determining species distribution. Alternative approaches have been proposed in which benthic associations are linked to

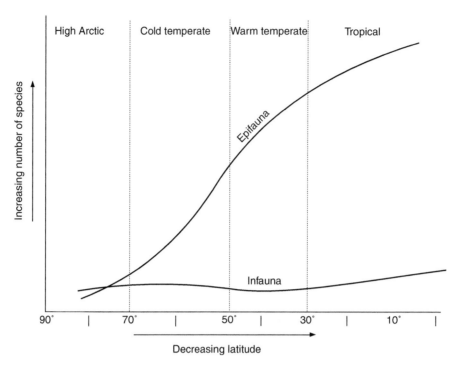

Figure 5 Relationship between numbers of epifaunal/infaunal species and latitude.

substratum (and latitude), suggesting that it is a common environmental requirement that forms the basis of the perceived community rather than an affinity of the members of the assemblage with one another.

Feeding Habits of Benthic Animals

Benthic animals, like all other animals, ultimately rely on the plant kingdom as their primary source of food. Most benthic animals are dependent on the rain of dead or partially decayed material (organic detritus) from above, and it is only in shallow coastal waters that macrophytes and phytoplankton are directly available for grazing or filtering by bottom-living forms. Much of this material consists of cellulose (dead plant cells) and chitin (crustacean exoskeletons). Few animals are able to digest these substances themselves and most rely on the action of bacteria to render them available, either as bacterial biomass or breakdown products.

Food particles are intercepted in the water before they reach the seabed, or are collected from the sediment surface after they have settled out, or are extracted from the sediment after becoming incorporated into it. These three scenarios reflect the three main types of detrivores: suspension feeders, selective deposit feeders, and direct deposit feeders.

Table 1 Petersen's benthic communities

Petersen's community	Typical species	Substratum
Macoma or Baltic community	Macoma balthica, Mya arenaria, Cerastoderma edule	Intertidal mud
Abra community	Abra alba, Macoma calcarea, Corbula gibba, Nephtys spp.	Inshore mud
Venus community	Venus striatula, Tellina fabula, Echinocardium cordatum	Offshore sand
Echninocardium–Filiformis community	E. cordatum, Amphiura filiformis, Cultellus pellucidus, Turritella communis	Offshore muddy sand
Brissopsis–Chiagei community	Brissopsis lyrifera, Amphiura chiagei, Calocaris macandreae	Offshore mud
Brissopsis–Sarsi community	B. lyrifera, Ophiura Sarsi, Abra nitida, Nucula tenuis	Deep mud
Amphilepis–Pecten community	Amphilepis norvegica, Chlamys (= Pecten) vitrea, Thyasira flexuosa	Deep mud
Haploops community	Haploops tubicola, Chlamys septemradiata, Lima loscombi	Offshore clay
Deep Venus community	Venus gallina, Spatangus purpureus, Abra prismatica	Deep sand

Table 2 Examples of parallel bottom communities identified by the Danish ecologist Gunnar Thorson

Species	Genera			
	NE Atlantic	NE Pacific	Arctic	NW Pacific
Macoma	baltica	nasuta	calcarea	incongrua
Cardium	edule	corbis	ciliatum	hungerfordi
Mya	arenaria	arenaria	truncata	
Arenicola	marina	claparedii	marina	

Table 3 Percentage representation of different feeding types from sandy and muddy sediments

Feeding types	Sandy sediment (< 15 silt/clay) (%)	Muddy sediment (> 90% silt/clay) (%)
Predators/omnivores	25.3	22.4
Suspension feeders	27.2	14.9
Selective deposit feeders	41.1	50.8
Direct deposit feeders	6.3	11.9

- Suspension feeders may be passive or active. Passive suspension feeders trap passing particles on extended appendages that are covered in sticky mucus and rely on natural water movements to bring the food to them (e.g., crinoid echinoderms). Active suspension feeders create a strong water current of their own, filtering out particles using specially modified organs (e.g., most bivalve mollusks).
- Selective deposit feeders either consume surface deposits in their immediate vicinity using unspecialized mouth parts, or, where food is less abundant, use extendable tentacles or siphons to pick up particles over a large area (e.g., terebellid polychaetes).
- Direct deposit feeders indiscriminately ingest sediment using organic matter and microbial organisms contained in the sediment as food. Polychaeta, such as the lugworm, *Arenicola*, construct L-shaped burrows and mine sediment from a horizontal gallery, reversing up the vertical shaft to defecate at the surface. Such animals play an important role in the physical turnover of the sediments.

Grazing or browsing animals are most common intertidally or in shallow waters. They are mobile consumers, cropping exposed tissues of sessile prey usually without killing the whole organism. They include animals that feed on macroalgae and those that feed on colonial cnidaria, bryozoans, and tunicates such as gastropod mollusks and echinoids. On the level bottom, the tips of tentacles and siphons of infaunal animals are grazed by demersal fish, providing a route for energy transfer from the seabed back into the pelagic system.

Predatory hunters are common among benthic epifauna and include crustaceans, asteroid echinoderms, and gastropod mollusks. These are mobile animals that seek out and consume individual prey items one at a time. Although less common, such predators are also found in the infauna, moving through the sediment, attacking their prey *in situ* (e.g., the polychaete *Glycera*).

Although, overall, deposit feeders form the largest single group of benthic infauna, the proportion of each trophic group is greatly influenced by the nature of the sediment. Thus in coarser sandy sediments, where water movement is relatively strong, the proportion of suspension feeders increases, while fine silts and muds are usually dominated by deposit feeders (see **Table 3**).

Spatial Distribution of Benthos

Competition between benthic infauna is usually for space. This is because most benthic animals are either suspension or deposit feeders and are competing for access to the same food source. In this respect, benthic communities are similar to those of terrestrial plants since, in both, competition between individuals is for an energy source that originates from above. Indeed, early approaches to the statistical analysis of benthic community structure were often rooted in principles originally developed by terrestrial botanists.

Suspension feeding may take place at more than one level. Some species, such as sabellid worms, extend their feeding organs several centimeters into the water column; some keep a lower profile, with short siphons projecting just a few millimeters above the sediment surface, while others have open burrows, drawing water into galleries below the surface. In addition, selective deposit feeders scour the sediment surface for food particles using palps or tentacles that in some species can extend up to a meter. The result is a contagious horizontal distribution of animals (i.e., neither random nor regular) that is maintained primarily by inter- and intra-specific competition for space.

Benthic infauna also show a marked vertical distribution. This is more a function of the subsurface physical conditions of the sediments and the need for the majority of species to be in communication with the surface than of competition between the animals.

Petersen chose physically large representatives to describe his benthic communities, primarily because they were easy to identify under field conditions. However, most benthic infaunal species are too small to be recognized with the naked eye, with burrows that penetrate no more than a few centimeters into the substratum. The result is that, for most level-bottom communities, some 95–99% of the animals are located within 5 cm of the sediment surface.

Reproduction in Benthic Animals

The act of reproduction offers benthic animals, the majority of which are either sessile or very restricted in their migratory powers, an opportunity to disperse and to colonize new ground. It is therefore not surprising that the majority of benthic species experience at least some sort of pelagic phase during their early development. Most invertebrates have larvae that swim for varying amounts of time before settlement and metamorphosis. The larvae, which develop freely in the surface waters of the ocean, either feed on planktonic organisms (planktotrophic larvae) or develop independently from a self-contained food supply or yolk (lecithotrophic larvae). Pelagic development in temperate waters can take several weeks, during which time developing larvae may be transported over great distances. Where it is within the interests of a particular species to ensure that its offspring are not dispersed (e.g., some intertidal habitats), a free-living larval phase may be dispensed with. In this case eggs may develop directly into miniature adults (oviparity) or may be retained within the body of the adult with the young being born fully developed (viviparity). Reproductive strategies such as these are also common in the deep-sea and polar regions where the supply of phytoplankton for feeding is unreliable or nonexistent. A good example of a latitudinal trend in this respect was demonstrated by Thorson. Analysing the developmental types of prosobranchs, he was able to show that the proportion of species with nonpelagic larvae decreases from the arctic to the tropics, while the proportion with pelagic larvae increases (**Figure 6**).

For many years deep-sea biologists believed that the energetic investment required to produce large numbers of planktotrophic larvae, and the huge distances required to be covered by such larvae in order to reach surface waters, would preclude such a reproductive strategy for deep-sea animals. However, it is now known that several species of ophiuroids living at depths of 2000–3000 m not only exhibit seasonal reproductive behavior but also produce larvae that feed in ocean surface waters.

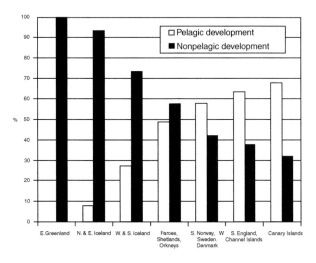

Figure 6 Percentage distribution of prosobranchs with pelagic and nonpelagic development in relation to latitude. (Adapted from Thorson (1950).)

Many benthic invertebrates are able to reproduce asexually. For example, polychaetes from the family the Syllidae are able to reproduce by budding; others, such as the cirratulid *Dodecaceria* or the ctenodrilid *Raphadrilus*, simply fragment, each fragment growing into a new individual. The ability to switch between sexual and vegetative means of propagation provides the potential for such species to rapidly colonize areas that have been disturbed. Where disturbance is accompanied by organic enrichment, for example, from sewage or paper pulp discharge, huge localized populations may result. These are the so-called opportunistic species that are sometimes used as indicators of pollution.

Although planktonic larvae are able to swim, they are very small and, for the most part, are obliged to go where ocean currents take them. The critical time arrives just before the larvae are about to settle. At one time it was thought that the process of settlement was random, with individuals that settled in unfavorable substrata perishing. Although this undoubtedly happens, most species seem to have some sort of behavioral pattern to increase their chances of finding a suitable substratum. The larvae usually pass through one or more stages of photopositive and photonegative behavior. These enable the larvae to remain near the sea surface to feed and then to drop to the bottom to seek a suitable substratum on which to settle. Depending on the species, larvae may cue on the mechanical attributes of the substratum or on its chemical nature. Chemical attraction is also important in gregarious species in which the young are attracted to settle at sites where adults of the same species are already present (e.g., oysters). Most larvae go through a period when, although able to settle permanently, they retain the ability to swim. This allows them to 'test' the substratum, rising back into the water and any prevailing currents should the nature of the ground be unsuitable. After settling, larvae may move a short distance, usually no more than a few centimeters. These early stages in the recruitment of benthic organisms are crucial in the maintenance of benthic community structure and it is now believed that it is at this time that the nature of the community is established.

It is clear that the vast majority of planktonic larvae never make it to adulthood. Mortality from predation and transport away from a suitable habitat are on a massive scale. To compensate, species with planktotrophic larvae produce huge numbers of eggs (e.g., the sea hare *Aplysia californiensis* spawns as many as 450 000 000 eggs at one time). This is possible because there is no need for a large, and energetically expensive, yolk; the larvae hatch at an early embryonic stage and rely almost entirely on plankton-derived food for their development. One consequence of this is that the recruitment varies depending on the success of the plankton production in a particular year and the vagaries of local currents. Thus, populations of benthic species that reproduce by means of planktotrophic larvae tend to fluctuate numerically from year to year, with the potential for heavy recruitment when the combination of environmental factors is favorable, or recruitment failure when they are not. Species reproducing by means of nonpelagic larvae or by direct development tend to produce fewer eggs, since there is a large yolk required to nourish the developing embryo. Although annual recruitment is relatively modest for these species, it is less variable between years, producing populations with a greater temporal stability (**Figure 7**). Because of this, these populations are

likely to be slow to recover from major natural environmental disturbances (e.g., unusual temperature extremes or physical disturbance) or major pollution events.

Deep-sea Benthos

Because of regional differences, it is difficult to define the exact upper limit of the deep-sea benthic environment. However, it is generally regarded as beginning beyond the 200 m depth contour. At this point, where the continental shelf gives way to the continental slope, there is often a marked change in the benthic fauna. In the past there have been many attempts to produce a scheme of zonation for the deep-sea environment. There are three major regions beyond 200 m depth – these are the bathyal region, the abyssal region, and the hadal region (see **Figure 1**). The bathyal region represents the transition region between the edge of the continental shelf and the true deep sea (the continental slope). Its boundary with the abyssal region has been variously defined by workers. It is believed that the 4 °C isotherm limits the depth at which the endemic abyssal organisms can survive. Since this varies in depth according to geographical and hydrographical conditions in the abyssal region, it follows that the upper depth limit of the abyssal fauna will also vary. This depth is usually between 1000 and 3000 m. The abyssal region is by far the most extensive, reaching down to 6000 m depth and accounting for over half the surface area of the planet. The hadal zone, sometimes called the ultra-abyssal zone, is largely restricted to the deep oceanic trenches. The composition of the benthos in these trenches differs from that of nearby abyssal areas. The trenches are geographically isolated from one another and the fauna exhibits a high degree of endemism.

The deep sea is aphotic and so has no primary production except in certain areas where chemosynthetic bacteria are found. Thus, the fauna of the deep sea is almost wholly reliant on organic material that has been generated in the surface layers of the oceans and has sunk to the seabed. Since the likelihood of a particle of food being consumed on its way down through the water column is related to time, the deeper the water the less food is available for the animals that live on the seafloor. This results in a relatively impoverished fauna, in which the density of organisms is low and the size of most is quite small. Paradoxically, the deep sea supports a few rare species that grow unusually large. This phenomenon is known as abyssal gigantism and is found primarily in crustacean species. The reasons for abyssal

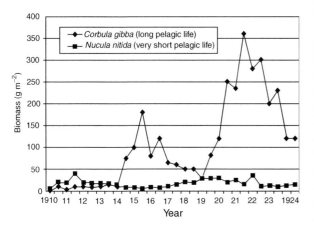

Figure 7 Example of two populations of bivalves showing the influence of type of larvae on population stability. (Adapted from Thorson (1950).)

gigantism are not clear, but it is believed to be the result either of a peculiar metabolism under conditions of high pressure or of the slow growth and time taken to reach sexual maturity.

Although the number of individuals in the deep sea is quite low, there are many species. This combination of many species represented by few individuals results in high calculated diversity values and has led to the suggestion by some that the biodiversity of the deep sea is comparable to that of tropical rain forests. Accepting that diversity is high in the deep sea, there are several theories that attempt an explanation. The earliest is the stability–time hypothesis put forward by Howard Sanders of the Woods Hole Oceanographic Institution in the late 1960s. This suggests that the highly stable environmental conditions of the deep sea that have persisted over geological time might have allowed many species to evolve that are highly specialized for a particular microhabitat or food source. Another theory, the cropper or disturbance theory, suggests that, as a result of the scarcity of food in the deep sea, none of the animals is a food specialist, the animals being forced to feed indiscriminately on anything living or dead that is smaller than itself. High diversity results from intense predation, which allows a large number of species to persist, eating the same food, but never becoming abundant enough to compete with one another. More recently, Grassle and Morse-Porteus have suggested that a combination of factors might be responsible, including the patchy distribution of organically enriched areas in a background of low productivity; the occurrence of discrete, small-scale disturbances (primarily biological) in an area of otherwise great constancy; and the lack of barriers to dispersal among species distributed over a very large area.

Although most animal groups are represented in the deep sea, the fauna is often dominated by Holothuroidea (sea cucumbers) or Ophiuroidea (brittle stars). Crustacea and polychaete worms are also important members of deep-sea communities. For many years it was believed that the deep seafloor was too remote from the surface, and the physical conditions were too constant, for the organisms living there to be influenced by the seasons. Although this may be the case for many deep-sea species, long-term time-lapse photography has shown that cyclical events, such as the accumulation of organic detritus on the deep seabed, do take place. Furthermore, in temperate waters, these appear to correspond with seasonally driven processes such as the spring plankton bloom. Where they occur, these pulses of organic input inevitably influence the deep-sea communities below with the consequence that seasonal life cycles are not uncommon in abyssal animals. Localized areas of organic enrichment can occur in the deep sea as the result of the sinking of large objects such as the carcass of a large sea mammal or fish or waterlogged tree trunk. The surprisingly quick response of deep-sea scavengers to large food-falls such as these, and the frequency with which they have been recorded, has led researchers to believe that they are important contributers to energy flow on the deep-sea floor. Hydrothermal vents are also thought to provide a significant input of energy into the benthic environment. These are a relatively recent discovery and, at first, were thought to be rare and isolated phenomena occurring only on the Galapagos Rift off Ecuador. It is now known that active vents are associated with nearly all areas of tectonic activity that have been investigated in the deep Pacific and Atlantic. These vents provide a nonphotosynthetic source of organic carbon through the medium of chemoautotrophic bacteria. Theses organisms use sulfur-containing inorganic compounds as an oxidizing substrate to synthesize organic carbon from carbon dioxide and methane without the need for sunlight. The chemicals come from hot water that originates deep within the Earth's crust. Some of the bacteria form dense white mats on the surface of the sediments similar to those of *Beggiatoa*, an anaerobic bacterium found in anoxic sediments in shallow water; others enter into symbiotic relationships with the bacteria, either hosting them on their gills (e.g., the mussel *Bathymodiolus*) or within a special internal sac, the trophosome (e.g., *Riftia*). The relationship is very complex as there has to be a compromise between the anaerobic needs of the bacteria and the aerobic needs of the animals. Nevertheless, the arrangement is very successful and the animals, which sometimes occur in huge numbers, often grow to gargantuan size. There is still much to be learned about these vent communities, which can support concentrations of biomass several orders of magnitude greater than that of the nearby seafloor. It remains a mystery how these vents become populated, since they are known to be transient and variable, probably lasting only decades or less. It has been suggested that pelagic larvae of many vent species may be able to delay settlement for months at a time so as to increase their chances of locating a suitable site.

Glossary

Abyssal region   That region of the seabed from between 1000 and 3000 m depth reaching down to 6000 m.

Atmosphere   Measure of pressure (1 atm = 101 kPa).

Bathyal region Region of the seabed that represents the transition region between the edge of the continental shelf and the true deep sea (the continental slope).

Benthal The benthic environment.

Benthos Those organisms that live on the seabed and the bottoms of rivers and lakes.

Continental shelf Region of the seabed from low-water mark to a depth of 200 m.

Detritivores Animals feeding on organic detritus.

Direct deposit feeders Animals indiscriminately ingesting sediment using organic matter and microbial organisms contained in the sediment as food.

Epifauna All animals living on or associated with the surface of the substratum.

Hadal zone Region of the seabed below 6000 m depth, largely restricted to the deep oceanic trenches.

Infauna All animals living within the substratum or moving freely through it.

Kilopascal (kPa) Measure of pressure (100 kPa = 1 bar = 0.987 atm).

Lecithotrophic larvae Pelagic larvae of marine animals that develop freely in the surface waters of the ocean, developing independently from a self-contained food supply or yolk.

Macrofauna Animals retained on a 0.5 mm aperture sieve.

Meiofauna Animals passing a 0.5 mm sieve but retained on a 0.06 mm sieve.

Microfauna Animals passing a 0.06 mm sieve.

Organic detritus Dead or partially decayed plant and animal material.

Oviparity Eggs laid by the adult develop directly into miniature adults.

Pelagic larvae Larvae of marine animals that swim freely in the water column.

Planktotrophic larvae Pelagic larvae of marine animals that develop freely in the surface waters of the ocean, feeding on planktonic organisms.

Selective deposit feeders Animals feeding on surface particles of organic matter or sediment particles supporting a rich bacterial flora.

Suspension feeders Animals feeding on organisms or organic detritus suspended in the water column.

Viviparity Young are born fully developed either from eggs retained within the body of the mother (oviparity) or after internal embryonic development.

See also

Benthic Boundary Layer Effects. Benthic Foraminifera. Deep-sea Fauna. Macrobenthos. Meiobenthos. Microphytobenthos. Phytobenthos.

Further Reading

Cushing DH and Walsh JJ (eds.) (1976) *The Ecology of the Seas.* Oxford: Blackwell Scientific Publications.

Friedrich H (1969) *Marine Biology: An Introduction to Its Problems and Results.* London: Sidgwick and Jackson.

Gage JD and Tyler PA (1991) *Deep-sea Biology: A Natural History of Organisms at the Deep-Sea Floor.* Cambridge: Cambridge University Press.

Hedgepeth JW (ed.) (1957) *Treatise on Marine Ecology and Paleoecology,* Vol. 1, *Ecology,* The Geological Society of America, Memoir 67. Washington, DC: Geological Society of America.

Jøgensen CB (1990) *Bivalve Filter-Feeding: Hydrodynamics, Bioenergetics, Physiology and Ecology.* Fredensborg, Denmark: Olsen and Olsen.

Levington JS (1995) *Marine Biology: Function, Biodiversity, Ecology.* Oxford: Oxford University Press.

Nybakken JW (1993) *Marine Biology.* New York: Harper-Collins College Publishers.

Thorson G (1950) Reproduction and larval ecology of marine bottom invertebrates. *Biological Reviews* 25: 1–25.

Webber HH and Thurman HV (1991) *Marine Biology.* New York: Harper Collins.

MICROPHYTOBENTHOS

G. J. C. Underwood, University of Essex, Colchester, UK

Introduction

Microphytobenthos is a descriptive term for the diverse assemblages of photosynthetic diatoms, cyanobacteria, flagellates, and green algae that inhabit the surface layer of sediments in marine systems. Microphytobenthos occur wherever light penetrates to the sediment's surface, and are abundant on intertidal mud and sandflats and in shallow subtidal regions. Microphytobenthic primary production may be high, matching that of phytoplankton in the overlying water column, yet this activity is compressed into a biofilm only a few millimeters thick. The relationship between irradiance and rates of microphytobenthic photosynthesis is fairly well understood, but new methods are revealing fine-scale effects of microspatial distribution within the vertical light profile and migration of cells throughout the diel illumination period. Patterns of biomass distribution and seasonal and spatial changes in species composition are well described, but studies differ on the relative importance of the factors influencing microphytobenthic biomass (irradiance, resuspension, nutrients, grazing, exposure, desiccation, etc.). Microphytobenthic biofilms play an important role in mediating the exchange of nutrients across the sediment–water interface, and microphytobenthos both stimulate and compete with various bacterial sediment processes. The presence of biofilms rich in extracellular polysaccharides alters the erosional properties of sediments, termed biostabilization.

Types of Microphytobenthos

Sediment properties play a major role in determining the type of microphytobenthic assemblage present in a particular environment. Sediments consisting of fine silts and clays (less that 63 µm) are termed cohesive sediments. The fine nature of such material and the lack of suitable attachment points result in assemblages dominated by motile microphytobenthic species. These are termed 'epipelic' biofilms (epipelic: living on mud), and the microphytobenthos are sometimes termed 'epipelon.' Sediments consisting of larger particles, silty sands, and sands are noncohesive, with greater pore space, and are generally more often disturbed. Growing attached to individual sand and silt particles are found 'epipsammic' taxa (epipsammic: living on sand). Epispammic assemblages usually contain a substantial proportion of epipelic taxa as well.

Epipelic biofilms The commonest epipelic microphytobenthos are biraphid diatoms, with the genera *Navicula*, *Gyrosigma*, *Nitzschia* and *Diploneis* usually well represented (**Table 1**). In fine sediment habitats, light penetration is very limited and, in order to photosynthesize, cells need to be able to position themselves at the sediment surface. Biraphid diatoms move by excreting extracellular polymeric substances (EPS) from the raphe slit present in each of the silica cell walls (valves) that make up the cell. Cyanobacterial filaments move by gliding and nonflagellated euglenids move by amoeboid movement. In dense biofilms of epipelic diatoms, the concentrations of EPS can become high (200–300 $\mu g\,g^{-1}$ dry weight of sediment), providing a carbon source to the sediment system. High concentration of EPS can increase the force needed to erode sediments, termed 'biostabilization.' Epipelic biofilms can be very extensive on intertidal estuarine mudflats, where they can contribute up to 50% of estuarine carbon budgets.

Epipsammic assemblages The 'epipsammon' are generally nonmotile, or only partially mobile. Diatoms are the major constituents, with araphid and monoraphid genera common (e.g., *Opephora*, *Achnanthes*, *Amphora*, and *Cocconeis*) (**Table 1**).

Table 1 Genera of photoautotrophs commonly found in microphytobenthic communities

Algal group	Epipelic	Epipsammic
Cyanobacteria	Oscillatoria	Oscillatoria
		Microcoleus
		Spirulina
Bacillariophyta	Navicula	Opephora
	Amphora	Raphoneis
	Fallacia	Achnanthes
	Staurophora	Cocconeis
	Gyrosigma	Fragilaria
	Pleurosigma	Navicula
	Nitzschia	Nitzschia
	Diploneis	Amphora
	Cylindrotheca	
Euglenophyta	Euglena	Euglena
Chlorophyta		Many

Epipsammic cells attach themselves to sand particles by a pad or short stalk of EPS, though many cells are also capable of movement. Filamentous and colonial cyanobacteria (*Oscillatoria*, *Microcoleus*), coccal green algae and motile flagellates and chlorophytes are common in epipsammic assemblages. Thus epipsammic assemblages often have a greater taxonomic diversity (at the level of algal groups). Light penetration is greater into sandy sediments, which are also disturbed by tidal and wind-induced currents. Cells are therefore frequently mixed within the sediment photic zone, and the requirement for motility is less. Indeed, in highly mixed systems, nonattached, motile taxa may be absent, and only attached species are found, often within depressions present on the surface of sand grains, where they receive protection from abrasion.

Primary Production

Photosynthesis

Microphytobenthos are photoautotrophic organisms. Hourly rates of primary production are high,

with annual primary production ranging between 0.3 and 234 g C m^{-2} a^{-1} (**Table 2**). Different techniques are used to measure microphytobenthic primary production: (1) oxygen exchange across the sediment–water interface; (2) (^{14}C)bicarbonate uptake in intact biofilms; (3) (^{14}C)bicarbonate uptake in slurries; (4) oxygen production within biofilms using oxygen microelectrodes; and more recently (5) modulated chlorophyll *a* fluorescence techniques (**Figure 1**). These techniques all measure slightly different aspects of photosynthesis, making intercomparisons difficult.

Oxygen exchange measurements on intact biofilms measure net community production, and if the oxygen uptake rate (negative) in the dark is subtracted from the net community production, then a measure of gross oxygen production is obtained (assuming that respiration in the light and dark do not differ). (^{14}C)Bicarbonate uptake into intact biofilms measures net photosynthesis, and tends to underestimate carbon fixation rates as it is not possible to measure accurately the specific ^{14}C activity within the thin photosynthetically active layer. In noncohesive sediments, percolation of sea water of known specific ^{14}C activity into the biofilm through the application of a slight vacuum to the bottom of a sediment core results in higher estimates of carbon fixation. Percolation techniques cannot be used with cohesive

Table 2 Daily and annual rates of primary production for epipelic and epipsammic microphytobenthos from a number of different habitats

Site	Daily production (mg C m^{-2} d^{-1})	Annual production (g C m^{-2} a^{-1})
Epipelon		
Ems-Dollard, Netherlands[a]	600–1370	62–276
Tagus Estuary, Portugal[b]	5–32 (h^{-1})	47–178
North Inlet, SC, USA[c]	–	56–234
Langebaan Lagoon, South Africa[d]	17–69	253 (mud)
Epipsammon		
Langebaan Lagoon, South Africa[d]	17–69	63 (sand)
Laholm Bay, Sweden[e]	10–200	0.3–20
Ria de Arosa, Spain[f]	–	54
Weeks Bay, AL, USA[g]	10–750	90.1

[a] Colijn, F De Jonge, V (1984) *Marine Ecology Progress Series* 14: 185–196.
[b] Brotas, V Catarino, F (1995) *Netherlands Journal of Aquatic Ecology*, 29: 333–339.
[c] Pinckney, JL (1994) In: *Biostabilization of Sediments* (ed. WE Krumbein, DM Paterson and LJ Stal). Universität Oldenburg, Oldenburg. pp. 55–84.
[d] Fielding P *et al.* (1988) *Estuarine Coastal shelf Science.* 27: 413–426.
[e] Sundbäck, K Jönsson, B (1988) *Journal of Experimental Marine Biology and Ecology* 122: 63–81.
[f] Varela, M Penas, E (1985) *Marine Ecologoy Progress Series* 25: 111–119.
[g] Schreiber, RA Pennock, JR (1995) *Ophelia* 42: 335–352.

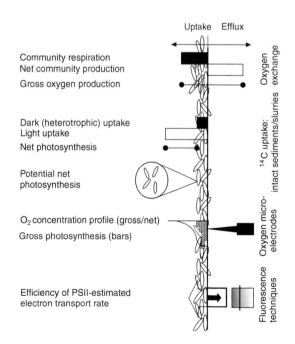

Figure 1 Techniques used to determine microphytobenthic primary production measure different aspects of photosynthesis in microphytobenthic biofilms, and have varying scales of vertical and horizontal resolution. Thus comparison of data needs to be made with care. PSII, photosystem II.

sediments. Oxygen exchange and ^{14}C methods require the microphytobenthic community to be submerged and this may underestimate intertidal primary production, where the majority of the photosynthesis occurs during low tide exposure. ^{14}C slurry techniques are a rapid method for measuring photosynthetic parameters, with photosynthesis versus light curves generated in a 'photosynthetron.' However, existing microgradients in the sediment are destroyed in slurries, and this technique therefore measures maximum potential primary production, in the absence of structure within the biofilm.

Oxygen microelectrodes measure gross primary production rates at small-scale (100–200 µm) depth intervals down a profile into the sediment. Construction of complete photosynthesis profile curves is time-consuming; the time taken to generate sufficient replicate production profiles is greater than some of the temporal properties of the biofilm (e.g., endogenous vertical migration). To avoid this, production rates can be calculated from the profile of oxygen concentration with depth under a fixed irradiance, assuming diffusion and porosity coefficients. Net oxygen production can be calculated from the slope in oxygen concentrations out of the sediment, but with exposed sediments this can be problematic. Significant amounts of variation in oxygen production profiles can be due to patchiness in the distribution of microphytobenthic biomass. Oxygen microelectrodes are an important tool for measuring the microspatial distribution of photosynthesis within sediments and response of photosynthesis to environmental variation, but scaling-up of these measurements to larger areal rates is contentious.

Variable fluorescent techniques measure the activity of the photosystem II (PSII) reaction centre, thus providing an estimate of the rate of production of electrons by the water-splitting system of PSII (electron transport rate). Being noninvasive, fluorescence techniques can be used to rapidly and repeatedly measure in situ activity. As oxygen is a product of the water-splitting process, there is a relationship between oxygen production and PSII electron transport rate (ETR), and also reasonable linearity between ^{14}C-fixation rates and ETR, especially in sediment slurries. Thus fluorescence techniques can provide an indirect (but nondestructive and rapid) measurement of microphytobenthic primary production. However, the relationship between ETR and oxygen evolution or ^{14}C fixation can become nonlinear at high irradiances, and vertical migration of cells within the biofilm can complicate the interpretation of results. Variable fluorescence measurements can also be made on single cells, using a modified fluorescence microscope and image analysis techniques, allowing the photosynthetic response of single cells within a mixed population to be measured in undisturbed biofilms (**Figure 2**).

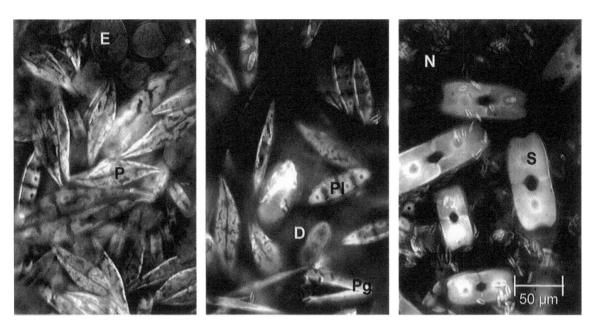

Figure 2 Fluorescence images of intact epipelic microphytobenthic biofilms, showing patchiness at a microscale in cell distribution, and differences in cell size (E = *Euglena* sp.; P = *Pleurosigma angulatum*; Pl = *Plagiotropis vitrea*; D = *Diploneis didyma*; Pg = *Petrodictyon gemma*; S = *Staurophora* sp.; N = small *Navicula* species). Fluorescence imaging techniques can calculate the photosynthetic efficiency of individual cells within the biofilm, allowing taxonomic differences to be determined. (Images courtesy of ARM Hanlon, University of Essex.)

Light Penetration and Photosynthesis

There are substantial spatial and temporal gradients in light availability in microphytobenthic habitats. Irradiance can exceed $2000\,\mu\text{mol}$ photons $\text{m}^{-2}\,\text{s}^{-1}$ on exposed intertidal sediments, while in clear shallow water sufficient light can penetrate to depths of 20–30 m, permitting microphytobenthic growth. Steep gradients of irradiance occur within sediments, where the attenuation of light is rapid (attenuation coefficients (k) between 1 and $3.5\,\text{mm}^{-1}$ for sandy and cohesive sediments, respectively). Thus the euphotic zone depth in sediments (1% of incident light) is usually much less than 1 cm (less than 2 mm in cohesive muddy sediments) (**Figure 3**). Light intensity just beneath the sediment surface, particularly at wavelengths $>700\,\text{nm}$ can be greater than the incident light, owing to backscatter effects within the sediment. The spectral quality of light also changes within sediments and is further modified by increased light attenuation of specific wavelengths (particularly blue and red) due to absorption by microalgal photopigments.

There is a fairly clear relationship between biomass-normalized primary production ($\mu\text{g}\,\text{C}(\mu g\,\text{Chl}a)^{-1}\,\text{h}^{-1}$, termed P^{B}) and irradiance in microphytobenthic systems, up to saturating irradiances (P^{B}_{max}). Irradiance accounts for between 30% and 60% of the variability in primary production, and biomass explains another 30–40%. Within cohesive sediments the majority of photosynthesis occurs within the top 200–400 μm of the sediment. In sandy sediments, where light penetration is greater, gross photosynthesis can occur deeper than this (up to 2 mm) (**Figure 3**) and may even show a biomodal distribution owing to distinct vertical separation of diatoms and cyanobacterial layers. Isolated microphytobenthos (i.e., in slurries, lens tissue preparations or cultures) reach P^{B}_{max} at light intensities between 100 and $800\,\mu\text{mol}$ photons $\text{m}^{-2}\,\text{s}^{-1}$ and show photoinhibition of P^{B} at higher light intensities. Depth-integrated rates of sediment photosynthesis obtained from *in situ* oxygen microelectrode measurements saturate at higher irradiances than slurries and show little or no evidence of photoinhibition. In undisturbed sediments, the peak of gross oxygen production occurs deeper in the sediment at high light intensities ($>1200\,\mu\text{mol}$ photons $\text{m}^{-2}\,\text{s}^{-1}$) than at lower light intensities, mainly because of migration of the bulk of the microalgal population down into the sediment away from high surface irradiance. Microphytobenthos are sensitive to light intensity and UVB radiation, with surface biomass varying with irradiance. Some subtidal assemblages are shade-adapted and migrate down into the sediment at midday to avoid high light levels. Taxonomic differences occur with regard to positioning with the light field. The euglenophyte *Euglena deses* commonly occurs on intertidal flats and at high irradiance occurs on the surface of sediments, with epipelic diatoms underneath. Mixed assemblages of filamentous cyanobacteria and epipelic diatoms also show vertical positioning, with cyanobacteria positioned beneath the diatom layer. The ability of cells to migrate away from high irradiance allows microphytobenthos to respond to the light climate and position themselves at optimal irradiances.

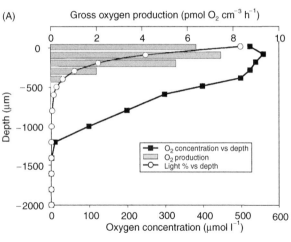

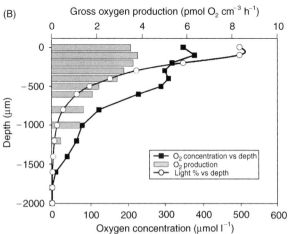

Figure 3 Typical oxygen concentration profiles and rates of gross oxygen production in an epipelic (A) and epipsammic (B) microphytobenthic biofilm. Light attenuation is less steep in sandier sediments, and thus oxygen production occurs to a greater depth.

Vertical migration Following disturbance and/or deposition of fresh sediment, microphytobenthos need to reposition themselves back within the euphotic zone to photosynthesize. In many intertidal habitats the microphytobenthos exhibit endogenous rhythms of vertical migration, with migration remaining in synchrony with the daily shift of the tidal cycle within the diel light frame. These endogenous rhythms can be maintained for between 3 and 4 days in the absence of any light or tidal stimuli. Intertidal sites are also subject to varying patterns of diel illumination periods. The shifting pattern of tidal exposure (≈ 55 min per day) within diel light curves, and the fortnightly cycle of spring and neap tides can result in periods when microphytobenthos are exposed to very high irradiance during low tide at solar noon (exceeding $2000 \, \mu mol \, photons \, m^{-2} s^{-1}$) and at other times (several days for some regions of the intertidal) when little or no light reaches the sediment surface. Thus cells need to be able to cope with periods of darkness, when they rely on intracellular carbon storage compounds (glucans) as an energy source.

Temperature effects on microphytobenthic photosynthesis The temperature of a mudflat can change rapidly during a tidal emersion period, at up to 2–$3 °C h^{-1}$, with daily ranges of $20 °C$ and seasonal ranges between 0 and $35 °C$. There is a clear relationship between P^B_{max} and temperature, with an optimal temperature for intertidal diatoms of $25 °C$, while at temperatures above $25 °C$ there can be significant inhibition (30%) of microphytobenthic photosynthesis, particularly on upper shore intertidal regions. This can lead to reductions of biomass on upper shores.

Extracellular Polysaccharide (EPS) Production and Sediment Biostabilization

Epipelic and epipsammic diatoms produce EPS either during motility or as an attachment structure. Microphytobenthos also excrete surplus photoassimilated carbon as carbohydrates when they are nutrient limited and subject to high irradiance. In diatom-rich biofilms, between 20% and 40% of the extracellular carbohydrate material present is polymeric, i.e., EPS. The remainder consists of nonpolymeric material, mainly simple sugars, leachates, and other photoassimilates. These low molecular weight exudates are rapidly utilized by bacteria, and may play a significant role in the ecology of cohesive sediments by providing bacteria with a readily available carbon source. Carbohydrate concentrations in sediments are much more a function of the epipelic rather than epipsammic (attached) diatom biomass, and within more mixed assemblages of photosynthetic microorganisms (cyanobacterial mats, high-saltmarsh algal assemblages), the close relationship between colloidal carbohydrate and chlorophyll *a* concentrations present in diatom-dominated sediments is not present.

In epipelic diatoms, production of EPS requires between 0.1% and 16% of photosynthetically fixed carbon. EPS is produced both during illuminated and darkened periods. In conditions of darkness, the relative amounts of EPS produced increase, possibly linked to increased cellular motility. Extrusion of EPS by pennate diatoms is an active metabolic process, as vesicles filled with polymeric material are transported from the Golgi body to the raphe. Motility is generated by the extrusion of polymers through the cell membrane within the raphe, and the polymer strand is moved along the raphe by actin fibres. In dark conditions, internal storage carbohydrates (glucans) are metabolized to provide the carbon and energy sources needed to produce EPS.

These mechanisms provide a route for the production of extracellular carbohydrate material into the surrounding sediments. The EPS produced by microphytobenthos diatoms binds together sediment particles and can form smooth surface layers. The binding strength of exopolymers varies with chemical composition and the degree of cross-linkage; and as polymers dehydrate during tidal exposure, their binding strength increases. Thus during tidal exposure there is an increase in concentrations (due to diatom photosynthesis and motility) and a reduction in sediment water content. This can significantly increase the critical shear required for sediment erosion (by up to 300%), when the tide covers the site. In epipsammic biofilms, sand particles can be stuck together by pads and fibers of EPS, as well as by filaments of cyanobacteria. These processes all result in biostabilization, and the presence of microphytobenthos significantly changes the sedimentological properties of their habitat.

Distribution and Biomass

Small-scale Heterogeneity in Microphytobenthos

Microphytobenthos show a high degree of spatial heterogeneity in biomass and species composition. This patchiness occurs on a scale of micrometers to many tens of meters. There are also patterns of vertical distribution within sediments, with the bulk of the active biomass (determined as chlorophyll *a*) found within the top few millimeters of cohesive sediments, and the top centimeter of sandy sediments. However, viable cells and chlorophyll *a* can be isolated from deeper layers, up to 10–15 cm. Given the shallow photic depth in most sediments, only the algae in the uppermost depths of the sediments will be photosynthetically active. Yet many microphytobenthos can survive prolonged periods (2–3 weeks) of darkness, and there is some limited evidence of heterotrophy. Thus buried cells may, if mixed back to the surface, resume photosynthesis.

Large-scale Heterogeneity

Sediment type is a major determining factor in the abundance and biomass of microphytobenthos. Sandy silts and sands support significantly lower concentrations of microalgal biomass than sites with fine cohesive sediments (chlorophyll *a* concentrations ranging from 1 to $560 \, mg \, m^{-2}$ or $0.1–460 \, \mu g \, g^{-1}$ sediment). As sediment grain size increases, the proportion of epipelic, motile taxa decreases and microphytobenthic assemblages in intertidal sands consist predominantly of smaller epipsammic taxa. Sands tend to have lower nutrient concentrations and are more frequently resuspended than are cohesive sediments, and all of these characteristics contribute toward lower microphytobenthic biomass.

On intertidal mudflats, microphytobenthic biomass tends to be greater toward the upper shore. Lower shore sediments have a higher water content and are less stable than sediments at the middle and upper shore, partly owing to the energy of tidal flow and regular resuspension of sediments. Periods of illuminated exposure are shorter on the low intertidal where light penetration is restricted by highly turbid estuarine waters. Thus low shore microphytobenthos are probably light limited (in terms of available photoperiod per 24 h), while biomass accumulation is prevented by frequent disturbance. At higher tidal heights on a shore, the pattern of illuminated emersion periods and reduced resuspension contribute to create conditions favorable for epipelic microphytobenthos. However, upper shore stations are also subject to greater desiccation and temperature effects, the effect of which can be increased by long periods of exposure during neap tide periods. These factors usually result in a unimodal distribution of biomass across an intertidal flat, with the peak somewhere between mid-tide level and mean high water neap tide level, and not necessarily at the highest bathymetric level. In subtidal habitats, microphytobenthic biomass tends to decrease with increasing water depth owing to increasing light limitation. However, very shallow ($<1 \, m$) sediments in exposed situations are more prone to mixing and disturbance due to wave action or tidal flows, and thus biomass decreases in such sites.

Temporal Variation

In temperate latitudes, increases in epipelic microphytobenthic biomass tend to occur during the summer months. However, peaks of biomass also occur frequently at other times of the year, and in many estuarine systems epipelic diatom assemblages are less seasonally influenced than are phytoplankton communities. High temporal variability in biomass is a common feature of epipelic microphytobenthos, with biomass dependent on local environmental changes such as erosion and deposition events, desiccation linked to tidal exposure and weather conditions, and periods of rapid growth. Rapid doubling times (1–2 days) permit microphytobenthos to increase rapidly in density during favorable conditions. Subtidal microphytobenthos are not subjected to the extremes of exposure present on the intertidal, with irradiances ranging from very high during exposure to virtually nil during immersion in turbid overlying water. Subtidal microphytobenthos tends to show greater degrees of seasonality, with peaks of biomass and activity following the annual pattern of irradiance.

Response of Microphytobenthos to Nutrients

Nutrient Limitation

The potential for nutrient limitation of microphytobenthos depends in part on the sediment type concerned. Fine cohesive sediments usually have high organic matter contents, with high rates of bacterial mineralization and high porewater concentrations of dissolved nutrients, while sand flats are more oligotrophic. There is therefore an increased possibility that microphytobenthos inhabiting sediments of a larger grain size will be nutrient limited. The spatial distribution of sediments within estuaries is also pertinent to whether nutrient limitation will occur, in that many estuaries exhibit significant nutrient gradients along their length and areas of extensive mudflats supporting microphytobenthos may coincide with regions of high nutrient concentration.

There are few experimental data showing nutrient effects on intertidal microphytobenthos independently of other covarying factors that also affect primary production and biomass (shelter, salinity, etc.). Nutrient enrichment experiments on mudflats have found no consistent short-term pattern of increased photosynthesis or biomass, though long-term reductions (over 16 years) in nutrient inputs in estuaries have been shown to result in declines in biomass. In contrast, enrichment experiments in subtidal epipsammic microphytobenthos and cyanobacterial mats in nutrient-poor habitats have shown varying degrees of stimulation of microphytobenthic photosynthesis and biomass. It is generally considered that epipelic microalgae are not nutrient limited and that they obtained nutrients both from within the sediment, particularly during migration during tidal immersion, and from the overlying water. However, the term 'nutrient limitation' includes both Liebig-type limitation (on final biomass) and short term (Monod type) effects on rates of photosynthesis/growth. To what extent short-term nutrient dynamics within biofilms influence the rate of photosynthesis and growth of microphytobenthos is not known. As porewater concentrations of many nutrients (e.g., ammonium, phosphate) increase with depth within the sediment, cells exhibiting vertical migration may obtain nutrients when they have migrated away from the surface. Although this seems logically sound, there is as yet no experimental evidence to support this hypothesis.

The nutrient environment is important in determining species composition. Ammonium concentrations in sediments influence the distribution of diatom species in both saltmarshes and mudflats. Concentrations of ammonium between 500 and 1000 $\mu mol\,l^{-1}$ are selective for some taxa of microalgae, with the toxic effects of ammonia being enhanced in high pH conditions. Sediment organic content and tolerance to sulfide also influence the species composition of microalgal biofilms. Given the steep gradients in pH, oxygen and sulfide within fine cohesive sediments, these are likely to be important selective factors determining the species diversity of microphytobenthic assemblages.

Interaction between Microphytobenthos and Nutrient Cycling

The activity of microphytobenthos biofilms can significantly affect the fluxes of nutrients across the sediment–water interface (**Figure 4**). This is due to assimilation of nutrients by the algae from the overlying water and underlying porewaters, and to

the high oxygen concentrations present in the surface sediments during photosynthesis. Photosynthetic production of oxygen increases the depth of the surface oxic layer, which increases the oxidation of vertically diffusing reduced molecules such as sulfide, ammonium, and phosphate. Thus the export fluxes of these compounds across the sediment–water interface can be significantly reduced compared to the fluxes under dark conditions (**Figure 4**). Assimilation by the algae of nutrients (ammonium, nitrite, nitrate, phosphate, CO_2, dissolved organic carbon) diffusing into the biofilm both from overlying water and from deeper layers within the sediments also alters the pattern of exchange fluxes. On intertidal mudflats, microphytobenthic biofilms develop an ammonium demand during periods of tidal exposure and photosynthesis that persists for up to 4 h after tidal cover.

Bacterial denitrification (the reduction of nitrate to nitrogen gas) is an important process in coastal sediments as it is the only mechanism by which nitrogen can be permanently removed from the marine environment. Microphytobenthos influence denitrification in a number of ways (**Figure 4**). The position of the microphytobenthos on the surface of the sediment allows them to assimilate nitrate from the overlying water column and reduce the amount diffusing into the sediment. Denitrification is an anaerobic process, and photosynthetic oxygen production increases the depth in the sediments at which it can occur, thereby increasing the diffusional path length for nitrate. By these processes, microphytobenthos reduce denitrification of nitrate from the water column. However, oxygen production can stimulate nitrification (production of nitrate from ammonium), and this nitrate can then be denitrified. Stimulation of this 'coupled nitrification–denitrification' pathway can be particularly significant,

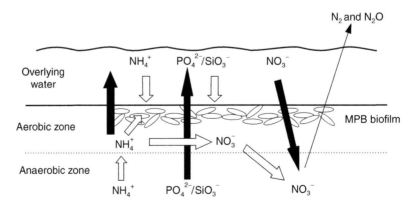

Figure 4 Interactions between microphytobenthos and nutrient fluxes across the sediment–water interface. Open arrows indicate fluxes/processes stimulated by the action of microphytobenthic primary production. Solid arrows represent processes reduced by microphytobenthic activity.

especially in low-nutrient environments. By these processes, microphytobenthos influence the nutrient dynamics of shallow water sediments (**Figure 4**). These processes will be affected by spatial and temporal (both diel and seasonal) differences in biofilm biomass, activity, and species composition. There is some evidence to suggest that differences in species composition effect the ability of biofilms to sequester C and N compounds from the overlying water.

See also

Benthic Boundary Layer Effects. Phytobenthos. Primary Production Methods.

Further Reading

Admiraal W (1984) The ecology of estuarine sediment-inhibiting diatoms. *Progress in Phycology Research* 3: 269–322.

Decho AW (1990) Microbial exopolymer secretions in ocean environments: their role(s) in food webs and marine processes. *Oceanography and Marine Biology Annual Reviews* 28: 73–153.

MacIntyre HL, Geider RJ, and Miller DC (1996) Microphytobenthos: the ecological role of the 'secret garden' of unvegetated, shallow-water marine habitats. I. Distribution, abundance and primary production. *Estuaries* 19: 186–201.

Miller DC, Geider RJ, and MacIntyre HL (1996) Microphytobenthos: the ecological role of the 'secret garden' of unvegetated, shallow-water marine habitats. II. Role in sediment stability and shallow water food webs. *Estuaries* 19: 202–212.

Paterson DM (1994) Microbial mediation of sediment structure and behaviour. In: Stal LJ Caumette P (eds.) NATO ASI Series vol. G35. *Microbial Mats* pp. 97–109. Berlin: Springer Verlag.

Round FE, Crawford RM, and Mann DG (1990) *The Diatoms, Biology and Morphology of the Genera.* Cambridge: Cambridge University Press.

Sullivan MJ (1999) Applied diatom studies in estuarine and shallow coastal environments. In: Stoermer EF and Smol JP (eds.) *The Diatoms: Applications for the Environmental and Earth Sciences*, pp. 334–351. Cambridge: Cambridge University Press.

Sundbäck K, Nilsson C, Nilsson P, and Jönsson B (1996) Balance between autotrophic and heterotrophic components and processes in microbenthic communities of sandy sediments: a field study. *Estuarine and Coastal Shelf Science* 43: 689–706.

Underwood GJC and Kromkamp J (1999) Primary production by phytoplankton and microphytobenthos in estuaries. *Advances in Ecological Research* 29: 93–153.

PHYTOBENTHOS

M. Wilkinson, Heriot-Watt University, Edinburgh, UK

What is Phytobenthos?

'Phytobenthos' means plants of the seabed, both intertidal and subtidal, and both sedimentary and hard. Such plants belong almost entirely to the algae although seagrasses, which form meadows on some subtidal and intertidal areas, are flowering plants or angiosperms. Algae of sedimentary shores are usually microscopic, unicellular or filamentous, and are known as the microphytobenthos or benthic microalgae. Marine algae on hard surfaces can range from microscopic single-celled forms to large cartilaginous plants. Some use the term 'seaweed' for macroscopic forms whereas others also include the smaller algae of rocky seashores. This article is concerned with the nature, diversity, ecology, and exploitation of the marine benthic algae. Other plants of the shore are dealt with elsewhere in this encyclopedia as saltmarshes, mangroves, and seagrasses.

What are Algae?

Algae were regarded as the least highly evolved members of the plant kingdom. Nowadays most classifications either regard the microscopic algae as protists, while leaving the macroscopic ones in the plant kingdom, or regard all algae as protists. This distinction is not important for an understanding of the ecological role of these organisms so they will all be called plants in this article. The fundamental feature that algae share in common with the rest of the plant kingdom is photoautotrophic nutrition. In photosynthesis they convert inorganic carbon (as carbon dioxide, carbonate, or bicarbonate) into organic carbon using light energy. Thus they are primary producers, which act as the route of entry of carbon and energy into food chains. They are not the only autotrophs in the sea. Besides the other nonalgal plant communities mentioned earlier, there are chemoautotrophs in hydrothermal vent communities, which use inorganic reactions rather than light as the energy source, and some bacteria are photosynthetic. However, algae are responsible for at least 95% of marine primary production. Algal photosynthesis uses chlorophyll *a* as the principal pigment that traps and converts light energy into chemical energy (although many accessory photosynthetic pigments may also be present) and water is the source of the hydrogen that is used to reduce inorganic carbon to carbohydrate. Oxygen is a byproduct so the process is called oxygenic photosynthesis. Broadly the same process occurs throughout the plant kingdom. Those true bacteria that are photosynthetic use alternative pathways, pigments, and hydrogen donors, e.g., hydrogen sulfide.

One group of organisms falls between the algae and the bacteria – the cyanobacteria, until recently regarded as blue–green algae. These perform oxygenic photosynthesis and have similar photosynthetic pigments to algae, but their cell structure is fundamentally different. In common with the bacteria they have the more primitive, prokaryotic cell structure, lacking membrane-bound organelles and organized nuclei. Algae have the more advanced and efficient eukaryotic cell structure, with membrane-bound organelles and defined nuclei, in common with all other plants and animals. Blue–greens also have some physiological affinities with bacteria, particularly nitrogen fixation. This means that they can use elemental nitrogen as a source of nitrogen for biosynthesis of various organic nitrogen compounds, starting from the simple organic compounds formed in photosynthesis. Algae and other plants have lost this ability and require to absorb fixed nitrogen, combined inorganic nitrogen as nitrate and ammonium ions, from solution in sea water. Despite internal cellular differences, blue–greens have similar overall morphology to smaller algae and live indistinguishably in algal communities as primary producers. They will therefore be included with algae in this review.

Algae are therefore similar to the 'true' plants in having oxygenic photosynthesis. They are distinguished from the rest of the plant kingdom only on a rather technical botanical point. Algae have simpler reproductive structures. In all the higher plant groups the reproductive organs are surrounded by walls of sterile cells (i.e., cells that are not gametes or spores) that are formed as a specific part of the reproductive organ. This does not occur in algae. Even in the highly complex large brown seaweeds with apparently complex reproductive structures, the gametangia, which produce eggs and sperm, enclosed within complex reproductive structures, have only membranous walls rather than cellular walls.

Diversity of Algae

Algae as defined in the previous section include a large diversity of organisms from microscopic single-celled ones, as little as about 2 μm in diameter, to the complex giant kelp nearly 70 m long, the largest plant on Earth. We can make sense of this diversity in three ways:

- Structural diversity
- Habitat diversity
- Taxonomic classification

Structural Diversity of Algae

The simplest algae are single cells, which can vary in size from about 2 μm to 1 mm. They can be non-motile, lacking flagella, or motile by means of flagella. An interesting intermediate situation is in the diatoms which lack flagella but are nonetheless motile by gliding over surfaces. Unicells can differ in the presence of external sculpturing and the number and orientation of flagella on each cell. Colonies are aggregations of single cells which can also be flagellate or nonflagellate. The simplest truly multi-cellular algae are filamentous, i.e., hair-like, chains of cells. These can be branched or unbranched and may be only one cell in thickness (uniseriate) or more than one cell in thickness (multiseriate). Heterotrichy is an advanced form of filamentous construction in which two separate branched systems of filaments may be present on one plant – a prostrate system which creeps along the substratum and an erect system which arises into the seawater medium from the prostrate system.

The larger more advanced types of seaweeds can be traced in origin to modifications of the hetero-trichous system. Reduction of one of the two fila-ment systems and elaboration of the other can give rise either to encrusting forms (erect reduced, prostrate elaborated) or to erect plants in which the prostrate system only forms the attachment organ or holdfast. Three further modifications give rise to a wide diversity of large cartilaginous (leathery), foliose (leaf-like) and complex filamentous seaweeds. These three modifications are:

- Presence of meristems, localized areas where cell division is concentrated, which may be apical, at the growing tips of branches, or may be inter-calary, located along the length of the plant (in simpler algae growth is diffuse with cell division occurring anywhere in the plant, not localized to meristems).
- Pseudoparenchyma formation – the aggregation of many separate filaments together to make a massive plant body, as opposed to true par-enchyma formation, where massive tissues result only from multiplication of adjacent cells. This is a different use of the term parenchyma from that in higher plants where it means an unspe-cialized type of cell which acts as packing tissue.
- The occurrence of two or more phases of growth, which may or may not differ in pattern (pseudo-parenchymatous or truly parenchymatous) and may involve formation of a secondary lateral meristem. Various phases of growth can give rise to the different tissues seen in cross-sections of seaweeds which may help in giving the ability to bend in response to water motion and wave action, without breaking.

Some seaweeds are able to secrete calcium car-bonate so that they appear solid. Red calcareous species appear pink and are common throughout the world whereas green calcareous forms are commoner in the tropics and subtropics. Calcareous encrusting red algae form a pink calcareous coating on the rock surface which can be mistaken by the nonspecialist for a geological feature.

The structure of a kelp plant illustrates the life of seaweeds. The plant is attached to the rock surface by a holdfast, which is branched and fits intimately to the microtopography of the rocks. From the holdfast arises the stipe, a stem-like structure which supports the frond in the water column. The frond is a wide flat area which gives a high surface area to volume ratio for light, carbon dioxide, and nutrient absorption. Superficially there is a resemblance to the roots, stem, and leaves of higher plants but there is no real equivalence because of the different life style. The holdfast is not an absorptive root system and does not penetrate the substratum, unlike roots penetrating the soil, since the seaweed can obtain all its requirements by direct absorption over its surface. The stipe is not a stem containing transport systems, as in higher plants, since these are not needed, again because of direct absorption. Similarly the frond does not have the complex structure of leaves with gas exchange and water retention organs such as stomata. Seaweeds do not have resistant phases such as seeds. Development is direct from spores or zygotes.

A seaweed, such as a kelp, can be viewed as a chemical factory taking in light, nutrients, and inorganic carbon from the water and converting them into organic matter. This production can be going on even when the plant does not seem to be increasing in size. The formation of new organic matter is then balanced by the loss of decaying tissue from the tip of the plant and by organic secretions from the frond. Both of these will be contributing to heterotrophic production in the kelp's ecosystem.

Habitat Diversity of Benthic Algae

Microphytobenthos in sedimentary shores can be distinguished according to whether they are epipsammic (attached to sand particles) or epipelic (between mud particles). They are mainly unicellular forms: diatoms, euglenoids, and blue–greens in estuarine muds; diatoms, dinoflagellates, and blue–greens in sand. Many show vertical migration within the top few millimeters of the sediment, photosynthesizing when the tide is out and burrowing before the return of the tide so that some escape being washed away. This is not 100% effective so that resuspended microphytobenthos can be a significant proportion of apparent phytoplankton in some estuaries. Diatoms migrate by gliding motility whereas euglenoids do so by alternate contraction and relaxation of the cell shape (metaboly). In estuaries, which may be turbid environments where photosynthesis by submerged plants may be reduced, and large expanses of intertidal mud flats may be available, microphytobenthos could be important primary producers which have been underestimated because they are not visually obvious. They may also help to stabilize sediments by the mucus secretions which keep them from desiccation when on the mud surface.

Seaweeds do not just grow attached to hard surfaces such as bedrock, boulders, and artificial structures. In Britain, a habitat-diverse, open coast shore is likely to have 70–100 species of seaweed present out of a British total of about 630 species. Such a high total on a shore is only realized because of many habitat variations that harbor the more microscopic species.

Most seaweeds have smaller species that grow attached to them as epiphytes. In turn they have even smaller species attached and this may continue for several orders down to very small microscopic plants. Endophytes are microscopic algae that grow between the cells within the tissues of larger ones. Epizoic algae grow attached to animals and endozoic forms grow inside animals, usually in skeletal parts. These include algae that penetrate calcareous substrata, i.e., shell-boring algae – red, green, and blue–green forms that bore through mollusk and barnacle shells and coral skeletons. They also include forms that inhabit proteinaceous animal skeletons – red and green filamentous species in skeletons of hydroids and bryozoans, and filamentous green seaweeds, mainly *Tellamia*, in the periostracum of periwinkles. Some of the shell-boring algae are also endolithic, boring through chalk rocks and so possibly aiding coastal erosion. Finally, there are endozoic algae that live in soft parts of benthic animals. Zooxanthellae are nonmotile dinoflagellate unicells in coral polyps, where they contribute to the high productivity of coral reefs, and unicellular blue–greens live in the tissues of some sea-slugs.

Taxonomic Classification of the Algae

Algae are classified into a number of divisions of the plant kingdom (equivalent to phylum), varying in number from about 8 to 16 depending on author. Distinction is based on fundamental cellular and biochemical features and so is independent of the form of the plant. Each division can contain a range of forms, from unicellular to complex multicellular, although in many divisions the unicellular and colonial forms predominate. General features used to distinguish the divisions are:

- The range of accessory photosynthetic pigments present in addition to chlorophyll *a* (other chlorophylls, carotenoids, and biloproteins);
- The secondary more soluble components of the cell wall present in addition to the main fibrillar component;
- The chemical nature of the insoluble storage products resulting from excess photosynthesis;
- The presence, fine structure, number, and position of flagella on vegetative cells of flagellate organisms or on flagellate reproductive bodies of larger species;
- Specialized aspects of cell structure, peculiar to particular divisions, such as the silica frustules, which encase diatom cells.

The three biochemical features above are relatively uniform in the higher plants, compared with the algae, and similar to those of one algal division, *Chlorophyta* (green algae). This suggested origin of land plants occurred from only this one division, although this is now contested. At a fundamental level the algae are therefore much more diverse than the higher plants. Characteristics for each division can be found in the Further Reading list.

The seaweeds are the macroscopic marine algae in the divisions *Chlorophyta* (green), *Phaeophyta* (brown) and *Rhodophyta* (red algae). Greens are mainly foliose and filamentous seaweeds. Browns have no unicellular forms and include the very large complex and leathery forms such as kelps and rockweeds. Reds include a wide range of heterotrichous forms forming a wide diversity of complex foliose and filamentous forms.

Seaweed Life Cycles

Most seaweeds have more than one phase in their life cycle. They have generally the same pattern as higher plants where a sexually reproducing gametophyte generation gives rise to an asexually reproducing sporophyte generation and vice versa. There is not always an obligate alternation as in higher plants and there is much more diversity in the nature of the different phases in algae, with some red algae even having a third generation. Some only have one generation. In some cases this may reflect lack of experimental culture which is necessary for life cycle determination.

The simplest life cycle with two phases is termed isomorphic where the gametophyte and sporophyte are morphologically identical. Many common green seaweeds are like this, e.g., *Ulva* and *Enteromorpha*. Life cycles with morphologically different phases are heteromorphic. An example is kelp plants which have a massive leathery sporophyte which alternates with a microscopic filamentous gametophyte.

In many cases the two generations in a heteromorphic life cycle may have been known since the nineteenth century by separate names from before the life cycle was determined in culture. For example, the various species of the red seaweed, *Porphyra*, alternate with a filamentous shell-boring sporophyte formerly known as *Conchocelis rosea*.

Life cycles can be under environmental control. In some *Porphyra* species the change between generations is controlled by daylength, bringing about an annual seasonal life cycle. In some simpler life cycles, individual generations may be able to propagate themselves so that the full life cycle is not seen. This can be environmentally controlled so that, for example, in Europe there is a change with latitude of the relative proportions of the sexual and asexual generations of the brown filamentous seaweed, *Ectocarpus*, connected with latitudinal variation of sea temperature.

The Validity of Laboratory Cultures of Phytobenthos

Experimental culture is needed to determine life cycles but it is difficult to simulate all environmental conditions. Wave action and water flow are not usually simulated in the numerous batch culture dishes needed for replicated ecological experiments. Artificial light sources are unlike daylight in spectral composition and intensity, yet light quality may control photomorphogenesis in plants. It is therefore possible that laboratory culture could give false results. Two examples are given below.

The red seaweeds *Asparagopsis armata* and *Falkenbergia rufulanosa*, originally described as separate species, are phases in the same life cycle. During the twentieth century they have been spreading their geographical limit northwards in Europe from the Mediterranean to northern Scotland. Populations in Britain are rarely seen with reproductive organs but have a mode of attachment to the rock surface that suggests they were produced by vegetative reproduction. The two phases seem to have spread independently in Britain although in culture they could be made to participate in the same life cycle. This reflects a wider phenomenon in red seaweeds where by going north and south from the center of geographical distribution, reproductive potential declines.

In ecological experiments aseptic conditions are often not used, to the surprise of microbiologists. This lack of sterility may be desirable. For example, the green foliose seaweed, *Monostroma*, develops abnormally as a filamentous form in aseptic culture because of the need for growth factors from contaminating bacteria. These examples are not meant to decry culture experiments but to counsel their critical interpretation.

Seaweed Ecology

Seaweeds are present on all rocky shores but are more obvious where wave action is less. On temperate shores intertidal zones are generally dominated by brown fucoid seaweeds (wracks or rockweeds), while the shallow subtidal area is occupied by a kelp forest formed of large laminarian seaweeds. A variety of red, green, and brown seaweeds forms an understorey.

Some general rules can be exemplified by consideration of shores in north-west Europe. Firstly, on intertidal rocky shores the number of species is greatest at the lower tidal levels and declines with increasing intertidal height. Shores sheltered from wave action show the greatest cover and biomass of seaweeds. Shores exposed to very strong wave action tend to be dominated by sessile animals rather than seaweeds. Most shores are intermediate in wave action and tend to have a mosaic distribution of organisms, at least on lower and mid-shore. This may include patches of grazing animals interspersed with patches of different seaweed communities. The mosaic may make it hard to see a zonation of organisms with height on the shore. On very sheltered shores there may be a very obvious zonation of large brown seaweeds, in order of descending height on the shore: *Pelvetia canalicaulata*, *Fucus spiralis*,

Fucus vesiculosus and *Ascophyllum nodosum, Fucus serratus, Laminaria digitata.* (Similar zonations, but with different species, may occur on temperate shores outside north-west Europe.) Clear-cut zonations with visually dominant species can give a false impression that discrete communities exist at different tidal heights. Understorey species also have zones but their boundaries do not coincide with the larger species so as not to give sharply delimited communities.

With increase in wave action there is a change of species, e.g., in fucoids *Fucus serratus* is replaced by *Himanthalia elongata* and in laminarians *Laminaria digitata* is replaced by *Alaria esculenta*. Some species change form with wave action, for example, the bladder wrack, *F. vesiculosus*, loses its bladders (such morphological plasticity is a common feature confusing seaweed identification). With increased wave action, zones increase in breadth and height on shore. All these features can be incorporated in biologically defined exposure scales for the comparative description of rocky shores, which place shores on a numerical exposure scale. This facilitates the comparison of similar shores in pollution-monitoring studies to ensure that differences along a pollution gradient are due to human disturbance rather than to wave action.

Rock pools interrupt the gradient of conditions with height on shore. They provide a constantly submerged environment, like the subtidal one, but which is of limited volume and so undergoes physicochemical fluctuations while the tide is out, unlike the open sea. There is a corresponding zonation of dominant seaweed types in rock pools. On the lower shore they are characterized by sublittoral species and on the mid-shore they include species restricted to pools such as *Halidrys siliquosa*. Upper shore pools, where salinity fluctuates, have few species and are characterized by euryhaline opportunists such as *Enteromorpha* spp.

Intertidal zonation is only partly due to the desiccation and salinity tolerance of the seaweeds. Such factor tolerance is particularly important on the upper shore where conditions are most harsh for a marine organism and so few species are present, with few biotic interactions. On mid- and lower-shores biotic factors are important in determining species boundaries. Grazing by limpets and periwinkles on smaller algae, and on the microscopic germlings of larger ones, is important as is biological competition, which narrows down species occurrence to less than their tolerance range.

Subtidally a zonation may also be seen with dense kelp forest, with a large variety of understorey species in the shallowest water, below which is a kelp park with only scattered plants. Below the kelp depth limit may be a red algal zone to the photic depth limit where light becomes insufficient for positive net photosynthesis. Important factors are again both physical and biotic. Light tolerance plays a role but in the shallowest waters, where plant density is greatest, competition is important and zone limits may be set by grazers, this time by sea urchins rather than limpets.

There is an old view that accessory pigment differences between green, brown, and red seaweeds equip them to dominate at different depths according to which spectral quality of light penetrates. This is an oversimplification. Deeper-growing plants are shade plants with lower overall light requirements and can be of any color group.

Distribution into estuaries along a generally decreasing salinity gradient is another modifying factor like wave action. Colonization of hard surfaces by phytobenthos in estuaries is largely by marine species with species number declining going upstream. This occurs by selective attenuation firstly of red, then of brown species. Estuaries have broadly two algal zones: an outer one with fucoid dominated shores, which are a species-poor version of a sheltered open coast shore; and an inner zone dominated by filamentous mat-forming algae, principally greens and blue–greens.

There can be a successional sequence in which a bare area of shore is successively colonized, starting with unicells, with increasingly larger and more complex algae. Patches in a mosaic distribution may be at different stages in such a sequence. The succession involves contrasting types of seaweed as shown in **Table 1**.

Although opportunists are good at colonizing bare rock, in a stable environment they are eventually replaced by more precisely adapted late successional species.

Various conditions may favor the unusual abundance of opportunists. Mistaken conclusions about effects of effluent discharges can be reached by environmentalists who do not realize that there are both natural and artificial causes. Opportunist domination of rocks is favored naturally by sand scour and they may have natural summer outbursts in temperate climates. Pollution, particularly by sewage-derived nutrients, may favor opportunist domination. On tidal flats, 'green tides' may occur where the mudflat is completely covered by thick opportunist mats. This can induce anoxia and ammonia release beneath the mat so inhibiting benthic invertebrate populations and interfering with bird-feeding on tidal flats.

Table 1   Types of seaweed involved in successional colonization

	Opportunist species early in succession, e.g., foliose and filamantous green algae	Late successional species, e.g., large cartilaginous plants such as fucoids
Morphology	Simple	Complex
Size	Smaller	Larger
Growth rate	Faster	Slower
Life span	Shorter	Longer
Reproduction	All year round	Likely to have seasons
Environmental tolerance	Very wide but not precise adaptation to a specific niche	Narrower but good adaptation to a specific niche

Uses of Seaweed

Benthic macroalgae are an important biological resource. Their principal uses are human food, animal feeds, fertilizer, and industrial chemicals.

In the West, human seaweed consumption is small compared with the east. One of the most valuable fisheries listed by the Food and Agriculture Organization (FAO) is a seaweed, *Porphyra*, used extensively for human consumption in Japan. In the West, human consumption is more a health food market with beneficial effects ascribed to high trace element content, because of the high bioaccumulation activity of many seaweeds, but this remains to be rigorously tested.

Aqueous extracts of seaweeds such as fucoids and kelps are used as commercial and domestic fertilizers. Beneficial effects have been claimed in horticulture such as increased growth, faster ripening and increased fruit yield. Effects are usually ascribed to trace element or plant hormone (usually cytokinin) content.

Various high value fine chemicals, such as pigments, can be obtained from seaweeds but the main seaweed chemical industry is extraction of phycocolloids. These are secondary cell wall components of red and brown seaweeds, which have gel-forming and emulsifying properties in aqueous solutions. This gives them hundreds of industrial applications ranging from textile printing to ice cream manufacture. The chemicals concerned are all macromolecular carbohydrates, principally alginates from brown seaweeds and agars and carrageenans from red seaweeds. Alginate is not a single substance but a biopolymer with considerable possibility for structural variation. It is a linear structure of repeating sugar units of two kinds, mannuronic acid (M units) and guluronic (G units). Different alginates vary in the ratio of M and G units and in chain length (total number of units). Structural differences confer different gel-forming and emulsifying properties making different alginates suitable for different industrial applications. In turn, different alginate structures are found in different species so that different seaweeds are required for different applications.

Ensuring Supplies of Commercial Seaweeds

Seaweeds can be harvested from natural populations or farmed in the sea. The approach may depend on the value of the product. In the West, harvesting is preferred for phycocolloids. Despite the wide industrial application they are low-value products and so will not support the high labor costs needed for cultivation. By contrast alginate is successfully produced from farmed kelps in China, where labor costs are lower, and *Porphyra* can be farmed for human consumption in Japan because of its high value.

Sustainable harvesting should consider the ability of the seaweed resource to recover. Mechanized *Macrocystis* (giant kelp) harvesting in California is sustainable yet productive because of the growth pattern of the plant. It grows to almost 70 m long and cropping of the distal few meters by barges floating over the forest canopy allows numerous meristems to remain intact so growth continues. By contrast *Laminaria hyperborea* harvesting in Norway is more destructive because the desired alginate is in the stipe (the supporting 'stem') of the plant. Harvesting by kelp dredges, i.e., large bags with cutters at the front end towed over the seabed on skis just above the rock surface, cuts plants off just above the holdfast, so that the meristematic area is harvested. Sustainability of the *Laminaria* forest is based on the vegetation structure. There are different layers of plants. The dredge takes only the large canopy-forming plants with tough, nonyielding stipes. The smaller plants bend rather than breaking and so survive harvesting. With the absence of the canopy they receive more light and so grow quickly to replace the harvested plants. The forest biomass is regenerated in 3 years so the Norwegian

Government licenses areas of the seabed for harvesting not less than every 4 years. The regenerated kelp is better for alginate extraction as it is not contaminated by epiphytes. However, the sustainability of the kelp forest biodiversity is separate from commercial yield. It is a diverse ecosystem with many invertebrates in the kelp holdfasts, as well as diverse seabed fauna and flora, and pelagic species sheltered by the forest. The diversity of this had not recovered after much more than 4 years so the licensing system does not conserve the full ecosystem.

There are biotic considerations in protecting the phytobenthic resource. In the early 1970s the decline of the sea otter population off California allowed its prey, a sea urchin which was a kelp grazer, to increase. This hindered natural regeneration of the forest, which was already declining due to sewage pollution from Los Angeles. Recovery required manual transplantation. Various species of sea urchin are among the most voracious subtidal grazers. Another one, *Strongylocentrotus*, was able to create barren areas of seabed off the Canadian coast where *Laminaria longicruris* had been harvested, thus preventing regeneration.

Farming of seaweed can require knowledge from laboratory culture of environmental tolerances and the factors controlling life cycles. This allows manipulation of stocks in seawater tanks on land under controlled conditions to produce reproductive bodies. These can be used to seed ropes, nets, canes or other substrata which are then planted out into sheltered sea areas for growing on. This increases the habitat area for attachment in the sea, and so increases yield. In the case of *Porphyra* in Japan, such manipulation of environmental conditions allows up to five generations per year from an area of coast where natural seasonal changes would only give one. It is not practical to grow the plants entirely on land in environmentally controlled seawater tanks because of the high cost of the facilities, but it is

feasible to retain a small stock of, for example, shells infected with the *Conchocelis*-phase of *Porphyra* from which spores can be obtained on demand by manipulation of daylength and temperature. The technology exists to cultivate seaweeds entirely in land-based tanks should the economics be favorable in the future. For example, there are different genetic strains of *Chondrus crispus*, whose carrageenan yield and type can be controlled by nutrient and salinity conditions. Such tank culture may be useful in the future to produce high value fine chemicals for medical applications.

Seaweeds are the most obvious type of plant in the sea and are the main component of the phytobenthos but other algae and other marine plants are described elsewhere in this Encyclopedia.

See also

Primary Production Distribution. Primary Production Methods. Primary Production Processes.

Further Reading

Guiry MD and Blunden G (eds.) (1991) *Seaweed Resources in Europe: Uses and Potential.* Chichester: Wiley.

Hoek C, van den, Mann DG, and Jahns HM (1995) *Algae: An Introduction to Phycology.* Cambridge: Cambridge University Press.

Lembi CA and Waaland JR (eds.) (1988) *Algae and Human Affairs.* Cambridge: Cambridge University Press.

Lobban CS and Harrison PJ (1997) *Seaweed Ecology and Physiology.* Cambridge: Cambridge University Press.

Luning K (1990) *Seaweeds: Their Environment, Biogeography and Ecophysiology.* New York: Wiley.

South GR and Whittick A (1987) *Introduction to Phycology.* Oxford: Blackwell Scientific Publications.

BENTHIC FORAMINIFERA

A. J. Gooday, Southampton Oceanography Centre, Southampton, UK

Introduction

Foraminifera are enormously successful organisms and a dominant deep-sea life form. These amoeboid protists are characterized by a netlike (granuloreticulate) system of pseudopodia and a life cycle that is often complex but typically involves an alternation of sexual and asexual generations. The most obvious characteristic of foraminifera is the presence of a shell or 'test' that largely encloses the cytoplasmic body and is composed of one or more chambers. In some groups, the test is constructed from foreign particles (e.g., mineral grains, sponge spicules, shells of other foraminifera) stuck together ('agglutinated') by an organic or calcareous/organic cement. In others, it is composed of calcium carbonate (usually calcite, occasionally aragonite) or organic material secreted by the organism itself.

Although the test forms the basis of foraminiferal classification, and is the only structure to survive fossilization, the cell body is equally remarkable and important. It gives rise to the complex, highly mobile, and pervasive network of granuloreticulose pseudopodia. These versatile organelles perform a variety of functions (locomotion, food gathering, test construction, and respiration) that are probably fundamental to the ecological success of foraminifera in marine environments.

As well as being an important component of modern deep-sea communities, foraminifera have an outstandingly good fossil record and are studied intensively by geologists. Much of their research uses knowledge of modern faunas to interpret fossil assemblages. The study of deep-sea benthic foraminifera, therefore, lies at the interface between biology and geology. This articles addresses both these facets.

History of Study

Benthic foraminifera attracted the attention of some pioneer deep-sea biologists in the late 1860s. The monograph of H.B. Brady, published in 1884 and based on material collected in the *Challenger* round-the-world expedition of 1872–76, still underpins our knowledge of the group. Later biological expeditions added to this knowledge. For much of the 1900s, however, the study of deep-sea foraminifera was conducted largely by geologists, notably J.A. Cushman, F.B. Phleger, and their students, who amassed an extensive literature dealing with the taxonomy and distribution of calcareous and other hard-shelled taxa. In recent decades, the emphasis has shifted toward the use of benthic species in paleoceanographic reconstructions. Interest in deep-sea foraminifera has also increased among biologists since the 1970s, stimulated in part by the description of the Komokiacea, a superfamily of delicate, soft-shelled foraminifera, by O.S. Tendal and R.R. Hessler. This exclusively deep-sea taxon is a dominant component of the macrofauna in some abyssal regions.

Morphological and Taxonomic Diversity

Foraminifera are relatively large protists. Their tests range from simple agglutinated spheres a few tens of micrometers in diameter to those of giant tubular species that reach lengths of 10 cm or more. However, most are a few hundred micrometers in size. They exhibit an extraordinary range of morphologies (**Figures 1** and **2**), including spheres, flasks, various types of branched or unbranched tubes, and chambers arranged in linear, biserial, triserial, or coiled (spiral) patterns. In most species, the test has an aperture that assumes a variety of forms and is sometimes associated with a toothlike structure. The komokiaceans display morphologies not traditionally associated with the foraminifera. The test forms a treelike, bushlike, spherical, or lumpish body that consists of a complex system of fine, branching tubules (**Figure 2A–C**).

The foraminifera (variously regarded as a subphylum, class, or order) are highly diverse with around 900 living genera and an estimated 10 000 described living species, in addition to large numbers of fossil taxa. Foraminiferal taxonomy is based very largely on test characteristics. Organic, agglutinated, and different kinds of calcareous wall structure serve to distinguish the main groupings (orders or suborders). At lower taxonomic levels, the nature and position of the aperture and the number, shape, and arrangement of the chambers are important.

Figure 1 Scanning electron micrographs of selected deep-sea foraminifera (maximum dimensions are given in parentheses). (A) *Epistominella exigua*; 4850 m water depth, Porcupine Abyssal Plain, NE Atlantic (190 μm). (B) *Nonionella iridea*; 1345 m depth, Porcupine Seabight, NE Atlantic (110 μm). (C) *Nonionella stella*; 550 m depth, Santa Barbara Basin, California Borderland (220 μm). (D) *Brizalina tumida*; 550 m depth, Santa Barbara Basin, California Borderland (680 μm). (E) *Melonis barleaanum*; 1345 m depth, Porcupine Seabight, NE Atlantic (450 μm). (F) *Hormosina* sp., 4495 m depth, Porcupine Abyssal Plain (1.5 mm). (G) *Pyrgoella* sp.; 4550 m depth, foothills of Mid-Atlantic Ridge (620 μm). (H) *Technitella legumen*; 997–1037 m depth, NW African margin (8 mm). (A)–(E) and (G) have calcareous tests, (F) and (H) have agglutinated tests. (C) and (D), photographs courtesy of Joan Bernhard.

Methodology

Qualitative deep-sea samples for foraminiferal studies are collected using nets (e.g., trawls) that are dragged across the seafloor. Much of the *Challenger* material studied by Brady was collected in this way. Modern quantitative studies, however, require the use of coring devices. The two most popular corers used in the deep sea are the box corer, which obtains a large (e.g., 0.25 m²) sample, and the multiple corer, which collects simultaneously a battery of up to 12 smaller cores. The main advantage of the multiple corer is that it obtains the sediment–water interface in a virtually undisturbed condition.

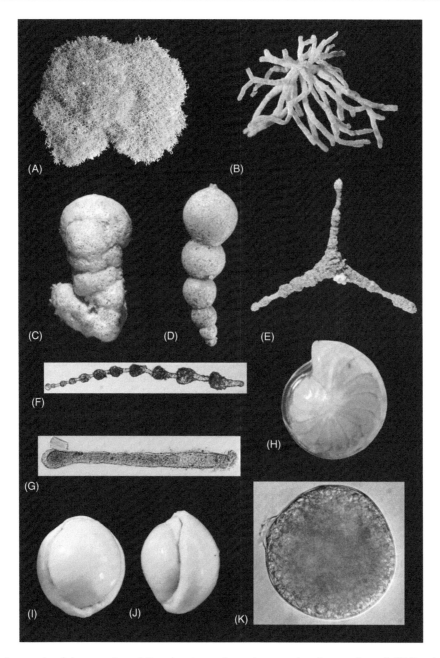

Figure 2 Light micrographs of deep-sea foraminifera (maximum dimensions are given in parentheses). (A) Species of *Lana* in which pad-like test consists of tightly meshed system of fine tubules; 5432 m water depth, Great Meteor East region, NE Atlantic (7.4 mm). (B) *Septuma* sp.; same locality (2 mm). (C) *Edgertonia* mudball; same locality (3.8 mm). (D) *Hormosina globulifera*; 4004 m depth, NW African margin (6.4 mm). (E) *Rhabdammina parabyssorum*; 3392 m depth, Oman margin, NW Arabian Sea (18 mm). (F) *Leptohalysis* sp.; 3400 m depth, Oman margin, NW Arabian Sea (520 μm). (G) Minute species of *Hyperammina*; 3400 m depth, Oman margin, NW Arabian Sea (400 μm). (H) *Lenticularia* sp.; 997–1037 m depth, NW African margin (2.5 mm). (I, J) *Biloculinella* sp.; 4004 m depth, NW African margin (3 mm). (K) Spherical allogromiid; 3400 m depth, Oman margin, NW Arabian Sea (105 μm). Specimens illustrated in (A)–(G) have agglutinated tests, in (H)–(J) calcareous tests and in (K) an organic test. (A)–(C) belong to the superfamily Komokiacea.

Foraminifera are extracted from sieved sediment residues. Studies are often based on dried residues and concern 'total' assemblages (i.e. including both live and dead individuals). To distinguish individuals that were living at the time of collection from dead tests, it is necessary to preserve sediment samples in either alcohol or formalin and then stain them with rose Bengal solution. This colors the cytoplasm red and is most obvious when residues are examined in water. Stained assemblages provide a snapshot of the foraminifera that were living when the samples were collected. Since the live assemblage varies in both time and space, it is also instructive to examine the dead assemblage that provides an averaged view of the foraminiferal fauna. Deep-sea foraminiferal assemblages are typically very diverse and therefore faunal data are often condensed mathematically by using multivariate approaches such as principal components or factor analysis.

The mesh size of the sieve strongly influences the species composition of the foraminiferal assemblage retained. Most deep-sea studies have been based on >63 μm, 125 μm, 150 μm, 250 μm, or even 500 μm meshes. In recent years, the use of a fine 63 μm mesh has become more prevalent with the realization that some small but important species are not adequately retained by coarser sieves. However, the additional information gained by examining fine fractions must be weighed against the considerable time and effort required to sort foraminifera from them.

Ecology

Abundance and Diversity

Foraminifera typically make up >50% of the soft-bottom, deep-sea meiofauna (**Table 1**). They are also often a major component of the macrofauna.

In the central North Pacific, for example, foraminifera (mainly komokiaceans) outnumber all metazoans combined by at least an order of magnitude. A few species are large enough to be easily visible to the unaided eye and constitute part of the megafauna. These include the tubular species *Bathysiphon filiformis*, which is sometimes abundant on continental slopes (**Figure 3**). Some xenophyophores, agglutinated protists that are probably closely related to the foraminifera, are even larger (up to 24 cm maximum dimension!). These giant protists may dominate the megafauna in regions of sloped topography (e.g., seamounts) or high surface productivity. In well-oxygenated areas of the deep-seafloor, foraminiferal assemblages are very species rich, with well over 100 species occurring in relatively small volumes of surface sediment (**Figure 4**). Many are undescribed delicate, soft-shelled forms. There is an urgent need to describe at least some of these species as a step toward estimating global levels of deep-sea species diversity. The common species are often widely distributed, particularly at abyssal depths, although endemic species undoubtedly also occur.

Table 1 The percentage contribution of foraminifera to the deep-sea meiofauna at sites where bottom water is well oxygenated

Area	Depth (m)	Percentage of foraminifera	Number of samples
NW Atlantic			
Off North Carolina	500–2500	11.0–90.4	14
Off North Carolina	400–4000	7.6–85.9	28
Off Martha's Vineyard	146–567	3.4–10.6	4
NE Atlantic			
Porcupine Seabight	1345	47.0–59.2	8
Porcupine Abyssal Plain[a]	4850	61.8–76.3	3
Madeira Abyssal Plain[a]	4950	61.4–76.1	3
Cape Verde Abyssal Plain[a]	4550	70.2	1
Off Mauretania	250–4250		26
		4–27	
46°N, 16–17°W	4000–4800	0.5–8.3	9
Indian Ocean			
NW Arabian Sea[b]	3350	54.4	1
Pacific			
Western Pacific	2000–6000	36.0–69.3	11
Central North Pacific	5821–5874	49.5	2
Arctic	1000–2600	14.5–84.1	74
Southern Ocean	1661–1680	2.2–23.7	2

[a]Data from Gooday AJ (1996) Epifaunal and shallow infaunal foraminiferal communities at three abyssal NE Atlantic sites subject to differing phytodetritus input regimes. *Deep-Sea Research I* 43: 1395–1421.
[b]Data from Gooday AJ, Bernhard JM, Levin LA and Suhr SB (2000) Foraminifera in the Arabian Sea oxygen minimum zone and other oxygen-deficient settings: taxonomic composition, diversity, and relation to metazoan faunas. *Deep-Sea Research II* 47: 25–54.
Based on Gooday AJ (1986) Meiofaunal foraminiferans from the bathyal Porcupine Seabight (northeast Atlantic): size structure, standing stock, taxonomic composition, species diversity and vertical distribution in the sediment. *Deep-Sea Research* 35: 1345–1373; with permission from Elsevier Science.

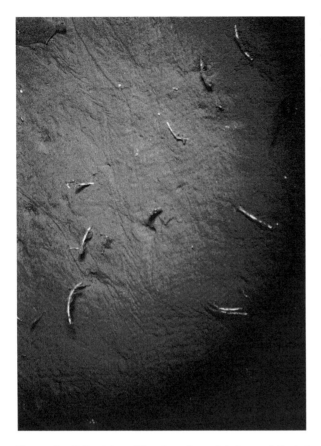

Figure 3 *Bathysiphon filiformis*, a large tubular agglutinated foraminifer, photographed from the Johnson Sealink submersible on the North Carolina continental slope (850 m water depth). The tubes reach a maximum length of about 10 cm. (Photograph courtesy of Lisa Levin.)

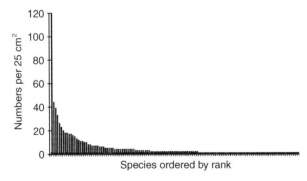

Figure 4 Deep-sea foraminiferal diversity: all species from a single multiple corer sample collected at the Porcupine Abyssal Plain, NE Atlantic (4850 m water depth), ranked by abundance. Each bar represents one 'live' (rose Bengal-stained) species. The sample was 25.5 cm^2 surface area, 0–1 cm depth, and sieved on a 63 μm mesh sieve. It contained 705 'live' specimens and 130 species.

Foraminifera are also a dominant constituent of deep-sea hard-substrate communities. Dense populations encrust the surfaces of manganese nodules as well as experimental settlement plates deployed on

the sea floor for periods of months. They include various undescribed matlike taxa and branched tubular forms, as well as a variety of small coiled agglutinated species (many in the superfamily Trochamminacea), and calcareous forms.

Role in Benthic Communities

The abundance of foraminifera suggests that they play an important ecological role in deep-sea communities, although many aspects of this role remain poorly understood. One of the defining features of these protists, their highly mobile and pervasive pseudopodial net, enables them to gather food particles very efficiently. As a group, foraminifera exhibit a wide variety of trophic mechanisms (e.g., suspension feeding, deposit feeding, parasitism, symbiosis) and diets (herbivory, carnivory, detritus feeding, use of dissolved organic matter). Many deep-sea species appear to feed at a low trophic level on organic detritus, sediment particles, and bacteria. Foraminifera are prey, in turn, for specialist deep-sea predators (scaphopod mollusks and certain asellote isopods), and also ingested (probably incidentally) in large numbers by surface deposit feeders such as holothurians. They may therefore provide a link between lower and higher levels of deep-sea food webs.

Some deep-sea foraminifera exhibit opportunistic characteristics – rapid reproduction and population growth responses to episodic food inputs. Well-known examples are *Epistominella exigua*, *Alabaminella weddellensis* and *Eponides pusillus*. These small (generally <200 μm), calcareous species feed on fresh algal detritus ('phytodetritus') that sinks through the water column to the deep-ocean floor after the spring bloom (a seasonal burst of phytoplankton primary production that occurs most strongly in temperate latitudes). Utilizing energy from this labile food source, they reproduce rapidly to build up large populations that then decline when their ephemeral food source has been consumed. Moreover, certain large foraminifera can reduce their metabolism or consume cytoplasmic reserves when food is scarce, and then rapidly increase their metabolic rate when food again becomes available. These characteristics, together with the sheer abundance of foraminifera, suggest that their role in the cycling of organic carbon on the deep-seafloor is very significant.

The tests of large foraminifera are an important source of environmental heterogeneity in the deep sea, providing habitats and attachment substrates for other foraminifera and metazoans. Mobile infaunal species bioturbate the sediment as they move through it. Conversely, the pseudopodial systems of foraminifera may help to bind together and stabilize deep-sea sediments, although this has not yet been clearly demonstrated.

Microhabitats and Temporal Variability

Like many smaller organisms, foraminifera reside above, on and within deep-sea sediments. Various factors influence their overall distribution pattern within the sediment profile, but food availability and geochemical (redox) gradients are probably the most important. In oligotrophic regions, the flux of organic matter (food) to the seafloor is low and most foraminifera live on or near the sediment surface where food is concentrated. At the other extreme, in eutrophic regions, the high organic-matter flux causes pore water oxygen concentrations to decrease rapidly with depth into the sediment, restricting access to the deeper layers to those species that can tolerate low oxygen levels. Foraminifera penetrate most deeply into the sediment where organic inputs are of intermediate intensity and the availability of food and oxygen within the sediment is well balanced.

Underlying these patterns are the distributions of individual species. Foraminifera occupy more or less distinct zones or microenvironments ('microhabitats'). For descriptive purposes, it is useful to recognize a number of different microhabitats: epifaunal and shallow infaunal for species living close to the sediment surface (upper 2 cm); intermediate infaunal for species living between about 1 cm and 4 cm (**Figure 5**); and deep infaunal for species that occur at depths

down to 10 cm or more (**Figure 6**). A few deep-water foraminifera, including the well-known calcareous species *Cibicidoides wuellerstorfi*, occur on hard substrates (e.g., stones) that are raised above the sediment–water interface (elevated epifaunal microhabitat). There is a general relation between test morphotypes and microhabitat preferences. Epifaunal and shallow infaunal species are often trochospiral with large pores opening on the spiral side of the test; infaunal species tend to be planispiral, spherical, or ovate with small, evenly distributed pores. It is important to appreciate that foraminiferal microhabitats are by no means fixed. They may vary between sites and over time and are modified by the burrowing activities of macrofauna. Foraminiferal microhabitats should therefore be regarded as dynamic rather than static. This tendency is most pronounced in shallow-water settings where environmental conditions are more changeable and macrofaunal activity is more intense than in the deep sea.

The microhabitats occupied by species reflect the same factors that constrain the overall distribution patterns of foraminifera within the sediment. Epifaunal and shallow infaunal species cannot tolerate low oxygen concentrations and also require a diet of relatively fresh organic matter. Deep infaunal foraminifera are less opportunistic but are more tolerant of oxygen depletion than are species living close to the sediment–water interface (**Figure 6**). It has been suggested that species of genera such as *Globobulimina* may consume either sulfate-reducing bacteria or labile organic matter released by the metabolic

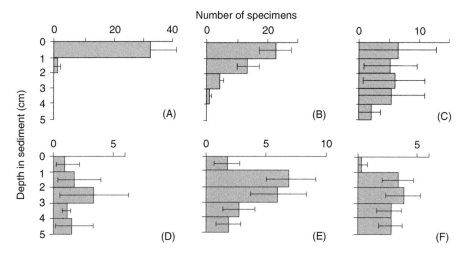

Figure 5 Vertical distribution patterns within the top 5 cm of sediment of common foraminiferal species ('live', rose Bengal-stained specimens) in the Porcupine Seabight, NW Atlantic (51°36′N, 13°00′W; 1345 m water depth). Based on >63 µm sieve fraction. (A) *Ovammina* sp. (mean of 20 samples). (B) *Nonionella iridea* (20 samples). (C) *Leptohalysis* aff. *catenata* (7 samples). (D) *Melonis barleeanum* (9 samples). (E) *Haplophragmoides bradyi* (19 samples). (F) *'Turritellella' laevigata* (21 samples). (Amended and reprinted from Gooday AJ (1986) Meiofaunal foraminiferans from the bathyal Porcupine Seabight (northeast Atlantic): size structure, standing stock, taxonomic composition, species diversity and vertical distribution in the sediment. *Deep-Sea Research* 35: 1345–1373; permission from Elsevier Science.)

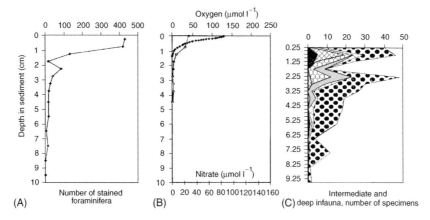

Figure 6 Vertical distribution of (A) total 'live' (rose Bengal-stained) foraminifera), (B) pore water oxygen and nitrate concentrations, and (C) intermediate and deep infaunal foraminiferal species within the top 10 cm of sediment on the north-west African margin (21°28.8'N, 17°57.2'W, 1195 m). All foraminiferal counts based on > 150 μm sieve fraction, standardized to a 34 cm³ volume. Species are indicated as follows: *Pullenia salisburyi* (black), *Melonis barleeanum* (crossed pattern), *Chilostomella oolina* (honeycomb pattern), *Fursenkoina mexicana* (grey), *Globobulimina pyrula* (diagonal lines), *Bulimina marginata* (large dotted pattern). (Adopted and reprinted from Jorissen FJ, Wittling I, Peypouquet JP, Rabouille C and Relexans JC (1998) Live benthic foraminiferal faunas off Cape Blanc, northwest Africa: community structure and microhabitats. *Deep-Sea Research I* 45: 2157–2158; with permission from Elsevier Science.)

activities of these bacteria. These species move closer to the sediment surface as redox zones shift upward in the sediment under conditions of extreme oxygen depletion. Although deep-infaunal foraminifera must endure a harsh microenvironment, they are exposed to less pressure from predators and competitors than those occupying the more densely populated surface sediments.

Deep-sea foraminifera may undergo temporal fluctuations that reflect cycles of food and oxygen availability. Changes over seasonal timescales in the abundance of species and entire assemblages have been described in continental slope settings (**Figure 7**). These changes are related to fluctuations in pore water oxygen concentrations resulting from episodic (seasonal) organic matter inputs to the seafloor. In some cases, the foraminifera migrate up and down in the sediment, tracking critical oxygen levels or redox fronts. Population fluctuations also occur in abyssal settings where food is a limiting ecological factor. In these cases, foraminiferal population dynamics reflect the seasonal availability of phytodetritus ('food'). As a result of these temporal processes, living foraminifera sampled during one season often provide an incomplete view of the live fauna as a whole.

Environmental Controls on Foraminiferal Distributions

Our understanding of the factors that control the distribution of foraminifera on the deep-ocean floor is very incomplete, yet lack of knowledge has not prevented the development of ideas. It is likely that foraminiferal distribution patterns reflect a combination of influences. The most important first-order factor is calcium carbonate dissolution. Above the carbonate compensation depth (CCD), faunas include calcareous, agglutinated, and allogromiid taxa. Below the CCD, calcareous species are almost entirely absent. At oceanwide or basinwide scales, the organic carbon flux to the seafloor (and its seasonality) and bottom-water hydrography appear to be particularly important, both above and below the CCD.

Studies conducted in the 1950s and 1960s emphasized bathymetry (water depth) as an important controlling factor. However, it soon became apparent that the bathymetric distribution of foraminiferal species beyond the shelf break is not consistent geographically. Analyses of modern assemblages in the North Atlantic, carried out in the 1970s, revealed a much closer correlation between the distribution of foraminiferal species and bottom-water masses. For example, *Cibicidoides wuellerstorfi* was linked to North Atlantic Deep Water (NADW) and *Nuttallides umbonifera* to Antarctic Bottom Water (AABW). At this time, it was difficult to explain how slight physical and chemical differences between water masses could influence foraminiferal distributions. However, recent work in the south-east Atlantic, where hydrographic contrasts are strongly developed, suggests that the distributions of certain foraminiferal species are controlled in part by the lateral advection of water masses. In the case of *N. umboniferus* there is good evidence that the main

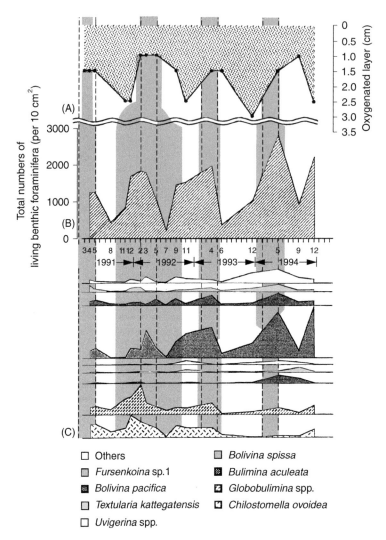

Figure 7 Seasonal changes over a 4-year period (March 1991 to December 1994) in (A) the thickness of the oxygenated layer, (B) the total population density of live benthic foraminifera, and (C) the abundances of the most common species at a 1450 m deep site in Sagami Bay, Japan. (Reprinted from Ohga T and Kitazato H (1997) Seasonal changes in bathyal foraminiferal populations in response to the flux of organic matter (Sagami Bay, Japan). *Terra Nova* 9: 33–37; with permission from Blackwell Science Ltd.)

factor is the degree of undersaturation of the bottom water in calcium carbonate. This abyssal species is found typically in the carbonate-corrosive (and highly oligotrophic) environment between the calcite lysocline and the CCD, a zone that may coincide approximately with AABW. Where water masses are more poorly delineated, as in the Indian and Pacific Oceans, links with faunal distributions are less clear.

During the past 15 years, attention has focused on the impact on foraminiferal ecology of organic matter fluxes to the seafloor. The abundance of dead foraminiferal shells >150 μm in size correlates well with flux values. There is also compelling evidence that the distributions of species and species associations are linked to flux intensity. Infaunal species, such as *Melonis barleeanum, Uvigerina peregrina,*

Chilostomella ovoidea and *Globobulimina affinis,* predominate in organically enriched areas, e.g. beneath upwelling zones. Epifaunal species such as *Cibicidoides wuellerstorfi* and *Nuttallides umbonifera* are common in oligotrophic areas, e.g. the central oceanic abyss. In addition to flux intensity, the degree of seasonality of the food supply (i.e., whether it is pulsed or continuous) is a significant factor. *Epistominella exigua,* one of the opportunists that exploit phytodetritus, occurs in relatively oligotrophic areas where phytodetritus is deposited seasonally.

Recent analysis of a large dataset relating the relative abundance of 'live' (stained) foraminiferal assemblages in the north-east Atlantic and Arctic Oceans to flux rates to the seafloor has provided a quantitative framework for these observations. Although species are associated with a wide flux range, this range diminishes as a species become relatively more abundant and conditions become increasingly optimum for it. When dominant occurrences (i.e., where species represent a high percentages of the fauna) are plotted against flux and water depth, species fall into fields bounded by particular flux and depth values (**Figure 8**). Despite a good deal of overlap, it is possible to distinguish a series of dominant species that succeed each other bathymetrically on relatively eutrophic continental slopes and other species that dominate on the more oligotrophic abyssal plains.

Other environmental attributes undoubtedly modify the species composition of foraminiferal assemblages in the deep sea. Agglutinated species with tubular or spherical tests are found in areas where the seafloor is periodically disturbed by strong currents capable of eroding sediments. Forms projecting into the water column may be abundant where steady flow rates convey a continuous supply of suspended food particles. Other species associations may be linked to sedimentary characteristics.

Low-Oxygen Environments

Oxygen availability is a particularly important ecological parameter. Since oxygen is consumed during the degradation of organic matter, concentrations of oxygen in bottom water and sediment pore water are inversely related to the organic flux derived from surface production. In the deep sea, persistent oxygen depletion ($O_2 < 1\,ml\,l^{-1}$) occurs at bathyal depths ($<1000\,m$) in basins (e.g., on the California Borderland) where circulation is restricted by a sill and in areas where high primary productivity resulting from the upwelling of nutrient-rich water leads to the development of an oxygen minimum zone (OMZ; e.g., north-west Arabian Sea and the Peru margin). Subsurface sediments also represent an oxygen-limited setting, although oxygen penetration is generally greater in oligotrophic deep-sea sediments than in fine-grained sediments on continental shelves.

On the whole, foraminifera exhibit greater tolerance of oxygen deficiency than most metazoan taxa, although the degree of tolerance varies among species. Oxygen probably only becomes an important limiting factor for foraminifera at concentrations well below $1\,ml\,l^{-1}$. Some species are abundant at levels of $0.1\,ml\,l^{-1}$ or less. A few apparently live in permanently anoxic sediments, although anoxia sooner or later results in death when accompanied by high concentrations of hydrogen sulfide. Oxygen-deficient areas are characterized by high foraminiferal densities but low, sometimes very low (<10), species numbers. This assemblage structure (high dominance, low species richness) arises because (i) low oxygen

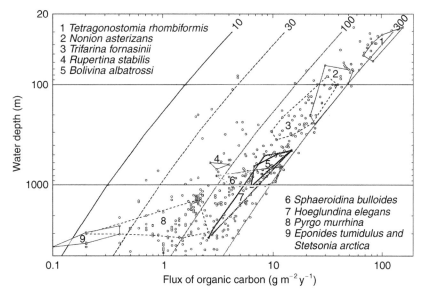

Figure 8 Dominant 'live' (rose Bengal-stained) occurrences of foraminiferal species in relation to water depth and flux or organic carbon to seafloor in the North Atlantic from the Guinea Basin to the Arctic Ocean. Each open circle corresponds to a data point. The polygonal areas indicate the combination of water depth and flux conditions under which nine different species are a dominant faunal component. The diagonal lines indicate levels of primary production (10, 30, 100, 300 g m^{-2} y^{-1}) that result in observed flux rates. Based on $>250\,\mu m$ sieve fraction plus $63–250\,\mu m$ fraction from Guinea Basin and Arctic Ocean. (Reprinted from Altenbach AV, Pflaumann U, Schiebel R et al. (1999) Scaling percentages and distribution patterns of benthic foraminifera with flux rates of organic carbon. *Journal of Foraminiferal Research* 29: 173–185; with permission from The Cushman Foundation.)

concentration acts as a filter that excludes non-tolerant species and (ii) the tolerant species that do survive are able to flourish because food is abundant and predation is reduced. Utrastructural studies of some species have revealed features, e.g., bacterial symbionts and unusually high abundances of peroxisomes, that may be adaptations to extreme oxygen depletion. In addition, mitachondria-laden pseudopodia have the potential to extend into overlying sediment layers where some oxygen may be present.

Many low-oxygen-tolerant foraminifera belong to the Orders Rotaliida and Buliminida. They often have thin-walled, calcareous tests with either flattened, elongate biserial or triserial morphologies (e.g., *Bolivina, Bulimina, Globobulimina, Fursenkoina, Loxotomum, Uvigerina*) or planispiral/lenticular morphologies (e.g., *Cassidulina, Chilostomella, Epistominella, Loxotomum, Nonion, Nonionella*). Some agglutinated foraminifera, e.g.,

Textularia, Trochammina (both multilocular), *Bathysiphon*, and *Psammosphaera* (both unilocular), are also abundant. However, miliolids, allogromiids, and other soft-shelled foraminifera are generally rare in low-oxygen environments. It is important to note that no foraminiferal taxon is currently known to be confined entirely to oxygen-depleted environments.

Deep-Sea Foraminifera in Paleo-Oceanography

Geologists require proxy indicators of important environmental variables in order to reconstruct ancient oceans. Benthic foraminifera provide good proxies for seafloor parameters because they are widely distributed, highly sensitive to environmental conditions, and abundant in Cenozoic and Cretaceous deep-sea sediments (note that deep-sea

Table 2 Benthic foraminiferal proxies or indicators (both faunal and chemical) useful in paleo-oceanographic reconstruction

Environmental parameter/property	Proxy or indicator	Remarks
Water depth	Bathymetric ranges of abundant species in modern oceans	Depth zonation largely local although broad distinction between shelf, slope and abyssal depth zones possible
Distribution of bottom water masses	Characteristic associations of epifaunal species	Relations between species and water masses may reflect lateral advection
Carbonate corrosiveness of bottom water	Abundance of *Nuttallides umbonifera*	Corrosive bottom water often broadly corresponds to Antarctic Bottom Water
Deep-ocean thermohaline circulation	Cd/Ca ratios and $\delta^{13}C$ values for calcareous tests	Proxies reflect 'age' of bottom watermasses; i.e., period of time elapsed since formation at ocean surface
Oxygen-deficient bottom-water and pore water	Characteristic species associations; high-dominance, low-diversity assemblages	Species not consistently associated with particular range of oxygen concentrations and also found in high-productivity areas
Primary productivity	Abundance of foraminiferal tests $> 150\,\mu m$	Transfer function links productivity to test abundance (corrected for differences in sedimentation rates between sites) in oxygenated sediments
Organic matter flux to seafloor	(i) Assemblages of high productivity taxa (e.g. *Globobulimina, Melonis barleeanum*) (ii) Ratio between infaunal and epifaunal morphotypes (iii) Ratio between planktonic and benthic tests	Assemblages indicate high organic matter flux to seafloor, with or without corresponding decrease in oxygen concentrations
Seasonality in organic matter flux	Relative abundance of 'phytodetritus species'	Reflects seasonally pulsed inputs of labile organic matter to seafloor
Methane release	Large decrease (2–3‰) in $\delta^{13}C$ values of benthic and planktonic tests	Inferred sudden release of ^{12}C enriched methane from clathrate deposits following temperature rise

sediments older than the middle Jurassic age have been destroyed by subduction, except where preserved in ophiolite complexes).

Foraminiferal faunas, and the chemical tracers preserved in the tests of calcitic species, can be used to reconstruct a variety of paleoenvironmental parameters and attributes. The main emphasis has been on organic matter fluxes and bottom-water/pore water oxygen concentrations (inversely related parameters), the distribution of bottom-water masses, and the development of thermohaline circulation (**Table 2**). Modern deep-sea faunas became established during the Middle Miocene (10–15 million years ago), and these assemblages can often be interpreted in terms of modern analogues. This approach is difficult or impossible to apply to sediments from the Cretaceous and earlier Cenozoic, which contain many foraminiferal species that are now extinct. In these cases, it can be useful to work with test morphotypes (e.g., trochospiral, cylindrical, biserial/triserial) rather than species. The relative abundance of infaunal morphotypes, for example, has been used as an index of bottom-water oxygenation or relative intensities of organic matter inputs. The trace element (e.g., cadmium) content and stable isotope (δ^{13}C; i.e., the deviation from a standard ^{12}C : ^{12}C ratio) chemistry of the calcium carbonate shells of benthic foraminifera provide powerful tools for making paleo-oceanographic reconstructions, particularly during the climatically unstable Quaternary period.

The cadmium/calcium ratio is a proxy for the nutrient (phosphate) content of sea water that reflects abyssal circulation patterns. Carbon isotope ratios also reflect deep-ocean circulation and the strength of organic matter fluxes to the seafloor.

It is important to appreciate that the accuracy with which fossil foraminifera can be used to reconstruct ancient deep-sea environments is often limited. These limitations reflect the complexities of deep-sea foraminiferal biology, many aspects of which remain poorly understood. Moreover, simple relationships between the composition of foraminiferal assemblages and environmental variables are elusive, and it is often difficult to identify faunal characteristics that can be used as precise proxies for paleo-oceanographic parameters. For example, geologists often wish to establish paleobathymetry. However, the bathymetric distributions of foraminiferal species are inconsistent and depend largely on the organic flux to the seafloor, which decreases with increasing depth (**Figure 8**) and is strongly influenced by surface productivity. Thus, foraminifera can be used only to discriminate in a general way between shelf, slope, and abyssal faunas, but not to estimate precise paleodepths. Oxygen concentrations and organic matter inputs are particularly problematic. Certain species and morphotypes dominate in low-oxygen habitats that also are usually characterized by high organic loadings. However, the same foraminifera may occur in organically enriched settings where oxygen levels are

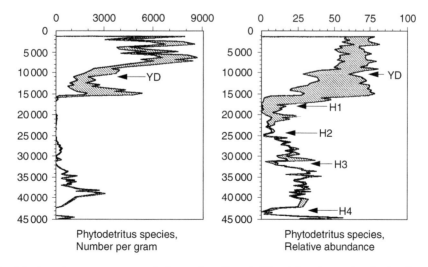

Figure 9 (A) Absolute (specimens per gram of dry sediment) and (B) relative (percentage) abundances of *Alabaminella weddellensis* and *Epistominella exigua* (>63 μm fraction) in a long-sediment core from the North Atlantic (50°41.3′N, 21°51.9′W, 3547 m water depth). In modern oceans, these two species respond to pulsed inputs of organic matter ('phytodetritus') derived from surface primary production. Note that they increased in abundance around 15 000 years ago, corresponding to the main Northern Hemisphere deglaciation and the retreat of the Polar Front. Short period climatic fluctuations (YD = Younger Dryas; H1–4 = Heinrich events, periods of very high meltwater production) are also evident in the record of these two species. (Reprinted from Thomas E, Booth L, Maslin M and Shackleton NJ (1995). Northeast Atlantic benthic foraminifera during the last 45 000 years: change in productivity seen from the bottom up. *Paleoceanography*. 10: 545–562; with permission from the American Geophysical Union.)

not severely depressed, making it difficult for paleo-oceanographers to disentangle the influence of these two variables. Finally, biological factors such as microhabitat preferences and the exploitation of phytodetrital aggregates ('floc') influence the stable isotope chemistry of foraminiferal tests.

There are many examples of the use of benthic foraminiferal faunas to interpret the geological history of the oceans. Only one is given here. Cores collected at 50°41'N, 21°52'W (3547 m water depth) and 58°37'N, 19°26'W (1756 m water depth) were used by E. Thomas and colleagues to study changes in the North Atlantic over the past 45 000 years. The cores yielded fossil specimens of two foraminiferal species, *Epistominella exigua* and *Alabaminella weddellensis*, both of which are associated with seasonal inputs of organic matter (phytodetritus) in modern oceans. In the core from 51°N, these 'phytodetritus species' were uncommon during the last glacial maximum but increased sharply in absolute and relative abundance during the period of deglaciation 15 000–16 000 years ago (**Figure 9**). At the same time there was a decrease in the abundance of *Neogloboquadrina pachyderma*, a planktonic foraminifer found in polar regions, and an increase in the abundance of *Globigerina bulloides*, a planktonic species characteristic of warmer water. These changes were interpreted as follows. Surface primary productivity was low at high latitudes in the glacial North Atlantic, but was much higher to the south of the Polar Front. At the end of the glacial period, the ice sheet shrank and the Polar Front retreated northwards. The 51°N site was now overlain by more productive surface water characterized by a strong spring bloom and a seasonal flux of phytodetritus to the seafloor. This episodic food source favored opportunistic species, particularly *E. exigua* and *A. weddellensis*, which became much more abundant both in absolute terms and as a proportion of the entire foraminiferal assemblage.

Conclusions

Benthic foraminifera are a major component of deep-sea communities, play an important role in ecosystem functioning and biogeochemical cycling, and are enormously diverse in terms of species numbers and test morphology. These testate (shell-bearing) protists are also the most abundant benthic organisms preserved in the deep-sea fossil record and provide powerful tools for making paleo-oceanographic reconstructions. Our understanding of their biology has advanced considerably during the last two decades, although much remains to be learnt.

See also

Benthic Foraminifera. Benthic Organisms Overview. Deep-sea Fauna. Macrobenthos. Meiobenthos. Microphytobenthos. Primary Production Processes.

Further Reading

Fischer G and Wefer G (1999) *Use of Proxies in Paleoceanography: Examples from the South Atlantic*. Berlin: Springer-Verlag.

Gooday AJ, Levin LA, Linke P, and Heeger T (1992) The role of benthic foraminifera in deep-sea food webs and carbon cycling. In: Rowe GT and Pariente V (eds.) *Deep-Sea Food Chains and the Global Carbon Cycle*, pp. 63–91. Dordrecht: Kluwer Academic.

Jones RW (1994) *The Challenger Foraminifera*. Oxford: Oxford University Press.

Loeblich AR and Tappan H (1987) *Foraminiferal Genera and their Classification*, vols 1, 2. New York: Van Nostrand Reinhold.

Murray JW (1991) *Ecology and Palaeoecology of Benthic Foraminifera*. New York: Wiley; Harlow: Longman Scientific and Technical.

SenGupta BK (ed.) (1999) *Modern Foraminifera*. Dordrecht: Kluwer Academic.

Tendal OS and Hessler RR (1977) An introduction to the biology and systematics of Komokiacea. *Galathea Report* 14: 165–194, plates 9–26.

Van der Zwan GJ, Duijnstee IAP, den Dulk M, *et al.* (1999) Benthic foraminifers:: proxies or problems? A review of paleoecological concepts. *Earth Sciences Reviews* 46: 213–236.

MEIOBENTHOS

B. C. Coull and G. T. Chandler, University of South Carolina, Columbia, SC, USA

Introduction

Meiobenthos live in all aquatic environments. They are important for the remineralization of organic matter, and they are crucial members of marine food chains. These small (less than 1 mm) invertebrates have representatives from 20 metazoan (multicellular) phyla and three protistan (unicellular) phyla. With their ubiquitous distribution in nature, high abundances (millions per square meter), intimate association with sediments, rapid reproduction and rapid life histories, the meiobenthos have also emerged as valuable sentinels of pollution.

Definitions and Included Taxa

Meio (Greek, pronounced 'myo') means smaller, thus meiobenthos are the smaller benthos. They are smaller than the more visually obvious macrobenthos (e.g., segmented worms, echinoderms, clams, snails, etc.). Conversely, they are larger than the microbenthos – a term restricted primarily to Protista, unicellular algae, and bacteria. Meiofauna are small invertebrate animals that live in or on sediments, or on structures attached to substrates in aquatic environments. Meiobenthos (*benthos* = bottom living) refers specifically to those meiofauna that live on or in sediments. Meiofauna is the more encompassing word. By size, meiofauna are traditionally defined as invertebrates less than 1 mm in size and able to be retained on sieve meshes of 31–64 µm.

Nineteen of the 34 multicellular animal phyla (**Table 1**) and three protistan (unicellular) phyla, i.e., Foraminifera, Rhizopoda, and Ciliophora, have meiofaunal representatives. Of these multicellular (metazoan) phyla, some are always meiofaunal in size (permanent meiofauna), whereas others are meiofaunal in size only during the early part of their life (temporary meiofauna) (**Table 2**). These are the larvae and/or juveniles of macrobenthic species (e.g., Annelida, Mollusca, Echinodermata). Members of the phylum Nematoda are the most abundant meiofaunal organisms, and copepods (Arthropoda, Crustacea) or Foraminifera are typically second in abundance worldwide. Representative meiofauna taxa are illustrated schematically in **Figure 1**.

The books listed under Further Reading by Higgins and Theil and by Giere, and any invertebrate zoology text, should allow one to identify field-collected meiofauna to phylum. Identification to the family, genus, and species level requires specialized literature. Good places to start are chapters on specific phyla in the two texts listed, and also the International Association of Meiobenthologists website: http://www.mtsu.edu/meio

Where Do Meiofauna Live?

Meiofauna occupy a variety of habitats from high-altitude lakes to the deepest ocean depths. In fresh water they occur in beaches, wetlands, streams, rivers, and even the bottoms of our deepest lakes. In marine habitats they occur from the intertidal splash zone to the deepest trenches. Wherever one looks in the aquatic environment, meiofauna are likely to be found. This holds true even in heavily polluted or anoxic sediments where the only living multicellular species are often a few meiofaunal taxa.

Sediment Habitats

Sediments, from the softest muds to the coarsest shell gravels and cobbles, harbor abundant meiofauna. Meiofauna associated with sediments live 'on' or 'in' the sediment. Those living on top of the sediment are epifaunal (or epibenthic) and are adapted to moving over sediment surfaces. Those living 'in' the sediment may burrow into the sediment (burrowing meiofauna), displacing sediment particles as they move, or they may move in the interstices between sediment grains and be called interstitial meiofauna (see **Table 2**). The interstitial fauna are restricted to sediments where there is sufficient space to move between the particles; typically sands and gravels. Sediments where the median particle diameter is below 125 µm provide little room for meiofauna to move between particles, and thus are inhabited by burrowing and epibenthic taxa. In those taxa having both interstitial and burrowing representatives (e.g., Nematoda, Copepoda, Turbellaria), there are often stark differences in the morphologies of the mud dwellers and sand dwellers. The sand fauna tend to be slender, since they must maneuver through narrow interstitial openings, whereas the mud fauna are not restricted to a particular morphology and are

Table 1 A list of meiobenthic taxa (Phyla of the Kingdom Animalia). Currently, 19 phyla (**bold**) from the 34 recognized phyla of the Kingdom Animalia have meiofaunal representatives. Of these 19 phyla, only five are exclusively meiofaunal (***bold italics***)

Phyla	Free-living			Symbiotic
	Marine	Freshwater	Terrestrial	
Porifera	Yes	Yes	No	No
Placozoa	Endemic	No	No	No
Cnidaria	Yes	Yes	No	Yes
Ctenophora	Endemic	No	No	No
Plathelminthes	Yes	Yes	Yes	Yes
Orthonectida	No	No	No	Endemic (marine)
Rhombozoa	No	No	No	Endemic (marine)
Cycliophora	No	No	No	Endemic (marine)
Acanthocephala	No	No	No	Endemic
Nemertea	Yes	Yes	Yes	Yes
Nematomorpha	No	No	No	Endemic
Gnathostomulida	Endemic	No	No	No
Kinorhyncha	Endemic	No	No	No
Loricifera	Endemic	No	No	No
Nematoda	Yes	Yes	Yes	Yes
Rotifera	Yes	Yes	Yes	Yes
Gastrotricha	Yes	Yes	No	No
Entoprocta	Yes	Yes	No	Yes
Pripaulida	Endemic	No	No	No
Pogonophora	Endemic	No	No	No
Echiura	Endemic	No	No	No
Sipuncula	Yes	No	Yes	No
Annelida	Yes	Yes	Yes	Yes
Arthropoda	Yes	Yes	Yes	Yes
Tardigrada	Yes	Yes	Yes	No
Onychophora	No	No	Endemic	No
Mollusca	Yes	Yes	Yes	Yes
Phoronida	Endemic	No	No	No
Bryozoa	Yes	Yes	No	No
Echinodermata	Endemic	No	No	No
Echinodermata	Endemic	No	No	No
Chaetognatha	Endemic	No	No	No
Hemichordata	Endemic	No	No	No
Chordata	Yes	Yes	Yes	Yes

(Modified from RP Higgins, unpublished, with permission.)

generally larger. Since sandy habitats often occur in areas with high wave and tidal action, most interstitial fauna have adhesive glands for attaching to sand grains so that they will not be washed away. They also tend to have a low number of eggs because their reduced body size cannot support large egg masses.

Table 2 Types of the meiofauna

Permanent meiofauna: always meiofaunal size
Interstitial: moves between sediment particles
Burrowing: displaces sediment particles
Epibenthic
On sediment surfaces
On plants or animals
Temporary meofauna: meiofaunal size in early life only
Larvae or juveniles of macrofauna: mostly bivalve molluscs and polychaete worms

Other Habitats

Meiofauna also occupy several 'above sediment' habitats, including rooted aquatic vegetation, moss, algae, sea ice, and various animal structures such as coral crevices, worm tubes, and echinoderm spines. Still other meiofauna are symbionts living commensally in animal tubes. Those meiofaunal assemblages living above the bottom, for example, in or on fouling communities, or on various animal

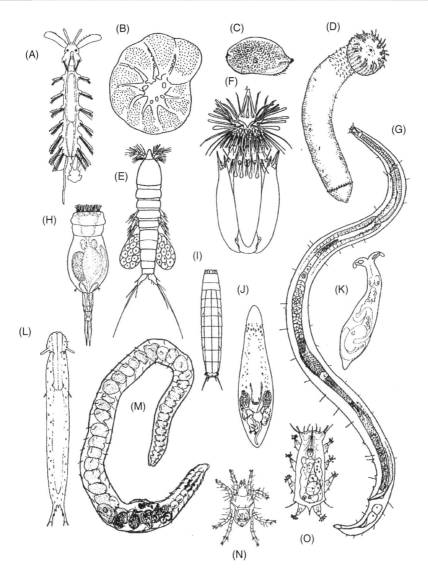

Figure 1 Schematic diagrams of representative meiofaunal animals: (A) Annelida, Polychaeta; (B) Foraminifera; (C) Crustacea, Ostracoda; (D) Priapulida; (E) Crustacea, Copepoda; (F) Loricifera; (G) Nematoda; (H) Rotifera; (I) Kinorhyncha; (J) Plathelminthes, Turbellaria; (K) Mollusca, Gastropoda; (L) Gastrotricha; (M) Annelida, Oligochaeta; (N) Arthropoda, Halacaroidea; (O) Tardigrada. The animals are not drawn to scale. (Modified from Higgins and Thiel (1988).)

structures, differ from sediment dwellers by having species composition and adaptive morphologies specific to particular epibenthic habitats.

Collection and Extraction of Meiofauna

Qualitative sampling of meiofauna will not allow estimation of abundance per unit area, but it is useful for a general assessment of faunal richness or to accumulate one or several species for experimental work. Qualitative samples of sediment are taken by scooping sediment arbitrarily with some device (shovel, hands, grab sampler, dredge), whereas qualitative samples of meiofauna living on structures are taken by collecting the structure itself. Such samples can be sieved live at the collection site, be taken to a laboratory for extraction of the fauna by physical or chemical means, or be preserved in their entirety for future examination. Quantitative meiofauna sampling requires that the sampling area be accurately known. For sediments, this typically involves pushing a core tube of known diameter into the sediment to a preselected depth, collecting all the sediment within the core, and ultimately counting all the fauna in the known area or volume.

Quantitative samples are typically preserved in formaldehyde or alcohol and subsequently counted and identified under a microscope. They are often stained with a protein stain (e.g., Rose Bengal) to help distinguish the animals from surrounding sediment and organic debris. Meiofaunal abundance values are preferably expressed as number per $10\,cm^2$, but also as number per m^2.

There are multiple ways of extracting meiofauna from sediments and surfaces. For live qualitative sediment samples, many species will be attracted to a focused directional light source (preferably cold fiberoptic light so as not to heat the sediments unduly) if sieved sediments are spread in a thin layer with a centimeter or so of overlying water. Sieved sediment can also be put into funnels, where established salinity and/or heat gradients will drive the fauna down the funnel and into a collecting dish. For animals clinging to surfaces, chemical relaxants – or fresh water for marine samples – will cause some fauna to release their purchase and be washed into overlying water where they can be collected onto sieves of appropriate size. For preserved quantitative samples, meiofauna can be separated from the sediments by decantation (swirling the sediment in a container and pouring off the less dense animals after the mineral particles have settled), elutriation (where water is passed through a sample continuously so that sediment is kept in suspension and the lighter animals come off with the flow), or by centrifugation in a density gradient solution so that the sediment (or debris) remains in one layer and the animals in another. All the products of extractions are sieved through a fine mesh (32–$100\,\mu m$ depending on the objective) and the portion retained on the mesh is observed, counted and identified under a microscope.

Distribution of Meiofauna

Geographic Distribution

Meiofauna inhabit some of the most dynamic environments imaginable (such as exposed high-energy shores) and these animals have traditionally been considered sedentary. Emphasis has centered on adaptations for remaining in close proximity to the substratum, particularly because pelagic larvae are almost nonexistent in the permanent meiofauna. Development, morphology, and biology all seem designed to ensure that the organism remains in or on the substratum. On the basis of such observations, one would expect limited worldwide distribution patterns for species. However, numerous species (identified by morphology, not by molecular genetic technologies) appear to be cosmopolitan. Plate tectonics has been invoked as a potential mechanism to describe pan-oceanic and worldwide meiofaunal distributions, as have dispersal via birds,

rafting on drifting materials, transport in the ballast of sailing vessels, and dispersal by suspension in the water column. On a local scale, meiofaunal dispersal is either a passive process of mechanical removal due to current scour or one in which the animals actively migrate to the water column. Animals occupying the sediment surface are obviously scoured much more easily than those living deeper in the sediment. The abundance of eroded species in the water column at any given time is a function of the magnitude of local current velocity and sediment erodability.

Large-scale Spatial Distribution

Meiofauna are rarely evenly distributed on, or in, a substrate. On the large scale (meters to kilometers) gradients in physical factors (e.g., salinity, tidal exposure, sediment grain size, oxygen concentrations) are primarily responsible for variances in abundance, whereas on smaller (centimeter) scales both physical and biological factors have been reported as important. Large-scale gradients lead to zonation of the fauna. For example, certain meiofauna species are confined to specific areas along salinity gradients in estuaries, across intertidal sandy and muddy habitats, and across the water depth gradient in lakes and in the ocean. With water depth, faunal changes are primarily a function of food availability (e.g., organic content of sediment), sediment type, temperature, and oxygen availability. Interestingly, the meiofauna at similar ocean depths are usually similar to each other all over the world. The same families and/or genera comprise a significant portion of the fauna at similar depths except in the Mediterranean and the Arctic, where many of the 'deep sea' genera also occur into shallower ($<500\,m$) depths.

Small-scale Spatial Distribution

Meiobenthos also exhibit spatial variation (patchiness) on a small (millimeter to centimeter) scale. A variety of factors have been suggested for the observed small-scale patchiness including: (1) microspatial variation in physical factors (oxygen, grain size); (2) food distribution; (3) physical structures in the habitat (worm tubes, algae, mud balls, etc.); (4) predation/disturbance, where a predator eats one patch of animals but not another; (5) interspecific competition, where species segregate themselves spatially to avoid competition for a resource; and (6) aggregations, where individuals come together for mating. While we presently lack a framework for experimentally testing how these factors effect microspatial distribution in the field, we know that species are aggregated more often than not. Small-scale zonation also takes place vertically in sediment.

Here the vertical distribution of the fauna is controlled primarily by the level of oxygen in the sediment layers. Most meiofauna require oxygen to survive, but certain adapted species can tolerate low oxygen or no oxygen. Species living in such sediments that can tolerate hydrogen sulfide, a known animal toxin, are called the 'sulfide fauna' or the thiobios. Copepods are typically the meiobenthic taxon most sensitive to decreased oxygen, and generally are confined to oxic sediments. Gnathostomulida primarily live in mild sulfidic and low oxic sediments, as do some Nematoda, Turbellaria, Ciliophora, Gastrotricha and Oligochaeta (see Table 1). While oxygen content is the ultimate factor controlling most meiofaunal vertical distribution, desiccation can also be important, particularly in intertidal marine beaches. As sand dries at low tide, the fauna face desiccation stress regardless of the oxygen content. Meiofauna therefore migrate downward on an ebbing tide and upward on a flooding tide, and this happens more at midday in the summer, when drying is greatest, than at midnight in the winter.

Abundance and Diversity of Meiofauna

On the average there are a million meiofaunal organisms per square meter of sediment surface, with a dry weight biomass of 0.75–2 g m^{-2} in shallow (<100 m) waters. Highest abundance values come from intertidal muddy estuarine habitats (6–12 million per m^2), lowest values from the deep sea (hundreds to thousands per m^2). In general, sediment grain size is the primary factor affecting the abundance and species composition of meiofaunal organisms within a given depth range. Different species occur in muddy versus sandy versus phytal habitats. In areas where temperature varies seasonally, meiobenthos abundance and species composition also vary seasonally. Typically, maximum abundances occur in the warmer months of the year, but individual species may reach maximum abundance at other times. Year-to-year variability in abundance also can be greater than within-year seasonal variability.

The highest known species diversity for a meiofaunal assemblage has been recorded for copepods from algal holdfast communities. Shallow-water algal frond assemblages and deep-sea sediments also yield high species diversities. Even though meiofaunal abundance in the deep sea is greatly reduced compared to shallow sediments, there are many different and exotic species. In shallow-water

sedimentary habitats, meiofaunal diversity appears similar worldwide, with ecologically equivalent species in different geographic regions. These communities usually have four to ten predominant species. While the database is limited and there are always difficulties interpreting diversity data, there appears to be a standard diversity range for most shallow-water meiofaunal assemblages. There is no evidence that meiofaunal species diversity increases toward the tropics. Pollution or other disturbances, such as hypoxia/anoxia, tend to decrease diversity.

Functional Role of Meiofauna

Meiofauna appear to have two major functional roles in aquatic ecosystems: to serve as food for organisms higher in the food web, and to facilitate mineralization of organic material and enhance nutrient regeneration. In addition, because they exhibit high sensitivity and rapid response to anthropogenic disturbance, they are excellent sentinels of pollution.

Food for Higher Trophic Levels

Meiofauna are very important nutritionally to a variety of animals that could not survive without them. Many predators go though an obligatory meiofaunal feeding stage, and copepods appear to be the major meiofauna prey item for most of these predators. These copepods primarily live in muddy sediment or on plants. Thus most predation on meiofauna takes place in muddy substrates or in areas with substantial sea grass or macroalgae. In muds, the meiofauna prey are restricted to the upper few millimeters or centimeters of oxidized sediment. Thus bottom-feeding predators only need to take a shallow bite to obtain abundant food. On aquatic plants, fish predation on meiofauna is analogous to birds eating insects on a tree. Over 90 species of juvenile fish are known to eat meiofauna, making them the major meiofaunal predators. Other predators are shrimp (prawns) and some bottom-feeding birds.

Mineralization and Nutrient Regeneration

Meiofauna are important in stimulating bacterial growth, which then enhances remineralization (the conversion of organic nitrogen, phosphorus, and carbon to their inorganic forms). Meiofauna package organic molecules and, because of their relatively short life span (months) and high metabolic rate, this packaged material is returned to the system rapidly (compare, for example, the carbon tied up in a clam that lives for 2–5 years). Meiofaunal nutrients then become part of the well-known microbial loop in which they are utilized by bacteria and can be converted into dissolved organic carbon for use by higher trophic levels and/or remineralized for primary producers. Meiofauna typically have less than 20% of the standing biomass of the larger, more visible, macrofauna, but they turn over as much or more carbon per year. These processes are important in all kinds of habitats, but they are probably most important in those sediments with high amounts of organic matter, i.e., muds.

Meiofauna and Pollution

Sediments are the ultimate repository for most of the persistent pollutants released to the ecosphere. Upon entering aquatic environments, most toxicants associate with dissolved organics, suspended silts, clays, and organic particulates and eventually accumulate in sediments. Meiofauna, of course, are intimately associated with this muddy-sediment geochemical soup, as they spend their entire life cycle there and have limited ability to leave. Because meiofauna reproduce very rapidly (often in 2–4 weeks), pollution effects on meiofaunal populations can be detected quickly and early in the history of contamination of a site. There have been three general approaches to using meiofauna to assess pollution: field studies, laboratory studies, and studies using replicas of the controlled natural environment (microcosm/mesocosm studies). In field studies, samples are typically collected from a polluted site and from a reference site, and differences in community (or genetic) structure between the sites are assessed. Laboratory studies usually examine the lethal effects (e.g., how many individuals die after exposure to specific dose levels of a contaminant) or sublethal effects (e.g., changes in egg production, embryonic development time, hatching success, or genetic diversity of contaminants singly or in mixture. Meiobenthic community responses to pollutants in micro/mesocosms are measurable and reproducible over reasonable time and spatial scales (owing to small organism size and rapid production/turnover), and are more effectively assessed than macrobenthos for toxicant-induced effects since meiofauna spend their entire life cycle in sediments and are not reliant on recruitment of a planktonic larval stage. After years of neglect, meiofauna are becoming more popular subjects of pollution studies.

See also

Benthic Foraminifera. Benthic Organisms Overview. Deep-sea Fauna. Fish Feeding and Foraging. Macrobenthos. Microphytobenthos.

Further Reading

Coull BC and Chandler GT (1992) Pollution and meiofauna: field, laboratory and mesocosm studies. *Oceanography and Marine Biology Annual Reviews* 30: 191–271.

Giere O (1993) *Meiobenthology. The Microscopic Fauna in Aquatic Sediments.* Berlin: Springer-Verlag.

Heip C, Vincx M, and Vranken G (1985) The ecology of marine nematodes. *Oceanography and Marine Biology Annual Reviews* 23: 399–489.

Hicks GRF and Coull BC (1983) The ecology of marine meiobenthic harpacticoid copepods. *Oceanography and Marine Biology Annual Reviews* 21: 67–175.

Higgins RP and Thiel H (eds.) (1988) *Introduction to the Study of Meiofauna.* Washington, DC: Smithsonian Institution Press.

International Association of Meiobenthologists web site: http://www.mtsu.edu/meio

McIntyre AD (1969) Ecology of the marine meiobenthos. *Biological Reviews of the Cambridge Philosophical Society* 44: 245–290.

Swedmark B (1964) The interstitial fauna of marine sand. *Biological Reviews of the Cambridge Philosophical Society* 39: 1–42.

MACROBENTHOS

J. D. Gage, Scottish Association for Marine Science, Oban, UK

Introduction

The macrobenthos is a size-based category that is the most taxonomically diverse section of the benthos. Only in shallow water does the macrobenthos include both plants and animals. Here, attached macrophytes (including various algae and green vascular plants) may make up a large part of the benthic biomass in coastal areas. Meadows of sea grass (which root into and stabilize coarser sediments) and forests of macroalgae (usually attached to hard bottoms) provide habitat for smaller plant and animal species. In warm, shallow water, stony corals, which flourish as a consequence of symbiotic algae living in their tissues, overgrow large areas and these reefs provide a biogenic habitat for a wealth of other species, and their broken-down skeletons provide much of the sediment in adjacent areas. But the importance of such areas declines rapidly with declining potential for photosynthesis as light quickly vanishes with increasing depth, and the macrobenthos becomes the exclusive domain of heterotrophic life (fueled by breakdown of complex organic material) in soft sediments. Only at deep-sea hydrothermal vents is there an exception to this. These support lush concentrations of benthic biomass that relies not on photosynthetic production ultimately derived from the surface but on the activity of chemoautotrophic bacteria exploiting the emissions of reduced sulfur-containing inorganic compounds.

The animal macrobenthos may be attached or may be able to move over hard surfaces provided by exposed bedrock, or may use as habitat the much larger and quantitatively important areas covered by soft sediment. Areas of rock, exposed as a consequence of water currents and turbulence (or, in the case of the newly formed sea floor at the spreading centers along the mid-ocean ridges, rock that has not had time to become covered in sediment settling from above) provide habitat for epibenthos, or epifauna. Even if epibenthos, both plant and animal, looks conspicuous between the tides (and is certainly important with respect to fouling of colonizers of submerged hard surfaces made by man, such as ships and jetties), it is only a vanishingly small proportion of the huge area of the benthic habitat covering more than half the globe that is not covered by soft sediment.

The term infauna has been used to categorize the organisms inhabiting soft sediment. But many epifauna, such as sea stars, are motile and can forage over the surface of sediments. The activities of the animals of the so-called infauna are usually focused on the sediment–water interface where their detrital food is concentrated, but this should not be taken to imply that sediment fauna is always burrowed out of sight. None the less, some species, particularly among larger crustacea, are capable of burrowing even a meter or more deep into the sediment. There are also many species closely related to epifaunal groups of hard substrata, such as sponges and Cnidaria (including sea pens, soft and stony corals and sea anemones), that anchor into the sediment for a sedentary life style, catching small particles from the bed flow.

Global Pattern in Macrobenthic Biomass

Extensive Russian sampling after World War II established the precipitous decline in benthic biomass with increasing depth into the abyss. This is caused largely by mid-water consumption of particles escaping from the euphotic zone. Globally, the amount leaving the euphotic zone should be equivalent to the so-called 'new production' of roughly $3.4–4.7 \times 10^9$ tonnes $C y^{-1}$, about 10% of total surface primary production. But this is distributed very unevenly. Perhaps 25–60% is exported in shallow seas, while in the deep ocean only 1–10% reaches the bottom; this is also influenced by latitude-related differences in depth of the mixed layer and intensity of seasonality in the upper ocean. **Figure 1** shows how influences such as upwelling and inshore surface productivity will also affect local values of benthic biomass. Overall, these range from highs of more than $500 \, \mathrm{g \, m^{-2}}$ in shallow, productive waters just tens of meters deep, to less than $0.05 \, \mathrm{g \, m^{-2}}$ (equivalent to $2 \, \mathrm{mg \, C \, m^2}$) on the abyssal plains. Trenches are deeper still but, by acting as sumps for material washed in from nearby island arcs and land mass, can support higher than expected biomass.

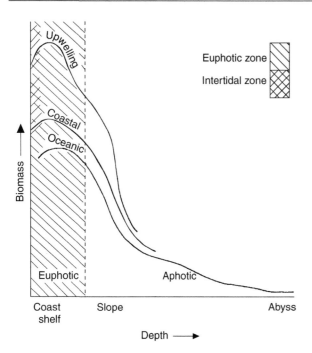

Euphotic zone

Intertidal zone

Figure 1 Conceptual model of food availability and benthic biomass in relation to depth. Upwelling areas provide nutrients for enhanced coastal productivity. The 'coastal' curve refers to shelf areas supporting high productivity (usually wider shelves with land inputs, e.g., rivers) compared to the 'oceanic' curve where oceanic effects prevail (usually narrow shelves with little land input). The mismatch in the intertidal between biomass and high food availability is explained by the co-occurrence with the latter of high hydrodynamic disturbance by waves and currents. (Modified from Pearson and Rosenberg (1987).)

History and Size Limits of Macrobenthos

The term macrobenthos dates from the early 1940s when Molly Mare published a study of an area of coastal soft sediment off Plymouth, England. In recognizing that the benthic ecosystem is fueled by a detrital rain of particles derived from photosynthetic production by macrophytes or phytoplankton, she identified the potential importance of the smaller size classes of metazoan and single-celled organism, right down to bacteria, in the decomposition cycle and food chains in the sediment. She differentiated the benthos into several subcategories based on size, or biomass. Before this time a distinct category for macrobenthos was unnecessary because the only part of the benthos generally thought to be worth studying was the animal life large enough to be eaten by fish. We now differentiate these from the very small metazoans and single-celled algae that Mare named meiobenthos. Another category is the hyperbenthos – small metazoan life that can swim off the bottom and form a distinct community in the benthic boundary layer.

The lower size limit of macrobenthos was determined by Mare as that part retained in a 1 mm sieve, but later became reduced downwards to just 0.5 mm as it was realized not only that small, juvenile sizes were being lost in numbers but also that smaller species of groups already being sampled and taken as part of the macrofauna were not always retained adequately. Mare recognized that the limit might depend on habitat and deep-sea benthic biologists found they had to use even finer-meshed screens to collect the same sorts of animals that characteristically make up the macrobenthos in shallow waters. In the 1970s, Hessler found that he had to use a 297 μm mesh sieve to catch sufficient animal macrobenthos for study in box cores from the abyssal central North Pacific (a very oligotrophic area – nutrient-poor and therefore thin in plankton). Sieves with meshes of just 250 μm are now standard in recent large European studies on the deep-sea macrobenthos.

Sources of Food and Feeding Types

Patterns in feeding of macrobenthos have often been used to distinguish ecological zones. Although exact definition of feeding category for individual organisms has been controversial, the simplest classification is into suspension and deposit feeders, carnivores, and herbivores. More detailed categorization has proved difficult because of overlap and behavioral flexibility. Although most macrobenthos feed on detrital particles settling from the water column, such as feces, molts, and dead bodies of plankton, this passive sinking is augmented by currents that may resuspend particles periodically from the bottom. Macrobenthos may gather these particles either by catching them from bottom flow or by ingesting the sediment itself as deposit feeders, either in bulk or more selectively for the most nutritious particles. Where currents vary periodically, some animals can feed on both suspended and settled particles by simply changing the way they use their feeding appendages. Just as the particles caught by suspension feeders may range from inert floating detritus up to small swimming organisms, deposit feeding shades into predation where the particles encountered include smaller living benthos. Whether macrofauna can utilize dissolved organic matter in the sediment porewaters to any great extent is still unclear.

Wildish has provided the most satisfactory classification of macrofaunal feeding types related to environment. This keeps all three categories but separates deposit feeders into surface and burrowing deposit feeders. Each of the five groups is subdivided in terms of motility and also in terms of food-gathering technique, such as use of jaws and particle-entangling structures. These may be arranged along an environmental gradient, such as that illustrated in **Figure 2**, to allow insight into the causal basis of previously described composition of macrofaunal communities.

The relation of feeding to small body size in deep-sea macrobenthos may be important. Thiel thought that small body size is a result of a balance between limited food and metabolic rate that makes larger size more efficient than smaller, and of the effects of small population size on reproductive success. Being small allows organisms to maintain higher population densities that increase the chance of encountering the opposite sex and so of reproducing and maintaining the population. The extent of faunal miniaturization is still debated and, surprisingly, not readily summarized by simply taking the total bulk of

the sample and dividing by the number of animals present. The exceptions seem be those organisms that have overcome the reproductive problem by being highly motile scavengers and that need also to be large enough to allow them to forage for the large food falls that occur very sporadically on the deep ocean bed. This scavenger community is quite well developed and includes close relatives of typical macrofaunal organisms in shallow water that in the deep sea grow to a relatively enormous size (**Figure 3**).

Size Spectra

If the sizes of all individuals from an area of sediment are measured and plotted as frequencies along a logarithmic size axis, a pattern of peaks shows up corresponding to the micro-, meio- and macrobenthic size classes (**Figure 4**). This supports practical intuition but does not explain why such peaks occur (no

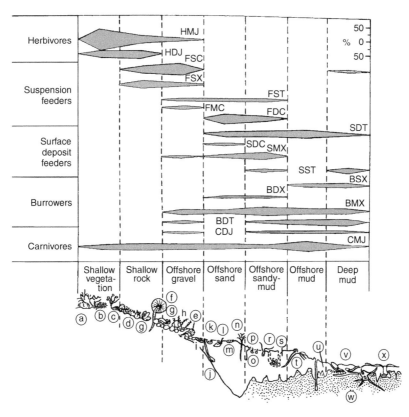

Figure 2 Distribution of functional groups in boreal coastal macrobenthos (compiled from genera listed by N. S. Jones for the Irish Sea) along an environmental gradient of decreasing food availability and water turbulence and increasing depth and sedimentation. Functional groups: H, herbivore; F, suspension feeder; S, surface deposit feeder; B, burrowing deposit feeder; C, carnivore. Motility: M, motile; D, semi-motile; S, sessile. Feeding habit: J, jawed; C, ciliary mechanisms; T, tentaculate; X, other types. In the upper panel, width of line representing each functional group along the gradient indicates proportional composition at that depth. The lower panel gives a diagrammatic representation of typical feeding position of taxa representative of various groups relative to the sediment–water interface along each gradient. Key to taxa: (a) macroalgae; (b) sea urchins, e.g., *Echinus* (HMJ); (c) limpets, e.g., *Patella* (HDL); (d) Barnacles, e.g., *Balanus* (FSX); (e), (f) serpulids, sabellids (FST); (g) epifaunal bivalves, e.g., *Mytilus* (FSC); (h) brittle stars, e.g., *Ophiothrix* (FMC); (i) *Venus* (FSX); (j) *Mya* (FSC); (k) *Cardium* (FDC); (l) *Tellinba* (FDT); (m) *Turritella* (SMX); (n) *Lanice* (SST); (o) *Abra* (SDC); (p) *Spio* (SST); (r) *Amphiura* (FDT); (s) *Echinocardium* (BMX); (t) *Ampharete* (SST); (u) *Maldane* (BSX); (v) *Glycera* (CDJ); (w) Thyasira (BDX); (x) *Amphiura* (SDT). From Pearson and Rosenberg (1987).

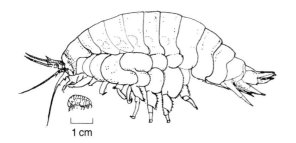

Figure 3 Gigantism in a scavenging amphipod (family Lysianassidae), the cosmopolitan deep-sea species *Eurythenes gryllus*, compared to the size of a typical shallow-water lysiannasid, *Orchomene nana* (bottom left), a northern European shallow-water species. (Redrawn from Gage and Tyler (1991) and Hayward PJ and Ryland JS (1995) *Handbook of the Marine Fauna of North-West Europe*. Oxford: Oxford University Press.)

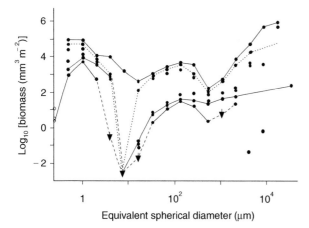

Figure 4 Macrobenthic size spectra measured from an intertidal inlet in Nova Scotia, subtidal Bay of Fundy, and abyssal sediment from the Nares Abyssal Plain south of Bermuda. The median lines (dotted for intertidal and inshore, and double-dashed for abyssal plain) and range (continuous or solid lines) show a coherent pattern with biomass peaks at 1256 and 8192 μm equivalent spherical diameter (ESD). Downward-pointing arrows indicate minimum detectable biomass. Abundance in the bacterial (leftmost) and meiofaunal (middle) peaks averages 5×10^3 mm^3 m^{-2}. The macrofaunal biomass peak is an order of magnitude higher, but shows greater variability. Biomass in the troughs is about 2–3 orders of magnitude less than adjacent peaks. Sediment was a fluid silt-clay. (From Schwinghamer P (1985) Observations on size-structure and pelagic coupling of some shelf and abyssal benthic communities. In: Gibbs PE (ed) Proceedings of the Nineteenth European Marine Biology Symposium, Plymouth, Devon, U.K. 16–21 September 1984, pp. 347–359. Cambridge: Cambridge University Press.)

such peaks occur, for example, in pelagic communities). Schwinghamer thought that these peaks reflect the way the organism perceives its sediment environment: macrofauna as a continuous medium on, or in, which to move and burrow; meiofauna as a series of interstices between sediment particles; while

to microbenthos and bacteria each particle is a little world on which to attach and grow. Warwick provided a complementary explanation that the peaks also reflect size-related adaptation in the way the life history of the organism is optimized to its environment. For example, larvae of macrofauna exploit the trough between macro- and meiofauna to escape from meiofaunal predators, and thereafter quickly grow into the size range of the 'macrofaunal' peak. This is generally lower and less defined than the meiofaunal peak, where organism longevity is just a few weeks at most and there is therefore a narrow range in size, while individual macrofauna might grow over several years so that population size distributions are wider. However, subsequent studies have not found that clear peaks in size spectra occur everywhere. In the deep sea, body size miniaturization does not destroy this pattern, even if the trough at 512–1024 μm between meio- and macrofauna may be less than in coastal sediment (**Figure 4**). It seems more likely that low food supply has become more important than anything else, so that macrofauna, although settling at roughly the same size, simply do not grow anything like as large as similar coastal species, rather than their somehow perceiving the sediment environment differently from typical macrofaunal organisms in shallow water.

We cannot therefore reject the idea that size-based differentiation of the benthos occurs; but is it sensible to stick rigidly to the strict size-based divisions that define the macrofauna as only those organisms within a given range of size, or is it better to compare like with like on the basis of higher taxa determining limits rather than size? With the former definition, the lower limit of the macrobenthos will be determined by size at 1.0 or 0.5 mm, even if this excludes smaller specimens belonging to the same higher taxon, or even much lower-level taxa. This assumes that a size-based functional distinction operates that for the purposes of the study (perhaps environmental impact assessment) will be more important in determining variability than taxonomic affinity. The former function-based definition has been referred to as macrofauna *sensu stricto*, while the latter, taxonomic one as macrofauna *sensu lato*.

Composition and Succession

Macrobenthos characteristically includes a huge range of phyla (the major divisions of the animal kingdom). In fact, most higher-level taxa are marine and benthic, with most of these part of the macrobenthos. Of the 35 or so known phyla (the major divisions of the animal kingdom), 22 are exclusively marine, with 11 restricted to the benthic environment. Virtually every known phylum is represented in the macrobenthos except for one, the Chaetognatha, or arrow worms (arguably found only in the plankton, although one bottom-living genus is known). This contrasts with the land (including freshwater environments) where only 12 phyla are found (there is only one small, obscure phylum of worm-like animals, the Onychophora, known only on land). This reflects the marine origins of life and the much shorter time of occupation for life on land (barely 400 million years), compared with 800 million years since metazoan organisms first appeared in the ancient ocean. Only five metazoan phyla are normally regarded as part of the next size group down, the meiofauna.

The proportional representation of major taxa is conservative, and seems to vary little worldwide with depth, latitude, or productivity regime. It is only in stressed soft-sediment environments, such as those with high organic loading and depleted oxygen (often occurring together), where major departures to this pattern are found (**Figure 5**). Inshore studies on effects of pollution have contributed to a concept whereby such stress leads to a modified macrobenthos with fewer species, and these characterized by opportunist forms, mainly polychaetes. In tracing recovery after pollution events, it is not clear to what extent a predictable succession occurs. The modern consensus is that there is a random component imposed on a facultative succession in which 'opportunist' species pioneer colonization and bring about amelioration in sediment conditions. This allows a more diverse set of species that are more highly tuned to particular habitats to become established through progressively deeper and more extensive bioturbation (**Figure 6**).

How Many Macrobenthic Species Are There?

Up to a few years ago it was thought that of the 1.4 to 1.8 million or so species recorded on earth there are perhaps only 160 000 or so known marine species, about 10% of the total. A large-scale sampling programme in deep water off the eastern United States has thrown this into doubt. Along a 180 km section of the continental slope at about 2000 m depth, Grassle and Maciolek found 58% of the species – especially among polychaete (bristle) worms and peracarids (small, sandhopper sized crustaceans) – new to science. The curve of the accumulation of species plotted against increasing area sampled showed no sign of tailing off; the steady increment of new (but rare) species encouraging an extrapolation that this will apply over the wider area of the deep ocean.

Depths below the shelf edge cover about 90% of the domain of macrobenthic infauna. But an area of only about 0.5 km^2 out of the almost 335×10^6 km^2 area below 200 m depth has yet been adequately sampled for macrofauna using grabs or corers. Because of this huge unexplored area, the actual number of marine macrobenthic species present today is unknown. It must be vastly greater than earlier estimates based on shallow seas, and according to Grassle and Maciolek is conservatively greater than one million, and more likely to rise to 10 million as more of the deep sea is sampled.

The overwhelming taxonomic challenge of describing these new species, the painstaking work of sorting samples from the sediment, and the difficulty of seabed experimentation have perhaps slowed progress in understanding the deep-water sediment community. In contrast, much more has been achieved in biological knowledge of hydrothermal vents (and to a lesser extent cold seep communities) since their discovery in 1977.

Large-scale Patterns in Macrobenthic Diversity

Large-scale patterns, other than that for biomass, remain controversial. On the basis largely of sampling of the continental shelf, Thorson pointed out that species richness of the epifauna, occupying less than 10% of the total area, and maximally developed intertidally, rises steeply from low levels in the ice-scoured shallows in the Arctic to high values in the tropics. In contrast, the sediment macrofauna he referred to as 'infauna,' usually found deeper and unaffected by ice and meltwater, show much less change. This lack of a latitudinal gradient is supported in some other studies. However, latitudinal comparisons by Sanders in the 1960s found depressed diversity in shallow boreal macrobenthos stressed by wide seasonal temperature change compared to the tropics. Thorson through there were about four times more epibenthic than infaunal species, the microhabitat complexity and consequently high species diversification of the epibenthic habitat being much less obvious in sediments. The sameness of this habitat regardless of latitude led Thorson to his concept of parallel level-bottom communities related to sediment type. But several recent studies indicate shallow tropical and deep-sea sediments do not follow this pattern, with much more species-rich communities developing, albeit including lots of 'rare' species, in these habitats.

Macrofaunal abundances

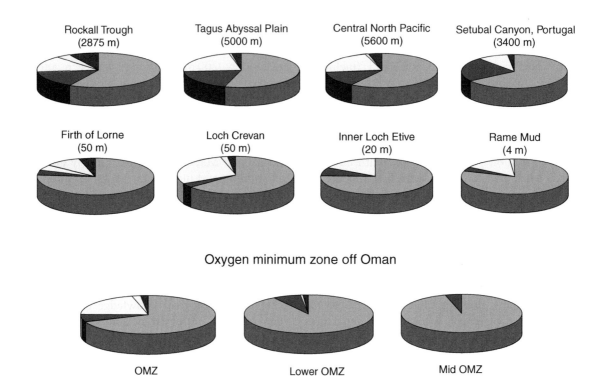

Oxygen minimum zone off Oman

Macrofaunal species

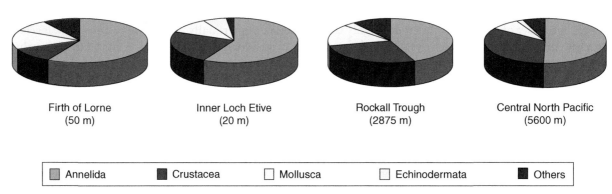

Figure 5 Proportional representation of the major taxonomic groups of macrofauna in differing soft sediment habitats and depths worldwide. The upper diagrams show this in terms of the relative abundance in these groups; the lower ones in terms of the number of species represented. A broadly similar pattern is shown between both sets although Crustacea are often more abundant in the deep sea rather than shallow-water macrobenthos. Representation of Annelida (mostly polychaete worms) shows most obvious variation in relation to organic carbon loading and oxygen, the three samples from the oxygen minimum zone (OMZ) in the Arabian Sea off Oman showing a pattern of increasing dominance by Annelida, and eventually complete loss of all other groups except Crustacea, with increasing oxygen depletion. (Data from Mare (1942); Gage (1972) Community structure of the benthos in Scottish sea-lochs. I. Introduction and species diversity. *Marine Biology* 14: 281–297, Gage J (1977) Structure of the abyssal macrobenthic community in the Rockall Trough. In: Keegan BF, O'Ceidigh P and Boaden PJS (eds) *Biology of Benthic Organisms* (11th European Marine Biology Symposium), pp. 247–260. Oxford: Pergamon Press; Levin LA, Gage JD, Martin C and Lamont PA (2000) Macrobenthic community structure within and beneath the oxygen minimum zone, NW Arabian Sea. *Deep-Sea Research II* 47: 189–226.)

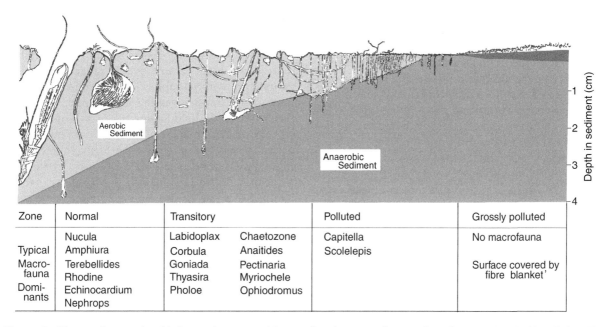

Zone	Normal	Transitory		Polluted	Grossly polluted
Typical Macro-fauna Domi-nants	Nucula Amphiura Terebellides Rhodine Echinocardium Nephrops	Labidoplax Corbula Goniada Thyasira Pholoe	Chaetozone Anaitides Pectinaria Myriochele Ophiodromus	Capitella Scolelepis	No macrofauna Surface covered by fibre blanket'

Figure 6 Changes in macrobenthic fauna along an enrichment–disturbance gradient, such as that associated with pollution. The gradient can be replaced by time in tracing recovery along the x-axis to a 'normal' community (left) after a severe pollution event (right). Note the shift in body size (reflecting change from quick-growing and fast-turnover 'pioneer' species to slower-growing, longer-lived population species) from right to left, as well as the increased depth and extent of bioturbation in the sediment. There is also a concomitant increase in macrobenthic diversity. (From Pearson TH and Rosenbeg R (1978).)

Depth-related patterns in macrobenthic composition have been studied, particularly for larger benthic invertebrates (technically megabenthos, see discussion earlier) and demersal (bottom-living) fish. Rates of macrofaunal turnover, and clinal variation in individual species, correspond to the rate of change in depth. It is highest in upper bathyal and slowest and most subtle in the abyssal. Many changes in species composition can be related to trophic strategies along a gradient in food and hydrodynamic energy (see **Figure 2**), but changes in the sort and intensity of biological interactions, such as predation, varying with depth, may also be important, as can life-history characteristics (such as the incidence of planktotropic larval development). Recent studies have also established a degree of pressure adaptation during early development that will further limit vertical range. Such ecological processes must be considered in concert with processes at the evolutionary timescale for understanding of zonation patterns. These processes have been summarized using multivariate statistics. While helping in formulating ideas on causal factors, these may obscure the underlying complexity, which is best understood as the sum of the range and adaptation and evolutionary history of individual species.

Rex postulated a mod-slope peak in macrobenthic diversity from studies of sled samples taken throughout the Atlantic by Sanders. However, there is high variability among individual sample values compared and there are conflicting results from other sites worked in the north-eastern Atlantic. That deep-sea macrobenthos has high species diversity seems well founded, but the extent to which this contrasts with shallow water is unclear. In his original study off the north-eastern United States, Sanders showed an impoverished species richness in samples of macrofauna compared to the adjacent slope and rise. Gray has pointed out that on the outer continental shelf off Norway macrofaunal diversity may be comparable to that found in Sanders' deep-sea samples, and it is considerably higher still off south-eastern Australia. This suggests not only that the inshore shelf off New England is rather poor in species richness but that the North Atlantic as a whole may be atypical, with perhaps historical factors operating there to restrict macrobenthic diversity compared to the southern hemisphere.

Such factors may determine the differing response shown in a comparison of deep-sea macrobenthic diversity among sites throughout the Atlantic where the depression at high latitudes is absent south of the equator. The reduced levels at high latitudes may simply reflect Quaternary glaciation so that the deep Norwegian Sea, isolated by shallow sills from deep water to the south, has a much more recently diverged and quite distinct macrofauna from that in the Atlantic.

Small-scale Pattern

In sampling macrobenthos from a ship it is easy to assume that the animals in this apparently homogeneous habitat are randomly distributed, but sample replicates may not always provide good estimates of the population mean and its sampling error. When samples are mapped over the sediment, or variability is analysed in large numbers of replicates, clumped distributions of some kind are commonplace (**Figure 7**). Nonrandomly even (regular) dispersions have been detected, but only at the centimeter scale, suggesting that they are actively defined by the ambit (such as the area swept by feeding tentacles) of individual animals. To describe rather than just detect such nonrandom spatial pattern has been a challenging task, not least because most macrofauna are not readily visible in seabed photographs and the very analysis of samples by sieving and mud will destroy fine-scale pattern. Spatial pattern is usually envisaged in the horizontal plane because macrobenthic organisms concentrate their activity on the sediment–water interface in feeding, movement, and reproduction. Pattern is a dynamic expression of this and consequently may change through time but marine sediments provide a three-dimensional habitat so that vertical as well as horizontal spatial patterns may occur. The latter may be best developed at the small scale where smaller macrofauna (and meiofauna) are concentrated around irrigatory or feeding burrows of larger species (**Figure 8**). The problem is that to analyze this pattern it is difficult not also to disrupt the habitat. Yet in order to understand the basis of such pattern it is vital to analyze and map pattern over a range of scales in conjunction with variability in the sediment habitat. Some of the most revealing studies have examined dispersions of individual species over plots measuring tens of meters square. These may reveal the two aspects of pattern, intensity and form. Intensity can relatively easily be measured by the ratio of variance to mean. This will distinguish distributions that are clumped, regular, or not statistically distinguishable from random. The form of pattern is an aspect that classical statistical tests of nonrandomness do not address. Yet a clumped pattern may be very different in form from that shown by another species that shows similar intensity of aggregation.

Although a nuisance for the easy interpretation of sample statistics, an understanding of the biological basis of patterns will provide important insight into the processes maintaining macrobenthic communities.

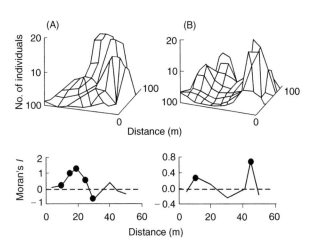

Figure 7 Upper, three-dimensional plots of abundance showing spatial dispersion patterns of two infaunal bivalve molluscs *Nucula hartivigiana* (A) and *Soletellina siliqua* (B) in a 9000 m² area of mid-tide sandflat with no obvious gradients in physicochemical conditions. Both species show quite different spatial patterns. The lower plots show spatial correlograms with significant autocorrelation coefficients (measured as Moran's *I*) denoted by filled circles (From Hall *et al.* (1984). Thrush *et al.*, (1989).)

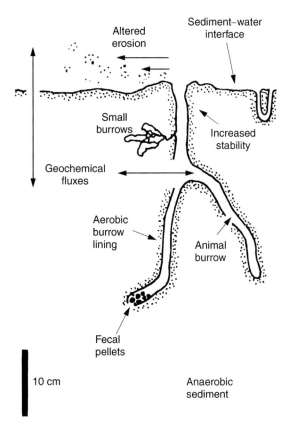

Figure 8 Benthic biological activity and seabed sediment structure. The diagram shows some of the ways macrobenthos, in conjunction with other size classes, influence sediment fabric, physicochemical properties and solute fluxes (see also **Table 1**). (After Meadows PS (1986) Biologica activity and seabedsediment structure. *Nature* 323: 207.)

Functional Importance of Macrobenthos

Using a grab as a quantitative sampler in the early years of the twentieth century, C. J. J. Petersen hoped to be able to work out from his thousands of samples taken in the North Sea how much food was available to fish such as flounder. Such links were well supported from finds of large numbers of benthic animals in fish guts, even if later work showed that fish are by no means as important predators of macrobenthos as are invertebrates such as sea stars. Petersen also noticed that characteristic and uniform assemblages of macrobenthos were found that could be related to sediment type and that provided him with statistical units that Thorson later used as his descriptive units in his concept of 'parallel level bottom communities.' Ecologists debated whether these existed as anything more than assemblages responding to similar conditions (as originally inferred by Petersen) or reflected functional units or 'biocoenoses' where biological interactions play an important, if unknown, role. However, the importance of biological interactions in the subtidal community is difficult to address experimentally, and most data are available from intertidal mudflats and sandflats where ecological gradients related to tidal exposure pose additional complexity. Manipulative experiments on the effects of predation by caging small areas show that predators like shore crabs can have big effects on prey densities, while other studies show competitive exclusion between worms with different burrowing styles that can be reflected in clumping patterns. This contrasts with the importance of grazers and predators in preventing dominance by fast-growing competitive superior species on rocky shores. Such biological interaction cascades down through the community – so-called 'top-down' control. But, in sediments, effects such as predation are not so marked overall. Perhaps the three-dimensional structure, the uneven distribution of food and irrigatory flows, and the often intense stratification of chemical processes reduce competition. Indirect effects, such an bioturbation, may take the place of competition. By bulk processing large quantities of sediment, large macrofaunal deposit feeders rework the sediment down to the greatest ocean depths and thus exert a major influence on benthic community structure. It has been suggested that this constant process of biogenic disturbance and alteration of the benthic environment by macrofauna (**Figure 8, Table 1**) encourages high species richness among smaller macrobenthos by reducing them to levels where competition is relaxed, so that more species can coexist. It is also argued that the constantly changing micro-landscape created by other, larger species provides a rich niche variety for macrofauna. It is difficult to see this process operating on the vast abyssal plains where faunal densities, and therefore such biogenic effects, are so low but species richness is high. Grassle's spatiotemporal mosaic theory sees the deep-sea bed having patchy and ephemeral food resources that create a relatively small, discrete, and widely separated patch structure promoting coexistence.

Effects of larger-scale disturbances are more difficult to detect let alone manipulate in experiments with the sediment community. Yet the evidence is that physical disturbance such as that caused by storm-driven sediment scour and resuspension may have an important effect on assemblage structure and species richness on the exposed continental shelf and margin. The expectation that, just as on an exposed sandy shore, only a relatively small suite of species will be able to adapt to such conditions is confirmed in the deep sea on the continental rise off Nova Scotia, where benthic storms occur with relatively high frequency. Benthic storms may occasionally occur on the abyssal plains, so it should not be assumed that biogenic structure is simply longer-lasting there because it takes so long to be covered by the very low rate of natural sedimentation.

Perhaps the most important determinant of the macrobenthic assemblage, or community, is the larval stage usually dispersed in the water column. Larvae can test the substratum and swim off until they find conditions suitable for settlement and metamorphosis. On rock this may involve a series of precise cues that can include presence of their own or other species. Less is known about settlement of infaunal species, but it is thought that positive cues such as microtopography may be much less important in sediment dwellers, while negative cues such as the presence of other species or unattractive sediment are more important. Nevertheless, a community will still be very largely constrained by supply of propagules. In a coastal area the access of larvae supply from adjacent breeding populations may be constrained by coastal topography and currents, not to mention barriers formed by features such as estuaries. The patch structure in the deep sea is maintained by water-borne dispersal stages with the resulting metapopulations spatially unautocorrelated (presence of an organism not dependent on other occurrences). In deep water the openness of the system may mean that the sediment is exposed to a much larger pool of species, even if they are at very low densities as larvae. In the tropics a similar effect results from the greater incidence of planktotrophy (larvae feeding in the plankton), when the longer

Table 1 Direct and indirect effects of macrofauna on soft sediments and their ecological consequences

Direct effect	Indirect effects and ecological feedback
Bioresuspension: pelletization of superficial sediment as egesta changes sediment granulometry and increases openness of sediment fabric	Bioresuspension and formation of bottom turbidity layers through decrease in threshold where bed-shear stress will mobilize sediment Alteration in benthic community by reducing suspension feeders (trophic group 'amenalism') Reduced numbers of epifauna, and total community diversity Disruption and rapid obliteration of biogenic traces
Biodeposition: increased stickiness of sediment surface, and sediment structure altering hydrodynamic conditions to trap particles in pits, etc.	Biodeposition increases sediment stability and trapping of fine particles utilizable as food Encourages suspension feeders, and hence increases trophic diversity Results in tighter sediment fabric; increased sediment microstructure, and habitat complexity causing greater niche variety, and thereby increasing community diversity Better preservation of biogenic traces
Sediment irrigation	Spatial and temporal variability in exchange of dissolved gases, dissolved or absorbed ions and complex chemical species nutrients across sediment water interface Alteration in vertical gradients in Eh, pH etc. Transfer of reduced compounds from reducing conditions below the interface to aerobic conditions above Enhances recycling of dissolved nutrients Creates chemical heterogeneity from hotspots of elements concentrated by larger organisms in the sediment
Vertical sediment transport by larger deposit feeders	Particle selection may cause vertical shifts, granulometry Alteration of chemical microenvironment and diagenesis of sediment and smearing of stratigraphic signal
Formation of surface traces, by movement, burrowing or defecation	Increases bottom roughness, affecting near-bed flow Increased complexity in microenvironment creates niche mosaic at sediment surface

larval life will ensure wider dispersal than that of the nonfeeding larvae prevalent in cooler waters. This may help to explain why so many, mostly rare, species can coexist in both environments.

Importance of Macrobenthos in Environmental Assessment

Because the benthic community (unlike fish or plankton) is stationary or at best slow-moving over a small area of bottom, it is useful in monitoring environmental change caused by eutrophication and chemical contamination. Macrobenthos studies have defined the generic effects of such sources of stress by changing representation of major taxa, reduction in diversity, and increasing numerical dominance by small-sized opportunist species causing a downward shift in size structure. This seems to be accompanied by greater patchiness, reflected by increased variability in species abundances in sample replicates. It is also seen as greater variability in local species diversity caused by greater heterogeneity in species identities. This reflects subtle changes in abundance

and, particularly in the more species-rich communities, changes in presence/absence of rare species that might be detected earlier at less severe levels of disturbance. It is claimed that in bioassessments comparing species richness using samples of macrobenthos rare species should receive greater attention by taking larger samples because they contribute relatively more to diversity than the abundant community dominants. Other workers argue that very many species, especially rare ones, are interchangeable in the way they characterize samples. This question requires investigation of the way stressors impact the community, and whether it is the dominant or the rare species that are most sensitive, and therefore most rewarding for study in detecting impacts.

Interpretation of impacts also has to proceed against a background of natural changes in benthic communities caused by little-understood, year-to-year differences in annual recruitment. In establishing a baseline there is a need also to take into account the little-understood effects of bottom trawling on coastal benthos. Such disturbance in parts of the North Sea may date back at least 100 years, and now means that virtually every square meter of bottom is trawled over at least once a year. Such monitoring has in the past entailed costly benthic survey and tedious analysis of samples to species level. Consequently, there has been effort to see whether the effects of stress can be detected at higher taxonomic levels, such as families. Higher taxonomic levels may more closely reflect gradients in contamination than they do abundance of individual species because of the statistical noise generated from natural recruitment variability and from seasonal cycles such as reproduction. This hierarchical structure of macrobenthic response means that, as stress increases, the adaptability of first individual animals, then the species, and then genus, family, and so on, is exceeded so that the stress is manifest at progressively higher taxonomic level.

Such new approaches, along with the nascent awareness of conservation of the rich benthic diversity, and with a need for improved environmental impact assessment on the deep continental margin, should ensure a continued active scientific interest in macrobenthos in the years to come.

See also

Benthic Boundary Layer Effects. Benthic Foraminifera. Benthic Organisms Overview. Deepsea Fauna. Meiobenthos. Microphytobenthos. Phytobenthos.

Further Reading

Gage JD and Tyler PA (1991) *Deep-sea Biology: A Natural History of Organisms at The Deep-sea Floor.* Cambridge: Cambridge University Press.

Graf G and Rosenberg R (1997) Bioresuspension and biodeposition: a review. *Journal of Marine Systems* 11: 269–278.

Gray JS (1981) *The Ecology of Marine Sediments.* Cambridge: Cambridge University Press.

Hall SJ, Raffaelli D, and Thrush SF (1986) Patchiness and disturbance in shallow water benthic assemblages. In: Gee JHR and Giller PS (eds.) *Organization of Communities: Past and Present*, pp. 333–375. Oxford: Blackwell.

Hall SJ, Raffaelli D, and Thrush SF (1994) Patchiness and disturbances in shallow water benthic assemblages. In: Giller PS, Hildrew HG, and Raffaelli DG (eds.) *Aqautic Ecology: Scale, Patterns and Processes*, pp. 333–375. Oxford: Blackwell Scientific Publications.

Mare MF (1942) A study of a marine benthic community with special reference to the micro-organisms. *Journal of the Marine Biological Association of the United Kingdom* 25: 517–554.

McLusky DS and McIntyre AD (1988) Characteristics of the benthic fauna. In: Postma H and Zijlstra JJ (eds.) *Ecosystems of the World 27, Continental Shelves*, pp. 131–154. Amsterdam: Elsevier.

Pearson TH and Rosenberg R (1978) Macrobenthic succession in relation to organic enrichment and pollution of the marine environment. *Oceanography and Marine Biology: an Annual Review* 16: 229–311.

Pearson TH and Rosenberg R (1987) Feast and famine: structuring factors in marine benthic communities. In: Gee JHR and Giller PS (eds.) *Organization of Communities: Past and Present*, pp. 373–395. Oxford: Blackwell Scientific Publications.

Rex MA (1997) Large-scale patterns of species diversity in the deep-sea benthos. In: Ormond RFG, Gage JD, and Angel MV (eds.) *Marine Biodiversity: Patterns and Processes*, pp. 94–121. Cambridge: Cambridge University Press.

Rhoads DC (1974) Organism–sediment relations on the muddy sea floor. *Oceanography and Marine Biology Annual Reviews* 12: 263–300.

Thorson G (1957) Bottom communities (sublittoral or shallow shelf). In: Hedgepeth JW (ed.) *Treatise on Marine Ecology and Paleoecology*, pp. 461–534. New York: Geological Society of America.

Thrush S (1991) Spatial pattern in soft-bottom communities. *Trends in Ecology and Evolution* 6: 75–79.

DEEP-SEA FAUNA

P. V. R. Snelgrove, Memorial University of
Newfoundland, St John's, Newfoundland, Canada
J. F. Grassle, Rutgers University, New Brunswick,
New Jersey, USA

Overview

The deep sea covers more of the Earth's surface than
any other habitat, but because of its remoteness and
the difficulty in sampling such great depths, our
sampling coverage and understanding of the environment have been limited. There has been a
common misperception that the deep sea is species
poor, and a commonly used 'desert' analogy is hardly
surprising given that early sampling found few organisms, and the first deep-sea photographs revealed
large plains of rolling hills covered in sediment with
little obvious life (**Figure 1**). Indeed, all lines of evidence suggested that the deep sea is a very inhospitable environment. Temperatures are low
($\sim 4°C$), ambient pressure is extremely high (hundreds of times greater than on land), light is completely absent, and food is generally in very low
abundance. But within the last few decades, quantitative samples have revealed what primitive sampling
gear and photographs could not – that sediments in
the deep sea are teeming with a rich diversity of tiny
invertebrates only a few millimeters in size or smaller. These benthic (bottom-dwelling) organisms may
reside just above the bottom but closely associated
with it (hyperbenthos), on the sediment surface
(epifauna), or among the sediment grains (infauna).

The change in perception regarding the species
richness of the deep sea has continued to evolve; we
now know that, on the basis of the combination of
species richness per unit area and total size, the deep
sea is the most species-rich habitat in the oceans and
among the richest on earth. A similar change in
perception has occurred with two other generalizations about the deep sea. First, the deep sea is
generally thought of as a food-limited environment,
where biomass of individuals and communities as a
whole are extremely low. Although this generalization usually holds, the surprising discovery of
hydrothermal vent communities in the late 1970s,
with meter-long tube worms and biomass that rivaled even the most productive shallow-water areas,
proved that clear exceptions exist. A second generalization is that the deep sea has been considered to
be an extraordinarily stable habitat, where variables
such as salinity, temperature, and food supply are
constant, and light and photosynthesis are uniformly
absent. Again, this generalization holds in some respects, in that temperature and salinity are often invariant and light is indeed absent. Studies in the last
two decades, however, have indicated that small-
scale patchiness is common, seasonal variation in
phytoplankton production in surface waters can be
directly reflected in the material that reaches deep-
sea sediments, and some deep-sea areas are very
dynamic in terms of currents and sediment
movement.

The recent discoveries in the deep sea raise several
interesting questions. First, how can a seemingly inhospitable and physically homogeneous habitat such
as the deep sea support a rich diversity of organisms?
Second, how can hydrothermal vents support such a
high biomass of organisms relative to most deep-sea
environments? We now have a firm understanding of
the latter question and some definite ideas on the
former.

Defining the Habitats

Some deep-sea biologists define deep-sea habitats
somewhat arbitrarily as those greater than 1000 m in
depth. For this review, the deep sea is defined as all
benthic habitats beyond the edge of the continental
shelf, including the continental slope, continental
rise, abyssal plains, ocean ridges (including hydrothermal vents), and deep-ocean trenches. Thus,
ocean bottom from ~ 200 to $10\,000$ m falls within
this definition. Most of these regions share the features described above, including low temperature,
dependence on organic production 1000s of meters
above, high pressure and a sedimentary bottom, but
each has unique characteristics as well (**Figure 2**).
The continental slope, because it is adjacent to the
continental shelf, generally receives a higher level of
organic input than abyssal areas and, with its $\sim 3°$
slope, it is the steepest of the deep-sea environments
other than trenches and seamounts and is subsequently subject to occasional sediment slides
(called turbidity currents) that may move large volumes of sediment down the slope to the rise and
abyssal plains. The continental slope is largely covered in sediments, which often derive from terrestrial
and riverine runoff but may also come from marine

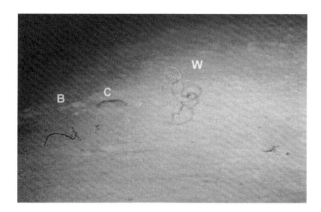

Figure 1 Photograph of a typical deep-sea landscape. This photo is from 750 m near St Croix, US Virgin Islands. Infaunal burrows (B), a sea cucumber (C) and a sea whip (W) are visible.

biological production. The continental rise occurs at the base of the slope and can exhibit elevated organic matter relative to lower slope areas because material moving down the slope may accumulate at the rise. By far, the abyssal plains cover the largest portion of the deep sea (∼40% of the Earth's surface), but they also represent the most benign of the deep-sea habitats. Because they are removed from land influence and very deep (≳4000 m), the amount of organic matter reaching the bottom is generally quite small, even in comparison with the slope and rise.

The deepest ocean habitats are trenches, which form in subduction areas and are characterized by relatively steep sides, poor circulation, and occasional mud slumping. The poor circulation and slumping make trenches particularly inhospitable to most organisms.

The sedimentary environment is vertically structured in terms of geochemistry and living organisms. Sediments have limited permeability and oxygen normally penetrates only a few millimeters by diffusion alone. Greater oxygen penetration occurs when bottom currents mix sediments or when organisms move pore water and sediments around (bioturbation). Sediments with active bioturbation are usually oxygenated within the top few centimeters, although strong bioturbation can lead to deeper pockets of penetration. Because light is absent from the deep sea, most productivity is provided by phytoplankton detritus and fecal pellets sinking from surface waters above, or closer to coastal habitats, from organic material transported seaward (e.g. kelps, seagrass etc.). Most of the available organic matter is concentrated near the sediment surface, although some species are capable of 'caching' food deeper in the sediment for later use. The combination of limited food and oxygen penetration at depth in sediments results in the vast majority of organisms being confined to the upper few centimeters of

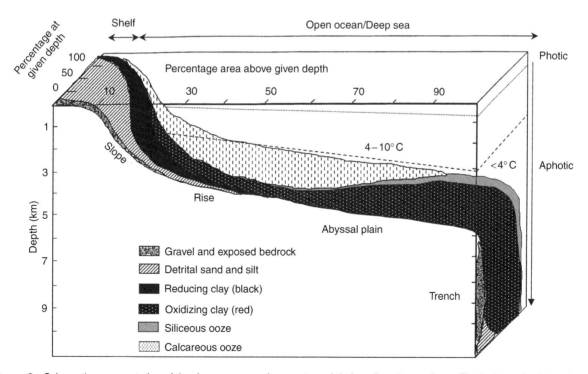

Figure 2 Schematic representation of the deep ocean environments and their sedimentary makeup. The horizontal axis has been greatly compressed, and the vertical axis is subsequently exaggerated. (Modified from Wright JE (ed.) (1977). *Introduction to the Oceans*. Milton Keynes, UK: The Open University.).

sediment near the sediment–water interface. Smaller organisms that can tolerate anoxia and larger organisms that maintain a burrow or appendage to the surface can live deeper, but even then distributions are usually only a few centimeters deeper.

Historically, the deep sea was perceived to be aseasonal because temperature is largely invariant at deep-sea depths, the absence of any light negates any day-length signal, and the habitat is so far removed from surface waters that it was thought that any signal from surface production would be completely dampened. Evidence in the last two decades has indicated that seasonality is a factor in many deep-sea environments. Samples from a number of different areas around the world and at a full range of depths have shown that the organic content of sediment does change seasonally. The strength of the seasonal signal, not surprisingly, varies with latitude and location; areas with very strong spring blooms are more likely to result in pulses of phytodetritus that sink to the seafloor than areas with weak production cycles. Where pulses are strong, the benthic fauna has been shown to respond quickly to organic input in terms of activity and biomass. Experimental patches of organic enrichment also generate a response by colonizing species.

The deep-sea floor lacks the large-scale physical heterogeneity of habitats such as forests and coral reefs, but it is nonetheless far from uniform. Biologically generated features such as burrows and feeding mounds create small-scale heterogeneity that persists for longer periods of time than in shallow water because physical redistribution of sediments by waves does not occur in the deep sea. As organic matter such as phytodetritus sinks to the seafloor and is carried horizontally by currents, small-scale bottom topography creates spatial variation in how that material settles. Depressions on the sea floor, for example, trap phytodetritus. Sessile species such as sea whips, glass sponges and protozoans called xenophyophores co-occur with mobile groups such as sea spiders and sea cucumbers, whose movements across the sediment can create tracks and topography. All of this small-scale heterogeneity acts in concert to create a mosaic of microhabitats for different organisms.

Seamounts

Seamounts, like volcanic islands, are mountains that are formed above the ocean floor near spreading centers, and they subsequently break up the landscape of abyssal plains. Because they are generally steep-sided, much of the substrate is volcanic rock, but sedimentary environments occur where sides are not steeply sloped or if the top of the seamount is flattened to form a guyot. Because seamounts can extend large distances above the bottom (thousands of meters), the fauna is often different from that found on the surrounding abyssal plains. The hard substrate, of course, supports very different species from sedimentary environments, and the relatively high flows that can occur over seamounts can support a higher proportion of organisms that feed on suspended particles than most deep-sea environments. In some cases the seamount may extend through the oxygen minimum layer of the Pacific Ocean; these low oxygen conditions favor a very different set of species than areas where oxygen is not limited.

Hydrothermal Vents

Hydrothermal vents represent a very specialized and unusual deep-sea environment, and prior to their discovery in 1977, the deep sea was thought to support very low densities of small invertebrates. There had been some debate until that time as to whether high pressure and low temperature constrain the size of benthic organisms, or whether deep-sea environments were simply food limited. The discovery of vents quickly answered that question, as researchers discovered dense concentrations of large organisms such as tube worms, clams, and crabs (**Figure 3**). Compared to the surrounding deep-sea environment, the vents supported extraordinary biomass of large organisms, and led to early analogies of 'oases in the desert'. From a biomass perspective this is an appropriate analogy, but from a species diversity perspective it is not.

Vents occur where tectonic spreading and subduction create fissures in the Earth's crust, allowing sea water to percolate through the crust and become heated by the mantle (**Figure 4**). When this water percolates out through the crust again, it is rich in minerals and reduced compounds such as hydrogen sulfide. Water temperature is extremely high (200–400 °C), and is prevented from boiling by the extreme pressure. It cools quickly, however, as it mixes with the ambient sea water that is typically ~4 °C. The mixture of sedimentary and hard substrate habitats that are characteristic of vent fields often support a large biomass of a very specialized fauna.

The key to the high productivity of hydrothermal vents are chemoautotrophic bacteria that live freely or form symbioses within specialized tube worms, clams, and mussels. These bacteria utilize hydrogen sulfide to synthesize organic compounds, which in

Figure 3   Photograph of tube worms (A), clams (B), and polychaete worms (C) from a hydrothermal vent.

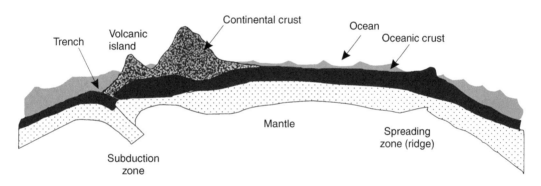

Figure 4   Hydrothermal vents occur where plate tectonic spreading and subduction occur. Water percolates through cracks that plate motion creates in the crust and is superheated in the mantle. The high temperatures cause chemical reactions, changing the chemistry of the sea water, and creating a fluid rich in hydrogen sulfide and various compounds.

the latter case may be passed on to the symbiont hosts. Not surprisingly, the hosts are characterized by reduced guts and little or no feeding structures. For both members of the symbiotic partnership, there are clear advantages. The host provides a physically stable habitat in the immediate proximity of hydrogen sulfide and the bacteria provide a rich food supply to the host. But relatively few species can utilize this symbiotic relationship. The extreme temperature gradients and toxic concentrations of hydrogen sulfide create a habitat that few organisms can tolerate. Moreover, the transient nature of vent environments, which generally persist for time scales of only decades, means that organisms must be able to colonize, grow quickly, and reproduce before vent flow ceases.

Defining the Organisms

Deep-sea biologists, like other benthic researchers, divide organisms based on size groupings. These groupings are not absolute in that the larval or juvenile stages of one group may be similar in size to adults from a smaller group, but this division is necessary because the sampling logistics that are appropriate for large organisms are inappropriate for small ones. Organisms that can be readily identified in bottom photographs, such as seastars and crabs, are commonly called megafauna (**Figure 5A–C**). Included here are characteristic deep-sea fish such as grenadiers, which cruise around near the bottom feeding on any falling carcass or disturbing the sediment and creating feeding pits as they feed on

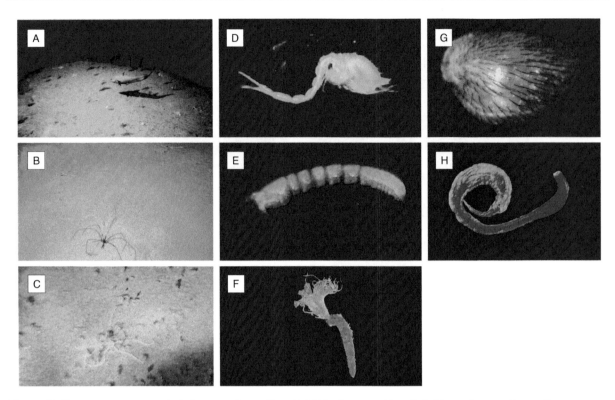

Figure 5 Deep-sea organisms, including megafauna (A, rattail fish; B, sea spider; C, brittle star), macrofauna (D, cumacean; E, tanaid; F, polychaete annelid) and meiofauna (G, ostracod; H, nematode). All taxa are from the northwest Atlantic except nematode from San Diego Trough (kindly provided by PJD Lambshead and D Thistle; G Hampson kindly provided photogrpahs D, E and G).

bottom invertebrates. Some of these taxa migrate up into the water column, providing an additional means by which energy may be cycled between the water column and the benthos. Macrofauna are organisms living on or in the sediments that are retained on a 300-µm sieve; this cutoff is in contrast to the coarser sieves used by shallow-water benthic researchers because the smaller size of deep-sea animals requires a finer sieve. This size grouping includes polychaete annelids, crustaceans, bivalves and many other phyla (**Figure 5D–F**). Meiofauna are organisms between 40 and 300 µm, and include nematodes, foraminiferans, tiny crustaceans, and many others (**Figure 5G–H**). Microorganisms pass through a 40-µm sieve and include bacteria and protists. Within any one of these groups, the taxonomic challenges are considerable in that few individuals are trained in taxonomy of deep-sea organisms and many species have yet to be described (see below). Not surprisingly, syntheses across groups are therefore rare in a single study.

Feeding modes in deep-sea sediments include omnivores, predators, scavengers and parasites. Most species are deposit feeders that ingest sediment grains and the organic particles and bacteria associated with them; through their feeding activity, these organisms are particularly important bioturbators.

In some areas, suspension feeders filter particles out of the water column above the bottom, but because they rely on suspended particles they are most abundant in energetic environments such as seamounts.

As bottom depth increases, both biomass and densities of organisms decrease (**Figure 6**). Because food resources become scarcer and scarcer as distance from surface water and primary producers increases, this pattern is not unexpected. What is less intuitive, however, is that as food becomes more limiting and densities of organisms decrease, species diversity does not decline.

Sampling the Fauna

Early efforts to sample the deep ocean floor used crude trawls towed from surface ships that were ineffective, and undoubtedly contributed to the idea that the environment is species poor. Modern trawls are now used in deep-sea fisheries, and some of these are effective for sampling megafauna living above the sediment or on bedrock. For smaller organisms, Howard Sanders and Robert Hessler, in their important work in the 1960s, used a semiquantitative device called an epibenthic sled comprising a metal frame surrounding a mesh bag. The sled is lowered

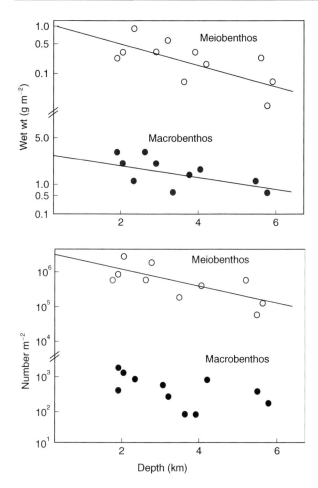

Figure 6 Based on Shirayama Y (1983) Size structure of deep-sea meio- and macrobenthos in the western Pacific. *International Revue der gesamten Hydrobiologie* 68: 799–810. Upper panel shows how densities of organisms decline with depth and lower panel shows a similar pattern for biomass.

to the bottom and towed behind a ship, where it skims off the surface sediment and associated fauna (**Figure 7C**). For hyperbenthos, this is still an instrument of choice, but for infauna it is only semi-quantitative and has been replaced by quantitative samplers.

In the late 1960s, biologists began using a large metal device called a spade or box corer that is commonly used today to sample macrofaunal organisms. Box corers range in size but typically sample $\sim 0.25\,\mathrm{m} \times 0.25\,\mathrm{m}$ of ocean bottom. A metal frame surrounds a metal box that is lowered into the sediment from a surface ship (**Figure 7A**). After the box, which is often subdivided by a grid of metal subcores, enters the sediment, the spade swings down and slices beneath the box, locks in place, and seals the sediment within the corer for the return trip to the surface ship. Although the box corer is effective for sampling macrofauna, it must be handled carefully to avoid a bow wave as it approaches the

sediment. For meiofauna and microbes living right at the sediment surface, even a slight bow wave is a problem because it can blow away the lightest sediments and organisms at the sediment–water interface. To circumvent this problem, a device called a multicorer was developed by the Scottish Marine Biological Association (now the Scottish Association of Marine Sciences). With this sampler, a frame is gently lowered onto the seafloor from a surface ship and individual acrylic cores slowly enter the sediment. Individual cores are typically $\sim 6\,\mathrm{cm}$ in diameter and a corer can have 4–12 individual cores or more.

Sampling gear that is deployed from surface ships has a distinct disadvantage; although the samples are quantitative, they are largely collected blindly, meaning that there is no frame of reference for the area where the sample was collected. Thus, the corer could land on or near some anomalous feature on the seafloor, and the investigating scientist could have trouble interpreting why the fauna was unusual. The development of research submersibles, such as ALVIN (**Figure 7B**) or the Johnson Sealink (**Figure 7D**) allows scientists to actually visit deep-sea environments and watch as their samples are collected at precisely the locations they request. To achieve this, a device called an ALVIN box corer, which is effectively a miniaturized box corer, has been developed. The manipulator arm of the submersible pushes the corer (typically $15\,\mathrm{cm} \times 15\,\mathrm{cm}$) into the sediment and then trips the doors that seal-in the sample, much as the spade does for the larger box corer. On board ship, individual subcores are processed over a sieve and then preserved in buffered 4% formaldehyde. Samples are kept in formaldehyde for at least 48 hours and then transferred to 70% ethanol.

Hard substrate environments present a different challenge in terms of quantitative sampling. Quantitative removal of hard substrate fauna is near impossible except where the substratum itself may be removed (e.g. manganese nodules). For these environments, visually based surveys, achieved through submersibles, remotely operated vehicles (ROVs), or towed cameras provide the most common means of evaluating fauna. These same approaches are also used for some megafaunal studies in sedimentary environments.

The instruments described above are the bases for evaluating faunal abundance, but studies of deep-sea fauna do not always focus on species composition. Physiologists have developed special respiration chambers to study metabolic processes, and ecologists have developed baited traps, colonization trays, and settlement tiles to study species response to

Figure 7 (A) The box corer is one of the basic sampling tools for deep-sea sediments. The corer is lowered from a surface vessel on a wire, and when the box penetrates into the sediment, the spade swings down and slices through the sediment, sealing the sample in the box. Fred Grassle is shown directing the deployment of the corer. (B) ALVIN was one of the first submersibles to be used for deep-sea research and played a vital role in the early characterization of hydrothermal vents. (C) An epibenthic sled, developed for deep-sea sampling in the 1960s, is towed so that it skims along the seafloor and samples sediments and near-bottom fauna. (D) The Johnson Sealink is a submersible that allows scientists to descend, with a pilot, to the deep-sea floor. ALVIN corers and colonization trays, seen on the front of the submersible, were to be deployed on this dive along with other gear.

resource availability. Most of these instruments are most effectively deployed by submersibles but other approaches have also been used, including free vehicles that are dropped to the bottom from surface ships, and later float to the surface when a release mechanism is triggered.

Patterns of Diversity

Although a few scientists before the turn of the century recognized that the deep sea was indeed a species-rich environment, general recognition of this fact did not occur until a series of papers were published by Howard Sanders and Robert Hessler. They collected a series of samples using an epibenthic sled; although this sampling approach is now known to significantly undersample, it represented a marked

improvement over previous gear. Comparison of the data with data from other environments (**Figure 8**) indicated that the deep sea was among the most diverse of marine sedimentary habitats. From data collected along a transect running from Martha's Vineyard, Massachusetts to Bermuda, they demonstrated that diversity in deep-sea sediments exceeded that in most shallow areas and rivaled that observed in shallow tropical areas. At the time, this finding represented a startling contrast to current thinking, and even now the desert analogy still persists in some textbooks. More recent work using a box corer (**Figure 7A**), has shown that the deep sea is not only species rich, but it may also rival tropical rain forests in terms of total species present.

Several studies have looked at broad scale pattern in the deep sea, and found that diversity is not uniform with depth, location, or latitude. Michael Rex analyzed patterns with depth in gastropods and other taxa, and found a parabolic pattern, with a peak in diversity between 2000 and 3000 m on the continental slope. Diversity in shallow-water sedimentary habitats, such as in estuaries and tidal flats, is relatively low, then increases somewhat on the continental shelf, peaks along the mid to lower slope and then declines to abyssal plains. A similar pattern has been noted by other researchers, although the slope depth of the diversity peak varies among studies. There have been exceptions noted to this pattern. The abyssal Pacific, for example, has higher diversity than shallower areas, and work in Australia suggests that some coastal environments are extremely diverse, perhaps even exceeding that observed in the deep sea.

Studies have examined latitudinal patterns and suggest that diversity in the deep sea may decrease with latitude, at least in the North Atlantic. This work has focused on North Atlantic macrofauna, but other studies on meiofaunal nematodes in the Atlantic and isopod crustaceans from the South Pacific do not support such a trend.

How Many Species Are There?

At higher taxonomic levels, the marine environment is inarguably more diverse than any other habitat on Earth, and most of the phyla and classes of organisms that are unique to the marine environment occur in sedimentary environments. Some 90% of all animal families and 28 of 29 nonsymbiont animal phyla occur in marine environments, and of these 29, 13 occur only in marine habitats. Only one animal phylum, the Onychophora, has no living representatives in marine habitats, but even then there are marine fossil forms now known.

A number of estimates have been made regarding the possible number of species in the oceans (**Figure 9**) and considerable debate has arisen from these estimates; much of this controversy arises from the general acceptance that the deep sea has many undescribed species but disagreement over how many. Of particular uncertainty is the validity of assumptions that have gone into various estimates. One approach has been to survey taxonomic specialists who work on different groups, and determine what proportion of their taxon remains undescribed (left panel in **Figure 9**). The famous tropical rainforest estimate by Terry Erwin (central panel in **Figure 9**) of 10 million insect species was generated by looking at numbers of beetle species associated with a given species of rainforest tree, estimating what portion of

insects comprises beetles, and then multiplying by estimated numbers of tropical tree species. Based on the rate at which species were added with increased area sampled along a 176-km long depth contour off the eastern United States, Grassle and Maciolek extrapolated to the total area of the deep sea and estimated that there are 10 million deep-sea macrofaunal species. Robert May suggested that because about half of Grassle and Maciolek's species were previously undescribed, one could extrapolate from the presently described 250 000 marine species to arrive at a projection of ~ 500 000 total species. Data from the Pacific suggests that only 1 in 20 species has been described; using this estimate, May's approach would yield ~ 5 million species. John Lambshead, a nematode ecologist, has noted that meiofaunal nematodes are more abundant and species rich in individual core samples than macrofauna, and his estimate for total nematodes in the deep sea is 100 million species. At present we know little about how widely distributed nematode species may be, making extrapolation even more tenuous than for macrofauna.

Species richness comparisons across different environments are complex. Within any given environment, estimating the total numbers of species present is difficult because it is impossible to fully sample the environment. The deep sea is particularly problematic. It has been estimated that of the ~ 3.25×10^8 km^2 of seafloor that is part of the deep sea, only about 2 km^2 has been sampled for macrofauna and 5 m^2 has been sampled for meiofauna. Sampling coverage for microbes is poorer still. In addition to the huge area involved, ship time and sample processing are expensive, and relatively few taxonomic specialists know the deep-sea fauna.

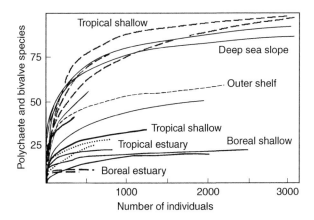

Figure 8 Comparison of deep-sea communities to other marine sedimentary communities based on plotting numbers of species versus numbers of individuals. (From Sanders HL (1969) Marine benthic diversity and the stability-time hypothesis. *Brookhaven Symposium on Biology* 22: 71–80.)

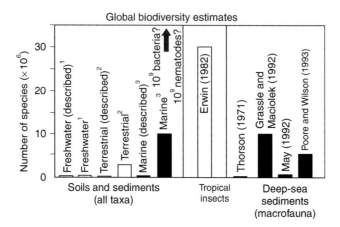

Figure 9 Various estimates of total species richness in marine sediments in comparison with estimates for other environments. Numbered sources in the left panel are: [1]Palmer MA, Covich AP, Finlay BJ *et al.* (1997) Biodiversity and ecosystem processes in freshwater sediments. *Ambio* 26: 571–577; [2]Brussaard L, Behan-Pelletier VM, Bignell DE *et al.* (1997) Biodiversity and ecosystem functioning in soil. *Ambio* 26: 563–570; [3]Snelgrove PVR, Blackburn TH, Hutchings PA *et al.* (1997). The importance of marine sedimentary biodiversity in ecosystem processes. *Ambio* 26: 578–583. For each of these sources, the numbers of described and projected species are given. The arrow in Snelgrove *et al.* indicates estimates if bacteria (10^{12}?) or nematodes (10^9?) are included; data on which estimates for these groups are based are very limited. Values in other panels are projected numbers of species. Numbers for marine systems are solid bars. Freshwater and terrestrial refer to species number for all global components of those environments pooled, although estimates for bacteria are not included in these numbers. Additional data sources are Thorson G (1971). *Life in the Sea.* New York; McGraw-Hill and Erwin TL (1982) Tropical forests: their richness in Coleoptera and other Arthropod species. *Coleopterists' Bulletin* 36: 74–75.

Thus, current conclusions on pattern and numbers are based on very limited spatial coverage.

Most of the data on species number and pattern has been based on data from the North Atlantic. Clearly the variability in patterns described above suggests that a wider database is needed to effectively test the generality of these concepts. Areas such as the southern oceans that have been poorly sampled in the past offer key pieces to the puzzle of deep-sea pattern. Another shortcoming with large-scale comparisons is that most focus on just one of the size groupings of organisms. Thus, between-group differences in pattern are difficult to attribute to differences in areas sampled or to real differences between the groups. In short, we need deep-sea studies that are broader in geographic coverage and taxonomic coverage before these questions can be resolved definitively.

Low Diversity Environments

Not all deep-sea communities are species rich. As described earlier, the hydrogen sulfide and heavy metals emitted at vents are toxic to most species, and species diversity is quite low. Moreover, the short life span of most individual vents makes them among the most unpredictable environments in the deep sea. Like the hard substrate environment around vents, the sediments that occur in hydrothermal vent areas

are inhospitable because of high concentrations of metals, sulfide, and hydrocarbons. There are, nonetheless, often mats of bacteria over these sediments that utilize the hydrogen sulfide as an energy source. Not surprisingly, an increase in hydrothermal flux can quickly 'cook' the bacteria, whereas a decrease in flux can starve them to death.

A number of other specialized deep-sea environments are species depauperate. Deep-sea trenches are subject to mud slumping and poor circulation as a result of the steep trench sides, and species diversity is very low. Sediments beneath upwelling regions and other highly productive areas, such as the upper slope off Cape Hatteras, are also generally low in diversity because large amounts of organic matter accumulate on the ocean floor and decompose, resulting in hypoxic (low oxygen) conditions that few organisms can tolerate. A few deep-sea areas, such as the 'HEBBLE' site on the continental slope off Nova Scotia, are subject to intensive 'storms', where currents become intense and sediment resuspension occurs. Evidence suggests that macrofaunal diversity is depressed in such areas, but surprisingly meiofaunal diversity is not. Presumably, the meiofauna are able to cope with the disturbance more effectively than the macrofauna.

Although ecological processes are undoubtedly important in the deep sea, evolutionary time scales and processes are also important. Defaunation of some deep-sea areas such as the Norwegian Sea over recent geological time scales is thought to have contributed to low diversity. Glaciation defaunated the area by blocking sunlight and reducing circulation, and the shallow sills that surround the basin have likely resulted in very slow reestablishment of deep-sea communities from adjacent basins.

Theories

One of the driving questions in deep-sea ecology since the 1960s is how an environment that appears so physically homogeneous is able to support a species-rich fauna. When Sanders documented the high diversity of deep-sea systems in the late 1960s, he proposed the stability–time hypothesis, in which the high level of stability afforded by the deep sea over evolutionary time has resulted in greater specialization and niche diversification in deep-sea fauna. Certainly the relative stability of deep-sea environments has contributed to the numbers of species present, but if stability were the lone explanation then it is inconsistent with observations of lower diversity on abyssal plains than on adjacent slope habitats. An additional problem is that most species are thought to be relatively nonselective deposit feeders, which is inconsistent with niche specialization.

In the early 1970s, several alternative theories were proposed. Predators could prevent competitive equilibria from being attained by infauna by cropping back individuals. But most deep-sea predators appear to be nonselective, and infauna are characterized by slow growth, late reproductive maturity and dominance by older age classes; none of these characteristics would be expected in a predator-controlled system. Evidence from shallow-water sediments suggests that at small scales at least, predators decrease sedimentary diversity, but the role that predators may play in maintaining deep-sea diversity remains largely unanswered at this point.

The increasing evidence for small-scale spatial and temporal heterogeneity in deep-sea systems has led to speculation that small-scale patches may be important. The patch mosaic model proposes that small patches create disequilibria habitats that promote different species and thus promote coexistence. Studies have tested the patch mosaic model by sampling natural patches or creating experimental patches; both types of study have found that species that occur in most patch types are usually rare or absent from nonpatch sediments. This pattern is consistent with the patch mosaic model, but experiments so far have demonstrated that patches promote only a modest number of species and the overall species richness in a given patch tends to be lower than in nonpatch areas. It is unclear whether we need to sample more patch types or invoke an altogether different explanation for high diversity.

Island biogeography has shown that larger areas tend to support more species, and it has been argued that the large area of the deep sea may be the primary reason for its species richness. To some extent this areal relationship must be a contributing factor, but if area were the only issue then abyssal plains would consistently exceed all other habitats in diversity. Moreover, the deep sea does not appear to add habitat heterogeneity with area as most habitats do; vast areas of sedimentary bottom may sometimes exhibit changes in sediment composition, but even compared to shallow areas the habitat heterogeneity is quite small.

Intermediate disturbance has been touted in a number of ecological systems as being important for high diversity. High levels of disturbance can eliminate sensitive species and invariant habitats may allow superior competitors to outcompete weaker species; both scenarios result in reduced diversity. Intermediate disturbance prevents competitive dominants from taking over but is not severe enough to eliminate sensitive species, resulting in high diversity. The strongest evidence for this hypothesis in the deep sea is the mid-slope peak in diversity described earlier, and the reduced diversity observed in disturbed deep-sea habitats such as hydrothermal vents, low oxygen areas, environments with benthic storms, and slumping areas.

In summary, we still lack a definitive explanation for which factors are most important in promoting diversity in deep-sea ecosystems. In all likelihood, no single explanation is correct and multiple factors will prove to be important. Efforts are currently underway to try to clarify this question using experimental approaches such as predator exclusion experiments and creation of artificial food patches, along with analytical approaches that analyze pattern with respect to environmental variables. As the available data increase, many of the questions about pattern and cause may become clearer, but there is considerable work to be done.

Threats and Benefits

The deep-sea environment has attributes that render it vulnerable to human disturbance, but impacts have nonetheless been modest compared to most marine habitats because of the distance from land, large size, and great depth. Pollutants and nutrients that have created severe problems in coastal areas via land runoff, river runoff or aerosol transport, are usually sufficiently diluted by the time they reach the open ocean that impacts are modest. There is biochemical evidence, however, that pollutants may occur in deep-sea organisms at low concentrations. Fishing, and the habitat destruction it causes, is less wide-spread in the deep sea than in shallow water because it is expensive and time consuming to fish at great depths. Moreover, the densities of organisms that are present are often insufficient to support commercial fishing. Having said that, there are a number of deep-sea fisheries, many of which utilize trawls and dredges that damage the integrity of the benthic habitat, injure organisms, and remove many non-target species as by-catch. The vulnerability of deep-sea species to human activities is well exemplified by fisheries such as that for the Australian orange roughy. Like many deep-sea fisheries, fishing effort has outpaced the capacity of the population to re-cover, raising the distinct possibility that a sustain-able deep-sea fishery may represent an oxymoron. This same vulnerability presumably applies to non-target species that are removed as by-catch or injured by fishing gear.

A third human impact is through waste disposal. Materials ranging from sewage sludge to radioactive waste have been dumped in the deep ocean, largely justified on the basis that the currents are weak so containment is more likely, food chains are far re-moved from most human harvesting activities, and the large area of the deep sea reduces the likelihood of a major impact. There is also a feeling among some that even if an impact occurs, the organisms that would be affected have little or no obvious economic value to humans. A final potential threat to deep-sea com-munities is deep-sea mining. There has been some interest in the mining of manganese nodules, small softball-sized nodules that are rich in manganese, nickel and other metals. These nodules occur on the seabed in abyssal plain areas of the oceans, but because of the depths involved, deep-sea mining is not com-mercially viable at present. Metals such as manganese are, however, of great strategic importance because many countries presently rely on foreign suppliers. Thus, there is a very real chance that deep-sea mining may someday occur. Different mining strategies will, of course, have different types of impact but habitat de-struction is likely to be the greatest problem.

Because many deep-sea organisms grow very slowly, reproduce at an older age than their shallow-water counterparts, and produce very few offspring per individual, they are thought to be extremely vulnerable to disturbance and habitat damage. In the past it has been assumed, however, that many deep-sea species are broadly distributed because there are few barriers to dispersal and little obvious habitat heterogeneity. If this assumption holds, then their vulnerability to extinction might be reduced. At present, our understanding of how quickly species turn over spatially is very limited, particularly for some of the groups like nematodes that have been least studied. With emerging molecular approaches, it is also becoming clear that species that have been treated as cosmopolitan may, in some instances, be species complexes.

Given that many deep-sea environments support species that are either of no commercial fishing interest or are not sustainable, is there any reason to exercise caution in how humans impact the deep sea? There are, in fact, several compelling reasons to be concerned. First, the deep sea represents one of the few remaining pristine habitats on Earth. We can say, with only a few exceptions, that deep-sea com-munities have not been compromised by human de-velopment. This attribute makes them one of the last natural laboratories on Earth where the 'chemicals' have not been tainted. Because it is so very diverse, the deep sea can provide a natural and uncompro-mised laboratory in which to test ideas on regulation of biodiversity. A second reason to exercise caution is that the deep sea may represent one of the largest species pools on Earth. From an ethical and esthetic perspective, it could be argued that this characteristic alone is sufficient motivation to limit human dis-turbance. But from an economic perspective, there is great interest among pharmaceutical companies in organisms with unusual physiologies; the thermo-philic bacteria that live at hydrothermal vents, for example, have generated tremendous interest for their bioactive compounds. A third concern is with respect to remediation. Although material dumped in the deep sea may be out of sight and mind, any de-cision at a later time to remediate (e.g. leaking radioactive waste) would be prohibitively expensive, if it was possible at all.

In summary, the deep sea is a vast and relatively undisturbed habitat that may be very vulnerable to human disturbances. Our current understanding of the deep sea and its immense diversity is very limited, but is nonetheless advancing steadily. A precaution-ary approach will ensure that the unusual attributes of the deep sea, including its rich biodiversity, will not be inadvertently destroyed by ignorance.

See also

Benthic Boundary Layer Effects. Benthic Foraminifera. Benthic Organisms Overview. Macrobenthos. Meiobenthos.

Further Reading

Gage JD and Tyler PA (1991) *Deep-Sea Biology. A Natural History of Organisms at the Deep-Sea Floor.* Cambridge: Cambridge University Press.

Grassle JF and Maciolek NJ (1992) Deep-sea species richness: regional and local diversity estimates from quantitative bottom samples. *American Naturalist* 139: 313–341.

Gray J, Poore G, and Ugland K (1992) Coastal and deep-sea benthic diversities compared. *Marine Ecology Progress Series* 159: 97–103.

Lambshead PJD (1993) Recent developments in marine benthic biodiversity research. *Oceanis* 19: 5–24.

May R (1992) Bottoms up for the oceans. *Nature* 357: 278–279.

Merrett NR and Haedrich RL (1997) *Deep-sea demersal fish and fisheries.* London: Chapman Hall.

Poore GBC and Wilson GDF (1993) Marine species richness. *Nature* 361: 597–598.

Rex MA (1983) Geographic patterns of species diversity in the deep-sea benthos. In: Rowe GT (ed.) *The Sea: Deep-Sea Biology*, vol. 8, pp. 453–472. New York: John Wiley.

Rex MA, Stuart CT, and Hessler RR (1993) Global-scale latitudinal patterns of species diversity in the deep-sea benthos. *Nature* 365: 636–639.

Sanders HL and Hessler RR (1969) Ecology of the deep-sea benthos. *Science* 163: 1419–1424.

COLD-WATER CORAL REEFS

J. M. Roberts, Scottish Association for Marine Science, Oban, UK

Introduction and Historical Background

Corals and coral reefs are not restricted to shallow, tropical waters. Deep-ocean exploration around the world is now revealing coral ecosystems at great depths in the cooler waters of the continental shelf, slope, and seamounts. Here, in permanent darkness and without the algal symbionts (zooxanthellae) of many tropical species, cold-water corals grow to form true deep-water scleractinian reefs or 'forests' of flexible gorgonian, black, gold, and bamboo corals.

Such corals have been known since the eighteenth century; the Reverend Pontoppidan, Bishop of Bergen, discussed corals as 'sea-vegetables' in his 1755 book *The Natural History of Norway* and Linnaeus subsequently described several cold-water coral species. The following century, the British naturalist Philip Henry Gosse summarized British corals in his 1860 book *Actinologica Britannica. A History of the British Sea-Anemones and Corals* (**Figure 1**). Further records and samples were obtained during the pioneering nineteenth century expeditions of HMS *Porcupine* (1869, 1870) and HMS *Challenger* (1872–76), and in the first half of the twentieth century scientific dredging studies by Dons and Le Danois identified sizeable coral banks off the Norwegian and Celtic margins, respectively. However, until relatively recently, cold-water coral banks remained best known to fishermen, especially trawlermen, who risked damaging their nets and marked coral areas on fishing charts. In the latter half of the twentieth century, advances in acoustic survey techniques and research submersibles allowed the first mapping and direct observations of coral colonies in deep water. Using the early research submersible *Pisces*, Wilson described cold-water coral patch development on the Rockall Bank west of the UK and was among the first to show the value of video surveys to document and understand these structurally complex habitats (**Figure 2**). In the last 20 years, there have been further advances in deep-ocean exploration, notably the development of multibeam echo sounders, remotely operated vehicles, and, most recently, autonomous underwater vehicles that are now beginning to reveal the true extent of cold-water coral ecosystems (**Figure 3**).

Cold-Water Corals

Cold-water corals are all members of the phylum Cnidaria and include species belonging to a number of lower taxonomic groups. Among these the colonial scleractinian or stony corals can develop sizable reef frameworks and are the focus of this article. *Lophelia pertusa* dominates reef frameworks in the Northeast Atlantic where *Madrepora oculata* is an important secondary species. *L. pertusa* is also abundant on the other side of the Atlantic Ocean in the US South Atlantic Bight where other framework corals

Figure 1 Color plate from Gosse's 1860 book *Actinologica Britannica*, showing a colony of *Lophohelia prolifera* (*Lophelia pertusa*) together with cup corals and zoanthids.

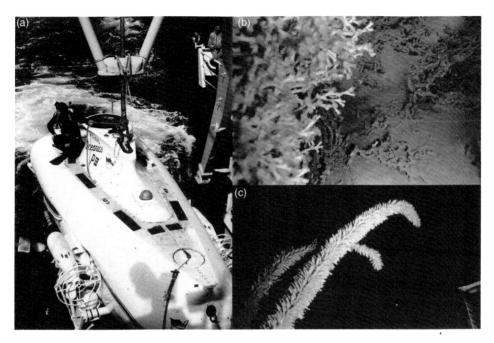

Figure 2 *Lophelia pertusa* reef patches on Rockall Bank in the Northeast Atlantic were among the first cold-water coral habitats observed from a manned submersible. (a) The *Pisces III* submersible in 1973. (b) Live coral framework and surrounding rubble. (c) Large antipatharian coral colony, possibly *Parantipathes hirondelle*. Images courtesy of Dr. John Wilson. (b) Reproduced from Wilson JB (1979) 'Patch' development of the deep-water coral *Lophelia pertusa* (L.) on Rockall Bank. *Journal of the Marine Biological Association of the UK* 59: 165–177, with permission of the Marine Biological Association of the UK.

include *Enallopsammia profunda*, *M. oculata*, and *Solenosmilia variabilis*. Along the eastern Florida shelf, the facultatively zooxanthellate coral *Oculina varicosa* forms banks up to 35 m in height at depths of 70–100 m. *S. variabilis* is also reported forming tightly branched frameworks on the Little Bahama Bank and Reykjanes Ridge in the Atlantic and on Tasmanian seamounts in the South Pacific. *Goniocorella dumosa* is only reported from the Southern Hemisphere, where it forms reef frameworks around New Zealand on the Chatham Plateau. Thus, there are just six cold-water scleractinian coral species currently known to form significant reef frameworks in deep water, compared to more than 800 species of shallow reef-building tropical corals. While the scleractinian reef framework-forming species form the basis of this article, other cold-water corals, notably gorgonians, antipatharians, and hydrocorals, can develop dense assemblages that also provide significant structural habitat for other species.

Reef Distribution and Development

The robust anastomosing skeletons of colonial cold-water scleractinians produce dense frameworks that over time can develop structures with significant topographic expression from the seafloor that alter local sedimentary conditions, are subject to the dynamic process of (bio)erosion, and provide habitat to many other species (**Figure 4**). By these criteria, cold-water scleractinian corals form reefs and these reefs can persist for tens of thousands of years.

Cold-water coral reef distribution is controlled by a suite of environmental factors. They are largely restricted to water masses with temperatures of 4–12 °C and salinities of 35 psu. Although they are often generically referred to as deep-sea corals, their wide depth distribution, from shallow fiordic sills at just 40 m to shelf and slope depths of 200–1000 m where the majority of cold-water coral reefs are found, reflects bathyal environments (200–2000 m) and not the abyssal depths (2000–6000 m) more often associated with the term 'deep sea'. On a global scale, the importance of seawater carbonate chemistry as a control on cold-water coral occurrence is now becoming apparent. Scleractinian corals secrete calcium carbonate skeletons in the form of aragonite. The boundary between saturated and undersaturated seawater, the aragonite saturation horizon (ASH), is relatively deep (>2000 m) in the Northeast Atlantic and relatively shallow (50–500 m) in the North Pacific. As shown in **Figure 5**, there are many records of reef framework-forming cold-water corals from the Northeast Atlantic and very few from the North Pacific where coral assemblages are dominated by octocorals and

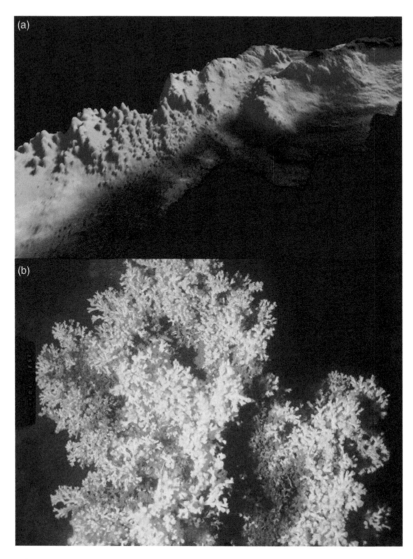

Figure 3 Multibeam echosounders are valuable tools for wide area mapping in deep waters. (a) Three-dimensional bathymetry, exaggerated sixfold in the vertical, showing the many seabed mounds formed by *Lophelia pertusa* reefs of the Mingulay Reef Complex, Northeast Atlantic. (b) Photograph of a *L. pertusa* colony from these mounds. Reproduced from Roberts JM, Brown CJ, Long D, and Bates CR (2005) Acoustic mapping using a multibeam echosounder reveals cold-water coral reefs and sourrounding habitats. *Coral Reefs* 24: 654–669.

hydrocorals that do not form robust aragonitic skeletons. Indeed, the scleractinians beneath the ASH from the Aleutian Islands in the North Pacific are dominated by species adapted to low-calcium-carbonate environments and the majority of the hydrocorals found there form calcitic rather than aragonitic coralla. Compellingly, calcite is *c.* 50% less soluble in seawater than aragonite.

In the 1990s, the hydraulic theory was advanced proposing that cold-water coral reefs, notably the well-developed *Lophelia* reefs of the Norwegian continental margin, were coupled to the geosphere via the seepage of light hydrocarbons. This intriguing idea led to a large research effort searching for evidence that the corals and associated fauna were reliant on local seepage. Despite proximity to seeps in certain areas, stable isotope analyses of coral tissue reflect material derived from surface productivity and recent ocean drilling through a coral carbonate mound in the Porcupine Seabight failed to find any evidence of gas accumulation or that mound growth had been initiated by local hydrocarbon seepage.

Thus while broad trends in the factors controlling cold-water coral reef distribution are becoming apparent, our understanding of their global distribution remains biased to those parts of the world, notably the Northeast and Northwest Atlantic, where most surveys have been carried out. Predictive modeling approaches suggest that suitable conditions for cold-water coral reef development may be found in areas of the continental slope that have not yet been surveyed and mapped in sufficient detail to test for their

Figure 4 Cold-water coral reefs form highly complex three-dimensional structural habitat. (a) The sloping flank of a giant coral carbonate mound in the Porcupine Seabight, Northeast Atlantic. (b) Dense scleractinian coral framework (*Lophelia pertusa* and *Madrepora oculata*) with purple octocorals (probably *Anthothelia grandiflora*) and glass sponges (probably *Aphrocallistes*). (c) Live polyps of *L. pertusa*. Images courtesy of *VICTOR-Polarstern* cruise ARKXIX/3a, Alfred-Wegener-Institut für Polar- und Meeresforschung and the Institut Français de Recherche pour l'Exploitation de la Mer.

occurrence. Similarly, only a small proportion of the tens to hundreds of thousands of seamounts that have been estimated to exist have ever been surveyed. Many of the seamounts that have been examined reveal abundant cold-water corals like the *S. variabilis* reefs on Tasmanian seamounts in the South Pacific or the assemblages of gorgonian, black, and bamboo corals on the Davidson Seamount off Monterey in the East Pacific.

Feeding, Growth, and Reproduction

Cold-water corals are typically reported from areas with locally accelerated currents or from regions offshore where internal tidal waves impinge on the slope and break, thus enhancing mixing and flux of food material from the surface to the seabed. Both productive surface waters and hydrographic conditions that transport this material to the benthos are needed to support cold-water coral growth. The reef framework-forming corals *L. pertusa* and *M. oculata* seem able to use both zoo- and phytoplankton and, as with other cnidarians, are likely to feed from a cosmopolitan mix of zooplankton prey, detritus, and dissolved organic matter.

Understanding cold-water coral growth has largely been limited to observations derived in two ways from dead coral skeletons. First, coral colonies found on man-made structures can be measured and used to estimate approximate extension rates based on the age of the man-made structure. Second, cycles in the stable isotopes of carbon and oxygen in the coral's skeleton can be used to infer annual extension rates, although recent work has shown this method is complicated by poor understanding of skeletal banding patterns. Using these approaches to study *L. pertusa*, annual extension rates of between 0.5 and 3 cm have been derived, with the faster growth rates associated with shallower colonies growing on North Sea oil platforms where enhanced food availability and competition for space may combine to accelerate linear extension rates. To date there have been no detailed reports on cold-water coral calcification rate or mode – a worrying deficiency in a time of predicted climate change and ocean acidification, discussed below.

While we lack detailed understanding of cold-water coral growth and calcification, some promising initial progress in unraveling the reproductive ecology of some species has been made. Histological studies show that *L. pertusa* polyps have separate sexes (gonochoristic) and that in the Northeast Atlantic they produce gametes over winter following phyto-detrital flux to the seafloor (**Figure 6**). Spawning has

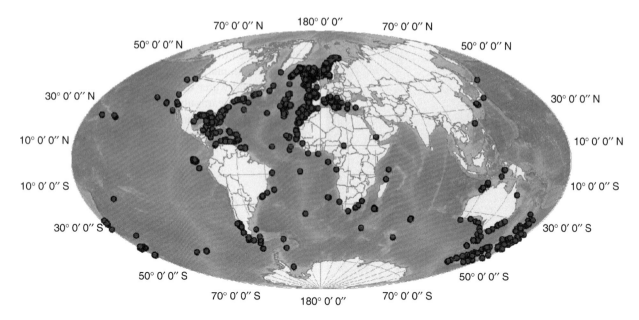

Figure 5 Global distribution of reef framework-forming cold-water corals. Map courtesy of Dr. Max Wisshak, University of Erlangen, and Dr. Andrew Davies, Scottish Association for Marine Science.

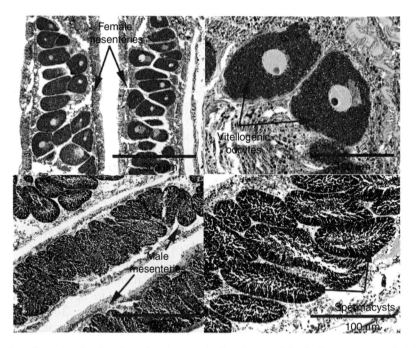

Figure 6 Histological sections showing female and male reproductive tissues of *Lophelia pertusa* sampled from a North Sea oil platform. Image courtesy of Dr. Rhian Waller, University of Hawaii.

not been observed yet and while no coral larvae have been sampled, the widespread occurrence and rapid colonization of man-made structures both point to a dispersive planula larva capable of remaining competent in the water column for several weeks.

Hidden Diversity and Molecular Genetics

Molecular genetic analysis has tremendous potential to reveal both species- and population-level information. Few molecular studies of cold-water corals have been completed and, as with most deep-sea biological research, those that have been attempted have been limited by small sample sizes. In spite of these constraints, interesting patterns are emerging. For example, at a species level, molecular analysis using partial sequences of the mitochondrial 16S ribosomal RNA encoding gene suggests that *M. oculata* may have been misclassified on the basis of its skeletal morphology. Rather than grouping with the Oculinidae, *M. oculata* may be more closely related to the Caryophyliidae and Pocilloporidae. On a population level, microsatellite and ribosomal internal transcribed spacer sequence analyses have shown that in the Northeast Atlantic some slope populations of *L. pertusa* are predominantly clonal and this species forms discrete fiord and shelf populations reflecting geographical isolation in fiord settings. Information like this is vital to develop management strategies to protect and conserve coral populations.

Habitats and Biodiversity

Coral reefs are renown for their structural complexity and cold-water coral reefs are no exception. Corals are ecological engineers, their skeletons forming complex three-dimensional structures that provide a multitude of surfaces for attached epifauna and shelter for mobile fauna. Coral frameworks trap resuspended sand and mud, forming sediment-clogged frameworks and providing further niches for infaunal species. On a larger scale, clear habitat zones develop around a cold-water coral reef. Live coral is largely unfouled by other organisms and supports relatively few species. Over time, as older polyp generations die back and exposed skeleton is (bio)eroded, coral frameworks degrade to a coral rubble apron that can extend for considerable distances downslope at the foot of the reef. The small-scale structural complexity of coral skeletal frameworks and the larger-scale diversity of habitat types combine to support highly diverse associated animal communities. For example, a recent compilation of European studies showed that cold-water coral reefs along the Northeast Atlantic margin supported 1317 other animal species (**Figure 7**).

However, significant gaps in our understanding remain. While it is clear that cold-water coral reefs sustain many species, we have very limited understanding of the functional relationships between these species. Largely because of the great expense and technical difficulties of working on structurally complex habitats in deep water, few examples of the natural history of these systems have been described. To date, only the most ubiquitous relationships have been examined in any detail. Of these the symbiosis formed between reef framework-forming scleractinians and eunicid polychaetes appears particularly significant. In the Northeast Atlantic, the large worm *Eunice norvegica* is very frequently found with both *L. pertusa* and *M. oculata*. The worm develops a fragile parchment tube through the coral

Figure 7 Examples of diverse fauna sampled from a giant coral carbonate mound in the Porcupine Seabight, Northeast Atlantic. (a) Isopod *Natatolana borealis*. (b) Gastropod *Boreotrophon clavatus*. (c) Brachiopod *Macandrevia cranium*. (d) Hydrocoral *Pliobothrus symmeticus*. Images courtesy of Dr. Lea-Anne Henry, Scottish Association for Marine Science.

framework that becomes calcified by the coral. As well as apparently strengthening the overall framework, recent aquarium observations have shown that the worms repeatedly aggregate small coral colonies. *In situ*, this behavior could be a significant factor in enhancing patch formation and accelerating reef growth.

We also know little of what organisms prey upon cold-water corals with only a few direct submersible observations showing asteroids apparently grazing on live coral colonies. However, some parasitic relationships have now been described from samples recovered from cold-water coral reefs. *L. pertusa* is parasitized by the foraminiferan *Hyrrokkin sarcophaga* and gall-forming copepods are sometimes associated with large gorgonian corals such as *Paragorgia arborea*. It is intriguing to think what other interactions and behaviors might remain to be described and *in situ* video and photographic records from unobtrusive benthic landers have great potential to provide these observations (**Figure 8**).

There has been great interest in whether cold-water coral reefs form essential fish habitat, for example, providing areas for spawning or nursery areas for juvenile fish. Investigations are again at an early stage but some broad trends are becoming apparent. The degree to which cold-water corals provide habitat to fish seems to depend on the coral habitat in question. Coral assemblages dominated by gorgonian 'forests' seem to sustain fish communities similar to those found near seabeds with other structural habitat such as large rocks. In contrast, there is reasonably

compelling evidence that the large, structurally complex reefs formed by scleractinians can provide important fish habitat. For example, higher numbers of gravid female redfish (*Sebastes marinus*) were found associated with large *Lophelia* reefs than neighboring off-reef areas. However, the studies to date rely on sparse data collected in different ways (e.g., long line catches vs. submersible observations) in different regions, making it hard to draw out clear patterns.

Thus although cold-water coral reefs form biodiversity hotspots on the continental slope, it is proving hard to understand their true significance in biogeographic and speciation terms. Studies are typically biased by the methodology used to collect samples and the taxonomic expertise applied to the fauna. Cold-water coral reef biodiversity studies rely on material gathered by trawl, dredge, box core, grab, or on megafaunal descriptions from photographs or lower-resolution video records. Each technique differs in the type of sample it can recover and further biases may be introduced by different sample-processing methods. Despite these frustrations, when the animal communities recovered from cold-water coral reefs are examined by taxonomic specialists, they frequently reveal undescribed species and often new records or range extensions of species known from other areas. For example, one study examining just 11 box core samples from coral carbonate mounds in the Porcupine Seabight reported 10 undescribed species and that coral-rich cores on-mound were 3 times more species-rich than off-mound cores.

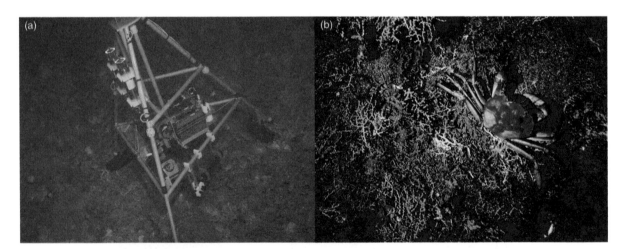

Figure 8 Developments in benthic landers and seafloor observatories will allow long-term environmental data recording and unobtrusive observations of cold-water coral reef fauna. (a) A 'photolander' deployed at 800 m water depth on a giant coral carbonate mound. (b) Lander photograph showing coral framework, glass sponges (probably *Aphrocallistes*), and deep-sea red crab (*Chaceon affinis*). (a) Courtesy of *VICTOR-Polarstern* cruise ARKXIX/3a, Alfred-Wegener-Institut für Polar- und Meeresforschung and the Institut Français de Recherche pour l'Exploitation de la Mer.

Timescales and Archives

As individual animals, corals can be extremely long-lived. [14]C dating has shown that one colony of the gold coral *Gerardia* (Zoanthidea) was approximately 1800 years old when collected, making it possibly the oldest marine animal known. Although scleractinian polyps are unlikely to live for more than 10–20 years, reef frameworks can persist for thousands to tens of thousands of years. In the Northeast Atlantic, the ages of scleractinian reefs clearly correspond to glacial history with relatively young reefs (8000 yr BP) found at latitudes affected by Pleistocene glaciation and far older reef frameworks (50 000 yr BP) further south well beyond glacial influence. Initial interpretation of the cores obtained by drilling through a large carbonate mound in the Porcupine Seabight suggest that the entire mound structure has developed over 1.5–2 My with periods of interglacial coral framework growth interspersed with periods of glacial sediments when conditions were unsuitable for corals.

An extensive literature now exists, showing the value of shallow, tropical corals as paleoenvironmental archives, and interest in studying historical patterns in ocean temperature and circulation has increased as evidence of anthropogenic climate change grows. Cores extracted from glacial ice sheets provide invaluable high-resolution climate archives but the potential of deep-ocean sediments to reveal similar high-resolution temporal patterns is limited by the mixing action of bioturbating infauna. Recent research has shown that cold-water coral skeletons not only provide a long-lasting archive but one that can be analyzed at high temporal resolution without confounding effects of bioturbation. Studies fall into two categories: those that use coral skeletal chemistry to estimate past seawater temperature and those that use combinations of dating techniques to trace ocean ventilation. The most convincing paleotemperature estimates come from gorgonian and bamboo corals, which, unlike reef framework-forming scleractinians, contain clear skeletal banding patterns that allow a good chronology to be developed (**Figure 9**).

We continue to learn more about the importance of deep-ocean circulation in regulating global climate and once again recent research demonstrates that cold-water corals provide a unique archive of ocean circulation patterns. By dating coral skeletons with both ^{14}C and U/Th techniques, it is possible to discriminate the coral's actual age from the age of the seawater in which it grew. This is possible because as corals calcify they use dissolved carbon from the ambient seawater, thus providing material that can later be ^{14}C-dated. This means that cold-water coral skeletons can record ocean ventilation history as water masses of differing ^{14}C age exchange. This approach has so far successfully followed ventilation patterns in the Southern Ocean and North Atlantic and offers great potential to study past ocean circulation at key oceanographic gateways.

Threats

As shallow-water fish stocks have diminished on continental shelves around the world, the fishing industry has expanded into deeper slope and even seamount waters. This move, made possible with larger, powerful refrigerated vessels and improved navigational technology (Global Positioning System), has seen the development of a series of boom-and-bust deep-water fisheries for species such as the orange roughy (*Hoplostethus atlanticus*) around New Zealand, the roundnose grenadier (*Coryphaenoides rupestris*) in the Northwest Atlantic, marbled rockcod (*Notothenia rossii*) around Antarctica, and pelagic armorheads (*Pseudopentoceros pectoralis*) on Pacific seamounts. Trawling for fish in deep waters requires heavier ground gear and large trawl doors that plough across the seafloor and can easily damage epifaunal communities dominated by corals and sponges (**Figure 10**). Visual and acoustic evidence of damage to cold-water coral habitats has now been recorded from the territorial seas of many nations including the Canada, Ireland, New Zealand, Norway, UK, and USA, and in each case measures to limit or ban trawling in some coral-rich areas have been instigated. However, on the High Seas beyond the jurisdiction of any one nation, deep-water trawling continues without regulation causing unknown damage to benthic communities. As well as direct physical impacts, deep-water trawling disturbs sediment producing a plume that could smother epifauna over a wider area. To date, few studies have examined this wider area impact.

While evidence for damage from deep-water trawling is visually clear and now known to be a concern in several regions, the impacts of other human activities are harder to pin down. As with the fishing industry, technological developments and reduced shallow-water reserves have made it economically viable for the oil industry to explore progressively deeper and deeper waters. At the time of writing, there are producing oil fields in the deep waters of the Atlantic Frontier (UK), Campos Basin (Brazil), and Gulf of Mexico (USA), where exploratory drilling has now taken place to depths over 3000 m. For example, in late 2003, the Toledo well was drilled at 3051 m water depth by Chevron Texaco in the Gulf of Mexico.

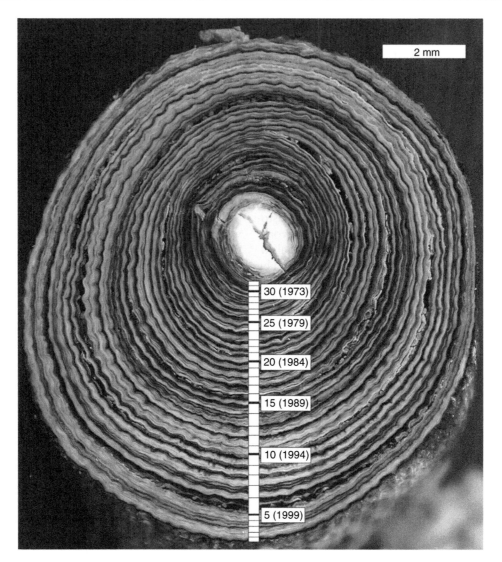

Figure 9 Cross section of the gorgonian *Primnoa resedaeformis* showing clear growth banding. Courtesy of Dr. Owen Sherwood, Memorial University of Newfoundland, reproduced from Sherwood OA, Scott DB, Risk MJ, and Guilderson TP (2005) Radiocarbon evidence for annual growth rings in the deep-sea octocoral *Primnoa resedaeformis*. *Marine Ecology Progress Series* 301: 129–134, with permission of Inter-Research.

As with all suspension-feeding invertebrate animals, cold-water corals are vulnerable to increased sediment loads that could smother polyps or even bury whole colonies. Tropical studies have shown corals are vulnerable to discharges, notably the muds and cuttings released during drilling operations, but it is frequently hard to disentangle physical effects of increased particle exposure from any toxic effects of exposure to drill muds or other additives. Intriguingly, both tropical and cold-water corals will settle and grow on producing oil platforms, sometimes close to drilling discharge points. Some of the older platforms in the northern North Sea support large colonies of *L. pertusa* that must have grown there continuously for over 20 years to have reached their present size and now form a reproductive population (**Figure 6**). Visual surveys have shown that coral polyps directly exposed to drill cuttings may be smothered but other polyps, even on the same colonies, that are not directly exposed can survive. This and the relatively small geographical extent of drilling suggest that if these activities are restricted in areas supporting cold-water coral reefs, their impacts will be considerably less than those of deep-water trawling now known to abrade large areas of the slope and to target seamount fish stocks. However, as tropical studies have shown, coral responses to drill discharges are hard to interpret and to date no detailed studies of reef framework-forming cold-water corals exposed to muds and cuttings have been carried out.

Figure 10 Fisheries damage to cold-water corals from coral carbonate mounds in the Porcupine Seabight and Porcupine Bank, Northeast Atlantic. (a) Nets and ropes with crushed coral rubble. (b) Closer view showing scavenging crabs. (c) Lost trawl net. (d) Abandoned trawl rope. (a), (b) Reproduced from Grehan *et al.* (2004) *Proceedings, ICES Annual Science Conference*, 22–25 Sep., Vigo, Spain, ICES CM 2004/AA07, with permission of International Council for the Exploration of the Sea (ICES). (c), (d) Courtesy of Dr. Anthony Grehan, National University of Ireland; *VICTOR-Polarstern* cruise ARKXIX/3a, Alfred-Wegener-Institut für Polar- und Meeresforschung and the Institut Français de Recherche pour l'Exploitation de la Mer.

Deep seabed mining for valuable materials such as manganese nodules or the rich mineral deposits of ridge and vent systems has yet to develop as a commercial proposition, although some forecasts suggest this could happen within the next 20 years. As with trawling and drilling, mining activities would have localized impacts on the area mined but could also disturb a sediment plume that would disperse affecting a wider area. Again, great care is needed to limit the impact of any developments in deep seabed mining on cold-water coral ecosystems.

Recent analysis suggests that rising atmospheric carbon dioxide levels will not only cause global warming, but also could cause the most rapid 'acidification' of the oceans seen in the last 300 My. Once again, no studies of cold-water coral response to ocean acidification have been carried out but tropical coral calcification could be reduced by over 50% if atmospheric carbon dioxide concentrations doubled. Perhaps of greatest concern are modeled predictions that the depth of the ASH could shallow by several hundred meters in as little as next 50–100 years, leading to concerns that regions currently suitable for cold-water coral reef development will become inhospitable. Corals living in lowered aragonite saturation states produce weaker skeletons more vulnerable to (bio)erosion; consequently, entire reef systems may shift from a phase of overall growth to erosion with severe implications for habitat integrity.

Conclusions

The last decade has seen an explosion of interest in cold-water coral reefs. This article attempts to summarize the many and exciting advances in our understanding of reef development, longevity, and diversity in deep waters while realizing that this work has only just begun. Work to map and characterize cold-water coral reefs is still needed, as shown by the geographical bias in studies so far, and efforts to unify sampling methodologies and species identification are vital. However, in the coming years, we will enter a phase in research on cold-water coral ecosystems that requires a shift away from baseline mapping to one centered on process-oriented questions that will examine the factors driving reef development in the deep ocean and allow us to really understand the significance and vulnerability of cold-water coral reefs and develop protected areas for their conservation.

See also

Deep-sea Fauna. Deep-sea Fishes.

Further Reading

Cairns SD (2007) Deep-water corals: An overview with special reference to diversity and distribution of deep-water Scleractinia. *Bulletin of Marine Science* 81: 311–322.

Expedition Scientists (2005) Modern carbonate mounds: Porcupine drilling. *Integrated Ocean Drilling Program Report Number 307*. Washington, DC: Integrated Ocean Drilling Program.

Freiwald A, Fosså JH, Grehan A, Koslow T, and Roberts JM (2004) *Cold-Water Coral Reefs*. Cambridge, UK: United Nations Environment Programme – World Conservation Monitoring Centre.

Freiwald A and Roberts JM (eds.) (2005) *Cold-Water Corals and Ecosystems*. Berlin: Springer.

Gage JD and Tyler PA (1991) *Deep-Sea Biology: A Natural History of Organisms at the Deep-Sea Floor*. Cambridge, UK: Cambridge University Press.

Grehan A, Unnithan V, Wheeler A, *et al.* (2004) *Proceedings, ICES Annual Science Conference*, 22–25 Sep., Vigo, Spain, ICES CM 2004/AA07.

Roberts JM, Brown CJ, Long D, and Bates CR (2005) Acoustic mapping using multibean echosounder reveals cold-water coral reefs and surrounding habitats. *Coral Reefs* 24: 654–669.

Roberts JM, Wheeler AJ, and Freiwald A (2006) Reefs of the deep: The biology and geology of cold-water coral ecosystems. *Science* 312: 543–547.

Rogers AD (1999) The biology of *Lophelia pertusa* (LINNAEUS 1758) and other deep-water reef-forming corals and impacts from human activities. *International Review of Hydrobiology* 4: 315–406.

Sherwood OA, Scott DB, Risk MJ, and Guilderson TP (2005) Radiocarbon evidence for annual growth rings in the deep-sea octocoral *Primnoa resedaeformis*. *Marine Ecology Progress Series* 301: 129–134.

Wilson JB (1979) 'Patch' development of the deep-water coral *Lophelia pertusa* (L.) on Rockall Bank. *Journal of the Marine Biological Association of the UK* 59: 165–177.

Relevant Website

http://www.lophelia.org
– Lophelia.org, an information resource on the cold-water coral ecosystems of the deep ocean.

BENTHIC BOUNDARY LAYER EFFECTS

D. J. Wildish, Fisheries and Oceans Canada,
St. Andrews, NB, Canada

Introduction

The benthic boundary layer (BBL) is a discrete layer of flowing sea water above a benthic substrate, delimited vertically by its contact with free stream flow. The degree of turbulence within the BBL and boundary shear forces exerted on the substrate are determined by the free stream velocity and the roughness characteristics at the substrate interface. Roughness elements may either be of geological origin, e.g. sand ripples of soft sediments, or of biological origin, e.g. tubes constructed by macrofauna that extend into the BBL.

Typical structure of a smooth BBL consists of a bed layer, inclusive of a viscous sublayer (laminar sublayer) closest to the substrate interface. Here, the flow is laminar and only a few millimeters thick, dictated by the free stream velocity. Next is the logarithmic layer, where mean velocity varies as the logarithm of the height above the substrate interface, and where the flow is often turbulent. In the outer layer of the BBL, turbulence decreases with distance from the substrate interface and is bounded by the free stream flow, situated immediately above it. In coastal and estuarine habitats, where many of the benthic animals discussed here live, BBL depth may vary from 10 cm to 5 m. In some conditions the BBL of coastal waters may extend throughout the water column. Such shallow environments are much influenced by tidal and wind forcing; the latter inducing oscillatory water movements in the bed layer.

A useful measure of the flow conditions for biologists is Reynold's number. It expresses the relative proportions of inertial and viscous forces within a flow as a dimensionless number. It is determined by measuring a characteristic length of a solid in flow measured in the same direction as the flow, multiplied by the velocity and divided by the kinematic viscosity of sea water. Other hydrodynamic measures useful for this presentation are lift and drag coefficients. A flat body resting on the substrate and in a flow field will experience lift due to Bernoulli's principle. This occurs because the velocity is locally higher on the upper than on the lower surface, due to the already mentioned effect of height on velocity within the BBL. The resultant pressure differences – the pressure is higher where flows are low – cause a lift force to be generated. The lift is resisted by the negative buoyancy of the body but, if exceeded as velocity increases, it is 'lifted' and carried downstream. The drag of a body in a flow field depends on the frontal area presented to the flow and the square of the velocity that it experiences.

Some examples of the wide range of taxa that can be found within the BBL are presented here. Concentrating on the epifauna and suprabenthos of the BBL macrofauna, a brief survey is made of their adaptations to the BBL environment; including examples from rocky shore, wave-exposed locations to soft sediment substrates where weak currents dominate.

Organisms of the Benthic Boundary Layer

Many of the organisms living within the BBL are commonly found elsewhere in the sea. They include microbiota, such as viral, bacterial, and planktonic life forms, and microalgae limited by light penetration to the shallow fringes of oceans. Of two distinctive life forms of macrofauna, the first one, epifauna, is usually sampled by grab or corer, inclusive of attached, free-living, and tube-living life forms. Epifauna characteristically protrude into the BBL where they feed. The second life form group is composed of near-bottom swimmers of the BBL macrofauna. They are sampled by drawing a plankton sampler through the BBL and are defined as those animals that are retained within a 0.5 mm mesh plankton net. Typical catches consist of zooplankters, common throughout the whole water column, eggs and larvae, from some of the epifauna and suprabenthos (also referred to as hyperbenthos). The suprabenthos are bottom-dependent animals that perform regular daily or seasonal vertical migrations above the bottom. The only life forms that are unique to the BBL are epifauna and suprabenthic animals.

Epifauna

Some examples of the major taxa of epifauna are shown in **Table 1**. Attached epifauna comprise a wide range of taxa from hydroids to cirripedes. Included among free-living epifauna are some echinoderms, decapods, and bivalves. Tube-living epifauna

Table 1 Life forms of epifauna

Life form	Examples
Attached epifauna	Cnidaria: Hydrozoa – hydroids
	Cnidaria: Anthozoa – gorgonian corals
	Cnidaria: Antipatharia – octocorals
	Bryozoa – bryozoans
	Brachiopoda – lampshells
	Echinodermata: Crinoidea – stalkless crinoids
	Porifera – sponges
	Ascidiacea – sea squirts
	Cirripedia – barnacles
Free-living epifauna	Some Echinodermata, e.g. feather stars (Crinoidea), sand dollars (Echinoidea)
	Some decapod Crustacea, e.g. sand crabs (Hippidae), porcelain crabs (Porcellanidae)
	Some Bivalvia, e.g. scallops (Pectinidae), mussels (Mytilidae)
Tube-living epifauna	
Tube normal to flow	Polychaeta, e.g. *Lanice conchilega, Streblospio benedicti, Eudistylia vancouveri*; Amphipoda, e.g. *Ampelisca abdita, A. vadorum, Haploops fundiensis*
Tube opposed to flow	Amphipoda, e.g. *Ampithoe valida*; Tanaidacea, e.g. *Tanais covolinni*
Truncated cone tube	Polychaeta, e.g. *Spio setosa, Fabricia limnicola, Mesochaetopterus sagittarius, Phyllochaetopterus verrilli*
Spar buoy tube	Polychaeta, e.g. *Potamilla neglecta*
Complex tube	Polychaeta, e.g. *Diopatra cuprea*

consist of species from a limited number of polychaete, amphipod, and tanaid families. Five types of tube builders can be distinguished on the basis of tube height (H) to tube diameter (D) ratio. The greatest difference between H/D ratios of the life forms shown in **Table 1** is between the truncated cone and spar buoy life forms. The former are short and fat (H/D < 1) and the latter are long and thin (H/D > 19). The name of the latter derives from the observation that the tube is articulated at its base so that it follows any change in flow direction, like a spar buoy, and its sabellid crown filtration surfaces are thus always downstream in the flow.

Suprabenthos

In contrast to epibenthic organisms, suprabenthic animals are represented by a narrower range of taxa,

Table 2 Densities and relative importance of suprabenthos from the Bay of Fundy

Taxon	Lower net[a]		Upper net[b]	
	%	Density Number per $100\,m^{-3}$ filtered	%	Density Number per $100\,m^{-3}$ filtered
Amphipoda, Gammaridea	44	420	23	30
Cumacea	29	279	11	14
Amphipoda, Caprelloidea	11	102	1	1
Mysidacea	8	74	11	15
Euphausiacea	5	50	52	66
Isopoda	2	15	< 1	< 1
Tanaidacea	1	11	< 1	< 1
Decapoda, Caridea	< 1	1	1	1

[a] 33–73 cm from sediment.
[b] 109–149 cm from sediment.

being limited to the Crustacea (**Table 2**). For example, in the Bay of Fundy, gammarid amphipods and euphausids are the dominant members representing the suprabenthos.

BBL Flow Adaptations

Over time, the word 'adaptation' has come to have conflicting meanings. These include the biological process of adjustment to environmental stresses that occur within an individual's lifetime, as well as the results of evolutionary changes that persist over more than one lifetime and produce individuals of superior survivability or reproductive capacity. To avoid confusion, the latter meaning of evolutionary adaptation is used throughout this article.

Evolutionary adaptation refers to the phenotypic characteristics of an organism, which have arisen by a process of Neo-Darwinian evolution. The adaptations of concern here are morphological, physiological, behavioral, or reproductive, in response to the physical processes characteristic of the BBL. Included in the definition are any extended phenotypic adaptations, e.g. worm tubes, that occur outside the individual body and persist as characteristic structures over many generations. An adaptation may have arisen in response to one environmental feature, and subsequently been modified for another purpose. A possible example is jet propulsive predator avoidance in scallops, modified as scallop swimming to seek a better feeding area. In the latter example, such

evolutionary change from predator avoidance to free swimming may be referred to as an exaptation. Since the adaptations and exaptations occurred in the past, it is only possible to use deductive methods and circumstantial evidence in their study, thus limiting confidence in the explanations proposed.

Life History Adaptations

Many epifaunal organisms have a complex life cycle in which the egg and larvae are the chief dispersal stages. Larval life is terminated by metamorphosis at the time of recruitment to a benthic substrate. Juvenile and adult are the main growth and gamete-producing phases of the life cycle (**Figure 1**). Because of the obviously different strategies of larval and juvenile/adult life, and therefore adaptations associated with them, they are dealt with separately.

An important function of larval life is to ensure that a sedentary adult, such as a cirripede barnacle, becomes spatially distributed to a suitable, unoccupied habitat where growth and reproduction can occur. Although larval dispersal is the major mechanism among epibenthic macrofauna, additional means of dispersal are also available during juvenile and adult life.

A characteristic feature of marine benthic larvae is their small size, in general < 1000 μm at all stages of larval life. Among a wide range of marine bivalve species the pediveliger larva (**Figure 1**) ranges in size from 160 to 350 μm. The pediveliger larva is the

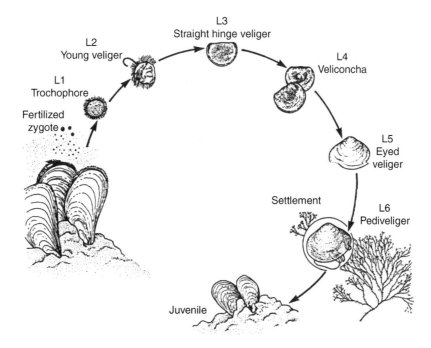

Figure 1 Life cycle of the blue mussel, *Mytilus edulis*.

Table 3 Comparison of larval adaptive strategies among marine epifauna

Characteristic	Lecithotrophic larva	Planktotrophic larva
Larval type	Larger, nonfeeding, food store	Smaller, feeding in plankton, no food store
Larval numbers	Few	Many
Larval period	Short	Long
Larval dispersal distance (maximum)	Short	Long
Inbreeding potential	High	Low
Biogeographic dispersal rate	Slow	Fast
Biogeographic range	Endemic	Pandemic (cosmopolitan)
Geological history of taxa	Short	Long
Speciation rate	High	Low

stage capable of finding and contacting a suitable substrate, and this property is referred to as competency. During larval life, small size is presumably an adaptation for passive dispersal by often large-scale physical oceanographic forces. Concomitant features resulting from small size are the feasibility of a large number of larvae per unit of reproductive effort and limitation of swimming locomotion to ciliary means.

The distance that a dispersing larva can travel will depend largely on the time that it spends in the plankton. This can vary from a few hours for the nonfeeding, lecithotrophic larva to up to 2 years for the planktotrophic larva. The latter suspension feeds at certain stages during its life cycle and consequently sustains its stay in the plankton. The two types of larvae shown in **Table 3** appear to use contrasting strategies in determining where to settle, metamorphose, and start growth as an attached, epifaunal organism. Larvae swim at speeds characteristic of ciliary beating and low Reynold's numbers, which is in the range of 0.02–$0.50\,\mathrm{cm\,s^{-1}}$. Such low speeds are sufficient for competent larvae to swim through the viscous sublayer of the BBL, so that they can contact the benthic surface guided by sensory cues, inclusive of hydrodynamics, chemicals, and/or surface roughness.

Lecithotrophic larvae (**Table 3**) spend a short time in the plankton and consequently are not dispersed far from the parental population. By contrast, planktotrophic larvae have a longer larval period and consequently can undertake longer dispersal pathways. This allows the competent larval stage to reach unoccupied habitats distant from the parental population.

Some larvae must settle on 'protruding bodies,' such as attached macroalgae, some branching corals, or man-made structures. This provides a challenge because of the characteristic flow around the protruding structures. At higher flows, the stream lines compress the laminar sublayer on the upstream and lateral surfaces, whereas wakes occur on the downstream surface. Some coral reef larvae have developed specialized adhesive mucous threads up to 100 larval body lengths in size, which aid the competent larva to contact and firmly attach to the protruding body. In the process, coral larvae overcome the shear forces that are locally high, as well as the accelerative reaction forces characteristic of flow around protruding bodies.

Suspension-feeding Adaptations

The function of juvenile life is to transform the recruited settlement stage to a large adult capable of producing gametes. To do this requires rapid growth, which must be fueled by a dilute diet of organic (inclusive of bacteria, phytoplankters, zooplankters, and organic carbon) and sedimentary particles in sea water, collectively called seston. The function of removal of the sestonic particles from sea water is achieved by suspension feeders in a variety of ways (**Table 4**). The characterization of **Table 4** depends on whether filtration is driven by ambient flow, as in passive forms, or by some form of energy supplied by the suspension feeder, as in active forms.

Passive suspension feeders must deflect the stream lines close enough to their filtration surfaces to enable them to capture sestonic particles by direct interception. Typical adaptations of passive forms are thus expansion of filtration surfaces and channeling flows, e.g. by polyps, tentacles, and pinnules.

By contrast, active suspension feeders provide a pump (ciliary, flagellar, muscular) or collecting device that requires muscular power to sweep and capture the seston. The operating characteristics of a representative range of pump types have been described, whereas the knowledge available regarding the precise filtration mechanism is incomplete. For taxa with mucous filter nets (see **Table 4**), the net is

Table 4 Classification of the filtration mechanisms of suspension-feeding epifauna

Classification	Mechanisms		Examples
	Pump	Filtration	
Passive	Absent	Direct interception	Sea pens, hydroids, sea whips, black coral, feather star, brittle star, sea cucumber
Active	Ciliary	?	Many bivalves
	Ciliary	?	Bryozoa
	Ciliary	Mucus net sieving	Sea squirts
Deposit/suspension feeder	Ciliary	?	Spionid polychaete worms
	Ciliary	?	Tellinid bivalves
Facultative passive/active	Cirral sweeping	Direct interception	Barnacles
	Maxilliped setal sweeping	Direct interception	Porcelain crabs (decapod Crustacea)

periodically secreted and then ingested, inclusive of the seston trapped by sieving. For taxa with a ciliary pump, such as many bivalves, the precise filtration mechanism is still unclear.

In deposit/suspension feeders, it is ambient flow, and/or seston concentration, that is the trigger to change the feeding mode. Thus, at flow velocities < 2–$5\,\mathrm{cm\,s^{-1}}$, spionid, tube-building polychaetes deposit feed by touching the elongated paired palps on the sediment surface. At faster flows, the palps become helically coiled in the downstream flow direction to optimize suspension feeding. In tellinid bivalves such as *Macoma balthica*, the extensible inhalant siphon is used to deposit or suspension feed in response to changing flow conditions. Thus, a combination of morphological/physiological and innate behavioral responses to flow changes allows these animals to feed in a wider range of flow conditions.

For barnacles of the facultative passive/active group, the cirri act passively at higher flows, but below a critical velocity of a few centimeters per second, begin active rhythmic sweeping so that the concave surface faces away in lower flows. Concave surfaces of cirri are optimum for passive suspension feeding in strong flows by direct interception and that is why the cirrus is rotated to face the oscillating flow by *Semibalanus balanoides* during passive feeding. Musculature and nervous control enables the barnacle to respond rapidly to waves by switching the cirrus to face the flow in flow oscillating conditions of wave surge and swash.

Adaptations to Resist Shear Stress

Rocky shores are home to epifaunal organisms such as barnacles and mussels, and experience the most extreme oscillatory movements due to wave forces. Here, 2–4 m breaking waves translate into a peak velocity of $8\,\mathrm{m\,s^{-1}}$, exerting considerable shear forces at the substrate interface. This results in drag, lift, and acceleration forces on the attached macrofauna. Such forces tend to dislodge sessile organisms and are resisted in a variety of different ways. Barnacles have developed adhesives that cement the animal firmly to the rock surface. In mussels, adhesion is achieved by byssus threads produced from the foot, which attach with secreted adhesive to nearby solid surfaces. If mussels are present as densely packed reefs, the drag and lift forces are shared by the group. A similar case occurs among South African sublittoral holothurians, but here two species occur together as a mixed group. One species has degenerate tube feet and insinuates its body beneath the other, thereby gaining a surrogate means of attachment. By this arrangement, the holothurian lacking tube feet is able to extend its range from a protected to a more exposed location.

Structural adaptations linked to shear stresses include the development of an elastic body that allows form changes proportional to velocity, resulting in less drag, reduced size where flows are energetic and body strengthening by secretion of skeletal materials such as the spicules in sponges or cnidarians.

Behavioral responses to resist drag and lift forces are also common among free-living epifauna. For example, many scallop species can recess into the sediment. They do this by jetting water at the sandy sediment until a pit is made and then they settle into it. Recessed scallops experience reduced drag forces because most of the body is situated beneath the BBL, although lift forces (resisted by the buoyant weight of the animal) are still present if the upper valve is within the logarithmic flow layer. Experimental observations have suggested that larger scallops, above a critical size threshold, are at risk from

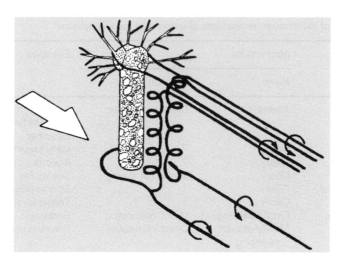

Figure 2 Flow stream lines around the polychaete tube of *Lanice conchilega.*

drag forces during valve opening, which is required for feeding, due to an increase of the frontal area. These results are consistent with field observations that show that it is the older scallops that are recessed.

Typical stream lines around an isolated, epifaunal tube normal to the flow are shown in **Figure 2**. The conical tube of the worm *Lanice conchilega* causes resuspension of the lighter organic particles in the wake of the tube and this material becomes available for downstream capture by the tentacles of the worm. Another worm studied in detail is *Spio setosa*, which makes a truncated cone tube. The 4–6 cm high tubes have an equally wide diameter and are functional in ensuring that the worm is able to suspension feed on good quality seston. This is because at 4–6 cm in height in the BBL there is a significantly greater proportion of nutritious organic particles than closer to the sediment surface where the seston carries more heavier, denser, and non-nutritious inorganic sedimentary particles.

The fact that tube-living worms rely on the trophic advantage provided by their tubes is evidenced by the fact that they are rapidly rebuilt if they become broken or eroded during storms. Another worm, *Spiochaetopterus oculatus*, is able to adjust the height of the tube so that at low flows, tube building occurs whilst tube cutting is initiated as energetic oscillating flows commence. Tube building in this worm is achieved by the fourth paropodia selecting and gluing suitable inorganic sedimentary particles in place, and cutting by the setae on this segment.

There is evidence that if densely packed aggregations of tube-living epifauna exceed a critical density, they will interact with flow to produce 'skimming flow'. This implies that the flow is diverted above the tubes, which act as roughness elements.

Skimming flow has important consequences for seston supply and particle settlement within the tube field, which needs further clarification.

Aggregation as an Adaptation

Possible reasons why epifaunal organisms may occur in aggregated groups include: to exploit an optimal niche for suspension feeding, to provide protection from extreme shear forces, to provide better opportunities for successful reproduction, and to provide better protection from predators. Examples could be quoted which appear to support each reason why some epifauna aggregate, but the work is considered to be at too early a stage in its development to give a comprehensive account.

The question as to how epifauna achieve their characteristically aggregated distribution may be solved either by larvae or adults. Thus, larvae may settle near adults of the same species, guided by pheromones from the adults or by the same chemical/physical cues from the habitat that originally attracted the earlier colonizers. Many juvenile and adult epifauna are attached, and hence do not usually take part in aggregation behavior, although there are exceptions, e.g. juvenile blue mussels that may move to a secondary site after first settling on red seaweeds. Free-living epifauna, such as scallops, can swim to occupy many different niches following initial settlement as the competent larva.

Table 5 shows examples of the population densities and biomasses of animals that aggregate as obtained by conventional grab sampling methods. The high biomasses evident in **Table 5** signify that the bivalve reefs, for example, are major producers in discrete local areas, sustained by enriched seston levels and/or optimal hydrodynamic conditions.

Table 5 Density and biomass of epifauna that aggregate

Taxa	Common name	Density (number m²)		Biomass (g dry m²)	
		Mean	Maximum	Mean	Maximum
Molluscs					
Modiolus modiolus	horse mussel	–	510	281	3038
Mesodesma (1)	Turton's	488	1550	6485	21 030
deuratum (2)	wedge clam	5890	12 010	939	4697
Crassostrea virginica	American oyster	–	4077	–	214
Mytilus edulis	blue mussel				
Rhode Is., USA		2139	–	10 962	–
Baltic Sea		36 000	158 000	101	–
Perna perna	green-lipped mussel	–	–	826	1285
Polychaetes					
Owenia fusiformis	a bamboo worm	500	15 000	–	–
Lanice conchilega	a fan worm	–	20 200	–	1094
Spio setosa	a mud worm	408	2002	61	109
Amphipod Crustacea					
Haploops fundiensis	ampelscid	376	923	0.35–0.53	0.96
Ampelisca abdita	ampelscid	1360	73 000	–	–
Ampelisca vadorum	ampelscid	1307	1885	–	–

Conclusions

This brief survey of the macrofauna most characteristic of the BBL, that is the epifauna and suprabenthos, leads to the conclusion that BBL physical processes have had far-reaching effects on their evolution. As the study of this subject is still so young, it is not possible to give a full account. Thus, almost nothing is known about the general biology of the suprabenthos. For example, it is not understood where they fit in the marine food web and their relative importance in it, let alone the adaptations which allow them to live in this niche. Many Crustacea, especially suprabenthic organisms, do not have a larval stage and hatch as a juvenile directly from the brooded egg.

For the larvae of epifauna, the most important adaptations appear to be associated with their small size, which permits sustained periods in the plankton and passive dispersal. The latter is made possible by adaptations to suspension feed during an extended period in the plankton. It is likely that most aggregated groups of epifaunal organisms are formed during larval settlement, but the precise behavioral mechanisms also need further study.

Benthic boundary layer adaptations of juvenile and adult life fall into the following groups:

- food collection from seston in sea water;
- resistance to the shear stresses that tend to dislodge macrofauna; and
- the building of epifaunal tubes.

Adaptations for collecting seston may have arisen in a local population to allow them to colonize or adapt to changing flow environments. Estimates of the physiological cost of operating the bivalve ciliary pump to suspension feed have been made and suggest that they are small – less than 2% of the overall energy budget. Nevertheless, the ontogenetic cost of constructing and maintaining the inline ciliary pump in the trophic fluid transport system of a bivalve must be high. To date, there do not appear to be any studies that have investigated this latter possibility.

Shear forces that try to dislodge epifauna are resisted by adaptations that include adhesives, skeletal strengthening, developing an elastic body deformation capability, and behavioral adaptations. The latter often involve changing the orientation of the body to minimize drag and lift forces.

For the few species of tube builders studied, the adaptations found seem linked to optimizing seston feeding where the quality is best. Tube building may be regarded as an extended phenotypic expression from genes that control the complex innate behavior involved in cementing sedimentary particles together. The type of sedimentary particle selected, as well as the shape and size of the tube constructed, can usually be used to identify the organism that created them.

Further Reading

Ackerman JD, Sim B, Nichols SJ, and Claud R (1994) A review of the early life history of zebra mussels

(*Dreissena polymorpha*): comparisons with marine bivalves. *Canadian Journal of Zoology* 72: 1169–1179.

Bayne BL (1976) The biology of mussel. In: Bayne BL (ed.) *Marine Mussels: Their Ecology and Physiology*, pp. 81–120. Cambridge: Cambridge University Press.

Carey DA (1983) Particle resuspension in the benthic boundary layer by a tube building polychaete. *Canadian Journal of Fisheries and Aquatic Science* 40: 301–308.

Chevrier A, Brunel P, and Wildish DJ (1991) Structure of a suprabenthic shelf sub-community of gammaridean Amphipoda in the Bay of Fundy compared with similar sub-communities in the Gulf of St Lawrence. *Hydrobiologia* 223: 81–101.

Denny MW (1998) *Biology and the Mechanics of the Wave-swept Environment*. Princeton: Princeton University Press.

Jørgensen CB (1990) *Bivalve Filter Feeding: Hydrodynamics, Bioenergetics, Physiology and Ecology*. Fredensborg: Olsen and Olsen.

Koehl MAR (1982) The interaction of moving water and sessile organisms. *Scientific American* 247: 124–134.

Sainte-Marie B and Brunel P (1985) Suprabenthic gradients of swimming activity by cold-water gammaridean amphipod Crustacea over a muddy shelf in the Gulf of St Lawrence. *Marine Ecology Progress Series* 23: 57–69.

Vogel S (1994) *Life in Moving Fluids: The Physical Biology of Flow*. Princeton: Princeton University Press.

Wildish DJ and Kristmanson DD (1997) *Benthic Suspension Feeders and Flow*. New York: Cambridge University Press.

FISH BIOLOGY

FISH: GENERAL REVIEW

Q. Bone, The Marine Biological Association of
the United Kingdom, Plymouth, UK

Introduction

Few other groups of animals are as diverse and so
abundant as fish. They live in a dense, relatively
oxygen-poor, conductive medium with a very wide
range of transparency and light regimes. Ambient
pressures can be up to some 450 times atmospheric,
and temperatures from 44 °C to below the freezing
point of seawater. Such varied aquatic surroundings
have led to some of the most remarkable adaptations
of all vertebrates. They have permitted fish to live in
almost every kind of habitat with water: in transient
puddles, in alkaline lakes, under the Antarctic and
Arctic ice sheets, in hypersaline lagoons, and in the
depths of ocean trenches. The adaptations are often
interlinked, so that, for example, the need to acquire
oxygen from the water has led to large exposed gill
surfaces, this in turn to varied osmotic ion pumps
and to extrarenal excretory mechanisms.

While most fish spend all their lives in waters of
one kind or another, some can fly, others climb trees
or sit about out of water, and a surprising number of
freshwater bony fishes spend part of their lives es-
tivating underground during droughts, like some
lungfish, mudfish, and the little salamander fish.

Abundance

There are huge numbers of fishes, far exceeding
any other vertebrate populations. Global populations
of the little bathypelagic minnow-sized *Cyclothone*
species *C. microdon* and *C. acclinidens*, living between 200 and 500 m, and ranging over three oceans,
must certainly consist of many thousands of millions
of fishes. Calculations based on marine data suggest
that the average number of individuals in a given
species is around 4×10^{10}, while some pelagic clupeid
species like anchovetas may have populations of 10^{12}
or more. Not only are populations of many fish huge,
but in 'numbers' of species fish are rivaled only by
insects and crustaceans, with crustaceans possibly
topping the list.

Freshwater fish populations are not quite so large,
average high and low values suggested by Horn in
1972 being 10^{10} and 10^7. Of course, both in fresh

water and (less commonly) in the marine environment, some species have very small populations.
The little desert pupfish (*Cyprinodon diabolus*) in
the Death Valley Devil's Hole is restricted to some
500 individuals. Several coral reef fish are restricted to single reefs, while the bythitid vent fish
(*Thermichthys hollisi*) is restricted to thermal vents
of the ocean floor rifts. Such small populations live in
small niches, but even pelagic and demersal fish
which have large populations over wide areas may be
at risk from overfishing if they have the misfortune to
be comestible. Despite increasing attempts at conservation measures, even the huge populations of
certain food fishes have been reduced to danger point
by the efficiency of modern fishing.

Diversity of Living Fish and Their Origins

Very large numbers of different fish species have been
described, all conventionally lumped together in the
class Pisces. This contains fish as diverse both in organization and mode of life as lampreys, lungfish,
and manta rays. More fish are being discovered and
named almost daily, so that the present total of
around 27 000 species may finally end up as a grand
total over 30 000. In comparison, the largest of the
other vertebrate classes, the birds (Aves), has only
some 9000 species, and all birds, reptiles, mammals
and amphibians together are today about the same
number as fish, and will certainly shortly be overtaken by fish. These other classes are very unlikely to
be added to significantly by the discovery of new
species. Much more likely, extinction (often our fault
by habitat reduction) will gradually reduce them
despite conservation efforts.

The huge array of living fish species was preceded
by several abundant and widespread fossil groups of
fish. The earliest found so far are the remarkable
fishlike fossils from the early Cambrian lägerstetten
(remarkable rich fossil localities with very unusually
complete preservation of soft parts) fields of China
(525–520 Ma), thus far (though clearly fish) not well
enough known to link directly as ancestral to any
living (or later fossil) fish groups. They are in some
ways rather like the little living amphioxus (*Bran-
chiostoma*), filtering food with its ciliated gills bars,
but some, like *Myllokunmingia*, have been linked
with hagfish. It is important to realize that there were
probably several essays at producing an early fish,

each sharing the chordate characters of an in-compressible notochord against which muscle fibers in V-shaped myotomes acted to oscillate the body, a nerve cord to control the muscles, and perhaps various anterior sense organs of different kinds.

Only one of these early fishlike forms seems to have given rise to all the living fish groups. We can be fairly sure that all living fish are derived from a common ancestor in one of these Cambrian groups, for not only do all today share such notable morphological features as myotomal innervation pattern, Mauthner hindbrain neurons, other brain structures, and so forth, but they also share DNA patterns in a well-studied molecular classification. The conodonts were another of these early fishlike groups, long known from small curious teeth-like fossils important strati-graphically, but of unknown affinity. Comparatively recently, fairly complete fossils were found, which showed that the later conodonts had several chordate characters like serial myotomes, a notochord, and large anterior eyes. The complex and bizarre structure of their feeding mechanism however probably debars the group from the ancestry of living fish.

After these tantalizing early glimpses of fishlike fossils, much better known and more complex fishes appear later in the fossil record, sometimes exqui-sitely preserved, so that they have been reconstructed in great detail, even down to the cranial nerves and impressions of gill filaments. Some of these different Silurian and Devonian fish groups lacked jaws and so have been suggested as ancestral to the modern jawless (Agnathan) lampreys and hagfish. But there is a large gap between these early agnathans and the first hagfish and lamprey fossils of the Upper Car-boniferous, which are similar to modern forms. In the Silurian as well as different groups of agnathan fish, there were several successful groups that had evolved jaws, and such gnathostomes rapidly di-versified, some of their descendants continue as the teleosts and elasmobranchs of today, and the two last lobe-finned coelacanth species (*Latimeria chalumnae* and *Latimeria menadoensis*).

The earliest ray-finned fishes (Actinopterygii) in-cluding the earliest bony teleosts have been shown very recently to have appeared 40 My earlier than previously supposed, in the Paleozoic.

Kinds of Living Fish and How They Are Related

There are a number of rather different groups of liv-ing fish, classified mainly according to their morphol-ogy, on such features as skull bones, scale and fin structure, and vertebral number and design. Most fish

taxonomists today set out the relationships of fishes according to 'cladistic' methods. These yield a branching tree or cladogram where a parent species gives rise to two daughter sister groups and the an-cestor disappears. The cladistic method depends upon discerning which characters are primitive and which are novel or derived, thus recognizing branch points where new sister groups arise. The cladogram re-sulting is then treated by various statistical ap-proaches to produce the minimum number of branch points and thus becomes a classification. A lack of space precludes a detailed discussion of cladistics, which results in a branching classification of the kind seen in **Figures 1** and **2**. The cladistic groupings of fish seen in the classification of **Figure 1** have been broadly supported by more recent molecular analyses.

Although not all fish taxonomists use the most rigorous form of the cladistic method, the great majority today does so. It works well where there is agreement about the status of the characters used, less well when (as for many fossil fish groups) it is uncertain or disputed whether a given character is ancestral or derived.

In **Figure 1**, the larger divisions are shown on the right: between the agnatha and all other fish; between the chimaeras, sharks and rays in the elasmobranchiomorpha; the remaining sturgeons, garpikes, and *Amia*; the lobe-finned lungfish and bichirs, and the vast array of bony teleosts.

The greatest difficulties that taxonomists face are those posed by the very numerous kinds of bony teleost fishes, and by their relationships with the living members of ancient fish lines, such as lungfish, sturgeons, or the bichir *Polypterus*. Bichirs (which so impressed Cuvier, that he would have regarded Napoleon's Egyptian expedition as a success had the discovery of the fish been its only result!) have long restlessly migrated in the classification scheme (see **Figure 2**) until recently, as too have the sturgeons, garpikes (*Lepisosteus*), and the bowfin *Amia*.

Teleosts and Elasmobranchs

The curious relict freshwater species of ancient lines are zoologically interesting, and sturgeons are eco-nomically important, but they are completely over-shadowed by the dominant bony teleosts and the cartilaginous elasmobranchs of both fresh and salt waters. In numbers of species, teleosts (96% of all living fish) far outweigh the 800 or so elasmo-branchs, and have diversified into all kinds of habi-tat, but elasmobranchs are by no means unsuccessful in other respects and apart from killer whales (*Orca*), large sharks are the top marine predators, as well as

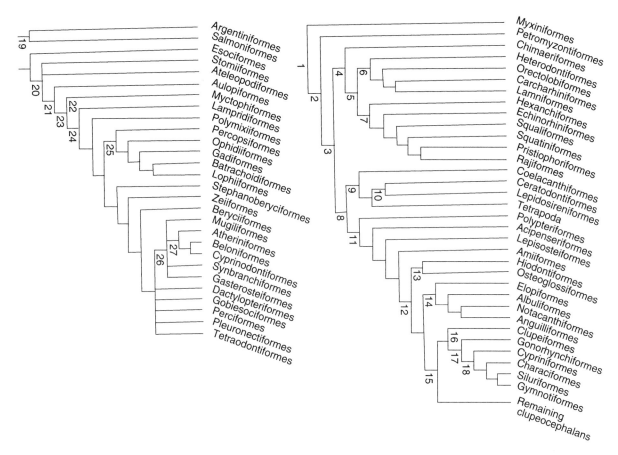

Figure 1 A cladistic classification of living fish. After Stiassny MLJ, Parenti LR, and Johnson GD (eds.) (1997) *Interrelationships of Fishes*. London: Academic Press.

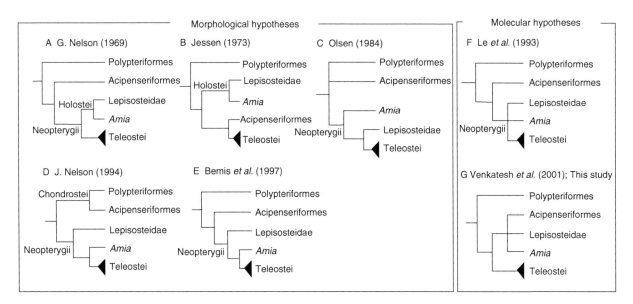

Figure 2 Alternative views about the relationships of *Polypterus* and other 'ancient' fishes. From Inoue JG, Miya M, Tsukamoto K, and Nishida M (2003) Basal actinopterygian relationships: A mitogenomic perspective on the phylogeny of the 'ancient fish'. *Molecular Phylogenetics and Evolution* 26: 110–120.

having relatively larger brains than almost all teleosts; and the plankton feeding whale sharks and basking sharks are the largest of all fish. Because cartilage appears earlier in ontogeny than bone, this suggested at one time that elasmobranchs are more 'primitive' than bony fish. However, not only do the two groups appear more or less simultaneously in the fossil record, but also some small amounts of bone are found in elasmobranchs, probably indicating reduction from earlier more ossified forms.

Teleosts

Living teleosts are the result of four main radiations (**Figure 3**), first clearly recognized some 50 years ago (on morphological grounds alone) in the classical paper published by several museum fish workers. One is very much larger than the others.

Of these four radiations, that in fresh water containing the electrolocating mormyrids and a few large fish like the arapaima (*Osteoglossum*) are regarded as the least advanced. The two other less advanced groups are the Clupeomorpha with the herrings, sprats, and anchovetas and the Elopomorpha with eels, bonefish, and some deep-sea forms. Clupeomorphs all share good hearing, due to a curious link between the swim bladder, the ear, and the lateral line. In some shads, this even enables them to hear the ultrasonic hunting cries of dolphins, and herrings are thought to avoid nets guarded against unintended capture of dolphins by ultrasonic devices. Elopomorphs, sometimes quite large fish, like the tarpon Megalops, all share the ribbon-like flattened leptocephalus larva, in some cases as in the notacanth *Aldrovandia* more than a meter long, but in most, as in *Anguilla* species, shorter and willow leaf-shaped.

On the whole, the progressive changes leading to the largest and most derived group, the euteleosts, from early teleost ancestors have been well summarized by W.B. Stout's famous advice in designing

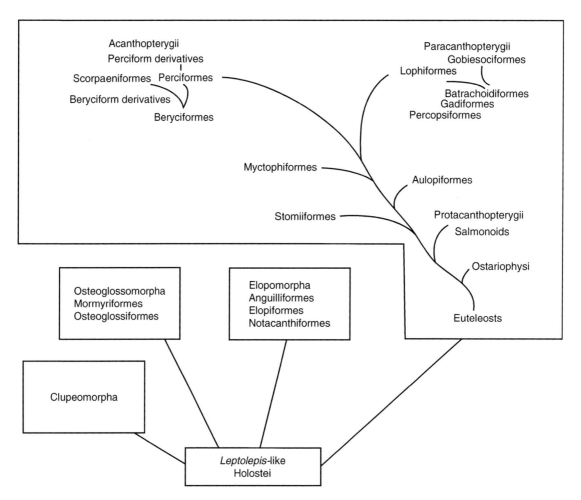

Figure 3 The four main teleost radiations. Reproduced from Bone Q and Marshall NB (1995) *Biology of Fishes*, 2nd edn. Glasgow, UK: Blackie. From Greenwood PH, Rosen SH, and Myers GS (1966) Phyletic studies of teleostean fishes with a provisional classification of living forms. *Bulletin of the American Museum of Natural History* 131: 339–456.

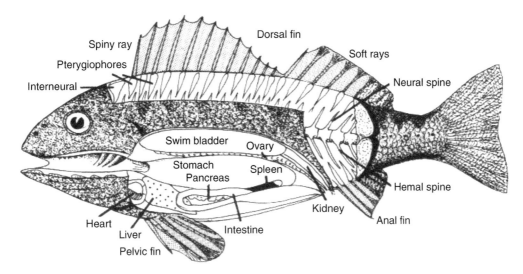

Figure 4 Main features of a typical acanthopterygian teleost, the carp *Cyprinus carpio*. Reproduced from Bone Q and Marshall NB (1995) *Biology of Fishes*, 2nd edn. Glasgow, UK: Blackie. After Dean B (1895) *Fishes, Living and Fossil. An Outline of their Forms and Probable Relationships.* New York, NY: Macmillan.

the 1930s Ford trimotor aeroplane "simplicate and add more lightness": from their paleoniscid ancestors, fishes with thick bony scales, similar to *Polypterus* or *Lepisosteus*, the modern euteleosts have thinner scales, lightened fin rays, skull bones braced by a fenestrated scaffolding structure, and fewer bones in the lower jaw.

At the apex of the euteleosts are the numerous acanthopterygian species, so diverse in habit and form as to include small gobies; the huge sunfish (*Mola*); scombroids like mackerel (*Scomber*) and barracudas (*Sphyraena*); and over 9000 perch-like fish.

Teleosts like carp (*Cyprinus carpio*; **Figure 4**), cod (*Gadus morrhua*), herrings (*Clupea harengus*), eels, and salmon are the great majority (96%) of all fish. Almost all the fish we usually eat are bony teleosts, though a few other groups like sturgeons are also valuable as food, and small dogfish sharks are used in fish (rock salmon) and chips. Because teleosts are extraordinarily varied in shape, size and color, they are not so easy to categorize, but all share bony skeletons with flexible fins supported by bony fin rays. Some are herbivorous, like the important cultivated grass carp (*Ctenopharyngodon*), others like herring are planktivorous, and many are carnivores. Some carnivores eat strangely restricted prey, for example, sunfish survive on a watery diet of jellyfish and salps, and some smaller fish bite off portions of fins of others, or even specialize in sucking the eyes out of other fish. The body is covered with protective scales; though these may be small and not obvious, they lie within epithelial pockets protecting them from the external medium. Teleosts in temperate waters show annual growth rings on their scales (and

in the bony otoliths of the inner ear), so are much easier to age than elasmobranchs. Only the least advanced teleosts have a spiral valve in the gut, and most have a gas-filled swim bladder making them close to neutral buoyancy.

Why have modern teleosts been so successful in radiating into almost all aquatic habitats and so greatly dominating in numbers of species? A significant part of the reason must be that the teleost design, refined in several ways from the ancestral forms, is capable of operating over a very large size range, and so can occupy many niches and trophic levels, particularly in fresh water, where *c.* 40% of all teleosts are found.

Elasmobranchs

Elasmobranchs like sharks and rays, and their relatives the chimaeras (rabbitfish or spookfish), come a very poor second in number of species to the teleosts. They differ from teleosts because they lack a swim bladder, contain large amounts of urea as an osmolyte, and have a mainly cartilaginous skeleton. All sharks and rays are covered in tooth-like denticles (much reduced in rabbit fish), sometimes with bony bases: all are either planktivorous or carnivorous, and have a spiral valve in the gut, and a special rectal gland which excretes NaCl. All have internal fertilization (seen in rather few teleosts), with large male intromittent organs (claspers). The young either leave the mother in tough egg cases, or are born alive, in some species after a period of cannibalism in the uterus, in others after being nourished by uterine 'milk'. In many shark species, the young gather in

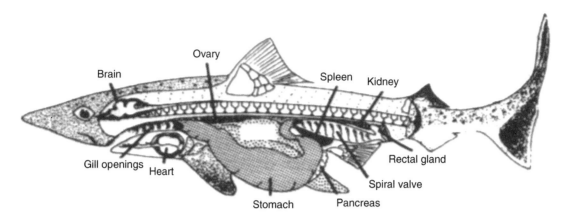

Figure 5 Main features of an elasmobranch, the spurdog *Squalus acanthias*. Reproduced from Bone Q and Marshall NB (1995) *Biology of Fishes*, 2nd edn. Glasgow, UK: Blackie. From Lagler KF, Bardach JE, Miller RR, and Passino DM (1977) *Ichthyology*, 2nd edn. New York: Wiley.

nursery areas where they are not at risk from predation by adults.

In contrast to teleosts, elasmobranchs have today a smaller size range, for although there are several very large filter-feeding elasmobranchs, there are none smaller as adults than the little cookie-cutter sharks (e.g., *Squaliolus*; **Figure 5**) around 30 cm. They have also almost entirely failed to invade fresh water, where, as seen above, *c*. 40% of teleosts live. At one time, it was supposed that small-sized elasmobranchs with a large surface/volume ratio might have osmoregulatory difficulties, owing to their blood containing a very easily diffusible osmolyte (urea). However, the small dogfish (*Scyliorhinus*) embryos in their 'mermaids purses' have no problem in osmoregulating. As to fresh water, sharks like the bull shark (*Carcharias leucas*) can swim up rivers into fresh water, and remain there for long periods, and some stingrays can only live in fresh water, so evidently it could have been possible for others in the group to adapt.

Unlike the bony teleosts and the few remnants of other groups like lungfish and bichirs or sturgeons, the modern elasmobranch skeleton is built from cartilage, often fairly thick, and strengthened at strategic points by calcifications (or even sometimes, as at scale bases, with bone).

Distribution of Marine Fishes

In oceanic waters, sufficient light for photosynthesis by phytoplankton penetrates down to at least 100–150 m. In this productive surface euphotic zone (especially in the clear warm waters of the Tropics), there are some 250 kinds of adult fish, including large predators like blue and white tip sharks, and scombroids like tunas and billfish. They feed on smaller fish like flying fish, gar, and halfbeaks. Many fish have buoyant eggs, and their larvae hatch and grow in this zone even if as adults they live deeper.

Below the euphotic zone, there are *c*. 850 fish species in the mesopelagic layers (200–1000 m). Most of these are quite small, 5–10 cm overall, either black or silvery, and most are myctophids, with sexually dimorphic lateral photophores. There are however some larger fish like the black trichiurid scabbard fishes (*Aphanopus carbo*) caught on deep long lines for the Madeiran and Portuguese fish markets, and the strange megamouth shark (*Megachasma*). Feeding on copepods which migrate upward daily from the deep scattering layer, myctophids, stomiatoids, and similar fish also undertake daily vertical migrations, while the few larger fishes like aleposaurids and giant swallowers remain at depth. The deep scattering layer is seen on echo sounders because of reflections from the swim bladders of the small fish feeding within it.

Many of the commonest bathypelagic fishes (living below 1000 m) are the small ubiquitous *Cyclothone* species (stomiatoids), though in numbers of species angler fishes dominate. This is a generally food-poor zone, and the fish within it have reduced muscles and skeletal tissues, though with large mouths to take opportunistic meals. Angler fish are floating traps, with 'parasitic' males fused to the larger females.

Deeper still, just off the bottom, are benthopelagic fish like the dark brown or black squaloid sharks floating on their oil-filled livers and many rat-tail (Macrourid) species. Other fish like the tripod fish (*Bathypterois*), lizard fish (Synodontids), and deep-sea rajids rest on the bottom itself.

These fishes of the deep sea live in a rather constant environment, and are cosmopolitan, but fish of shallower seas are less widely distributed. The richest fish fauna, as we might guess, is found around coral reefs and atolls, where many different families of the most advanced teleosts live, most diverse in the Phillipines. But even on the temperate seasonally changing shores of the (richer) N. Pacific and N. Atlantic, there are nearly 1000 coastal fishes.

Special Adaptations of Fish

Probably the most interesting aspect of the two major fish groups is the way that the two have taken different approaches to solving the same problems dictated by life in water. Sometimes, indeed, the two have independently arrived at very similar solutions.

Buoyancy

Most teleosts possess a gas-filled swim bladder of sufficient volume to counterbalance the dense components of their bodies and render them neutrally buoyant. For most fish this requires a swim bladder around 5% of body volume in seawater, and 7% in fresh water, but there are fish with dense scales and heavier than normal bones that exceed these values: for instance, gurnards need a swim bladder around 9% of body volume for neutral buoyancy, while in the gar *Lepisosteus* with its dense scales, the swim bladder occupies 12% of body volume. Neutral buoyancy is advantageous for several reasons, and is found in fish from the surface to the depths of the sea, but the use of swim bladder gas to achieve it demands remarkable physiological adaptations, including special properties of the blood, ingenious counter-current flow rete mirabilia serving the swim bladder gas gland, and a simple method of preventing gas diffusing out of the bladder. Since swim bladders obey Boyle's law nearly perfectly, depth changes can pose difficult problems. Ambient pressure changes by 1 atm (101.3 Pa) for every 10-m depth change, so it is small wonder that many fish like mackerel (*Scomber*) or tunas, which hunt up and down from the surface, either reduce or lose their swim bladders. Midwater and deep-sea fish suffer little from small depth changes, since a 10-m depth change at say 400 m will hardly affect swim bladder volume. Nevertheless, numbers of fish like mycto-phids undergo a daily vertical migration of hundreds of meters, following their vertically migrating co-pepod prey, and it remains unclear whether they can retain neutral buoyancy over this wide depth range. That this is a real problem is suggested by the way that many adult myctophids replace the gas in their swim bladders with lipids like wax esters where the static lift provided alters little with depth.

Elasmobranchs lack swim bladders, and avoid the complexities of gas regulation within them. Instead, sharks like basking sharks or the deep-sea squaloids gain static lift instead from low-density oils like squalene stored in their livers, which offers lift that changes little as the fish changes depth. However, oil storage has its problems too, for there have to be complex biochemical controls to maintain buoyancy lipid separate from that used for other purposes, including fuel for locomotion.

Locomotion

All fish swim by contracting body or fin muscles to affect the surrounding water in such a way as to move the fish forward (and sometimes backward). It is hardly astonishing that almost all kinds of living fish from hagfish and lampreys to elasmobranchs and teleosts have basically the same muscular swim-ming machinery. The constraints that so dense a medium imposes are severe, curtailing alternatives. Speed in water is energetically very costly, so most fish cruise slowly using muscles that operate aerobically. Only in emergencies do they swim rapidly for short bursts using muscles with a much less efficient anaerobic use of fuel. Streamlining is mandatory for fast-swimming fish, and several have ingenious drag-reducing devices of one kind or another aimed at maintaining the boundary layer. For instance, both teleosts and elasmobranchs have independently produced microridging to delay boundary layer separation.

Body Fluids and Osmoregulation

Apart from hagfish, all marine fish have blood and body fluids quite different in composition to the surrounding seawater. Teleosts have blood that is more dilute than seawater, and so face loss of ions and ingress of water. To overcome osmotic water loss, marine teleosts have to drink seawater. The sea bass *Serranus* drinks around 12% of its body wt. daily, and absorbs about 75% of this water in the gut. But this replaces an osmotic problem with an ionic one! The introduction of radioactive tracers like Na^{24} showed very surprisingly that almost 10 times the amount of NaCl entered the marine teleost via the gills than that drunk to avoid water loss. The gradual accumulation of evidence showing that special 'chloride cells' on the gills (first discovered in 1932) regulated NaCl excretion from the blood (as was then suggested) lasted for almost 50 years! The final proof came (in the 1990s) from a vibrating probe showing negative current peaks over the apices of actively secreting immunolabeled chloride cells.

Elasmobranchs on the other hand have blood that is osmotically similar to seawater, because they store nitrogenous osmolytes, mainly urea (at just over 0.5 M). The problems are that NaCl diffuses into the blood down the concentration gradient between 0.5 and 1.0 M, and that urea is small and easily diffusible. Though there are chloride cells in elasmobranch gills, the greater part of the NaCl entering is excreted by a gland opening into the rectum, which secretes M NaCl! The elasmobranch and teleost approaches to the osmotic problem might be very roughly compared by the requirement in terms of ATP required for Cl$^-$ secretion, when the elasmobranch solution seems energetically much less costly. But this omits considerations of the costs of urea transporters and special gill membrane lipids which are hard to estimate.

Warm Blood

Some scombroid teleosts and lamnid and carcarhinid sharks warm different regions of the body above ambient. Blue fin tuna (*Thunnus thynnus*) cannot only maintain body temperature above ambient, but can regulate their body temperature in waters of different temperature. Salmon sharks (*Lamna ditropis*) need to keep their red cruising muscles around 25 °C to enable them to operate properly. In these cases, cool blood passing from the heart to the cruising musculature is warmed by blood passing from it, in parallel counter-current retia.

Warming locomotor muscles may increase their output, but probably the main reason for accepting the increased capillary resistance of the retia is that it enables these fish to swim efficiently in waters of very different temperature. Similar retia are arranged to warm the viscera in these fish, and the brain in other fishes. In the swordfish (*Xiphias gladius*), there are special arrangements to warm the eye, and so increase its acuity. Usually, normal muscular contractions provide the heat to warm different regions (for instance, a special vein passes from the warm muscle of the trunk to the brain retia in *L. ditropis*), but sometimes modified muscle tissue is designed solely for heat production.

Eyes, Visual Pigments, and Photophores

Fish in the oceans live in a variety of visual environments. Excellent vision is a prerequisite for most fish, and even in the stygian depths of the oceanic trenches, most fish have large sensitive eyes to see the glows and flashes of photophores. No oceanic fish has divided its eye like the freshwater *Anableps*, but clinids and flying fish have corneal flattened windows to scan their surroundings during aerial excursions.

In the oceans, except just near the land where runoff tinges the water, the wavelength (λ) best transmitted is around 470 nm. For most fish, the optimal visual pigment is one that absorbs maximally (λ_{max}) around 470 nm – for with this, they can see farthest. There are some dragonfishes (stomiatoids) however, which not only have a visual pigment absorbing maximally in the red, around 575 nm, but also a red headlight photophore, which shines invisibly onto their prey, whose visual pigments do not absorb at this wavelength. Most photophores, as we should expect, glow at around 470 nm, to be seen as far as possible to attract mates and lure prey.

On land, the light regime is irregular and varies a good deal, but in the oceans the polar distribution of light is constant, and symmetrical. This symmetric light distribution enables a fish such as a herring to resemble a mirror in the ocean by silvering its scales with guanine. Predators looking at a herring from the side will see a mirror reflecting light identical to that seen if the herring was absent, so it is well camouflaged. The only difficulty is that predators looking upward will see the ventral surface of a silvery fish darkly silhouetted against downwelling light. Herring and sprat more or less evade this problem by making their ventral surface knife-like, that is, they have virtually no ventral surface, but other fish are even more ingenious.

They shine their own light downward from ventrally directed photophores, so that seen from below they match downwelling light. Evidently, for the fish to be completely invisible from below, its photophores have to emit light of the appropriate intensity, wavelength, and direction. This complex problem is solved by the extraordinary laterally flattened sternoptychids like *Argyropelecus*. In such fish, there are not only special color filters and hemispherical reflectors, but also half-silvered mirrors!

Electroreception, Magnetic Fields, and Navigation

There are no marine fish which have developed the extraordinary electrolocating and 'communication' systems of the freshwater mormyrids. Only two kinds of marine fish, the stargazers (*Astroscopus* and *Uranoscopus*) and many (perhaps all) rays, generate electric pulses. The stargazers stun small fish, while rays seem to use their electric organs to communicate with conspecifics.

However, all elasmobranchs have exquisitely sensitive electroreceptors, the ampullae of Lorenzini, with which small sharks have been shown by experiment to detect the electrical signals from their buried prey. The electroreceptors involved respond to minute voltage gradients of $0.01\,\mu V\,cm^{-1}$, corresponding to $1\,mV\,km^{-1}$! It seems likely that the large basking shark and perhaps manta rays detect swarms of their prey copepods by the muscular electrical activity of the copepod filtering limbs.

In addition, these ampullary electroreceptors are sufficiently sensitive to detect changes in the electrical field of the Earth as the fish swim through it. Although the only type of experimental attack on how the fish may use this information has come from small-scale orientation experiments on stingrays, it seems probable that wide-ranging oceanic sharks navigate in this way.

There is also evidence that some teleost fish can use the magnetic field of the Earth to navigate with magnetoreceptors containing magnetite crystals. After a considerable search, it was finally shown conclusively that bluefin tuna have magnetoreceptors associated with olfactory receptors in the nasal epithelium. Archival tags have shown that these fish regularly make annual journeys around the N. Atlantic to Mediterranean spawning sites, presumably using magnetic navigation.

Value to Man and Relation to Other Disciplines

Of course, many kinds of fish are of direct or indirect economic value, but it is less obvious that studies on fish have led to important scientific discoveries that have advanced our knowledge of human physiology. Perhaps the most rewarding of such studies have been endocrinological. Pituitary neurosecretion was demonstrated first in teleosts over 70 years ago, and a recent review of various forms of a reproductive hormone (GnRH) pointed out that fish in general and teleosts in particular have often played a leading part in changing established concepts. Indeed, several hormones first found in fish, such as the urotensins and stanniocalcin, were later found to be important in humans.

An interesting feature of the relation of work on fish to other disciplines has been that of the ways in which each has contributed to the other. Recent studies of fish locomotion have been transformed by the advent of digital particle velocimetry from engineering flow visualization work, but it is at present unclear whether the studies on polymer drag reduction (apparently initiated by work on barracuda mucus) have been practically valuable. Since both the fish mucus studies and work on longitudinal grooving of shark scales seem to have led to banishment of both

on racing yachts, they apparently have had a definite (if forbidden) effect on yacht hydrodynamic design. Vortex generators, however, were independently invented by Boeing engineers before it was realized that they were to be found on many fish species.

See also

Antarctic Fishes. Coral Reef Fishes. Fish Ecophysiology. Fish Locomotion. Fish Migration, Horizontal. Fish Migration, Vertical. Fish Reproduction. Marine Mammal Evolution and Taxonomy. Marine Mammal Migrations and Movement Patterns. Marine Mammal Overview. Marine Mammal Trophic Levels and Interactions. Mesopelagic Fishes.

Further Reading

Bone Q and Marshall NB (1995) *Biology of Fishes*, 2nd edn. Glasgow: Blackie.

Dean B (1895) *Fishes, Living and Fossil. An Outline of their Forms and Probable Relationships.* New York, NY: Macmillan.

Denton EJ (1970) On the organization of reflecting surfaces in some marine animals. *Philosophical Transactions of the Royal Society of London, B* 258: 285–513.

Greenwood PH, Rosen SH, and Myers GS (1966) Phyletic studies of teleostean fishes with a provisional classification of living forms. *Bulletin of the American Museum of Natural History* 131: 339–456.

Horn M (1972) The amount of space available for marine and freshwater fishes. *Fishery Bulletin* 70: 1295–1297.

Hurley I, Mueller RL, Dunn KA, *et al.* (2007) A new time-scale for ray-finned fish evolution. *Proceedings of the Royal Society of London, B* 274: 489–498.

Inoue JG, Miya M, Tsukamoto K, and Nishida M (2003) Basal actinopterygian relationships: A mitogenomic perspective on the phylogeny of the 'ancient fish'. *Molecular Phylogenetics and Evolution* 26: 110–120.

Lagler KF, Bardach JE, Miller RR, and Passino DM (1977) *Ichthyology*, 2nd edn. New York: Wiley.

Lethimonnier C, Madigou T, Muñoz-Cueto J-A, Lareyre J-J, and Kah O (2004) Evolutionary aspects of GnRHs, GnRH neuronal systems and GnRH receptors in teleost fish. *General and Comparative Endocrinology* 135: 1–16.

Nelson JS (2006) *Fishes of the World*, 4th edn. New York: Wiley.

Stiassny MLJ, Parenti LR, and Johnson GD (eds.) (1997) *Interrelationships of Fishes.* London: Academic Press.

ANTARCTIC FISHES

I. Everson, Anglia Ruskin University, Cambridge, UK

Introduction

Antarctica is a continental landmass much of which is covered by an ice cap, consequently the ichthyofauna is totally marine. Surrounding the continent is the Southern Ocean, approximately 36 million km^2, continuous with the Atlantic, Indian, and Pacific Ocean basins to the north and whose northern limit is generally taken as the Antarctic Polar Frontal Zone (APFZ). There is a clear separation between the Antarctic and the Southern Hemisphere continents, the nearest connection being with South America via the Scotia Arc, a series of islands separated from each other by deep water.

The Antarctic Circumpolar Current (ACC) and the general oceanographic regime mean that marine isotherms are more or less concentric around the continent. Close to the continent the seasonal variation in temperature is rarely more than 1 °C while even at the northern limit, as for example at South Georgia, the range is little more than 4 °C.

These two factors, geographical isolation and constant low temperature, have a major effect on Antarctic fish.

Fish Fauna

The Southern Ocean ichthyofauna is relatively sparse and unusual in composition, consisting of 213 species belonging to only 18 families (**Table 1** and **Figure 1**). Nearly half the species belong to one group, the perciform notothenioids, which make up 45% of the fish fauna. Restricting consideration to the shelf, and particularly in the highest latitudes, notothenioids make up 77% of the species and 90–95% of the biomass of fish. Notothenioids are morphologically and ecologically diverse, and have variegated into a wide variety of niches, mainly demersal, and also in the water column and even within sea ice. As a group, this makes them more diverse than, for example, the finches of the Galápagos archipelago. The concept of species flocks has been developed for freshwater fish to identify groups that have a close affinity; typically such flocks are to be found in ancient lake systems and it is extremely unusual for such a flock to be identified from a large marine environment. Antarctic notothenioids with their high species diversity and endemism form a species flock comparable to that of Lake Baikal.

Early taxonomic studies were based on the traditional methods of morphometric and meristic analyses. Recent studies have used molecular biological analyses not only of nuclear material but also of antifreeze compounds to indicate phylogeny.

Adaptations

Cold Adaptation

Some of the earliest studies on the physiology of Antarctic fish concerned the measurement of oxygen uptake rates. Initially it had been assumed that, since many biochemical processes are temperature-dependent, the metabolic rates of Antarctic fish might be very low. The initial experiments indicated that rates were substantially higher than those of temperate fish when studied at low temperature and the degree of elevation of the metabolic rate in Antarctic fish was attributed to a phenomenon termed 'cold adaptation'. Subsequent studies demonstrated that the greater part of this elevation was caused by handling stress and the extended recovery

Table 1 Composition of Southern Ocean ichthyofauna

Taxon	Benthic	Benthopelagic	Pelagic
Agnatha	2		
Chondrichthyes	8	2	1
Osteichthyes			
Notacanthiformes		2	
Anguilliformes		2	
Salmoniformes	4		5
Stomiiformes			12
Aulopiformes			9
Myctophiforms			35
Gadiformes	9	11	1
Ophidiiformes	1		1
Lophiiformes			3
Lampriformes			2
Beryciformes			6
Zeiformes		1	
Scorpaeniformes	32		
Perciformes			
Zoarcidei	22		
Notothenioidei	95		
Blennioidei	1		
Scombroidei			2
Stromateoidei			1
Pleuronectiformes	4		

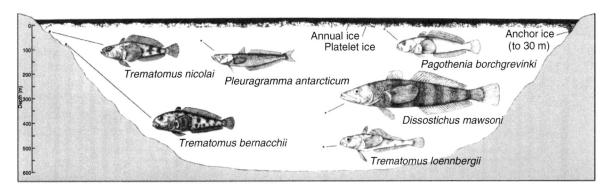

Figure 1 These six species from McMurdo Sound demonstrate some of the life-history types included in the Nototheniidae. Pelagic, cryopelagic, epibenthic, and benthic species are illustrated. Dots indicate typical habitat, although most species have considerable depth ranges. Modified from Eastman and DeVries (1986).

time, of the order of 24 h or more, following introduction into respirometers. In spite of this, it is now accepted that some slight elevation of metabolic rate remains that cannot be explained wholly by experimental technique. Consideration of the phenomenon has raised some controversy between different workers. The existence of the phenomenon has been demonstrated experimentally, although it does not appear to confer any evolutionary advantage because it implies a higher energy requirement on the part of the fish. All these studies have been undertaken on whole fish; the overall oxygen uptake rate being the balance between all the component metabolic pathways that are present. As such it has been argued that the term 'cold adaptation' has little meaning and that it is more sensible to consider each metabolic component separately to provide an overall balance.

Antifreeze

Pure water freezes at $0\,°C$, but the presence of salts causes the freezing point to be depressed such that normal seawater freezes at around $-1.85\,°C$. At McMurdo Sound the annual mean water temperature is $-1.87\,°C$ and varies within the range -1.40 to $-2.15\,°C$. Body fluids, such as the blood plasma, of most teleost fish have a freezing point of $c.\ -0.7\,°C$. Even though this difference is small it is important, because living in waters close to the freezing point of seawater, Antarctic fish require some mechanism to prevent their body fluids from freezing.

In the absence of ice, fish could live in a supercooled state. Unfortunately this is not a stable state because very few ice crystals are required to cause a supercooled liquid to freeze. An alternative adaptation is required. The ionic concentration of the blood of most marine teleosts is $320–380\,\mathrm{mOsm\,kg^{-1}}$, only about one-third of that of Antarctic seawater ($1050\,\mathrm{mOsm\,kg^{-1}}$). The freezing

point depression of some notothenioids at McMurdo Sound is $-2.2\,°C$, although their blood osmolality is $550–625\,\mathrm{mOsm\,kg^{-1}}$, equivalent to a freezing point depression of -1.02 to $-1.16\,°C$. Thus although there appears to be some compensation as measured by the osmolality, it is insufficient to explain all of the depression in freezing point. Compensation for this difference comes in the form of antifreeze glycopeptides (AFGPs) which exert their effect by a mechanism known as adsorption-inhibition (**Figures 2 and 3**). Even though ice crystals can form, their further growth is prevented when AFGPs are adsorbed onto them because the AFGP molecule prevents growth of the ice crystal along its main axis. Thus the AFGPs have an antifreeze function, lowering the freezing point beyond that which would be expected from the osmolality. The AFGPs however do not lower the melting point.

The AFGP molecules are of such a size that they would be lost through the glomeruli of normal teleost kidneys. In glomerular nephrons of normal teleosts, molecules with a molecular weight of $<68\,000\,\mathrm{Da}$ pass through the filtration barrier. As the urine passes through the different parts of the nephron, it is modified by reabsorption of nonwaste products and secretion of waste products. The AFGP molecules are of such a size that they would pass through the glomeruli but would need to be reabsorbed later on in the nephron. The kidneys of all Antarctic fish which possess AFGPs are aglomerular, obviating this requirement. Thus the evolution of the aglomerular trait in Antarctic fish complements that of the presence of antifreeze.

Cardiovascular

A continuous low water temperature means that the oxygen-carrying potential of seawater is high. Thus, as long as the partial pressure of oxygen in the seawater remains high, so will the available oxygen. It is against this background that further cardio-vascular adaptations have evolved.

Early taxonomic studies relied on specimens preserved in alcohol or formalin, both of which affect the color of the fish. Because fish typically possess red

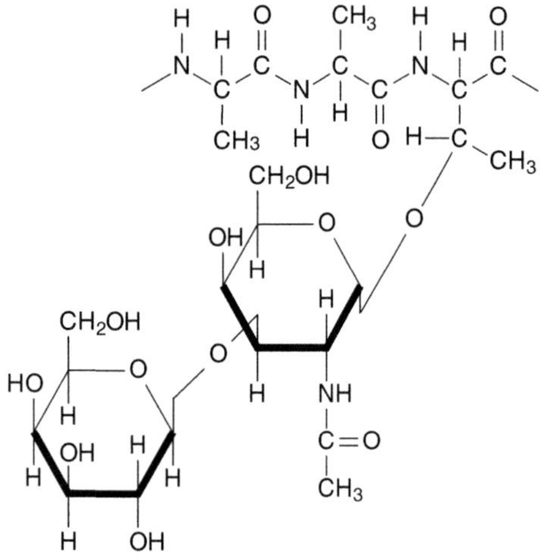

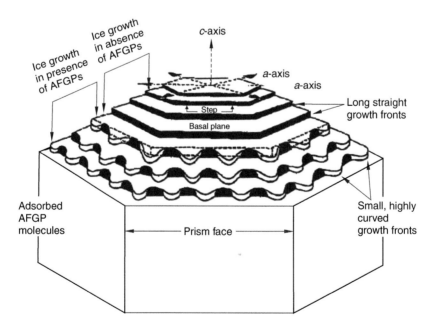

Figure 2 Basic repeating structural unit of the AFGPs of notothenoids. The peptide consists of amino acids in the sequence [alanyl-alanyl-threonine]$_n$. Each threonine is joined to a disaccharide through a glycosidic linkage. In low-molecular-weight AFGPs 6–8, proline is periodically substituted for alanine at position 1 of the tripeptide. Reproduced from Eastman JT (1993) *Antarctic Fish Biology: Evolution in a Unique Environment*. London: Academic Press.

blood, until the 1950s no mention was made of the anemic appearance of the gills of some species of Antarctic fish. At that time, it was noticed that members of the Channichthyidae (at that time called Chaenichthyidae) were white, as a result of which they were called 'white-blooded fish' or 'icefish'. The blood of channichthyids is devoid of hemoglobin, although small numbers of nonfunctional erythrocytes have been described in a few species.

Initial consideration was given to determine whether, because channichthyids do not possess scales, cutaneous respiration might be a major factor in oxygen uptake. However, the absorptive area and vascularization relative to the gills mitigated against that mechanism. Alternatively it was thought possible that channichthyids possessed either a more efficient oxygen utilization mechanism or else lowered oxygen requirement. This second consideration was being examined at a time when the concept of metabolic cold adaptation was under discussion.

The viscosities of the plasma of red-blooded notothenioids and channichthyid fish are very close, although the blood of the former is approximately 25% higher than the latter. Studies on oxygen uptake rates indicated that channichthyids utilized oxygen at a slightly lower rate as compared with equivalent red-blooded notothenioids. In the absence of hemoglobin, the oxygen-carrying capacity of channichthyid blood is only about one-tenth that of

Figure 3 Model of an ice crystal depicting adsorption-inhibition as a mechanism for the freezing point depression of water by antifreezes. In the absence of AFGPs, ice crystal growth occurs as water molecules are added to the crystal in a regular fashion at steps on the basal planes. When the AFGPs are adsorbed, ice cannot propagate over them and long straight fronts become divided into many small curved fronts. Reproduced from Eastman JT (1993) *Antarctic Fish Biology: Evolution in a Unique Environment*. London: Academic Press.

red-blooded fish. Two mechanisms are possible to compensate for this effect: either channichthyid blood is circulated at a much faster rate or there is much more of it in the system. The latter has proven to be the case and channichthyid blood takes up 8–9% of the total volume of the fish (2–4 times that of other teleosts); the heart rate and blood pressure are low but the stroke volume and resultant cardiac output are large. To reduce the resistance to flow, the capillaries are larger than in other teleosts and the blood is less viscous.

Even though the hemoglobin-less condition is clearly effective, it is a feature that confines the fish to areas of high oxygen tension such as those present in Antarctic waters. Only one channichthyid species, *Champsocephalus esox*, is found outside of the Antarctic zone. Experimental studies have demonstrated that channichthyids are particularly sensitive to hypoxia, indicating that in their natural habitat the oxygen saturation is always consistently high.

See also

Fish Ecophysiology. Fish Feeding and Foraging. Fish Reproduction.

Further Reading

Cheng C-HC, Cziko PA, and Evans CW (2006) Nonhepatic origin of notothenioid antifreeze reveals pancreatic synthesis as common mechanism in polar fish freezing avoidance. *Proceedings of the National Academy of Sciences of the United States of America* 103: 10491–10496.

Clarke A (1991) What is cold adaptation and how should we measure it? *American Zoologist* 31: 81–92.

Di Prisco G, Pisano E, and Clarke A (eds.) (1998) *Fishes of Antarctica; A Biological Overview.* Milan: Springer.

Eastman JT (1993) *Antarctic Fish Biology: Evolution in a Unique Environment.* London: Academic Press.

Gon O and Heemstra PC (eds.) (1990) *Fishes of the Southern Ocean.* Grahamstown: JLB Smith Institute of Ichthyology.

Kock K-H (1992) *Antarctic Fish and Fisheries.* Cambridge, MA: Cambridge University Press.

CORAL REEF FISHES

M. A. Hixon, Oregon State University, Corvallis, OR, USA

Introduction: Diversity, Distribution, and Conservation

Coral reef fishes comprise the most speciose assemblages of vertebrates on the Earth. The variety of shapes, sizes, colors, behavior, and ecology exhibited by reef fishes is amazing. Adult body sizes range from gobies (Gobiidae) less than 1 cm in length to tiger sharks (Carcharhinidae) reportedly over 9 m long. It has been estimated that about 30% of the some 15 000 described species of marine fishes inhabit coral reefs worldwide, and hundreds of species can coexist on the same reef. Taxonomically, reef fishes are dominated by about 30 families, mostly the perciform chaetodontoids (butterflyfish and angelfish families), labroids (damselfish, wrasse, and parrotfish families), gobioids (gobies and related families), and acanthuroids (surgeonfishes and related families).

The latitudinal distribution of reef fishes follows that of reef-building corals, which are usually limited to shallow tropical waters bounded by the 20 °C isotherms (roughly between the latitudes of 30° N and S). The longitudinal center of diversity is the Indo-Australasian archipelago of the Indo-Pacific region. Local patterns of diversity are correlated with those of corals, which provide shelter and harbor prey. There is a high degree of endemism in reef fishes, especially on more isolated reefs, and many species (about 9%) have highly restricted geographical ranges.

The major human activities that threaten reef fishes include overfishing (especially by destructive fishing practices and live collections for restaurants and aquariums), and habitat destruction, which includes both local effects near human population centers and the ongoing worldwide decline of reefs due to coral bleaching and ocean acidification caused by anthropogenic carbon emissions and global warming. Worldwide, about 31% of coral reef fishes are now considered critically endangered and 24% threatened. The major solution for local conservation is fully protected marine reserves, which have proven effective in replenishing depleted populations.

Fisheries

Where unexploited by humans, coral reef fishes typically exhibit high standing stocks, the maximum being about $240 \, t \, km^{-2}$ (about $24 \, t \, C \, km^{-2}$). High standing crops reflect the high primary productivity of coral reefs, often exceeding $10^3 \, g \, C \, m^{-2} \, yr^{-1}$, much of which is consumed directly or indirectly by fishes. Correspondingly, reported fishery yields have reached $44 \, t \, km^{-2} \, yr^{-1}$, with an estimated global potential of $6 \, Mt \, yr^{-1}$. These fisheries provide food, bait, and live fish for the restaurant and aquarium trades. However, the estimated maximum sustainable yield from shallow areas of actively growing coral reefs is around $20–30 \, t \, km^{-2} \, yr^{-1}$, so many reefs are clearly overexploited. Indeed, overfishing of coral reefs occurs worldwide, due primarily to unregulated multispecies exploitation in developing nations. Few and inadequate stock assessments or other quantitative fishery analyses, susceptibility of fish at spawning aggregations (see below), and destructive fishing practices (including the use of dynamite, cyanide, and bleach) are contributing factors. In the Pacific, some 200–300 reef fish species are taken by fisheries, about 20 of which comprise some 75% of the catch by weight. As fishing intensifies in a given locality, large fishes, especially piscivores (see below), are typically depleted first, followed by less-preferred, smaller, and more-productive planktivores and benthivores. (Note that fishing of some piscivores is naturally inhibited in some regions by ciguatera fish poisoning, caused by dinoflagellate toxins concentrated in the tissues of some species.) The indirect effects of overfishing include the demise of piscivores, perhaps enhancing local populations of prey species or causing a trophic cascade. Overfishing of urchin-eating species (such as triggerfishes, Balistidae) and various herbivorous fishes may provide sea urchins predatory and competitive release, respectively. Although urchin grazing may help maintain benthic dominance by corals, overabundant urchins may overgraze and bioerode reefs.

Morphology

A typical perciform reef fish (virtually an oxymoron) is laterally compressed, with a closed swimbladder and fins positioned in a way that facilitate highly maneuverable slow-speed swimming. Compared to more generalized relatives, reef fishes have a greater proportion of musculature devoted to both locomotion and feeding. Their jaws and pharyngeal apparatus are complex and typically well developed for suction feeding of smaller invertebrate prey, with tremendous variation reflecting a wide variety of diets. For example, most butterflyfish (Chaetodontidae) have forceps-like jaws that extract individual polyps from corals, many damselfish (Pomacentridae: e.g., genus *Chromis*) have highly protrusible jaws that facilitate pipette-like suction feeding of zooplankton, and parrotfishes (Scaridae) have fused beak-like jaw teeth and molar-like pharyngeal teeth enabling some species to excavate algae from dead reef surfaces. (This excavation and subsequent defecation of coral sand can bioerode up to 9 kg of calcium carbonate per square meter annually.) Tetraodontiform reef fishes typically swim relatively slowly with their dorsal and anal fins, and consequently are morphologically well defended from predation by large dorsal–ventral spines (triggerfishes, Balistidae), toxins (puffers, Tetraodontidae), or quill-like scales (porucupinefishes, Diodontidae). The latter two families have fused dentition which is well adapted for consuming hard-shelled invertebrates.

Diurnal reef fishes are primarily visual predators. Visual acuity is high and retinal structure indicates color vision. At least some planktivorous damselfishes have ultraviolet-sensitive cones, which may assist in detecting zooplankton by enhancing contrast against background light. Coloration is highly variable (including ultraviolet reflectance), ranging from cryptic to dazzling. Bright 'poster' colors are hypothesized to serve as visual signals in aggression, courtship, and other social interactions. Sexually dimorphic coloration is associated with haremic social systems (see below). Nocturnal reef fishes are either visually oriented, having relatively large eyes (e.g., squirrelfishes, Holocentridae), or rely on olfaction (e.g., moray eels, Muraenidae).

Behavior

Overt behavioral interactions between coral reef fishes include mutualism (when both species benefit), interference competition (often manifested as territoriality), and predator–prey relationships.

Mutualism

Three of the best-documented cases of marine mutualism occur in reef fishes. 'Cleaning symbiosis' occurs when small microcarnivorous fish consume ectoparasites or necrotic tissue off larger host fish, which often allow cleaners to feed within their mouths and gill cavities. The major cleaners are various gobies (Gobiidae) and wrasses (Labridae). Some of the cleaner wrasses are specialists that maintain fixed cleaning stations regularly visited by hosts, which assume solicitous postures. The interaction is not always mutualistic in that cleaners occasionally bite their hosts, and some saber-tooth blennies (Blenniidae) mimic cleaner wrasse and thereby parasitize host fish. Anemonefishes (Pomacentridae, especially the genus *Amphiprion*) live in a mutualistic association with several genera of large anemones. By circumventing discharge of the cnidarian's nematocysts, the fish gain protection from predators by hiding in the stinging tentacles of the anemone. In turn, the fish defend their host from butterflyfishes and other predators that attack anemones. However, some host anemones survive well without anemonefish, in which case the relationship is commensal rather than mutualistic. Finally, some gobies cohabit the burrows of digging shrimp. The shrimp provides shared shelter and the goby alerts the shrimp to the presence of predators.

Territoriality

The most overt form of competition involves territoriality or defense of all or part of an individual's home range. Many reef fishes behave aggressively toward members of both their own and other species, but the most obviously territorial species are benthic-feeding damselfishes (Pomacentridae: e.g., genus *Stegastes*). By pugnaciously defending areas about a meter square from herbivorous fishes, damselfish prevent overgrazing and can thus maintain dense patches of seaweeds. These algal mats serve as a food source for the damselfish as well as habitat for small juvenile fish of various species that manage to avoid eviction. At a local spatial scale, the algal mats can both smother corals as well as maintain high species diversity of seaweeds. By forming dense schools, nonterritorial herbivores (parrotfishes and surgeonfishes) can successfully invade damselfish territories.

Piscivory and Defense

Predation is a major factor affecting the behavior and ecology of reef fishes. There are three main modes of piscivory. Open-water pursuers, such as reef sharks (Carcharhinidae) and jacks (Carangidae), simply overtake their prey with bursts of speed. Bottom-oriented stalkers, such as grouper (Serranidae) and trumpetfishes (Aulostomidae), slowly approach their prey before a sudden attack. Bottom-sitting ambushers, such as lizardfishes (Synodontidae) and anglerfishes (Antennariidae), sit and wait cryptically for prey to approach them. The vision of piscivores is often suited for crepuscular twilight, when the vision of their prey is least acute (being adapted for either diurnal or nocturnal foraging). Hence, many prey species are inactive during dawn and dusk, resulting in crepuscular 'quiet periods' when both diurnal and nocturnal species shelter in the reef framework. (Parrotfishes may further secrete mucous cocoons around themselves at night, and small wrasses may bury in the sand.) Otherwise, prey defensive behavior when foraging or resting typically involves remaining warily near structural shelter and shoaling either within or among species. Associated with day–night shifts in activity are daily migrations between safe resting areas and relatively exposed feeding areas. Caribbean grunts (Haemulidae) spend the day schooling inactively on reefs, and after dusk migrate to nearby seagrass beds and feed. Reproducing reef fishes may avoid predation by spawning (in some combination) offshore, in midwater, or at night. Spawning during ebbing spring tides that carry eggs offshore or guarding broods of demersal eggs further defends propagules from reef-based predators. Subsequent settlement of larvae back to the reef, which occurs mostly at night, is also an apparent antipredatory adaptation.

Reproduction

Social Systems and Sex Reversal

The best-studied examples of highly structured social systems in reef fishes are the harems of wrasses and parrotfishes. Typically, these fish are born as females that defend individual territories or occupy a shared home range. A larger male defends a group of females from other males, thereby sequestering matings. When the male dies, the dominant (typically largest) female changes sex (protogyny) and becomes the new harem master. At high population sizes, some fish may be born as males, develop huge testes, resemble females, infiltrate harems, and sneak spawnings with the resident females. Spatially isolated at their home anemones, anemonefishes have monogamous social systems in which the largest individual is female, the second largest is male, and the remaining fish are immature. Upon the death of the female, the male changes sex (protandry) and the behaviorally dominant juvenile fish matures into a male. Simultaneous hermaphroditism occurs among a few sparids and serranine sea basses. These fish have elaborate courtship behaviors during which individuals switch male and female roles between successive pair spawnings. Regardless of the broad variety of mating systems found in reef fishes, each individual behaves in a way that tends to maximize lifetime reproductive success.

Life Cycle

The typical bony reef fish has a bipartite life cycle: a pelagic egg and larval stage followed by a demersal (seafloor-oriented) juvenile and adult stage. Most bony reef fishes broadcast spawn, releasing gametes directly into the water column where they are swept to the open ocean. Smaller species spawn at their home reefs and some larger species, such as some grouper (Serranidae) and snapper (Lutjanidae), migrate to traditional sites and form massive spawning aggregations. Gametes are released during a paired or group 'spawning rush' followed by rapid return to the seafloor. Exceptions to broadcast spawning include demersal spawners that brood eggs until they hatch, either externally (e.g., egg masses defended by damselfishes) or internally (e.g., mouthbrooding cardinalfishes, Apogonidae), and a few ovoviviparous or viviparous species that give birth to well-developed juveniles (including reef sharks and rays). Annual fecundity of broadcast spawners ranges from about 10 000 to over a million eggs per female. Spawning is weakly seasonal compared to temperate species, typically peaking during summer months but not strongly related to any particular environmental variable. Lunar and semilunar spawning cycles are common. These are presumably adaptations that transport larvae offshore away from reef-based predation, maximize the number of settlement-stage larvae returning during favorable conditions that vary on lunar cycles, and/or benefit spawning adults in some way.

Little is known about the behavior and ecology of reef-fish larvae. Duration of the pelagic larval stage ranges from about 9 to well over 100 days, averaging about a month. Larval prey include a variety of small zooplankters. Comparisons of fecundity at spawning to subsequent larval settlement back to the reef suggest that larval mortality is both extremely high and extremely variable, apparently due mostly to predation. Patterns of endemism, settlement to isolated islands, and limited data tracking larvae directly suggest that there is considerable larval retention at the scale of large islands, yet substantial larval dispersal nonetheless. Later-stage larvae are active swimmers and may control their dispersal by selecting currents among depths. The overall reproductive strategy is apparently to disperse the larvae offshore from reef-based predators, but then to retain offspring close enough to shore for subsequent settlement in suitable habitat.

Settlement, the transition from pelagic larva to life on the reef (or nearby nursery habitat), occurs at a total length of *c.* 8 to *c.* 200 mm. Larger larvae are either morphologically distinct (e.g., the acronurus of surgeonfishes) or essentially pelagic juveniles (e.g., squirrelfishes and porcupinefishes). Choice of settlement habitat is apparent in some species, and both seagrass beds and mangroves can serve as nursery habitats. Some wrasse larvae bury in the sand for several days before emerging as new juveniles. There is typically weak metamorphosis during settlement involving the growth of scales and onset of pigmentation. Estimates of settlement are generally called 'recruitment' and are based on counts of the smallest juveniles that can be found by divers some time after settlement. Once settled, most reef fish are thought to live less than a decade, although some small damselfish live at least 15 years.

Ecology

Coral reef fishes are superb model systems for studying population dynamics and community structure of demersal marine fishes because they are eminently observable and experimentally manipulable *in situ*. These characteristics make studies of reef fishes conceptually relevant to demersal fisheries and ecology in general.

Population Dynamics

Because reefs are patchy at all spatial scales and reef fish are largely sedentary, coral reef fishes form metapopulations: groups of local populations linked by larval dispersal. Many local populations are demographically open, such that reproductive output drifts away and is thus unrelated to subsequent larval settlement originating from elsewhere. Ultimately, the degree of openness depends on the spatial scale examined. For example, anemonefish populations are completely open at the scale of each anemone, may be partially closed at the scale of an oceanic island, and mostly closed at the scale of an archipelago. It is clear that variability in population size is driven by variation in recruitment due to larval mortality (and perhaps spawning success). Input to local populations via recruitment varies considerably at virtually every spatial and temporal scale examined. Increasing evidence indicates that two mechanisms predominate in regulating reef fish populations. First, given that density-dependent growth is common and that there is a general exponential relationship between body size and egg production in fish, density-dependent fecundity is likely. Second, early postsettlement mortality is often density-dependent, and has been demonstrated experimentally to be caused by predation in a variety of species.

Community Structure

Due to high local species diversity, reef fish communities are complex. There are about five major feeding guilds, each containing dozens of species

locally (with approximate percentage of total fish biomass): zooplanktivores (up to 70%), herbivores (up to 25%), and piscivores (up to 55%), with the remainder being benthic invertebrate eaters or detritivores. The benthivores can be further subdivided based on prey taxa (e.g., corallivores) or other categories (e.g., consumers of hard-shelled invertebrates). Grunts that migrate from reefs at night and feed in surrounding seagrass beds subsequently return nutrients to the reef as feces. There is also considerable consumption of fish feces (coprophagy) by other fish on the reef. Fishes thus contribute substantially to nutrient trapping (via planktivory and nocturnal migration) and recycling (via coprophagy and detritivory) on coral reefs. Within each feeding guild, there is typically resource partitioning: each species consumes a particular subset of the available prey or forages in a distinct microhabitat. Communities are also structured temporally, with a diurnal assemblage being replaced by a nocturnal assemblage (the resting assemblage sheltering in the reef framework). The diurnal assemblage is dominated by perciform and tetraodontiform fishes, whereas the nocturnal assemblage is dominated by beryciform fishes (evolutionary relics apparently relegated to the night by more recently evolved fishes).

Maintenance of Species Diversity

Four major hypotheses have been proposed to explain how many species of ecologically similar coral reef fishes can coexist locally. There are data that both corroborate and falsify each hypothesis in various systems, suggesting that no universal generalization is possible. The first two hypotheses are based on the assumption that local populations are not only saturated with settlement-stage larvae, but also regularly reach densities where resources become limiting. First, the 'competition hypothesis', borrowed from terrestrial vertebrate ecology, suggests that coexistence is maintained despite ongoing interspecific competition by fine-scale resource partitioning (or niche diversification) among species. Second, the 'lottery hypothesis', derived to explain coexistence among similar territorial damselfishes that did not appear to partition resources, is based on the assumptions that, in the long run, competing species are approximately equal in larval supply, settlement rates, habitat and other resource requirements, and competitive ability. Thus, settling larvae are likened to lottery tickets, and it becomes unpredictable which species will replace which following the random appearance of open space due to the death of a territory holder or the creation of new habitat. The relatively restrictive assumptions of this hypothesis can be relaxed if one considers the 'storage effect', which is based on the multiyear life span of reef fishes and the fact that settlement varies through time. Even though a species is at times an inferior competitor, as long as adults can persist until the next substantial settlement event, that species can persist in the community indefinitely. The third hypothesis, 'recruitment limitation', assumes that larval supply is so low that populations seldom reach levels where competition for limiting resources occurs, so that postsettlement mortality is density-independent and coexistence among species is guaranteed (assuming a storage effect). Finally, the 'predation hypothesis' predicts that early postsettlement predation, rather than limited larval supply, keeps populations from reaching levels where competition occurs, thereby ensuring coexistence.

See also

Fish Feeding and Foraging. Fish Predation and Mortality. Fish Reproduction. Fish Vision.

Further Reading

Böhlke JE and Chaplin CCG (1993) *Fishes of the Bahamas and Adjacent Tropical Waters*. Austin, TX: University of Texas Press.

Caley MJ (ed.) (1998) Recruitment and population dynamics of coral-reef fishes: An international workshop. *Australian Journal of Ecology* 23(3).

Lieske E and Myers R (1996) *Coral Reef Fishes: Indo-Pacific and Caribbean*. Princeton, NJ: Princeton University Press.

Randall J (1998) *Shore Fishes of Hawai'i*. Honolulu, HI: University of Hawai'i Press.

Randall JE, Allen GR, and Steene RC (1997) *Fishes of the Great Barrier Reef and Coral Sea*, 2nd edn. Honolulu, HI: University of Hawai'i Press.

Sale PF (ed.) (1991) *The Ecology of Fishes on Coral Reefs*. San Diego, CA: Academic Press.

Sale PF (ed.) (2002) *Coral Reef Fishes: Dynamics and Diversity in a Complex Ecosystem*. San Diego, CA: Academic Press.

DEEP-SEA FISHES

J. D. M. Gordon, Scottish Association for Marine Science, Argyll, UK

Introduction

For the purpose of this article a deep-sea fish is one that lives, at least for most of its life, at depths greater than 400 m. The fishes of the continental shelves are usually classified as either pelagic or demersal. These categories are often further subdivided in the deep sea. The pelagic component is comprised of the mesopelagic and bathypelagic fishes that live entirely in the water column and are generally of small adult size. Mesopelagic fishes, e.g. lantern fishes (family Myctophidae) and cyclothonids (family Gonostomatidae), live beneath the photic zone to a depth of approximately 1000 m. Bathypelagic fishes live below 1000 m and are usually highly adapted to life in a food-poor environment. Examples are the deep-water angler fishes (family Ceratidae) and the gulper eels (family Eurypharyngidae).

Although the term demersal, referring to fishes living on or close to the bottom, is equally appropriate to the deep sea it has become customary to divide these fish into benthic and benthopelagic species. There is also a trend to refer to these as 'deep-water' rather than 'deep-sea' fish, thus avoiding the nautical use of deep-sea meaning distance from land. Benthic fishes are those that spend most of their time on the bottom and include the rays (family Rajidae), the flatfishes (e.g. family Pleuronectidae) and the tripod fishes (family Chlorophthalmidae). The benthopelagic fishes are those that swim freely and habitually near the ocean floor and examples include the squalid sharks (family Squalidae), the macrourid fishes (family Macrouridae) and the smoothheads (family Alepocephalidae) (see **Figure 1**).

The long-held belief that deep-sea fish belonged to old (in evolutionary terms) and archaic groups is no longer tenable. The deep sea has been invaded many times and there is little doubt that the specialized pelagic fauna have undergone much of their evolution in the deep sea. On the other hand, the demersal fishes probably evolved mainly in the shallower waters and secondarily invaded the deep sea and therefore, although well-adapted for life at depth, their special morphological features are usually less well-developed than those found in the meso- and bathy-pelagic fishes. In some areas, such as the deep Norwegian Sea, the colonization by demersal fishes appears to be of fairly recent origin.

Although there are very marked regional differences in the deep-water demersal fish faunas there is also a degree of global similarity, and certain families such as the macrourids and smoothheads are dominant. Some species such as several deep-water sharks and the orange roughy (*Hoplostethus atlanticus*) are widely distributed on continental slopes. Many abyssal species, of which the armed grenadier (*Coryphaenoides armatus*), the blue hake (*Antimora rostrato*), and the lizardfishes (*Bathysaurus* spp.) are good examples, are cosmopolitan in their distribution.

Our knowledge of the deep-water demersal fishes ultimately depends on the sampling techniques. The bottom trawl, beam or otter, has been the most widely used sampling method. Each design of trawl is selective for a particular spectrum of fishes, and on the upper continental slope – especially in areas where there is commercial exploitation – a wide range of trawls have been used that has resulted in a better understanding of the fish assemblages. On the lower slope and at abyssal depths fewer, more specialized sampling gears have been utilized, and the assemblages may not be adequately described. Where the seabed is unsuitable for bottom trawling, sampling by longlines and traps has been successful. Advances in submersible technology continue to provide information on the distribution and behavior of deep-water fishes.

Depth-related Changes in Abundance and Biomass

The abundance, biomass, and usually the number of species, decreases with increasing depth. This in turn is related to the food supply which, with the exception of a very small contribution of probably <1% from chemosynthesis, is ultimately derived from photosynthesis at the surface, see **Figure 2**.

Surface production is not uniform throughout the world's oceans, as it depends on the factors necessary for photosynthesis such as light, temperature, and nutrients. In general, it tends to be greatest at higher latitudes where there is strong winter mixing and in areas of upwelling. On the basis of this production,

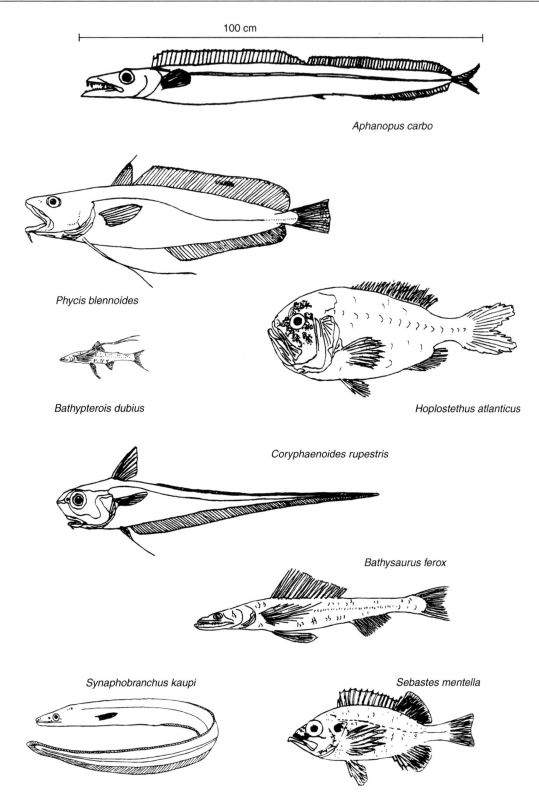

100 cm

Aphanopus carbo

Phycis blennoides

Bathypterois dubius

Hoplostethus atlanticus

Coryphaenoides rupestris

Bathysaurus ferox

Synaphobranchus kaupi

Sebastes mentella

Figure 1 Some deep-water species to show different morphologies.

the upper layer of the world's oceans can be divided into a series of fairly well-defined faunal provinces, which extend into the mesopelagic. These provinces have been well-defined for some groups such as the myctophid fishes. Since the demersal fishes ultimately depend on the same energy supply it is not unreasonable to suppose that such faunal provinces exist also for the demersal fishes, but with a few

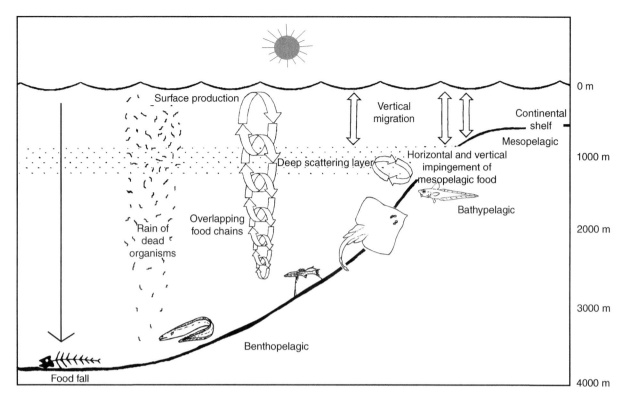

Figure 2 A diagram showing some of the pathways by which food reaches the deep-sea fish. Vertical migration is probably an important food source for bottom fish at around 1000 m and could account for the increased biomass at this depth.

exceptions the level of sampling has been insufficient to identify their presence. An indication that such faunal provinces may exist and that they might be directly related to surface production comes from some work on the deep-water, abyssal fishes of the north-eastern Atlantic. The investigations were carried out in three main areas, the Porcupine Abyssal Plain (*c.* 48°N; 16°W), the Madeiran Abyssal Plain (*c.* 31°N; 25°W) and off the west African coast (20°N). All the stations were at depths between about 4000 and 5550 m. Striking differences in species composition, morphology, maximum fish size, feeding pattern, and reproductive strategies were demonstrated between the fish catches of these areas. In the area of the Madeiran Abyssal Plain, where there is a well-established seasonal thermocline and surface productivity is relatively low, the abyssal fish fauna is most diverse. The individual fish tend to be of small body size and adapted to life where the food supply is dispersed and limited. The stations off the African coast have a distinct fauna, which is probably attributable to higher productivity caused by upwelling along the continental margin. Further north, where there is a marked seasonal cycle of productivity, the fauna is less diverse and the individual fish have larger body sizes and are adapted to exploiting food sources that tend to be patchily

distributed. There appears to be a general trend for decreasing diversity and increasing body size from low to high latitudes. Overall, the biomass at abyssal depths is low compared with the continental slope and shelf. Another trend that has been reported in the Pacific is for a decrease in fish biomass with increasing distance from the continental land mass. This probably results from increased surface productivity attributable to terrestrial inputs.

The abyssal fishes are dependent on the 'rain' of detritus and associated bacteria and occasional large food falls for their food supply and as these decrease exponentially from the surface to the seabed, the low biomass is easy to explain. However, at depths of around 1000 m on the continental slopes or around seamounts there is often an increase in demersal fish abundance and biomass. It is this increase in biomass that forms the basis of the developing deep-water fisheries. Most of the fishes at this depth are benthopelagic and studies of their diets (see below) have shown that pelagic and benthopelagic organisms dominate their diet. Many of the prey organisms are vertical migrators, ascending towards the surface at night to feed and descending to a depth of about 1000 m during the day, where they form a deep-scattering layer. Where this scattering layer impinges either vertically or horizontally onto the slope, it provides a rich source of food for the benthopelagic fishes.

The Diet of Deep-water Demersal Fishes

In situ investigations in the deep sea are difficult and inevitably this means that much of our knowledge on feeding is derived from stomach content analysis. This has many inherent difficulties, such as the very high percentage of empty or everted stomachs due to the expansion of the gas during recovery from depth in those species that have gas-filled swim bladders. Very often all that remains are hard parts such as vertebrae, squid beaks, and crustacean appendages that become trapped in the lining of the stomach and which can lead to an overestimation of these prey types. The problem of identifying food items, often from fragments, can lead to bias especially when some prey taxa are poorly known, as is often the case in the deep sea. Indeed in some studies, the contents of the fish stomachs have been considered as yet another method of sampling the deep-sea fauna. For example, a significant part of the mesopelagic fauna of Madeira was described from the stomachs of deep-water species such as the black scabbardfish (*Aphanopus carbo*) landed by the commercial fishery. Net feeding is also considered to be a problem because of the long time the fish spend in the net after initial capture. In some fishes, such as the Alepocephalidae, there is often a high percentage of unidentified soft tissue that may result from feeding on gelatinous plankton. The gelatinous plankton is poorly sampled by nets but the deployment of cameras has shown that it can be abundant and therefore it should not be neglected as a potentially significant food source. Indirect evidence of feeding modes can be obtained from parasite loadings, presence of sediment in the gut, the morphology of the fish, and its associated sensory systems. Stable isotope ratios are beginning to be used to determine the level of the different fishes in the food chain. Direct observation from manned submersibles, remotely operated vehicles, and baited camera systems is a useful tool for understanding feeding behavior.

The feeding strategies of the deep-sea fishes cover as wide a range as their shelf counterparts and, as one might expect, reflect the habitat differences (e.g. depth, bottom topography). The piscivorous fishes can broadly be divided into those that adopt a sit-and-wait strategy, such as the lizardfish (*Bathysaurus*) and those that are active predators such as some deep-water sharks and the black scabbardfish. Many deep-water species, including the macrourids,

feed on a mixed diet of the larger pelagic and benthopelagic crustaceans, cephalopods, and small fishes. Others feed on a mixed diet of the smaller benthopelagic organisms and epifauna including amphipods, mysids, and copepods. Again this group can be divided into those that actively forage, such as some of the smaller macrourid fishes, and those that sit and wait, such as the tripod fish (*Bathypterois* spp.). Feeding directly on the benthos, whether it is browsing at the surface, sifting the sediment for infauna, cropping or even scavenging is relatively unimportant and reflects the relatively low amount of energy that reaches the deep-sea floor.

Sensory Systems

Olfaction is well developed in some groups such as the Gadiformes (families Macrouridae and Moridae), some of the sharks, and the synaphobranchid eels. It is probably mostly used for the detection of food, but because it can be sexually dimorphic in some species it may also be used for mate recognition. For example, many male macrourids have larger nostrils than the female.

In general the eyes of the benthopelagic fishes do not have the wide range of adaptations to life in the deep sea that are found in the meso- and bathy-pelagic fishes. The eyes tend to remain large and functional and are probably used mainly to detect bioluminescence. Relatively few of the benthopelagic fishes have photophores. Some families have adaptations to maximize the incoming light. In the Alepocephalidae there is a large aphakic space which allows more light to reach the retina from around the edge of the lens, even if it is less focussed. In the squalid sharks and in some teleosts there is a reflective tapetum behind the retina which maximizes the stimulation resulting from the light entering the eye. Some species such as the tripod fish and the forkbeards (*Phycis* spp.) have developed long fin rays that are sensitive to touch and are used for detecting prey. As may be expected in a dark environment, the lateral line system for detecting movements is particularly well developed in deep-water species and in some species it is particularly elaborate and extends onto the head as a series of canals. The elongate body form of many species is probably an adaptation to increase the sensitivity of the lateral line.

In some deep-water species, including many macrourids and gadids of the slope, the males have drumming muscles on their swim bladders for producing sound. This adaptation is absent in the abyssal species.

Buoyancy

Fishes in general have evolved many different methods of reducing the energy required to maintain them in the water column and deep-water fishes are no exception. Gas-filled swim bladders are widely used by the benthopelagic fishes of the continental slopes and also in some abyssal fishes, such as the macrourid fishes, where it has been shown that there is a direct relationship between the length of the retia mirabila (the blood supply to the gas gland) and the depth of occurrence. In some species the swim bladder has become filled with low-density lipid, such as wax esters as in the orange roughy. Some gas-filled swim bladders also have considerable amounts of phospholipid and/or cholesterol, although their role in buoyancy control is uncertain. Reduction in body density can be achieved by having lipids distributed throughout the body. The orange roughy has wax esters in a layer beneath the skin and in vacuoles on the head. Density can also be decreased by reducing the ossification of the skeleton and by having a high water content in the tissues, such as in the alepocephalid fishes. The deep-water sharks have very large livers and also generate hydrodynamic lift by their pectoral fins during swimming, as in their shallow-water counterparts.

Longevity

The otoliths (earbones) of most deep-water fishes have well defined opaque and transparent zones typical of those found in shallow water where, at least in temperate latitudes, they correspond to seasonal changes in growth and hence can be used to age the fish (**Figure 3**). In shallow water the broader opaque zones are associated with faster summer growth, but it is not so obvious why fish living in the aseasonal deep sea should have changes in growth rate, unless they are linked to seasonal changes in food availability and/or quality. Although otoliths, and sometimes scales, have been used to estimate age of deep-water species on the assumption that the rings are laid down annually it has seldom been possible to directly validate these age estimates except in juvenile specimens. Radiometric aging, although controversial, has tended to confirm the generally held view that many deep-water species are long-lived. The commercially exploited grenadiers (Macrouridae) can live to about 50 years and the orange roughy in New Zealand waters lives to more than 100 years.

Reproduction

The benthopelagic fishes have relatively few of the more extreme reproductive adaptations, such as

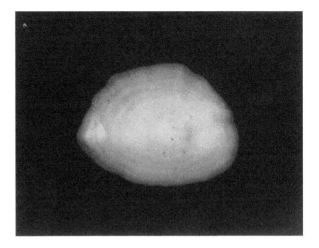

Figure 3 An otolith (ear bone) of an abyssal macrourid (*Coryphaenoides rupestris*) showing growth zones similar to the annual rings in shallow-water fish. If these rings are annual then this fish will be 6 years old.

parasitic males and hermaphroditism that are more frequently found in the meso- and bathy-pelagic fishes. However, hermaphroditism does occur in some groups such as the tripod fishes and live-bearing occurs in the Scorpaenidae (e.g., *Sebastes* spp.) and in the deep-water squalid sharks. It was a long held belief that the lack of seasonality in the deep sea would result in year-round reproduction. However, it has become increasingly apparent that many of the deep-water species of the continental slopes, especially in areas of the oceans where there are marked seasonal cycles of production at the surface, have well-defined spawning seasons. At greater depths year-round spawning or asynchronous spawning is more common. It is also possible that some abyssal species, such as the armed grenadier, are semelparous (spawn once in a lifetime). Many shallow-water fishes begin spawning before they reach full adult size, however there is some evidence that some deep-water fishes may not mature until they reach adult size, thus partitioning the energy supply first for somatic growth and then for reproduction.

There is a lack of information on the egg and larval stages of many of the benthopelagic fish species. For example, the eggs and larvae of the commercially important black scabbardfish are unknown and the eggs of the roundnose grenadier (*Coryphaenoides rupestris*) another widespread and exploited species in the Atlantic, have only been described from the Skagerrak. On the other hand, the abundance of eggs of the orange roughy in the South Pacific has been used for stock assessments. It is probable that the eggs of most species are pelagic, but although there has been speculation about the eggs rising and hatching in the food-rich waters associated with the thermocline, there is little evidence to substantiate this. Indeed, recent investigations suggest that the ornamentation of the surface of the eggs of some macrourid species might be an adaptation to restrict the ascent of the eggs through the water column and avoid too wide a dispersal.

Life Histories

The relatively low level of sampling in the deep sea, its restriction to a small number of areas and/or depths, and a general lack of seasonal sampling have all resulted in incomplete life history information. Except where there are special physical features, such as extreme temperature changes with depth (e.g. Norwegian Sea), there is little evidence of zonation in the deep-sea fishes. Instead each species has a depth range which can extend over several thousand meters, as in the cut-throat eel (*Synaphobranchus kaupi*) or over a few hundred meters, as in the tripod fish (*Bathypterois dubius*) (both examples from the north-east Atlantic). The 'bigger-deeper' phenomenon is a common feature among the deep-water demersal fishes, although it might be more correctly referred to as 'smaller-shallower'. The juveniles of many of the demersal fishes of the continental slopes live at shallower depths than the adults. While in some regions the horizontal distribution of a species can be well documented there is little information on stock discrimination. With present technology it is difficult to tag, release, and recapture deep-water fishes and therefore there is very little information on the movements of deep-water fishes. Some of the commercial deep-water fisheries exist because they often target spawning aggregations, such as orange roughy in the South Pacific and blue ling (*Molva dypterygia*) in the North Atlantic. Some of the shark species are often found in single sex shoals, and in the exploited leafscale gulper shark (*Centrophorus squamosus*) of the North Atlantic the gravid females have never been found. The juveniles of many demersal species have never been found in trawl surveys, which suggests that there are separate nursery grounds or that they occur higher in the water column and are not sampled by bottom trawls.

See also

Bioluminescence. Deep-sea Fishes. Fish Migration, Vertical. Fish Reproduction. Fish Schooling. Mesopelagic Fishes.

Further Reading

Garter JV Jr, Crabtree RE, and Sulak KJ (1997) Feeding at depth. In: Randall DJ and Farrell AP (eds.) *Deep-sea Fishes*, pp. 115–193. San Diego: Academic Press.

Gordon JDM and Duncan JAR (1985) The ecology of the deep-sea benthic and benthopelagic fish on the slopes of the Rockall Trough, northeastern Atlantic. *Progress in Oceanography* 15: 37–69.

Gordon JDM, Merrett NR, and Haedrich RL (1995) Environmental and biological aspects of slope-dwelling fishes of the North Atlantic Slope. In: Hopper AG (ed.) *Deep-water Fisheries of the North Atlantic Oceanic Slope*, pp. 1–26. Dordrecht: Kluwer Academic Publishers.

Haedrich RL (1997) Distribution and population ecology. In: Randall DJ and Farrell AP (eds.) *Deep-sea Fishes*, pp. 79–114. San Diego: Academic Press.

Marshall NB (1979) *Developments in deep-sea biology.* Poole: Blandford Press.

Mauchline J and Gordon JDM (1991) Oceanic pelagic prey of benthopelagic fish in the benthic boundary layer of a marginal oceanic region. *Marine Ecology Progress Series* 74: 109–115.

Merrett NR (1987) A zone of faunal change in assemblages of abyssal demersal fish in the eastern North Atlantic; a response to seasonality in production? *Biological Oceanography* 5: 137–151.

Merrett NR and Haedrich RL (1997) *Deep-sea Demersal Fish and Fisheries.* London: Chapman and Hall.

Montgomery J and Pankhurst N (1997) Sensory physiology. In: Randall DJ and Farrell AP (eds.) *Deep-sea Fishes*, pp. 325–349. San Diego: Academic Press.

Pelster B (1997) Buoyancy at depth. In: Randall DJ and Farrell AP (eds.) *Deep-sea Fishes*, pp. 195–237. San Diego: Academic Press.

Randall DJ and Farrell AP (eds.) (1997) *Deep-sea Fishes.* San Diego: Academic Press.

INTERTIDAL FISHES

R. N. Gibson, Scottish Association for Marine Science, Argyll, Scotland, UK

Introduction and Classification of Intertidal Fishes

The intertidal zone is the most temporally and spatially variable of all marine habitats. It ranges from sand and mud flats to rocky reefs and allows the development of a wide variety of plant and animal communities. The members of these communities are subject to the many and frequent changes imposed by wave action and the ebb and flow of the tide. Consequently, animals living permanently in the intertidal zone have evolved a variety of anatomical, physiological and behavioral adaptations that enable them to survive in this challenging habitat. The greater motility of fishes compared with most other intertidal animals allows them greater flexibility in combating these stresses and they adopt one of two basic strategies. The first is to remain in the zone at low tide. This strategy used by the 'residents', requires the availability of some form of shelter to alleviate the dangers of exposure to air and to predators. 'Visitors' or 'transients', that is species not adapted to cope with large changes in environmental conditions, only enter the intertidal zone when it is submerged and leave as the tide ebbs. The extent to which particular species employ either of these strategies varies widely. Many species found in the intertidal zone spend most of their lives there and are integral parts of the intertidal ecosystem. At the other extreme, others simply use the intertidal zone at high tide as an extension of their normal subtidal living space. In between these extremes are species that spend seasons of the year or parts of their life history in the intertidal zone and use it principally as a nursery or spawning ground. The different behavior patterns used by residents and visitors mean that few fishes are accidentally stranded by the outgoing tide.

Habitats, Abundance and Systematics

Fishes can be found in almost all intertidal habitats and in all nonpolar regions. Most shelter is found on rocky shores in the form of weed, pools, crevices and spaces beneath boulders and it is on rocky shores that resident intertidal fishes are usually most numerous. If, however, fishes are capable of constructing their own shelter in the form of burrows in the sediment, as in the tropical mudskippers (Gobiidae), they may be abundant in such habitats. Visiting species may also be extremely numerous on occasions and are particularly common in habitats such as sandy beaches, mudflats and saltmarshes. Few fishes occupy gravel beaches although species like the Pacific herring (*Clupea pallasii*), capelin (*Mallotus villosus*, Osmeridae) and some pufferfishes (Tetraodontidae) may spawn on such beaches. Estimating abundance in terms of numbers per unit area can be difficult because of the cryptic nature of the fishes and the patchiness of the habitat. The difficulty is particularly acute on rocky shores where fishes may be highly concentrated in areas such as rock pools but absent elsewhere. Nevertheless, fish densities can be relatively high, particularly at the time of recruitment from the plankton, and on occasions may exceed 10 individuals per m^2.

Over 700 species of fishes from 110 families have so far been recorded in the intertidal zone worldwide. This figure represents less than 3% of known fish species but is certainly an underestimate because it is based only on species recorded on rocky shores. Species on soft sediment shores are not included and many areas of the world have yet to be studied in detail. The final count is therefore likely to be much higher. Intertidal fish faunas are frequently dominated by members of a few families (**Table 1**). Generally speaking, more species are found in the tropics than in temperate zones and each area of the world tends to have its own characteristic fauna. The Atlantic coast of South Africa, for example, is characterized by large numbers of clinid species, New Zealand by triplefins, the northeast Pacific by sculpins and pricklebacks (Stichaeidae) and many other areas by blennies, gobies and clingfishes.

Characteristics of Intertidal Fishes as Adaptations to Intertidal Life

Resident intertidal fishes are probably descended from subtidal ancestors and have few, if any, characters that are truly unique. They are thus representatives of families that have convergently evolved morphological, behavioral and physiological traits that enable them to survive in shallow turbulent

Table 1 Analysis of 47 worldwide collections of rocky shore intertidal fishes to show the 10 families with the largest number of species. Based on Prochazka *et al.* in *Horn et al.* (1999). Note that abundance of species does not necessarily imply that families are also numerically abundant

Family	Common name	Number of species
Blenniidae	Blennies and rockskippers	55
Gobiidae	Gobies and mudskippers	54
Labridae	Wrasses	44
Clinidae	Clinids, kelpfishes, klipfishes	33
Pomacentridae	Damselfishes	30
Tripterygiidae	Triplefin blennies	30
Cottidae	Sculpins	26
Labrisomidae	Labrisomids	26
Scorpaenidae	Scorpionfishes	25
Gobiesocidae	Clingfishes	24

habitats. The distribution of many resident intertidal species also extends below low water mark but they mainly differ from their fully subtidal relatives in the degree to which they can withstand exposure to air and are capable of terrestrial locomotion. Nevertheless, a few species can be considered truly intertidal in their distribution because they never occur below low water mark. Examples are the

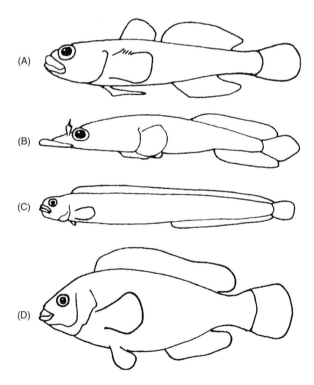

Figure 1 Sketches of four intertidal fish species to demonstrate the basic body shapes. (A) Terete (*Gobius paganellus*, Gobiidae); (B) dorsoventrally flattened (*Lepadogaster lepadogaster*, Gobiesocidae); (C) elongate (*Pholis gunnellus*, Pholidae); (D) laterally compressed (*Symphodus melops*, Labridae).

mudskippers of the genus *Periophthalmus* and blennies of the genera *Alticus* and *Coryphoblennius*.

Morphology

All the common families of intertidal fishes possess the characteristic morphological features of fishes adapted for benthic life in turbulent waters. They are cryptically colored, are rarely more than 15 cm long and are negatively buoyant because they lack a swimbladder or possess one that is reduced in volume. Four basic body shapes can be recognized: elongate, dorsoventrally flattened (depressed), smoothly cylindrical (terete), or laterally compressed (**Figure 1**). In many species the fins are modified to act as attachment devices to prevent dislodgement by turbulence or to assist movement over rough surfaces. In most blennies, the rays of the paired and ventral fins are hooked at their distal ends and may be covered with a thick cuticle to minimize wear. In the gobies and clingfishes, the pelvic fins are fused to form suction cups and allow the fish to attach themselves firmly to the substratum. There seem to be few 'typical' sensory adaptations to intertidal life although many species have reduced olfactory and lateral line systems. These sensory systems would be of limited use for species living in turbulent waters or in those that frequently emerge from the water.

Behavior

Intertidal fishes also show characteristic behavior that enables them to cope with the rigors of intertidal life. Their modified fins and relatively high density allow them to remain on or close to the bottom with the minimum of effort and to resist displacement by surge. Most are also thigmotactic, a behavior that keeps as much of their body touching the substratum as possible and ensures that they come to rest in contact with solid objects when inactive. Their mode of locomotion also reflects this bottom-dwelling lifestyle. Few excursions are made into open water and those species with large pectoral fins use them as much as the tail for forward movement and swim in a series of short hops. Clingfishes and gobies can progress slowly over horizontal and vertical surfaces using their sucker. Elongate forms, which usually have reduced paired fins, creep along the bottom using sinuous movements of their body or alternate lateral flexions of the tail. Strong lateral flips of the tail are also used by some blennies and gobies that can jump between rock pools at low tide and by the semiterrestrial mudskippers in their characteristic 'skipping' movements over the surface of the mud. When progressing more slowly mudskippers use the muscular pectoral fins as 'crutches' to move over the substratum.

Resident species are generally active at a particular state of the tide, although these tidally phased movements can be modulated by the day/night cycle so that some species may only be active, for example, on high tides that occur during the night. Visitors present a more complicated picture because, although their movements are also basically of tidal frequency, they are modulated by a wider range of cycles of lower frequency. Individuals may migrate intertidally on each tide, on every other tide depending on whether they are diurnal or nocturnal, or only on day or night spring tides. They may enter the zone as juveniles in spring or summer, stay there for several months, during which time they are tidally active, and then leave when conditions become unsuitable in winter or as they grow and mature.

The distances over which fishes move in the course of their intertidal movements are dependent on several factors. At one end of the scale are relatively gradual but ultimately extensive shifts in position related to the seasons. At the other end are local, short-term, tidally related foraging excursions. Residents tend to be very restricted in their movements and are often territorial, whereas visitors regularly enter and leave the intertidal zone and may cover considerable distances in each tide. Body size also determines the scale of movement. Residents are small and have limited powers of locomotion whereas visitors usually possess good locomotory abilities and can travel greater distances more rapidly.

The small size and poor swimming abilities of resident species partially account for the restricted extent of their movements but there is good evidence that some species also possess good homing abilities. Most evidence comes from experiments in which individuals are experimentally displaced short distances from their 'home' pools and subsequently reappear in these pools a short time later. Experiments with the goby *Bathygobius soporator* suggest that this species acquires a knowledge of its surroundings by swimming over them at high tide. It can remember this knowledge for several weeks and use its knowledge to return to its pool of origin. Homing is also known in some species of blennies and sculpins and is presumably based on the use of visual clues in the environment. Displacement experiments with the sculpin *Oligocottus maculosus*, however, suggest that this species at least may also use olfactory clues to find its way back to its home pool.

The energy expended in these movements at the various temporal and spatial scales described suggest that they play an important part in the ecology of both resident and visiting species alike. Several functions have been proposed for these movement patterns of which the most obvious is feeding. Visitors move into the intertidal zone on the rising tide to take advantage of the food resources that are only accessible at high water and move out again as the tide ebbs. Residents, on the other hand, simply move out of their low-tide refuge, forage while the tide is high, and return to the refuge before low tide. Following the flooding tide into the intertidal zone may have the added benefit of providing protection from larger predators in deeper water. Movements at both short and long time scales may also be in response to changing environmental conditions. Visitors avoid being stranded above low water mark at low tide because of the lack of refuges or because they are not adapted for the low tide conditions that may arise in possible refuges such as rock pools. Longer term seasonal movements into deeper water can be viewed as responses to changes in such physical factors as temperature, salinity and turbulence. Finally, several species whose distribution is basically subtidal move into the intertidal zone to spawn (see Life histories and reproduction below).

In order to synchronize their behavior with the constantly changing environment fishes must be able to detect and respond to the cues produced by these changes. The cues that fish actually use in timing and directing their tidally synchronized movements are mostly unknown. Synchronization could be achieved by a direct response to change. The flooding of a tide pool or the changing pressure associated with the rising tide, for example, could be used to signal the start of activity. In addition, behavior may be synchronized with the external environment by reference to an internal timing mechanism. The possession of such a 'biological clock' that is phased with, but operates independently of, external conditions is a feature common to all intertidal fishes in which it has been investigated. In the laboratory the presence of the 'clock' can be demonstrated by recording the activity of fish in the absence of external cues. Under these conditions fish show a rhythm of swimming activity in which periods of activity alternate with periods of rest (**Figure 2**). In most cases the period of greatest activity appears at the time of predicted high tide on the shore from which the fish originated. These 'circatidal' activity rhythms, so called because in constant conditions the period of the rhythm only approximates the average period of the natural tidal cycle (12.4 h), can persist in the laboratory for several days without reinforcement by external cues. After this time activity becomes random but in the blenny *Lipophrys pholis* the rhythm can be restarted (entrained) by exposing fish to

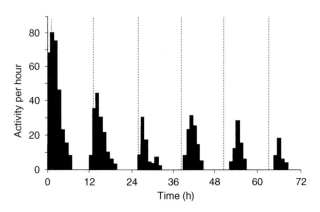

Figure 2 The 'circatidal' activity pattern shown by the blenny *Lipohrys pholis* in constant laboratory conditions. Peaks of activity correspond initially to the predicted times of high tide (dotted lines) but gradually occur later because the 'biological clock' has a period greater than that of the natural tidal cycle (12.4 h). In the sea the clock would be continually synchronized by local tidal conditions.

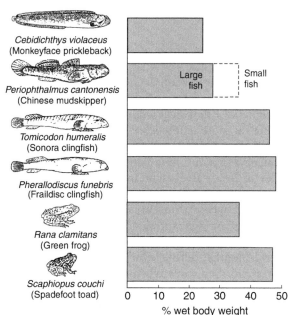

Figure 3 Tolerance of water loss as a percentage of wet body weight in four intertidal fishes compared with two amphibians. (Reproduced with permission from Horn and Gibson, 1988.)

experimental cycles of wave action or hydrostatic pressure or by replacing them in the sea. The probable function of this circatidal 'clock' is that it enables a fish to anticipate future changes in tidal state and regulate its activity and physiology accordingly. If activity is prevented in the wild, by storms for example, then the persistent nature of the clock allows fish to resume activity that is appropriate to the tidal state at the next opportunity.

Physiology

The ebbing tide exposes the intertidal zone to air and so the location of a fish is critical to its survival. Over the low tide period resident fish face not only the danger of exposure to air but also to marked changes in other physical and chemical conditions. On rocky shores, pools act as low-tide refuges but even here temperature, salinity, pH and oxygen content of the water can change markedly for the few hours that the pool is isolated from the sea. Consequently, many resident species are usually more tolerant of changes in these factors than subtidal species. Exposure to air could result in desiccation but, surprisingly, resident intertidal fishes show no major physiological or anatomical adaptations for resisting desiccation. Instead, desiccation is minimized by behavior patterns that ensure fish hide in pools, or in wet areas under stones and clumps of weed at low tide. Nevertheless, many species can survive out of water in moist conditions for many hours and tolerate water losses of more than 20% of their body weight, equivalent to that of some amphibians (**Figure 3**). The ability to tolerate water loss is generally correlated with position on the shore; those fish that occupy higher

levels are the most resistant. Prolonged emersion could also present fish with problems of nitrogen excretion and osmoregulation but those species that have been investigated seem to be able to cope with any changes in their internal medium caused by the absence of water surrounding them.

A further consequence of emersion is the change in the availability of oxygen. Although air contains a greater percentage of oxygen than water its density is much lower causing the gill filaments to collapse and reducing the area of the primary respiratory surface. Unlike some freshwater fishes, intertidal fishes that leave the water or inhabit regions where the water is likely to become hypoxic have no specialized air-breathing organs but maximize aerial gas exchange in other ways. Some species have reduced secondary gill lamellae, thickened gill epithelia and their gills are stiffened with cartilaginous rods. Such features reduce the likelihood of gill collapse when the fish is out of water. The skin is also used as an efficient respiratory surface because it is in contact with air and is close to surface blood vessels. In order to be effective, however, the skin must be kept moist and fish that are active out of water frequently roll on their sides or return to the sea to wet the skin. It is probable that some species use vascularized linings of the mouth, opercular cavities, and, possibly, the esophagus as respiratory surfaces.

The ability to respire in air is currently known for at least 60 species from 12 families and many of these voluntarily leave the water. Fish capable of respiring out of water have been classified into three main types. 'Skippers' are commonly seen out of water and actively feed, display and defend territories on land. They are typified by the tropical and subtropical mudskippers (Gobiidae) and rock-skippers (Blenniidae). The second much less terrestrially active group, the 'tidepool emergers', crawl or jump out of tide pools mainly in response to hypoxia. Species from several families show this behavior but it has been best studied in the sculpins. The third group, the 'remainers', comprises many species that sit out the low-tide period emersed beneath rocks and in crevices or weed clumps. Some may be guarding egg masses and they are not active out of water unless disturbed.

Feeding Ecology and Predation Impact

Intertidal fishes are no more specialized or generalized in their diets and feeding ecology than subtidal fishes. Most are carnivorous or omnivorous and a few are herbivores. Herbivores appear to be less common in higher latitudes but a satisfactory explanation of this phenomenon is still awaited. In some cases the diet changes with size so that the youngest stages are carnivorous but larger amounts of algae are included in the diet as the fish grow. The small size of most resident fishes also means that their diet is composed of small items such as copepods and amphipods. In common with many other small fishes, they may also feed on parts of larger animals such as the cirri of barnacles or the siphons of bivalve molluscs, a form of browsing that does not destroy the prey. The extent to which intertidal fishes have an impact on the abundance of their prey and on the structure of intertidal communities is not clear. In some areas no impact has been detected whereas in others fish may have a marked effect, particularly on the size and species composition of intertidal algae. Those species that only enter the intertidal zone to feed contribute to the export of energy from this area into deeper water.

Life Histories and Reproduction

The majority of resident intertidal fishes rarely live longer than two to three years although some temperate gobies and blennies have a maximum life span of up to 10 years. Maturity is achieved in the first or second year of life and the females of longer-lived species may spawn several times a year for each year thereafter. Representatives of 25 teleost families are known to spawn in the intertidal zone. Of these,

residents and visitors make up about equal proportions. Intertidal spawning has both costs and benefits. For resident species, intertidal spawning reduces the likelihood of dispersal of the offspring from the adult habitat. It also obviates the need for movement to distant spawning grounds; a process that would be energetically costly for small-bodied demersal fishes with limited powers of locomotion and would at the same time expose them to greater risks of predation. These advantages do not apply to subtidal species many of which are good swimmers and whose adult habitat is offshore. For these species, the benefits are considered to be reduced egg predation rates and possibly faster development if the eggs become emersed. In both residents and visitors alike eggs spawned intertidally may be subject to the costs of increased mortality caused by desiccation and temperature stress. In addition, visitors may be vulnerable to avian and terrestrial predators during the spawning process.

The rugose topography of rocky shores and the cryptic sites chosen for spawning has led to the development of complex mating behavior in many species, particularly the blennies and gobies. In these species, courtship displays by the male include elements of mate attraction and a demonstration of the location of the chosen spawning site. Observing these displays is difficult in most groups but field observations have been made on several Mediterranean blennies that live in holes in rock walls and on mudskippers that perform their mating behavior out of water on the surface of the mud.

All species spawn relatively few (range approximately 10^2–10^5) large eggs (~ 1 mm diameter) that are laid on or buried in the substratum. Large eggs produce large larvae which may reduce dispersal from shallow water by minimizing the amount of time spent in the planktonic stage. On hard substrata the eggs are laid under stones, in holes and crevices and in or under weed. They may be attached individually to the substratum surface in a single layer (blennies, gobies, clingfishes) or in a clump (sculpins). In the gunnels and pricklebacks the eggs adhere to each other in balls but not to the surface. In soft sediments the eggs may be buried by the female as in the grunions (Atherinidae), or laid in burrows as in the mudskippers and some gobies. Killifishes (*Fundulus*) lay their eggs in salt-marsh vegetation.

In temperate latitudes spawning usually takes place in the spring and early summer, but spawning rhythms of shorter frequency may be superimposed on this annual seasonality. Subtidal species such as the grunions, capelin and pufferfishes that use the intertidal zone as a spawning ground mostly take advantage of spring tides to deposit their eggs in the sediment high on the shore, usually at night. During the reproductive season such fish, therefore, spawn at fortnightly intervals at the times of the new and full moons. The larvae develop over the intervening weeks and hatch when the eggs are next immersed.

Buried eggs are left unattended but eggs laid in layers, clumps or balls are always cared for by the parent. The sex of the individual that undertakes this parental care varies between species. In some it is the male, in others the female and in yet others both sexes participate. In some members of the families Embiotocidae, Clinidae and Zoarcidae fertilization is internal and the young are produced live. Parental care by oviparous species takes a variety of forms but all have the function of increasing egg survival rates and removal of the guardian parent greatly increases mortality. Most species guard the eggs against predators but some also clean and fan the eggs to maintain a good supply of oxygen and reduce the possibility of attack by pathogens.

Development time of the larvae depends on species and temperature but when fully formed the eggs hatch to release free-swimming planktonic larvae. In only one case, the plainfin midshipman (*Porichthys notatus*, Batrachoididae) does the male parent also care for the larvae, which in this species remain attached to the substratum near the nest site. The factors stimulating hatching are mostly unknown although it has been suggested that wave shock and temperature change associated with the rising tide may be involved. Until recently it was assumed that the hatched larvae were dispersed randomly by currents and turbulence. It has been shown, however, that at least some species minimize this dispersion by forming schools close to the bottom and are rarely found offshore (**Figure 4**). On completion of the larval phase the larvae metamorphose into the benthic juvenile phase and settle on the bottom. The clues used by settling larvae to select the appropriate substratum are poorly known but there is some

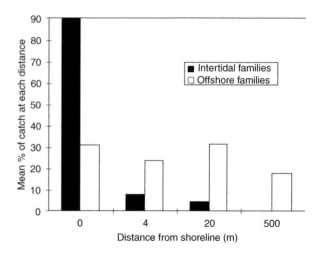

Figure 4 Comparison of the distribution of fish larvae from four intertidal and five offshore families caught in plankton nets at four distances from the shoreline in Vancouver harbour, British Columbia. (Based on data in Marliave JB (1986) *Transactions of the American Fisheries Society* 115: 149–154.)

evidence to suggest that individuals can discriminate between substratum types and settle on their preferred type.

See also

Fish Ecophysiology. Fish Larvae.

Further Reading

Gibson RN (1996) Intertidal fishes: life in a fluctuating environment. In: Pitcher TJ (ed.) *The Behaviour of Teleost Fishes*, 2nd edn, pp. 513–586. London: Chapman Hall.

Horn MH and Gibson RN (1988) Intertidal fishes. *Scientific American* 256: 64–70.

Horn MH, Martin KLM, and Chotkowski MA (eds.) (1999) *Intertidal Fishes: Life in Two Worlds*. San Diego: Academic Press.

MESOPELAGIC FISHES

A. G. V. Salvanes and J. B. Kristoffersen,
University of Bergen, Bergen, Norway

Introduction

'Meso' meaning intermediate and mesopelagic (or midwater) fish refers to fish that live in the intermediate pelagic water masses between the euphotic zone at 100 m depth and the deep bathypelagic zone where no light is visible at 1000 m. Most mesopelagic species make extensive vertical migrations into the epipelagic zone at night, where they prey on plankton and each other, and thereafter migrate down several hundred meters to their daytime depths. Some species are distributed worldwide, and many are circumpolar, especially in the Southern Hemisphere.

Much research on distribution and natural history of mesopelagic fish was conducted in the 1970s, when FAO (Food and Agriculture Organization) searched for new unexplored commercial resources. The total biomass was at that time estimated to be around one billion tonnes with highest abundance in the Indian Ocean (about 300 million tonnes) approximately 10 times the biomass of the world's total fish catch. No large fisheries were, however, developed on mesopelagic fish resources, perhaps due

to the combination of technology limitations and a high proportion of wax-esters, of limited nutritional value, in many species. From 1990 there was renewed interest in these species in connection with interdisciplinary ecosystem studies, when vertically and diel migrating sound-scattering layers (SSLs) turned out to be high densities of mesopelagic fish. These findings formed the basis for studies of the life history and adaptations of mesopelagic fish in the context of general ecological theory.

The thirty identified families of mesopelagic fish are listed in **Table 1** and typical morphologies are shown on **Figure 1**. The taxonomic arrangements of the families differ between various classification systems. In terms of the number of genera per family, the families Gonostomatidae, Melanostomiatidae, Myctophidae, and Gempylidae are the most diverse.

Mesopelagic fish are abundant along the continental shelf in the Atlantic, Pacific, and Indian Oceans and in deep fiords, but have lower abundance offshore and in Arctic and sub-Arctic waters. Most populations have their daytime depths somewhere between 200 and 1000 m. They show several adaptations to a life under low light intensity: sensitive eyes, dark backs, silvery sides, ventral light organs that emit light of a spectrum similar to ambient light, and reduced metabolic rates for deeper-living fish. Vertically migrating species have muscular bodies, well-ossified skeletons, scales, well-developed central nervous systems, well-developed gills, large hearts, large kidneys, and usually a swim bladder. The

Table 1 Families of mesopelagic fish with corresponding number of genera

Family	Number of genera	Family	Number of genera
Argentinidae	2	Alepisauridae	1
Bathylagidae	2	Scopelarchidae	5
Opisthoproctidae	4	Evermannellidae	3
Gonostomatidae	20	Giganturidae	2
Sternoptychidae	3	Nemichthyidae	ca.5
Stomiatidae	2	Trachypteridae	3
Chauliodontidae	1	Regalecidae	2
Astronesthidae	6	Lophotidae	2
Melanostomiatidae	ca.15	Melamphaeidae	2
Malacosteidae	4	Anoplogasteridae	2
Idiacanthidae	1	Chiasmodontidae	5
Myctophidae	ca.30	Gempylidae	20
Paralepididae	5	Trichiuridae	8
Omosudidae	1	Centrolophidae	1
Anotopteridae	1	Tetragonuridae	1

Adapted from Gjøsæter and Kawaguchi (1980).

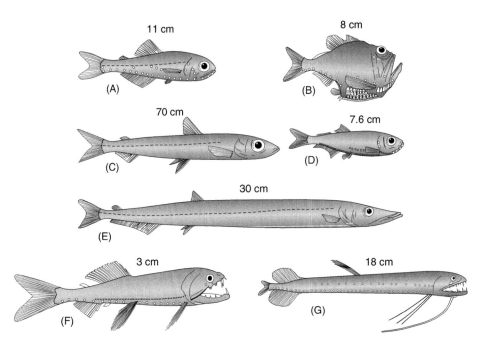

Figure 1 Mesopelagic fish. (A) *Benthosema glaciale* (Myctophidae). (B) *Argyropelecus olfersii* (Sternoptychidae). (C) *Argentina silus* (Argentinidae). (D) *Maurolicus muelleri* (Sternoptychidae). (E) *Notolepis rissoi kroyeri* (Paralepidae). (F) *Astronesthes cyclophotus* (Astronesthidae). (G) *Bathophilus vaillanti* (Melanostomiidae).

ventral light organs are species specific in some families, such as the Myctophidae. The deeper-living species have reduced skeletons, a higher water content in their muscles, lower oxygen consumption, and probably reduced swimming activity compared with species that live at shallower depths.

Life Histories

Most mesopelagic fish species are small, usually 2–15 cm long, and have short life spans covering one or a few years. Some species, especially at higher latitudes, become larger and older. A few larger species such as the blue whiting *Micromesistius poutassou* also live in the mesopelagic habitat, but have the characteristics of epipelagic species. Because of a generally small size, mesopelagic fish have low fecundity, ranging from hundreds to a few thousand eggs. This implies a low mortality in the early life stages, whereas adult mortality is high compared with many epipelagic species. Despite their low fecundity mesopelagic fish have a higher reproductive rate than long-lived epipelagic species which have higher fecundity but a much longer generation time. Neither eggs nor larvae from mesopelagic fish appear to have fundamentally different morphology from those of epipelagic fish, and the larvae all inhabit the epipelagic zone and have growth rates comparable with larvae of epipelagic fish. The higher survival

among the early life stages of mesopelagic fish than of epipelagic species has not yet been quantified. One possible explanation could be different advective loss. The early life stages of large epipelagic populations are passively transported long-distances which means high advective loss. No particular long-distance drift pattern is yet known for mesopelagic fish and this may reflect lower advective loss and lower mortality.

Generally, mesopelagic species that live at high latitudes or at shallow depths have more defined spawning seasons than those that live deeper or at lower latitudes. Some species (e.g., *Maurolicus muelleri*, *Gonostoma ebelingi*, *Cyclothone pseudopallida*) exhibit batch spawning, with repeated spawning throughout a prolonged season of several months. Egg diameters do not differ from those of other fish with pelagic eggs and range between 0.5 and 1.65 mm. The eggs are released either in the daytime in deep water, or epipelagically at night. Eggs and larvae have a dilute internal milieu which makes them buoyant. In some species these buoyancy chambers are later replaced by a swim bladder. Other species, especially among the deepest-living forms, do not have a swim bladder. Those with a gas-filled swim bladder often deposit increased amounts of fat in the swim bladder as the fish become older. Before metamorphosis the larvae inhabit the productive epipelagic zone. During metamorphosis the skin becomes pigmented, light organs develop, and the young start to move down towards the adult habitat. Among some myctophids this ontogenetic shift in habitat is believed to be recorded in the otoliths as accessory primordia, that is, structures that appear as extra nuclei outside the true nucleus of the otoliths.

Growth and age composition of some species have been studied by counting presumed annuli or daily increments in the otoliths. In cold and temperate waters both annual and daily increments may be found. In tropical waters only daily increments can usually be detected, partly because of a shorter longevity in these waters and partly because of the lack of seasonality that fish from temperate regions experience. When there are seasonal changes in the environment this is usually registered as annuli in the otoliths. Only seldom has the periodicity of the increments been validated for mesopelagic fish, Nevertheless, studies have verified the daily basis of microincrements in, for example, *Maurolicus muelleri*, *Benthosema suborbitale*, *B. pterotum*, *B. fibulatum*, *Lepidophanes guentheri*, *Diaphus dumerilii*, *D. diademophilus*, *Lampanyctus* sp., and *Myctophum spinosum*. Annual increments have been partially validated for *M. muelleri*, *B. glaciale*, *Notoscopelus kroyeri*, and *Stenobrachius leucopsaurus*.

The usual pattern of growth towards an asymptotic size (usually expressed by fitting the von Bertalanffy growth equation to empirical data of length versus age), which is common in fish, may not occur in all mesopelagic species. Some show almost linear length increase with age and tend not to reach any asymptotic length in their lifetime. Others slow down their length increase as they become older but do reach an asymptotic length.

Among widely distributed mesopelagic species, geographical variation has been found in morphology, life history or genetics. Based on morphology, 15 subspecies of *M. muelleri* have been identified worldwide. Meristic characters of *B. glaciale* and *Notoscopelus elongatus* differ between the Mediterranean and the North Atlantic, which suggest genetic heterogeneity. Furthermore, populations of *B. glaciale* in west Norwegian fiords are genetically different from each other and from the Norwegian Sea population, and their life histories also vary, with a faster growth towards a lower maximum length in the fiord populations. Genetic isolation is probably possible because of the generally deep distribution of *B. glaciale* combined with relatively shallow sills at the mouth of the fiords. *Maurolicus muelleri* in Norwegian fiords have lower mortality than those in oceanic water masses. The estimated light level at the depth occupied by *M. muelleri* is also lower in the fiords than off the shelf, and this may give the fiord fish improved protection from visually oriented predators. The growth rate, reproductive strategy and predation risk also tend to differ between fiords.

Sexual size dimorphism is observed in many mesopelagic species as well as in numerous other fish species. In such dimorphic species the average size of females is larger than for males. Possible explanations for such differences are lower mortality and/or higher growth among females. In some species (e.g., *Cyclothone microdon*, *Gonostoma gracile*) sex change occurs; they change from male to female as they grow older. That females are larger than males indicates that a large body size is of greater benefit for females than males, possibly because large females are more fecund than small females. Secondary sexual characters are also found in some species. Among myctophids, males have a supracaudal light organ, whereas females have an infracaudal light organ. These light organs are perhaps structures that could be associated with courtship behavior.

Behavior

The behavior of mesopelagic fish has mostly been studied indirectly through monitoring of sound-scattering layers (SSLs) by echosounder and by pelagic trawling to obtain samples with some *in situ* sightings from submersibles. These show that mesopelagic fish are often oriented obliquely or vertically in the water column and it is thought that they may be in a dormant state during daytime. Fish with extensive vertical migrations are not good animals for laboratory experiments. Attempts to keep such light-sensitive mesopelagic fish in aquaria have failed because the fish attempted to migrate downwards, or battered themselves against the walls of the container until they became lifeless. In specially designed spherical containers with water jets, captured myctophid fish have survived a maximum of 72 h.

Although no long-distance horizontal spawning or feeding migrations are known for small mesopelagic fish, many species (particularly the myctophids and some stomiatoids) undertake nightly vertical feeding migrations into the productive surface layer. Species with gas-filled swim bladders are most prominent on the records of echosounders, and populations may appear as distinct layers. Such sound-scattering layers move upward after sunset and downward before dawn to their daytime depths. Vertical migration speeds up to $90 \, \text{m h}^{-1}$ have been measured. The entire population does not necessarily migrate to the surface every night. For instance, a considerable proportion of the adult population of *Benthosema glaciale* is present at daytime depths during the night, whereas juveniles are most numerous in the surface layers. Ontogenetic differences in daytime levels have been observed for *Maurolicus muelleri*. In winter and spring, juveniles are found in a separate scattering layer above the adults. Some evidence for depth segregation between the sexes is also reported; female *M. muelleri* tend to stay deeper than males at daytime during the spring in temperate regions. Depending on the season, females of *Lampanyctodes hectoris* have been reported to stay either shallower or deeper than males.

During daytime mesopelagic fish can adjust their vertical position to accommodate fluctuating light intensities caused by changes in cloudiness and precipitation. The adjustment of the daytime depth levels of the scattering layer thus suggests that vertically migrating mesopelagic fish tend to follow isolumes, at least over short time periods. However, during a 24 h cycle in the summer the estimated light intensity at the depth of *M. muelleri* has been observed to change by three orders of magnitude.

Light is a common stimulus for the vertical displacements and acts as a controlling, initiating and orientation cue during migration. It has been suggested that the ratio between mortality risk and feeding rate in fish, which locate their prey and predators by sight, tend to be at minimum at intermediate light levels. Thus, migration during dawn and dusk may extend the time available for visual feeding while minimizing the predation risk (so-called 'antipredator-window'). At high latitudes in summer the nights become less dark, and the optimal vertical distribution for catching prey and avoiding predators is altered. For example, *Maurolicus muelleri* in the northern Norwegian Sea changes between winter to summer months from a daily vertical migration behavior to schooling. Schooling serves as an alternative antipredator behavior during feeding bouts in the upper illuminated productive water masses.

Adaptations

Mesopelagic fish experience vertical gradients in light intensity, temperature, pressure, rate of circulation, oxygen content, food availability and predation risk. Species of mesopelagic fish have adapted morphologically and physiologically to a midwater life in various ways. Mouth morphologies are generally large horizontal mouths with numerous small teeth, typical of fish that feed on large prey, combined with fine gill rakers, typical of fish that feed on small prey. This arrangement may partially explain their success, since it enables the fish to feed on whatever prey comes along, regardless of size.

Considered broadly, three main groups of mesopelagic fish can be identified based on the morphology: (1) small-jawed plankton eaters, mostly equipped with swim bladders; (2) large-jawed piscivorous predators with a swim bladder; (3) large-jawed piscivorous predators without a swim bladder. **Table 2** lists most of the traits that are typical for these groups.

The physiological and morphological adaptations in mesopelagic fish can be regarded as indirect or direct responses to light stimuli. For example, except for Omosudidae, all mesopelagic fish have pure-rod retinas which are characterized with a high density of the photosensitive pigment, rhodopsin. Eyes of mesopelagic fish tend to be large. The larger the absolute size of the eye and the greater the relative size of its pupils and lens, the better it is for gathering and registering the light from small bioluminescent flashes emitted by photophores. At times, such flashes may be frequent enough to merge into a nearly continuous background of light. Some mesopelagic fish (members of the families Gonostomatidea, Sternoptychidae, Argentinidae, Opistoproctidae, Scopelarchidae, Evermannellidae, Myctophidae, and Giganturidae) have even evolved tubular eyes with large lenses and a larger field of binocular vision, which improve resolution and the ability to judge distances of nearby objects. Coupled with short snouts such eyes enable the individuals to pick out small planktonic organisms in dim light. Tubular eye design for improved binocular vision is achieved at the cost of lateral vision. Many species have modified the eyes further with an accessory retina or even accessory lenses that also allow lateral vision.

The possibility of protection for mesopelagic fish lies in camouflage coloration which matches the light conditions in their habitat. Most of them lack spines or other protrusions that may serve as a defence against predators. In the deep ocean they find protection in twilight and darkness, where dark-skinned predatory fish are also well camouflaged. In shallower waters good camouflage is provided by transparency, by reflecting light to match the background perceived by a visual predator, or in certain surroundings by having a very low reflectance. The shallow-living larvae of mesopelagic fish are generally transparent to light. During and after metamorphosis, when the mature coloration is developing, the young fish move down to the dim or dark depths of their adult habitat. The adult coloration of a large proportion of the mesopelagic fish consists of silvery sides, a silvery iris, and a dark back. Most kinds of silvery-sided fish live at the upper mesopelagic levels, where, to the eyes of a visual predator, uncamouflaged prey will stand out against the background of light, except when viewed from above.

Table 2 Organization of the major groups of mesopelagic fish. The comparisons of the predators are relative to the plankton consuming group

Features	Mesopelagic planktivores[a]	Mesopelagic piscivores with swim-bladder[b]	Mesopelagic piscivores without swim-bladder[c]
Colour	Often silvery sides	Black skin	Black and brassy
Photophores	Numerous and well developed in most species		
Jaws and teeth	Relatively short	Large jaws and teeth	
Eyes	Fairly large to very large, with large and sensitive pure-rod retina		Eyes and optic centers of the brain are moderately developed
Olfactory organs	Moderately developed in both sexes of most species		
Central nervous system	Well-developed in all parts		
Myotomes	Well-developed myotomes and large red muscle		Some reduction in myotomes, but still large red muscle system
Skeleton	Well ossified, including scales		Reduction in skeleton
Swim bladder	Usually present, highly developed		Regressed and invested by adipose tissue
Gill system	Gill filaments numerous, bearing very many lamellae to increase oxygen extraction at low ambient oxygen partial pressure		Not so well-developed
Kidneys	Relatively large with numerous tubules		
Heart	Large		
Metabolic rate	Decrease with increasing minimum depth		Low
Vertical position	Most of diel migratory fish belong to this group		Centered at lower rather than upper mesopelagic levels
Maturation age	Low		Higher
Fecundity	Low		Higher

[a]Families Gonostomatidae, Argentinidae, Sternoptychidae, Myctophidae and Melamphaeidae;
[b]Families Astronesthidae, Trichiuridae and Chiasmodontidae;
[c]Families Melanostomiatidae, Stomiatidae, Chauliodontidae, Malacosteidae, Scopelarchidae, Evermannellidae, Omosudidae, Alepisauridae, Anotopteridae and Paralepididae.
Families underlined have vertically migrating members.
Adapted from Marshall (1971) and Childress (1995).

Silvery-sided fish are very vulnerable to attacks from below, particularly from black-skinned visual predators. When a visual predator looks upwards it will see its prey in silhouette. It has been argued that the ventral light organs in mesopelagic fish are an adaptation to emit light that matches the background of downwelling ambient light, in order to break up its silhouette and so to make attack from below more difficult.

The ability of mesopelagic fish, which are nearly neutrally buoyant, to undertake vertical migration is related to the structure of their myotomes. They have a muscular organization for sustained efforts with a large proportion of red muscle fibers. These are rich in fat, contain lots of glycogen, myoglobin, and many mitochondria and are richly supplied with blood and thus oxygen. White fibers that dominate the muscles of epipelagic fish, hold little or no fat, little glycogen, no myoglobin, few mitochondria and are more sparsely supplied with blood. White muscles work anaerobically in short bursts, such as rapid escape responses towards predators. The

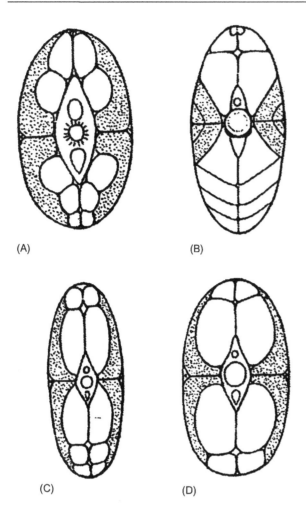

(A)   (B)

(C)   (D)

Figure 2 Transverse sections through the tails of three mesopelagic fish showing the extent of the red muscles (stippled). (A) *Notolepis coasti*, an Antarctic paralepidid. (B) *Electrona antarctica*, a myctophid. (C) *Maurolicus*, a Sternoptychidae. (D) *Astronesthes lucifer*. From Marshall (1977).

metabolic cost is related mostly to the requirements of the red muscle in moving the fish upward at a cruising speed in order to search for food. Little energy would be needed during descent when mesopelagic fish, whether they have a swim bladder or not, are likely to be negatively buoyant. The comparative development of red muscle in the tail of selected species is shown in **Figure 2**. Those with highest proportions of red muscle fibers undertake the most pronounced vertical migrations.

Originally the adaptive value of daily vertical migration was related to factors such as reduced competition among species through resource partitioning; minimizing horizontal displacement through advection; and bioenergetic benefits by foraging in warm surface waters and digesting in cooler deep waters. It was also suggested that mesopelagic fish could use vertical current gradients as a way of being transported to new feeding areas. Subsequently there has been more focus on the balance between predation risk and food demand and how this affects vertical distribution patterns. Emphasis has been laid on how vertical migration during dawn and dusk extends the time available for visual feeding, while minimizing the visibility towards predators. The earlier view relied on research in tropical and subtropical regions of the ocean where the fish always experience a change in temperature of about 10°C between the daytime depth and surface, and where also the daily light changes are similar all year round. Although similar temperature differences also exist during the summer in temperate regions, there is hardly any temperature difference between shallow and deep water in winter, and occasionally shallow water may be colder than the deeper water. The observations that mesopelagic fish also undertake daily vertical migration during the winter in west Norwegian fiords suggest that there are other explanations than bioenergetics. It is more likely that vertical migration extends the time available for visual feeding while minimizing the visibility towards predators. This is also consistent with the camouflage coloration in mesopelagic fish and that juveniles can stay in shallower water than adults because they are smaller, often transparent and thus less visible than adults.

There is a difference of a factor of 15 in metabolic rates between species that live at the surface and those that come no shallower than 800 m. This difference is found to be too great to be explained by decreases in temperature, oxygen content, decrease in food availability or increase in pressure. Comparative analyses of fish from different regions show similar depth trends even in isothermal regions (e.g., the Antarctic) for species which live at similar depths but at different oxygen concentrations. Several lines of research indicate that the metabolic decline is related to a reduction in locomotory abilities with increasing depth. It is suggested that the higher metabolic rates at shallower depths in groups with image-forming eyes is the result of selection action to favor the use of information on predators or prey at long distances when ambient light is sufficient. Hence, good locomotory abilities will be beneficial in order to escape predators. This idea is supported by the fact that major gelatinous groups that lack image-forming eyes do not show a decline in metabolic rate with depth. Thus, the lower metabolic rates found in fish living deeper where visibility is lower, result from the relaxation of selection for locomotory abilities, and is not a specific adaptation to environmental factors at great depths. If so, high metabolic rate in the surface waters indicates a metabolic cost of predation risk because good locomotory abilities require high metabolism. At greater depths the predation risk is much lower and the need for locomotory abilities decreases.

See also

Fish Feeding and Foraging. Fish Locomotion. Fish Migration, Vertical. Fish Predation and Mortality. Fish Reproduction. Fish Schooling. Fish Vision.

Further Reading

Andersen NR and Zahuranec BJ (eds.) (1977) *Oceanic Sound Scattering Prediction*. New York: Plenum Press.

Baliño B and Aksnes DL (1993) Winter distribution and migration of the sound scattering layers, zooplankton and micronecton in Masfjorden, western Norway. *Marine Ecology Progress Series* 102: 35–50.

Childress JJ (1995) Are there physiological and biochemical adaptations of metabolism in deep-sea animals. *Trends in Ecology and Evolution* 10: 30–36.

Farquhar GB (1970) *Proceedings of an International Symposium on Biological Sound Scattering in the Ocean*. MC Report 005. Maury Center for Ocean Science. Washington, DC: Dept. of the Navy.

Giske J, Aksnes DL, Baliño B, *et al.* (1990) Vertical distribution and trophic interactions of zooplankton and fish in Masfjorden, Norway. *Sarsia* 75: 65–81.

Gjøsæter J and Kawaguchi K (1980) *A Review of the World Resources of Mesopelagic Fish*. FAO Fish. Tech. Paper No. 193. Rome: FAO.

Kaartvedt S, Knutsen T, and Holst JC (1998) Schooling of the vertically migrating mesopelagic fish *Maurolicus muelleri* in light summer nights. *Marine Ecology Progress Series* 170: 287–290.

Kristoffersen JB and Salvanes AGV (1998) Life history of *Maurolicus muelleri* in fjordic and oceanic environments. *Journal of Fish Biology* 53: 1324–1341.

Marshall NB (1971) *Exploration in the Life of Fishes*. Cambridge, MA: Harvard University Press.

Rosland R (1997) Optimal responses to environmental and physiological constraints: evaluation of a model for a planktivore. *Sarsia* 82: 113–128.

PELAGIC FISHES

D. H. Cushing, Lowestoft, Suffolk, UK

Introduction

In the economy of the sea pelagic fish play a central part. The simplest food chain comprises phytoplankton, copepods and pelagic fish. They spend their lives in the midwater of coastal seas and oceans. Much of our knowledge about them comes from the fisheries that exploit them, which yield very large catches. The three groups, clupeoids, tunas and mackerels live in all parts of the ocean and many of them migrate across the seas for considerable distances. The clupeoids include such fishes as herring, sardines and sprats which have supported large fisheries since the earliest times. Anchovies are widespread and the Peruvian anchoveta once supported a very large fishery. Herring are caught in the North Atlantic and North Pacific; sardines are taken in the upwelling areas and sprats are mainly caught in the North Sea. Herring can migrate for up to 2000 km each year and their stocks (or populations) have yielded annual catches of as much as one million tonnes, wet weight. They are smallish fish ranging in length from about 12 cm (sprats) to 20 cm (sardines) and 25–30 cm or longer (herring).

The tunas are larger fishes, one or two meters in length and each year they make transoceanic migrations. They include yellowfin, albacore, bigeye and bluefin. Catches amount to about a million tonnes each year and they are taken at many places in the world ocean, but particularly in upwelling areas and at fronts. Mackerels are larger than herring (up to 40 cm in length) and they also migrate across considerable distances. The common mackerel in the North Atlantic and the cosmopolitan Spanish mackerel are typical examples; horse mackerels (in a different suborder) are also widely distributed. The pelagic fish spend their lives in the near-surface layers of the ocean and they swim steadily for long periods. From their central position in the marine ecosystem they control the passage of energy up the food webs from the algal cells and copepods to fish.

Clupeoids

Herring: a Case Study of a Pelagic Fish

Herring live in the North Atlantic (*Clupea harengus* L.) and in the North Pacific (*Clupea pallasii* Val.). The edges of the scales on the belly of the Atlantic herring are rough, but in the Pacific herring they are smooth. Atlantic herring are found in the east from the Bay of Biscay to Spitzbergen and the Murman coast and in the west from Greenland and Labrador to Cape Hatteras on the eastern seaboard of the United States. The Pacific herring is found off British Columbia, off Hokkaido in Japan and off Kamchatka. Although they live in the midwater, both species lay their eggs on the seabed; the Atlantic herring usually lay their eggs at depths of 40–200 m and in some cases intertidally on gravel and small stones and the Pacific herring spawns on seaweeds between tidemarks. Both species spawn on narrow strips; for example that near the Sandettié Bank not far from Dover is 3000 m long and 300–360 m wide.

Herring are small fish, the adults being about 25–30 cm in length; Baltic herring (*Clupea harengus/membras*) are smaller and the Norwegian herring are somewhat larger. North Sea herring live for about twelve years and Norwegian herring for about twenty years. They mature at about three or four years of age and their annual fecundity amounts to 40 000–100 000 eggs. Throughout life the total fecundity would be about ten times greater and only two survivors are needed to replace the stock. Hence the annual mortality each year of the eggs, larvae and juveniles is high; indeed, it is approximately the inverse of the annual fecundity. The natural mortality of the adults is that sustained under predation in the absence of fishing and, as might be expected, it is rather difficult to establish; that of the herring might be about 10–20% of numbers per year. They live and swim in large shoals, each of which may be many kilometers across. They migrate towards the surface at night and then the shoals tend to disperse. They make long migrations each year from spawning ground to feeding ground and back, distances of up to 2000 km. To do this they usually swim down tide or current. Herring feed on *Calanus* among other plankton animals. They grow quite quickly particularly during the first year of life, but as adults they grow relatively slowly, as more energy is devoted to reproduction. The herring of any spawning group spawn at the same season each year to within about a week. It is likely that they return to the grounds on which they were spawned, although this cannot be shown decisively.

In the Northeast Atlantic there are a number of spring spawning stocks: Norwegian, Murman, Shetland and Faroe Islands. The Icelandic spring spawners may no longer exist, but the Icelandic summer spawners flourish. There are some small stocks in the Skagerak and there are very small local stocks that spawn at the mouths of certain rivers such as the Elbe. In the North Sea there are two autumn spawning stocks, Buchan and Dogger in the northern and central North Sea and a winter spawning stock, the Downs, in the southern North Sea. The Buchan group spawns off the northeast coast of Scotland; the Dogger group spawns off Whitby and around the Dogger Bank. The Downs stock spawns in the English Channel off Dover and in the Baie de la Seine off northern France. There is another winter spawning stock in the western English Channel. Icelandic summer spawners lay their eggs on three grounds off the north coast of the island. The Norwegian herring spawn in spring between Egersund and Bergen, between Bergen and Kristiansund and off Lofoten and Westeralen. The

North Sea herring migrate in a clockwise circuit in the North Sea (**Figure 1**). The Norwegian herring migrate from the Norwegian coast across to Iceland and then northwards to Jan Mayen and the Barents Sea before returning to their spawning grounds on the Norwegian coast (**Figure 2**). The Icelandic herring stocks migrate round the island in a clockwise direction.

In the Northwest Atlantic there are a number of herring stocks: off Labrador, the southern and western coasts of Newfoundland, the Scotian Shelf, Georges Bank, and to the south of Georges Bank. Most of these stocks are spring spawners, but after 1950, off the south and west of Newfoundland the migrations of the stock on the Scotian Shelf are quite limited but the reason for this difference is not known.

In the Pacific there are two main groups of stocks, that off British Columbia and that in the Far East. The British Columbian fishery is established off Vancouver Island, and off Queen Charlotte Is. There are four major stocks in the east, the Sakhalin-Hokkaido (a spring spawner), the Kamchatka herring, the Kora Karazynsk herring and the Okhotsk herring. Each stock migrates from feeding ground to spawning ground over distances of about 1200 km, for the first three stocks and about 500 km for the Okhotsk herring (**Figure 3**).

In the North Sea herring have been caught from the earliest times by drift net off Shetland, off the northeast coast of Scotland, off northeast England and off East Anglia. Later, they were caught by trawl on the Fladen Ground northeast of Aberdeen, off the Dogger Bank and on the narrow grounds not far from Dover and Boulogne. In the Norwegian Sea herring were originally caught by drift net in the fiords and later by purse seine. Finally, when the purse seine could be worked in the open sea, herring were caught there off southern Norway and also north of the Lofoten Is. Off Iceland, drift nets were also replaced by purse seines worked in the open sea. Nowadays most herring are caught by purse seine and midwater trawl.

In the Northwest Atlantic the traditional fishery took place in the Gulf of Maine and in the Bay of Fundy. In recent years, trawl fisheries have been developed on Georges Bank and purse seine fisheries off Nova Scotia. There were fisheries also off Newfoundland and in the Gulf of St Lawrence. Catches were converted into fish meal and oil. In the Far East, the major fishery was that for the Hokkaido spring herring which were caught by offshore gill nets. The fishery peaked between 1895 and 1905 after which it declined to low levels. Since 1950, the offshore gill nets have been replaced by trawlers. Off Sakhalin

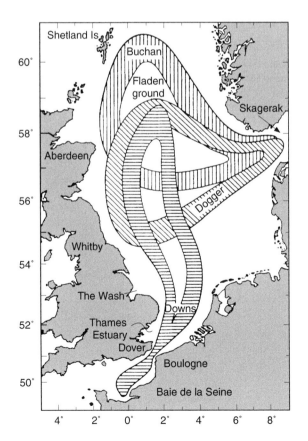

Figure 1 The migration circuits of the three groups of spawners in the North Sea: Buchan, Dogger and Downs.

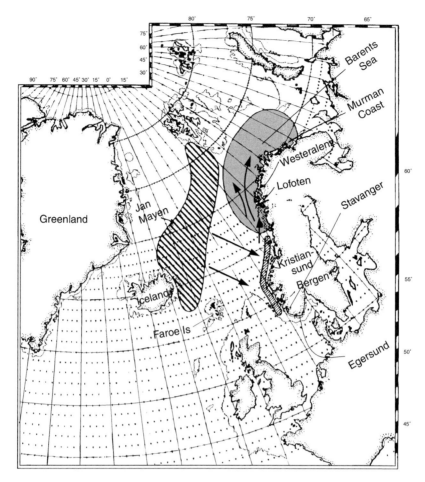

Figure 2 The migration circuit of the spring spawning Norwegian herring; the spawning ground lies off the Norwegian coast, the nursery ground spreads north into the Barents Sea and the feeding ground lies between Iceland and Spitzbergen.

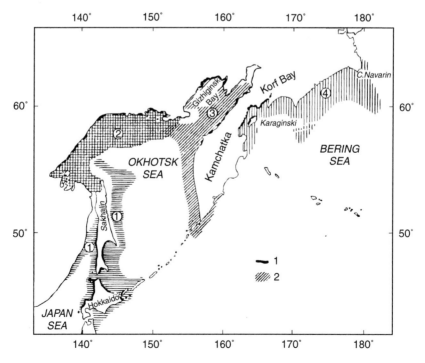

Figure 3 The stocks of *Clupea pallasi* in the Far East. 1, Sakhalin-Hokkaido herring; 2, Okhotsk herring; 3, Gyzhigynsk-kamchatka herring; 4, Korto-karagynsk herring. Bold line, spawning grounds; Hatched areas, feeding grounds.

and in the Sea of Okhotsk herring have been caught by gill nets for a long time.

The global catches of herring have amounted to several million tonnes. In the North Sea the Downs stock was overexploited in the late 1950s by trawling on the spawning grounds and the two other groups, Buchan and Dogger, suffered from overfishing by purse seiners in the late 1960s. The Norwegian spring spawning stock of herring was overexploited by the offshore purse seiners in the early 1970s. In the Northwest Atlantic the fishery on Georges Bank became too heavy in the 1970s. It is possible that the Hokkaido stock suffered from overfishing but this cannot be shown decisively.

Sprats

Sprats, *Clupea sprattus* (L.), are small clupeoids about 10–13 cm in length. They live for about five to seven years, but the most abundant age group is the third. Like other clupeoids, they feed on copepods and other plankton animals. They spawn in spring and summer, but in the northern Adriatic they spawn between December and March. They mature from the end of their second year to the end of their third year and they lay between 10 000 and 40 000 eggs each year. The natural mortality is probably high. There are major spawning grounds in the southern North Sea and in the southern Norwegian fiords. The eggs, larvae and juveniles are fully pelagic.

Sprats are found in the Baltic, in the North Sea, in the northern Adriatic and off Romania in the Black Sea. Fisheries were established in the Scottish firths, between Bergen and Stavanger in Norway, off Brittany in France and around the English coast, particularly in the Wash and in the Thames estuary. In the main, these are winter fisheries worked with drift nets and stow nets, i.e. bag nets hung from fixed poles in the tidal streams.

Sardines

There are a number of sardines and related species: *Sardina pilchardus* Walb in European waters, *Sardinops caeruleus* (Girard) off California, *Sardinops ocellatus* (Pappé) off South Africa, *Sardinops melanostictus* (Schlegel) off Japan and *Sardinella aurita* Val. off West Africa. The pilchard (*S. pilchardus*) lives in the English Channel, in the Bay of Biscay and off the Iberian peninsula. Elsewhere, sardines live mainly in the upwelling areas in subtropical oceans. An upwelling area is an extensive region along western coasts in the subtropical ocean both north and south of the equator. The four main regions lie off California, off Peru and Chile, off southern Africa and off northwest Africa. There are also lesser upwellings off the west coast of the Iberian peninsula, off Ghana, off southern Arabia and off the Malabar coast of India. In an upwelling area the wind blows towards the equator. Water advects offshore and the nutrient-rich replacement is drawn up from below and in it production starts. In these rich areas sardines flourish in large populations.

Sardines are smaller than herring with an average length of just over 20 cm. They live up to ten years but the abundant year classes peak at about four or five years of age. An average year class of *Sardinops caeruleus* off California might comprise about a billion fish. The natural mortality of the sardine is high because they are eaten by a wide range of predators. They spawn in spring and fall off Baja California; within the Baja they spawn in late winter. Their fecundity may amount to about 40 000 eggs year^{-1}. Shoals lie along the line of the tide or current and, like herring shoals, they are much longer than they are wide. Sardines do migrate within an upwelling area, but their movements are restricted. They feed on copepods; the larvae eat copepod nauplii and the larger fish feed on adult copepods. They tend to spawn in winter and spring in both hemispheres, perhaps a little before the upwelling strengthens (but off southern California, they can also spawn in the fall). Eggs, larvae and juveniles are all pelagic.

Records of sardine catches off Japan go back to the fifteenth century. The rich periods lasted from twenty to seventy years, for example, from 1660 to 1730, 1818 to 1864 and 1917 to 1939. From five to seven local groups were recognized but they tended to flourish or decline together. One of the most remarkable events is the trend in catches peaking between the 1930s and the 1950s, off Japan, off California, off Spain and in the northern Adriatic. Catches of the Japanese sardine (*Sardinops melanostictus*) rose to a peak of about 2 500 000 tonnes in 1935–36 and then declined to very low levels between 1945 and 1972. With the year classes, 1977 and 1980, catches recovered to over 4 000 000 tonnes each year by 1985. Subsequently catches again declined. These events occurred at places far apart and there is, as yet, no explanation for the fact that the catches rose and declined together.

Pilchards were caught by drift net off Brittany and by ring net off Cornwall. In the upwelling regions purse seines were used. Catches reached a peak quite quickly and then declined and the fishermen would switch to another pelagic stock. Off South Africa sardines were replaced by anchovies and off Peru, the anchoveta was replaced by sardines. The sardine stock off California collapsed and was replaced by the stock of the northern anchovy (*Engraulis mordax*

Girard) which was subsequently exploited further south by Mexican fishermen.

Off northwest Africa sardine catches remained fairly steady in the north off Cap Ghir but further south between Cap Blanc and Cap Bojador catches reached very high levels before a sudden collapse. The fisheries for sardines and anchovies were very large and they supplied a market for fish meal and oil. Anchovies (*Engraulis* spp.) are small fishes and they live not far from the coast. They are caught in small fisheries in European waters. Off India, a rather larger fishery was practiced on *Rastrelliger kanagurta* (Cuvier) and a much bigger fishery has been that for the Peruvian anchoveta (*Engraulis ringens* Jenyns). At its peak it was the largest fishery in the world, yielding about eleven million tonnes each year.

Tuna

The tunas are large pelagic fish which grow to up to 2 m in length. The principal species are the yellowfin (*Thunnus albacares* Bonaterre), bigeye (*Thunnus obesus* Lowe), skipjack (*Katsuwonus pelamis* L.), albacore (*Thunnus alalunga* Bonaterre) and bluefin (*Thunnus thynnus* L.). They are, in the main, subtropical animals and are distributed across each of the oceans making transoceanic migrations. The natural mortality of the tunas is probably rather high because they do not appear to live very long. Catches of each species amount to about 100 000 or 200 000 tonnes each year; for all tuna species total catches amount to about one million tonnes each year.

Tuna larvae are found all over the subtropical ocean at nearly all seasons which means that spawning is widespread and continuous, but they are also found in the upwelling areas. In three years the bluefin tuna grows to 50 kg. The yellowfin is most abundant in the divergences and convergences of the equatorial system. There is a spawning migration into the Mediterranean and tuna have been caught in small traps and sighted from vedette (look-out posts) in Dalmatia, on the Roussillon coast of France and off Algeria. Bluefin are caught in the tonnare di corsa (spawning migration) in Spain, Portugal, France, Sardinia and Sicily. The tonnara is a complex structure; that off Trapani in Sicily was manned some years ago by ninety-three men. Porters, stevedores, cooks, boxmakers and coopers supported such operations. The fish were herded into the mattanza or 'death room' with white palm leaves and as the nets were brought towards the surface the tuna were killed with lances.

Off Brittany, albacore are caught with hooks on long rods. The fish are most abundant at the shelf edge, associated with 'heavy swells' perhaps where internal waves ride from south to north. Off California, tuna are caught either by pole and line with live bait or with purse seine in the equatorial region and there is also a sport fishery off the coasts of the United States. Yellowfin are prominent in the catches perhaps because they tend to live in stocklets just north of the North Equatorial Current. Albacore tagged off California can be recovered off Japan for the fish may live right across the North Pacific gyre. The larger fish live in the North Equatorial Current and spawn east of the Philippines.

Mackerels

There are a number of mackerels together with their relatives. The Atlantic or common mackerel, *Scomber scombrus* L., lives in the Atlantic off the North American coast and in European waters. The chub mackerel, *Scomber japonicus* Houttuyn, is cosmopolitan. The Indian mackerel is *Rastrelliger kanagurta* (Cuvier). The common horse mackerel, *Trachurus trachurus* L., lives in the Atlantic; the Pacific form is *Trachurus japonicus*. The Atka mackerel of the North Pacific is a scombrid, *Pleurogrammus azonus* Jordan and Metz. The Spanish mackerel, *Scomberomorus commerson* Lacépède, is a larger animal.

Mackerels are larger than herring, reaching as much as 40 cm in length or more. They spawn in the midwater in productive seas where the larvae and juveniles grow up, and they spend the rest of their lives there. They feed on copepods and other plankton animals. They swim quickly and some of the larger ones are predatory. The natural mortality of the mackerels is perhaps high, up to 30% of numbers per year. They do not shoal as herring do but there may be small and transient aggregations.

In the Northeast Atlantic annual catches of mackerel have reached as much as one million tonnes from two stocks, one in the North Sea and the other to the west of the British Isles. They are found in the Mediterranean but catches have only amounted to about 30 000 tonnes each year. Considerable catches of the Atlantic mackerel have been made off the eastern seaboard of the United States. The stock of Atka mackerel in the Northwest Pacific yielded annually about 100 000 tonnes. The Pacific mackerel yielded large catches for a period, after which they declined. Tens of thousands of tonnes of *Rastrelliger* are taken each year off the Indian coast.

The horse mackerel (*Trachurus trachurus* L.) has usually been caught in trawls. It is somewhat larger than the common mackerel, is pelagic and feeds on small animals in the plankton. It probably does not shoal very much and lives rather deeper than the common mackerel or the herring. In the South Atlantic about 100 000 tonnes of maasbanker (*Trachurus trachurus* L.) are caught. Some tens of thousands of tonnes are caught in the open ocean.

Conclusion

The pelagic fish occupy a central position in the marine ecosystem as they harvest food from lower trophic levels and support the predators in higher ones. The stocks respond to climatic changes and provide near stability to the tuna and fishes like them. Catches are very large, indeed the largest sector in the world harvest.

See also

Fish Migration, Horizontal. Fish Reproduction.

Further Reading

Cushing DH (1982) *Fisheries Biology.* Madison: University of Wisconsin Press.

Cushing DH (1996) *Towards the Science of Recruitment in Fish Populations.* Oldendorf Luhe, Germany: Ecology Institute.

FAO (1983) *Species Catalogues.* No. 125, Vol. 2 *Scombrids of the World*; No. 125, Vol. 7(1) *Clupeid Fishes of the World*; No. 125, Vol. 7(2) *Clupeid Fishes of the World.* Rome: FAO.

Graham M (1943) *The Fish Gate.* London: Faber.

Gulland JA (1974) *The Management of Marine Fisheries.* Bristol: Scientica.

Hardy A (1959) *The Open Sea: Fish and Fisheries.* London: Collins.

Rothschild BJ (1986) *The Dynamics of Marine Fish Populations.* Cambridge, MA: Harvard University Press.

SALMONIDS

D. Mills, Atlantic Salmon Trust, UK

Introduction

The Atlantic and Pacific salmon and related members of the Salmonidae are anadromous fish, breeding in fresh water and migrating to sea as juveniles at various ages where they feed voraciously and grow fast. Survival at sea is dependent on exploitation, sea surface temperature, ocean climate and predation. Their return migration to breed reveals a remarkable homing instinct based on various guidance mechanisms. Some members of the family, however, are either not anadromous or have both anadromous and nonanadromous forms.

Taxonomy

The Atlantic salmon (*Salmo salar*) and the seven species of Pacific salmon (*Oncorhynchus*) are members of one of the most primitive superorders of the teleosts, namely the Protacanthopterygii. The family Salmonidae includes the Atlantic and Pacific salmon, the trout (*Salmo* spp.), the charr (*Salvelinus* spp.) and huchen (*Hucho* spp.). The anatomical features that separate the genera *Salmo* and *Oncorhynchus* from the genus *Salvelinus* are the positioning of the teeth. In the former two genera the teeth form a double or zigzag series over the whole of the vomer bone, which is flat and not boat-shaped, whereas in the latter the teeth are restricted to the front of a boat-shaped vomer. In the genus *Salmo* there is only a small gap between the vomerine and palatine teeth but this gap is wide in adult *Oncorhynchus* and not in *Salmo*. A specialization occurring in *Oncorhynchus* and not in *Salmo* is the simultaneous ripening of all the germ cells so that these fish can only spawn once (semelparity). There are a number of anatomical features that help in the identification of the various species of *Salmo* and *Oncorhynchus*; these include scale and fin ray counts, the number and shape of the gill rakers on the first arch and the length of the maxilla in relation to the eye.

Origin

There has been much debate as to whether the Salmonidae had their origins in the sea or fresh water. Some scientists considered that the Salmonidae had a marine origin with an ancestor similar to the Argentinidae (argentines) which are entirely marine and, like the salmonids and smelts (Osmeridae), bear an adipose fin. Other scientists considered the salmonids to have had a freshwater origin, supporting their case by suggesting that since the group has both freshwater-resident and migratory forms within certain species there has been recent divergence. Furthermore, there are no entirely marine forms among modern salmonids so they can not have had a marine origin. The Salmonidae have been revised as relatively primitive teleosts of probable marine pelagic origin whose specializations are associated with reproduction and early development in fresh water. The hypothesis of the evolution of salmonid life histories through penetration of fresh water by a pelagic marine fish, and progressive restrictions of life history to the freshwater habitat, involves adaptations permitting survival, growth and reproduction there. The salmonid genera show several ranges of evolutionary progression in this direction, with generally greater flexibility among *Salmo* and *Salvelinus* than among *Oncorhynchus* species (**Table 1**). Evidence for this evolutionary progression is perhaps even greater if one starts with the Argentinidae and Osmeridae which are basically marine coastal fishes. Some enter the rivers to breed, some live in fresh water permanently, and others such as the capelin (*Mallotus villosus*) spawn in the gravel of the seashore.

Life Histories

Members of the Salmonidae have a similar life history pattern but with varying degrees of complexity. A typical life history involves the female excavating a hollow in the river gravel into which the large yolky eggs are deposited and fertilized by the male. Because of egg size and the protection afforded them in the gravel the fecundity of the Salmonidae is low when compared with species such as the herring and the cod which are very fecund but whose eggs have no protection. On hatching the salmonid young (alevins) live on their yolk sac within the gravel for some weeks depending on water temperature. On emerging the fry may remain in the freshwater environment for a varying length of time (**Table 2**) changing as they grow into the later stages of parr and then smolt, at which stage they go to sea (**Figure 1**). Not all the Salmonidae have a prolonged freshwater life before entering the sea, and the juveniles of some

Table 1   Examples of flexibility of life history patterns in salmonid genera: anadromy implies emigration from fresh water to the marine environment as juveniles and return to fresh water as adults; nonanadromy implies a completion of the life cycle without leaving fresh water, although this may involve migration between a river and a lake habitat

Genus	Species		
	Anadromous form only	Both anadromous and nonanadromous forms	Nonanadromous form only
Oncorhynchus	gorbuscha (pink salmon) keta (chum salmon) tschawytscha (chinook salmon)	nerka (sockeye salmon) kisutch (coho salmon) masou (cherry salmon) rhodurus (amago salmon) mykiss (steelhead/rainbow trout) clarki (cut throat trout)	aguabonito (golden trout)
Salmo	none	salar (Atlantic salmon) trutta (sea/brown trout)	
Salvelinus	none	alpinus (arctic charr) fontinalis (brook trout) malma (Dolly Varden) leucomanis	namaycush (lake trout)
Hucho	none	perryi	hucho (Danube salmon)

Adapted from Thorpe (1988).

species such as the pink salmon (*Oncorhynchus gorbuscha*) and chum salmon (*O. keta*) migrate to sea on emerging from the gravel. Others such as the sockeye salmon (*O. nerka*) have specialized freshwater requirements, namely the need for a lacustrine environment to which the young migrate on emergence (**Table 2**).

Distribution

Atlantic Salmon

The Atlantic salmon occurs throughout the northern Atlantic Ocean and is found in most countries whose rivers discharge into the North Atlantic Ocean and Baltic Sea from rivers as far south as Spain and Portugal to northern Norway and Russia and one river in Greenland. It has been introduced to some countries in the Southern Hemisphere, including New Zealand where they only survive as a land-locked form.

Sea Trout

The marine distribution of the anadromous form of *S. trutta* is confined to coastal and near-offshore waters and is not found in the open ocean. It has a more limited distribution than the salmon being confined mainly to the eastern seaboard of the North Atlantic, although it has been introduced includes Iceland, and the Faroe Islands, Scandinavia, the Cheshkaya Gulf in the north, throughout the Baltic and down the coast of Europe to northern Portugal. It occurs as a subspecies in the Black Sea (*S. trutta labrax*) and Caspian Sea (*S. trutta caspius*). It has been introduced to countries in the Southern Hemisphere including Chile and the Falkland Islands.

Sockeye Salmon

The natural range of the sockeye, as other species of Pacific salmon, is the temperate and subarctic waters of the North Pacific Ocean and the northern adjoining Bering Sea and Sea of Okhotsk. However, because sockeye usually spawn in areas associated with lakes,

Table 2 Life histories

Species	Length of freshwater life (years)	Particular features
Oncorhynchus nerka	1–3	Lake environment required for juveniles
O. gorbuscha	Migrate to estuarine waters on emergence	Tend to spawn closer to sea than other oncorhynchids, and may frequent smaller river systems
O. keta	Migrate to estuarine waters on emergence	Spawning takes place in lower reaches of rivers
O. tschawytscha	Some migrate to estuary on emergence, others remain in fresh water for one or more years	Two races: *anadromous*, long freshwater residence; *semelparous*, short freshwater residence
O. kisutch	1–2	Tend to utilize small coastal streams
O. masou	1–2	Large parr become smolts and go to sea; Small to medium-sized parr remain in fresh water
Salmo salar	1–7	A small percentage spawn more than once. 'Land-locked' forms live in lakes and spawn in afferent or efferent rivers
S. trutta	1–4 (anadromous form)	May spawn frequently, the anadromous form after repeat spawning migrations
S. alpinus	2–6	Anadromous form only occurs in rivers lying north of 60°N

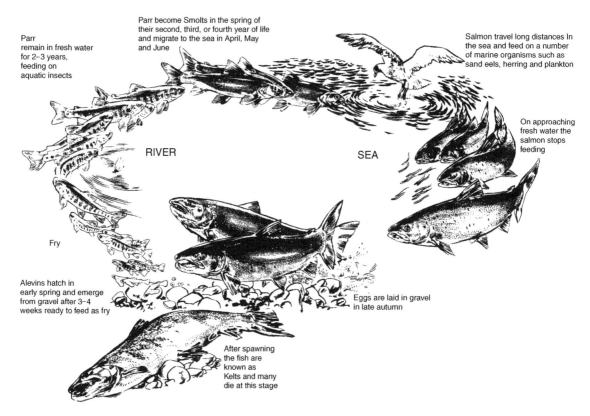

Figure 1 Life cycle of the Atlantic salmon. (Reproduced from Mills, 1989.)

where their juveniles spend their freshwater existence before going to sea, their spawning distribution is related to north temperate rivers with lakes in their systems. The Bristol Bay watershed in southwestern Alaska and the Fraser River system are therefore the major spawning areas for North American sockeye.

Pink Salmon

The natural freshwater range of pink salmon embraces the Pacific coast of Asia and North America north of 40°N and during the ocean feeding and maturation phase they are found throughout the

North Pacific north of 40°N. The pink is the most abundant of the Pacific salmon species followed by the chum and sockeye in that order. Pink salmon have been transplanted outside their natural range to the State of Maine, Newfoundland, Hudson Bay, the North Kola Peninsula and southern Chile.

Chum Salmon

The chum occurs throughout the North Pacific Ocean in both Asian and North American waters north of 40°N to the Arctic Ocean and along the western and eastern arctic seaboard to the Lena River in Russia and the Mackenzie River in Canada.

Chinook Salmon

The chinook has a more southerly distribution than the other Pacific salmon species, extending as far south as the Sacramento–San Joaquin River system in California as well as into the northerly waters of the Arctic Ocean and Beaufort Sea. It has been transplanted to the east coast of North America, the Great Lakes, south Chile and New Zealand.

Coho Salmon

The coho is the least abundant of the Pacific salmon species but has a similar distribution as the other species. It has been introduced to the Great Lakes, some eastern states of North America and Korea and Chile.

Masu Salmon

This species only occurs in Far East Asia, in the Sea of Japan and Sea of Okhotsk. Some have been transported to Chile. The closely related amago salmon mainly remain in fresh water but some do migrate to coastal waters but few to the open sea.

Arctic Charr

The anadromous form of this species has a wide distribution throughout the subarctic and arctic regions of the North Atlantic north of 40°N. It enters rivers in the late summer and may remain in fresh water for some months.

Migrations

Once Atlantic salmon smolts enter the sea and become postsmolts they move relatively quickly into the ocean close to the water surface. Patterns of movement are strongly influenced by surface water currents, wind direction and tidal cycle. In some years postsmolts have been caught in the near-shore zone of the northern Gulf of St Lawrence throughout

their first summer at sea, and in Iceland some postsmolts, mainly maturing males, forage along the shore following release from salmon-ranching stations. These results indicate that the migratory behavior of postsmolts can vary among populations. Pelagic trawl surveys conducted in the Iceland/Scotland/Shetland area during May and June and in the Norwegian Sea from 62°N to 73°N in July and August, have shown that postsmolts are widely distributed throughout the sampled area, although they do not reach the Norwegian Sea until July and August. Over much of the study area, catches of postsmolts are closely linked to the main surface currents, although north of about 64°N, where the current systems are less pronounced, postsmolts appear to be more diffusely distributed.

Young salmon from British Columbia rivers descend in discontinuous waves, and it has been suggested that this temporal pattern has evolved in response to short-term fluctuations in the availability of zooplankton prey. Although zooplankton production along the British Columbia coast is adequate to meet this seasonal demand, there is debate about the adequacy of the Alaska coastal current to support the vast populations of growing salmon in the summer. Density-dependent growth has been shown for the sockeye salmon populations at this time, suggesting that food can be limiting.

The movements of salmon in offshore waters are complex and affected by physical factors such as season, temperature and salinity and biological factors such as maturity, age, size and food availability and distribution of food organisms and stock-of-origin (i.e. genetic disposition to specific migratory patterns). Through sampling of stocks of the various Pacific salmon at various times of the year over many years scientists have been able to construct oceanic migration patterns of some of the major stocks of North American sockeye, chum and pink salmon (**Figure 2**) as well as for stocks of chinook and masu salmon. Similarly, as a result of ocean surveys and tagging experiments it has been possible to determine the migration routes of North American Atlantic salmon very fully (**Figures 3** and **4**). A picture of the approximate migration routes of Atlantic salmon in the North Atlantic area as a whole has been achieved from tag recaptures (**Figure 5**).

Movements

Vertical and horizontal movements of Atlantic salmon have been investigated using depth-sensitive tags and data storage tags. Depth records show that salmon migrate mostly in the uppermost few meters

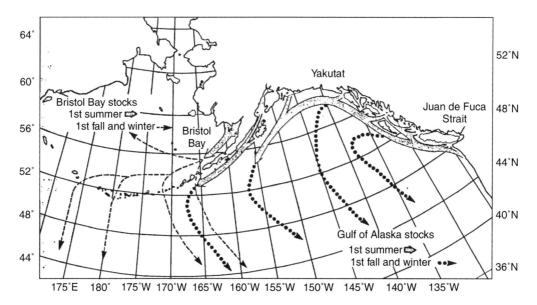

Figure 2 Diagram of oceanic migration patterns of some major stocks of North American sockeye, chum and pink salmon during their first summer at sea, plus probable migrations during their first fall and winter. (Reproduced from Burgner, 1991.)

and often show a diel rhythm in vertical movements. The salmon are closest to the surface at mid-day and go deeper at night. They have been recorded diving to a depth of 110 m.

Available information from research vessels and operation of commercial Pacific salmon fisheries suggest that Pacific salmon generally occur in near-surface waters.

Details of rates of travel are given in **Table 3**.

Food

The marine diet of both Pacific and Atlantic salmon comprises fish and zooplankton. The proportion of fish and zooplankton in their diet varies with season, availability and area. Among the fish species sockeye eat capelin (*Mallotus villosus*), sand eels (*Ammodytes hexapterus*), herring (*Clupea harengus pallasi*) and pollock (*Theragra chalcogramma*). Zooplankton organisms include euphausiids, squid, copepods and pteropods.

The diet of pink salmon includes fish eggs and larvae, squid, amphipods, euphausiids and copepods, whereas chum salmon were found to take pteropods, salps, euphausiids and amphipods. Chinook salmon eat herring, sand eels, pilchards and anchovies and in some areas zooplankton never exceeds 6% of the diet. However, the diet varies considerably from area to area and up to 21 different taxonomic groups have been recorded in this species' diet. Similarly, the diet of coho salmon is a varied one, with capelin, sardines, lantern fish (myctophids), other coho salmon, being eaten along with euphausiids, squid, goose barnacles and jellyfish.

Masu salmon eat mainly small-sized fish and squid and large zooplankton such as amphipods and euphausiids. Fish species taken include capelin, herring, Dolly Varden charr, Japanese pearlside (*Maurolicus japonicus*), saury (*Cololabis saira*), sand eels, anchovies, greenlings (*Hexagrammos otakii*) and sculpins (*Hemilepidotus* spp.).

There is no evidence of selective feeding among sockeye, pink and chum salmon.

The food of Atlantic salmon postsmolts is mainly invertebrate consisting of chironomids and gammarids in the early summer in inshore waters and in the late summer and autumn the diet changes to one of small fish such as sand eels and herring larvae. A major dietary study of 4000 maturing Atlantic salmon was undertaken off the Faroes and it confirmed the view that salmon forage opportunistically, but that they demonstrate a preference for fish rather than crustaceans when both are available. They are also selective when feeding on crustaceans, preferring hyperiid amphipods to euphausiids. Feeding intensity and feeding rate of Atlantic salmon north of the Faroes have been shown to be lower in the autumn than in the spring, which might suggest that limited food is available at this time of year. Similar results have been found for Atlantic salmon in the Labrador Sea and in the Baltic. Salmon in the north Atlantic also rely on amphipods in the diet in the autumn, whereas at other times fish are the major item. Fish taken include capelin, herring, sprat (*Clupea sprattus*), lantern fishes, barracudinas and pearlside (*Maurolicus muelleri*).

Predation

Marine mammals recorded predating on Pacific salmon include harbour seal (*Phoca vitulina*), fur seal (*Callorhinus uresinus*), Californian sea lion (*Zalophus grypus*), humpbacked whale (*Megaptera novaeangliae*) and Pacific white-sided dolphin (*Lagenorhynchus obliquideus*). Pinniped scar wounds on sockeye salmon, caused by the Californian sea lion and harbor

seal, increased from 2.8% in 1991 to 25.9% in 1996 and on spring-run chinook salmon they increased from 10.5% in 1991 to 31.8% in 1994.

Predators of Atlantic salmon postsmolts include gadoids, bass (*Dicentrarchus labrax*), gannets (*Sula bassana*), cormorants (*Phalacrocorax carbo*) and Caspian terns (*Hydroprogne tschegrava*). Adult fish are taken by a number of predators including the grey seal (*Halichoerus grypus*), the common seal (*Phoca*

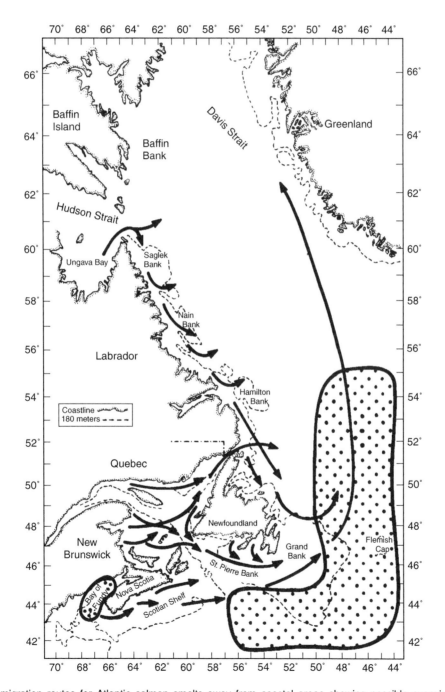

Figure 3 The migration routes for Atlantic salmon smolts away from coastal areas showing possible overwintering areas and movement of multi-sea winter salmon into the West Greenland area. Arrows indicate the path of movement of the salmon, and dotted area indicates the overwintering area. (Reproduced from Reddin, 1988.)

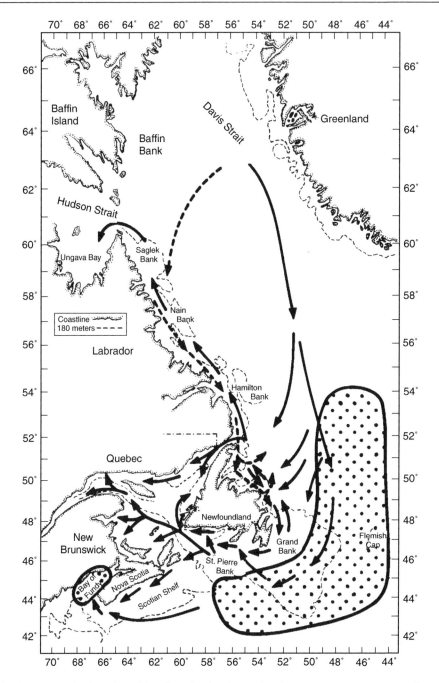

Figure 4 The migration routes of salmon from West Greenland and overwintering areas on return routes to rivers in North America. Solid arrows indicate migration in mid-summer and earlier; broken arrows indicate movement in late summer and fall; dotted areas are the wintering areas. (Reproduced from Reddin, 1988.)

vitulina), the bottle-nosed dolphin (*Tursiops truncatus*), porbeagle shark (*Lamna cornubica*), Greenland shark (*Somniosus microcephalus*) and ling (*Molva molva*).

Environmental Factors

Surface Salinity

Salinity may have an effect on fish stocks and it has been shown that the great salinity anomaly

of the 1970s in the North Atlantic adversely affected the spawning success of eleven of fifteen stocks of fish whose breeding grounds were traversed by the anomaly. In the North Pacific there was found to be little relation between the high seas distribution of sockeye salmon and surface salinity. Sockeye are distributed across a wide variety of salinities, with low salinities characterizing 'salmon waters' of the Subarctic Pacific Region.

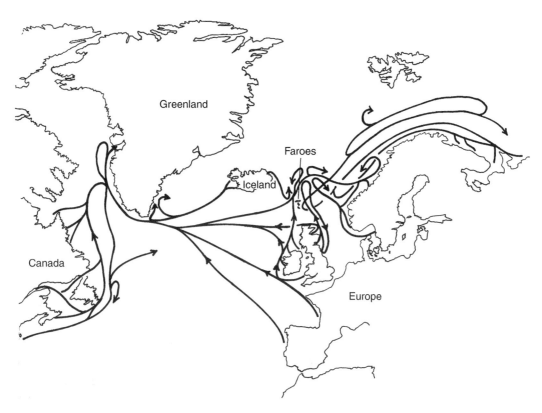

Figure 5 Approximate migration routes of Atlantic salmon in the North Atlantic area. (Reproduced from Mills, 1989.)

Table 3 Rates of travel in the sea of Atlantic and Pacific salmon

Species	Rate of travel (km day^{-1})	Conditions
Atlantic salmon	19.5–24	Icelandic postsmolts from ranching stations
	15.2–20.7	Postsmolts based on smolt tag recaptures
	22–52	Grilse and large salmon
	32	Average for maturing salmon of all sea ages
	26	For previous spawners migrating to Newfoundland and Greenland
	28.8–43.2	Icelandic coastal waters
	10–50	Baltic Sea
Sockeye salmon	46–56	During their final 30–60 days at sea
Pink salmon	17.2–19.8	Juveniles during first 2–3 months at sea
	45–54.3	Fish recovered at sea and in coastal waters
	43.3–60.2	For eastern Kamchatka stocks tagged in
		Aleutian Island passes or the Bering Sea
Coho salmon	30	
	55	Could be maintained over long distances

Sea Surface Temperature

Sockeye salmon are found over a wide variety of conditions. Sea surface temperature (SST) has not been found to be a strong and consistent determinant of sockeye distribution, but it definitely influences distribution and timing of migrations. Sockeye tend to prefer cooler water than the other Pacific salmon species.

Temperature ranges of waters yielding catches of various salmon species in the northwest Pacific in winter are: sockeye, 1.5–6.0°C; pink, 3.5–8.5°C; chum, 1.5–10.0°C; coho, 5.5–9.0°C.

A significant relationship was found for SST and Atlantic salmon catch rates in the Labrador Sea, Irminger sea and Grand Banks (**Figure 6**), with the greatest abundance of salmon being found in SSTs between 4 and 10°C. This significant relationship suggests that salmon may modify their movements at sea depending on SST. It has also been suggested that the number of returning salmon is linked to environmental change and that the abundance of salmon off Newfoundland and Labrador was linked to the amount of water of <0°C on the shelf in summer. In years when the amount of cold water was large and the marine climate tended to be cold, there were fewer salmon returning to coastal waters. A statistically significant relationship has been found between the area of ice off Labrador and northern Newfoundland and the number of returning salmon in one of the major rivers on the Atlantic coast of Nova Scotia. The timing and geographical distribution of Atlantic salmon along the Newfoundland and Labrador coasts have been shown to be dependent on the arrival of the 4°C water, salmon arriving earlier during warmer years.

In two Atlantic salmon stocks that inhabit rivers confluent with the North Sea a positive correlation was found between the area of 8–10°C water in May and the survival of salmon. An analysis of SST distribution for periods of good versus poor salmon survival showed that when cool surface waters dominated the Norwegian coast and the North Sea during May salmon survival has been poor. Conversely, when the 8°C isotherm has extended northward along the Norwegian coast during May, survival has been good.

Temperature may also be linked to sea age at maturity. An increase in temperature in the northeast Atlantic subarctic was found to be associated with large numbers of older (multi-sea winter) salmon and fewer grilse (one-sea winter salmon) returning to the Aberdeenshire Dee in Scotland.

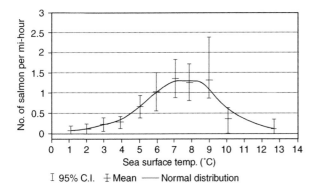

Figure 6 The relationship between sea surface temperatures and salmon catch rates from Labrador Sea, Irminger Sea and Grand Banks, 1965–91. (Reproduced from Reddin and Friedland, 1993.)

Ocean Climate

Ocean climate appears to have a major influence on mortality and maturation mechanisms in salmon. Maturation, as evidenced by returns and survival of salmon of varying age, has been correlated with a number of environmental factors. The climate over the North Pacific is dominated by the Aleutian Low Pressure System. The long-term pattern of the Aleutian Low Pressure System corresponds with trends in salmon catch, with copepod production and with other climatic indices, indicating that climate and the marine environment may play an important role in salmon production. Survival of Pacific salmon species varies with fluctuations in large-scale circulation patterns such as El Niño Southern Oscillation (ENSO) events and more localized upwelling circulation that would be expected to affect local productivity and juvenile salmon growth. Runs of Pacific salmon in rivers along the western margin of North America stretching from Alaska to California vary on a decadal scale. When catches of a species are high in one region (e.g. Oregon) they may be low in another (e.g. Alaska). These changes are in part caused by marked interdecadal changes in the size and distribution of salmon stocks in the northeast Pacific, which are in turn associated with important ecosystem shifts forced by hydroclimatic changes linked to El Niño and possibly climate change.

Similarly, in the North Atlantic there have been annual and decadal changes in the North Atlantic Oscillation (NAO) index. Associated with these changes is a wide range of physical and biological responses, including effects on wind speed, ocean circulation, sea surface temperature, prevalence and intensity of Atlantic storms and changes in zooplankton production. For example, during years of positive NAO index, the eastern and western North Atlantic display increases in temperature while temperatures in central North Atlantic and in the Labradon Basin decline. The extremely low temperatures in the Labrador Sea during the 1980s coincided with a decline in salmon abundance. Similarly, in Europe, years of low NAO index were associated with high catches, whereas stocks have declined dramatically during high index years.

In the northeast Atlantic significant correlations were obtained between the variations in climate and hydrography with declines in primary production, standing crop of zooplankton and with reduced abundance and altered distribution of pelagic forage fishes and salmon catches.

By analogy with the North Pacific it is likely that changes in salmon abundance in the northeast Atlantic are linked to alterations in plankton productivity and/

or structural changes in trophic transfer, each forced by hydrometeorological variability and possibly climate change as a result of the North Atlantic Oscillation and the Gulf Stream indexes.

Homing

It is suggested that homeward movement involves directed navigation. Fish can obtain directional information from the sun, polarized light, the earth's magnetic field and olfactory clues. Chinook and sockeye salmon have been shown capable of detecting changes in magnetic field. It has been suggested that migration from the feeding areas at sea to the natal stream must be accomplished without strictly retracing the outward migration using a variety of geopositioning mechanisms including magnetic and celestial navigation coordinated by an endogenous clock. Olfactory and salinity clues take over from geopositioning (bi-coordinate navigation) once in proximity of the natal stream because even small changes in declination during the time at sea correspond to a large search area for the home stream.

See also

Fish Feeding and Foraging. Fish Larvae. Fish Migration, Horizontal.

Further Reading

Burgner RL (1991) Life history of sockeye salmon (*Oncorhynchus nerka*). In: Groot C and Margolis L (eds.) *Pacific Salmon Life Histories*, pp. 2–117. Vancouver: UBC Press.

Foerster RE (1968) The sockeye salmon, Oncorhynchus nerka. *Bulletin of the Fisheries Research Board of Canada* 162: 422.

Groot C and Margolis L (eds.) (1991) *Pacific Salmon Life Histories*. Vancouver: UBC Press.

Healey MC (1991) Life history of chinook salmon (*Oncorhynchus tschawytscha*). In: Groot C and Margolis L (eds.) *Pacific Salmon Life Histories*, pp. 312–393. Vancouver: UBC Press.

Heard WR (1991) Life history of pink salmon (*Oncorhynchus gorbuscha*). In: Groot C and Margolis L (eds.) *Pacific Salmon Life Histories*, pp. 120–230. Vancouver: UBC Press.

Kals F (1991) Life histories of masu and amago salmon (*Oncorhynchus masou and Oncorhynchus rhodurus*). In: Groot C and Margolis L (eds.) *Pacific Salmon Life Histories*, pp. 448–520. Vancouver: UBC Press.

McDowell RM (1988) *Diadromy in Fishes*. London: Croom Helm.

Mills DH (ed.) (1989) *Ecology and Management of Atlantic Salmon*. London: Chapman and Hall.

Mills DH (ed.) (1993) *Salmon in the Sea and New Enhancement Strategies*. Oxford: Fishing News Books.

Mills DH (ed.) (1999) *The Ocean Life of Atlantic Salmon*. Oxford: Fishing News Books.

Reddin D (1988) Ocean Life of Atlantic Salmon (*Salmo salar* L.) in the northwest Atlantic. In: Mills D and Piggins D (eds.) *Atlantic Salmon: Planning for the Future*, pp. 483–511. London and Sydney: Croom Helm.

Reddin D and Friedland K (1993) Marine environmental factors influencing the movement and survival of Atlantic salmon. In: Mills D (ed.) *Salmon in the Sea and New Enhancement Strategies*, pp. 79–103. Oxford: Fishing News Books.

Salo EO (1991) Life history of chum salmon (*Oncorhynchus keta*). In: Groot C and Margolis L (eds.) *Pacific Salmon Life Histories*, pp. 232–309. Vancouver: UBC Press.

Sandererock FK (1991) Life history of coho salmon (*Oncorhynchus kisutch*). In: Groot C and Margolis L (eds.) *Pacific Salmon Life Histories*, pp. 396–445. Vancouver: UBC Press.

Thorpe JE (1988) Salmon migration. *Science Progress Oxford* 72: 345–370.

EELS

J. D. McCleave, University of Maine, Orono, ME, USA

Introduction

Migratory, catadromous eels of the genus *Anguilla* have among the most fascinating life cycles of all fishes. Catadromous fishes breed in the ocean but feed and grow in fresh waters or estuaries. This cycle includes a larval or juvenile migration from breeding to feeding area and an adult migration back. These two migrations have been termed denatant and contranatant to imply that the larval migration is accomplished largely by drift with currents and the adult migration by swimming against the currents. These terms get the essence of eel migration, but they fail to capture the complexity and mystery of eel migrations. Anguillid eels are sexually, ecologically, and behaviorally highly adaptive. They occur naturally in a greater diversity of habitats than any other fishes. These statements apply mainly to the feeding and growth stages in continental waters. In contrast, successful spawning and larval survival seems to depend on rather specific oceanic conditions for all *Anguilla*.

Juveniles and adults of *Anguilla* are elongate, rather cylindrical, darkly pigmented fishes, reaching about 30 cm to nearly 200 cm at maturity depending on species and gender. They have long continuous median fins extending from the anus around the tail and well forward on the back. The contrast with the larvae, termed leptocephali, is extreme. Leptocephali are laterally compressed and deep bodied. They are nearly transparent, with pigment restricted to the retina of the eye. A major metamorphosis occurs between larva and juvenile.

This article considers eels of the Family Anguillidae, which is one of about 22 families of eels. Eels are primitive bony fishes. Within the bony fishes (Class Osteichthyes), there are two orders of eels. The Anguilliformes contains 15 families, including spaghetti eels, morays, cutthroat eels, worm eels, snipe eels, and conger eels. The Saccopharyngiformes contains seven families, including deep-sea gulper eels. These two orders are unified by the presence of leptocephali with continuous dorsal, caudal, and anal fins. W. Hulet and R. Robins have argued that the evolution of the leptocephalus, which is in ionic equilibrium with, and nearly isosmotic with sea water, in these primitive fishes was one solution allowing fishes to complete their life cycle in the sea.

Only species in the genus *Anguilla* are truly catadromous, though not all individuals enter fresh water. The other families are marine, ranging from abyssalpelagic to epipelagic to coastal, with juveniles of some species being estuarine. The easily viewed stages in the life cycle of *Anguilla* occur in fresh water, and they are the only group of eels to enter fresh waters, so they are often called freshwater eels, an unfortunate name, given their extensive migrations at sea.

Taxonomic and Geographic Diversity of *Anguilla*

Fifteen species of *Anguilla* comprise the Anguillidae (**Table 1**). These can be grouped into tropical and temperate species on the basis of coastal and freshwater distribution, and of proximity of those distributions in continental waters to the spawning areas. Two temperate species occur in the North Atlantic Ocean, one in the North Pacific, and two in the South Pacific. Eight tropical species are all distributed in the western Pacific and Indian Oceans. Two species extend from the tropics into temperate zones, one in the South Pacific and one in the Indian Ocean.

All species require warm, saline, offshore water for successful reproduction. Appropriate currents must be present to transport the larvae toward continental waters. The widely distributed, temperate species use anticyclonic, subtropical gyres for spawning and use associated western boundary currents for distribution of larvae. The European eel is unusual in having its continental distribution on the eastern side of an ocean basin.

Life Cycle

Life cycles are best known for the temperate species, but they are undoubtedly similar for the tropical species (**Figure 1**).

Spawning Areas and Times

Spawning of adult eels has never been observed in nature. Spawning areas and times are inferred from the distribution of small leptocephali. A fascinating case was the discovery of the spawning area of the European eel by Danish fishery biologist, J. Schmidt.

Table 1 Taxonomic diversity and continental distribution of the Family Anguillidae with its single genus Anguilla, 15 species and four subspecies

Scientific name	English common name	Continental distribution
A. anguilla (Linnaeus, 1758)	European eel	(Te)[a] Iceland, Europe and North Africa from Norway to Morocco, Mediterranean basin to Black Sea, Canary Islands, Azores
A. australis[b] Richardson, 1841	Shortfin eel	(Te) Lord Howe Island, east coast of Australia, Tasmania, New Zealand, and Auckland, Chatham, and Norfolk Islands, New Caledonia
A. bengalensis bengalensis (Gray, 1831)	Indian mottled eel	(Tr) India, Sri Lanka, Myanmar, Andaman Islands, northern Sumatra
A. bengalensis labiata (Peters, 1852)	African mottled eel	(Tr) Eastern Africa from Kenya to South Africa
A. bicolor bicolor McClelland, 1844	Indonesian shortfin eel	(Tr) Eastern Africa, Madagascar, India, Myanmar, Sumatra, Java, Timor, north-western Australia
A. bicolor pacifica Schmidt, 1928	—	(Tr) Eastern Indonesia, New Guinea, Taiwan
A. celebesensis Kaup, 1856	Celebes longfin eel	(Tr) Sumatra, Java, Timor, the Philippines, Celebes, western New Guinea, smaller islands of eastern Indonesia
A. dieffenbachii Gray, 1842	New Zealand longfin eel	(Te) New Zealand, and Auckland and Chatham Islands
A. interioris Whitley, 1938	New Guinea eel	(Tr) Eastern New Guinea
A. japonica Temminck and Schlegel, 1846	Japanese eel	(Te) Northern Vietnam, northern Philippines, Taiwan, China, Korea, and Japan
A. malgumora Popta, 1924	—	(Tr) Eastern Borneo
A. marmorata Quoy and Gaimard, 1824	Giant mottled eel	(Tr) South Africa, Madagascar, Indonesia, the Philippines, Japan, southern China, Taiwan, eastward through Pacific islands to the Marquesas
A. megastoma Kaup, 1856	Polynesian longfin eel	(Tr) Solomon Islands and New Caledonia eastward to Pitcairn Island
A. mossambica (Peters, 1852)	African longfin eel	(Tr–Te) Eastern Africa from Kenya to South Africa, Madagascar, Mascarenes
A. obscura Günther, 1871	South Pacific eel	(Tr) New Guinea to the Society Islands
A. reinhardtii Steindachner, 1867	Speckled longfin eel	(Te–Tr) Australia from Victoria to Cape York (north tip), New Caledonia, Lord Howe Island
A. rostrata (Lesueur, 1817)	American eel	(Te) Southern Greenland, eastern North America from Labrador through the Gulf of Mexico and the West Indies to Venezuela and Guyana, Bermuda

[a]Te, temperate species; Tr, tropical species.
[b]Australian and New Zealand subspecies are sometimes recognized.

When Schmidt in 1904 caught a 7.5-cm long leptocephalus of the European eel west of the Faroes in the north-eastern Atlantic Ocean, the chase was on. Between 1913 and 1922, Schmidt made four research cruises in the North Atlantic, and commercial vessels collected plankton samples for him. Only in the western North Atlantic were European leptocephali less than 1 cm long captured, in an area 20–30°N and 50–65°W. Larger leptocephali were found to the north and east across the Atlantic toward Europe. Schmidt also caught a few small American eel leptocephali west of the locus of small European leptocephali.

Recent research by F.-W. Tesch in Germany and at the University of Maine has confirmed that the south-western portion of the North Atlantic, the Sargasso Sea, is a primary spawning area for both Atlantic *Anguilla*. From the distribution in space and time of leptocephali less than 1 cm long, the European eel spawns from February through June, primarily March and April, at about 50–75°W in a narrow zonal band. The American eel spawns from February through April, at about 53–78°W. The two species overlap in space and time. The northern limit to spawning seems to be near-surface frontal zones in the subtropical convergence, which may serve as a cue for the adults to cease migrating and to spawn. Not all areas of the North Atlantic, especially south of 20°N, have been adequately sampled to rule out other spawning areas.

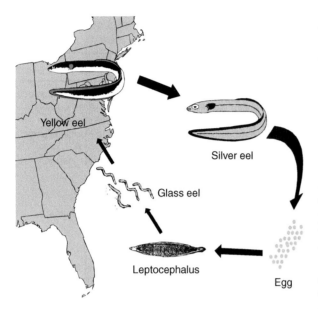

Figure 1 Catadromous life cycle of a temperate species of *Anguilla*, exemplified here by the American eel, *A. rostrata*.

Small leptocephali of the Japanese eel have been collected in July in a salinity frontal zone at the northern edge of the North Equatorial Current between 131° and 147°E, west of the Mariana Islands in the western North Pacific. Spawning areas for the Australian and New Zealand temperate species and for most of the tropical species are speculative. Spawning areas of some tropical species may not be far offshore of the areas of continental distribution.

Eggs

In the sea, eggs are presumed to be broadcast into the plankton and fertilized externally. Mature eggs of American, European, and Japanese eels, obtained by hormone injections of females, are transparent and about 1 mm in diameter. Eggs of the Japanese eel hatch in 38–45 h at 23°C. Japanese leptocephali are about 2.9 mm long at hatching.

Leptocephali

Leptocephali of *Anguilla* are part of the epipelagic plankton. Their bodies are transparent and laterally compressed, with body height being about 20% of length. A series of W-shaped myomeres (muscle segments) extends from head to tail. Inside the myomeres is an acellular, mucus-like matrix. The myomeres are overlain by a thin epithelium. The head is small (leptocephalus means slender head) and bears a set of fang-like teeth projecting forward. Eyes and olfactory organs are well developed. Dorsal, caudal, and anal fins are continuous around the posterior of the body. Small pectoral fins are present. Leptocephali of

Anguilla sp. are separated from one another primarily by the number of myomeres, e.g. 103–111 for the American eel and 112–119 for the European eel.

The mode of nutrition is in debate. Most workers have reported the absence of food in the simple guts of leptocephali, but the ingestion of bacteria, microzooplankton, or gelatinous organisms might go undetected. E. Pfeiler proposed that leptocephali absorb dissolved organic matter from the sea water across their epithelium. N. Mochioka and T. Otake showed that guts of leptocephali other than *Anguilla* collected at sea do contain larvacean houses, zooplankton fecal pellets, and detrital aggregates (particulate organic matter). Leptocephali may absorb dissolved organic matter across the gut.

Larval life lasts from a few months for the tropical species to debatably less than one to more than two years for the European eel. Estimates of larval duration are based on counting putative daily rings in otoliths (ear stones), but otoliths as indicators of age in days have not been validated. Leptocephali of the temperate species grow to about 6–8 cm, whereas those of the tropical species grow to 5–6 cm. Concurrently, leptocephali are gradually transported toward continental waters.

Glass Eels

At some time, a dramatic metamorphosis of form and physiology occurs. The body loses height and becomes rounder in cross-section, the larval teeth are resorbed, and the mucus matrix is catabolized. Newly transformed eels, still lacking pigment, are termed glass eels. Initially, they are still pelagic, and they move across the wide or narrow continental shelf into coastal waters. Along the way, they lose their strict pelagic habit and move on and off the bottom. Glass eels entering coastal waters and estuaries may become resident there, or they may continue into fresh waters. The invasion of estuaries and fresh waters is seasonal in the temperate species. At a particular location, the immigration usually peaks over two months, earlier at lower latitudes than at higher latitudes. For the Japanese and American eels and the two New Zealand species, glass eel immigration is in late winter, spring, and early summer. Immigration occurs throughout the year in some tropical species.

Elvers and Yellow Eels

Pigmentation develops rapidly after entry into estuarine or fresh waters, and the eels are known as elvers and then yellow eels. Elver refers loosely to small pigmented eels. Yellow eels are named because their ventral surfaces are yellow to white. The dorsal surfaces are usually shades of green to brown, plain or mottled depending on species. Yellow eels are in the juvenile growth phase of life.

Plasticity and adaptation

Habitat selection   Gradual upstream movement of a segment of the yellow eel population occurs for several years, so older yellow eels become distributed from coastal waters to far inland. Eels occur in various habitats including saline coastal waters, estuaries, marshes, large rivers, small streams, lakes, ponds, and even subterranean springs. They occur in highly productive to highly unproductive waters. Temperate species invade waters with near tropical temperatures and waters that are seasonally ice covered.

Diet   Yellow eels are opportunistic, consuming nearly any live prey that can be captured. Benthic invertebrates predominate in the diets, but fish, including eels, become important to larger eels. Eels respond to local abundances of appropriately sized prey through the seasons. Insect larvae may predominate in early summer and young-of-the-year fishes in late summer. Yellow eels are nocturnal, feeding mostly during the early hours of the night.

Sex determination and differentiation   Sex is partially or wholly determined by environmental conditions. There are no morphologically differentiated sex chromosomes. Differentiation of the gonads does not occur until the yellow eel phase, with considerable variation in age size at differentiation. For American and European eels, a high population density of small eels seems to result in a high proportion of males and vice versa. Within a river basin, lakes generally have a higher proportion of females than riverine sections, which may reflect a population density or productivity effect.

Earlier hypotheses that there was a cline of increasing proportion of female American eels with increasing latitude, and that this was due to longer larval life of females, are not supported by current knowledge. Male-dominated rivers occur in northern as well as southern latitudes, and widely varying sex ratios occur in neighboring rivers. This is indicative of some mechanism for sex determination that acts at the river or habitat level.

Growth rate and sexual dimorphism   Growth rate varies with length of the growing season and with productivity of the habitat. Growth rate also varies among individuals in the same habitat. For a given age, growth rate is greater for females than males, and females attain greater size at maturity than

males. Annual growth rate decreases with age. Because the average age at maturity of females is greater than males, the annual average growth rate to maturity may sometimes be greater for males than females.

Determination of age of eels is by counting annual growth rings in the otoliths, but preparation of otoliths has varied among investigators. Accuracy and precision are low, with a tendency to overestimate the age of young eels and underestimate the age of older eels. Therefore, calculated growth rates must be interpreted cautiously.

Size and age at maturity   Within a species and gender, size is more characteristic than age for when a yellow eel will metamorphose into a silver eel and migrate to sea (**Table 2**).

Studies of ages and sizes at migration of the European eel at 44 sites from Tunisia to Sweden allows generalization. Faster-growing eels of both sexes mature at an earlier age but not a larger size than slower-growing eels. The productivity of the habitat influences growth rate and, therefore, size and age at maturation. Thus, variation in length and age at maturity can occur in different habitats within a restricted geographic range. The length of the growing season and the temperature are negatively correlated with latitude, so age at maturity is strongly correlated with latitude. Both sexes display a time-minimizing strategy, i.e., they mature at the earliest opportunity. However, females mature at a larger size than males because they require sufficient energy to migrate to the spawning area and to produce eggs. Length at migration increases with distance to the spawning area for both sexes.

Metamorphosis   Toward the end of the yellow phase, many morphological and physiological changes occur, transforming a bottom-oriented yellow eel into an oceanic, pelagic, migratory silver eel. Lipid in the muscle increases to 20–35% of muscle mass in European eels. The gut degenerates, suggesting that silver eels do not feed on migration. The eye increases in diameter by approximately 50%, the number of rod cells in the retina increases, and the spectral absorption maxima shifts more toward blue. This results in increased sensitivity for conditions in the oceanic mesopelagic zone by day and the epipelagic zone by night. The ventral surface of the body becomes whiter and more reflective, increasing the countershading. The swim bladder retial capillaries increase in length from yellow to silver eel, by a factor of 2.5 in American eels, increasing swim-bladder gas deposition rate. Additional guanine is deposited in the swim-bladder

Table 2 Range of mean ages and mean lengths at maturity of silver eels of five species of *Anguilla*; the European eel is by far the most well studied in this regard

Species	Sex	Age at maturity			Length at maturity		
		Mean age range (y)	Factor[a]	N[b]	Mean length range (cm)	Factor	N
A. anguilla	Male	2.3–15.0	6.5	21	31.6–46.0	1.5	33
	Female	3.4–20.0	5.9	28	44.9–86.8	1.9	38
A. rostrata	Male	3.0–12.7	4.2	6	27.7–39.2	1.4	9
	Female	7.1–19.3	2.7	6	41.7–95.7	1.9	15
A. australis	Male	14.2–14.4	1.0	3	43.2–46.5	1.1	3
	Female	19.4–23.6	1.2	4	60.9–94.0	1.5	4
A. dieffenbachii	Male	23.2	—	1	62.3–66.6	1.1	2
	Female	34.3–49.4	1.4	2	106.3–115.6	1.1	2
A. japonica	Male	6.4	—	1	48.3	—	1
	Female	8.3	—	1	61.4	—	1

[a]Factor, largest value divided by smallest value.
[b]N, number of geographic locations studied.

Table 3 Fecundity (number of eggs) of females of four species of *Anguilla* over the size range of eels studied; estimates are probably least accurate for *A. anguilla*

Species	Smaller eels		Larger eels	
	Length (cm)	Fecundity	Length (cm)	Fecundity
A. anguilla	65	775 000	85	1 956 000
A. australis	50	410 000	95	3 901 000
A. dieffenbachii	70	1 009 000	145	21 374 187
A. rostrata[a]	45	1 447 000	115	23 357 000
A. rostrata[b]	50	646 000	75	2 949 000

[a]Estimate from Maine, USA, 45°N.
[b]Estimate from Chesapeake Bay, USA, 37°N.

wall, reducing diffusive loss of gas. Premigratory silver American eels maintained swim-bladder inflation at a simulated depth of 150 m compared with 60 m for yellow eels.

Silver Eels

Silver eels return to the open ocean, migrate to the spawning area, spawn, and presumably die. The temperate species typically leave fresh and coastal waters in mid–late summer or autumn, earlier at higher latitudes than at lower latitudes. They are presumed to spawn at the next spawning season. However, the journey and spawning have not been witnessed for any species, so the biology of this oceanic stage is speculative.

The fecundity of eels increases exponentially with length, ranging from about 0.4 to 25 million eggs depending on species and size (**Table 3**).

Migrations in the Ocean

Silver Eels

In a telemetric study in an estuary, silver American eels migrated seaward by selective tidal stream transport. By ascending into the water column when the tide was ebbing and descending to the bottom when the tide was flooding, eels moved seaward in a saltatory fashion. Directed swimming may also be important in less strongly tidal estuaries.

In shallow waters of the North Sea, silver European eels have been shown by telemetry to maintain travel in a given direction regardless of tidal direction, without direct contact with the sea bottom. They also have the ability to move along the tidal axis by selective tidal stream transport. How these mechanisms are used in actual migration is unknown. In deeper waters of the western Mediterranean Sea, silver eels also maintained approximately

Table 4 Estimated lengths of migration of silver eels of three species of *Anguilla* from various locations to the approximate center of the spawning areas[a], assuming travel on a great circle route. Data arranged from south to north.

Species	Location	Distance (km)
A. anguilla	Tejo River, Portugal	4980
	Po River, Italy	8200
	Loire River, France	5600
	River Shannon, Ireland	4965
	IJsselmeer, The Netherlands	7025
	Lake Vidgan, Sweden	8300
	Thjórsá River, Iceland	5150
A. japonica	Pearl River, China	2840
	Shih-Ting River, Taiwan	2205
	Yangtze River, China	2605
	Hamana Lake, Japan	2135
	Naktong-gang River, South Korea	2480
A. rostrata	Mississippi River, Louisiana, USA	2265
	Cooper River, South Carolina, USA	1440
	Chesapeake Bay, Virginia, USA	1550
	Penobscot Bay, Maine, USA	2165
	St John's, Newfoundland, Canada	2840
	St Lawrence River, Quebec, Canada	3820

[a] Assumed spawning locations: *A. anguilla* 25°N 60°W; *A. japonica* 15°N 140°E; *A. rostrata* 25°N 68°W.

unidirectional movement for hours to days. They also made daily vertical migrations, moving upward at dusk to about 160 m and down at dawn to about 320 m.

Routes and rates of silver eel migrations in the open oceans are unknown. I infer from the morphological and physiological changes in the eye, the swim bladder, and the skin that occur at metamorphosis to the silver stage that migration occurs in the epipelagic and upper mesopelagic zones. A model of European eel migration, which combined oriented swimming toward the Sargasso Sea with modeled surface currents, predicted an arrival in the Sargasso Sea somewhat south (15–20°N) and east of where the smallest leptocephali have been captured. A second simulated arrival occurred later at about 28–30°N.

Four migrating females of the New Zealand longfin eel were tagged with archival 'pop-up' satellite transmitters, programmed to release after two or three months. All four moved eastward from the New Zealand coast as much as 1000 km. This technology may allow rapid advances in knowledge of silver eel migrations at sea, at least for females of the larger species.

Because the spawning areas of many of the species are ill defined, the lengths of migrations of many are unknown. Apparently, many of the tropical species spawn over deep water just off the edge of the continental shelves, e.g., the Celebes, Molucca, and Banda seas in the western Pacific. In contrast, the migrations of the temperate species are lengthy (**Table 4**) or presumed so.

Leptocephali

Leptocephali of American and European eels < 5 mm long are distributed between 50 and 300 m deep both day and night, perhaps indicative of the spawning depth of eels. Larger leptocephali perform daily vertical migrations, which increase in magnitude with increasing body size. Those 5–20 mm long descend from 50–100 m deep by night to 100–150 m deep by day. Those > 20 mm long are concentrated at 30–50 m by night and 125–250 m by day.

The classical account of the horizontal migration of leptocephali of American, European, and Japanese eels is gradual westward transport south of the subtropical convergences of the Atlantic and Pacific Oceans. The westward transport of the Japanese leptocephali is by the North Equatorial Current. The vertical migration of the leptocephali moves them into the Ekman layer at night, so wind drift influences the trajectories of leptocephali in addition to influence of deeper geostrophic currents. The leptocephali then become entrained in the strong western boundary currents, where they are transported rapidly northward. The western species metamorphose and detrain from the Gulf Stream and Kuroshio, whereas the European leptocephali, not yet ready to metamorphose, continue eastward in the North Atlantic Current.

Some have recently claimed that European leptocephali swim toward Europe by a more direct route from the Sargasso Sea. Their arguments are spatial and temporal. First, leptocephali off the continent of Europe increase in mean length from south to north, suggesting a migration in that direction. However, European leptocephali are found in the Gulf Stream and the North Atlantic Current. There could also be other drift routes eastward, such as an eastward countercurrent associated with the subtropical convergence zone. Secondly, the length of larval life is claimed to be only about 6–7 months on the basis of presumed daily rings in otoliths, versus 2 years or more in the classical account. Whether the migration of European leptocephali is passive or at least partially active cannot currently be resolved. However, back calculation of birth dates on the basis of presumed daily rings of the leptocephali implies that spawning occurs throughout the year. Yet, small leptocephali of the European eel only occur in the Sargasso Sea during part of the year. Unless there is another spawning area and time, the oceanic data are incompatible with the otolith data.

Glass Eels

Somewhere, perhaps the edge of the continental shelf, metamorphosis from leptocephalus to glass eel occurs. My speculation is that the diurnal vertical migration of leptocephali brings them into contact with the sea bottom on the shelf, perhaps triggering both metamorphosis and a change in behavior.

In near-shore waters and in estuaries, glass eels use selective tidal stream transport to migrate against a net seaward flow. The circatidal vertical migration is probably phased to the local tidal cycle through olfactory cues. Whether the cross-shelf migration is drift on residual currents, selective tidal stream transport, or oriented horizontal swimming is unknown.

Genetics and Panmixia

How closely related the eel species are is controversial. Debate has focused primarily on the distinctness of the European and American eels, but applies to all *Anguilla*. European and American eels separate morphologically on number of vertebrae. The mean vertebral numbers are 114–115 and 107–108, respectively, with little overlap in ranges. J. Schmidt considered them separate during his extensive research in the North Atlantic in the early 1900s.

In 1959, D. Tucker offered the bold hypotheses that: (1) eels from Europe do not return to breed but die without spawning; (2) the two eels are not separate species but are ecophenotypes of the American eel, their distinguishing characters being environmentally determined during egg and early larval stages; and (3) European populations are maintained by offspring of American eels. A north–south temperature gradient in the Sargasso Sea was proposed as the environmental influence on number of myomeres and vertebrae. Tucker's hypotheses, though criticized immediately, called attention to the lack of knowledge of the breeding and oceanic biology of eels.

Today, Tucker's hypotheses fail on oceanographic and genetic grounds. The spawning areas of both species overlap considerably in space and time, and they stretch primarily zonally not meridionally. The northern limit of spawning of the two species seems to be the very feature that Tucker invoked to provide the environmental difference. The two are not subject to systematically differing temperature conditions. Small leptocephali captured in single plankton net tows in the spawning area segregate bimodally on myomere numbers.

Analyses of nucleotide sequences of mitochondrial DNA (mtDNA) and nuclear DNA from European and American eels show the two to be closely related but distinct species. There are small but consistent differences in cytological characteristics of at least 9 of the 19 pairs of chromosomes. Hybrids of the two species are found in low frequency in Iceland, indicating that genetic isolation is not complete.

From Schmidt came the idea that European and American eels were each panmictic, i.e., a single breeding population of each species. Examination of nucleotide sequences of both mitochondrial and nuclear DNA has been used to address the question, with mixed results. European eels with particular sequences collected over wide geographic areas from Morocco and Greece to Sweden and Ireland did or did not cluster geographically in different studies. One study suggested weak structuring of the population, with a southern group (North Africa), a western–northern European group, and an Icelandic group. Another suggested no geographic genetic differentiation.

The single study of the American eel using mtDNA showed no geographic structuring among samples collected from the Gulf of Mexico to the Gulf of Maine, 4000 km of coastline. Japanese eels collected at seven sites in Taiwan, two in mainland China, and one in Japan, showed no geographic structuring. MtDNA analysis showed genetic similarity between *Anguilla australis* from Australia and New Zealand, suggesting they not be treated as subspecies.

Evolution and Paleoceanography

Two studies of evolutionary relationships among nine key species (seven in common) were based on sequence analysis of mtDNA. Both considered that the genus evolved originally in the eastern Indian Ocean, then the Tethys Sea, or Indonesian area, with one suggesting *A. celebesensis* and one suggesting *A. marmorata* as the ancestral species *A. japonica* and *A. obscura* branched off early. More recently evolved and in the same clade are *A. australis*, *A. mossambica*, *A. reinhardtii*, *A. anguilla*, and *A. rostrata*. *A. mossambica* and *A. reinhardtii* are sufficiently related that further molecular analysis may suggest a single species. The molecular phylogenies match well with groupings of V. Ege from 1939 based primarily on dentition. The exception is that Ege believed the Japanese eel was closely related to the Atlantic species, a relationship difficult to envision zoogeographically.

Anguilla is known from fossils in the early Eocene Epoch, perhaps 50 million years ago (50 Ma), and the family may date from 100 Ma. The evolutionary dispersal of the Austral-Asian species occurred during the time when Australia was moving closer to the Indo-Pacific islands and when the Indian subcontinent had broken from Africa and was drifting toward Asia. Invasion of the Atlantic by the ancestor of the two Atlantic species probably was through the Tethys Sea and its connection to the Atlantic between Africa and Asia. Closing of the Tethys Sea about 30 Ma isolated them. Separation of the closely related *A. australis* from the Atlantic ancestor must have occurred prior to the closing. The timing of the separation of *A. anguilla* and *A. rostrata* is problematic.

Fisheries and Aquaculture

Fisheries occur for glass eels, yellow eels, and silver eels in continental waters. All fisheries are for pre-reproductive stages. World catches of *Anguilla* reported to the Food and Agriculture Organization averaged 234 000 t (metric tonnes) in 1995–1998. Asia accounted for about 90% and Europe nearly 7%. Under-reporting of catches is probably widespread. These values are misleading in terms of impact because they combine fisheries for glass eels (a few thousand per kilogram) with fisheries for large silver female eels (a kilogram per eel).

Glass eel fisheries are heaviest on the Japanese eel and the European eel, with some commercial harvest of other species. At a peak in the 1970s, glass eel catch in the estuary of the River Loire, France, alone averaged more than 500 t annually. Glass eels are used for human consumption (e.g., in Spain and Portugal), for restocking rivers (e.g., in the Baltic Sea area), and primarily for aquaculture (e.g., in Asia and Europe).

Eel culture operations in Asia produce about 120 000 t annually, almost all in China, Taiwan, and Japan. In Europe, eel farms produced about 10 000 t of eels in 1998, mostly in Italy, The Netherlands, and Denmark. Wild and cultured eels are prepared as fresh fish, smoked fish, or kabayaki, traditional Japanese grilled eel with a soy sauce.

International trade in live eels and the development of large-scale eel culture has negative as well as positive consequences. European and American eels were accidentally or purposely introduced into Japanese rivers, where in some cases they dominate the eel fauna. In Taiwan, wells drilled to supply water to eel farms resulted in aquifer depletion and land subsidence. The nematode, *Anguillicola crassus*, which naturally coexists with the Japanese eel, was introduced into the populations of European and American eels. Larval and adult worms infest the swim bladder wall and lumen, with unknown effects on swimming ability of silver eels migrating at sea.

Status of Eel Populations

The status of the stocks is best known for Atlantic *Anguilla*. That of the European eel has been assessed frequently and that of the American eel recently by working groups of the International Council for the Exploration of the Sea. For yellow and silver European eels, fishery-dependent and fishery-independent trends have been largely downward in the last 20–50 years. For American eels, trends have been downward in the last 20 years from peaks in the late 1970s or early 1980s, to very low levels in many cases. It is unknown if those peaks represented unusually high population levels. However, loss of habitat and commercial fishing have contributed to declines.

Recruitment of young Atlantic *Anguilla* from the sea has declined. Long-term data on glass eel recruitment are available in The Netherlands, where the trend was dramatically downward in the 1980s to low levels through the 1990s. This is paralleled by the trend in commercial catch of glass eels in the estuary of the Loire River. The numbers of older yellow American eels moving upstream in the St Lawrence River at an eel ladder declined by three orders of magnitude from the early 1980s to 1999. Two to three order declines in upstream-migrating yellow European eels have occurred in Sweden over the last 50 years.

Whether or not recruitment declines are the result of lowered spawning stock is not known. It is possible that regime shifts in North Atlantic oceanic conditions have resulted in decreased survival of leptocephali at sea or in altered transport pathways for the leptocephali. There are negative correlations between the North Atlantic Oscillation Index, indicative of northern North Atlantic circulation patterns and productivity, and recruitment of glass European eels in The Netherlands (dating to 1938), and recruitment of yellow American eels in the St Lawrence River lagged by four years (dating to 1974).

Glossary

Abyssalpelagic Zone of the water column of the ocean encompassing depths between 2000 and 6000 m.

Anticyclonic Direction of atmospheric or oceanic circulation around an area of high pressure, clockwise in the Northern Hemisphere and anticlockwise in the Southern Hemisphere.

Benthic Organisms dwelling near, on, or in the bottom of the sea.

Catadromy Life cycle in which a species of fish breeds in the ocean, migrates into fresh water for growth, and returns to the sea at maturity.

Circatidal vertical migration Vertical movement by an organism in phase with the tidal cycle of 12.5 h, with vertical movement occurring during slack tides.

Clade A group of organisms, e.g., a group of species, sharing characteristics derived from a common ancestor.

Cline Geographic trend in some characteristic of a species or population.

Contranatant A migration against the direction of prevailing water current flow.

Countershading Color pattern of a pelagic species with a dark, sometimes mottled dorsal surface grading to a silvery ventral surface to reduce the contrast between the animal and its background.

Daily (diurnal) vertical migration Vertical movement by an organism in phase with the solar cycle of 24 h, with vertical movement occurring during dusk and dawn, usually upward at dusk and downward at dawn.

Denatant A migration along the direction of prevailing water current flow, usually involving drift of young stages.

Detritus Dead organic matter.

Ecophenotype The expressed characteristics of a particular subset of a genetically similar population caused by environmental conditions.

Ekman layer The near-surface layer, approximately 100 m deep, where wind-generated currents prevail.

Epipelagic Zone of the ocean encompassing approximately the upper 200 m.

Frontal zone Zone of the ocean in which the gradient (usually horizontal) in features of interest is steep, e.g., temperature, salinity, productivity, fauna.

Geostrophic current Currents in balance between a pressure-gradient force (gravity) and the Coriolis deflection.

Guanine A double-ringed nitrogenous compound forming part of a nucleotide found in DNA, but here also a crystalline substance deposited in the wall of a fish's swim bladder to reduce gaseous diffusion.

Ionic equilibrium Exhibiting equal concentrations of the same ions inside and outside an organism.

Isosmotic Exhibiting equal osmotic pressure inside an outside an organism.

Larvacean houses Delicate gelatinous cases secreted, and periodically discarded, by planktonic individuals of the Class Larvacea, Subphylum Urochordata, Phylum Chordata.

Leptocephalus The larval stage of eels of the Orders Anguilliformes and Saccopharyngiformes and of tarpons Order Elopiformes and bonefishes and spiny eels Order Albuliformes.

Meridional Distribution of oceanic characteristics or organisms on a north–south (longitudinal) axis.

Mesopelagic Zone of the water column of the ocean encompassing depths of 200–1000 m.

Mitochondrial DNA Genetic material of the mitochondria, the organelles that generate energy for animal cells, which is passed from female to offspring in eggs.

Molecular phylogeny Evolutionary relationships among taxonomic groups based on molecular techniques, especially the analysis of nucleotide sequences in DNA or amino acid sequences in proteins.

Myomere Muscle segment along the flank of a fish, here a larval eel.

North Atlantic Oscillation Index Difference in sea-level atmospheric pressure between Lisbon, Portugal (or Ponta Delgada, Azores) and Reykjavik, Iceland.

Nuclear DNA Genetic material of the nucleus of cells, a component of which is passed to offspring from both female and male parents.

Nucleotide Fundamental structural unit of the nucleic acid group of organic macromolecules, here those involved in information storage (in DNA: units containing adenine, cytosine, guanine or thymine), the sequences of which form codes for protein synthesis.

Otolith Concretions of calcium carbonate in a protein matrix deposited in the inner ears of bony fishes, frequently sectioned to examine annual or daily growth.

Panmixia The characteristic of a species being composed of a single breeding population.

Pelagic Areas of the ocean, or organisms dwelling, well away from the bottom of the sea.

Plankton Organisms in the water column of oceans and lakes that have weak swimming abilities, and are wafted by currents.

Recruitment The entry of organisms, typically fishes, into the next stage of a life cycle or into a fishery by virtue of growth or migration.

Regime shift A transition in oceanic conditions from one quasi-stable state to another.

Selective tidal stream transport Mechanism whereby an organism migrates along the tidal axis by ascending into the water column when the tide is flowing in the appropriate direction and descending to hold position near the bottom when the tide is flowing in the inappropriate direction.

Subtropical gyre Anticyclonic circular pattern of circulation in each of the major ocean basins between the equator and the temperate zone.

Subtropical convergence An area of the ocean where waters of differing characteristics come together, in this case equatorial and temperate waters meeting in the subtropical regions of the Northern and Southern Hemispheres.

Swim-bladder retia Countercurrent network of capillaries on the surface of a fish's swim bladder allowing gas pressure to increase and causing diffusion of gas into the swim bladder.

Western boundary current Western geostrophic currents of subtropical circulation gyres which are swift, deep, and narrow.

Zonal Distribution of oceanic characteristics or organisms on an east–west (latitudinal) axis.

See also

Fish Larvae. Fish Migration, Horizontal. Fish Migration, Vertical.

Further Reading

Bruun AF (1963) The breeding of the North Atlantic freshwater-eels. *Advances in Marine Biology* 1: 137–169.

FishBase, A Global Information System on Fishes. http://www.fishbase.org/search.cfm.

Hulet WH and Robins CR (1989) The evolutionary significance of the leptocephalus larva. In: Böhlke EB (ed.) *Fishes of the Western North Atlantic, part 9*, vol. 2, pp. 669–677. New Haven, CT: Sears Foundation for Marine Research, Yale University.

Schmidt J (1922) The breeding places of the eel. *Philosophical Transactions of the Royal Society of London, Series B* 211: 179–210.

Smith DG (1989) Order Anguilliformes, Family Anguillidae, freshwater eels. In: Böhlke EB (ed.) *Fishes of the Western North Atlantic*, part 9, vol. 1, pp. 25–47. New Haven, CT: Sears Foundation for Marine Research, Yale University.

Tesch F-W (1977) *The Eel. Biology and Management of Anguillid Eels*. London: Chapman and Hall.

Tesch F-W (1999) *Der Aal*, 3rd edn. Berlin: Parey. (In German).

Tucker DW (1959) A new solution to the Atlantic eel problem. *Nature* 183: 495–501.

Usui A (1991) *Eel Culture*. Oxford: Fishing News Books.

FISH ECOPHYSIOLOGY

J. Davenport, University College Cork, Cork, Ireland

Introduction

The earliest known jawless fish (Agnathans) date from the Cambrian period 500–600 million years ago. It is generally agreed among paleontologists that they evolved from sessile, filter-feeding ancestors in shallow fresh or brackish water, not the sea. Their environment was characterized by low salt levels and was probably turbid and at least intermittently oxygen-depleted (hypoxic). They have since spread to virtually all aquatic habitats (**Table 1**), but freshwater is still a stronghold, with a third of all fish species living in rivers, lakes and streams even though such habitats only contain a minute proportion ($<0.01\%$) of the total water on earth. Fish make up the most numerous vertebrate class, both in terms of species' numbers and biomass. They form a remarkably plastic group, exhibiting a diversity of sizes (from 8 mm Philippine gobies, *Mistichythys*

Table 1 Relative proportions of *c.* 30 000 living fish species living in different habitats

Major category	Subcategory	% of total
Marine fish (58.2%)	Shallow warm water	39.9
	Shallow cold water	5.6
	Deep benthic	6.4
	Deep pelagic	5.0
	Epipelagic	1.3
	Diadromous	0.6
Freshwater fish (41.2%)	Primary inhabitants	33.1
	Secondary inhabitants	8.1

- A third of species are primary inhabitants of freshwater, though freshwater makes up only 0.0093% of the total water of the earth.
- Possibilities of isolation and allopatric speciation are greater in freshwater than in seawater.
- Many marine species have re-invaded freshwater, or are at home in both media.
- Several intertidal and freshwater fish species spend some of their time on land.

After Cohen DM (1970) How many recent fishes are there? *Proceedings of the California Academy of Science* 38: 341–345.

luzonensis, to 18 m whale sharks *Rhincodon typicus*), shapes and life histories. Living forms include the numerically dominant teleost bony fish, the elasmobranchs (sharks and rays) and smaller groups such as lungfish, coelocanths (*Latimeria* spp.) and the surviving jawless lampreys and hagfish.

Despite this great diversity, fish are monophyletic, i.e. they have a single origin. All living fish share anatomical and physiological features of their earliest ancestors and are still constrained by them to a greater or lesser extent. Basic fish features include the following:

- Vertebral column with associated myotomal (segmented) muscles. The vertebral column sets fish length and prevents shortening of the animal when muscle contraction powers undulatory swimming.
- Head with food capture apparatus and bilaterally symmetrical sense organs.
- Gills for respiratory exchange, excretion of nitrogenous waste, plus regulation of body fluid ion content and pH.
- Closed vascular system with chambered heart, circulating red blood cells and (in teleost fish) dilute blood plasma (250–600 mosmol kg^{-1}, compared with 1000 mosmol kg^{-1} of seawater).
- Elasmobranch fish and coelocanths have similar plasma ionic levels to teleost fish (i.e. much lower than the ionic concentration of seawater), but have high urea and trimethylamine oxide (TMAO) levels that result in total plasma osmolarity being similar to that of seawater (**Figure 1**).

Biotic and Abiotic Factors in Distribution of Marine Fish

Distribution in fish is always controlled by a mixture of biotic and abiotic factors. For example, herbivorous marine fish are limited to shallow water where photosynthesis by microalgae, seagrasses or seaweeds is possible. Carnivorous fish (particularly postlarval and juvenile forms) may also be limited to such areas because they specialize in feeding on herbivorous invertebrates. Young gadoid fish are found living for protection beneath the bells of stinging jellyfish – a resource limited to near-surface waters at specific times of the year. These are all biotic constraints. Abiotic influences are those imposed by physical or chemical factors such as temperature, salinity or oxygen tension, and they are the main

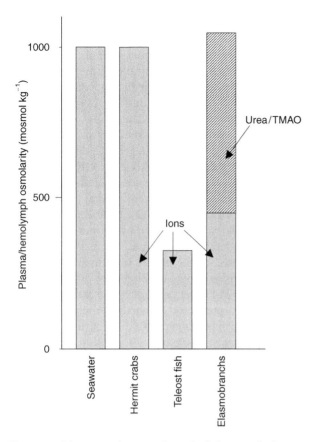

Figure 1 Diagrammatic comparison of relative contributions to total plasma/hemolymph osmolarity of ions and urea/trimethylamine oxide (TMAO) in teleost fish, elasmobranchs and hermit crabs (*Pagurus bernhardus*). Data for seawater given as reference.

focus of this article since they profoundly affect fish physiology and tolerances and therefore distribution. There are other features that are strictly abiotic, in particular physical habitat type (e.g. mud, sand, coral reefs, tangled mangrove roots), but they do not impinge directly on fish physiology, so will not be considered here.

Tolerances and Limits to Distribution

Presence of Water

Limitation to an aqueous habitat is the most fundamental physiological constraint imposed on fish. No elasmobranch or agnathan species can survive out of water, and only a few dozen amphibious teleost species plus the three surviving lungfish species have the ability to live out of water for significant periods. Most of these species are freshwater, so outside the scope of this encyclopedia. Amphibious intertidal fish such as mudskippers (*Periophthalmus* sp.), shannies (*Lipophrys pholis*) or butterfish (*Pholis gunnellus*) are rarely more than a few meters from

seawater and are emersed only for a few hours at a time. Eels (*Anguilla* sp.) are catadromous and move between freshwater and seawater, these migrations often involving movement on damp/wet land between water bodies.

When fish are emersed they have problems of respiration, excretion, acid–base balance and locomotion. Fish respiration involves the passage of an incompressible medium (water) over the gills. If an unadapted fish is taken out of water it becomes short of oxygen (hypoxic), accumulates CO_2 (hypercapnia) and cannot excrete H^+ – all because the gills collapse and the animal cannot circulate compressible air over them. It soon dies because the blood becomes acid, not because of lack of oxygen, and usually death will occur long before dehydration becomes a factor. Mudskippers and shannies have strengthened gills with relatively few filaments that are less prone to collapse. They also have scaleless, well-vascularized skins that allow respiratory exchange; the gills are relatively less important for respiration, though they can still take up oxygen from water held within the buccal cavity, and are involved in H^+ regulation. Nitrogen excretion is also a problem for amphibious fish. Most fish excrete nitrogen as NH_3 or NH_4^+, with 50–70% of excretion being by diffusion across the gills in marine fish. Although NH_3/NH_4^+ is metabolically inexpensive to produce, it is toxic and therefore cannot be accumulated within the body. Amphibious marine fish such as mudskippers reduce protein breakdown when in air, and also accumulate a proportion of N_2 as urea or trimethylamine oxide (TMAO), both of which are less toxic than ammonia. However, for marine fish, nitrogenous excretion is a fundamental constraint on survival on land; they have to return to the sea to get rid of accumulated nitrogen and metabolites. Basic fish anatomy is unsuitable for terrestrial locomotion, though butterfish and eels 'swim' through three-dimensional habitats, such as pebbles and thick grass, relying on secretion of mucus for lubrication. Shannies and mudskippers have strengthened prop-like pectoral fins that stop them falling over and raise the belly off the ground to some extent. They are propelled in a series of hops by the tail.

Depth of Water

Four decades ago Jacques Piccard took photographs from the bathyscaphe *Trieste* of tripod fish (Chlorophthalmidae), resting on the bed of an oceanic trench at a depth of over 10 000 m. In doing so he demonstrated that fish could live at all depths, despite their shallow-water origin. Trawling, cameras and submersible observations have confirmed that a diverse ichthyofauna may be found at all depths, in all seas. However, increasing depth may control the sort of fish that are found. Some depth-related constraints are biotic; deep water generally has a restricted energy supply due to absence of light and distance from the productive surface layers. However, there are two major physical problems imposed by depth: increasing pressure and low temperature (particularly at depths greater than 1000 m). In addition, there is an important related chemical problem. In warm surface waters, the sea is practically a saturated solution of calcium carbonate and relatively little energy is needed to maintain calcareous materials (e.g. bone, shells) in solid form. However, solubility rises with increasing pressure and decreasing temperature. In consequence, building and sustaining solid calcareous materials becomes more expensive, particularly at depths beyond 3000 m (below which calcareous sediments are unknown). The problems of pressure and calcium carbonate solubility interact in the physiology of fish buoyancy. A high proportion of shallow-water teleost fish have swim bladders, which develop from gut diverticula. They have sophisticated volume regulatory mechanisms (employing the lactate-secreting gas gland and its associated countercurrent multiplier system), particularly in physoclistous fish in which the swim bladder is isolated from the gut. The original adaptive value of swim bladders probably lay in offsetting the burden of dense scales and armoured skulls, so that shallow-water benthic fish could swim into the water column without undue effort. At depths down to about 1000 m some 75% of fish have swim bladders, but at greater depths pelagic fish usually have no swim bladder or a swim bladder filled with fat; they also show progressively reduced musculature and ossification, plus a very high water content. Lack of ossification has been interpreted as an energy-saving strategy in a food-poor environment where dissolution of calcium carbonate takes place. Loss of swim bladders has been attributed to the metabolic expense at great depth of pumping gas into swim bladders. Interestingly, benthopelagic fish (i.e. those living close to the sea bed) from deep water can possess working swim bladders even at depths of 7000 m; they are also of robust skeleton and musculature. The near-sea bed environment is now known to be much more energy-rich than the water column above, particularly due to the fall of large carrion items. This suggests that benthopelagic fish can 'afford' to expend energy counteracting abiotic depth-related factors on fish form, because of the biotic influence of a good food supply.

Temperature

Fish are almost all ectothermic animals with no significant production or retention of metabolic heat. A few tuna species can keep their locomotory muscles warm, and some big lamnid sharks (e.g. the great white shark, *Carcharodon carcharias*) are partially endothermic with core body temperatures being held at around 25°C in waters of 15°C. However, the body temperature of most fish is directly determined by environmental temperature. Metabolic rate in ectotherms is strongly affected by temperature, with a useful rule of thumb (the Q_{10} relationship) stating that metabolic rate is doubled by an increase of 10°C in environmental temperature. There is a huge literature devoted to the effects of temperature on aspects of the physiology, development and ecology of marine fish. However, despite this wealth of information, the question of whether thermal physiological constraints control fish distribution is difficult to answer, since marine fish are found in all available habitats, from Antarctic ice tunnels at −2.5°C to Saudi reef pools at over 50°C. Antarctic fish usually die at around +5°C, whereas most temperate fish are stressed severely at temperatures above 30°C. Tropical fish reach their thermal limits at about 45°C (also the upper thermal limit for most tropical marine invertebrates).

The position and breadth of a species' thermal niche is determined by a variety of factors. At the biochemical level, fish must have enzymes with appropriate thermal optima. Cell membranes are effectively liquid crystals that must remain fluid if the cell is to survive. Maintaining fluidity over the full environmental temperature range requires modulation of the fatty acid composition of membrane lipids (homeoviscous adaptation). In many fish these characteristics can vary geographically and seasonally, but within overall limits which are species specific. These limits do constrain fish distribution. In the northern hemisphere capelin (*Mallotus villosus*) do not penetrate much further south than the 5°C summer surface isotherm, whereas the corresponding limit for cod (*Gadus morhua*) is about 20°C. In marked contrast, deep-water fish living at very stable low temperatures (*c.* 2°C) can have extremely wide distributions despite limited thermal tolerance. For example, the orange roughy (*Hoplostethus atlanticus*) living at 800–1200 m has been fished off New Zealand, Namibia and northern Europe.

A specialized warm-water example of thermal constraints on distribution is provided by flying fish (Exocoetidae). Flying fish take off through or from the sea surface at very high speed (15–30 body lengths s^{-1}). This in turn demands extremely high rates of tail beat (*c.* 50 beats s^{-1}). Calculations demonstrate that take-off is unlikely to be possible at water temperatures below 20°C, and surface water of this temperature approximates to the northern and southern limits of this essentially tropical group.

Physiologically, fish need greater gill areas per unit mass at high temperature because their respiratory requirements are greater and because the solubility of oxygen decreases with rising temperature, so less environmental oxygen is available. At the ecological level, high temperatures imply high metabolic rates and elevated food consumption. A typical tropical fish needs roughly six times the oxygen to support resting metabolism as does a typical polar fish. High temperature, high activity lifestyles can only be sustained in energy-rich environments. However, it should not automatically be assumed that low-temperature environments (e.g. the deep sea, polar waters) are inexpensive to live in. Low temperature is associated with high viscocity so swimming demands proportionally more energy, as does pumping of blood around the body. Antarctic fish in particular tend to show viscocity-related modifications; they (and some deep-water bathypelagic fish) generally have low hematocrits (i.e. have relatively few red blood corpuscles) so that effective blood viscosity is reduced. Icefish (Channichthyidae) provide extreme examples of this, having no red blood corpuscles or hemoglobin and possessing wide-bore blood vessels through which viscous plasma flows more easily. The trade-off is that they are sluggish fish with extremely low endurance, demonstrating again that lifestyle can be constrained by temperature.

Oceanic seawater of salinity 34‰ (osmolarity 1000 mosmol kg^{-1}) freezes at about -1.9°C. Elasmobranch fish are at little risk of freezing because their blood has a similar osmolarity (**Figure 1**). Because teleost fish have relatively dilute body fluids (300–600 mosmol kg^{-1}) they are potentially liable to freezing at temperatures of -0.6 to -1.0°C, when seawater is still fluid. Lower temperatures than this occur in the winter in the Arctic and throughout the year in the Antarctic. This contrasts with the situation for freshwater fish in which freezing cannot take place unless the water around them is itself frozen. Intertidal pools can become even colder, unfrozen high salinity water beneath ice sheets reaching -3 to -8°C. High latitude marine fish show various adaptations to deal with freezing risk. Anadromous arctic charr (*Salvelinus alpinus*) migrate from the sea to freshwater in the winter, whereas other arctic fish migrate to low latitudes or deep water. There appears to be a fundamental constraint on the distribution of resident intertidal fish, which are absent from northern Norway, Russia and Canada throughout the year – presumably because they cannot compete effectively in deeper water in winter, are too small to migrate significant distances southwards, and cannot avoid freezing.

Antarctic fish and permanent surface-water residents of the winter Arctic all exhibit physiological adaptations that permit freezing avoidance even when pack ice is present. All have relatively high blood osmolarities, depressing their potential freezing points, and most can produce so-called antifreezes: peptides or glycopeptides. These molecules do not actually stop ice formation in body fluids, instead they adsorb onto the crystal surfaces of minute ice nuclei and prevent these nuclei from growing or propagating. In icefish these mechanisms are effective to -2.5°C, allowing them to live in ice tunnels in Antarctic ice shelves. Several arctic species overwinter in deep water which is colder than their potential blood freezing point, but does not contain ice nuclei that can initiate freezing. This supercooled state is precarious and there are records of schools of capelin (for instance) straying into shallow water containing ice during cold weather and freezing instantly.

Salinity

The great majority of marine fish species live under very stable salinity conditions (34–35‰; osmolarity *c.* 1000 mosmol kg^{-1}). This medium, though stable, is much more concentrated than the freshwater or brackish media encountered by ancestral fish. As a result, the osmotic and ionic physiology of marine teleost fish is very different from that of freshwater fish, and relatively few species can live in both media (mainly anadromous and catadromous fish such as salmonids and eels). Elasmobranch fish (sharks, skates and rays) are adapted to seawater by virtue of high blood urea levels that make the blood slightly hyperosmotic to seawater so that they have little or no osmotic problem. Although this group does have a few brackish water species, it is generally limited to fully marine habitats and will not be discussed further.

Briefly, marine teleost fish have to drink seawater to replace water lost osmotically (mainly across the gills) because the blood is much more dilute than seawater (**Figure 2**). To gain access to the water taken into the foregut, salt pumps (actually ATP-ase enzymes embedded in cell membranes) sited on the intestinal wall pump salts from gut fluid to blood until (in the posterior parts of the gut) the osmolarity of the gut fluid is below that of the blood, so that water flows into the blood osmotically. The combination of drinking seawater and active desalting of the gut fluid results in much salt uptake, augmented by diffusion from the

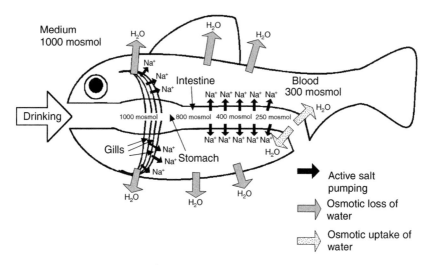

Figure 2 Diagrammatic representation of osmotic and ionic regulation in a marine teleost fish.

salty external medium. To counteract this, salt pumps located mainly in 'chloride cells' on the gills actively pump salt outwards.

In the bulk of the world's oceans, salinity does not constrain distribution. Difficulties only occur at the seas' margins, in estuaries, lagoons and pools where salinities can be much higher or lower. Generally, there appears to be an upper limit of survivable salinity for fish of around 80–90‰, with one or two specialists (such as the killifish *Fundulus heteroclitus*) tolerating up to 128‰. Impressive though this performance is, it is much poorer than that exhibited by many invertebrates, particularly crustaceans, some of which can tolerate up to 300‰ (e.g. the brine shrimp, *Artemia*). Fish are constrained by what is known as the 'osmorespiratory compromise'. Fish gills are necessarily large in surface area to support gaseous exchange. Gill epithelia are also thin to permit ready diffusion. Unfortunately, these two characteristics also favor rapid osmotic loss of water that has to be replaced by drinking, plus salt diffusion that must be opposed by salt pumps. There comes a point where the balance breaks down; killifish are unusual in that they tolerate a 30% rise in plasma osmolarity at 128‰, most fish would succumb.

The majority of marine fish have blood osmolarities of around $300 \, mosmol \, kg^{-1}$, equivalent to about 10‰. If unadapted marine fish are placed in media less concentrated than this, they die because of blood dilution caused by osmotic uptake of water and diffusional loss of salts. To survive in dilute media, euryhaline fish of marine origin have to stop drinking the medium and pump salts inwards at the gills, effectively reversing the osmotic physiology exhibited in seawater. Many fish of this type acclimate slowly over a period of days to dilute media, since profound microanatomical and biochemical changes have to

take place; migratory salmonids fall into this category. Some fish that forage regularly into brackish water (flounders (*Pleuronectes flesus*), mullets (*Mugilidae*), ròvolos (*Eleginops* sp.)) respond more quickly, and a few species such as the shanny are capable of reacting to extreme salinity changes in a matter of minutes. Shannies inhabit the intertidal zone and may be found in crevices that are fed by freshwater runoff when the tide falls. They exhibit the constrained features of such highly euryhaline fish, i.e. small size, thickened gills and a low skin permeability to salts and water, that slow changes in blood concentration and reduce the energetic costs of regulation. Most marine fish that are regularly exposed to low salinity are benthic and slow moving, another consequence of the osmorespiratory compromise.

Oxygen tension

Fish in general, and marine fish in particular are intolerant of oxygen-depleted (hypoxic) conditions. Many freshwater habitats are hypoxic and fish have multiple hemoglobins to deal with this situation and extract oxygen from hypoxic water. Broadly speaking, marine habitats are mostly close to equilibration with the atmosphere (air-saturated; normoxic), so oxygen tension poses no constraints and most fish have few or single hemoglobins. Pollution incidents have often revealed the sensitivity of marine fish to hypoxia, with diatom bloom formation sometimes resulting in massive fish kills at night when plant respiration has dramatically reduced water column oxygen tension (Po_2). In general marine fish avoid hypoxic areas rather than tolerating them, though large schools of clupeoids (e.g. herring, *Clupea harengus*) create their own hypoxic environments in the heart of the shoals. Shannies will even leave nocturnally hypoxic intertidal pools and respire in air, particularly in summer when their respiratory demands are high. Marine fish unable to avoid transient hypoxic conditions (e.g. sole, *Solea solea* in shallow organic-rich water) usually respond by reduced activity, depressed basal metabolic rate and activation of anaerobic metabolism. This is a short-term response and is supplemented by behavioral responses such as burst-swimming to the surface where higher oxygen tensions prevail.

pH

Fish have a plasma pH of 7.4–8.1. Unlike higher vertebrates they have limited internal buffering mechanisms and the viscocity of water means that ventilatory pH control is much less effective than in air-breathers. Excretion rather than respiration controls fish acid–base balance. In many freshwater habitats, whether natural or affected by acid rain, environmental pH poses considerable physiological costs and constraints. This is not the case for marine fish because the sea is a slightly alkaline environment (pH 7.8–8.0) that has enormous buffering capacity for H^+ and CO_2 and poses no problems whatsoever. Only in rock pools are great pH fluctuations known (7.2 at night; 10.6 by day), and even here there are no documented problems for rock pool fish.

Optima

Optimal abiotic environmental conditions for fish species have been studied from two perspectives. First, there are now numerous commercially valuable marine fish species in culture, principally for human food, but increasingly in tropical countries for the aquarist trade. An extensive literature reporting on the ideal conditions for survival, rapid growth and effective reproduction has arisen. Much of the work done has involved multifactorial experimental approaches (e.g. combinations of temperature, salinity and oxygen tension) and these have often revealed changes in optima at different stages of the life history. Rearing densities are high and have biotic effects on optimal rearing conditions, but these in turn can create constraining abiotic problems (e.g. NH_4 and nitrite accumulation) that do not arise in nature. Such studies have revealed constraints on where economic acquaculture can be performed. For example, halibut (*Hippoglossus hippoglossus*) farming is practiced only in cold areas of northern Europe (e.g. Norway, Scotland) because *Hippoglossus* is a deep-water species intolerant of elevated temperature. Conversely, turbot (*Scophthalmus maximus*) farming, once tried in such areas, proved to be

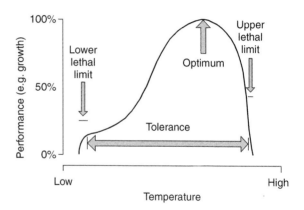

Figure 3 Diagram to illustrate tolerance range, optimum performance and lethal limits (using temperature as an example).

uneconomic because extra (expensive) heat was needed to secure fast growth; production is now centered in France and Spain.

Secondly, abiotic optima have been considered from a biogeographical perspective, almost exclusively in thermal terms (**Figure 3**). This is linked with the constraints on distribution already discussed, but is presently a rather poorly developed research area, ripe for development. For temperature, there is evidence that fish seek out preferred (assumed optimal) temperature regimes, though much of this work has been conducted on freshwater species. Widely distributed marine species have been assumed to have optima either in the center of distribution, or close to the warmer limits, but some work on fish living at the edge of distributions show no evidence of maladaptation or of particular population instability. Difficulties lie in deciding what optimal performance means, and in disentangling biotic and abiotic influences. Fish in temperate zones usually grow more quickly in the warmer areas of their distribution. They may reproduce at an earlier age, often at smaller size. How much of this is due to biotic influences (e.g. quantity and quality of food supply, trophic structure of the local ecosystem) is usually unclear. Optimal performance essentially involves maximizing contribution to the gene pool; establishing the ecophysiological conditions that deliver this is still a challenge.

See also

Antarctic Fishes. Deep-sea Fishes. Eels. Fish Larvae. Intertidal Fishes. Salmonids.

Further Reading

Bone Q, Marshall NB, and Blaxter JHS (1999) *Biology of Fishes* 2nd edn. Glasgow: Stanley Thomas.

Dalla Via D, Van Den Thillart G, Cattani O, and ortesi P (1998) Behavioural responses and biochemical correlates in *Solea solea* to gradual hypoxic exposure. *Canadian Journal of Zoology* 76: 2108–2113.

Davenport J and Sayer MDJ (1993) Physiological determinants of distribution in fish. *Journal of Fish Biology* 43(supplement A): 121–145.

DeVries AL (1982) Antifreeze agents in coldwater fish. *Comparative Biochemistry and Physiology* 73A: 627–640.

Kinne O (1970) *Marine Ecology*, vol. 1. New York: Wiley Interscience.

Kinne O (1971) *Marine Ecology*, vol. 2. New York: Wiley Interscience.

Marshall NB (1979) *Developments in Deep Sea Biology*. Poole: Blandford Press.

Rankin JC and Davenport J (1981) *Animal Osmoregulation*. Glasgow: Blackie.

Sayer MDJ and Davenport J (1991) Amphibious fish: why do they leave the water? *Reviews in Fish Biology and Fisheries* 1: 159–181.

FISH FEEDING AND FORAGING

P. J. B. Hart, University of Leicester, Leicester, UK

Introduction

There are approximately 24 600 species of fish of which 58% or 14 268 live in the sea. The sea covers *c.* 71% of the surface area of the Earth and has an average depth of around 3800 m, so that the total volume of the marine environment is about $1370 \times 10^6 \, km^3$. Much of this volume, removed from the influence of the sun's rays, is an inhospitable place to live, being dark, cold, and very low in available food. With such a volume for living, it is no surprise to learn that there are some 15 basic ways in which fish can gain food from the environment (**Table 1**). Because the open ocean and the deep ocean have low productivity compared with the shallow seas, most of the fish diversity is found in waters less than 200-m depth with the highest concentrations being found in tropical waters over coral reefs. The fish in these areas also have the greatest diversity of ways of making a living. Coral reefs and other inshore areas also have the most complex food chains with many links between fish and their prey.

Modes of Feeding in Fishes

During the course of evolution, fish in the marine environment have developed a diverse array of behavioral, morphological, and physiological adaptations to cope with the food they most commonly eat. Although fish feeding habits can be classified into a relatively few groups, the diversity within each group is significant. The different modes of feeding are shown in **Table 1** together with a selection of illustrative species. With this table in mind, it becomes possible to examine in more detail the principles of behavioral adaptations used to cope with different conditions.

Feeding mode can be classified by the type of food eaten. A species adopting a particular type of food, say that of a piscivore, will develop a body form and a set of foraging tactics suiting it to the particular types of prey taken and the habitat in which the piscivore lives. For an example, a whiting (*Merlangius merlangus*) living in shallow areas of the North Sea uses vision to locate prey and has a larger mouth than an invertebrate feeder such as the haddock

(*Melanogrammus aeglefinus*). The related rat-tail macrourid living at 3000-m depth is more likely to use olfaction or the lateral line to find prey, and many deep-sea fish have very large mouths to allow them to take whatever prey they encounter. The categories in **Table 1** are a useful way to illustrate how form, function, and behavioral habits influence the characteristics of different feeding types.

Fish biologists also use the term 'guild' when classifying feeding modes of fish. In contrast to a niche, a guild refers to a collection of fish species, which can be from different taxonomic groups that feed on the same type of prey and show convergent adaptations to the food. For example, herring and mackerel are pelagic planktivores and show many similar adaptations of body form that are designed to cope with living in the open sea, yet the two species belong to very different taxonomic groups.

Table 1 Feeding modes of fishes: major trophic categories in fishes

Category	Examples
Detritivore	Mullet, *Mugil*
Scavengers	Dogfish, *Squalus*; hagfishes, Myxinidae
Herbivores	
Grazers	Parrotfishes, Scaridae
Browsers	Surgeon fishes, Acanthuridae
Phytoplanktivores	Peruvian anchoveta, *Engraulis ringens*
Carnivores	
Benthivores	
Picking at relatively small prey	Lemon sole, *Microstomus kitt*
Disturbing then picking at prey	Gurnards, Triglidae
Picking up substratum and sorting prey	Black surfperch, *Embiotica jacksoni*
Grasping relatively large prey	Triggerfish, *Balistes fuscus*
Zooplanktivores	
Filter feeders	Menhaden, *Brevoortia tynrannus*
Particulate feeders	Anchovy, *Engraulis mordax*
Piscivores	
Ambush hunters	Megrim, *Lepidorhombus wiff-iagonis*
Lurers	Angler fish, *Lophius piscatorius*
Stalkers	Trumpet fish, *Aulostomus maculatus*
Chasers	Bluefin tuna, *Thunnus thunnus*

Most shallow-water environments contain varying amounts of detritus derived from dead plants and animals. This can provide a source of food for some fish, although this mode does not comprise a major feeding type in the sea when compared with fresh waters. In **Table 1**, mullets (Mugilidae) are given as an example and this highlights a problem with the classification: the categories are not exclusive. Very few fish specialize to such a degree that they never eat anything but the prey type classed as their principal food. So, even though mullet species do eat detritus, they also graze on plant material and capture invertebrate food. It is probably true of all species that eat some detritus that this food source is a supplement to their diet, resorted to when other items are scarce.

A large range of fish types feed on the dead remains of other fish, marine mammals, or invertebrate species. Hagfish (Myxinidae) are primitive and have no jaws. Inside their buccal cavity they have teeth that are used to rasp flesh once they have attached themselves with the suckerlike mouth. Although they often feed on dead fish, they also consume living fish if they can first obtain a good hold on them. This is facilitated by the presence of an irregularity on the skin of the prey, such as a wound. Other species, such as the spur dog, *Squalus acanthias*, take dead material if it is available, although they mainly eat fish and larger invertebrates. As with many carnivores from other animal groups, dead meat is rarely ignored.

Herbivores in the sea are limited in their choice. They can either frequent the shallow waters and consume macroalgae or algae encrusting rocks, or they can live in the open water near the surface and eat phytoplankton. If they choose the second option, they are most likely to eat zooplankton also. Grazers and browsers are most common on coral reefs (**Table 2**), where there are numerous species feeding on algae. Many herbivores are very selective in the species they eat and the grazing effect has a strong influence on competition for space between algal species. Herbivorous fish on coral reefs adopt one of three feeding strategies: they defend a territory, with some species of damsel fish (Pomacentridae) 'tending' gardens; they can adopt a home range within which all feeding occurs, as exemplified by some species of pomacanthid angelfishes; or they feed in mixed species groups as in some surgeon fishes. Herbivores on a reef feed only during the day and hide in crevices during the night.

Although they are separated in **Table 1**, phytoplanktivores and zooplanktivores will be dealt with together. Fish that feed on plankton can adopt one of two tactics: either they can sieve the water to extract the plankton or they can pick off items individually. The two are presented as individual tactics in **Table 1**, but in reality species will switch between the two depending on the density and size of food. For species focusing on phytoplankton, sieving is the only alternative, as the plants are too small to take individually. The Peruvian anchoveta (*Engraulis ringens*) takes a mixture of phytoplankton and zooplankton but, because phytoplankton is so rich, the bulk of what they eat could be of plant origin. Most other planktivores eat a mixture of both, with zooplankton predominating. The classic planktivores are species such as the herring (*Clupea harengus*), mackerel (*Scomber scombrus*), pilchard (*Sardina pilchardus*), and sprat (*Sprattus sprattus*). They live in the epipelagic region of the ocean, have fusiform streamlined bodies, and most often live in large shoals. Many of them make significant migrations to reach feeding areas that are seasonally worth exploiting. An example is the mackerel stock that spawns off southwest England in the spring and then migrates into the North Sea either via the west coast of Britain or up through the English Channel.

A wonderful example of a plankton feeder is the basking shark (*Cetorhinus maximus*). It is remarkable that such large animals, up to 10 m long, can be sustained by their microscopic prey. To survive, these 3000-kg fish have to filter very large volumes of water and do so by swimming for long periods with mouths wide open. Fine rakers on the gill arches act

Table 2 Proportions of different types of feeders in a temperate and a tropical marine system

Feeding category	Gulf of Maine, Atlantic		Marshall Island, Pacific (coral reef)	
	%	N	%	N
Herbivores				
Phytoplankton	0.7	1	0	0
Benthic diatoms	0	0	1.5	3
Filamentous algae	0	0	16.0	33
Vascular plants and seaweeds	0	0	8.7	18
Detritivores	0.7	1	3.9	8
Carnivores				
Zooplanktivores	16.9	25	6.3	13
Benthic invertebrates	41.2	61	54.9	113
Piscivores	39.2	58	a	
Omnivores	2.0	3	8.9	18

[a]Category absent.

as filters removing plankton from the stream of water leaving the gill slits.

In planktivores such as the herring, prey items are mostly selected and the frequency of prey species found in the stomach is not the same as their frequency in the environment. A famous study of the diet of herring by Sir Alister Hardy, made in the early 1920s, showed how complex the feeding habits of a fish are. Like all species of teleost fish, herring grow throughout their lives, starting as microscopic larvae and finally reaching a size of around 30–40 cm. As revealed by Hardy, the diet of the fish changes dramatically as the fish increases in size, and the figure that Hardy produced to show this (**Figure 1**) has become a classic of the marine biology literature. As larvae, the herring feed on very small planktonic prey such as the early stages of copepods, larval mollusks, tintinnids, and dinoflagellates. At this stage of their lives, herring are as much food for other fish as they are predators themselves. With growth, the young herring can begin to take larger planktonic prey such as the copepods *Pseudocalanus*, *Temora*, and *Acartia*, common in inshore waters off the British Isles. Juvenile and adult herring feed extensively on *Calanus finmarchicus*, one of the most common copepods, euphausiids (krill), amphipods, and fish. By changing their diet through their life history, the herring are moving niche too, and this also has a spatial component as the young herring live in nursery areas close inshore.

Carnivorous fish come in a wide range of forms (**Table 1**). A basic division is between species that feed mainly on prey dwelling in or on the bottom and those that take prey from the water column. Benthic feeding fish show a range of adaptations reflecting the differences in lifestyle of the species. For example, the lemon sole (*Microstomus kitt*) is a visual feeder swimming over the bottom searching for annelid worms, which make up the bulk of its diet. In contrast, the sole (*Solea solea*) lies buried in the sand or mud during the day and forages only at night or when light conditions are very low during the day. It searches for food by touch. Both species could be categorized as 'benthivores' that pick at relatively small prey, but their differences are not insignificant. These differences are largely behavioral as both species are bottom dwellers superbly designed for their habitat, having flattened bodies and the habit of burying themselves in sediment.

As with herring, bottom feeders show life history changes in feeding behavior. For example, the black surf perch (*Embiotica jacksoni*), living on reefs off

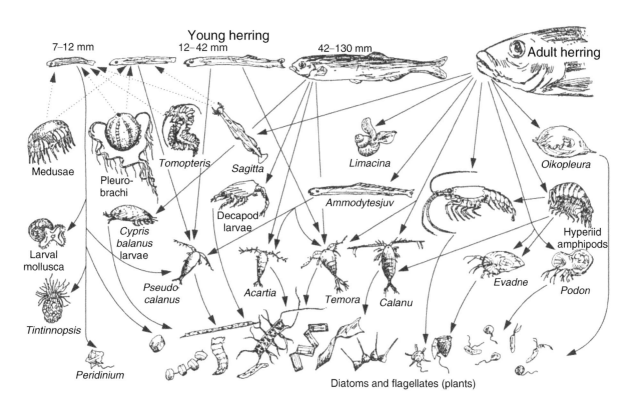

Figure 1 The food of herring from larval stages to adults. Also shown are the connections between the prey of herring and the food they eat. Redrawn from Hardy AC (1924) The herring in relation to its animate environment. Part I: The food and feeding habits of the herring. *Fishery Investigations, London, Series II* 7(3): 53.

southern California, is named as an example in **Table 1** of a species that picks up substratum and, in the mouth, sorts prey from gravel and sand. These fish use this 'winnowing' tactic only once they have grown above a certain size. When they are small they fall into category 'Picking at relatively small prey' (under 'Benthivores') of **Table 1** as they feed by picking up each prey item. Their diet is limited, by the gape of the mouth, to food particles below a certain size.

The greatest problem for a piscivore is that its prey is mostly mobile and well able to see the predator coming. Two basic tactics are used by piscivores to capture prey: either they use stealth in various forms or they try to outswim the prey in a chase. These tactics have had profound influences on the selective forces influencing the fish found in each group. Those that use stealth have developed along two routes. Many species have special adaptations to lure prey close; the classic example of this is the angler fishes (Lophiidae) with their dorsal fin ray modified to form a movable rod with a lure on the end. These fish have also developed body coloration and shapes that camouflage them as they sit and wait on the bottom. Ambush hunters have used other means of getting close to prey. They hide in weed or adopt coloration that makes them inconspicuous. For example, the megrim (*Lepidorhombus wiffiagonis*) eats mainly fish and shows the characteristic morphology of a piscivore despite its flattened body form. It has a slim body relative to other flatfish and has large eyes and mouth. To catch fish it lies half-buried on the bottom until a fish comes close, when it springs forward to make a capture.

Stalkers are also well camouflaged but get close to their prey by either using cover to disguise their intentions or by moving so slowly that the prey do not notice the advance until too late. For example, the trumpet fish (*Aulostomus maculatus*) living over coral reefs join shoals of nonpredatory fish as a way of coming close to their prey. The remarkable aspect of this tactic is that the trumpet fish changes its head color to match the color of the fish in the shoal.

At sea, fish such as tuna (*Thunnus* spp.), the larger sailfish (*Istiophorous albicans*), and salmon (*Salmo salar*) hunt their prey at speed. These fish have bodies designed for fast swimming and also have large mouths, often with backward-pointing teeth, to grab the prey securely once caught. Many prey fish that are attacked in this way have developed behavioral tactics to reduce the risk of predation. They can shoal or school, they can develop camouflage, or they can avoid contact with predators by appearing only at night.

A few species of fish have specialized in exploiting others for food. At its most aggressive, this mode includes fish that eat scales or fins of other species, although most of the examples of these modes are from fresh water. Some species have adopted the role of cleaners who specialize in picking ectoparasites off other fish. This mode is not exploitative in that both sides of the interaction benefit. There are wrasse species (genus *Labroides*) in the Indo-Pacific that specialize entirely on cleaning and have a characteristic color scheme – blue with a longitudinal black stripe – that allows their 'clients' to identify them as cleaners. Any such system can be exploited: and the saber-toothed blenny, *Aspridonotus taeniatus*, adopts the same color scheme as the labrids but when it gets close to the client it tears pieces of flesh out rather than picking off ectoparasites.

In describing the various modes of fish feeding we have seen that the success of individuals at capturing prey is a consequence of having the right morphology and behavior. It is also important to realize that many species are not confined to just one mode of feeding. The tactics adopted by a species can change with age or size, time of day, and geographical location. On a moment-to-moment basis the behavior adopted by a fish is critical in determining the diet taken and the energetic consequences of food intake. It is assumed in modern studies of foraging that behaviors have been molded by natural selection in the same way as has morphology.

Foraging Strategies

Given the assumption that behavior can be molded by natural selection, it becomes possible to analyze behaviors from an economic viewpoint. If a fish behaves so as to maximize its lifetime fitness, then in the short term it will choose to do things that maximize short-term gains such as increased growth rate or egg production and to minimize costs such as energy consumed or risk of predation during foraging. It then becomes possible to ask what behavioral strategy will maximize short-term gain or, in the jargon of foraging theory, optimize the behavior. With this approach it has been possible to predict what the optimal foraging strategy is for a species selecting prey from an environment with particular characteristics.

The way in which tunas behave while foraging near ocean fronts can be understood with the aid of optimality arguments. Tuna in the bluefin group (genus *Thunnus*) travel widely in search of prey. They have often been observed aggregating at ocean fronts where warm water is separated from cooler water by a narrow transitional zone. It is characteristic of these fronts that the productivity of tuna food is highest on the cool side of the front. This may be because the cooler water has recently upwelled and has higher plant nutrient levels. The dilemma facing the foraging tuna is that it prefers to be in the warmer water from a thermoregulation point of view but its best feeding opportunity is in the cooler water. Unlike many smaller fish, tuna have some control over their core body temperature. The vascular system has a heat exchange process by which blood moving from the center of the body outward passes vessels taking blood from the outside in. In this way, the core temperature of a tuna can be maintained significantly above ambient and controlled at a relatively constant level. For the tuna, this regulation becomes harder and more energetically costly in cool water, so that a prolonged stay in cool water could lead to death.

The question for the tuna then is to decide how long it should stay foraging in the cool water where food availability is higher than in the more 'comfortable' warm water on the other side of the front. Using optimality methods borrowed from engineering, it is possible to model the physiology and behavior of the fish and to calculate the energetic costs and benefits of the fish being in either the undesirable cold water with high food or the desirable warm water with low food. The model predicts that the fish will behave optimally, that is, maximize its net energy gain, if it spends all of its nonfeeding time in the warm water, making quick sorties into the cold water area to fill its stomach. As soon as this has been achieved the fish withdraws again to the warm water to digest its meal. How long it has to stay in the cool water is a function of the abundance of food and the clarity of the water. The tuna is a visual predator, so the encounter rate with prey (prey met per unit time) will be a function of these two variables. Adopting this strategy will lead to the fish hovering around the boundary and, when applied to a school of tuna, may provide a mechanism for the observed aggregation behavior.

For many species, food acquisition takes place in a competitive environment. As already mentioned, some species shoal together in an attempt to reduce the individual risk of predation. One cost of this behavior is that all the individuals in the group will be searching in the same area for the same type of food, although group foraging often means that food is found faster. The optimal behavior for an individual will then depend on what others choose to do. Individuals cope with this type of competition in a number of different ways. Experiments with groups of cod (*Gadus morhua*) in large aquaria show that access to food items delivered one at a time is determined largely by the visual acuity, swimming speed, and hunger of each fish. The individuals that take the first few prey that are offered tend to be bigger than the others and hungrier, and may have a genetically determined higher basic metabolic rate. This type of competition by cod is usually termed 'scramble competition'.

Other species handle group competition in different ways. For example, the omnivorous damsel fish, *Eupomacentrus planifrons*, defends a territory against conspecifics, so ensuring for itself a private supply of food. A further method of coping with intraspecific competition is to develop a hierarchy so that individuals can recognize the status of others from behavioral signals. When confronted with a dominant, a subdominant will give way without a fight. In this way, the cost of contests is reduced, although the subdominant might be forced to feed as an opportunistic forager while the dominant takes a more selective diet. However, in the context of foraging in a group, the subdominant is doing the best it can.

If competing individuals are genetically related, or live together for a long time, individuals might be prepared to give way to a competitor in any particular interaction over food. In this way, familiar or related competitors might operate on a tit-for-tat basis, thus sharing the resource. There is some evidence from three-spined sticklebacks (*Gasterosteus aculeatus*) that this occurs. Fish that have been living together spend less time chasing a partner that has caught first a prey offered simultaneously to them both than do fish that have met for the first time in the competitive arena. This outcome implies that individual fish recognize each other and can remember who did what when. Later work has shown that sticklebacks differentiate between familiar and unfamiliar partners using smell. Vision alone is not enough and it is likely that fish cannot recognize each other as known individuals.

Situations in which the optimal behavior depends on how others behave are best handled theoretically using aspects of game theory. This predicts how rational decision makers should behave to maximize their payoffs in the face of competition. In fish behavioral studies, aspects of game theory have been used to predict how groups of sticklebacks should divide themselves when exploiting patches of food with different profitabilities and how individuals of the same species should behave when two or more are approaching a predator to undertake what is called 'predator inspection'. Here individual prey fish suddenly leave their shoal and swim deliberately toward a predator before turning back and rushing back to the safety of the shoal. Such individuals are often accompanied by one or more conspecifics that lag behind the leader. This behavior has been modeled as a cooperative interaction between the inspectors using a branch of game theory called the 'prisoners' dilemma'.

Food Chains

Everything that has been said so far emphasizes links between fish at various levels in the ecosystem, as shown in **Figure 1**. A similar diagram could be drawn for any species of fish, meaning that the dynamics of marine ecosystems is a function of the relationships established through feeding. Certain fish species have key roles to play in that they are prey for a wide range of species. One such species in the North Sea is the sand eel (*Ammodytes marinus*), which is a major food item for herring, mackerel, cod, whiting, pollack (*Pollachius pollachius*), saithe (*Pollachius virens*), haddock, bass (*Dicentrarchus labrax*), turbot (*Scophthalmus maximus*), brill (*Stemmatodus rhombus*), megrim, plaice (*Pleuronectes platessa*), halibut (*Hippoglossus hippoglossus*), and sole. In addition, the sand eel is an important food item for many seabirds, particularly during the nesting season when, for example, the survival of puffin chicks (*Fratercula arctica*) depends on their parents bringing sufficient numbers of sand eels back to the nest.

This one example shows how critical the links between species are in a marine ecosystem. Early attempts at fisheries management in the North Sea, and in most other areas of the world, ignored the interconnectedness of species through trophic interactions. Since the late 1980s, there has been an effort to take note of the interactions when fish stock assessment is carried out. One of the major effects of sustained fishing pressure on marine ecosystems has been the gradual reduction of abundance of the larger fish within a species and of the larger species. This has had the consequence of reducing the predation pressure on lower levels of the trophic web, so that species that have traditionally been given a low commercial value have increased in abundance and are all that is available. The sand eel illustrates this well. Until the early 1970s, there was no significant fishery for sand eels in the North Sea. The growing demands for fish meal, generated by the poultry and pig production industries, created a market for previously unused species such as sand eels. Coupled with reduced catches of higher-valued species such as herring and cod, this stimulated fishermen to focus on sand eels and this, together with the continued sustained high levels of effort on the predators of sand eels, is hastening the demise of the whole system. There have also been serious consequences for

the seabird populations that have suffered a number of years with little or no fledging of young.

Feeding Behavior and Climate Change

In the first section it was made clear that many marine fish species start life as microscopic individuals that grow to be several orders of magnitude larger in length and weight. During the animal's life it occupies a series of feeding niches, and this process has come to be known as an otogenetic niche shift. The process has important consequences for the ecology of fish.

An ontogenetic niche shift is well illustrated by the North Atlantic cod where the eggs are numerous and small, larvae only a few millimeters long, and a 20-year-old adult can be over 1 m long. During its life history, the cod is most vulnerable when at its smallest size. This is partly because the larvae are prey to many planktonic organisms, but also because the survival and growth of the fish are dependent on them finding the right kind of food at the right time. In the North Sea, cod larvae appear in the plankton in the spring and one of their favourite foods is the copepod *C. finmarchicus*, which is also most abundant in the first part of the year. Over the past 20 years, data from the Continuous Plankton Recorder Survey, which has been mapping plankton distributions in the Northeast Atlantic for the past 70 years, has shown that *C. finmarchicus* abundance has fallen. *C. finmarchicus* favors cool water and is most abundant in northerly regions. Its place has been taken in the North Sea by *Calanus helgolandicus*, which is a warmer-water species and not so favored as food by cod. The timing of this species' highest abundance is also not favorable to cod larvae as it is most abundant in the second part of the summer season. By this time, cod larvae have dropped out of the plankton and have become juveniles in inshore nursery areas. The change in plankton composition is thought to be due to the warming of the surface waters of the North Sea brought about by global climate change. As a result of this change in the plankton community, cod larvae are obtaining less food and dying in greater numbers, thus reducing the recruitment of new young fish to the adult population. This, coupled with continued overfishing, has brought the North Sea cod stock to a perilous state where it is almost at the point of commercial extinction.

This example illustrates how critical feeding is to the well-being of a fish population. It also illustrates the complex interactions that link what happens to individuals, large-scale events such as climate change, and events at the population level.

See also

Benthic Organisms Overview. Coral Reef Fishes. Gelatinous Zooplankton. Mesopelagic Fishes. Pelagic Fishes. Seabird Foraging Ecology.

Further Reading

Beaugrand G, Brander KM, Lindley JA, Souisi S, and Reid PC (2003) Plankton effect on cod recruitment in the North Sea. *Nature* 426: 661–664.

Brill RW (1994) A review of temperature and oxygen tolerance studies of tunas pertinent to fisheries oceanography, movement models and stock assessments. *Fisheries Oceanography* 3: 204–216.

Gerking SD (1994) *Feeding Ecology of Fish*. San Diego, CA: Academic Press.

Giraldeau L-A and Caraco T (2000) *Social Foraging Theory*. Princeton, NJ: Princeton University Press.

Hardy AC (1924) The herring in relation to its animate environment. Part I: The food and feeding habits of the herring. *Fishery Investigations, London, Series II* 7(3): 53.

Hart PJB (1993) Teleost foraging: Facts and theories. In: Pitcher TJ (ed.) *Behavior of Teleost Fishes*, ch. 8. London: Chapman and Hall.

Hart PJB (1997) Foraging tactics. In: Godin J-G (ed.) *Behavioral Ecology of Teleost Fishes*, ch. 5. Oxford, UK: Oxford University Press.

Hart PJB (1998) Enlarging the shadow of the future: Avoiding conflict and conserving fish. In: Pitcher TJ, Hart PJB, and Pauly D (eds.) *Reinventing Fisheries Management*, ch. 17. Dordrecht: Kluwer.

Hart PJB and Reynolds JD (eds.) (2002) *Handbook of Fish Biology and Fisheries, Vol. I: Fish Biology*, chs. 11–14 and 16. Oxford, UK: Blackwell.

Helfman GS, Collette BB, and Facey DE (1997) *The Diversity of Fishes*. Oxford, UK: Blackwell.

Lowe-McConnell RH (1987) *Ecological Studies in Tropical Fish Communities*. Cambridge, MA: Cambridge University Press.

Pauly D, Christensen V, Dalsgaard J, Froese R, and Torres F, Jr. (1998) Fishing down marine food webs. *Science* 279: 860–863.

Stephens DW and Krebs JR (1986) *Foraging Theory*. Princeton, NJ: Princeton University Press.

Wootton RJ (1998) *The Ecology of Teleost Fishes*. Dordrecht: Kluwer.

FISH LARVAE

E. D. Houde, University of Maryland, Solomons, MD, USA

Introduction

Most marine teleost (bony) fishes produce thousands to millions of planktonic eggs and larvae. Newly hatched larvae, usually 1–5 mm in length, are delicate, poorly developed, and retain many embryonic characteristics. They usually hatch with undeveloped mouth parts, fins, and eyes. Larvae drift and disperse in the sea. Most die before transforming into the juvenile stage. The larval stage originates as a non-feeding, yolk-sac larva that derives nutrition from stored yolk and develops into an actively feeding larva that eventually transforms to a juvenile morphologically resembling a small adult. Transformation, which occurs from a few days to more than a year after hatching, often involves major changes in morphology (e.g. eels, flounders, herring) or may be less dramatic (e.g. cods, basses, sea breams). Lengths at metamorphosis are usually <25 mm, but can range from a few millimeters to many centimeters (some eels). The diverse, sometimes bizarre, suite of larval types that are collected represents a range of adaptations that promote survival and fitness in marine environments ranging from estuaries to the deep sea (**Figure 1**).

Fisheries scientists and managers are concerned about growth and survival during the earliest life stages of fishes because variability in those processes can lead to 10-fold or greater differences in numbers of recruits that survive to catchable size. Causes of mortality are seldom evaluated. Since early in the twentieth century, scientists have realized that ocean circulation, frontal systems, and turbulence might be key physical factors controlling larval survival. Biological factors, especially larval nutrition and predation also are major controllers of survival and growth. The physical and biological factors combine to determine success of the reproductive effort, termed 'recruitment' by fisheries scientists. Recruitment processes are not confined to the larval stage but act on earlier (eggs) and later (juveniles) stages. The larval stage is important but does not stand alone as a 'critical stage', and its relative importance in determining recruitment success can vary annually and seasonally.

Fish Larvae and the Plankton

Larvae of marine fishes, termed ichthyoplankton, usually are pelagic, drifting in the sea and interacting with pelagic predators and planktonic prey. Most fish larvae, even of species that ultimately are herbivores as juveniles or adults, are primarily carnivorous during the larval stage, feeding upon smaller planktonic organisms. In turn, larval fishes are the prey of larger nektonic and planktonic organisms. Escape from the precarious larval stage is accomplished via growth and ontogeny. Only a few individuals from thousands of newly hatched larvae survive the ever-present threats of starvation and predation during planktonic life.

Eggs and larvae of marine fishes are collected in fine-meshed plankton nets or specially designed traps. Surveys at sea estimate distributions, abundance, diversity, and structure of 'ichthyoplankton' communities, including associations of larvae with their predators and prey. Such surveys sometimes are a component of stock assessments used in fisheries management.

Larval distributions within nursery areas, including vertical distributions, differ among species. Most larvae are in the upper 200 m of the water column, although buoyant eggs, which gradually rise towards surface after spawning, may be deeper if adults spawned at depth. Accurate estimates of abundances and production of eggs and small larvae are possible because these stages usually cannot avoid samplers. Declines in abundances of older (and larger) larvae in plankton collections provide the means to estimate mortality rates if dispersal losses by water currents can be determined, and if the probability of sampler avoidance is known.

Larval Survival

The larval stage may be most important in controlling levels of subsequent recruitment. Early in the twentieth century, Johann Hjort (1914) offered the 'critical period' hypothesis, proposing that incidence of starvation at the time of yolk-sac exhaustion, when larvae first required plankton as food, was the primary factor determining variability in year-class recruitment success. In combination, probabilities of starvation and transport to unsuitable nursery habitats constitute the two elements of Hjort's hypothesis. Although the hypothesis cannot be rejected after 85 years, it is now apparent that recruitment variability is generated by a multitude of processes acting throughout the early-life stages in fishes.

Hjort's hypothesis provided a foundation for subsequent attempts to explain recruitment variability. The potential for high and variable larval stage mortality led many scientists to hypothesize that coarse controls on the magnitude of recruitment are set during the larval stage rather than later in life. The 'match-mismatch' hypothesis proposed by Cushing builds upon Hjort's ideas, emphasizing the importance of temporal coincidence in spawning and bloom dynamics of plankton, the primary food of larval fish. There is support for this hypothesis,

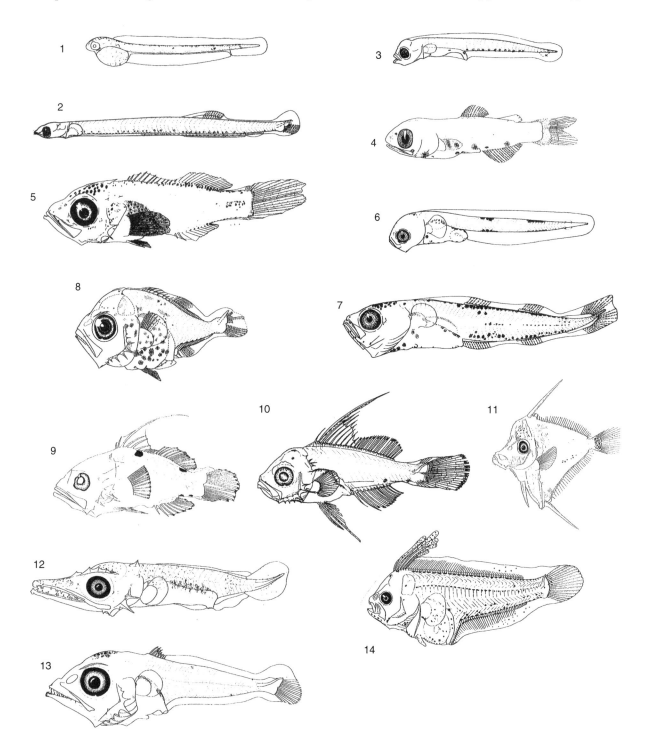

especially for species with short spawning seasons in high-latitude seas. A 'match' between spawning and spring plankton blooms ensures larval growth and survival, while a 'mismatch' results in high mortality. The 'stable ocean' hypothesis, proposed by Lasker in the 1970s, is another nutrition-related explanation for variability in larval survival. Lasker hypothesized that relaxation of storm winds and intense upwelling resulted in a stable, vertically stratified ocean in which strata of fish larvae and their prey coincide, promoting larval nutrition and survival. The hypothesis is strongly supported for species inhabiting coastal upwelling regimes.

Contrasting with the match-mismatch hypothesis is the proposal by Sinclair and Iles that gyre-like circulation features in well-mixed coastal waters define spawning areas while retaining eggs and larvae. This hypothesis emphasizes physics and circulation features, rather than nutritional factors, as the controller of recruitment variability. The hypothesis gains support from evidence on recruitment of numerous herring (*Clupea harengus*) and some cod (*Gadus morhua*) stocks in the North Atlantic. Sinclair and Iles argued that high larval retention promotes strong recruitment while failed retention diminishes success. Retention not only reduces larval losses to dispersal, but ensures genetic integrity of a stock by confining reproduction within the retention area. Although this hypothesis differs from the 'match-mismatch' hypothesis, there is evidence from some cod stocks that 'match-mismatch' and retention mechanisms operate together to promote larval survival.

In the tropics and especially on coral reefs, scientists have debated whether supply of larvae or post-settlement mortality of juveniles is the primary factor generating variability in recruitment levels. The 'lottery' hypothesis, of Sale, proposes that fish larvae are delivered by ocean currents to reefs where they settle onto structure that, if by chance is free of predators or competing fish, will lead to successful recruitment. Recent evidence supports both larval supply and post-settlement controls as important for recruitment success. Also, there is accumulating evidence that long-range dispersal, once assumed to dominate early-life processes in tropical seas, may in fact be limited and that retention near islands and reefs is common as larvae develop from hatching to settlement stages.

Foods and Feeding

Fish larvae must feed frequently to ensure fast growth, a prerequisite for high survival. Successful initiation of feeding depends upon availability of suitable kinds and sizes of prey, which usually is zooplankton 50–100 μm in width. Sizes of prey that are consumed are strongly related to larval size (specifically mouth size). Nauplii of copepods are perhaps the most common prey of small fish larvae.

Figure 1 (Left) Larval fishes. (1) *Sardinella zunasi* (Clupeidae), 4.8 mm. (Reproduced with permission from Takita T (1966).) Egg development and larval stages of the small clupeoid fish *Harengula zunasi* Bleeker and some information about the spawning and nursery in Ariake Sound. *Bulletin of the Faculty of Fisheries, Nagasaki University* 21: 171–179. (2) *Clupea pallasi* (Clupeidae), 10.4 mm. (Reproduced with permission from Matarese AC, Kendall AW Jr, Blood DM and Vinter BM (1989).) *Laboratory Guide to Early Life History Stages of Northeast Pacific Fishes. NOAA Technical Report, NMFS 80* Seattle, WA: National Marine Fisheries Service.) (3) *Diaphus theta* (Myctophidae), 4.6 mm. (Reproduced with permission from Matarese *et al.*, 1989.) (4) *Benthosema fibulatum* (Myctophidae), 8.7 mm. (Reproduced with permission from Moser HG and Ahlstrom EH (1974).) Role of larval stages in systematic investigations of marine teleosts: the Myctophidae, a case study. *Fishery Bulletin, US* 72: 391–413. (5) *Sebastes melanops* (Scorpaenidae), 10.6 mm. (Reproduced with permission from Matarese *et al.*, 1989.) (6, 7) *Theragra chalcogramma* (Gadidae), 6.2 and 13.7 mm. (Reproduced with permission from Matarese *et al.*, 1989.) (8) *Carangoides* sp. (Carangidae), 4.4 mm. (Reproduced with permission from Leis JM and Trnski T (1989).) *The Larvae of Indo-Pacific Shorefishes.* Sydney: University of New South Wales Press.) (9) *Plectranthias garupellus* (Serranidae), 5.5 mm. (Reproduced with permission from Kendall AW Jr (1979).) *Morphological Comparisons of North American Sea Bass Larvae (Pisces: Serranidae). NOAA Technical Report NMFS Circular 428.* Rockville, MD: National Marine Fisheries Service. (10) *Lutjanus campechanus* (Lutjanidae), 7.3 mm. (Reproduced with permission from Collins LA, Finucane JH and Barger LE (1980).) Description of larval and juvenile red snapper, *Lutjanus campechanus. Fishery Bulletin, US* 77: 965–974. (11) *Naso unicornis* (Acanthuridae), 5.9 mm. (Reproduced with permission from Leis JM and Richards WJ (1984).) Acanthuroidei; development and relationships. In: Moser HG, Richards WJ, Cohen DM, Fahay MP, Kendall AW Jr, and Richardson SL (eds) *Ontogeny and Systematics of Fishes*, pp. 547–551. American Society for Ichthyologists and Herpetologists, Special Publication 1. (12) *Xiphias gladius* (Xiphiidae), 6.1 mm. (Reproduced with permission from Collette BBT, Potthoff T, Richards WJ, Ueyanagi S, Russo JL and Nishikawa Y (1984)) Scombroidei: development and relationships. In: Moser HG *et al.* (eds) *Ontogeny and Systematics of Fishes*, pp. 591–620. American Society of Ichthyologists and Herpetologists, Special Publication 1. (13) *Thunnus thynnus* (Scombridae), 6.0 mm. (Reproduced with permission from Collette BB *et al.*, 1984.) In: Moser HG *et al.* (eds) *Ontogeny and Systematics of Fishes*, 591–620. American Society of Ichthyologists and Herpetologists Special Publication 1.) (14) *Paralichthys californicus* (Paralichthyidae), 7.0 mm. (Reproduced with permission from Ahlstrom EH, Amaoka K, Hensley DA, Moser HG and Sumida BY (1984).) In: Moser HG *et al.* (eds) *Ontogeny and Systematics of Fishes*, pp. 640–670. American Society of Ichthyologists and Herpetologists, Special Publication 1.)

Concentrations of nauplii and other zooplankton have often been analyzed to evaluate feeding conditions for larvae in the sea. Recent research has demonstrated that many species of fish larvae initiate feeding on a diverse spectrum of small planktonic organisms, including some phytoplankton and protozoa that are often more abundant than copepod nauplii. Laboratory experiments indicate that concentrations of copepod nauplii exceeding 100 per liter may be required to support feeding and growth of larval fish, leading to hypotheses and simulation models that implicate patchiness of prey and small-scale turbulence as mechanisms promoting prey encounter, successful feeding, and growth. Such enhancing mechanisms clearly are important, but if diverse diets are the norm, as recent evidence suggests, then availability of suitable larval prey in the sea may be higher than once believed.

As development and growth proceed, feeding success increases. Maximum prey sizes in larval diets increase substantially, but the mean size increases only slowly because larvae continue to consume small prey while adding larger prey to the diet. This ontogenetic shift occurs gradually in most fishes, but in some mackerels, tunas, and other species the shift is dramatic, and piscivory on larval fish becomes their primary source of nutrition during the earliest larval stage. Large amounts of prey must be consumed.

Marine fish larvae typically consume >50% of their body weight daily to achieve average growth. Some larvae from warm seas, e.g. tunas and anchovies, consume >100% of their body weight each day to grow at average rates. Such high food requirements have reinforced the belief that poor feeding conditions, slow growth, and starvation are major causes of larval mortality in the sea.

Starving fish larvae are seldom observed in the sea because such larvae are believed to be selectively preyed upon, then disintegrating rapidly in stomachs of predators. Thus, dead or starving larvae are infrequent in ichthyoplankton collections. Nutritional condition of fish larvae can be evaluated by morphological, histological, and biochemical approaches. Nucleic acid analysis, when properly conducted, shows that poorly fed larvae in environments with low prey have low RNA/DNA ratios, indicating poor potential for growth. Fish larvae that fail to initiate feeding soon after yolk depletion will become nutritionally compromised and poorly conditioned, reaching a 'point-of-no-return' within 2–10 days, from which they cannot recover even if adequate food is provided.

Little is known about specific feeding behaviors of fish larvae. Larvae mostly feed visually and must be in close proximity (e.g. less than one body length) to successfully encounter and capture prey. Most feeding occurs during daylight, although threshold light levels

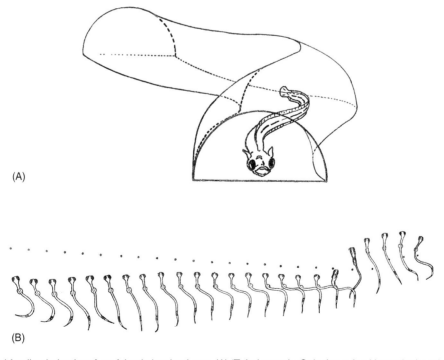

(A)

(B)

Figure 2 Typical feeding behavior of an Atlantic herring larva. (A) 'Tube' search. Only those food items in the tube are available to the larva. (B) S-flex feeding. Sequence of stages during an unsuccessful feeding attempt by a larva. (Reproduced with permission from Rosenthal H and Hempel G (1970).) Experimental studies in feeding and food requirements of herring larvae (*Clupea harengus* L.) In: Steele JH (ed.) *Marine food chains*, pp. 344–364. Berkeley: University of California Press.)

are low, in the range 0.01–0.10 lux, levels equivalent to dawn and dusk periods. In elongate, herring-like larvae, a typical 'S-flex' behavior has been described in which a larva, upon encountering a prey, flexes its body into an S-shape before striking at the prey (**Figure 2**). Other kinds of larvae use a modified 'S-flex' behavior or a 'cruise and strike' searching behavior. As in other life stages, successful feeding by larvae depends on the relationship: $P = E \cdot A \cdot C$, where E, A, and C are probabilities of encountering, attacking, and capturing a prey. Modeling and understanding this relationship, and changes in it during larval development, are important to evaluate larval survival potential.

Predation

Much evidence indicates that predation is the major direct cause of mortality to early-life stages of fish. This conclusion does not exclude starvation and nutritional deficiencies as causes of, or contributors to, larval mortality. Because predation in the sea is strongly size-dependent and predators are size-selective, fish larvae that are slow-growing or starving will remain in small, highly vulnerable size classes longer and may have higher predation risk. Thus, larval growth rate and its dependence on nutrition links the predation and starvation processes. Larval growth rate under many circumstances is an important measure of susceptibility to predation.

There are many kinds of predators on fish eggs and larvae. Predators include jellyfishes (ctenophores and medusae), fish, and crustacea (e.g. euphausiids, large copepods). Cannibalism on eggs and larvae occurs and can be an important population regulatory mechanism. The relative importance of the various predators is poorly known. Although a suite of predators of different taxa and sizes eat larvae of preferred sizes, the integrated effects of complex, multispecific predation on recruitment outcomes have hardly been evaluated in the sea or in laboratory experiments. Models of predation also generally have focused on single predators of near-uniform size preying on fish larvae of a single species. In natural ecosystems a community of predators of varying kinds and sizes, which are distributed patchily in time and space, interacts with an assemblage of fish early-life stages. Outcomes are difficult to predict and will be modified by availability of alternative prey resources.

Vulnerability of larval fish to a predator of defined size can often be represented by dome-shaped responses in which vulnerability peaks at an intermediate larval size or by a consistent decline in vulnerability as larval size increases (**Figure 3**). Predation rate on larvae is frequently a dome-shaped function of the ratio of prey size : predator size, where larval fish (the prey), are consumed most efficiently when they are 5–15% of a predator's length, a ratio that is consistent for predators as diverse as jellyfish or fish. Vulnerability of larval fish to predation is the product of encounter rate (dependent upon abundances, sizes, and swimming speeds of both predators and prey) and susceptibility (attack probability × capture probability) of larvae. Thus, sizes and swimming speeds of predators and larvae determine vulnerability of an individual larva. At the population level, size-specific abundances of the predator clearly will be important in determining whether a particular predator can control or inflict significant mortality on a population of larval fish. Many common predators, e.g. some medusae and fish, span a broad range of sizes and may be predators on populations of fish larvae over a wide size range.

Although size is the dominant factor determining vulnerability of fish larvae to predation, ontogeny plays an important role. Fish larvae can react to predators by sensing vibrations via free neuromasts that are present at hatching. As growth occurs, neuromasts proliferate, the lateral line develops, and vision may become increasingly important to detect and escape predators. Development of musculature and fins, especially the caudal fin complex, increases the swimming power and escape ability of larval fishes. Ontogenetic improvements in larval ability to avoid predators are counteracted to an extent by the increased sizes of predators targeting larger larvae and juvenile fish. Mortality rates of large larvae and juveniles generally decline, indicating that survival advantages attained through ontogeny and growth outweigh increased suitability of these fish to a suite of larger predators.

Temperature and Salinity

Temperature exerts strong control over rate processes associated with metabolism and growth in larval fishes. Temperature and body size may be the two most important controlling variables in the early lives of fishes. Temperature also controls rate processes in predators and thus indirectly controls rates at which larvae die from predation. Furthermore, seasonal development of planktonic prey of fish larvae depends in part upon temperature, and is especially important in seasonally variable mid and high latitudes where combined effects of temperature and prey levels control larval growth potential. Salinity

Encounter rate Susceptibility Vulnerability

Ambush raptorial invertebrate

(A)

Cruising invertebrate

(B)

Filter-feeding fishes

(C)

Raptorial fishes

(D)

Relative scale

Relative larval size

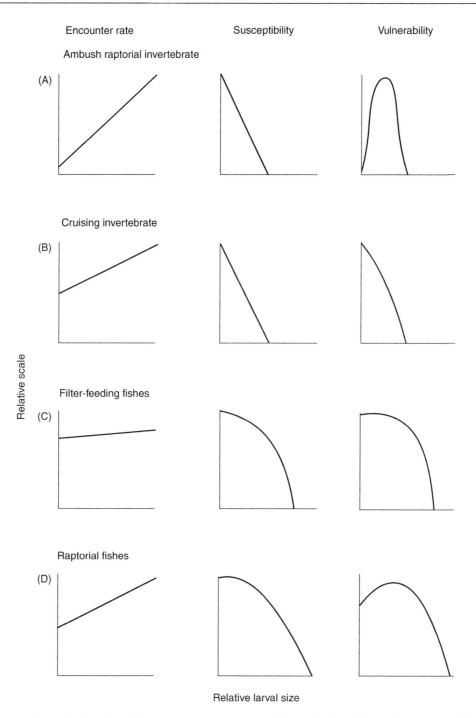

Figure 3 Conceptual models showing relative encounter rates, susceptibility, and vulnerability of fish larvae to different predator types. (Reproduced with permission from Bailey KM and Houde ED, 1989.)

generally is of secondary importance, but it too can be critical for some anadromous fishes whose larvae inhabit estuarine zones along the salinity gradient or at the interface between fresh water and salt waters.

Growth rates of marine fish larvae are surprisingly plastic. Weight-specific growth rates may vary by more than fourfold in response to temperature differences within tolerable ranges, which imparts high variability

to larval stage durations. Although temperature can exercise physiological control over growth, prevailing temperatures in many continental shelf and oceanic nursery areas differ only slightly during a spawning season or between years, and larval growth may in fact be controlled more by prey availability in those ecosystems. However, in shallow estuaries and neritic habitats, fluctuations in temperature induced by weather fronts or shifts in circulation patterns may be sufficiently strong to influence larval growth rates or to directly cause mortality. Ranges of temperature tolerated by larvae are generally lower than for juveniles and adults.

Temperature and body size control bioenergetics relationships in marine poikilotherms, including larval fish. Growth rates, metabolic rates, and possibly growth efficiencies and assimilation rates of fish larvae, are sensitive to changes in temperature. Each fish species has a temperature at which it performs optimally. In meta-analyses across species and ecosystems, larval growth rates were demonstrated to increase as temperature increased. Species from high latitudes grow slowly (often <10% body weight per day) while tropical species grow fast (>30% per day). In the meta-analysis, mortality rates are positively correlated with growth rates, and thus also are temperature-dependent.

Salinity controls water balance in osmotic regulation. Marine fish larvae often have surprisingly broad tolerances of salinity. Even anadromous fishes, whose larvae normally live in salinities <1 psu, may perform well in laboratory experiments at salinities of 5–10 psu. At high salinities marine fish larvae drink sea water to regulate water balance. Drinking rates increase as either salinity or temperature increases. Salinity and temperature often act together to affect the physiological performance of larvae. In particular, optimum salinity–temperature combinations determine hatching success and larval performance. Moreover, temperature and salinity combinations in the sea determine the density field (σ_t) in which larvae are located. The density field controls pycnocline and thermocline depths, which define strata, discontinuities, shear zones, and transition depths that may aggregate prey and predators of fish larvae as well as the larvae themselves.

Behavior

Except for larval feeding behavior (see above), relatively little is known about the behavior by individuals with respect to environmental cues or other stimuli. Swimming behaviors are documented for some species. Larvae with elongate bodies (e.g. sardines) swim with an anguilliform motion, while larvae with more compact bodies (e.g. cod-like or bass-like larvae) adopt a modified carangiform, 'cruise-pause' swimming behavior. When searching for food, swimming typically is slow, one or two body lengths per second, and is the predominant behavior of feeding-stage larvae during daylight hours. Feeding-stage larvae generally can cruise at sustained swimming speeds of 3–10 body lengths per second, but newly-hatched, yolk-sac larvae are incapable of strong, directed swimming in a horizontal plane. Burst swimming to escape predators, at >10 body lengths per second is observed. Predator detection and avoidance behaviors become increasingly effective during ontogeny as sensory systems develop.

Fish larvae may migrate vertically over tens of meters on a diel basis. This ability allows larvae to track food by adjusting depth distributions to coincide with depths where prey is abundant. In addition, larvae can potentially avoid parts of the water column where predators prevail or where environmental conditions and water quality (e.g. low dissolved oxygen) are unfavorable. Dispersal of larvae or, alternatively, retention on nursery grounds can depend on depth selection that promotes 'selective tidal stream transport.' This behavioral mechanism requires that larvae coordinate vertical migrations with tides and currents to ensure transport to, or retention within, favorable nursery areas. This mechanism is particularly important in estuaries, but may also regulate larval drift or retention on continental shelf nursery areas.

Integration

Larval populations of many marine fishes are distributed over tens to thousands of kilometers and have early-life durations that range from days to months. However, the fates of individual larvae depend upon interactions with their environment, prey, and predators measured on spatial scales of micrometers to meters and on timescales of fractions of seconds to hours. Events at all spatial and temporal scales are potentially important and can generate significant variability in larval survival.

The potential for recruitment success is indicated by both survival and proliferation of cohort biomass during early life. While numbers decline steadily, biomass of successful cohorts increases. Stage-specific survival during early life depends upon relative rates of mortality (M) that reduce a cohort's numbers and growth (G) that controls accumulation of individual mass. Biomasses of most marine fish cohorts decline during the larval stage to levels <1% of their biomasses at hatching because M/G > 1.0. In many marine and estuarine fish larvae (e.g. walleye pollock *Theragra chalcogramma*, American shad *Alosa sapidissima*, bay anchovy *Anchoa mitchilli*), the body size at which the transition from M/G > 1.0 to <1.0 occurs, after which cohort biomass increases, varies annually. Cohorts that are strong contributors to recruitment make the transition at smaller size than unsuccessful cohorts (**Figure 4**).

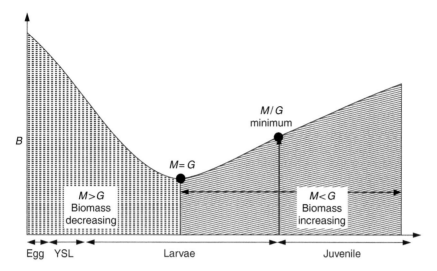

Figure 4 Conceptual diagram of trend in cohort biomass (B) during early-life stages. Cohort biomasses decrease during egg and yolk-sac larva (YSL) stages but may increase during the larval and juvenile stages. M is the instantaneous mortality rate and G is the weight-specific growth rate. A 'transition size' at which M = G is the size of an individual at which cohort biomass begins to increase. (Reproduced with permission from Houde ED (1997) Patterns and consequences of selective processes in teleost early life histories. In: Chambers RC and Trippel EA (eds) *Early life history and recruitment in fish populations*, pp. 173–196. Dordrecht: Kluwer Academic Publishers.)

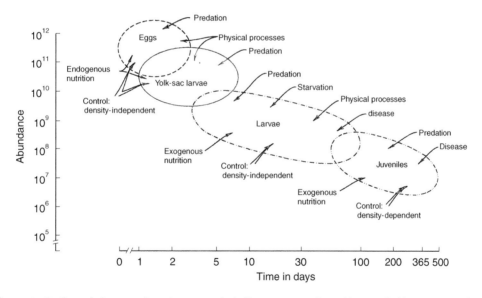

Figure 5 Conceptualization of the recruitment process including sources of nutrition, probable sources of mortality, and hypothesized mechanisms of control for four early-life stages. (Reproduced with permission from Houde ED, 1987.)

Important processes may act sequentially during development (**Figure 5**). Initially, density-independent processes are believed to dominate as controllers of growth and mortality.

As growth proceeds and the young fish are less dominated by physical conditions of their immediate environment, density-dependent processes related to competition for prey or predation pressure will become increasingly important in regulating recruitment.

The ability to assign ages to sampled fish larvae from daily increment counts in otolith microstructure and thus to estimate sizes-at-age, growth rates, and mortality rates provides a powerful tool to retrospectively analyze cohort dynamics and relate variability in larval-stage survival to environmental factors. Observed mortality rates of larval populations usually decline as size increases from hatching to metamorphosis, but patterns in growth rates are less predictable, differing among species and among cohorts within species. While the potential advantages of fast larval growth were cited by Cushing many years ago, convincing evidence supporting this hypothesis has been provided only recently in studies on larval cod, plaice *Pleuronectes platessa*, and other species. In these species, cohorts that grow fast often contribute disproportionately to recruitment. Although growth rate differences alone can generate strong variability in early-life survivorship, variability in larval mortality rates probably contributes most to fluctuations in recruitment.

Models to explore and simulate larval stage dynamics have proliferated in recent years. Individual-based models have proven to be useful in simulations of survival and growth responses of individual larvae, and to track the status and trajectory of cohorts over time with respect to variability in physical and biological factors. Coupled physical–biological models are making major contributions to understanding how circulation features transport or retain larvae and support larval production.

Research Needs

It remains difficult to accurately estimate abundances of fish early-life stages. As larvae grow their abundances decline, while they become increasingly more mobile and difficult to sample with conventional nets. This problem continues to plague attempts to estimate larval mortality rates and production, which require accurate abundance-at-age estimates. Improved samplers that allow depth-specific collections have been available since the 1970s and a new generation of collecting and sensing instruments (e.g. optics and acoustics) may partly resolve the problem. Distribution patterns and patchiness of early-life stages must be appreciated to design surveys that can adequately account for the temporal and spatial variability in distributions of early-life stages. The issues here are complex because critical spatial scales are poorly known and may change as larvae grow. Obtaining better abundance estimates ultimately depends upon an integrated knowledge of biotic and abiotic factors, especially larval behavior and ocean physics, and the interactions of the two at many scales.

There are few extensive time series of abundances of early-life stages of fish. Such series, while costly to obtain, provide information on annual variability in abundances with respect to long-term changes in climate and ocean environment. Fundamental advances in understanding causes of variability in larval production would emerge from surveys repeated over many years. It is quite probable that, in addition to annual variability, decadal and other low-frequency

patterns of variability in ocean productivity also exercise controls over survival of fish early-life stages and levels of recruitment. Long time series of larval abundances and dynamics may not be essential to manage fisheries in a tactical sense, but they would be important strategically to document long-term trends and causes of variability in recruitment success.

It is essential to develop better knowledge of trophic relationships in the plankton because they largely determine the fate of larval fish cohorts. Better conceived, interdisciplinary, and integrated research programs that emphasize how physics at appropriate scales (e.g. microturbulence) accounts for interactions between larvae and prey and between larvae and predators are needed. Modeling research must play a major role to help explain and predict variability in larval survival. Existing models range from simple statistical models that describe allometric relationships of physiological responses during larval development to models of ecosystem-level complexity that include many environmental factors. Forecasting larval stage survival and recruitment are worthy, but elusive, goals of such modeling . It is important to build upon progress in development of coupled biological–physical models, including individual-based, spatially explicit models that predict larval survival (e.g. walleye pollock, Atlantic cod, Atlantic herring, Atlantic menhaden *Brevoortia tyrannus)*. Modeling physics at appropriate scales has progressed more rapidly than biological modeling. Improved models to elucidate biological variability in the larval stage are needed before forecasts of recruitment can be achieved.

Development of new technologies and adoption of technologies from other disciplines have advanced knowledge of larval stage ecology. Since the 1970s, analysis of microstructure in larval otoliths has provided the breakthrough that allows more or less routine aging of fish larvae from daily increments in otoliths. Biochemical methods, such as analysis of RNA/DNA and lipids, have proved useful to evaluate growth potential and nutritional condition of larvae. New biochemical indicators to assess larval condition or to categorize habitats (water quality) in which larvae were collected are being developed. Recent research combining analyses of otolith chemistry and microstructure holds great promise for hindcasting environmental conditions (temperature, salinity) that larvae experienced earlier in life. New genetic tools have potential to identify stocks, track larval dispersal, and differentiate survival potentials among closely related, but genetically distinct, stock units. The problem of obtaining adequate and representative samples of fish larvae is being addressed by applying new sensing technologies, e.g. *in situ* optical and acoustical samplers aboard ship or moored on buoy systems and designed to sense larval fish as well as their zooplankton prey and predators. Advances in remote sensing capability will allow better assessments of environmental conditions, hydrographic variability, and sea state, which will permit more critical evaluations of larval habitats and water quality.

Conclusions

Coarse controls on reproductive success and establishment of year-class abundances are exercised primarily during the embryonic and larval stages of marine fish during their drift in the plankton. The larval stage of marine fish is on average >35 days long, during which time much of the variability in survival to recruitment may be generated. Control may be mostly density-independent during these stages, although theoretical and modeling research demonstrates that density-dependence, even at modest levels, could be a powerful regulator of early-life dynamics. Finer controls operate during the juvenile stage when a shift towards density-dependence may occur. Year-class size and recruitment level can be set in either the larval or juvenile stage. Importantly, effects of poor larval survival seldom can be compensated by high juvenile survival. The converse is not true, i.e. high larval stage survival does not guarantee recruitment success; year classes may fail during the juvenile stage.

Cushing (1975) referred to 'the single process' during early lives of marine fishes. Growth and mortality, nutrition and predation, and physical conditions in nursery areas all act together to control recruitment and mold a year class. It is not 'starvation, predation or advective losses,' but the net result of those combined processes that controls recruitment. The subtle, difficult-to-measure small differences in mortality or growth rates over weeks or months, rather than episodic events generating massive mortalities, usually dominate in determining recruitment success. Initially, eggs and larvae are very abundant, and their growth and mortality rates are high and variable. Even small differences in those rates, if not compensated in the juvenile stage, assure large fluctuations in cohort survival.

Survival of marine fish larvae appears to be strongly size-dependent. Compared with freshwater fish larvae, marine larvae tend to be small. Their hatch weights average nearly 10 times less than weights of freshwater larvae (Table 1). The difference in size is believed to be the primary factor determining the higher average mortality rates, longer

Table 1 Dynamic properties of marine and freshwater fish larvae

Variable	Marine		Freshwater	
	Mean	Standard error	Mean	Standard error
W_o	37.6	6.4	359.7	72.8
W_{met}	10 846	953	9277	1604
G	0.200	0.011	0.177	0.019
M	0.239	0.021	0.160	0.040
D	36.1	1.1	20.7	1.1

Temperature-adjusted mean values are presented. W_o = dry weight at hatch (µg); W_{met} = dry weight at metamorphosis (µg); G = weight-specific growth rate (d^{-1}); M = mortality rate (d^{-1}); D = larval stage duration (d). (Modified from **Table 1** in Houde ED (1994). Differences between marine and freshwater fish larvae; implications for recruitment. *ICES Journal of Marine Science* 51: 91–97.)

stage duration, higher weight-specific respiration rates, and higher weight-specific food consumption by marine larvae. As a consequence and generalization, control over recruitment probably is exercised primarily in the relatively long larval stage of marine fish but in the juvenile stage of freshwater fish.

An 'average' fish larva has little chance for survival. Typically, numbers are reduced by three orders of magnitude during the larval stage. Thus, larval survivors may have special qualities, in addition to extraordinary luck. Individuals that behave appropriately or grow faster will survive because they are proficient at feeding and avoiding predators, and perhaps better at selecting favorable habitat (e.g. by vertically migrating). Favored or selected characteristics may be phenotypic or inherited. As an example of the former, there is evidence that in many species bigger larvae hatch from eggs spawned by larger and older females. Such larvae have higher survival potential. Thus, the fates of fish larvae may be partly programmed by genes from their parents, but also will depend on selective fishing practices that target the largest adults, in addition to the myriad environmental factors that normally act on larval populations.

See also

Fish Ecophysiology. Fish Feeding and Foraging. Fish Locomotion. Fish Migration, Horizontal. Fish Migration, Vertical. Fish Predation and Mortality. Fish Reproduction. Pelagic Fishes.

Further Reading

Bailey KM and Houde ED (1989) Predation on eggs and larvae of marine fishes and the recruitment problem. *Advances in Marine Biology* 25: 1–83.

Beyer JE (1989) Recruitment stability and survival – simple size-specific theory with examples from the early life dynamics of marine fish. *Dana* 7: 45–147.

Blaxter JHS (1986) Development of sense organs and behavior of teleost larvae with special reference to feeding and predator avoidance. *Transactions of the American Fisheries Society* 115: 98–114.

Cushing DH (1975) *Marine Ecology and Fisheries.* Cambridge: Cambridge University Press.

Cushing DH (1990) Plankton production and year-class strength in fish populations: an update of the match/mismatch hypothesis. *Advances in Marine Biology* 26: 250–293.

Hjort J (1914) Fluctuations in the great fisheries of northern Europe. *Rapport et Procés-Verbaux des Réunions, Conseil International pour l'Exploration de la Mer* 20: 1–228.

Houde ED (1987) Fish early life dynamics and recruitment variability. *American Fisheries Society Symposium* 2: 17–29.

Houde ED (1997) Patterns and consequences of selective processes in teleost early life histories. In: Chambers RC and Trippel EA (eds.) *Early Life History and Recruitment in Fish Populations.* Dordrecht: Kluwer Academic Publishers.

Lasker R (ed.) (1981) *Marine Fish Larvae. Morphology, Ecology, and Relation to Fisheries.* Seattle: University of Washington Press.

Sale PF (ed.) (1991) *The Ecology of Fishes on Coral Reefs.* San Diego: Academic Press.

Sinclair M (1988) *Marine Populations. An Essay on Population Regulation and Speciation.* Seattle: University of Washington Press.

FISH LOCOMOTION

J. J. Videler, Groningen University, Haren,
The Netherlands

Introduction

After 500 million years of evolution, fish are extremely well adapted to various constraints set by the aquatic environment in which they live. In the dense fluid medium they are usually neutrally buoyant and use movements of the body to induce reactive forces from the water to propel themselves. Propulsive forces overcome drag, are required for acceleration, and disturb the water. Animal movements are powered by contracting muscles and these consume energy. These basic principles of fish locomotion are used by approximately 25 000 extant species. The variation in swimming styles, within the limits of these principles, is extremely large.

The Swimming Apparatus of Fish

Fish are aquatic vertebrates with a skull, a vertebral column supporting a medial septum dividing the fish into two lateral halves, and lateral longitudinal muscles segmentally arranged in blocks, or myotomes. The vertebral column is laterally highly flexible and virtually incompressible longitudinally. Consequently, contraction of the muscles on one side of the body bends the fish, and waves of curvature along the body can be generated by series of alternating contractions on the left and right side.

Fish vertebrae are concave fore and aft (amphicoelous) and fitted with a neural arch and spine on the dorsal side. In the abdominal region, lateral projections are connected with the ribs enclosing the abdominal cavity. The vertebrae in the caudal region bear a hemal arch and spine. Neural and hemal spines point obliquely backward. The number of vertebrae varies greatly among species: European eels have 114 vertebrae, and the numbers in the large perciform order vary between 23 and 40. The number is not necessarily constant within a species. Herring, for example, may have between 54 and 58 vertebrae. The end of the vertebral column is commonly adapted to accommodate the attachment of the tail fin. Several vertebrae and their arches and spines are partly rudimentary and have changed shape to contribute to the formation of platelike

structures providing support for the finrays of the caudal fin. Most fish species have unpaired dorsal, caudal, and anal fins and paired pectoral and pelvic fins.

In fish, relatively short lateral muscle fibers are packed into myotomes between sheets of collagenous myosepts. The myotomes are cone-shaped and stacked in a segmental arrangement on both sides of the median septum (**Figure 1(a)**). In cross sections through the caudal region, the muscles are arranged in four compartments. On each side there is a dorsal and a ventral compartment, in some groups separated by a horizontal septum. The left and right halves and the dorsal and ventral moieties are mirror images of each other. In cross sections the myosepts are visible as more or less concentric circles of collagen. The color of the muscle fibers may be red, white, or intermediate in different locations in the myotomes (**Figure 1(b)**). Red fibers are usually situated under the skin. The deeper white fibers form the bulk of lateral muscles, and in some species intermediately colored pink fibers are found between the two. The red fibers are slow but virtually inexhaustible and their metabolism is aerobic. They react to a single stimulus owing to the high density of nerve terminals on the fibers. The white fibers are fast, exhaust quickly, and use anaerobic metabolic pathways. White fibers are either

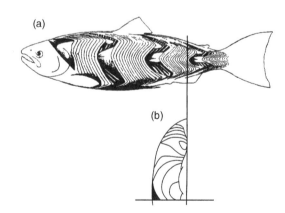

Figure 1 (a) The myotomes and myosepts on the left side of the king salmon. Myotomes have been removed at four places to reveal the complex three-dimensional configuration of the lateral muscles. (b) A cross section through the upper left quarter of the caudal region of a salmon. Red muscle fibers are situated in the dark area near the outside. The lines represent the myosepts between the complex myotomes. (a) Redrawn from Greene CW and Greene CH (1913) The skeletal musculature of the king salmon. *Bulletin of US Bureau of Fisheries* 33: 25–59. (b) Based on Shann EW (1914) On the nature of lateral muscle in teleostei. *Proceedings of the Zoological Society of London* 22: 319–337.

focally or multiply innervated. Pink fibers are intermediate in most aspects.

The red muscles of tuna and mackerel sharks are positioned well inside the white muscle mass, an arrangement that increases the muscle temperature by about 10 °C during swimming. This halves the twitch contraction time of the white muscles and doubles the swimming speed.

The orientation of the muscle fibers in the myotomes is roughly parallel to the longitudinal axis of the fish. However, a more precise analysis of the fiber direction shows a complex helical pattern. Fiber directions may make angles as large as 80° with the main axis of the fish. A possible function of this arrangement could be that it ensures equal strain rates during lateral bending for fibers closer to and farther away from the median septum. The purpose of the complex architecture of fish lateral muscles is not yet fully understood.

Fish fins are folds of skin, usually supported by fin rays connected to supporting skeletal elements inside the main body of the fish. Intrinsic fin muscles find their origin usually on the supporting skeleton and insert on the fin rays. The fins of elasmobranchs (sharks and rays) are permanently extended and rather rigid compared to those of teleosts (bony fish). The fin rays consist of rows of small pieces of cartilage. Muscles on both sides of the rows running from the fin base to the edge bend these fins. Teleost fins can be spread, closed, and folded against the body. There are two kinds of teleost fin rays: spiny, stiff unsegmented rays, and flexible segmented ones. Spiny rays stiffen the fin and are commonly used for defense. The flexible rays (**Figure 2**) play an important role in adjusting the stiffness and camber of the fins during locomotion. They consist of mirror image halves, each of which has a skeleton of bony elements interconnected by collagenous fibers. Muscles inserting on

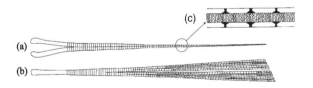

Figure 2 The structure of a typical teleost fin ray. (a) (Dorsal or ventral view) The left and right fin ray halves are each other's mirror image. (b) (Lateral view) The size of the bony elements decreases to the right after each of the bifurcations. The position of the bifurcations in the various branches does not show a geometrically regular pattern. (c) Longitudinal section through the bony elements of the fin ray at a position indicated in (a). Note that a joint with densely packed collagenous fibers connects the elements. The collagenous fibers connecting the fin ray halves have a curly, serpentine appearance. Reproduced from Videler JJ (1993) *Fish Swimming*. London: Chapman and Hall.

the fin ray heads can change the bending of the rays or the stiffness against bending forces.

The body shape of fish may vary greatly among species, but the best pelagic swimmers have a common form. Their bodies are streamlined with gradually increasing thickness from the point of the snout to the thickest part at about one-third of the length. From that point the thickness gradually decreases toward the narrow caudal peduncle. A moving body in water encounters friction and pressure drag. Friction drag is proportional to the surface area, pressure drag to the area of the largest cross section. A spherical body has the lowest friction for a given volume; a needle-shaped body encounters minimal amounts of pressure drag. An optimally streamlined body is a hybrid between a sphere and a needle and offers the smallest total drag for the largest volume. It has a diameter-to-length ratio between 0.22 and 0.24. The best pelagic swimmers have near-optimal thickness-to-length ratios.

The mechanically important part of fish skin is the tissue (the stratum compactum) underneath the scales, which consists of layers of parallel collagenous fibers. The fibers in adjacent layers are oriented in different directions, the angles between the layers vary between 50° and 90°, but the direction in every second layer is the same. The packing of layers resembles the structure of plywood, except that in the fish stratum compactum there are also radial bundles of collagen connecting the layers; the number of layers varies between 10 and 50. In each layer, the fibers follow left- and right-handed helices over the body surface. The angle between the fibers and the longitudinal axis of the fish decreases toward the tail. In some species the stratum compactum is firmly connected to the myosepts in the zone occupied by the red muscle fibers; in other fish there is no such connection. The strongest fish skins tested are those of eel and shark. Values of Young's modulus (the force per unit cross-sectional area that would be required to double the length) of up to 0.43 GPa (1 GPa = $10^9 \, \mathrm{N \, m^{-2}}$) have been measured. This is about one-third of the strength of mammalian tendon, for which values of 1.5 GPa have been measured.

Scales are usually found at the interface between fish and water. Several swimming-related functions have been suggested. Scales might serve to prevent transverse folds on the sides of strongly undulating fish, keeping the outer surface smooth. Spines, dents, and tubercles on scales are usually arranged to form grooves in a direction of the flow along the fish. Roughness due to microstructures on scales in general creates small-scale turbulence, which could delay or prevent the development of drag-increasing large-scale turbulence.

Fish mucus is supposed to reduce friction with the water during swimming. This assumption is based on the idea that mucus shows the 'Toms effect', which implies that small amounts of polymers are released that preclude sudden pressure drops in the passing fluid. Measurements of the effects of fish mucus on the flow show contradictory results varying from a drag reduction of almost 66% (Pacific barracuda) to no effect at all (California bonito). Experiments with rainbow trout showed that mucus increases the thickness of the boundary layer, which implies that viscous friction is reduced. However, the penalty for a thicker boundary layer is that the fish has to drag along a larger amount of water. The conclusion might be that the effect of mucus is beneficial during slow-speed cruising but detrimental during fast swimming and acceleration.

Swimming-Related Adaptations

Various fish species are adapted to perform some aspect of locomotion extremely well, whereas others have a more general ability to move about and are specialized for different traits not related to swimming. Generalists can be expected to have bodies that give them moderately good performance in various special functions. Specialists perform exceptionally well in particular skills. Fast accelerating, braking, high-speed cruising, and complex maneuvering are obvious examples. Swimming economically, top-speed sprinting, making use of ground effect, backward swimming, swimming in sandy bottoms, precise position keeping, flying, and straight acceleration by recoil reduction are less apparent. The special swimming adaptations shown in **Figure 3** are only a few out of a wealth of possible examples. A closer study of the swimming habits of a large number of species will show many more specialist groups than the dozen or so described here.

Styles of Swimming

Most fish species swim with lateral body undulations running from head to tail; a minority use the movements of appendages to propel themselves.

The waves of curvature on the bodies of undulatory swimmers are caused by waves of muscle activations running toward the tail with a 180° phase shift between the left and right side. The muscular waves run faster than the waves of curvature, reflecting the interaction between the fish's body and the reactive forces from the water. The swimming speed varies between 0.5 and 0.9 times the backward speed of the waves of curvature during steady swimming. The wavelength of the body curvature of slender eel-like fish is about $0.6L$ ($L=$ body length), indicating that there is more than one wave on the body at any time. Fast-swimming fish such as mackerel and saithe have almost exactly one complete wave on the body and on short-bodied fish as carp and scup there is less than one wave on the length of the body during steady swimming. The maximum amplitude (defined as half the total lateral excursion) may increase toward the tail linearly, as in eels and lampreys, or according to a power function in other species. The increase in maximum amplitude is concentrated in the rearmost part of the body in fast fish like tuna. The maximum amplitude at the tail is usually in the order of $0.1L$ with considerable variation around that value. The period of the waves of curvature determines the tail beat frequency, which is linearly related to the swimming speed. The distance covered per tail beat is the 'stride length' of a fish. It varies greatly between species but also for each individual fish. Maximum values of more than one body length have been measured for mackerel; the least distance covered per beat of the same individual was $0.7L$. Many species reach values between $0.5L$ and $0.6L$ during steady swimming bouts.

Swimming with appendages includes pectoral fin swimming and median fin propulsion. Pectoral fin movements of, for example, labrids, shiner perches, and surfperches make an elegant impression. The beat cycle usually consists of three phases. During the abduction phase, the dorsal rays lead the movement away from the body and downward. The adduction phase brings the fin back to the body surface led by horizontal movement of the dorsal rays. During the third phase, the dorsal rays rotate close to the body back to their initial position. Stride lengths vary with speed and may reach more than one body length at optimal speeds.

Undulations of long dorsal and anal fins can propel fish forward and backward and are used in combination with movements of the pectoral fins and the tail. In triggerfish, for example, tail strokes aid during fast swimming and the pectorals help while maneuvering. There is usually more than one wave on each fin (up to 2.5 waves on the long dorsal fin of the African electric eel).

Interactions between Fish and Water: Fish Wakes

Every action of the fins or the body of a fish will, according to Newton's third law, result in an equal but opposite reaction from the surrounding water. A swimming fish produces forces in interaction with

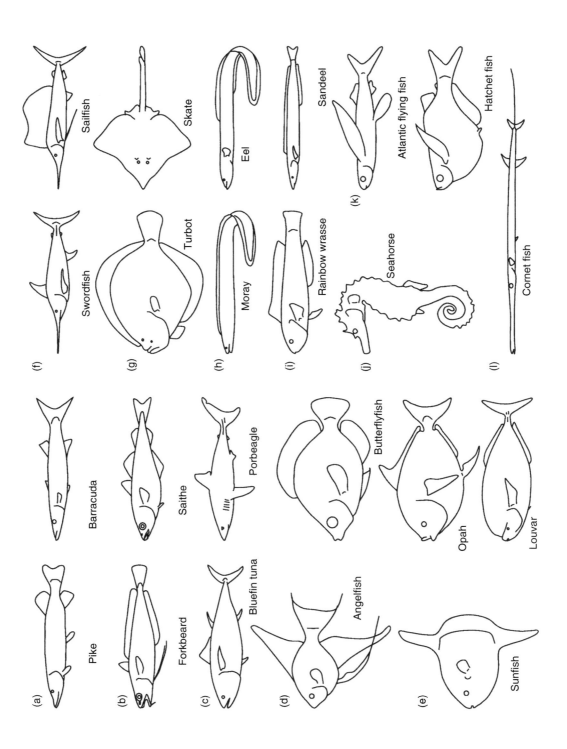

Figure 3 (a) Specialists in accelerating, such as the pike and the barracuda, are often ambush predators. They remain stationary or swim very slowly until a potential prey occurs within striking distance. Color patterns adapted to each specific environment provide the essential camouflage to make them hardly visible against the background. These species have a reasonably streamlined body and large dorsal and anal fins positioned extremely rearward, close to the caudal fin. Acceleration during the strike is caused by the first two beats of the tail, which is in effect enlarged by the rearward position of the dorsal and anal fins. The relative skin mass of the pike is reduced, compared with other fish, increasing the relative amount of muscles and decreasing the dead mass that has to be accelerated with the fish at each strike. Maximum acceleration rates measured for pike vary between 40 and 150 m s^{-2}, which equals 4–15 times the acceleration due to gravity ($g = 9.8$ m s^{-2}). The highest peak acceleration value reported for pike is 25 g. (b) Braking is difficult while moving in a fluid medium. Gadoids with multiple or long unpaired fins are good at it. The forkbeard swims fast and close to the bottom with elongated pelvic fins extended laterally for the detection of bottom-dwelling shrimp. The fish instantly spreads the long dorsal and anal fins and throws its body into an S-shape when a prey item is touched. Braking is so effective that the shrimp has not yet reached the caudal peduncle before the fish has stopped and turned to catch it. Fish use the unpaired fins and tail, usually in combination with the pectoral and pelvic fins, for braking. In the process, the fin rays of the tail fin are actively bent forward. The highest deceleration rate measured is 8.7 m s^{-2} for saithe. The contribution of the pectorals to the braking force is about 30%; the rest comes from the curved body and extended median fins. (c) Cruising specialists migrate over long distances, swimming continuously at a fair speed. Many are found among scombrids and pelagic sharks, for example. Cruisers have highly streamlined bodies, narrow caudal peduncles with keels, and high-aspect-ratio tails (aspect ratio being the tail height squared divided by tail surface area). The bluefin tuna, for example, crosses the Atlantic twice a year. The body dimensions are very close to the optimum values, with a thickness-to-length ratio near 0.25. Cruising speeds of 3-m-long bluefin tunas measured in large enclosures reached 1.2L per second (260 km d^{-1}). (d) Angelfish and butterflyfish are maneuvering experts with short bodies with high dorsal and anal fins. Species of this guild live in spatially complex environments. Coral reefs and freshwater systems with dense vegetation require precise maneuvers at low speed. Short, high bodies make very short turning circles possible. Angelfish make turns with a radius of 0.065L. For comparison, the turning radius of a cruising specialist is in the order of 0.5L, an order of magnitude larger. (e) Sunfish, opah, and louvar are among the most peculiar fish in the ocean. They look very different but have large body sizes in common. The sunfish reaches 4 m and 1500 kg, the opah may weigh up to 270 kg, and the louvar is relatively small with a maximum length of 1.9 m and weight of 140 kg. Little is known about the mechanics of their locomotion. They all seem to swim slowly over large distances. The opah will use its wing-shaped pectorals predominantly and the louvar has a narrow caudal peduncle and an elegant high-aspect-ratio tail similar to those of the tunas. The sunfish has no proper tail but the dorsal and ventral fins together are an extremely high-aspect-ratio propeller. Sunfish swim very steadily, moving the dorsal and ventral fins simultaneously to the left and, half a cycle later, to the right side. The dorsal and ventral fins have an aerodynamic profile in cross section. The intrinsic fin muscles fill the main part of the body and insert on separate fin rays, enabling the sunfish to control the movements, camber, and profile of its fins with great precision. Although there are no measurements to prove this as yet, it looks as though these heavy species specialize in slow steady swimming at low cost. Inertia helps them to keep up a uniform speed, while their well-designed propulsive fins generate just enough thrust to balance the drag as efficiently as possible. (f) Swordfishes (Xiphiidae) and billfishes (Istiophoridae) show bodily features that no other fish has: the extensions of the upper jaws, the swords, and the shape of the head. They are probably able to swim briefly at speeds exceeding those of all other nektonic animals, reaching values of well over 100 km h^{-1}. The sword of swordfish is dorsoventrally flattened to form a long blade (up to 45% of the body length). The billfish (including sailfish, spearfish, and marlin) swords are pointed spikes, round in cross section and shorter (between 14% and 30% in adult fish, depending on species) than those of swordfish. All the swords have a rough surface, especially close to the point. The roughness decreases toward the head. One other unique bodily feature of the sword-bearing fishes is the concave head. At the base of the sword the thickness of the body increases rapidly with a hollow profile up to the point of greatest thickness of the body. The rough surface on the sword reduces the thickness of the boundary layer of water dragged along with the fish. This reduces drag. The concave head probably serves to avoid drag-enhancing large-scale turbulence. The caudal peduncle is dorsoventrally flattened, fitted with keels on both sides. These features and the extremely high-aspect-ratio tail blades with rearward-curved leading edges are hallmarks of very fast swimmers. (g) The shape of the body of flatfish and rays offers the opportunity of hiding in the boundary layer close to the seabed where speeds of currents are reduced. There is another possible advantage connected with a flat body shape. Both flatfish and rays can be observed swimming close to the bottom. These fish are negatively buoyant and like flying animals, must generate lift (a downwash in the flow) at the cost of induced drag to remain 'waterborne'. Swimming close to the ground could reduce the induced drag considerably, depending on the ratio between height off the ground and the span of the 'wings'. (h) Only a few species can swim both forward and backward. Eels, moray eels, and congers can quickly reverse the direction of the propulsive wave on the body and swim backward. The common feature of these fish is the extremely elongated flexible body. Swimming is usually not very fast; they prefer to swim close to the bottom and operate in muddy or maze-type environments. (i) Sand eels and rainbow wrasses sleep under a layer of sand, sand eels in daytime and rainbow wrasses during the night. Both species swim head-down into the sand using high-frequency low-amplitude oscillations of the tail. If the layer of sand is thick enough, the speed is not noticeably reduced. Body shapes are similar, that is, slender with a well-developed tail. The wrasses use their pectoral fins for routine swimming and move body and tail fin during escapes and to swim into the sand. (j) Most neutrally buoyant fish species are capable of hovering at one spot in the water column. Some species can hardly do anything else. Sea horses and pipefishes (Syngnathidae) rely on camouflage for protection from predators. They are capable of minute adjustments of the orientation of their body using high-frequency, low-amplitude movements of the pectoral and dorsal fins. (Sea horses are the only fish with a prehensile tail.) (k) Flying fish have exceptionally large pectoral fins to make gliding flights out of the water when chased by predators. Some species are four-winged because they use enlarged pelvic fins as well. The lower lobe of the caudal fin is elongated and remains beating the water during takeoff. Hatchet fishes (Gasteropelecidae) actually beat their pectoral fins in powered flight. The pectoral fins have extremely large intrinsic muscles originating on a greatly expanded pectoral girdle. The flying gurnards (Dactylopteridae) have tremendously enlarged pectoral fins, usually strikingly colored. There is still some dispute over their ability to use these large pectorals when airborne. Many more species occasionally or regularly leap out of the water but are not specially adapted to fly. (l) Cornet fishes (Fistulariidae) are predators of small fish in the littoral of tropical seas, most often seen above seagrass beds or sandy patches between coral reefs. They seem to have two tails. The first one is formed by a dorsal and anal fin and the second one is the real tail. Beyond the tail there is a long thin caudal filament. They hunt by dashing forward in one straight line without any side movements of the head, using large-amplitude strokes of the two tail fins and the trailing filament. It looks as though the double tail fin configuration with the trailing filament serves to allow fast acceleration without recoil movements of the head. Precise kinematic measurements are needed to provide evidence for this assumption. Based on Videler JJ (1993) *Fish Swimming*. London: Chapman and Hall with figures (b) added.

(a)

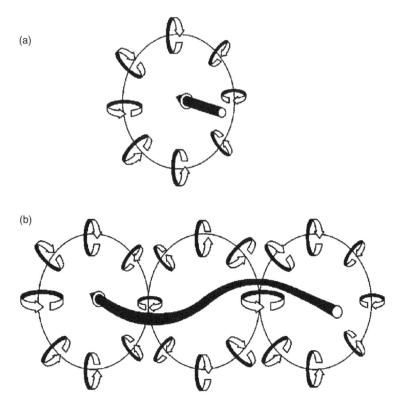

(b)

Figure 4 Schematic drawings of vortex ring structures. (a) A single vortex ring. The ring-shaped center of rotation is drawn as a line. The rotations of the vortex ring structure draw a jet of water through the center of the ring, indicated by the arrow. (b) A three-dimensional reconstruction of a chain of three connected vortex rings. A resulting jet of fluid undulates through the center of the vortex rings building the chain. From Videler JJ, Stamhuis EJ, Müller UK, and van Duren LA (2002) The scaling and structure of aquatic animal wakes. *Integrative and Comparative Biology* 42: 988–996.

the water by changing water velocities locally. The velocity gradients induce vortices, being rotational movements of the fluid. Vortices may either end at the boundary of the fluid or may form closed loops or vortex rings with a jet of water through the center. Vortex rings can merge to form chains (**Figure 4**).

Quantitative flow visualization techniques have been successfully applied to reveal the flow patterns near fish using body undulations to propel themselves. The interaction between undulating bodies and moving fins and the water results in complex flow patterns along and behind the swimming animals. A schematic three-dimensional impression of the wake generated by the tail behind a steadily swimming fish is shown in **Figure 5**. This shows the dorsal and ventral tip vortices generated during the tail beat as well as the vertical stop–start vortices left behind by the trailing edge of the tail at the end of each half-stroke. During a half-stroke, there is a pressure difference between the leading side of the fin and the trailing side. Dorsal and ventral tip vortices represent the water escaping at the fin tips from the leading side, with high pressure to the trailing side where the pressure is low. At the end of the

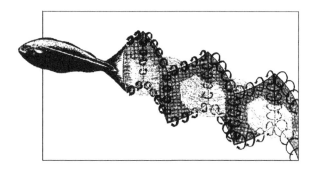

Figure 5 Artist's impression of the flow behind a steadily swimming saithe. The tail blade is moving to the left and in the middle of the stroke. At the end of each half-stroke, a column vortex is left behind when the tail blade changes direction. Tail-tip vortices are shed dorsally and ventrally when the tail moves from side to side. Together the vortices form a chain of vortex rings (as shown schematically in **Figure 4**) with a jet of water winding through the centers of the rings in the opposite swimming direction. Reproduced from Videler JJ (1993) *Fish Swimming*. London: Chapman and Hall.

half-stroke the tail changes direction and builds up high pressure on the opposite side of the fin, leaving the previous pressure difference behind as a vertical vortex column. These vertical, dorsal, and ventral

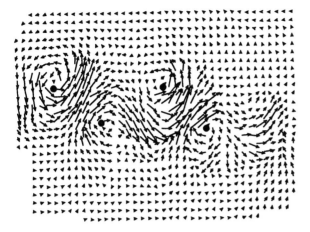

Figure 6 The wake of a continuously swimming mullet. The arrows represent the flow velocity in $mm\,s^{-1}$ scaled relative to the field of view of 195 mm × 175 mm. The shaded circles indicate the centers of the column vortices. The picture represents a horizontal cross section halfway down the tail through the wake drawn in **Figure 5**. Based on Müller UK, van den Heuvel BLE, Stamhuis EJ, and Videler JJ (1997) Fish foot prints: Morphology and energetics of the wake behind a continuously swimming mullet (*Chelon labrosus* Risso). *Journal of Experimental Biology* 200: 2893–2906.

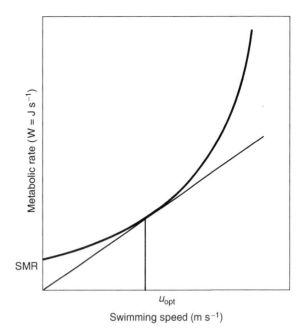

Figure 7 A theoretical curve of the rate of work as a function of swimming speed. SMR is the standard or resting metabolic rate at speed 0. The amount of work per unit distance covered ($J\,m^{-1}$) is at a minimum at u_{opt}. Reproduced from Videler JJ (1993) *Fish Swimming*. London: Chapman and Hall.

The Energy Required for Swimming

Swimming fish use oxygen to burn fuel to power their muscles. Carbohydrates, fat, and proteins are the common substrates. A mixture of these provides about 20 J per ml oxygen used. Measurements of energy consumption during swimming are mainly based on records of oxygen depletion in a water tunnel respirometer. Respiration increases with swimming speed, body mass, and temperature and varies considerably between species. The highest levels of energy consumption measured in fish are about $4\,W\,kg^{-1}$. Fast streamlined fish can increase their metabolic rates up to 10 times resting levels during swimming at the highest sustainable speeds. Short-burst speeds powered by anaerobic white muscles can cost as much as 100 times resting rates. Most of the energy during swimming at a constant speed is required to generate enough thrust to overcome drag. The drag on a steadily swimming fish is proportional to the square of the swimming speed; the energy required increases as the cube of the speed. In other words, if a fish wants to swim twice as fast it will have to overcome four times as much drag and use eight times as much energy. A fair comparison of the energy used requires standardization of the speed at which the comparison is made. The energetic cost of swimming is the sum of the resting or standard metabolic rate and the energy required to produce thrust. Expressed in watts (joules per second), it increases as a J-shaped curve with speed in meters per second (**Figure 7**). The exact shape of the curve depends mainly on the species, size, temperature, and condition of the fish. Owing to the shape of the curve there is one optimum speed at which the ratio of metabolic rate over speed reaches a minimum. This ratio represents the amount of work a fish has to do to cover 1 m ($J\,s^{-1}$ divided by $m\,s^{-1}$). To make fair comparisons possible, the optimum speed (u_{opt}), where the amount of energy used per unit distance covered is at a minimum, is used as a benchmark. Series of measurements of oxygen consumption at a range of speeds provide the parameters needed to calculate u_{opt} and the energy used at that speed. The energy values are normalized by dividing the active metabolic rate at u_{opt} (in $W = Js^{-1} = wt\,m\,s^{-1}$) by the weight of the fish (in newtons (N)) times u_{opt} (in $m\,s^{-1}$), to reach a dimensionless number for the cost of transport (COT expressed in $JN^{-1}m^{-1}$). Hence, COT represents the cost to transport 1 unit of weight over 1 unit of distance.

Available data show that u_{opt} is positively correlated with mass (proportional to mass$^{0.17}$); u_{opt} decreases with mass if it is expressed in L per second (proportional to mass$^{-0.14}$). The variation measured is large but $2L$ per second can serve as a reasonable first estimate of the optimum speed in fish. At u_{opt} the COT values are negatively correlated with body mass with an exponent of -0.38 (**Figure 8**). Fish use on average $0.07\,JN^{-1}$ to swim their body length at

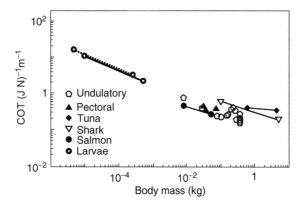

Figure 8 Doubly logarithmic plot of dimensionless COT, being the energy needed to transport 1 unit of mass over 1 unit of distance ($JN^{-1}m^{-1}$) during swimming at u_{opt}, related to body mass. The connected points indicate series of measurements of animal groups indicated separately; 'undulatory' and 'pectoral' refer to measurements of fish using body plus tail and pectoral fins, respectively, for propulsion. Reproduced from Videler JJ (1993) *Fish Swimming*. London: Chapman and Hall.

u_{opt}. If the weight and the size of the animals are taken into account as well by calculating the energy needed to transport the body weight over the length, the amount of energy used to swim at u_{opt} increases in proportion to body mass with an exponent of 0.93 (**Figure 9**).

Energy-Saving Swimming Behaviors

Burst-and-coast (or kick-and-glide) swimming behavior is commonly used by several species. It consists of cyclic bursts of swimming movements followed by a coast phase in which the body is kept motionless and straight. The velocity curve in **Figure 10** shows how the burst phase starts off at an initial velocity (u_i) lower than the average velocity (u_c). During a burst the fish accelerates to a final velocity (u_f), higher than u_c. The cycle is completed when velocity u_i is reached at the end of the deceleration during the coast phase. Energy savings in the order of 50% are predicted if burst-and-coast swimming is used during slow and high swimming speeds instead of steady swimming at the same average speed. The model predictions are based on a threefold difference in drag between a rigid body and an actively moving fish.

Schooling behavior probably has energy-saving effects. **Figure 5** – the wake of a steadily swimming – fish shows an undulating jet of water in the opposite swimming direction through a chain of vortex rings. Just outside this system, water will move in the swimming direction. Theoretically, following fish could make use of this forward component to facilitate their propulsive efforts. One would expect fish in a school to swim in a distinct three-dimensional spatial configuration in which bearing and

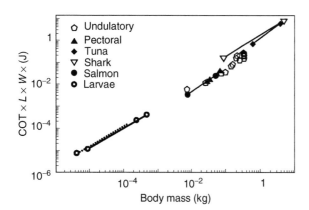

Figure 9 Doubly logarithmic plot of the energy needed by a swimming fish to transport its body weight over its body length as a function of body mass. Symbols as in **Figure 7**. Based on Videler JJ (1993) *Fish Swimming*. London: Chapman and Hall.

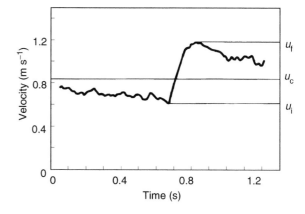

Figure 10 Part of a velocity curve during burst-and-coast swimming of cod. The average speed u_c was $3.2L$ per second. The initial speed and the final velocity of the acceleration phase are indicated as u_i and u_f, respectively. Reproduced from Videler JJ (1993) *Fish Swimming*. London: Chapman and Hall.

distance among school members showed a distinct constant diamond lattice pattern and a fixed phase relationship among tail beat frequencies. This has not been confirmed by actual observations. However, energetic benefits for school members have been confirmed by indirect evidence. It has been observed that the tail beat frequency of schooling Pacific mackerel is reduced compared with solitary mackerel swimming at the same speed. In schools of sea bass, trailing individuals used 9–14% lower tail beat frequencies than fish in leading position. There is also some evidence showing that fast swimming fish in a school use less oxygen than the same number of individuals would use in total in solitary swimming at the same speed.

Swimming Speed and Endurance

Maximum swimming speeds of fish are ecologically important for obvious reasons. However, slower swimming speeds and the stamina at these speeds represent equally important survival values for a fish. **Figure 11** relates swimming speed, endurance, and the cost of swimming for a 0.18-m sockeye salmon at 15 °C. At low speeds, this fish can swim continuously without showing any signs of fatigue. The optimum speed u_{opt} is between 1 and $2L$ per second. Limited endurance can be measured at speeds higher than the maximum sustained speed (u_{ms}) of somewhat less than $3L$ per second. For these prolonged speeds, the logarithm of the time to fatigue decreases linearly with increasing velocity up to the maximum prolonged speed (u_{mp}) where the endurance is reduced to

a fraction of a minute. Along this endurance trajectory, the fish will switch gradually from partly aerobic to totally anaerobic metabolism. The maximum burst speed in this case is in the order of $7L$ per second.

A comparison of published data reveals that u_{ms} for fish varying in size between 5 and 54 cm is in the order of $3L$ per second and that u_{mp} is about twice that value. Fifteen-centimeter carp are capable to swim at $4.6L$ per second for 60 min and at $7.8L$ per second during 0.2 min. Within a single species, u_{ms} and u_{mp} expressed in L per second decline with increasing length, and so does endurance.

Demersal fish living in complex environments have shallower endurance curves than pelagic long-distance swimmers, which fatigue more quickly when they break the limit of the maximum sustained speed. Endurance in fish swimming at prolonged speeds is limited by the oxygen uptake capacity. Higher speeds cause serious oxygen debts.

The maximum burst speed in meters per second increases linearly with body length; the slope is about 3 times as steep as that of the maximum sustained and prolonged speeds. The average relative value is about $10L$ per second, a figure that turns out to be a fairly good estimate for fish between 10 and 20 cm long. Small fish larvae swim at up to $60L$ per second during startle response bursts. Speed record holders in meters per second are to be found among the largest fish. Unfortunately, reliable measurements are not usually available. The maximum burst speed of fish depends on the fastest twitch contraction time of the white lateral muscles. For each tail beat, the muscles on the right and on the left have to contract once. Hence the maximum tail beat frequency is the inverse of twice the minimum contraction time. The burst speed is found by multiplying the stride length and the maximum tail beat frequency. Muscle twitch contraction times halve for each 10 °C temperature rise and the burst speed doubles. Larger fish of the same species have slower white muscles than smaller individuals. The burst swimming speed decreases with size with a factor of on average 0.89 for each 10-cm length increase. Estimates based on muscle twitch contraction times and measured stride length data for 226-cm-long bluefin tuna vary between 15 and $23\,\mathrm{m\,s^{-1}}$ (54–83 km h^{-1}). Estimates for 3-m-long swordfish exceed $30\,\mathrm{m\,s^{-1}}$ (108 km h^{-1}). Measured values for burst speeds are difficult to find. The maximum swimming speed ever recorded in captivity is that of a 30-cm mackerel swimming at $5.5\,\mathrm{m\,s^{-1}}$ (20 km h^{-1}). Its tail beat frequency was 18 Hz and the stride length $1L$.

The relationship between swimming speed and endurance is not straightforward due to the separate

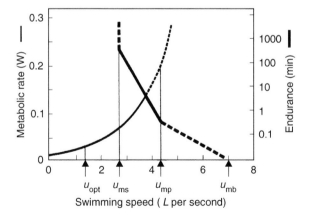

Figure 11 The metabolic rate (linear scale) and the endurance (logarithmic scale) of a 0.18-m, 0.05-kg sockeye salmon as functions of swimming speed in L per second. The water temperature was 15 °C. The optimum swimming speed (u_{opt}), the maximum sustained speed (u_{ms}), the maximum prolonged speed (u_{mp}), and an estimate of the maximum burst speed (u_{mb}) are indicated. Reproduced from Videler JJ (1993) *Fish Swimming*. London: Chapman and Hall.

use of red, intermediate, and white muscle. Virtually inexhaustible red muscles drive slow cruising speeds, while burst speeds require all-out contraction of white muscles lasting only a few seconds. Endurance decreases rapidly when speeds above cruising speeds are swum at. The variation in performance between species is large, and details have been hardly investigated so far.

See also

Fish Migration, Horizontal. Fish Migration, Vertical. Fish Predation and Mortality.

Further Reading

Domenici P and Blake RW (1997) The kinematics and performance of fish fast-start swimming. *Journal of Experimental Biology* 200: 1165–1178.

Greene CW and Greene CH (1913) The skeletal musculature of the king salmon. *Bulletin of US Bureau of Fisheries* 33: 25–59.

Müller UK, van den Heuvel BLE, Stamhuis EJ, and Videler JJ (1997) Fish foot prints: Morphology and energetics of the wake behind a continuously swimming mullet (*Chelon labrosus* Risso). *Journal of Experimental Biology* 200: 2893–2906.

Sakakibara J, Nakagawa M, and Yoshida M (2004) Stereo-PIV study of flow around a maneuvering fish. *Experiments in Fluids* 36: 282–293.

Shadwick RE and Lauder GV (eds.) *Fish Biomechanics*. San Diego, CA: Academic Press.

Shann EW (1914) On the nature of lateral muscle in teleostei. *Proceedings of the Zoological Society of London* 22: 319–337.

Videler JJ (1993) *Fish Swimming*. London: Chapman and Hall.

Videler JJ (1995) Body surface adaptations to boundary-layer dynamics. In: Ellington CP and Pedley TJ (eds.) *Biological Fluid Dynamics*, pp. 1–20. Cambridge, UK: The Society of Biologists.

Videler JJ, Stamhuis EJ, Müller UK, and van Duren LA (2002) The scaling and structure of aquatic animal wakes. *Integrative and Comparative Biology* 42: 988–996.

Wardle CS, Videler JJ, and Altringham JD (1995) Tuning in to fish swimming waves: Body form, swimming mode and muscle function. *Journal of Experimental Biology* 198: 1629–1636.

Relevant Website

http://www.fishbase.org
– FishBase.

FISH MIGRATION, HORIZONTAL

G. P. Arnold, Centre for Environment, Fisheries & Aquaculture Science, Suffolk, UK

Introduction

Unlike migration in insects, for which migration usually entails emigration in response to unfavorable environmental conditions, fish migration usually involves an annual migration circuit, which each individual sustains for several years, sometimes decades. In temperate and arctic waters, where fish commonly occupy separate feeding and spawning areas at different seasons, migration usually involves an annual two-way journey, often over considerable distances.

Occurrence of Migration

There are about 800 species of freshwater fish. A further 12 000 species live in the sea and about 120 species move regularly between the two environments. There are only about 200–300 species in polar latitudes but many more in temperate waters. The majority (>75%) occur in the tropics, with as many as 500 species on a single reef. Most species are confined to fairly limited areas and their movements are generally limited to distances of less than 50 km. Several hundred species, however, include populations that migrate between widely separated areas. Migrations are often spectacular, covering distances of several hundred to several thousand kilometres; the annual circuits of some oceanic species (e.g., bluefin tuna) extend to 10 000 km. Migratory species sustain many of the world's large commercial fisheries. In 1998 (the most recent year for which FAO data are available) the total world catch of marine finfish was nearly 65 million tonnes from over 740 species. Forty percent of this catch was derived from 17 species (**Table 1**), all of which are considered to be migratory.

Terminology

Diadromous species migrate between the sea and fresh water. Anadromous species feed at sea but return to fresh water to spawn. Catadromous species, in contrast, spawn in the sea, after completing most of their postlarval feeding and growth in fresh water. Amphidromous species spawn in fresh water but feed in both environments. The larvae of amphidromous species emigrate soon after hatching and early feeding takes place in the sea; post-larvae or juveniles

Table 1 Nominal world catches of the 17 most valuable species of marine finfish in 1998. Catches of these species comprised 40% of the total world catch of all marine finfish, which was 64.8×10^6 tonnes. Nominal world catches of diadromous species amounted to a further 1.8×10^6 tonnes

Species	Catch (mt)	Catch (%)	Cumulative catch (%)
Alaska (walleye) pollock, *Theragra chalcogramma*	4049317	15.62	15.62
Atlantic herring, *Clupea harengus*	2419117	9.33	24.95
Japanese anchovy, *Engraulis japonicus*	2093888	8.08	33.02
Chilean jack mackerel, *Trachurus murphyi*	2025758	7.81	40.83
Chub mackerel, *Scomber japonicus*	1910254	7.37	48.20
Skipjack tuna, *Katsuwonus pelamis*	1850487	7.14	55.34
Anchoveta (Peruvian anchovy), *Engraulis ringens*	1729064	6.67	62.01
Largehead hairtail, *Trichiurus lepturus*	1409704	5.44	67.44
Atlantic cod, *Gadus morhua*	1214470	4.68	72.13
Blue whiting, *Micromesistius poutassou*	1191184	4.59	76.72
Yellowfin tuna, *Thunnus albacares*	1152586	4.45	81.17
Capelin, *Mallotus villosus*	988033	3.81	84.98
European pilchard (sardine), *sardina pilchardus*	940727	3.63	88.61
South American pilchard, *Sardinops sagax*	937269	3.61	92.22
European sprat, *Sprattus sprattus*	696243	2.69	94.91
Round sardinella *Sardinella aurita*	663578	2.56	97.47
Atlantic mackerel *Scomber scombrus*	657278	2.53	100
Total catch of 17 most valuable species	**25928957**		

Data from Tables A-1(a) and A-1(e) of *FAO Yearbook*, Fishery Statistics, vol. 86/1, Rome, 2000).

subsequently return to fresh water, where they feed, grow, and mature. The migrations of oceanodromous and potamodromous species are limited to the sea and fresh water, respectively.

Ecology

Life histories of migratory fish are geared to regional production cycles. Most bony fishes (teleosts) produce large numbers of pelagic eggs, which are carried passively by the prevailing current, until they hatch as yolk-sac larvae. The yolk sac provides an initial reserve of food, but the larvae must find a sufficient density of planktonic food if they are to survive once the yolk is exhausted. In tropical latitudes, where standing stocks are low and production is continuous, there is probably always enough food to ensure survival of moderate broods of larval fish and spawning can occur more or less continuously. In temperate and polar latitudes, however, production is discontinuous and, although standing stocks are much larger and capable of sustaining very large populations of fish, they are also much shorter-lived. In these regions, spawning is most successful when the larvae hatch at times of high food abundance. Eggs and early larvae are carried passively by the current, so reproduction is also most successful when spawning takes place upstream of a suitable nursery ground. Productive upwelling areas provide similar feeding opportunities for large stocks of fish in tropical and subtropical waters. The season is longer but production moves poleward during the summer, so here, too, reproductive success favors those species whose eggs hatch in the right place at the right time.

In many species spawning areas and spawning grounds are well defined and persist for many decades, possibly centuries. After recruiting to the adult stock, fish generally home to the same spawning ground, even though this may not be where they themselves were spawned. Most fish spawn annually through adult life, which may last several decades in unfished populations. In homing, many fish follow regular migration routes, which take them between their feeding and overwintering grounds and back to their spawning grounds. In some species (e.g., cod and herring), new recruits probably learn the migration route by accompanying the older fish on their way to the spawning grounds. Homing ensures that the adults return to a location from which it is probable that eggs and larvae will be carried to favourable nursery grounds at the start of each new generation. Spawning migrations compensate for the drift of eggs and larvae, and migratory fish stocks tend therefore to be contained within oceanic gyres. Fish that stray outside the gyre are generally lost to the parent stock.

Typical Life Histories

Anadromous Species

Anadromous species are found in a wide range of families that includes the Petromyzontidae (lampreys), Acipenseridae (sturgeons), Osmeridae (northern smelts), Retropinnidae (southern smelts), and Clupeidae (shads but not herrings). Salmon are by far the best known, with examples in both Atlantic (*Salmo*) and Pacific (*Oncoryhnchus*) genera.

Salmon typically lay their eggs in redds excavated in the gravel of a headwater stream. The alevins that hatch from the eggs remain in the gravel for some weeks before emerging as parr. Most Atlantic salmon (*S. salar*) parr spend up to 3 years in fresh water, before emigrating to sea as smolts, as do most Pacific species. In pink (*O. gorbuscha*) and chum (*O. keta*) salmon, however, the juveniles migrate to sea almost immediately after hatching. Most species spend several years feeding and growing at sea before they return to fresh water to spawn. Atlantic salmon spend 1–5 years at sea and fish from some European stocks range as far afield as West Greenland, where they mix with fish from the eastern seaboard of Canada and the United States; others go to the Norwegian Sea. Pacific salmon from North America migrate north to Alaska and then westward along the Aleutian Chain before moving out into the open ocean, where they mix with fish from Asia. Chinook salmon (*O. tshawytscha*) spend 5–6 years at sea, during which they probably make several circuits of the Gulf of Alaska before returning to US and Canadian rivers. Most adult salmon return to the river and stream in which they were themselves spawned. Once in coastal or estuarine waters, they find their way back by olfaction, following a specific homestream odor, which appears to be produced by other fish of the same species. Most Pacific salmon die immediately after spawning; some Atlantic salmon return to the sea as kelts to spawn again after a further period of feeding in the sea.

Catadromous Species

Catadromy also occurs in a wide range of families that includes the Anguillidae (eels), Mugilidae (mullets), Galaxiidae (galaxiids and southern whitebaits), Cottidae (sculpins), Gobiidae (gobies) and Pleuronectidae (flounders). The life histories of anguillid eels, of which there are 15 species, are probably the best known.

Anguillid eels spawn at sea, usually in tropical to subtropical waters and often over great oceanic depths; they die after spawning. The eggs hatch as distinctive, leaf-shaped (leptocephalus) larvae, which have ferocious teeth and feed for one or more years as they are transported by ocean currents back to the continent from which their parents originated. The leptocephali metamorphose when they reach coastal waters and turn first into active, transparent, eel-shaped glass eels and then (up to a year later) into pigmented bottom-living elvers. Like salmon, elvers seek out fresh water by olfaction, although, unlike salmon, they appear to be attracted by the mix of odors produced by decaying detritus and associated microorganisms, rather than the odor of their conspecifics. Although there are substantial coastal eel populations, elvers enter fresh water in vast numbers. They move large distances upstream, disperse, grow, and become yellow eels, which spend up to 20 years in fresh water before reaching sexual maturity. A series of physical and physiological changes then turns the resident yellow eels into migratory silver eels, which move downstream to sea at night in the autumn, primarily during the first and third lunar quarters (the 'dark of the moon'). Feeding ceases, the gut atrophies, and silver eels rely on high fat reserves during their oceanic migration. Eye pigmentation also changes, so that peak spectral sensitivity shifts to the blue part of the spectrum, typical of clear oceanic waters. There are also changes in gas gland morphology, which allow the swimbladder to function in deep water during the oceanic spawning migration.

European (*Anguilla anguilla*) and American (*A. rostrata*) eels spawn in the Sargasso Sea within the Subtropical Convergence Zone, but in different locations. Spawning appears to be associated with areas of thermal density fronts. Some feature of southern Sargasso Sea water, possibly odor, may serve as a signal for returning silver eels to stop migrating and start spawning. Leptocephali of both species appear to be carried away from the spawning areas by gyres in the south-western Sargasso Sea, an Antilles Current, and the Florida Current north of the Bahamas. Thereafter they drift in the Gulf Stream and North Atlantic Current until they reach the American or European continental shelf. American eel leptocephali are found mostly in the west; European eel leptocephali are found all across the North Atlantic Current and its branches. The distribution of the larvae by size indicates that leptocephali take between 1.5 and 2.5 years to cross the North Atlantic. Despite some suggestions to the contrary, there are no reliable data to suggest that anything other than passive drift is involved. There may also be a more southerly migration route, originating in jetlike currents at the fronts in the western Sargasso Sea, which transport leptocephali toward the northern Canary Basin.

Amphidromous Species

The Japanese ayu (*Plecoglossus altivelis*), a small freshwater salmoniform, is a typical amphidromous species of the Northern Hemisphere. Demersal eggs are spawned in fresh water during autumn. On hatching, the larvae emigrate to sea, where they live for several months, before returning to fresh water during spring at a size of 50–60 mm. The fish grow and mature over the summer, before spawning and dying the following autumn. A similar life history is evident in a number of galaxiids in the Southern Hemisphere.

Oceanodromous Species

Life histories vary between oceanic and continental shelf species. The migration circuits of several species of tuna span whole ocean basins in both Northern and Southern hemispheres. The albacore tuna (*Thunnus alalunga*), for example, ranges over the entire North Pacific during its life (**Figure 1**); the southern bluefin tuna (*T. maccoyii*) occurs in both Pacific and Indian Oceans and may possibly make circumpolar migrations. Scombrids, which are closely related to tunas, are also migratory. The Western Stock of the Atlantic mackerel (*Scomber scombrus*), for example, which spawns over a large area of the Bay of Biscay along the edge of the continental slope, migrates to and from the Norwegian Sea, where it feeds in summer (**Figure 2**). Its migrations are possibly related to the northward-flowing European Ocean Margin Continental Slope Current.

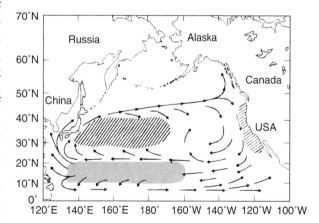

Figure 1 Distribution of albacore (*Thunnus alalunga*) in the North Pacific. The areas of the American fishery on young albacore and the Japanese fishery on older albacore are cross-hatched; the area in which the adults are believed to spawn is indicated by stippling. The main features of the subtropical gyre (Kuroshio, California, and North Equatorial Currents) and the North Equatorial Counter Current are shown by arrows. (From Harden Jones in Aidley (1980), with permission of Blackwell Science.)

During the 1980s large, dense shoals of mackerel overwintered off the coast of Cornwall, where they formed the basis of a large fishery. Subsequently this behavior has changed and for unknown reasons the mackerel now overwinter much farther north.

The herring (*Clupea harengus*), another pelagic species, has many migratory populations. The Atlanto-Scandian herring, for example, which spawns off the southern coast of Norway in the spring, used to range through the Norwegian, Greenland, and Icelandic Seas, feeding at the productive polar front and wintering between Iceland and the Faeroes, along the boundary of the East Icelandic current. Following overfishing, the migration circuit changed and the stock now over-winters in Ofotfjord and Tysfjord, near Lofoten, where fish collect in immense, dense, and largely inactive shoals. The three stocks of North Sea herring – Buchan, Dogger and Downs – make similar, although

much less extensive, migrations on the European continental shelf (**Figure 3**).

Many demersal species also have migratory populations. The Alaskan (walleye) pollack (*Theragra chalcogramma*), for example, the second most valuable species in the world (**Table 1**), occurs along the continental shelves of Asia and North America. In the Bering Sea, spawning occurs close to the Aleutian Islands in spring and summer, when dense shoals accumulate to the north of Unimak Pass (**Figure 4**). Spawning occurs between March and June, with a peak in May. Pelagic eggs, larvae, and young fish are distributed across the continental shelf by the local current system. At the end of the autumn the distribution contracts and with the onset of winter the pollack return to the deeper waters on the edge of the shelf. The migrations of the adult fish appear to be related to the Bering Sea Slope Current (**Figure 4**),

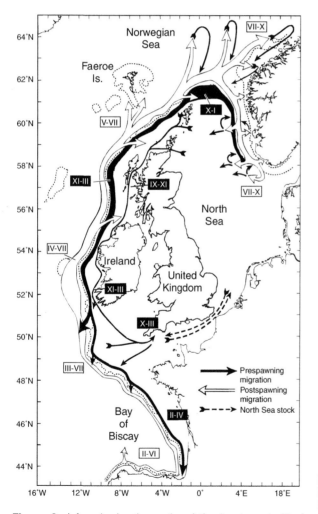

Figure 2 Inferred migration paths of the 'western stock' of Atlantic mackerel (*Scomber scombrus*). The Roman numerals indicate the months during which the fish are present in each area. (From SEFOS, Final Report to EU AIR Programme, May 1997.)

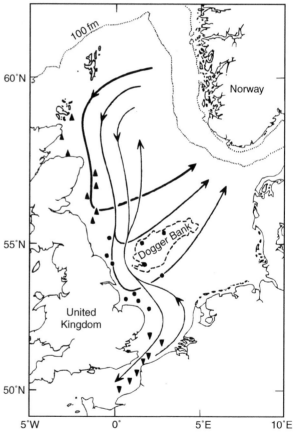

Figure 3 Migration routes and spawning grounds of three herring (*Clupea harengus*) stocks in the North Sea. (▲) northern North Sea (Buchan) summer spawners; (●) middle North Sea (Dogger) autumn spawners; (▼) Southern Bight and English Channel (Downs) winter spawners. The three groups share a common feeding ground in the northern North Sea. The Dogger Bank is shown by the dashed line. (From Harden Jones (1968, 1980), with permission.)

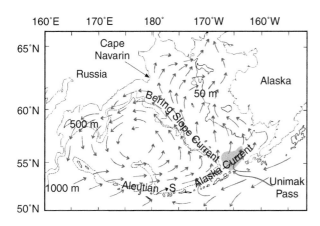

Figure 4 The spawning area (hatched) of Alaskan (walleye) pollack (*Theragra chalcogramma*) and the main surface currents in the Bering Sea. The migrations of older fish are related to the Bering Sea Slope Current and the distribution of larvae to the West Alaska Current. (From Harden Jones in Aidley (1981), with permission of Cambridge University Press.)

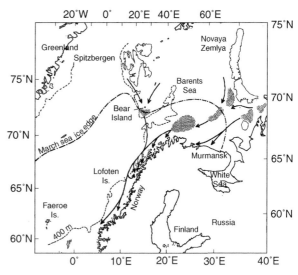

Figure 5 Autumn concentrations (stippled) and winter migrations of Arcto-Norwegian cod (*Gadus morhua*). (From Harden Jones (1968), with permission.)

which flows north-west from the Aleutian Islands toward Cape Navarin in Russia.

The cod (*Gadus morhua*) has a comparable distribution in the North Atlantic, with separate stocks off the coasts of New England, Newfoundland, Labrador, West Greenland, East Greenland, Iceland, Faeroe. There are also stocks at Faeroe Bank and in the Irish Sea, North Sea, Baltic Sea, and Barents Sea. Greenland stocks are at the northern limit of their range and only produce large year-classes during the negative phase of the North Atlantic Oscillation (NAO) when eggs and larvae are carried to Greenland from spawning grounds off south-east Iceland. Juvenile cod can then reach Greenland in large numbers because of the increased flow of the Irminger Current produced by consistent easterly winds, a feature of the negative phase of the NAO at these latitudes. Large year-classes of cod at Greenland can result in dramatic increases in the fishery; in 1989, for example, catches at West Greenland exceeded 100 000 tonnes.

The Arcto-Norwegian cod stock feeds in summer over a large area of the Barents Sea between Spitsbergen and Novaya Zemlya (**Figure 5**). Large shoals collect in winter off northern Norway and at Bear Island, before migrating to coastal spawning grounds between the Murman coast of Russia and Romsdal in southern Norway. The West Fjord in the Lofoten Islands is the most important. After spawning, the spent fish return to the north, followed by the eggs and larvae (**Figure 6**), which are carried to the Barents Sea by the Norwegian Coastal Current (NCC). The depth at which the adult fish swim is unknown, but it is suggested that spent cod may occur near the surface, while maturing fish may

possibly occur in the south-going countercurrent that underlies the NCC.

The plaice (*Pleuronectes platessa*) is a typical flatfish, which is found throughout the shallow coastal waters of Northern Europe from the White Sea to the Mediterranean. In the North Sea there are four stocks, which spawn off the Scottish East Coast, Flamborough Head, and in the Southern and German Bights, respectively. Fish in the southern North Sea are highly migratory and a proportion of the Southern Bight stock migrates through the Dover Strait to spawn in the eastern English Channel,

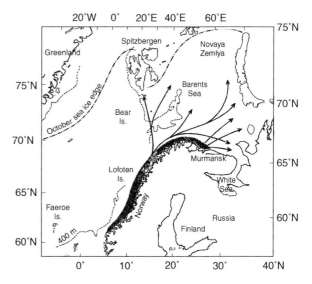

Figure 6 Spawning grounds (hatched) and postspawning migrations of Arcto-Norwegian cod (*Gadus morhua*). (From Harden Jones (1968), with permission.)

where about 60% of spawning fish are of North Sea origin. Mature plaice move 200–300 km south in November and December to spawn in the Southern Bight and eastern English Channel. Peak spawning occurs in January and spent fish return north in January, February, and March. Meanwhile, the pelagic eggs and larvae drift north-east with the residual current flowing from the Atlantic through the Channel to the North Sea. The larvae take about 5–8 weeks to reach metamorphosis, when they become miniature flatfish. Most take to the bottom along the coasts of Belgium and Holland and many enter the Dutch Wadden Sea. Juvenile plaice leave their coastal nursery grounds and move to deeper water from their second year of life. Some larger males reach first maturity in their third year and join mature fish of earlier year-classes on the spawning ground during the latter part of the spawning season. Most first-time spawners migrate down the eastern side of the Southern Bight; most repeat spawners migrate up and down the English coast.

Mechanisms of Migration

Water is a fluid environment that is much more resistant to movement than air and contains very much less oxygen (3–5%) in the same volume. As a result, fish generally swim rather slowly, except when feeding or avoiding predators, when they can attain quite high speeds. Because cruising speeds are low – usually less than 1 fish length per second – water currents can have a significant effect on the ground track of the fish. They are a major factor for eggs and newly hatched larvae and are often important for adult fish too. Migrating fish may swim downstream in the direction of the current and it is then not clear whether the current provides directional information as well as transport, or whether the fish has an independent system of navigation. To resolve this fundamental question, it is necessary to measure the speed and direction of the water and the fish at the same depth, a virtually impossible task until the advent of electronic tags, modern sonars, and split-beam echo sounders.

The Role of Ocean Currents

There is a prima facie case that the migration circuits of many fish follow the oceanic gyres. Several species of Pacific salmon, for example, spend several years in the Alaskan Gyre and the route they follow on returning to the Fraser River is affected by the position of the Sitka Eddy, which governs whether they pass north or south of Vancouver Island. Southern bluefin tuna may follow the West Wind Drift around the Southern Ocean. Albacore tuna, which spawn

between the Hawaiian Islands and the Philippines, appear to follow the clockwise movements of the Kuroshio, California, and North Equatorial Currents that make up the subtropical gyre in the North Pacific. The migrations of bluefin tuna, which spawn in the Gulf of Mexico and the Mediterranean, may also be linked to the subtropical gyre in the North Atlantic, although it is not yet known whether complete circuits of the basin are common. Two tagged bluefin, which in 1961 traveled a minimum distance of 4832 km from the Bahamas to Norway at an average speed of 40 km d^{-1}, almost certainly received substantial assistance from the Gulf Stream and its extension, the North Atlantic Drift. There are, however, as yet too few observations of the detailed movements of highly migratory fish to know whether they routinely follow the oceanic gyres, or only take advantage of the currents when it is energetically favorable.

The Role of Tidal Streams

Extensive fish tracking and midwater trawling studies in the North Sea have shown that several species of demersal fish – plaice, sole, flounder, cod, silver eels, and dogfish – make sophisticated use of the tidal streams (**Figure 7**) during their annual spawning migrations. Adult migrants show a 12-hour pattern of vertical movement related to the tides and juveniles of some species (e.g., plaice) show the same behavior. The fish leaves the seabed at one slackwater and spends about 6 h off the bottom, usually swimming in the upper half of the water column and often near the surface (**Figure 8**). It returns to the seabed around the time of the next slackwater and spends the ensuing tide on the bottom, not moving significantly (**Figure 9**). Flatfish bury in the sand during the adverse tide; roundfish probably refuge from the flow behind topographical features, where these are available. Fish using selective tidal stream transport progress rapidly in a consistent direction, determined by the choice of transporting tide. Prespawning and postspawning fish move in opposite directions and in the Dover Strait, for example; maturing and spent plaice can be caught in midwater on opposing tides in January.

Migration speed is determined by a number of factors, including the average speed of the tidal stream (commonly between 1 and 2 m s^{-1}), the size of the fish, and whether it uses all available tides. At various times, some fish use the transporting tide at night only; at other times, the same fish uses the transporting tide during the day as well. Fish using all transporting tides can move very rapidly (**Figure 10**) and visit more than one spawning area within a period of weeks. Vertical movements are usually well synchronized with times of local slackwater, although

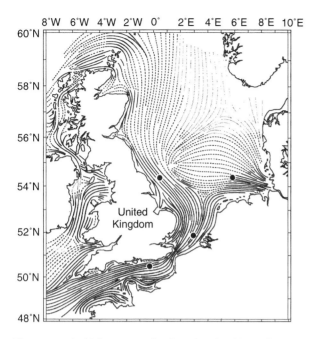

Figure 7 A tidal streampath chart for the North Sea and adjacent areas. The black circles indicate the centres of spawning of plaice off Flamborough and in the German Bight, Southern Bight, and eastern English Channel. (From Arnold in Aidley (1981) with permission of Cambridge University Press.)

occasionally the fish remains in midwater over the turn of tide and is carried back some distance in the opposite direction. Recent studies have shown that migrating plaice save energy by heading downtide within $\pm 60°$ of the tidal stream axis and swimming through the water at a speed of approximately 0.6 fish length per second. For a 35 cm female plaice the saving on a typical annual migration circuit of 560 km is equivalent to 30% of the energy content of the eggs it spawns each year. Other recent studies indicate that

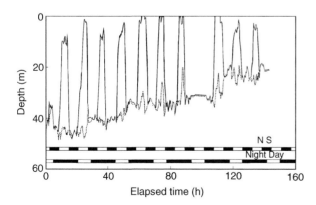

Figure 8 Vertical movements (solid line) of a 44 cm maturing female plaice tracked in the southern North Sea in January 1991. The dotted line indicates the depth of the seabed. The upper bar and lower bars indicate the direction of the tidal stream (north and south) and times of day and night, respectively.

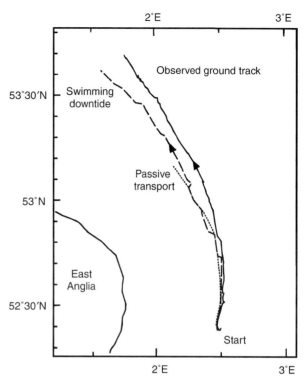

Figure 9 Geographical track of a 44 cm female plaice tracked in the southern North Sea in January 1991 (**Figure 8**). Also shown are computer simulations of how far the fish would have moved if it had been transported passively with the tide or had swum downtide at a speed of 1 fish length per second. The fish moved 152 km to the north in 6 days by selective tidal stream transport (British Crown Copyright).

plaice do not use tidal stream transport in areas of the North Sea where the speed of the tidal stream is insufficient for them to save energy. Instead they resort to other patterns of movement, which appear to be independent of the tidal streams.

Orientation and Navigation

There is ample evidence that fish can maintain a consistent heading over deep water in the open sea at night and thus obtain guidance from an external reference. The clue may well be geophysical in origin, possibly geomagnetic, although no fish has yet been shown to truly navigate, in the sense of knowing its geographical coordinates and adjusting its course to compensate for lateral displacement by a current. This ability has, however, recently been demonstrated in green turtles (*Chelonia mydas*), during their postspawning migration from Ascension Island to Brazil. Individual turtles maintained remarkably similar courses for the first 1000 km, following the South Atlantic Equatorial Current in a west-south-westerly direction, possibly using chemical clues to remain in it. Subsequently, however, they began to compensate for southward

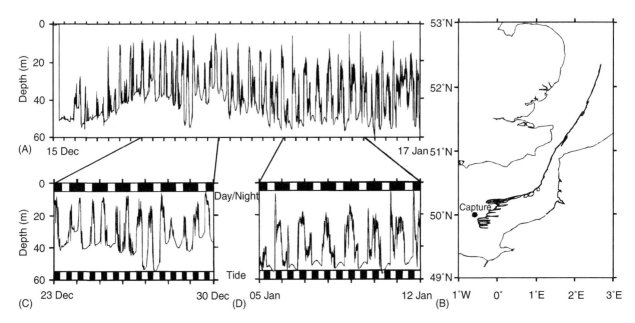

Figure 10   The reconstructed track of a 48 cm maturing female plaice fitted with a data storage tag, which recorded depth (pressure) at 10-minute and water temperature at 24-hour intervals. The track (B) was reconstructed from the vertical movements of the fish (A) and a computer simulation model of the tidal streams (**Figure 7**). During the first 7 days the fish showed selective tidal stream transport (C) and moved rapidly from the Southern Bight into the English Channel. During the remainder of the track the fish showed a diel pattern of vertical migration (D) and swam in midwater at night. It remained within the plaice spawning area in the eastern English Channel, where it was caught by a French trawler (British Crown Copyright).

displacement by the current and progressively corrected their course to head toward the easternmost part of the coast of Brazil. The data were obtained with radio tags, whose positions were determined by the Argos satellite system, which allows the movements of individual animals to be followed in detail over large distances with an indication of the accuracy of each estimated position. While ideal for marine animals like turtles, which spend most of the time at shallow depths and surface frequently, the Argos system cannot be used for fish, which, with the exception of a few unusual species, spend all their time submerged.

Discussion

Until the advent of electronic tags in the early 1970s, fisheries scientists had to rely on indirect information, such as catch rates and seasonal changes in distribution of fisheries, to deduce patterns of migration. Useful data could also be obtained from returns of conventional external tags, such as disks and streamers. Conventional tag data were, however, heavily biased by the distribution of fishing effort, which was often not well described. The returns could also not yield any information about the actual track of the fish between the release and recapture positions. In consequence, our knowledge about the mechanisms of fish migration and underlying sensory systems is based primarily on tracks of fish marked with acoustic tags and followed by research vessels. The technique has produced much useful information about migration on the continental shelves, but is not very suitable for the open ocean. It is also expensive, cannot be maintained for long periods, and does not readily permit the replication of observations. Increasingly, therefore, fisheries biologists are turning to archival (data storage) tags, which can record simple data such as depth, temperature, and light intensity for many months and retain them for several years. These data can be used to reconstruct the track of the fish in a variety of ways, depending on the behavior of the fish and the environment in which it lives. Archival tags can be deployed in large numbers and recovered through commercial or recreational fisheries. Alternatively, they can be programmed to detach from the fish after a pre-set interval and float to the surface, where they transmit data to a low-orbit satellite system, such as Argos. Although the data transmission capability of such systems is limited at present, the accelerating use of electronic tags is likely to lead to rapid advances in descriptions of the behavior of migrating fish and also an increased understanding of how they find their way around the oceans. Good descriptions of the migration circuits of the commercially exploited species should help to improve assessment, management, and conservation.

Conclusions

Fish spawn so that their eggs and larvae are carried to good feeding grounds for their juvenile stages. The adults must subsequently compensate for this drift, if the population is to sustain itself. As they grow, fish become less dependent on the environment and some may migrate without any reference to ocean currents at all. Adults of some species, however, still use the current if there is an energetic advantage in 'hitching a ride,' especially in shallow tidal seas. How fish find their way around the oceans and whether they can truly navigate, or only obtain guidance from local clues, is not yet known. There is, however, physiological evidence that fish do have a magnetic compass sense and behavioral evidence to suggest the involvement of geophysical, and perhaps also topographical clues. Rapid advances in understanding are to be expected in the near future with the increasing use of sophisticated electronic tags that allow the tracks of migrating fish to be described in detail over seasonal timescales and long distances.

See also

Eels. Fish Larvae. Fish Locomotion. Fish Migration, Vertical. Fish Predation and Mortality. Fish Reproduction. Fish Schooling. Pelagic Fishes. Salmonids.

Further Reading

Aidley DJ (ed.) (1981) *Animal Migration*. Cambridge: Cambridge University Press.

Harden Jones FR (1968) *Fish Migration*. London: Edward Arnold.

Harden Jones FR (1980) The Nekton: production and migration patterns. In: Barnes RK and Mann KH (eds.) *Fundamentals of Aquatic Ecosystems*, pp. 119–142. Oxford: Blackwell Science.

Lockwood SJ (1988) *The Mackerel. Its Biology, Assessment and the Management of a Fishery*. Farnham, Surrey: Fishing News Books Ltd.

McCleave JD (1993) Physical and behavioral controls on the oceanic distribution and migration of leptocephali. *Journal of Fish Biology* 43 (Supplement A): 243–273.

McDowall RM (1988) *Diadromy in Fishes*. London: Croom Helm.

Metcalfe JD and Arnold GP (1997) Tracking fish with electronic tags. *Nature London* 387: 665–666.

Metcalfe JD, Arnold GP, and McDowall RM (2001) Migration. In: Hart PJB and Reynolds JD (eds.) *Handbook of Fish and Fisheries*, Chap 9. Oxford: Blackwell Science.

Smith RJF (1985) *The Control of Fish Migration*. Berlin: Springer–Verlag.

Walker MM (1997) Structure and function of the vertebrate magnetic sense. *Nature. London* 390: 371–376.

FISH MIGRATION, VERTICAL

J. D. Neilson, Department of Fisheries and Oceans, New Brunswick, Canada
R. I. Perry, Department of Fisheries and Oceans, British Columbia, Canada

Introduction

While the often spectacular long-distance migrations of fish have been well described by the scientific community and are the source of considerable wonder to the general public, fish can also undertake migrations in the vertical dimension. Such migrations can occur at the earliest life history stages when fish are free-swimming, and can continue throughout their lives. Some species exhibit vertical migration behavior at certain stages of their life history, but not in others. In general, vertical migrations occur with 24-h periodicity and, to a lesser degree, display constancy in phase and amplitude. Attempts to explain the regularity of changes in depth distributions of fish have often involved the notion of circadian or endogenous rhythmicity. The term 'circadian' refers to a self-sustained rhythm of 24-h periodicity, either synchronized to a natural cyclic phenomenon or free-running. When migrations (or other events) occur with 24-h periodicity, they are said to be 'diel' in nature, in contrast to 'diurnal' and 'nocturnal', meaning day- and night-active, respectively.

Why study vertical migrations? From the ecological perspective, knowledge of vertical migrations advances our understanding of the interactions among fish, their predators and prey, and the abiotic environment. Vertical migrations are also mechanisms of energy transfer among the various depths of the ocean. Vertical migrations can also modify the horizontal distributions of fish by exposing populations to depth-specific variation in current strength. From the perspective of determining the number of fish in a population, ignoring vertical migrations can bias the interpretations of survey results. This is a particular problem for acoustic surveys that are unable to detect fish close to the bottom. If fish are periodically unable to be caught by the sampling gear, a potential bias may occur in the survey. Finally, from the perspective of a fishery, knowledge of vertical migration is critical in the choice and design of appropriate fishing gear and strategies.

Examples of Vertical Migration Patterns

Diel vertical migrations have been classified into two types. The first (type I, **Figure 1**) refers to the most common situation where fish move up in the water column at dusk, and down at dawn. As is the case in the example shown in **Figure 1**, some authors model the distribution of the fish in the water column using sinusoidal functions.

Type II vertical migrations are the converse, with the fish being found higher in the water column during daylight hours (**Figure 2**) and closer to the bottom at night. Of the two, examples of type II vertical migration have been less frequently documented (see Neilson and Perry (1990) for a tabulation of literature reporting these types of vertical migration patterns).

Variation in Patterns of Vertical Migration

Any classification of vertical migration runs the risk of oversimplification. Variation in patterns of vertical migrations can occur among individuals, locations, species, and seasons. Some examples of the rich diversity in patterns of vertical migration are provided below.

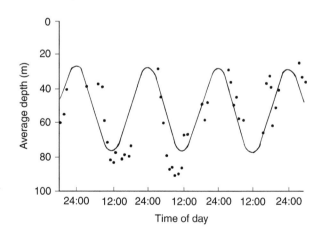

Figure 1 Vertical distribution of the sardine (*Sardina pilchardus*) in the Thracian Sea. The dots show the observed average depths, and the solid line shows the predicted average depth of the distribution according to a cosine function model based on the time of day. (Modified from Giannoulaki M, Machias A and Tsimenides N (1999) Ambient luminance and vertical migration of the sardine *Sardina pilchardus*. *Marine Ecology Progress Series* 178: 29–38.)

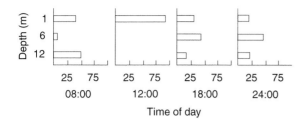

Figure 2 Vertical distribution of 15–21 mm gulf menhaden (*Brevoortia patronus*) larvae in the Gulf of Mexico expressed as percentages at four time intervals. (Modified from Sogard SM, Hoss DE and Govini JJ *et al.* (1987) Density and depth distribution of larval Gulf menhaden, Brevoortia patronus, Atlantic croaker, *Micropogonias undulatus*, and spot *Leiostomus xanthurus*, in the Northern Gulf of Mexico. *Fishery Bulletin*, 85: 601–609.)

Individual

Within a population, individuals that are larger than average often display a greater amplitude or range of depths during vertical migrations. In the case of Atlantic cod (*Gadus morhua*), for example, fish in their first year of life initially live in the pelagic zone. During this early stage, fish exhibit size-related vertical migrations, with larger individuals undertaking more extensive diel vertical migrations. Once these young fish reach a size of 80–100 mm, they become much more closely associated with the bottom. However, even when cod occur mostly on the bottom, they may still make vertical migrations whose amplitude appears positively related to their size. Many other species demonstrate this pattern of increasing range of vertical migration with increasing size, but few show the persistence across different life history stages illustrated by some Atlantic cod populations.

Population/Site

Variations in the pattern of vertical migration can occur because of differences in environmental conditions between locations. Site-specific variation in the pattern of vertical migrations occurs in populations of walleye pollock (*Theragra chalcogramma*) in the eastern Bering Sea, with differences related to the availability of their prey. For juvenile haddock on both sides of the Atlantic, site-specific differences in patterns of vertical migration are observed. In the western Atlantic, differences in vertical migration patterns are again related to the relative abundance of prey. In many cases, it has been shown that the presence of thermoclines or pycnoclines can modify patterns of vertical migration. Meteorological conditions at a site may also influence vertical migration, particularly for those species such as redfish (*Sebastes* spp.) for which light is an important controlling factor.

Seasonal

Seasonal changes in patterns of vertical migration are shown by adult Atlantic cod in the Gulf of St. Lawrence. This population shows a type I vertical migration from mid-July to September. However, earlier in the year, a type II vertical migration is present. Similarly, larval herring in the Gulf of Finland occur in relatively deep water by day and closer to the surface at night during early summer, but the behavior is reversed later in the summer. The effects of increasing size and age are related to seasonal effects, and there are numerous studies demonstrating changes in patterns of vertical migration as fish develop.

Genus/Species

There is considerable intergeneric, interspecific, and even individual variation in patterns of vertical migration within a Pacific community of co-occurring species of myctophids (lanternfishes; **Figure 3**). In this example, most of the members of the myctophid community exhibit a constant type I vertical migration, with variation in the vertical extent of the migration. One species (referred to as a semimigrant) shows a facultative vertical migration, with part of the population remaining relatively deep in the water column regardless of the time of day. In this community, vertical migration appears to be a nighttime feeding strategy, with most species undertaking movement into the more productive upper portions of the water column. The semimigrants may be exhibiting an energy-conserving behavior, in which satiated fish remain in the relatively cool (deeper) water at night. This behavior is a further example of variation in patterns of vertical migration at the individual level.

Potential Factors Influencing Vertical Migrations of Fish

From the examples discussed above, it is clear that variations in amplitude of vertical migrations occur frequently. Variations in phase and period are much less common, however, leading some workers to suggest that vertical migrations are an example of an endogenous circadian rhythm that is entrained by some cyclic natural cue. Natural processes that have been proposed to influence the patterns of vertical migration include light, tide, food, and other species.

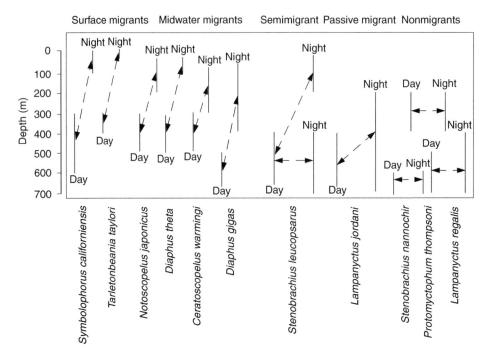

Figure 3 Changes in the day–night vertical distribution of a community of myctophids (lanternfishes) in the western North Pacific. (Modified after Watanabe H, Moku M, Kawaguchi *et al.* (1999) Diel vertical migration of myctophid fishes (Family Myctophidze) in the transitional waters of the western North Pacific. *Fisheries Oceanography* 8: 115–127.)

Light

For animals in the photic zone of the world's oceans, the daily cycle of light and dark is perhaps the most powerful environmental signal available. One of the earliest reports indicating light as a primary factor in vertical migration was by Russell in 1927, who suggested that zooplankton occupied depths that had an optimum light intensity. He further suggested that, as animals moved away from this optimum intensity, physiological reactions controlled by photochemical mechanisms stimulated them to return to the zone of optimum light intensity. Later reviewers concluded there was abundant (albeit circumstantial) evidence implicating light as a significant factor regulating the vertical migration of fish. This evidence included the timing of vertical migration with respect to dusk and dawn, a lack of vertical migration at high latitudes, and the results from experiments with artificial lights and during solar eclipses.

Researchers have elaborated on the concept of a preferred level of illumination by developing mathematical models to describe the diel vertical migration of fish based on light, and have developed expressions to predict the depth range over which vertical migration should occur. However, other studies have suggested that differences in water transparency will have a stronger influence on diel vertical migrations of fish than the surface illumination, since the attenuation coefficient determines the light intensity at depth.

The interpretation of these studies is that light has an important role in mediating vertical migrations of fish. Compared among species, however, light may be more important for some fish such as the herrings and sardines, but it is not the sole determinant of vertical migration.

Tides

For animals that live in near-shore and continental shelf regions, the daily variation of the tide is another powerful rhythmic environmental signal that could influence the behavior of fish. Several species of intertidal fish show strong rhythms of locomotor activity at near-tidal periods. For example, fish living on sandy shores (e.g., the sand goby *Pomatoschistus minutus*) are more active in the laboratory at times corresponding to ebb tide in their natural environment, which may prevent them from becoming stranded as the tide recedes. In contrast, fish on rocky shores (e.g., Blenniidae) become more active in the laboratory at times corresponding to high tide in their natural environment, which may relate to their habitat being tide pools and the fact that high tide provides an opportunity for them to move about more widely.

The larvae of some estuarine fish have developed behaviors that appear to enhance their retention within the estuary, or enhance their movement from offshore spawning grounds into estuarine nursery areas. For example, hogchoakers (*Trinectes maculatus*) in a Maryland estuary show a persistent rhythmic activity with maximum activity corresponding to slack tide, even under conditions of continuous light, which may help them maintain their position. The larvae of other species of estuarine fish have been observed to remain on or near the bottom during the day and during ebb tides, but to move off the bottom at night and during flood tides, which would tend to transport them towards shore or into the estuary.

Mechanisms that have been suggested to enable fish to detect tidal currents include physical factors such as changes in light, turbidity, turbulence, temperature, and salinity. Many of these have been criticized as tidal cues, however, because they will also vary as a result of nontidal factors such as clouds and storms. Other suggested factors include olfactory cues indicating currents toward or away from the coast, direct detection of flow direction or reversal, and induction of electric fields in the water.

Food

Another obvious possible driver of vertical migration in fish is the vertical migration of their prey, which includes both zooplankton and other fish. The presence or absence of prey can restrict or promote vertical migrations of fish, in particular of their early stages. For example, downward migrations of young walleye pollock (*Theragra chalcogramma*) in the Bering Sea are delayed when food abundances are high, but enhanced when food is low in the upper waters. Other species of fish, e.g., capelin (*Mallotus* sp.), Atlantic herring (*Clupea harengus*), and mesopelagic species, have been observed to follow closely the vertical distributions of their zooplankton prey. Adult cod follow a type I vertical migration pattern when feeding on pelagic zooplankton, but switch to a type II vertical migration pattern when feeding on benthic animals. When it occurs, the less common type II pattern appears to be related to the movements of the preferred prey.

Commensal Species

In open pelagic waters, the young stages of certain species of fish are often found closely associated with gelatinous zooplankton (e.g., jellyfishes). Some of these jellyfish undertake vertical migrations, and so the fish associated with them follow along. In a plankton sample this may appear as independent vertical migration by these fish, unless the sample is carefully observed for gelatinous zooplankton (or their remains).

Theories to Explain Diel Vertical Migration Behavior

A number of theories have been proposed to explain the occurrence and persistence of vertical migration across a broad range of species.

Bioenergetics

Considering the widespread occurrence but highly variable patterns of vertical migratory behaviors in fish, the question arises as to what advantages they might provide. For a behavior to persist and become widely distributed among different species, it must confer an evolutionary advantage that increases the survival (fitness) of an individual. One way that fitness is enhanced is by increasing individual growth rates. This implies that fish should feed where prey are most abundant and are readily available to the fish. In 1993, Bevelhimer and Adams, building upon work by Brett in 1971, noted that the most advantageous locations for feeding (and subsequent growth) are not necessarily those with the greatest food density. Other factors that are significant include the potential feeding rate, stomach capacity, and temperature effects on stomach evacuation rate, respiration rate, and other metabolic processes. These observations lead to the bioenergetic advantage hypothesis for vertical migration of fish, which states that fish should feed where the net intake of energy is maximized, then spend their nonfeeding time where energetic costs (e.g., respiration) are minimized, providing that the energetic costs of migrating between these two locations are minimal.

In most marine systems and when prey are not limiting, this bioenergetic hypothesis implies that vertical migrations will occur between deeper, cooler waters with less prey and shallower, warmer waters with more prey – a type I pattern. It can also explain the type II migration pattern if the fish are acclimated to warmer temperatures but make foraging excursions into colder water, where prey may be more abundant in some situations. But what happens when the vertical thermal gradient breaks down, either temporarily owing to storms and a deeper mixed layer, or to seasonal changes in stratification? For some fish at least, such as juvenile salmon in lakes, their vertical migratory behavior appears to stop.

If food is not abundant, the vertical migration patterns predicted by the bioenergetic hypothesis can be changed. Studies show that juvenile walleye pollock remain in warmer upper layers when food is easily available, but migrate into deeper, colder waters when prey densities are reduced. Further experimental studies show that juvenile walleye pollock switch to an energy-conserving behavior when food is limiting, by moving increasingly into colder water so that the optimal temperature for growth also decreases. Therefore, prey abundance and its availability to a fish species can be an important modifier of diel vertical migrations by that fish species.

Predation

The potential for the death of fish due to predation is another strong modifier of the vertical distribution of fish. Japanese sand eels (*Ammodytes* spp.) and northern anchovy larvae (*Engraulis mordax*) exhibit a type II vertical migration pattern in which they cease swimming at night and sink. This inactivity may reduce attacks by predators that detect prey through motion and vibrations. In 1978, Eggers suggested a predator avoidance hypothesis to explain the vertical migratory behavior of juvenile sockeye salmon. Although these studies were conducted on fish in fresh water, salmon are often used as models for pelagic marine predators because they are relatively easy to study. In this hypothesis, fish become distributed at depths where the searching ability of visual predators is reduced (i.e., at the minimum irradiance level that maximizes the reaction distance of a predator to its prey). This was subsequently combined with the bioenergetic hypothesis to suggest that the characteristic patterns of diel vertical migration are a three-way trade-off (optimization) between predator avoidance (minimizing predation mortality), maximizing food intake, and minimizing metabolic losses. In 1988, Clark and Levy developed a model of this trade-off, based on juvenile sockeye salmon and their visual predators. The model predicts that the ratio of risk-of-mortality to feeding rate for juvenile sockeye should be minimized at the low and intermediate light levels that occur at dawn and dusk, which they termed the lsquo;antipredation window'. The optimal behavior for survival, therefore, would be for the pelagic planktivorous fish to migrate into surface waters to feed during dawn and dusk and to migrate to cooler less-illuminated deeper waters during the day (**Figure 4**).

Optimization Models

Rosland and Giske have developed a similar optimization model to describe the vertical behaviors of the mesopelagic planktivore *Maurolicus muelleri*, a common species in the fiords and continental shelf of Norway. In winter, the juveniles of this species migrate over the upper 100 m, rising to the surface at dawn and dusk. The adults, in contrast, remain below 100 m, with no distinct vertical migration

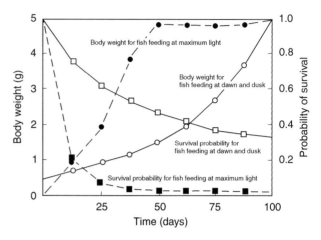

Figure 4 Results from a computer simulation model of two feeding behaviors by a planktivorous fish. Dashed lines represent the growth (●) and survival (■) of a fish feeding during the maximum light intensity, whereas the solid lines represent growth (○) and survival (□) for a fish feeding at dawn and dusk and migrating below the upper water layer during the day. Although growth is slower, survival is clearly enhanced for the fish that feed at dawn and dusk-feeding. (Modified after Clark CW and Levy DA (1988). Diel vertical migrations by juvenile sockeye salmon and the antipredation window. *American Naturalist* 131: 271–290.)

pattern until summer, when they migrate to the surface at dawn and dusk. The optimal depth position was calculated in their model as a balance between feeding opportunity and risk of mortality from predation, which correctly predicted the dawn and dusk migration and feeding pattern. The observed differences in vertical migratory behaviors of this species, as an example, depend on individual differences in age, size, energetic state, variations in the seasonal environment, and an optimization between minimizing predation losses, maximizing food intake, and minimizing metabolic losses that may depend on different life history requirements.

Conclusions

Diel vertical migrations of marine fish are relatively common phenomena that occur in many species and at different life history stages. The relative constancy of their diel periods is consistent with the notion of an underlying circadian rhythmicity. The process of vertical migration also appears to be a facultative one in many cases, as the pattern of vertical migration can be changed by a number of factors. An example of a hypothetical system of multiple controls on diel vertical migrations is shown in **Figure 5**. In this model, an endogenous rhythm of vertical migration is determined initially by photoperiod. Under certain circumstances, the vertical migration pattern of the fish switches from being entrained by a light–dark cycle to entrainment by the tidal cycle, for

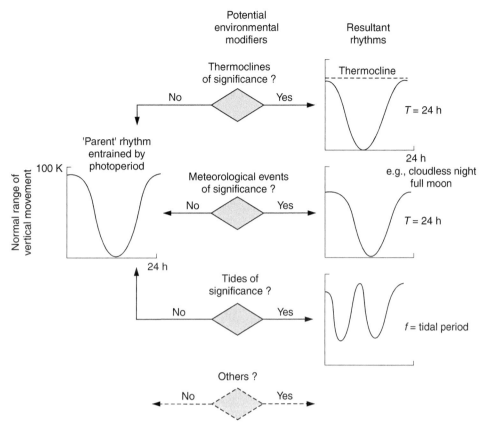

Figure 5 Diagrammatic figure showing how some environmental features could modify a cyclic vertical migration that is otherwise entrained by photoperiod. (After Neilson JD and Perry RI (1990) Diel vertical migrations of marine fishes: an obligate or facultative process? *Advances in Marine Biology* 26: 115–168.)

example, with the result that the period of vertical migration activity is modified. Likewise, events such as a full moon on a cloudless night might act to modify the rhythm by suppressing the amplitude of the vertical migration.

Such variations in vertical migration pose profound difficulties for surveys of the abundance of fish. To reduce these problems, researchers may study different life history stages or use different sampling techniques that include the range of vertical migration, when this is known.

See also

Fish Feeding and Foraging. Fish Larvae. Fish Locomotion. Fish Migration, Horizontal. Intertidal Fishes.

Further Reading

Bevelhimer MS and Adams SM (1993) A bioenergetics analysis of diel vertical migration by kokanee salmon. *Onchorhynchus nerka. Canadian Journal of Fisheries and Aquatic Sciences* 50: 2336–2349.

Blaxter JHS (1975) The role of light in the vertical migration of fish – a review. In: Evans GC, Bainbridge R, and Rackham O (eds.) *Light as an Ecological Factor II*, pp. 189–210. Oxford: Blackwell Scientific.

Brett JR (1971) Energetic responses of salmon to temperature. A study of some thermal relations in the physiology and freshwater ecology of sockeye salmon (Onchorhynchus nerka). *American Zoologist* 11: 99–113.

Clark CW and Levy DA (1988) Diel vertical migrations by juvenile sockeye salmon and the antipredation window. *American Naturalist* 131: 271–290.

Eggers DM (1978) Limnetic feeding behaviour of juvenile sockeye salmon in Lake Washington and predator avoidance. *Limnology and Oceanography* 23: 1114–1125.

Neilson JD and Perry RI (1990) Diel vertical migrations of marine fishes: an obligate or facultative process? *Advances in Marine Biology* 26: 115–168.

Rosland R and Giske J (1997) A dynamic model for the life history of Maurolicus muelleri, a pelagic planktivorous fish. *Fisheries Oceanography* 6: 19–34.

Russell FS (1927) The vertical distribution of plankton in the sea. *Reviews of the Cambridge Philosophical Society* 2: 213–262.

FISH PREDATION AND MORTALITY

K. M. Bailey and J. T. Duffy-Anderson
Alaska Fisheries Science Center, Seattle, WA, USA

Overview

Not only do fish prey on one another, but almost every other type of animal in the sea from jellyfish to whales and seabirds eat enormous quantities of fish. Apart from some less usual conditions (such as outbreaks of disease, mass starvations, harmful algal blooms, or extreme over-fishing) predation by other animals is the largest source of mortality of fishes in the sea. Among the most voracious of these predator groups, other fishes consume the lion's share, but in some seas marine mammals also consume large amounts (**Figure 1**). Predation mortality is generally highest on juvenile fishes, but fishing mortality increases as fish mature and grow, so in some areas commercial harvesting is the greatest source of mortality of the adult stage.

As opposed to fishing mortality, natural mortality rates (which include predation) of most marine fishes decline throughout their life span, resulting from a narrowing scope of the field of potential predators. For example, in the case of walleye pollock, the source of one of the world's most important commercial fisheries, mortality rates for eggs and larvae decline from an average of about 10% loss per day to 1% loss per day for 6-month-old juveniles to 0.05% loss per day for adults. However, because fish are increasing in size with age, the loss of biomass due to mortality peaks during the juvenile stage (**Figure 2**). As a consequence, it is juvenile fish that are most important as prey for providing energy to higher trophic levels.

The Diversity of Predators

A broad variety of predator types and sizes feed on marine fishes. These predators vary from near-microscopic organisms such as *Noctiluca* which feed

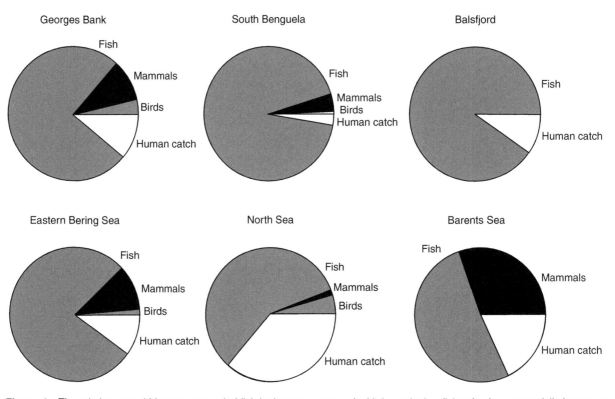

Figure 1 The relative annual biomass removal of fish by humans, mammals, birds, and other fishes in six commercially important marine ecosystems. (Adapted with permission; from Bax NJ (1991). A comparison of the fish biomass flow to fish, fisheries, and mammals in six marine ecosystems. *ICES Journal of Marine Science* 193: 217–224. As noted by that author, these estimates of relative biomass removals can be regarded as gross approximations.)

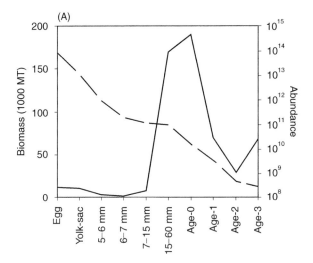

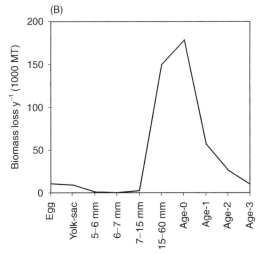

Figure 2 (A) Trends in abundance (dashed line) and biomass (solid line) of different life stages of a typical cohort of walleye pollock (*Theragra chalcogramma*) in the Gulf of Alaska. Numerical abundances of eggs and larvae are much higher than juveniles (note log scale), but total cohort biomass peaks during the juvenile period due to rapid growth in weight. (Adapted with permission from Brodeur RD and Wilson MT (1996). A review of the distribution, ecology and population dynamics of age-0 walleye pollock in the Gulf of Alaska. *Fisheries Oceanography* 5 (suppl. 1): 148–166.) (B) Annual loss of biomass due to natural mortality. Assuming that most of the loss is due to predation, the annual removal of biomass by predators peaks in the early juvenile stages due to the longer stage durations of juveniles.

on fish eggs, to invertebrates such as jellyfish which feed on fish larvae, to whales which feed on juvenile and adult fishes. Fish as prey may be attacked from above by birds or from below by benthic crabs, shrimp, and bottom fishes.

Invertebrate predators employ diverse ways to detect and secure their prey (**Table 1**). Ambush raptorial invertebrates include ctenophores and chaetognaths. Siphonophores may use lures to attract and attack their larval fish prey. Many invertebrate predators such as copepods and chaetognaths detect their prey by mechanoreception of larval fish swimming activity. Jellyfish may depend on random encounters with fish larvae, ensnare them with mucus, and/or then immobilize them with stinging nematocysts. Euphausiids are actively cruising-contact predators, which probably sweep fish eggs and larvae into their mouth parts with feeding currents generated by thoracic legs. Other crustacean predators may use their chemosensory abilities to detect prey, and some may use vision.

Fish also use a variety of methods to capture prey (**Table 1**). Herring either filter-feed or actively bite individual prey, depending on the prey's size and relative density. Other fishes are obligate filter (e.g. menhaden) or raptorial feeders (e.g walleye pollock). Filter-feeding fishes generally feed on small, abundant particles, while raptorial feeding fish generally pursue larger, less abundant prey.

How to Study Predation

There are several methods (**Table 2**) and steps to assessing the impact of predation on marine fishes. First, is the identification of their predators and prey. This has been done by direct observation of predator stomach contents, either looking at whole prey or parts of prey (for example, otoliths or scales), using immunochemical techniques, or looking at the presence of prey DNA in predator stomachs. After prey are identified, and their relative presence is determined, then the amount eaten is estimated from models that utilize the quantity of prey in guts and prey digestion rates to calculate predator daily rations. Alternatively, daily rations are calculated from energetic-demand models and then the amounts of specific prey consumed are estimated from their percent composition in gut contents. The amount of prey consumed by the overall predator population is extrapolated from the abundance of the predators. Predation is sometimes inferred from inverse oscillations in predator and prey populations, although other factors may also be involved, such as competition for resources or distribution shifts.

The dynamics and impacts of who is eating how much of what were examined by ecosystem modeling in the late 1970s. Ecosystem models range from simple to complex. In the past few years, partly due to enhanced computing power, ecosystem modeling has made a comeback. More complex models require that a large number of parameters be estimated. For example, in such models it is assumed that the input data are representative (i.e. that a relatively small sample of stomachs collected over a limited temporal

Table 1 The diversity of fish predators, prey capture strategies, methods of detection, life stages consumed, and predation rates

Predator type	Hunting strategy	Modes of detection	Manner of capture	Stages consumed	Predation rates (from various sources)
Ctenophores	Cruising	Contact	Entanglement	Eggs, larvae	0.4–8% d^{-1}
Jellyfish	Cruising	Contact	Entanglement	Eggs, larvae	2–5% d^{-1}
Chaetognaths	Ambush	Mechanoreception	Raptorial	Larvae	Negligible
Copepods	Ambush	Mechanoreception	Raptorial	Larvae	6–100% d^{-1}
Amphipods	Cruising, ambush	Vision, chemoreception	Grasping	Eggs, larvae	0.1–45% d^{-1}
Euphausiids	Cruising	Contact	Raptorial	Eggs, larvae	1.7–2.8% d^{-1}
Shrimp	Ambush	Chemoreception, mechanoreception	Raptorial	Larvae	16% d^{-1}
Filter-feeding fishes	Cruising	Vision	Filtering	Eggs, larvae	0.15–42% d^{-1}
Biting fishes	Ambush	Vision, mechanoreception	Biting	Juveniles, adults	20–80% d^{-1}
Birds	Ambush	Vision	Biting	Juveniles, adults	10% $month^{-1}$
Otarids	Cruising	Vision	Biting	Juveniles, adults	10–20% y^{-1}
Baleen whales	Cruising	Sonar	Filtering	Juveniles, adults	0.3–2.6% $month^{-1}$

Table 2 Methods of studying predation

Approach	Advantages	Disadvantages
Direct counts from stomach contents	Widely available application; little technical skill required	Fish larvae are digested rapidly, predation rates may be underestimated
DNA analyses	Provide conclusive identification to the species level	Difficult to quantify; time-consuming
Immunochemistry	Results can be obtained rapidly	Time-consuming development of antibodies, non-specific reactivity
Laboratory studies of predator–prey behavior	Provide fine control of multiple variables	True field dynamics are not well represented
Mesocosm studies	Simulate physical and chemical characteristics of water column	Container effects may elevate contact rates between predator and prey
Predator/prey abundance estimates	Provide estimates of mortality due to specific predators	Correlations in predator–prey abundances are not equivalent to causation
Models	Provide a systematic approach to testing system function	Estimation of mathematical function is limited by biological data

and spatial scale represents what the whole population is eating annually), and that they include the correct terms that account for both seasonal movements and interactions between predators and prey. Another approach is to incorporate predation as a component in catch-at-age fisheries models – so-called multispecies virtual population analyses. These models share many of the same data problems and assumptions as ecosystem models.

The Predation Equation

The act of predation consists of a sequence of events that either lead to a successful feeding bout or failure (**Figure 3**). An encounter begins when the prey enters the volume within which a predator can detect it. The rate of encounter is a function of population densities and swimming speeds. Detection occurs when a predator locates the prey and is a function of prey 'visibility' and predator acuity, which depends on the sensory system utilized. Encounter and detection are followed, or perhaps not, by pursuit, strike, and capture.

Predation rates on fish are size-, age-, stage-, and species-specific. It is known that on a gross-scale, size is the most important characteristic of individuals that determines predation rates, as it is associated with escape abilities, swimming speeds,

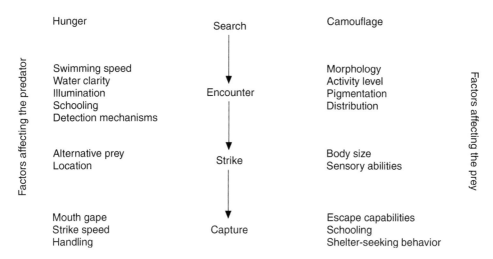

Figure 3 A simplistic conceptual model of the predation process. Left: factors affecting the predator at each stage; right: factors affecting the prey. More complex versions of the model include other steps such as detection, pursuit, contact, satiation, and digestive pause.

and encounter rates. In particular, jellyfish and crustaceans show decreasing rates of predation on fish larvae of increasing size. In the case of raptorial fishes, predation rates may increase with larval prey sizes due to increased visibility and encounter rates, and then decrease when a critical predator size to prey size ratio is attained.

Often the young stages of marine fishes are located in patches or schools, and predators that forage in such aggregations display definite changes in behavior. Swimming speeds change, predominant directions of swimming shift, and changes in the number of turns have been demonstrated for predators encountering high concentrations of prey. All of these behaviors may serve to increase encounters and/or keep predators within prey patches.

From the prey's perspective, once it has been encountered it may detect the oncoming predator (although the sensory systems to detect oncoming predators are poorly developed in the early larval stage of most species), or it may escape after contact with the predator. The success of a larva's escape response depends on its development, startle response, burst swimming performance, and the capture tactics and capability of the pursuing predator.

The Changing Predator Field

As fish grow and develop, their vulnerability to specific predators changes, as does the suite of predators that may consume them (**Figure 4**). Predators of fish eggs include many types of invertebrates and other fishes. Generally, egg loss by invertebrate predation may result in predation mortalities ranging from 1 to 10% loss of the population per day. These are high but not devastating mortalities. Pelagic fishes may also be important egg predators. It has been calculated that herring and sprat could theoretically consume the total standing crop of cod eggs in regions of the Baltic Sea.

Predators on fish larvae include most species that feed on eggs, as well as numerous others that are either not able to detect eggs because of transparency and immobility, or cannot grasp them and puncture them. Fish larvae are quickly digested, so it has proven difficult to get reliable estimates of predation rates from stomach content analyses.

As fish reach the juvenile stage, many of the smaller invertebrate predators and fishes are no longer able to consume fish of a larger size, but they are replaced by a new array of predators including larger raptorial fishes, birds, and marine mammals. Predation by benthic crustaceans such as shrimp may be particularly important for demersal fishes that undergo a transition from planktonic larva to benthic juvenile.

Adult fishes are prey to other larger fishes, sea-birds and marine mammals. Natural mortality rates of adults probably fluctuate with the abundance of predators, and may be influenced by migrations. Adult forage fishes can be severely impacted by migrations of predators, examples being capelin which are consumed by migratory cod in the Barents Sea, and sandlance being eaten by migrating schools of mackerel on Georges Bank (north-western Atlantic Ocean).

Life Transitions and Predation

Predation seems to be highest during or immediately after important life history transitions. For example, newly hatched larvae are vulnerable to a variety of new predator types. Newly hatched larvae are no

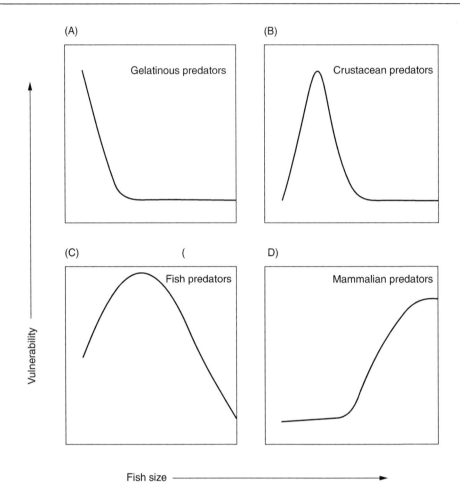

Figure 4 Schematic representation of changes in the relative vulnerability of fishes to different types of predators with increasing fish size. Vulnerability is the product of encounter rates and susceptibility or capture success, as determined by the prey's ability to detect and respond to a predator's presence or attack. (A) Very small fish are most vulnerable to cruising gelatinous predators because they have a poorly developed escape response and encounter rates between predator and prey are high. (B) Vulnerability to some ambush crustacean predators might be dome-shaped because encounter rates increase as developing fish increase their swimming speeds, but vulnerability declines as larger fishes become more adept at escaping. (C) Visual fish predators may not see very small fish prey and larger fish prey have improving escape and avoidance abilities that eventually reduce their vulnerability to fish predators. (D) Mammals may select larger fish.

longer protected by an egg shell and they now attract predators with their swimming motion. In addition, they have yet to learn predator avoidance and escape responses (or their innate abilities are still poorly developed).

Metamorphosis from the larval to juvenile stage has also been described as a period of high predation. For many species, metamorphosis represents a shift in habitat. The transition of flatfish larvae (such as plaice and Japanese flounder) as they settle from the plankton to the epibenthos has been found to be a period of high mortality that may modify recruitment levels. Likewise for coral reef fishes, the transition from freely floating planktonic larvae to their association with reef structures is a period of high mortality. This period of high mortality associated with a change in habitat may result from a lack of recognition of

potential predators, or delays in acquiring camouflage, finding refugia, or learning behaviors that may protect them from predators (such as burying and hiding). In addition, competition for refugia or shelter may leave the losers exposed.

Predation and Recruitment

Recruitment is the abundance of an annual cohort at an age just prior to their joining the adult or harvestable population. At this age, the cohort strength has been determined, but it has not been impacted by the fishery. In marine fishes, the level of recruitment is highly variable but is critical in establishing population levels. For many species, there is ample evidence that recruitment is established by the end of the larval period (e.g. Arctic cod and Pacific hake).

For other species, heavy predation on juveniles can influence relative recruitment levels (e.g. northern anchovy and walleye pollock); this is very likely to be the case for many forage fish species or in geographic areas where predators of juveniles are typically abundant. Predation on early stages is one factor that is believed, in some cases, to increase recruitment variability, and in other cases to dampen variability.

The abundance of predators on eggs and larvae, and in some cases juvenile fishes, has been correlated with poor recruitment success of a number of species. For example, large numbers of adult herring are believed to consume vast quantities of cod and plaice eggs in the North Sea, thus depressing recruitment of these species. The variability in recruitment caused by predation is probably associated with the abundance of predators and the degree of their spatial and temporal overlap with prey, which can be related to environmental conditions. In particular, there are several studies demonstrating the deleterious impact of high consumption rates of gelatinous predators of fish eggs and larvae (see **Table 1**). The heavy predation of pelagic fishes on fish eggs is also well documented. Anchovies are known consumers of fish eggs, and in fact, several studies indicate that cannibalism by anchovies on their own eggs accounts for a large proportion of the total anchovy egg mortality. When pelagic predators are very abundant, they may consume a large portion of a year class. A strong year class can occur when predators decline or conditions dissociate the distribution of predators and prey in time or space, causing a release from predation pressure. Alternatively, a strong year class can result when larval production is so high that it swamps the predatory capacity of the ecosystem. There is also an interplay between growth and predation, such that the longer that fish remain in stages vulnerable to heavy predation pressure, the higher the cumulative mortality.

Density-dependent predation mortality has been shown to occur for juvenile fishes, but not for larvae. Density-dependent mortality can result from individual predators feeding disproportionately on prey of increasing abundance, so-called switch-feeding, and from predator swarming on abundant prey. In theory, other mechanisms include density-dependent growth interacting with size-dependent mortality, density-dependent condition of prey, and limited refugia from predators that become filled at a threshold density (leaving the excess unprotected from predation).

Predation and Community Structure

The effects of predation on structuring marine rocky intertidal and freshwater communities is well known, but more recently, predation has become more widely recognized as a force that shapes the structure of marine communities. This is especially the case for coral reef communities where fish densities are high, predators are varied and abundant, and prey refuges are limited. Ecological disturbances such as hurricanes, El Niño events and long-term climate changes may shift conditions that favor certain species and thus reorganize community structure. For example, a climate shift in the late 1970s in the North Pacific Ocean favored increasing abundances of long-lived piscivorous flatfish species, which has altered the pattern of recruitment of their prey species. Trophic cascades are important indirect forces that play a role in the organization of marine communities. These cascades describe the repercussions of predator–prey interactions throughout the food web. In the Baltic Sea, for example, the mortality of sprat and herring has declined as the biomass of its major predator, cod, has decreased. Since, as mentioned above, herring and sprat also consume large quantities of cod eggs, different interactions between predators and prey at different life stages may reinforce (or sometimes counteract) shifts in community structure.

Predation and Evolution

Evolution is driven, in part, by the ability to survive long enough to contribute to the gene pool. As such, predation is a strong determinant of natural selection and fish have evolved an amazing variety of tactics in nearly all aspects of their life history to avoid predation mortality. For example, many tropical species, such as sciaenids and engraulids, demonstrate crepuscular or night-time spawning activity that minimizes the proficiency of visually foraging predators to locate newly spawned eggs, potentially reducing total egg mortality. Moreover, reproduction during periods of reduced light intensity may also minimize the vulnerability of spawning adults to predation by larger animals, leaving iteroparous spawners an opportunity to spawn again. Live-bearing strategies, such as viviparity and ovoviviparity, are common among elasmobranchs and may minimize predation on early life stages. Gross changes in fish distribution are also evolutionary adaptations to predation pressure. Current hypotheses suggest that diel vertical migrations are an adaptive behavior to avoid daytime predation by visual predators and permit feeding at night when predation risks are presumably reduced (**Figure 5**). Studies of fish aggregations, particularly of small-sized fish, indicate that synchronized schooling is an important 'safety in

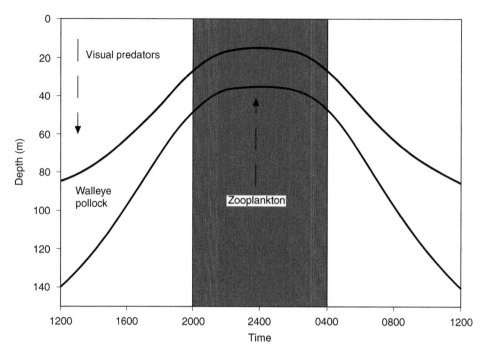

Figure 5 Spatial–temporal variations of young-of-the-year walleye pollock (*Theragra chalcogramma*) in the eastern Bering Sea. Walleye pollock occur at depth during daylight hours when visual predators in the surface waters are most active. At night (shaded area) when the risk of attack from visual predators is reduced, walleye pollock move to the upper water column to feed on available zooplankton.

numbers' evolutionary strategy to confuse and evade a pursuing attacker.

Feeding in fish is often a trade-off between foraging success and predator avoidance. A variety of studies have demonstrated that fishes can assess predation risk and modify their behavior to maximize fitness. This ability provides a strong, selective advantage to fishes that in the long run, increases their likelihood of reproducing. As a final example, predation has had profound impacts on fish morphology, ranging from adaptations that aid in disguising the fish from its predators to those that make the fish highly conspicuous. In the former case, fishes have evolved to resemble seaweeds, sponges, sticks, detritus, and sand. These cryptic strategies are common in blennies, pipefish, seahorses, and flatfishes. Fishes also have evolved patterns that disrupt their outlines, including silvery sides, lateral bands, and countershading. These strategies are common among silversides, killifish, and a variety of pelagic species. In the case of increased conspicuousness, fishes have evolved elaborate spines, bright warning colours, distinct eyespots, and mimicry of inedible species. An interesting example of mimicry has been observed in coral reef systems around Florida, USA. The blenny, *Hemiemblemaria simulus*, is very similar in both appearance and behavior to the cleaner wrasse, *Thalassoma bifasciatum*. The mimic blenny has a comparable body shape and color pattern to the wrasse and it adopts a similar swimming strategy. The blenny apparently benefits not only from the protection from predation the cleaner wrasse receives from other fishes, but also by consuming ectoparasites on host fishes that come to be groomed by the wrasse.

Conclusions

Predation plays a significant role in the recruitment and population dynamics of marine fishes. The broad variety of predators that consume fishes, coupled with the potential for the removal of large portions of the available population, make it likely that predation is an important part of observed fluctuations of fish populations. Integrated studies of the physical and biological processes that influence predation, and especially the spatial overlap of predators and prey, coupled with long-term observations of the consequences, can provide useful information for evaluating the role of predation to overall recruitment success.

See also

Fish Feeding and Foraging, Fish Larvae, Seabird Foraging Ecology.

Further Reading

Bailey KM and Houde ED (1989) Predation on eggs and larvae of marine fishes and the recruitment problem. *Advances in Marine Biology* 25: 1–83.

Bailey KM (1994) Predation on juvenile flatfish and recruitment variability. *Netherlands Journal of Sea Research* 32: 175–189.

Blaxter JHS (1986) Development of sense organs and behavior of teleost larvae with special reference to feeding and predator avoidance. *Transactions of the American Fisheries Society* 115: 98–114.

Fuiman L and Magurran AE (1994) Development of predator defenses in fishes. *Reviews in Fish Biology and Fisheries* 4: 145–183.

Hixon MA (1991) Predation as a process structuring coral reef fish communities. In: Sale P (ed.) *The ecology of fishes on coral reefs*, pp. 475–508. New York: Academic Press.

Leggett WG and Deblois E (1994) Recruitment in marine fishes: is it regulated by starvation and predation in the egg and larval stages? *Netherlands Journal of Sea Research* 32: 119–134.

FISH REPRODUCTION

J. H. S. Blaxter, Scottish Association for Marine Science, Argyll, UK

Introduction

Two characteristics of fish – that they are cold blooded and that many species continue to grow after reaching sexual maturity and throughout most of their life span – have profound effects on their reproductive ability. The ambient temperature influences the rate of development and growth, the time to reach sexual maturity and the life span. Fish living at high latitudes or in deep water usually grow more slowly, reach sexual maturity later and have longer life spans than their low-latitude counterparts. The seasonality of temperature change in high latitudes causes seasonal patterns of growth and has a role (with daylength) in determining spawning time. Some of these influences may be less clear in species such as tuna that have ocean-wide migrations during their life histories.

General Life Histories

In most fish species fertilization is external, there is no sexual dimorphism (the sexes look alike) and the large number of eggs produced annually by a female (its fecundity) develop and hatch and the larvae grow without parental protection. Sexual dimorphism is more likely to occur when fertilization is internal or elaborate courtship behavior takes place. In some species the fish are hermaphrodite, usually changing from one sex to another during their lifetime. Self-fertilization is rare. The generation time (age at first spawning) varies from a few weeks in small tropical species to several years in large temperate and deep-water species. Most species spawn several times during their lifespan (iteroparity), a few such as the Pacific salmon spawn once and then die (semelparity).

Typically, teleost (bony fish) eggs are 1–2 mm in diameter (**Figure 1**) and hatch after a few days, the time for incubation being dependent on ambient temperature (**Figure 2**). The larvae at hatching are a few mm long (**Figure 3**) and have a prominent yolk sac containing an endogenous supply of nutrients that they utilize until they reach the first-feeding stage. They are then large and well-developed enough to search for an exogenous supply of live food in the

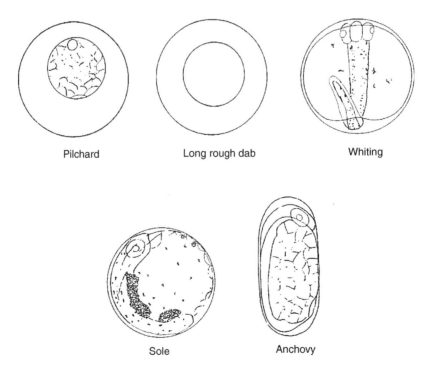

Pilchard   Long rough dab   Whiting

Sole   Anchovy

Figure 1 Free-living (oviparous) fish eggs of various species at different stages of development (diameter about 1 mm). (Reproduced with permission from Bone *et al.*, 1999.)

microzooplankton of the surrounding water. After a period of days to months, depending on species and temperature, the larvae metamorphose into juveniles, i.e. immature stages of adult-like form.

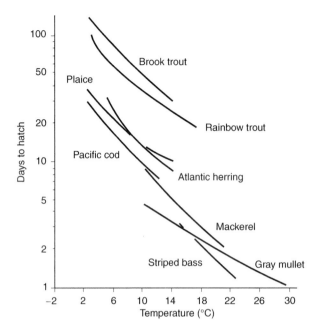

Figure 2 Time from fertilization to hatching of various species depending on temperature. (Reproduced with permission from Bone *et al.*, 1999.)

Such a typical early life history has many exceptions. Some teleosts and all elasmobranchs (cartilaginous sharks and rays) have internal fertilization and the eggs of many but not all species develop within the female, which produces live young. Other species guard their eggs or young in one way or another.

This article elaborates on the themes of spawning season, fecundity and egg size, spawning behavior and parental care, egg and larval development and behavior, the underlying theme being the range of 'reproductive strategies' that have been adopted by fishes.

Spawning Season

The time of spawning is synchronized with ambient conditions by temperature and daylength. In high latitudes spawning is most often associated with cycles of productivity in the plankton and the presence of food of appropriate size for larvae when they first start to feed. A 'match' between the presence of first-feeding larvae and their microzooplanktonic food is one key to their survival, the other being a 'mismatch' with their predators. Spring spawning is most common and some species such as cod produce all their eggs in one batch over a few days (single-batch spawning) whereas others such as the plaice produce several batches of fewer eggs over several

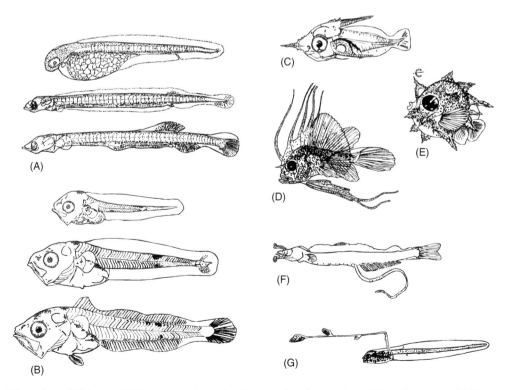

Figure 3 Teleost (bony fish) larvae. (A) Three stages in the development of northern anchovy (*Engraulis mordax*). (B) Three stages in the development of hake (*Merluccius productus*). (C)–(G) more unusual larvae: (C) *Holocentrus vexillarius*; (D) *Lophius piscatorius*; (E) *Razania laevis*; (F) *Myctophum aurolaternatum*; (G) *Carapus acus*. Drawings not to scale. (Reproduced with permission from Bone *et al.*, 1999.)

weeks. The reproductive strategy is clear. The single-batch spawner takes a 'gamble' on all its offspring reaching first-feeding when the food supply is of suitable size and quantity; the multiple-batch spawner 'hedges its bets' with the prospect of at least one (smaller) batch matching the available food, whose abundance can vary in space and time.

In low latitudes, the seasonality of temperature (and daylength) is less marked, but other environmental factors such as monsoon conditions or wind-induced upwelling may influence spawning season. Some tropical species spawn year round whereas others follow the limited seasonality, both types producing many small batches of eggs over the year. In deep water with a stable temperature regime, seasonality may be imposed by the descent of nutritive particles (fecal material and dead bodies) from the surface, where seasonal cycles of production prevail.

Fecundity and Egg Size

Their relationship can be summarized as follows:

1. Given that there is limited space within the body cavity of a female, species with high fecundities have small eggs and those with low fecundities have large eggs.
2. Fecundity within a species increases with age (size) of the female (**Figure 4**). This has important implications for overfishing since the loss of large

fish means a disproportionate reduction in total egg production by a stock. In this way fecundity can also be looked on as the lifetime production of eggs by a female.
3. Fecundity is highest in species that release their eggs into open water and lowest in those species that bear their young alive or show parental care.
4. Within a species, egg size decreases in multiple-batch spawners as the spawning season progresses (**Figure 4**). In the particular and unique case of the herring, which has many physiological races with characteristic spawning times throughout the year, fecundity and egg size are inversely related on a seasonal basis (**Figure 5**).
5. Larger eggs produce larger larvae at hatching with more yolk and a longer period of endogenous feeding, larger body size at first-feeding and probably a better chance of survival in relation to feeding and predation.

Reproductive strategies such as generation time, breeding once or several times in the lifespan, spawning season, single vs. multiple-batch spawning, egg size vs. fecundity and parental vs. no parental care lead to the concept of k and r strategies. A k-strategy is where low fecundity, large eggs and a long development period are favored in stable but crowded environments; r-strategists have high fecundity, small eggs and a short developmental period suited to less stable, less crowded environments in order to exploit the opportunities for expansion.

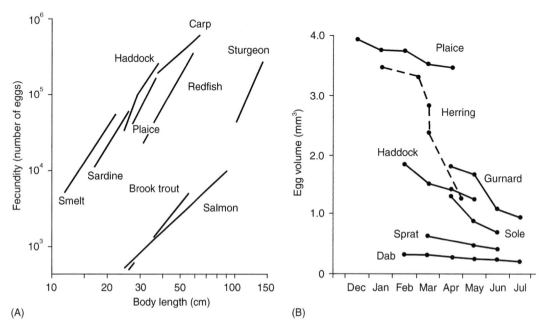

Figure 4 (A) The fecundity of various species depending on their length. (B) The size (volume) of fish eggs as the spawning season progresses. (Reproduced with permission from Bone *et al.*, 1999.)

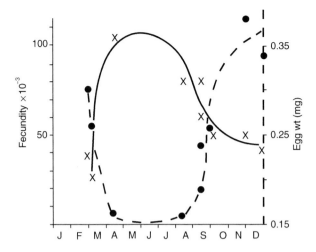

Figure 5 The inverse relationship between egg size (dry weight, ●) and fecundity (×) in the herring from various seasonal spawning races. (Reproduced with permission from Bone *et al.*, 1999.)

Spawning Behavior and Parental Care

Although inshore or littoral species restrict their ranges to small areas over their life span, many marine species make seasonal migrations for spawning, feeding and overwintering. Often the juveniles occupy nursery grounds inshore and move into deeper water as they grow.

The annual concentrations of many commercial species make them vulnerable to exploitation. They are typically r-stategists having oviparous (free-living) eggs that are released in large numbers into the water and left to fend for themselves. It is important that this release should not be too random, in order to maximize fertilization. Some degree of courtship or forms of signaling between males and females is known to occur in many species. For example, vocalization occurs when haddock are spawning and visual signals are known to play a part in the spawning of herring. Because it is difficult to maintain and breed many of the larger oceanic species in captivity, we are ignorant of much of their behavior

The k-strategists often have some form of courtship or mating interaction that may range from the elaborate behavior of the male stickleback to ensure the female lays her eggs in his nest, to internal fertilization of some live-bearing teleosts and all elasmobranchs. The teleosts have modified fins as an intromittent organ and the male elasmobranchs have claspers. Some elasmobranchs then lay and abandon their large horny eggs (e.g. the oviparous mermaid's purses of rays and sharks, **Figure 6**) attached to the substratum. Incubation can take several months. More commonly ovoviviparity occurs in which the

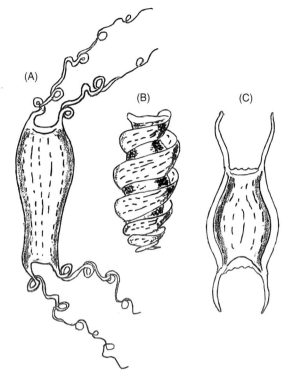

Figure 6 Egg cases of oviparous elasmobranchs. (A) spotted dogfish (*Scyliorhinus* sp.) 11 cm long without tendrils; (B) Port Jackson shark (*Heterodontus phillippi*) 9 cm long; (C) ray (*Raia* sp.) 12 cm long. (Adapted from Norman JR (1931) *A History of Fishes*. London: Ernest Benn.)

large eggs develop independently within the female's 'uterus' and are produced alive. More common still is viviparity where the developing young derive nourishment directly from the mother, as in the smooth dogfish and many shark species. In the ovoviviparous species, the embryos usually depend on their own yolk supply or they eat other eggs (intrauterine cannibalism or oophagy). In some ovoviviparous rays a nutritive uterine milk is secreted and protrusions from the uterine wall extend into the mouth cavity of the embryos. The viviparous sharks have a form of placenta to transfer nutrient to the young. All these species produce quite small numbers of young, less than 50 or so, but some are as long as 30 cm at birth. The viviparous basking shark produces young 150–180 cm long.

The advantage of both ovoviviparity and viviparity is obvious – the young are large and well-formed at birth and so more independent of their environment. The disadvantage lies in the fact that if the mother is eaten, all the young are lost.

Parental protection takes several forms: the nests of sticklebacks already mentioned, the pouches of seahorses and pipefish and the mouth-brooding of the male catfish. Many intertidal species protect their young. These survival mechanisms all suffer from the concentration and so the increased vulnerability of the young. It has to be said that the most successful species in terms of numbers or biomass are the r-strategists.

Egg and Larval Development

The eggs of r-strategists are small, round and transparent with a fairly tough egg membrane or chorion which protects them against wave action. However, anchovies have ellipsoidal eggs and in some gobies they are pear-shaped. With the exception of a few species such as the herring, capelin and most littoral species the eggs are buoyant and move passively with the currents. Development is meroblastic, the embryo developing at the animal pole as a blastodermal cap that overgrows the yolk and invaginates to form a gastrula (**Figure 7**). As the neurula develops the embryonic axis becomes visible as a head with prominent eye cups and a body of segmented muscle (the myotomes). As the body grows a tail is formed, the heart starts to beat and the embryo can be seen to move within the chorion. At hatching, part of the chorion is softened by hatching enzymes. In most oviparous teleosts the larva is a few millimeters long at hatching; it is very transparent and neither the mouth nor gut is fully formed. During yolk resorption over the next few days the mouth and gut develop and become functional. At first-feeding the larvae are predators of the microzooplankton (young stages of many invertebrates and especially crustacea) and use vision to feed. The larvae must obtain adequate food within a few hours to days if they are not to die of starvation. Feeding behavior usually takes the form of stalking, the larvae approaching the prey, bending into an s-shape and darting forward to engulf it. Some degree of suction may also be involved. The locomotor organs comprise a primordial finfold. Being so small the larvae are in a 'viscous' locomotor regime and swim in an eel-like or anguilliform style.

As the larvae grow the median (anal and dorsal) fins appear, but the most significant event is flexion when the posterior tip of the body turns upward and a caudal or tail fin develops. This greatly improves locomotor performance, which includes both capture of prey and escape from predators. Flexion and increasing size tend to bring the larvae into the 'inertial' locomotor regime occupied by the adults.

The sensory systems are fairly well known. The embryo may move within the chorion when stimulated by light, the eye cups being visible early on in the development of the neurula stage. At hatching the eyes may or may not be pigmented but are always functional at first feeding. At this time they usually have a pure-cone retina giving good acuity for perceiving food particles. The rods, which are designed to perceive movement, develop later. Large nasal pits containing a high density of cilia are present on the upper surface of the snout and taste buds can be observed in the pharyngeal area. Prominent over the head and body are free neuromast organs, consisting of bundles of hair cells each with cilia at the tip invested by a gelatinous cupula. The neuromasts, which are vibration receptors, proliferate during development and some of them become arranged in linear series to form the head and trunk lateral line. Little is known about the development of the inner ear, but there is a great interest in the otoliths which contain daily growth rings that can be used for determining the age of the larva. The development of the sense organs is of particular interest in terms of survival because the early larvae are poorly equipped and only later does the full suite of sense organs become available to make the larva maximally aware of its environment.

In the youngest larvae a liver and pronephric kidney is present but the gut is often a straight or coiled tube and develops into different regions later. The circulation may include an extensive blood supply to the yolk sac to help mobilize the yolk in endogenous feeding. The blood is a colorless fluid, the red pigmentation of hemoglobin appearing later often near metamorphosis. In the earliest stages respiration is cutaneous and gills appear later in the larval stage. It seems likely that development occurs in such a way as to keep the larva as transparent as possible as an antipredator device, but increasing size and thickness of the muscle tissue eventually makes the larvae more conspicuous, the eye being the most difficult organ to camouflage. Undigested food also makes the larva more conspicuous.

After a larval period of days to weeks, metamorphosis takes place as the larvae assume the juvenile stage, which is usually in the form of a

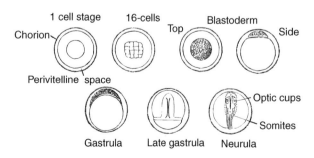

Figure 7 Development of a typical teleost egg (diameter of egg about 1 mm). (Reproduced with permission from Bone *et al.*, 1999.)

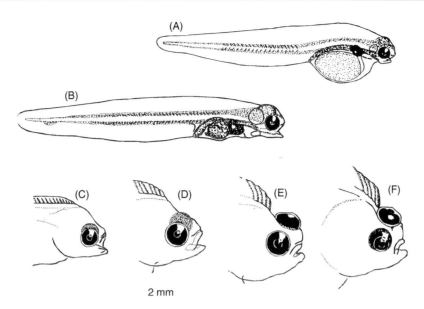

Figure 8 Larval stages and metamorphosis in the plaice: (A) yolk-sac larva; (B) first-feeding larva; (C)–(F) eye movements during metamorphosis. (Reproduced with permission from Bone *et al.*, 1999.)

miniature adult. The body thickens, pigment and scales appear in the skin and the blood turns red; behavior also changes. In flatfish, such as plaice, the larva is bilaterally symmetrical; at metamorphosis the eye on one side moves across the head and the juvenile lies on one side (**Figure 8**). Metamorphosis is not so clear in elasmobranchs and viviparous teleosts; the young hatch or are born at an advanced state of development, although some may still retain a yolk sac.

Behavior

While searching for food the larvae may move continuously or have periods of rest. Having a higher specific gravity than seawater, they then sink. Larvae are at the mercy of small-scale turbulence as well as larger scale wind-induced, tidal and residual currents that influence their distribution. These are important dispersion mechanisms that often bring the larvae inshore to nursery grounds. Within this imposed movement the larvae search for food and make predatory feeding movements. As the larvae grow the gape of the jaw increases, allowing larger food to be eaten; the searching speed and volume of water searched also increase. Light has some influence on behavior and larvae show changes in their vertical distribution by day and night especially as they get bigger. This may also be a dispersion mechanism allowing them to exploit currents at different depth horizons.

Antipredator mechanisms of small larvae seem to consist usually of showing little or no response to predatory attacks so depending on their transparency to remain inconspicuous; as they grow they respond by a fast-start mechanism that is mediated via giant nerve fibers in the spinal cord (Mauthner cells). The evasion strategy is to make a fast-start escape response, taking a few hundred milliseconds, at a very close response distance to the predator, so preventing it from reprogramming its attack.

At and after metamorphosis the juveniles adopt different defense mechanisms. Many species come together in schools or aggregations that make it difficult for the predator to select a target. Others, such as flatfish (**Figure 9**), seek cover on the sea bed where they adapt to the color of the background or bury in the substratum. Camouflage is achieved by the deployment of pigment cells (chromatophores) in the skin that adapt the fish to its background using

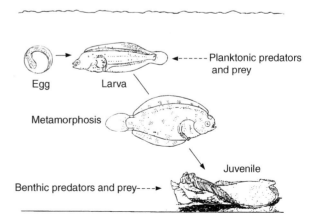

Figure 9 Life history of the plaice with a planktonic larva and bottom-living (benthic) juvenile and adult. (Reproduced with permission from Bone *et al.*, 1999.)

countershading or color-matching mechanisms. Silvery-sided fish have reflecting layers of guanine crystals on their flanks that act as mirrors.

Despite millions of years of evolution, the mortality of fish during the early life history is enormous. Small, soft-bodied organisms are ready prey for a vast suite of predators. There is a premium on fast growth to reduce predation by smaller predators. High mortality is compensated by various reproductive strategies as described above. It has even been suggested that the production of large numbers of young provides an additional food supply, the faster growers cannibalizing their slower-growing siblings.

See also

Fish Larvae. Fish Predation and Mortality.

Further Reading

Bailey KM and Houde ED (1989) Predation on eggs and larvae of marine fishes and the recruitment problem. *Advances in Marine Biology*, vol. 25, pp. 1–83. London: Academic Press.

Blaxter JHS (1988) Pattern and variety in development. *Fish Physiology*, vol. 11A, pp. 1–5. London: Academic Press.

Bone Q, Marshall NB, and Blaxter JHS (1999) *Biology of Fishes*, 2nd edn. Cheltenham: Stanley Thornes.

Ferron A and Leggett WC (1994) An appraisal of condition measures for marine fish larvae. *Advances in Marine Biology*, vol. 30, pp. 217–303. London: Academic Press.

Hoar WS and Randall DJ (eds.) (1998) *Fish Physiology*, vol. 11A, p. 546. London: Academic Press.

Kamler E (1992) *Early Life History of Fishes: an Energetic Approach*. London: Chapman and Hall.

Lasker R (ed.) (1981) *Marine Fish Larvae. Washington Sea Grant Program*. Washington, Seattle: University of Washington.

Moser HG (ed.) (1984) Ontogeny and Systematics of Fishes. *American Society of Ichthyologists and Herpetologists*.

Russell FS (1976) *The Eggs and Planktonic Stages of British Fishes*. London: Academic Press.

Wootton RJ (1990) *Ecology of Teleost Fishes*. London: Chapman and Hall.

FISH SCHOOLING

T. J. Pitcher, University of British Columbia,
Vancouver, Canada

Introduction

Wheeling and turning in synchrony, flashing iridescent silver flanks, fish in a school have been a source of inspiration to poets and naturalists since ancient times. But, to understand schooling behavior, scientists ask 'How?' and 'Why?' questions to address both form and function (**Table 1**). Schooling ('form'), is brought about by an integrated physiological system of muscles, nerves, and senses ('how?') that has evolved under natural selection ('why?') because of benefits to survival ('function'). This article surveys our knowledge of the physiological mechanisms that cause schooling behavior, the behavioral and ecological rules that govern its evolution, the implications of schooling for scale, pattern, and process in the ocean, and the impacts of schooling on human fisheries.

Definitions

Most of the 24 000 known species of bony fish form cohesive social groups known as 'shoals' at some stage of their life history. Social groups occur because animals choose to stay with their own kind to gain individual benefit, whereas grouping for extrinsic reasons such as food, shelter from water currents or oxygen availability is known as aggregation. The term 'school' is restricted to coordinated swimming groups, so schooling is one of the behaviors shown by fish in a shoal; there can be others, such as feeding or mating (**Figure 1**). The tendency to form shoals or schools varies both between and within species, depending on their ecological niche and motivational state respectively. For example, many species of fish shoal for part of the time (e.g. mullet, squirrelfish, cod), while other species adapted to fast swimming (e.g. mackerel, tuna, saithe), or rapid maneuvering around a reef (barracuda, seabream), generally school most of the time. Some species (e.g. minnows and perch in fresh water, herring and snappers in the sea) opportunistically switch between shoaling and schooling to maximize survival, or – strictly speaking – evolutionary fitness. At one time, species that schooled a lot were termed 'obligate' schoolers while those that schooled part-time were termed 'facultative' schoolers, but these terms have been replaced by 'frequent' and 'occasional' schoolers.

The School Rules and School Size

In contrast to early work on fish shoals that emphasized the collective actions of the whole group as though they were some kind of super-individual, insight into fish shoaling and schooling has come from examining the costs and benefits to individuals. Constantly, from second to second, shoaling fish take decisions to join, leave, or stay with the group (JLS). This provides a flexible 'online' response to the environment, which can change rapidly, for example when a food source is found or when a potential predator appears. Because of differences among individuals in opportunity and motivation, the tensions and conflicts underlying such a system are evident even in the most impressive phalanx of mackerel, which will break ranks to feed, or among schooling herring, which segregate by hunger level.

The size of the group is one important elective adjustment that is made to adjust individual pay-offs in a shifting regime. But adjustment of shoal size to the prevailing food/predation regime is possible only if fish shoals both split and meet so that they have the opportunity to merge and exchange members. Such shoal meetings, long observed in laboratory experiments, have recently been measured in the wild.

Are there rules that govern how fish pack in a school? Mathematicians can prove that the maximum packing of spheres in 3-D is in layers of offset hexagons, but fish do not do this. As fish join a group, they adopt a roughly equal distance from neighbors. Hence three fish form a triangle, four a pyramid and so on. Experiments with minnows, herring, saithe, and cod show that schools are like roughly stacked pyramids, about 15% less dense than the maximum. However, fish in schools do not behave like rigid crystal lattices, and the most useful finding from this work is that fish in a school tend to occupy a water volume of approximately one body length cubed, with neighbors about 0.7 of a body length apart. This value, recently validated using sonar on wild herring schools, shifts with swimming speed and timidity, but generally does not change with school size.

There are many reports that fish of similar size shoal together. For example, wild mackerel and

Table 1 Matrix, pioneered by the Nobel Laureate ethologist Niko Tinbergen, showing 'How?' and 'Why?' questions in relation to form and function in fish schooling behavior

	How?	*Why?*
Form	Swimming hardware, sensory inputs, neural decision mechanisms	Evolutionary shaping of schooling hardware (linked to below)
Function	Finding and eating food; hiding from or escaping from a predator	Trade-offs between schooling costs and benefits to evolutionary fitness (linked to above)

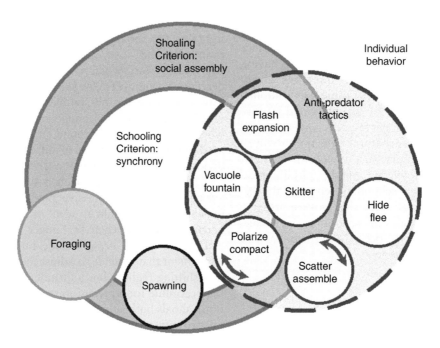

Figure 1 Fish schooling and shoaling behavior (Venn diagram). Criteria for these two behaviors are indicated. Three other behavior categories are superimposed: feeding, spawning and anti-predator behavior; some examples of behaviors in the latter category are also shown. (Concept from Pitcher, 1993.)

herring choose school neighbors within a 15% band of their own length. Recent observations from the wild support laboratory experiments showing that this choice is not merely because smaller fish cruise more slowly and fall behind, but because competition for food and coordinated escape from predators favor size-sorting.

Senses and Schooling

Vision is the predominant sense used in schooling, but swimming fish also use the lateral line, the 'distant touch' sense. Olfaction and hearing are more important in assembly of shoals, or their break-up when food is detected.

Experiments show that schooling fish use vision to join other fish, mirror images of themselves, or fish behind transparent barriers, although their behavior shows that they can distinguish these different kinds of visual images from the real thing – perhaps this is where other senses become important, and perception is mediated through central integration by the brain. In fact, many schooling fish species have visual display signals such as colored spots, longitudinal stripes along the flank, or even spots on fins or gill covers that can be raised or lowered at will. These schooling signals act as visual cues for JLS decisions.

In herring, which have a particularly well-developed lateral line system over the head, velocity changes are communicated by very rapid pressure waves from the accelerations of neighbors. Surgically cut lateral line nerves hamper minnows in their coordinated escape maneuvers from predatory pike. The roles of vision and lateral line were teased out in experiments where saithe were temporarily blinded with opaque blinkers. Blind fish were able to join and swim with a school when repeatedly passed by intact saithe swimming in a large annular tank. Saithe with cut lateral line nerves were also able to do this. The two sensory-deprived types of fish schooled differently; blind fish kept more precise distances to neighbors, while fish with no lateral line kept neighbors at 90°, where they could better detect velocity changes. Not surprisingly, cutting the lateral lines of blind saithe eliminated schooling. The work shows that the lateral line sense is critical to synchronization of acceleration and turning in schools.

Food, Predators, and Schooling

Predators and food are the keys to understanding what shoaling is for and why it has evolved. The functions of shoaling in foraging and in providing anti-predator advantage have been investigated with carefully controlled and replicated experiments in large laboratory aquaria, and in recent years these investigations have been extended to the wild using high resolution sonar or scuba diving.

One of the advantages of foraging in a larger group is that randomly located food items are located more rapidly. Moreover, in larger groups, fish spend more time feeding and are less timid. Furthermore, when the amount or quality of food changes, shoaling fish switch to a better location more efficiently in larger groups. All these effects are achieved by modifying JLS decisions after subtle observation of the behavior of other fish. If some individuals succeed in finding food, other fish copy their moves, including sampling new feeding patches. The benefits of foraging in a shoal get larger as the numbers in the group increase to a few dozen, but improvement becomes progressively less as shoal numbers get larger and the law of diminishing returns comes into play. Costs of competition for food get larger as shoal size increases, and such intra-school competition seems to help segregate size classes of fish in the wild, since large fish win in contests for food items.

As well as switching among food patches, some species of shoaling fish like clupeids and some cichlids can actually switch feeding methods. For example, schooling herring can filter-feed using their gill rakers, swimming with mouths and gills open, or alternatively, can bite at larger food organisms. The switch between the two feeding methods occurs when the density of small food is high enough to sustain the faster swimming speed and energy consumption of filter feeding. When food density is close to the threshold, individuals make different estimates of the switch point so both types of feeding can occur in a school.

Often used in ecology, the theory of the 'ideal free distribution' predicts that individuals will distribute themselves among food patches in proportion to the reward encountered, so that all individuals have the same average intake rate. But alternative strategies in competition for food or differences in perceived predation risk affect the JLS decisions of individual fish and result in distributions that differ from this theory. More modern theory, supported by experiments, is based on trade-offs between predation risk and food reward.

Although food is vital to survival and breeding, and hence is an important component of fitness, avoiding being eaten is even more critical: the Life/Dinner Principle. Shoaling fish try to reduce the success of predator attack through tactics of avoidance, dilution, abatement, detection, dodging, mitigation, confusion, inhibition, inspection, and anticipation. Predation events occur rapidly, and are not very frequent for a human observer. In the wild they are hard to observe at all, whereas in the laboratory, there is a worry about introducing artifacts. Fortunately, many laboratory and field experiments investigating anti-predator functions in fish shoals have successfully employed protocols in which dummy predators approach test shoals. This protocol has the advantage of being replicable, and most shoal responses appear realistic in the early stages of a simulated attack.

Fish shoals tend to have an oblate spheroid shape that may reduce their envelope of visibility but, unlike the situation in air, there is only a minor advantage of shoaling as a defense against detection by a searching underwater predator. The scattering of light under water means that the distance at which a shoal may be detected is almost the same as for an individual fish. And in fact, detection is not so important, because fish in shoals are often accompanied closely by many of their predators, like big game herds in the Serengeti.

Apparently, fish in a group have a clear advantage over singletons through being less likely to be the one selected as a victim by an attacking predator. Logically, this benefit should be in proportion to the reciprocal of group size, the 'attack dilution effect'. But to check if dilution may cause shoaling to evolve, we must compare the risk to individual fish that adopt solitary or grouping strategies. In both cases, the joint probability of being in the group attacked by a predator, and of being the victim picked out of the group is identical (because the dilution probability within groups is exactly balanced by the attack probability among groups). However, a singleton joining a larger group decreases its risk. But having joined, individuals in the now enlarged group bear an increased risk, but have no way of reducing it by leaving – unless they exclude newcomers, a behavior not observed in fish schools. This is termed the 'attack abatement' effect, and is an example of an evolutionarily stable strategy (ESS).

In larger shoals, experiments show vigilance to be a major anti-predator advantage. Fish detect a threat earlier because of the many eyes in the group. It is important to observe fish closely in these experiments; recording alarm or flight is not sufficient, because fish may choose to stay feeding when a predator approaches as they are less nervous, or more confident of successful escape. In experiments with minnows, counting two subtle behaviors were the key: 'skitters' are rapid alarm signals, and 'inspections' are approaches towards the predator (discussed in more detail below). In larger shoals, both of these behaviors were more frequent earlier in a predator attack. In another experiment, a clever protocol demonstrated a faster reaction of neon tetras to a randomly located light flash.

In a wide range of fish shoals, experiments have generally shown a declining success of attacks by fish, cephalopod, mammal, and bird predators as shoal size increases. Moreover, fish separated from the shoal are more likely to be eaten, and predators may learn not to attack larger groups. All of these phenomena derive from a large repertoire of anti-predator tactics performed by fish in shoals ranging from sandlance to tuna. Fish select tactics from the repertoire partly at random, to counteract predator learning, and partly according to the likelihood of an attack. Many fish can tell by olfaction when predators are nearby, and can pick up subtle visual cues as to their state of attack readiness and hunger.

Compaction, where fish reduce distance to neighbors and become more polarized, allows fish to take advantage of coordinated escape tactics. Compact groups may glide slowly out of predator range (**Figure 1**), taking advantage of cover provided by weed or rock. A 'pseudopodium' of fish may join two subschools like a thin neck along which individual fish may travel, so that one potential target next to the predator shrinks while the other enlarges surreptitiously. In the 'fountain maneuver' fish initially flee in front of the predator, turn, pass alongside in the opposite direction, and then turn again to reassemble behind the predator. This serves to relocate a target out of attack range. Tightly packed balls of fish, seen in response to severe attacks by cetacean, bird, and fish predators on schools, inhibit or deflect attack, like the 'silver wall' caused by highly polarized schooling fish suddenly changing direction in unison. And there are a few reports of 'mobbing' in fish; inhibition of predator attack by physically pushing it away. Information about approaching danger travels rapidly across compact polarized schools, termed the

'Trafalgar effect' because of its resemblance to the flag signaling system invented early in the nineteenth century by the British Navy. Impressively, the message can move among schooling fish two to seven times faster than the approach speed of the predator.

Predators attacking prey, like humans operating radar screens, become less accurate as the number of potential targets increases. This is known as the 'confusion effect', and probably results from overloading the peripheral visual analysis channels of the brain (the midbrain optic tectum in fishes). Confusion could also be cognitive, as in a dog unable to choose between several juicy bones. Two tactics in the anti-predator repertoire of shoaling appear to be designed specifically to exploit predator confusion. First, 'skittering' (see above) may confuse predators attempting to lock-on to a target. Secondly, 'flash expansion' (**Figure 1**) occurs when fish in a polarized compact school rapidly accelerate away from the center, like an exploding grenade (a behavior brought about by the 'Mauthner system' of rapid nerve fibers). One disadvantage of flash expansion is being found alone by the predator afterwards, and so there is a premium on rapid reassembly, or hiding if refuges are nearby.

One of the most interesting discoveries among anti-predator tactics in fish shoal is 'predator inspection' behavior. 'Inspecting' fish leave the shoal and swim towards an approaching predator, halt for a moment and then return to the group. It is clear that inspection carries a real risk of being eaten, and inspectors behave to try to minimize this risk. How can such evidently dangerous behavior have evolved? Clever experiments, where fish can see school fellows but not an attacking predator, have demonstrated transfer of information from inspecting fish about an impending attack. Strikes by pike on minnows are anticipated after inspectors return to a shoal, and predators may be inhibited in their attack by seeing inspections. (A counter-intuitive suggestion that inspection invites attack, giving prey an advantage in controlling how attack occurs, has received no experimental support.) The repetition rate of inspections may code for the degree of danger. Moreover, fish perform inspections in larger groups as risk increases, so dilution of danger during inspection may be a way of mitigating the costs. It seems that the sheer advantages of information about the predator derived from inspection may outweigh the risk of getting eaten. Under this theory, inspection behavior has evolved because, although noninspectors will get some of the benefit through transfer of information concerning risk, fish that inspect can act on more accurate information about the predator.

An alternative view states that, although inspectors are more likely to die than noninspectors, the behavior has evolved because genes coding for inspection increase in the population through kinship or in some other way. If fish in shoals were genetically related (see below), inspection behavior could evolve to save kin in the shoal. A second way in which altruistic inspection behavior might evolve is revealed by game theory. 'Tit-for-tat' is helping another at cost to oneself immediately after receiving benefit from the same move by the other player. A series of elegant experiments involving mirrors and companion inspectors revealed that tit-for-tat may be implicated in the evolution of inspection behavior. But, at present, it is not clear which of the two competing theories for the evolution of inspection behavior is correct.

Recent studies into the ways in which fish trade-off feeding and predator risk have led to productive insights of the evolution of shoaling behavior. For example, shoaling fish foraging on patches which were either safe or where predators might appear, altered feeding in almost perfect proportion to risk and food: a 'risk balancing' trade-off (see **Figure 2**). Elegant experiments demonstrate that hunger increases risky behavior. Even more complex sets of trade-offs occur in the wild, and involve motivational factors such as mating, hunger, food availability, competition, and perceived predator risk. As yet, few such complex circumstances have been investigated with experiments that test theoretical expectations.

Successful predators on fish schools employ a number of clever devices to counteract the schooling prey's defenses. For example, predators may attack

shoals from below at dawn and dusk, when the prey are silhouetted and dim light gives predators' eyes an advantage. This 'twilight hypothesis' was confirmed in experiments with a dummy pike and shoaling minnows. Nevertheless, minnows can compensate using inspection behavior to shift feeding to a safer location.

Many predators on shoaling fish are considerably larger than their fish prey and one common anti-schooling technique employed by tuna, sawfish, bluefish, marlin, swordfish, thresher sharks, and dolphins is to disrupt a prey school and split off individuals that may subsequently be pursued without confusion costs. Central positions in the school might be safer, simply because they are not on the edge where predators arrive first. Nevertheless, specialized predators like jacks attack the centre of schools at high velocity. Stripe-and-patch patterns on bird and cetacean bodies and flippers may serve to disrupt schools through the 'optomotor response'. One ingenious experiment showed that rotating striped penguin models depolarized anchovy schools. Some predators herd their prey in one way or another. For example, humpback whales blow bubble rings and rise to the surface to engulf entire schools of capelin. Other fish shoal predators themselves school and hunt in packs. For example, schools of sailfish may herd prey in rings formed by their large raised dorsal fins. Barracuda, jacks, tuna, yellowtail, and perch are species that hunt in schools.

A Genetic Basis for Behavior in Fish Schools

What are the origins of JLS dynamics and the impressive switches among the spectrum of behaviors seen in individual fish that shoal? Do these adaptive behaviors have a genetic basis or are they learned in some way from experiences in early life? Either of these mechanisms can produce adult animals with adaptive behavior.

This question has been addressed in elegant experiments that raised groups of fish from the egg. Minnows were collected from two locations in Britain; one a river in England where minnows lived with pike, the other a river in Wales where pike were absent. Previously, it had been found that fish from the wild population living with pike had more effective anti-predator behavior. The fish were spawned in aquaria, and the eggs from each location were divided into two batches. From each location, one batch experienced a test with a model pike at 3 months old, while the other batch had a sham test. When adult at 2 years old, all four batches of

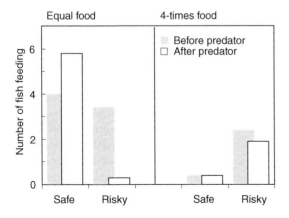

Figure 2 Results of an experiment demonstrating a risk balancing trade-off between food and risk of predation in schooling minnows offered food on two feeding patches. Number of fish feeding were counted before and after fish saw a diving bird at the risky feeding patch. When four times as much food was present, the fish accepted the risk and fed. (Data from Pitcher *et al.* 1988.)

minnows were tested with model pike. Adult minnows grown from the population that lived with pike, and had seen the pike model when juvenile, performed better than individuals which had not seen the pike, suggesting that they had learned from their early encounter. Conversely, minnows from the non-pike population were not able to learn from early experience. These results suggest that both genes and learning are important, but that there is a genetic basis to what can be learned, an example of the 'innate schoolmistress' – the genetic programming of animals' learning agenda.

Some Other Functions of Schooling

A number of other advantages of shoaling have been documented. For example, shoals of sticklebacks have a lower per-individual incidence of ectoparasites and the fish formed larger shoals in the presence of the ectoparasite.

Fish swimming in schools may make better estimates of the right direction in which to swim. For example, directional changes appropriate to either good or poor conditions of food, salinity, temperature or oxygen, spread through schools of migrating herring. A wave of turns passes through the school to fish that have yet to encounter the new good or bad conditions themselves: a behavior termed 'synchrokinesis'.

Swimming in a school may bring energy saving through some sort of hydrodynamic advantage involving the chain of rotating vortices set up by fishes' tails (these vortices are the main mechanism producing thrust in fish swimming.) Experiments with saithe, cod, and herring produced no support for hydrodynamics when quantitative predictions from theory were tested using tens of thousands of frames of film. And there is a more serious objection to the theory. Since only fish behind the leaders get energy savings, leaders would choose to fall back to get it, and so we would see continuous jostling for position, something that is not observed. Many experiments report lower oxygen consumption in larger shoals, but on its own this is not sufficient to prove hydrodynamic advantage, because fish are calmer in larger groups. However, recent experiments reporting energy saving from slower tail beats in fish at the rear of a school lend support to the hydrodynamic theory of fish schools.

Schooling, Fisheries and Pattern in the Ocean

The application of knowledge about fish shoaling lies in its impact on human fisheries. More than 60%

of the world's fisheries are for species that are frequent schoolers, and nearly all species shoal to some extent. Modern commercial fishing gear, such as mid-water trawls and mechanized purse seines, have been designed to exploit schooling fish; entire schools of tuna, mackerel, or herring may be caught by a purse seine, which may be over a kilometer in diameter. Purse seine technology has itself replaced a clever device for catching schools of giant bluefin tuna that was in use in the Mediterranean since the time of the ancient Greeks. The 'tonnare' fishery consists of kilometers of long guide fences leading to traps constructed from sisal rope, representing a preindustrial technological solution to catching schools of giant 3 m long fish migrating along the coastline at 20 knots. Today there is only one 'tonnare' fishery left, operated annually for tourists in Sicily.

The most important applied aspects of shoaling behavior are population collapse and range reduction. In both of these phenomena, shoaling can cause spatial problems that are hard to correct by intervention from management. The behavioral adaptations of pelagic fish in feeding, spawning, migration, and schooling are driven by the opportunity to exploit transient high levels of planktonic production: the highest plankton production levels are intrinsically patchy. Planktivorous fish constantly move in groups to minimize predation risk and get foraging advantages, to seek out these ephemeral food sources. This is an oceanographic perspective on why schooling and JLS dynamics have evolved, fitting pelagic fishes to their niche by determining their ocean distribution.

Unfortunately, there is a mismatch between human fisheries and this behavior. On an annual timescale, the mismatch generates volatility, range, and stock collapse. On a timescale of decades, the human response to uncertainty in pelagic fisheries has been to develop ever more effective levels of fish catching technology (see **Figure 3**).

When fish populations collapse from natural changes in habitat, from unsustainable levels of human harvest, or from an unholy alliance of both of these factors, two linked phenomena generally occur; stock collapse and range collapse. Stock collapse is defined as a rapid reduction in stock abundance, and is distinguished from short-term natural fluctuations. Range collapse is a progressive reduction in spatial range. Although some seek environmental correlates of collapse, sufficiently powerful mechanisms driving stock collapse can be generated by the impact of harvest on fish population dynamics and fish behavior. Range collapse makes a stock collapse more serious because the concentration of remaining fish into a reducing area

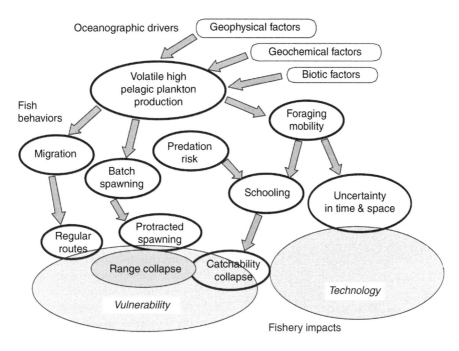

Figure 3 Diagram showing the factors that may have brought about the evolution of fish schooling, and the mismatch of these factors with human fisheries. For further details, see text. (Redrawn from Pitcher 1995.)

makes fish easier to locate and concentrates the fishing power of a fleet built with profits from a previous era of higher abundance (see **Figure 4**).

Catchability is the proportionality coefficient between fishing effort and population abundance. In a collapse event, schools stay the same size as before and so the catch rate within a school stays high. Also, schools, rather than individual fish, are located by fishing vessels. These two factors mean that catchability increases as range decreases, until the last school is caught. For schooling fish, the catch rate, conventionally used to predict abundance, stays constant as population abundance declines, bring about a rapid collapse (see **Figure 5**). Acting together, these forces can cause a great reduction in abundance and this is thought to be the mechanism behind disastrous collapses in many fisheries, such as the Monterey sardine in the 1950s, the NE Atlantic herring in the 1970s, and the Newfoundland cod in the 1980s, all of which had profound economic consequences.

In one of the most powerful theories underlying range collapse, known as the 'basin' model, spatial collapse is driven by environmental forces through competition among fish for optimal habitat. An alternative model is based on the shoaling behavior of fish where leaving and joining behavior adjusts school size to local conditions (see **Figure 6**). Schools need to be proximate for such meetings. When a population is greatly reduced by fishing, schools that

do not have encounters with others move faster until they do, and this process concentrates schools in an area of ocean. The process proceeds until the spatial collapse is complete. In practice, both basin and school size adjustment mechanisms may operate. The size-adjustment hypothesis raises the prospect of obtaining cheap diagnostics of impending collapse by monitoring the behavioral and spatial parameters of shoaling fish.

The model that best describes schooling over a fish's life history is analogous to that of the metapopulation, where groups comprised of essentially random individuals assemble for periods of their life history and then split up. School formation and dissolution is on a more rapid timescale, occurring within each phase of the life history; termed 'metasociality'. A pelagic schooling species like herring is made up of meta-populations that assemble to breed on spawning grounds, are advected by ocean currents as larvae, and then, as juveniles and adults, adopt a dynamic schooling regime during their feeding migrations. Individuals join and leave schools according to their perception of an ever-changing mix of predator and feeding trade-offs. So at each stage and on each timescale, individuals are shuffled by the behaviors that have evolved to maximize fitness.

But, at each level, these processes are not totally random. Three types of ocean processes (upwellings, gyres, and fronts) act as retention zones for fish

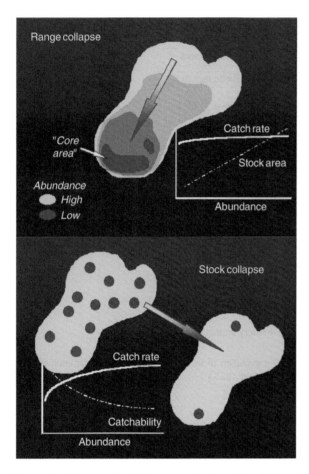

Figure 4 Diagrams illustrating range (top) and stock (lower) collapse in schooling fishes. Inset graphs plot catch-per-unit-effort (= catch rate), population range, and fish catchability against fish population abundance.

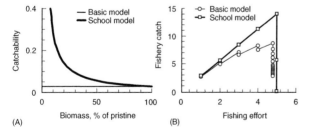

Figure 5 Graphs showing two alternative models of fisheries for schooling species subjected to severe depletion causing a collapse. 'Basic model' is the classic 'surplus production' model used in assessing fisheries; 'school model' takes account of school dynamics as described in the text. (A): Catchability plotted against stock biomass. 'Basic model' assumes constant fish catchability; in 'school model' catchability increases as biomass falls. (B): Simulated annual fishery catch plotted against fishing effort showing a rapid collapse in schooling fish. (Data from Pitcher 1995.)

larvae and allow long-term persistence of ocean fish populations. Moreover, homing to the natal area is a widespread basic trait in bony fishes (the obvious examples such as salmon are but extreme cases). In herring, there is a genetic basis for spawning at a particular time of year and for homing to a general spawning area, but in both Atlantic and Pacific herring it is unclear how much homing there is to precise spawning localities. These have very important implications for fishery management, since locally based populations require more precautionary management to sustain local fisheries and preserve genetic diversity. Genetic studies have generally failed to show much evidence for local stocks, but many argue that local populations were wiped out long ago by inshore herring fisheries.

Evidence is emerging to support the idea of fidelity of fish to a shoal. Early work on freshwater perch and reef grunts in the wild suggested this, but experiments in many laboratories never proved consistent allegiance to particular schools. Recoveries of tagged tuna from fisheries did not support school fidelity either. However, recent work on yellowfin and bluefin tuna (and white sharks) in Hawaii and Australia using sophisticated archival and acoustic tags has demonstrated regular homing to very precise coastal locations after days and even months elsewhere. Here, it looks like schooling predators repeatedly cruise a huge range looking for food, while minimizing their own predation risk, so high school fidelity may be more a consequence of this behavior than any active choice of particular individuals to swim with, which is what would be required for fidelity to be regarded as a trait intrinsic to fish schools.

There is almost no evidence of genetic relatedness in fish schools. Isozymes and mitochondrial DNA among individual fish sampled have been sampled from the same schools in the wild several times, but no close genetic ties were found. In fresh water, minnows and sticklebacks in the same watershed often have closer genetic affinity, but there is no link to schools. The lack of kinship is not surprising theoretically as it would have to provide a major selective advantage to outweigh the benefits of having a flexible school size that can respond rapidly to local predator/food trade-offs through JLS decisions. This makes some unpublished work on anchovy and sardine schools sampled with purse seines at night in the Adriatic Sea even more intriguing. Fish were taken when two fishing vessels were about 5 km apart and had set their purse net within 20 minutes, so the same fish school could not have swum to the other vessel. Comparisons based on DNA fingerprinting showed that anchovies within each school were more closely related than between schools. This was not the case

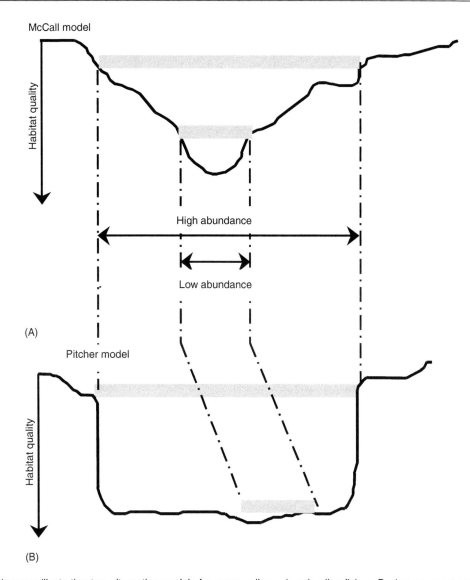

Figure 6 Diagrams illustrating two alternative models for range collapse in schooling fishes. Basins represent spatial distance in horizontal dimension, habitat quality (reversed) in vertical dimension. Thick shaded bars represent the fish population at high and low abundances. (A) Basin model proposed by Alex McCall, where the depleted population collapses to the best quality habitat as individal fish compete for resources. (B) School size adjustment model proposed by the author, where the depleted population collapses to a random location as schools maintain their size in relation to local conditions by packing in a mesocale pattern that allows exchange of members. (After Pitcher 1997.)

for sardines, perhaps reflecting the higher mobility and range adopted by this species.

The dynamics of schooling decisions of young herring on their spring feeding migration in Norwegian waters were studied with a very high resolution scanning sonar originally designed to detect small floating nonferrous mines. The machine could track and resolve individual herring in schools at 300 m range. Herring schools could be sampled using a precisely controlled mid-water trawl so that ages, stomach samples, and other fish swimming with the herring were measured. The findings were dramatic.

Herring schools were found to be accompanied by a mix of predators, rather like game herds on the Serengeti. Cod and haddock swam with the herring, picking off prey from time to time, and causing minor changes to school structure, but not dispersal of the school. Saithe swept in to attack as a fast-moving school, causing the herring to bring their last line of defense into play – a rapid dive to 200 m. Driving the research vessel at them causes herring to dive like this. Moreover, there was a dynamic regime of school splitting and joining as herring schools distributed throughout this region of ocean encountered each

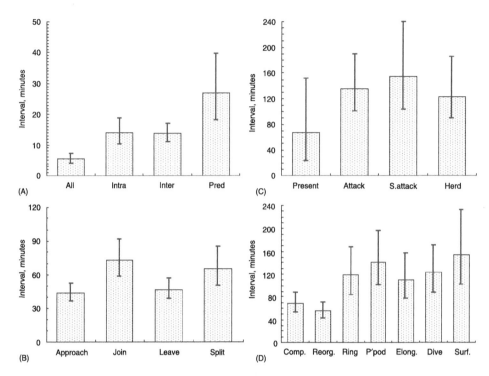

Figure 7 Mean frequency intervals (columns), and 95% confidence limits (bars), for 15 recorded behavioral events scored for herring schools observed with high-resolution sonar in the Norwegian Sea. (A): total of all events, and for intra-school, inter-school, and predator categories of behavior. (B): Four inter-school behaviors. (C): Four putative predator interaction behaviors. (D): Seven intra-school behaviors. (Data from Pitcher *et al.* 1996.)

other. On average an 'event' in a herring school occurred every 5 minutes; school encounters occurred every 15 minutes and splitting and joining events occurred every 30 minutes (see **Figure 7**). Events that were tentatively distinguished as predation happened every 25 minutes. Moreover, the sonar enabled the visualization of school formations such as 'rings' and 'pseudopodia' previously only studied with light. The overall conclusion was that herring school decisions were shaped by trying to minimize predation.

The distribution of older herring migrating northward in much deeper water 200 miles offshore appears to be limited by the southern edge of the polar front. They feed initially on copepods overwintering with eggs in deep layers, and then on euphausids near the surface at night as the spring bloom begins. Here there are no fish predators swimming with the herring schools. The herring exhibit a marked nightly vertical migration, often dispersing into loose shoals when they reach the surface to feed, whereas in the day they are found in exceptionally deep, dense, nonfeeding schools at around 300 m or more (see **Figure 8**). The sonar revealed night attacks on the herring by fin whales, causing great school compaction and rapid diving. On the surface, about a dozen fin whales were seen in

the area. It was calculated that even a few fin whales cruising the Norwegian Sea might have a major impact on the evolution of herring behavior. For example, 12 fin whales could easily search the whole Norwegian Sea during a 6 month season. In this population, herring live to 12–15 years of age, each year taking part in the spring and summer feeding migration after spawning, and then assembling in a fiord in northern Norway to overwinter. Now, an individual herring has to meet a feeding fin whale only once in this life history to die. In fact, the chances are that it will meet fin whales at least once per year, and hence 10 times during its life. Such selection pressure seems sufficient to shape the behavioral schooling decisions that drive the ocean movements of herring.

In fact, ecologists have recently come to believe that much of the spatial behavior of fish is driven in very profound fashion by attempts to minimize predation: schooling represents one of these strategies. Manipulation of cover and food in experimental lakes has revealed that fish choose habitats as refuges and feed only when hunger and reward provide a beneficial trade-off with the risk of being eaten themselves. Where cover is absent, as in pelagic and open ocean habitats, schooling is the best defense. Modeling of

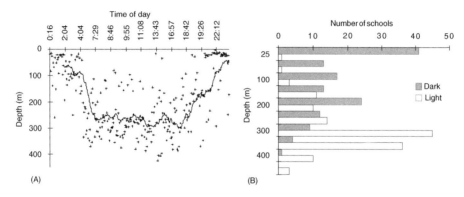

Figure 8 Diurnal changes in vertical distribution of herring recorded by echosounder in the Norwegian sea. (A) Depth distribution of herring schools – line is running average. (B) Number of schools with depth zone at day and night. (Data from Mackinson *et al.* 1999.)

Table 2 Analysis of the implications of various fish shoaling traits for distribution dynamics on the ocean, especially range collapse and resilience to fishing

Social behavior regime		Shoal	Cannibal	Fidelity	Relatedness
(schema)	Undepleted				
(schema)	Depleted				
Shoaling		Shoaling	No shoaling	Smaller shoals	Smaller shoals
Ecology	Range	Collapse	Patchier	Reduced	Patchier
	Refugia	Few	Many	Some	Many
Behavior	Join	High	Low	High	Low
	Stay	Low	Low	High	High
	Leave	High	Low	High	Low
Fishery	Resilience	Fragile	Resilient	Less fragile	More resilient
	Rebuild time	Slow	Fast	Medium	Faster

Rows 1 and 2 show schema representing school size and distribution before and after a severe depletion event. For further details of the four columns, see text.

predator–prey interactions in ecosystem simulation models has taken advantage of this finding. Refuge behavior produces more realistic and stable dynamics than classical Lotka-Volterra equations.

The spatial pattern of fish in the ocean depends on the type of behavior associated with shoaling. The following analysis assumes that frequent challenges to school membership arise from the actions of predators, the detection of food, and physical process in the ocean. It assumes that depletion is brought about mainly by overfishing, although population may also be reduced by environmental changes.

With normal shoaling, as described above, range collapse can occur, so that rebuilding of populations is from a small number of refugia (**Table 2**). The probabilities of 'join' and 'leave', decisions of individual fish probabilities are high while 'stay' is low, in order to adjust group size to local conditions. This leads to populations fragile to overfishing and slow to rebuild. On the other hand, nonshoaling piscivorous fish, like

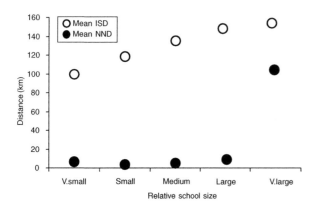

Figure 9 Relationships between average distance to nearest neighbor school distance (NND), average inter-school distance (ISD), and school size for echosounder data on herring in the Norwegian sea. (Data from Mackinson et al., 1995.) The results show a mesoscale pattern of school patches in the ocean for all except very large herring schools.

cannibalistic hake, space out rather than school, and 'join', 'leave', and 'stay' decisions are all of low probability. Hence such fish tend not to exhibit range collapse, serious depletion results in patchy abundance throughout the range, and rebuilding is fast because it can occur from many refugia. In other words, these species exhibit resilience in the face of depletion or environmental perturbations. In support of this idea, hake species differ in their degree of cannibalism, and this seems to be reflected in their relative resilience. The two strategies are summarized in the first two columns of **Table 2**.

Genetic relatedness (column 4 in **Table 2**) within schools implies high fidelity, so that only the 'stay' decision probability is high. Under this behavioral regime, schools shrink with population depletion and there is no reason for schools to be near each other, so that range collapse is less likely and there are more refugia from which the population may rebuild. This implies that resilience is higher, while the opportunity to adjust to local conditions is lower. A documented collapse and rebuilding of anchovies in the Adriatic Sea in the 1980s may fit this scenario.

A behavioral regime of high intrinsic fidelity within schools (column 3 in **Table 2**) would shift 'join', 'stay', and 'leave' decisions to high probability, so that, as the number of fish reduced, schools would shrink. However, schools might stay close together so that members could reassemble with their former schoolmates. Under this regime, fragility and rebuilding time would be intermediate. Measurements of the mesoscale distributions of fish shoals can distinguish among the hypotheses above. One attempt at measuring the patchy ocean distribution of herring schools is shown in **Figure 9**.

See also

Fish Feeding and Foraging. Fish Locomotion. Fish Predation and Mortality. Mesopelagic Fishes.

Further Reading

Fernö A, Pitcher TJ, Melle W, et al. (1998) The challenge of the herring in the Norwegian Sea: making optimal collective spatial decisions. Sarsia 83: 149–167.

Guthrie DM and Muntz WRA (1993) Role of vision in fish behavior. In: Pitcher TJ (ed.) Behavior of Teleost Fishes, pp. 89–128. London: Chapman and Hall.

Klimley P and Holloway CF (1999) School fidelity and homing synchronicity of yellowfin tuna. Marine Biology 133: 307–317.

Krause J, Ruxton GD, and Rubenstein D (1998) Is there always an influence of shoal size on predator hunting success? Journal of Fish Biology 52: 494–501.

Krause J, Hoare D, Croft C, et al. (2000) Fish shoal composition: mechanisms and constraints. Proceedings of the Royal Society B 267: 2011–2017.

Krause J, Hoare D, Krause S, Hemelrijk, and Rubenstein D (2000) Leadership in fish shoals. Fish and Fisheries 1: 82–89.

Mackinson S, Sumaila R, and Pitcher TJ (1997) Bioeconomics, and catchability: fish and fishers behavior during stock collapse. Fisheries Research 31: 11–17.

Mackinson S, Nøttestad L, Guénette S, et al. (1999) Cross-scale observations on distribution and behavioral dynamics of ocean feeding Norwegian spring spawning herring (Clupea harengus L.). ICES Journal of Marine Science 56: 613–626.

Murphy KE and Pitcher TJ (1997) Predator attack motivation influences the inspection behavior of European minnows. Journal of Fish Biology 50: 407–417.

Naish K-A, Carvalho GR, and Pitcher TJ (1993) The genetic structure and microdistribution of shoals of Phoxinus phoxinus, the European minnow. Journal of Fish Biology 43(Suppl A): 75–89.

Pitcher TJ (1983) Heuristic definitions of shoaling behavior. Animal Behavior 31: 611–613.

Pitcher TJ (1992) Who dares wins: the function and evolution of predator inspection behavior in shoaling fish. Netherlands Journal of Zoology 42: 371–391.

Pitcher TJ (1995) The impact of pelagic fish behavior on fisheries. Scientia Marina 59: 295–306.

Pitcher TJ (1997) Fish shoaling behavior as a key factor in the resilience offisheries: shoaling behavior alone can generate range collapse in fisheries. In: Hancock DA, Smith DC, Grant A, and Beumer JP (eds.) Developing and Sustaining World Fisheries Resources: The State of Science and Management, pp. 143–148. Collingwood, Australia: CSIRO.

Pitcher TJ and Alheit J (1995) What makes a hake? A review of the critical biological features that sustain global hake fisheries. In: Pitcher TJ and Alheit J (eds.) Hake: Fisheries, Ecology and Markets, pp. 1–15. London: Chapman and Hall.

Pitcher TJ and Parrish J (1993) The functions of shoaling behavior, In: Pitcher TJ (ed.) *The Behavior of Teleost Fishes*, 2nd edn. pp. 363–439. London: Chapman and Hall.

Pitcher TJ, Lang SH, and Turner JR (1988) A risk-balancing trade-off between foraging rewards and predation hazard in shoaling fish. *Behavioral Ecology and Sociobiology* 22: 225–228.

Pitcher TJ, Misund OA, Ferno A, Totland B, and Melle V (1996) Adaptive behavior of herring schools in the Norwegian Sea as revealed by high-resolution sonar. *ICES Journal of Marine Science* 53: 449–452.

FISH VISION

R. H. Douglas, City University, London, UK

Introduction

The eyes of all vertebrates are built to a common plan, with a single optical system focusing radiation on a light-sensitive retina lining the posterior part of the eye (**Figure 1**). It is, none the less, impossible to describe a general fish eye. Fish inhabit almost every conceivable optical environment, from the deep sea where darkness is punctuated only by brief bioluminescent flashes to the sunlit surface waters, from the red peat lochs of Scotland to the green coastal waters of the English channel and the blue waters of a tropical lagoon (**Figure 2**). Fish therefore live at all levels of illumination and are exposed to many different spectral environments. The fish visual system has adapted superbly to these various conditions.

Image Formation

Although most fish eyes are approximately symmetrical, with spherical lenses (**Figure 1**), there are several exceptions. Some bottom-dwelling flatfish and rays, for instance, have an asymmetrical eye so that the retina is at varying distances behind the lens (**Figure 3**). This so called 'ramp retina' results in objects at different distances appearing in focus at different points on the retina. Odd-shaped eyes are also observed in some deep-sea species. Since the last vestiges of sunlight come from above the animal, their eyes are often positioned on the top of their head. In order to accommodate as large an eye as possible – to maximize sensitivity – within a head of reasonable size, the sides of a normal eye have been removed during the course of evolution, resulting in a tubular-shaped, upward-pointing, eye (**Figures 4** and **8B–D**).

While in terrestrial vertebrates the major refractive surface of the eye is the cornea, the lens serving primarily to adjust the focus, in aquatic animals the only important refractive element in most fish is the lens because the similarity of the refractive indices of water, cornea, and aqueous humor neutralize the optical power of the cornea. Thus, while the lenses of diurnal terrestrial animals are generally flattened and less powerful (nocturnal animals tend to have rounder lenses, as well as wider pupils, as their shorter focal length results in a brighter image), the lenses of most teleost fish are spherical. However, some larger-eyed teleosts, as well as many elasmobranchs, do possess somewhat flattened lenses, and the surface-dwelling 'four-eyed' fish, *Anableps anableps*, has an asymmetrical lens to allow simultaneous vision in air and water (**Figure 5**).

Since the cornea is optically ineffective underwater, the pupil of a fish eye is usually immobile (see below), and the iris does not cover significant parts of the lens, the quality of the lens dictates the optical quality of the whole eye. Two major optical aberrations could theoretically affect the quality of the image produced. First, light entering the lens at different points might be focused at different distances behind it. Such 'spherical aberration' is minimized in fish by a refractive index gradient within the lens (high refractive index in the centre, decreasing toward the edge). Second, since the refractive index of a substance depends on the wavelength of light, different wavelengths will be focused at different points ('chromatic aberration'). Although fish lenses do suffer small amounts of both spherical and chromatic aberration, in general they are of high optical quality. In most species the image quality, and hence visual performance, will be unaffected by these small imperfections, since the lens provides a better image than can be resolved by the retina given its photoreceptor spacing. At rest an eye can be focused in one of 3 ways. Parallel light rays entering the eye either focus on the retina (emmetropia), behind it (hyperopia) or in front of it (myopia). Although determining the refractive state of animals is fraught with difficulty, most are likely to be emmetropic, resulting in a resting eye focused from an intermediate distance to infinity. Some teleosts, however, are myopic, with an eye focused for close objects, which, given the often limited visibility underwater, appears appropriate. The suggestion that some elasmobranchs may be hyperopic, resulting in an unfocussed image at rest, on the other hand, seems unlikely.

Whatever the refractive state of the resting eye, most animals have the ability to change the point at which it is focused. While mammals, birds, and reptiles perform such accommodation largely by deforming the shape of their soft lens, and in some instances the cornea, amphibia and fish accommodate by repositioning their hard spherical lenses. In teleosts, one or two retractor lentis muscles

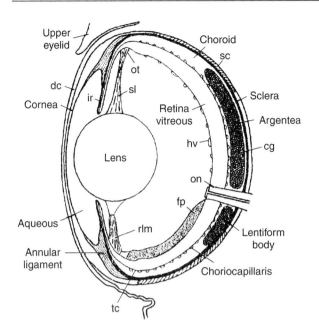

Figure 1 Schematic representation of a vertical median section through a teleost eye: cg = cgchoroid gland; dc = dermal component of cornea; fp = falciform process; hv = hyaloid vessel; ir = iris; on = optic nerve; ot = ora terminalis; rl = mretractor lentis muscle; sc = scleral cartilage; sl = suspensory ligament; tc = position of tensor choroidea. (From Nicol (1989).)

(**Figure 1**) contract to pull the lens backward, focusing the previously myopic eye for distance, in some instances by as much as 40 diopters (**Figure 6**), while in at least some elasmobranchs retractor lentis muscles move the lens forward, closer to the cornea.

Pupil

While constriction of the pupil in response to increased illumination is widespread among other vertebrates, in fish it is largely restricted to elasmobranchs. The very few teleosts that have mobile irises are mainly cryptic bottom-dwelling species among whom pupillary closure may serve to camouflage the otherwise very visible pupil. In all such species the dilated pupil is more or less round. However, in many species, when constricted the pupil consists of either a vertical or horizontal slit, a crescent-moon-shaped opening, or even two small pinholes (**Figure 7**). Such irregular pupils may serve a number of functions, for example, the maintenance of a small depth of field to facilitate distance judgment using accommodative cues, reduction of chromatic aberration, and spatial filtering of the image.

Ocular Filters

Both water molecules and small particles, such as plankton, suspended in the water column, tend to

(A)

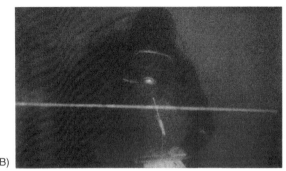

(B)

(C)

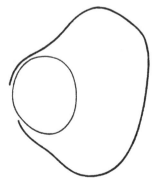

Figure 2 The author in three different bodies of water, illustrating the diversity of the underwater light environment: (A) Cayman Islands; (B) English Channel; (C) Loch Turret (Scotland).

Figure 3 Outline tracing of the lens and retina–choroid border in a stingray (*Dasyatis sayi*), showing the asymmetrical nature of the eye. (From Sivak JG (1980 Accommodation in vertebrates: a contemporary survey. *Current Topics in eye Research* 3: 281–330.))

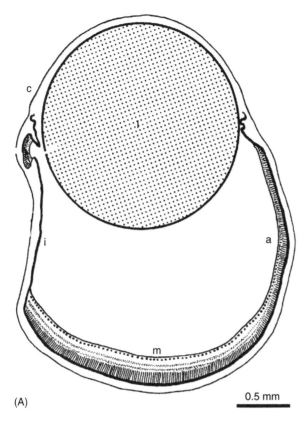

(A)

0.5 mm

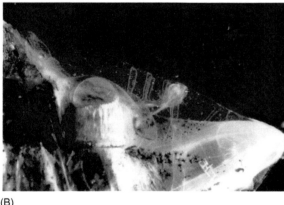

(B)

Figure 4 (A) Camera lucida drawings of resin sections of the tubular eye of *Opisthoproctus grimaldii*: c = cornea; l = lens; m = main retina; a = accessory retina; i = reflective iris. (From Collin SP, Hoskins RV and Partridge JC (1997) Tubular eyes of deep-sea fishes: a comparative study of retinal topography. *Brain Behaviour* and Evolution 50: 335–357.) (B) Lateral view of *Opisthoproctus* sp. showing its tubular eye. (Photograph by N. J. Marshall.)

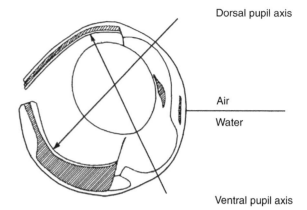

Dorsal pupil axis

Air

Water

Ventral pupil axis

Figure 5 The eye of *Anableps anableps*, half of whose eye lies above water surface and half beneath when the fish is at rest. It is able to see simultaneously in both media by the possession of an aspheric lens positioned such that light entering the eye from the air, and hence going through the refractive cornea, passes through the short axis of the lens and therefore is refracted less. Light from below the animal, passing through the optically ineffective cornea goes through the more powerful long axis of the lens. (From Sivak JG (1976) Optics of the eye of the "four-eyed" fish (Amableps anableps). *Vision Research* 16: 531–534.)

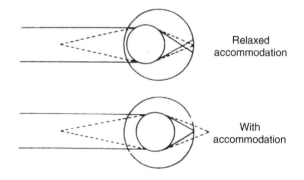

Relaxed accommodation

With accommodation

Figure 6 Schematic representation of a 'relaxed' (myopic) teleost eye in which close objects are imaged on the retina, and an accommodated eye in which the lens has moved toward the retina, focusing more distant objects. (From (1980) Accommodation in vertebrates: a contemporary survey. *Current Topics in eye Research* 3: 281–330.)

scatter light, resulting in a greatly reduced range at which fish will be able to see objects. Such (Rayleigh) scatter is particularly severe at short wavelengths. This part of the spectrum is also most prone to chromatic aberration (see above) further degrading the retinal image, and comprises those wavelengths most likely to damage the eye. It is therefore not surprising that short-wave absorbing filters, which often appear yellow to the human observer, are widespread within the corneas and lenses of fish (**Figure 8**). A similar function is performed by reflective layers in the corneas of some fish resulting in corneal iridescence.

The drawback of such pigmentation is that the overall level of illumination reaching the retina is reduced significantly. Short-wave filters are therefore generally found in diurnal animals inhabiting high light levels. Furthermore, both corneal pigmentation and irridescence are often restricted to, or are densest in, a dorsal region above the pupil. Such 'eyeshades' will selectively reduce the amount of

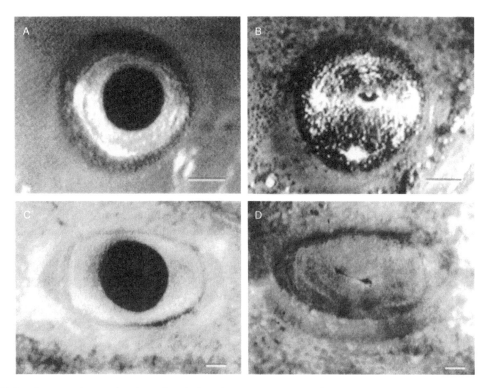

Figure 7 Fully dilated and constricted pupils of the plainfin midshipman *Porichthys notatus* (A and B) and swell shark *Cephaloscyllium ventriosum* (C and D). The scale bar represents 1 mm. (From Douglas RH, Harper RD and Case JF (1998) The pupil response of a teleost fish, *Porichthys notatus*: descrption and comparison to other species. *Vision Research* 38(18): 2697–2710.)

bright downwelling light, while leaving the less intense light, entering the eye along the optic axis, relatively unaffected. Some species have negated the undesirable loss of sensitivity in low light levels by developing 'occlusable' corneas, in which pigment is aggregated around the edge of the cornea in dim light and only migrates to cover the central cornea on exposure to higher light levels. In general animals inhabiting lower light levels lack short-wave absorbing pigmentation and have ocular media that transmit all wavelengths below the infrared down to around 320 nm (**Figure 8**).

Surprisingly, however, yellow lenses are found in some species inhabiting the deep sea (**Figure 8A–D**). This is because there are two sources of illumination here; dim residual sunlight, consisting mainly of a narrow band of radiation between 450–480 nm, and bioluminescence produced by most animals in the deep sea. In the mesopelagic zone (200–1000 m), where both bioluminescence and downwelling sunlight may be present, there is a potential conflict in the perception of these two sources. Although both tend to be most intense around 450–500 nm, bioluminescent emissions often contain more long-wave radiation than the surrounding spacelight. Short-wave absorbing filters will decrease the intensity of downwelling sunlight more than the relatively

long-wave rich bioluminescence, thereby enhancing the contrast of the bioluminescence and making it more visible (see below). At depths below which sunlight has become visually irrelevant (1000 m in the clearest waters but much shallower in most water bodies) animals no longer possess pigmented lenses.

Retinal Structure

As in other vertebrates, the fish retina is a layered structure composed primarily of six different nerve cell types (**Figure 9**). Within this basic framework, however, different species have undergone extensive 'adaptive radiation.'

Species Variation

The majority of vertebrate retinas, including those of most fish, contain two classes of photoreceptor: rods and cones. Cones are employed at higher (photopic) light levels, such as are experienced during the day near the water surface, and in most instances mediate high-acuity color vision. Rods are utilized at lower (scotopic) light levels to maximize sensitivity, usually at the expense of chromatic and spatial detail. Not surprisingly, animals habitually exposed to low light levels have increased their sensitivity by developing

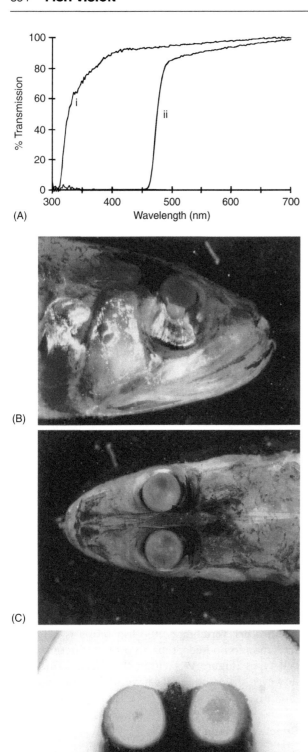

(A)

(B)

(C)

(D)

Figure 8 (A) Spectral transmission curves of the lenses of *Opisthoproctus soleatus* (i) whose tubular eye with a 'colorless' lens is shown in **Figure 4B** and *Scopelarchus analis* (ii) whose tubular eyes with 'yellow' lenses are shown in (B) lateral view, (C) dorsal view, and (D) transverse section. (Photographs by N. J. Marshall.)

more extensive rod systems than those living in brighter sunlit waters. For example, freshwater catfish have reduced cones and much enlarged rod outer segments compared to other species (**Figure 10**). Such specialization is taken to the extreme in deep-sea fish, whose retinas are usually cone-free and in which the rod outer segments have become either extremely long (**Figure 11A**) or banked into several layers (**Figure 11B**).

Sensitivity is further increased in some nocturnal or deep-dwelling species by a reflective tapetum behind the photoreceptors, giving rise to the phenomenon of eyeshine (**Figure 12**). Such tapeta can be located in either the retinal pigment epithelium or the choroid, and may contain a number of different reflecting materials, most commonly guanine. Tapeta are often restricted to the dorsal half of the eye, which receives the lowest-intensity radiation. Since tapeta are usually only beneficial in low light levels and cause image degradation in brighter light, some elasmobranchs have occlusable tapeta in which melanin migrates to cover the reflective material in brighter light.

Most fish, like other vertebrates, contain more than one spectral class of 'single' cone in their retinas (see below). Many fish, like some birds, also have 'double' (**Figures 9** and **13**) and sometimes even triple or quadruple cones. While cones are generally thought to subserve color vision, it is likely that double cones enhance sensitivity. Evidence of a role for the involvement of double cones in increasing sensitivity is provided, for instance, by the observation that some surface-dwelling fish have retinas containing primarily single cones, but that these are replaced by double cones when the animals migrate to deeper water later in their life cycle. Double cones may also be important in the ability of a fish to see the plane of polarized light (see below).

While the photoreceptors of most animals appear to form no consistent pattern, the cones of many fish are arranged in regular mosaics (**Figure 13**). Although the function of such arrangements is not completely clear, they tend to be observed in species where vision is the dominant sense and have a role to play in polarization sensitivity (see below).

Many fish show extensive regional variations in retinal structure. Some have a well-developed fovea, comparable to that of humans, where a high density of cones is located in a retinal depression. In other species, a distinct retinal depression may be lacking, but one still finds 'areas' of high photoreceptor or ganglion cell density. Such regional specializations, which in surface-dwelling species are areas subserving high spatial acuity, can have a variety of different shapes (**Figure 14**) and be located in quite

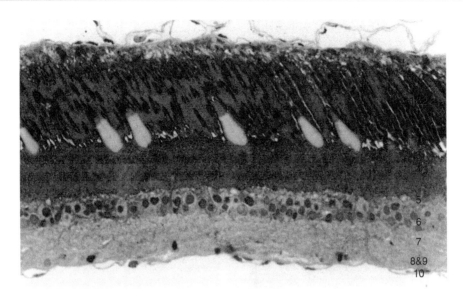

Figure 9 Transverse section of the retina of *Porichthys notatus*. This species displays the 'standard' design of the vertebrate retina divided into 10 layers. (1) Retinal pigment epithelium (RPE). (2) Layer of rod and cone inner and outer segments. As this retina is light-adapted, the melanin within the RPE is dispersed and there is no clear division between layers (1) and (2). (3) External limiting membrane formed by tight junctions between Müller cells and photoreceptors. (4) Outer nuclear layer composed of rod and cone nuclei. (5) Outer plexiform layer composed of synapses between photoreceptors, bipolar, horizontal, and interplexiform cells. (6) Inner nuclear layer comprising cell bodies of horizontal, bipolar, amacrine, interplexiform, and Müller cells. (7) Inner plexiform layer consisting of synapses between bipolar, amacrine, interplexiform, and ganglion cells. This is also the layer in which retinal efferents synapse. (8) Ganglion cell layer composed primarily of ganglion cell bodies but probably also containing some interplexiform and 'displaced' amacrine cell nuclei. (9) Nerve fiber layer containing ganglion cell axons running toward the optic disk. (10) Internal limiting membrane. The fact that *Porichthys notatus* is adapted to life at low light levels is shown by the preponderance of rods and double cones and thin ganglion cell and nerve fiber layers (indicative of convergence and therefore increased sensitivity). The nerve fiber layer is in fact so thin that it is not distinguishable at this magnification. Species inhabiting well-lit environments would tend to have more morphological (and hence spectral) cone types, including single cones, and a thicker nerve fiber layer.

different parts of the retina depending on the most relevant line of sight and habitat structure for a particular species.

Light and Dark Adaptation

During the course of a normal 24-hour period many fish can be exposed to a wide range of light levels, and those living near the surface experience fluctuations in brightness of up to 10 log units during the course of a day. Despite the fact that the dynamic range of photoreceptors is only about 3 log units at any one time, most species are able to maintain some form of vision throughout all of the light–dark cycle. This is achieved through adaptive processes that involve a variety of biochemical, physiological, and morphological changes within the retina. In fish the most prominent structural changes that accompany the diurnal light–dark cycle are retinomotor (photomechanical) movements of the rods, cones, and melanosomes within the cells of the retinal pigment epithelium (RPE) (**Figure 15**). On light exposure, cone myoids contract to position the cone outer segments near the external limiting membrane (ELM) while the rod myoids elongate burying the

rod outer segments behind a dense screen of melanosomes that are dispersed within the apical processes of the RPE (**Figures 15A, 9**, and **10A**). In lower light levels these positions are reversed and the rods now lie closest to the ELM, while the cone outer segments are positioned near the choroid along with the melanosomes aggregated at the base of the RPE cells (**Figures 15B** and **10B**). These movements serve primarily to make optimal use of the retinal space and to protect the rods from being bleached in high light levels.

Visual Pigments

Structure

All visual pigments, which are located within the membranes of the photoreceptor outer segment disks, consist of two components: the chromophore, an aldehyde of vitamin A, which absorbs the light; and a protein, opsin, which determines the spectral absorption characteristics of the chromophore. The chromophore in most vertebrates is retinal, a derivative of vitamin A_1. However, some fish possess an additional chromophore, 3,4-dehydroretinal, derived

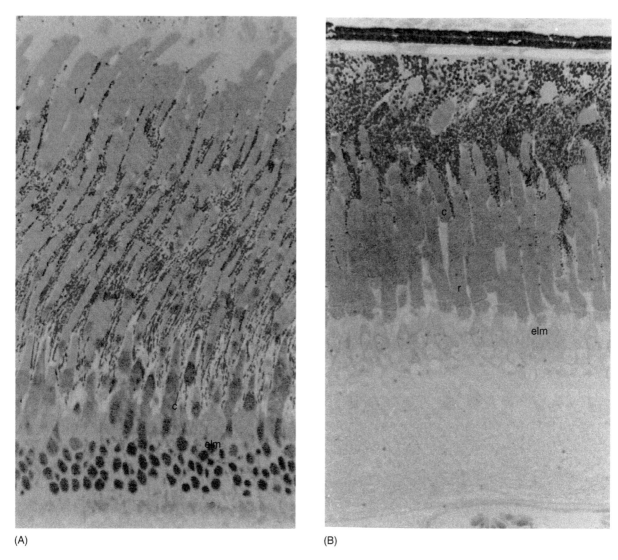

Figure 10 Transverse sections through the retina of (A) the tench (*Tinca tinca*) and (B) a glass catfish (*Kryptopterus bichirris*). Note the relatively larger rod outer segments in the catfish. c = cone; r = rod; elm = external limiting membrane. The rods and cones appear in different positions in the two figures as the tench is light-adapted and the catfish is dark-adapted (see below **Figure 15**).

from vitamin A_2. Visual pigments with retinal as their chromophore are known as rhodopsins, while vitamin A_2-based pigments are referred to as porphyropsins. The retinal photoreceptors of some animals contain 'pigment pairs' in which the same opsin is bound either to retinal or to 3,4-dehydroretinal. A visual pigment consisting of a given opsin and using retinal as the chromophore will have a narrower absorption spectrum peaking at shorter wavelengths than a pigment composed of the same opsin bound to 3,4-dehydroretinal (**Figure 16**).

All opsins have a similar structure, consisting of a chain of around 350 amino acids that crosses the outer segment disk membrane seven times in the form of α-helices. Isolated retinal and 3,4-dehydroretinal absorb at $\sim$380 and 400 nm, respectively. When bound to the opsin, the amino acids 'tune' the

chromophore to absorb at longer wavelengths. Thus, the absorption spectrum of a visual pigment depends both on the identity of the chromophore and on the amino acid composition of the opsin surrounding that chromophore.

Spectral absorption

Visual pigments have bell-shaped absorption spectra (**Figure 16**) that are most easily characterized by their wavelength of maximum absorption (λ_{max}). Of all classes of vertebrate, fish exhibit the greatest range of visual pigments, with λ_{max} values from around 350 nm in the ultraviolet to 635 nm in the red, adapting them to the optical diversity of the aquatic habitat, which can vary enormously in its spectral composition depending on the quantity and identity

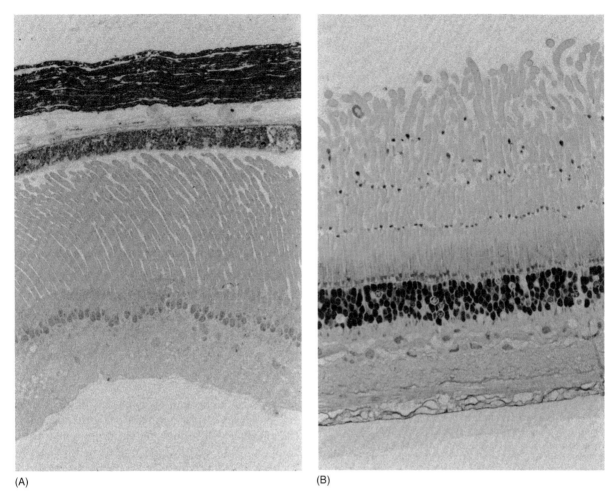

Figure 11 Transverse sections through the retinas of two species of deep-sea fish: (A) *Xenodermichthys copie* displaying a single bank of long rods; and (B) *Notacanthus chemnitzii* with four banks of shorter rods. The RPE has been removed in this section. (Photograph by H.-J. Wagner.)

Figure 12 Freshly caught specimens of the benthopelagic *Chimaera monstrosa* photographed with an electronic flash that is reflected by the tapetum lucidum, giving rise to the appearance of 'eyeshine.'

of material dissolved or suspended within it (see **Figure 2**).

Fresh water, for example, transmits primarily long-wave radiation. Consequently, fish living within it tend to have red-shifted visual pigments compared to those inhabiting oceanic environments where shorter wavelengths penetrate the water most easily. This difference in spectral sensitivity is partly caused by animals in the two environments possessing different opsins, but is also often due to differences in the chromophore. In fresh water many animals use 3,4-dehydroretinal, which, as noted above, forms porphyropsin pigments absorbing at longer wavelengths than rhodopsins, which are the pigments generally used by oceanic animals. Some species, in both the marine and fresh water environment, contain both rhodopsins and porphyropsins in their photoreceptors, and the relative proportions of these two pigment types, for reasons that are both metabolically and ecologically unclear, are influenced by factors such as season, day length, temperature, reproductive state, age, and diet. Interestingly, animals that migrate between fresh water and oceanic environments, such as eels and some salmonids, 're-tune' their visual pigments by changing their chromophore accordingly. Eels additionally begin to manufacture a different rod opsin when migrating

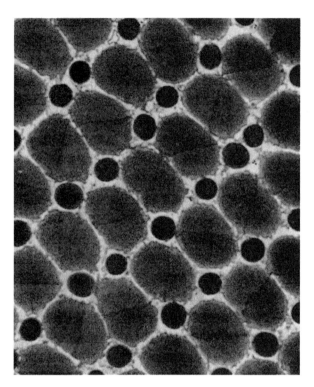

Figure 13 Tangential section through the retina of a rainbow trout (*Oncorhynchus mykiss*) at the level of the cone inner segments. The two members of the double cones (large lighter circles) contain visual pigments with λ_{max} values 531 and 576 nm, while the two types of single cone (small dark circles) contain pigments with λ_{max} 365 and 434 nm. The spatial resolving power (acuity) of the animal is to a large extent determined by the spacing of cones within this 'mosaic'. (From Hawryshyn CW (1992) Polarisation vision in fish. *American Scientist*, March-April: 164–175.)

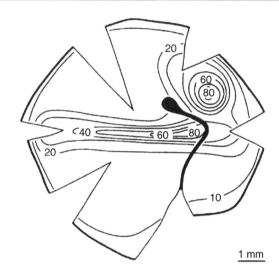

Figure 14 Regional specialization of the blue tuskfish (*Choerodon albigena*) retina. Lines represent isodensity contours (10^3 cells mm^{-2}) of retinal ganglion cells. The optic nerve head and falciform process are shown in black. This species has both a horizontal 'visual streak,' which could be used to enhance the perception of the 'horizon,' and a temporal 'area centralis,' possibly utilized for imaging prey approaching from the front. (From Collin (1997).)

from rivers to the sea. This opsin produces a short-wave-sensitive visual pigment typical, in spectral absorption, of a pigment in the eyes of deep-sea fish.

The relationship between visual pigment absorption and optical environment can most easily be seen by considering rod visual pigments, whose λ_{max} values show a gradual shift from the long to shorter wavelength end of the spectrum in animals inhabiting fresh water, coastal, and deep marine environments (**Figure 17**). While surface-dwelling oceanic species have rod visual pigments with λ_{max} values close to 500 nm, the vast majority of deep-sea fish have their λ_{max} within the range 470–490 nm. This hypsochromic shift is partly an adaptation to the detection of the residual downwelling sunlight, but in most species most probably serves primarily to enhance the perception of bioluminescence. Both residual sunlight and bioluminescence in the deep sea are most intense around 450–500 nm.

The most notable example of visual pigment adaptation to detect bioluminescent emissions is provided by three species of stomiiform dragonfish,

which have suborbital photophores producing far-red bioluminescence with peak emissions above 700 nm, that will be invisible to most animals in the deep sea. Two of these genera, *Aristostomias* and *Pachystomias*, are able to perceive their own far-red bioluminescence through the possession of at least three long-wave-shifted visual pigments (**Figure 18**), while the third, *Malacosteus*, has less red-shifted visual pigments, but long-wave sensitivity is enhanced through the use of a unique chlorophyll-derived, long-wave-absorbing photosensitizer (**Figure 19**). These dragonfish therefore have a 'private' wavelength enabling them to communicate with one another and illuminate their prey without detection by either predators or prey.

The above discussion has centered around rod photoreceptors. However, with the exception of most deep-sea species, the majority of fish have retinas also containing cones. A few species, such as freshwater catfish (**Figure 10B**) and some elasmobranchs, have just a single spectral class of cone (**Figure 20A**). Other fish, such as adult pollack, are dichromats and possess two spectral classes of cone similarly to most mammals (**Figure 20B**). Many fish, however, like humans, have three cone visual pigments (**Figure 20C**). The possession of a fourth visual pigment, absorbing in the ultraviolet is also not uncommon (**Figure 20D**). In general, like rod pigments, freshwater animals have more long-wave-sensitive cone pigments than those inhabiting oceanic environments. Marine fish thus lack the long-wave-sensitive cones frequently seen in

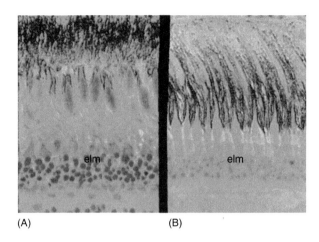

(A)   (B)

Figure 15 (A) Light-adapted and (B) dark-adapted outer retina of the perch (*Perca fluviatilis*). elm = external limiting membrane.

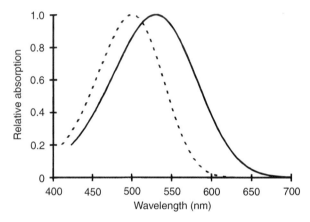

Figure 16 Absorption spectrum of a rhodopsin visual pigment with λ_{max} 500 nm (dotted) and a porphyropsin (solid) based on the same opsin (λ_{max} 530 nm) forming a pigment 'pair.'

freshwater animals. However, extreme shortwave (ultraviolet) sensitivity occurs among both freshwater and marine teleosts.

All of the above arguments assume that optimum advantage is bestowed upon the animal if the λ_{max} of the visual pigment matches the predominant wavelength in the surrounding environment. However, such a 'sensitivity hypothesis' may not always be appropriate. A visual pigment with a λ_{max} 'offset' from the spacelight would enhance the contrast of relatively bright objects with a different spectral radiance to the background (see below).

Visual Abilities

Vision has a major role to play in many aspects of fish behavior, from locating and attracting a mate to ensuring an adequate food supply and orientating in space. The importance of vision, and the constraints it may place on a behavior, can only be assessed by examining the visual capabilities of the animal.

Detection

The most fundamental task of a visual system is to detect the stimulus. In certain conditions, such as a deep-sea fish viewing a bioluminescent source against a dark sunless background, this means increasing absolute sensitivity by catching as many photons as possible. However, in most cases an object is viewed against an illuminated background and the task of the visual system is to detect the contrast between stimulus and background.

Absolute sensitivity The absolute sensitivity of a fish depends on the level of background illumination, which in most instances correlates most directly with the time of day. Sensitivity (one/threshold) is high when ambient illumination is low, while in high light levels, when capturing photons is not at a premium, sensitivity is reduced. In animals that possess both rods and cones, this involves switching between one receptor type and another, which, as already outlined above, in fish involves retinomotor relocation of the photoreceptors. Several studies have shown that fish as diverse as goldfish and lemon sharks have a dark-adapted sensitivity exceeding that of humans. In deep-sea fish, with their larger outer segments, reflecting tapeta and higher convergence ratios (pooling the output of rods together), this sensitivity is likely to be enhanced even further. Thus, in ideal conditions humans can see sunlight down to a depth of around 850 m in the clearest oceans, while some fish may perceive vestiges of sunlight down to around 1000 m. As the ontogenetic formation of rods lags behind that of cones in fish, many larval retinas possess only cones, and sensitivity therefore increases during development.

Contrast Given a sufficient amount of light to see by, the next most important task for a fish is probably to detect the stimulus at the greatest possible distance. An object only becomes visible when the observer can detect a difference between it and the background. Such contrast detection is especially problematical under water as contrast is severely degraded by both light absorption and scattering; the visual environment underwater has been likened to a 'colored fog'. Some species, when the spectral composition of the background and image-forming light differ, improve contrast by filtering out the background using either intraocular filters or visual pigments with λ_{max} values 'offset' from the background radiance.

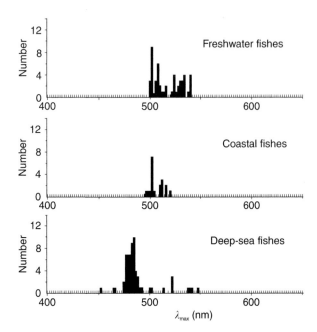

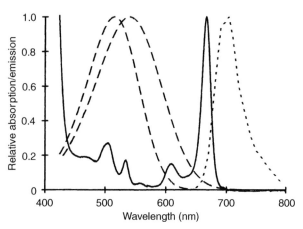

Figure 17 Histogram of λ_{max} values of rod visual pigments from fish inhabiting three different habitats. (From Partridge JC (1990) The colour sensitivity and vision of fish. In (Henring PJ, Campbell AK, Whitfield M and Maddock L eds) *Light and Life in the sea*. Cambridge University Press.)

Figure 19 Absorption spectra of the two visual pigments of *Malacosteus niger* (a rhodopsin with $\lambda_{max} = 515$ nm and a porphyropsin with $\lambda_{max} = 540$ nm; dashed curves) and of a chlorophyll-derived photosensitizing pigment located within the outer segments of this species (solid line), as well as the bioluminescent emission spectrum of a long-wave-emitting suborbital photophore (dotted line) The bioluminescence, which is unlikely to be absorbed by the visual pigments directly, is initially absorbed by the photosensitizer, which then activates the visual pigments through some as yet undefined mechanism. (Douglas RH, Partridge JC, Dulai KS, Hunt DM, Mullineaux CW and Hynninen PH (1999) Enhanced retinal longwave sensitivity using a chlorophyll-derived photosensitiser in *Malacosteu niger*, a deep-sea dragon fish with far red bioluminescence. *Vision Research* 39: 2817–2832.)

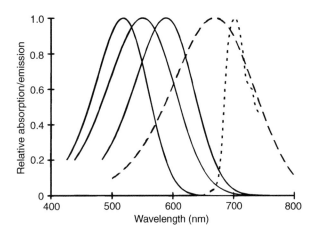

Figure 18 Bioluminescence of the *Aristostomias tittmanni* suborbital photophore (dotted line), and absorption spectra of the three visual pigments (solid lines) so far identified in its retina (a rhodopsin/porphyropsin pigment pair with λ_{max} values 520 nm and 551 nm and a rhodopsin with $\lambda_{max} = 588$ nm). The dashed curve represents a fourth visual pigment ($\lambda_{max} = 669$ nm) believed for theoretical reasons to exist within the retina of this species. (After Douglas *et al.* (1998).)

A measure of a fish's sensitivity to contrast is the minimum difference between the radiance of an object (R_o) and its background (R_b) that the animal can detect as a function of background radiance (R_b). This measure $((R_o - R_b)/R_b)$, the Weber fraction, ranges between 0.3% and 7.0% in various teleosts at threshold, which is somewhat higher than values

($\sim 1\%$) noted for humans under comparable conditions. The Weber fraction generally decreases at higher levels of background irradiation. Therefore, small percentage changes in brightness are easier to discriminate at higher light levels.

Spatial Resolution

The ability to resolve fine detail may be important to a fish for any number of reasons; for instance to recognize an animal by its body markings, or to react to a prey item at a maximum distance. The simplest way to define the ability of a fish to resolve detail is to determine the angle formed at the eye by two objects that the animal can just recognize as being separate. The value of this minimum resolvable angle varies greatly between species. Thus, the yellowfin tuna (*Thunnus albacares*) has a minimum resolvable angle of 3.7 minutes of arc, while other species can have values as high as 20 minutes of arc. Such variations are closely related to the density of photoreceptors (**Figure 13**) and ganglion cell spacing. Humans, under comparable conditions, have a minimum separable angle of around 1 minute of arc. In most species acuity will increase with age, as increased eye size and continued addition of new photoreceptors results in the image being sampled by more cells in older

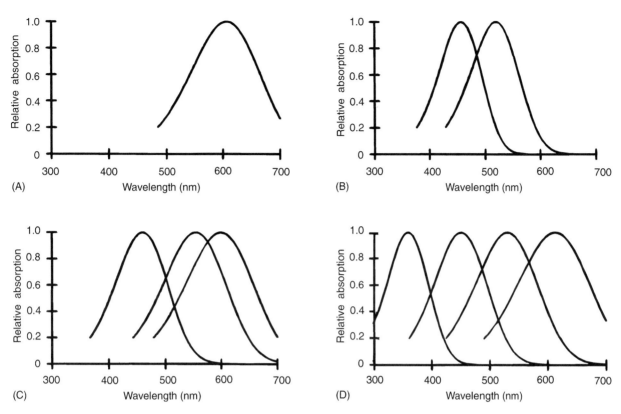

Figure 20 Cone visual pigment absorbance spectra of various teleost fish. (A) Glass catfish (*Kryptopterus bicirrhis*), which posseses only one spectral class of cone ($\lambda_{max} = 607$ nm). (B) A 'dichromatic' adult pollack (*Pollachius pollachius*) ($\lambda_{max} = 458$ nm, and 521 nm). (C) The 'trichromatic' cichlid *Nannacara anomala* ($\lambda_{max} = 460$ nm, 555 nm, and 600 nm). (D) The 'tetrachromatic' goldfish (*Carassius auratus*) ($\lambda_{max} = 360$ nm, 452 nm, 532 nm, and 614 nm). (After Douglas Douglas RH (2001) The ecology of teleost visual pigments: a good example of sensory adaptation to the environment. In (FG Barth and Schmid A, eds) Ecology.)

animals, who will therefore, for example, be able to see smaller prey and be able to detect them at greater distances than younger animals.

To achieve high acuity, not only must different parts of the retinal image be sampled by different photoreceptors but the signals generated by these cells have to be kept apart in subsequent neural processing. Consequently, at high light levels, when cones are being used, acuity is high because the cone system displays relatively little neural convergence. However, one of the factors that ensures enhanced sensitivity of the system of rods is their high convergence ratio with postreceptoral neurons. This inevitably leads to a decrease in spatial acuity at low light levels. Although individual photoreceptors are stimulated by different parts of the visual image, this spatial information is lost as the output of all these cells is combined.

Spectral Responses

As noted above, fish have the broadest range of visual pigments of any animal and many behavioral and electrophysiological studies have shown fish to be sensitive from the near ultraviolet (~ 320 nm) to the far red (~ 800 nm). The lower limit is set by the absorption of the ocular media (see above) and the upper limit by the sensitivity of the visual pigment.

To prove that an animal possesses color vision, one has to show that it can discriminate stimuli of different colors irrespective of their brightness. Such behavioral experiments are difficult to perform and the only fish shown to discriminate color in this way has been the goldfish. However, since most fish possess more than one spectral class of cone, and many have been shown to respond to a wide variety of monochromatic stimuli, it is very likely that the vast majority of fish do perceive color.

Distance Perception

While spatial acuity and color sensitivity are perhaps the most important attributes of vision that allow identification of potential predators, prey and conspecifics, in order to respond appropriately to them it is also important to know both their size and distance.

The simplest way to judge the distance of an object is to measure the angle it subtends on the retina, closer objects appearing larger. However, this is only appropriate if the size of the object is known. Thus, some fish, when using a constant-sized landmark to locate a food source rely solely on visual angle for judging its distance. However, in most situations, if animals estimated distance based purely on visual angle, confusions would continuously arise. A small animal close by, which might provide a potential meal, could be confused with a larger, more distant, predator, because both would subtend the same visual angle. The ability to judge true size, taking distance into account is known as size constancy. Although size constancy has only been demonstrated in goldfish, it is hard to believe that it is not an ability common to most fish.

Polarization

Light has three variable attributes: intensity, frequency (wavelength), and polarization. While the first two of these have been extensively studied (see above), the plane of polarization, like ultraviolet sensitivity, is often forgotten, largely one assumes because humans are unable to detect it. However, there is now considerable evidence that several species of teleost display polarization sensitivity. This appears to rely on the use of double and ultraviolet-sensitive single cones within a regular photoreceptor mosaic (e.g., **Figure 13**), and fish are able to use this information for, among other things, spatial orientation and navigation.

See also

Bioluminescence.

Further Reading

Ali MA (1975) *Vision in Fishes.* New York: Plenum Press.

Ali MA and Antcil M (1976) *Retinas of Fishes: An Atlas.* Berlin: Springer-Verlag.

Archer SN, Djamgoz MBA, Loew ER, Patridge JC, and Vallerga S (1999) *Adaptive Mechanisms in the Ecology.* Dordrecht: Kluwer Academic.

Bowmaker JK (1996) The visual pigments of fish. *Progress in Retinal and Eye Research* 15(1): 1–31.

Collin SP (1997) Specialisations of the telecost visual system: adaptive diversity from shallow-water to deep-sea. *Acta Physiologica Scandinavica* 161(supplement 638): 5–24.

Collin SP and Marshall NJ (2000) Sensory processing of the aquatic environment. *Philosophical Transactions of the Royal Society, Biological Sciences* 355(1401): 1103–1327.

Douglas RH and Djamgoz MBA (1990) *The Visual System of Fish.* London: Chapman and Hall.

Douglas RH, Patridge JC, and Marshall NJ (1998) The eyes of deep-sea fish I: lens pigmentation, tapeta and visual pigments. *Progress in Retinal and Eye Research* 17(4): 597–636.

Nicol JAC (1989) *The Eyes of Fishes.* Oxford: Clarendon Press.

FISH: DEMERSAL FISH (LIFE HISTORIES, BEHAVIOR, ADAPTATIONS)

O. A. Bergstad, Institute of Marine Research, Flødevigen His, Norway

Introduction

Fishes of many families, shapes, and sizes obtain their food in the near-bottom zone and show morphological and behavioral adaptations for life on or near the seabed. These are the demersal fishes, comprising both benthic and benthopelagic species, the latter usually performing vertical migrations to feed.

Demersal fishes occur at all depths and in all near-bottom habitats of the oceans. In this article, the emphasis will be on species inhabiting the continental shelves, that is, mainly waters shallower than about 200 m but deeper than the littoral and shallower part of the sublittoral. Some 7.5% of the marine environment belongs to this category, which is very little when compared with oceanic waters (91.9%), but considerably more than estuaries, algal beds, and reefs that together cover only 0.6%. More emphasis will be placed on fishes living in offshore waters than those inhabiting typical shallow coastal environments although it is recognized that very shallow habitats may constitute highly significant nursery areas for many shelf species.

Worldwide, about 85% of the total continental shelf area has sandy or muddy substrate. Only about 6% is rocky or gravelly, and the remaining areas are coral reefs or shellbeds (e.g., mollusk shells). There are, however, both latitudinal and depth-related patterns. Corals are almost entirely confined to low latitudes where organically enriched muddy areas are also most extensive, particularly near the mouths of major rivers or below highly productive upwelling areas. Sandy, rocky, and gravelly sediments are more common at high latitudes. Regional and local modification of the distribution and character of soft sediments is common, for example, due to water currents flushing the shelves.

In addition to offering a range of physical habitats to fishes, continental shelves are usually highly productive, and especially in temperate and boreal regions, demersal species of the shelf waters are very abundant and support some of the world's major fisheries. Of the c. 13 500 marine fish species, 1000–2000 inhabit continental shelf water of temperate and boreal zones. The majority of these are demersal species. In the subtropical and tropical zones, the richness and diversity are much greater.

Taxonomic Diversity, Geographical Patterns, and Assemblages

A wide range of families, including both cartilaginous and bony fishes, have demersal representatives in shelf waters. There are some rather consistent geographical patterns, however, both on a worldwide and regional scale. The taxonomical diversity tends to be higher at low than at high latitudes. Both species richness and evenness are normally highest in tropical and subtropical waters. An example is the Gulf of Thailand, a rather shallow soft-substrate shelf sea, where 850–900 fish species from around 125 families occur, of which at least 300 demersal species are commercially important. By comparison, the number of species in the boreal North Sea is only 160–170 belonging to about 70 families. Of these roughly 70 may be caught regularly in bottom trawl surveys offshore and only 10–15 are commercially important demersal species. These numbers decline even further in subarctic shelf waters.

The abundance and biomass of demersal fish are, however, considerably higher at high latitudes, that is, in temperate and boreal waters. The latter is normally due to high abundances of a few species, especially gadiform fishes, such as hakes (Merlucciidae) and cod-like fishes (Gadidae), and also flatfishes (Pleuronectiformes) and rockfishes (Scorpaenidae) (**Figure 1**). Moreover, consistent differences between the oceans have evolved: for example, gadiform fishes are more diverse on temperate Atlantic shelves than on North Pacific shelves where scorpaenids are very diverse and few gadiforms occur. The evolutionary history forms the background for present-day species composition patterns that are maintained by the rather strong structuring influence of regional and local environmental conditions, including the patterns of biological production in the surface layers.

Upwelling areas, well-mixed shallow-shelf seas or shoals, and hydrographical frontal zones along the shelf break are typical highly productive areas. Each of these environments is inhabited by subsets of the

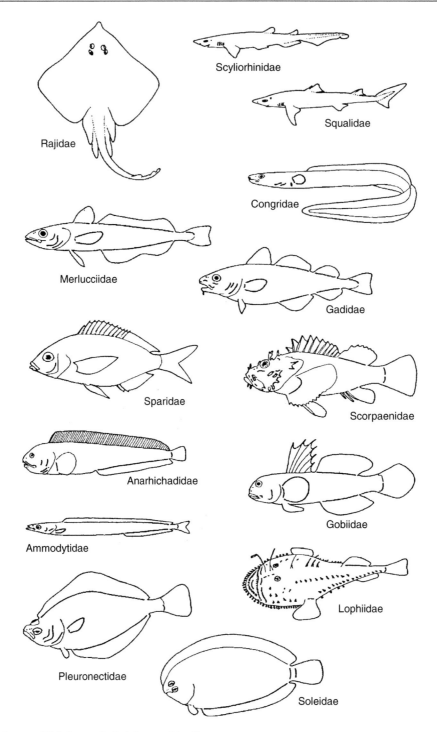

Figure 1 Body shapes of fish from selected demersal families.

fish species found in the zoogeographical province to which they belong. Also, within such environments there may be a range of demersal habitats characterized by their hydrographical regime, currents, substrate quality, depth range, demersal invertebrate fauna, etc., each favoring particular assemblages of fish species. Underlying this structuring is each species' preference for certain environmental conditions, for example, not only a limited range of depth, temperature, sediment type, prey type and size, etc., but also biotic processes among the fishes, such as predator–prey relations and competition. Such assemblage patterns have been found in a variety of coastal and shelf regions. The distribution of such assemblages on and around the Georges Bank off the east coast of North America is an example (**Figure 2**).

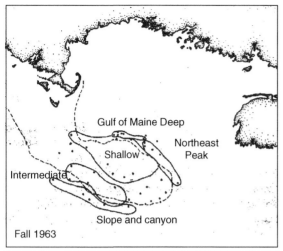

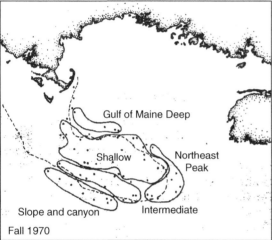

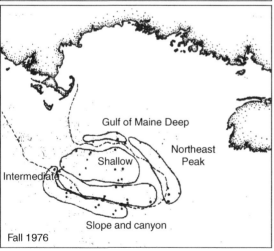

Figure 2 The geographical distribution of demersal fish assemblages on the Georges Bank of the Northwest Atlantic in 1963, 1970, and 1976. Each of the encircled areas has a distinctive assemblage of fishes characterized by the relative abundance or biomass structure of its member species. Broken line is the 100-m isobath. From Overholtz WJ and Tyler AV (1985) Long-term responses of the demersal fish assemblages of Georges Bank. *Fishery Bulletin* 83: 507–520.

Narrow shelves tend to have fewer demersal species than wide shelves and shelf seas. The highly productive upwelling areas associated with eastern boundary currents have few demersal species, and most are benthopelagic. Hakes (Merluccidae) are very well adapted to these environments and live both demersally and pelagically, mainly feeding on other fish. In contrast, demersal assemblages in major shelf seas and on offshore shoals have tens to hundreds of species with a wide range of adaptations. Even in the subarctic shelf seas, such as the Barents Sea and Bering Sea, there are high numbers of demersal species, yet only a few are very abundant.

Appearance and Behavior

In contrast with the slender and often torpedo-shaped pelagic fishes, typical demersal fishes are not so well adapted in terms of body shape or physiology for sustained fast swimming. Many shapes have, however, proven successful (**Figure 1**); ranging from the eel-shaped (e.g., Congridae and Ammodytidae) through the more classical fish shapes of the cods (Gadidae), sparids (Sparidae), and rockfishes (Scorpaenidae), to the laterally compressed flatfishes (Pleuronectiformes) and dorsoventrally compressed rays (Rajiformes). Eels and other elongated fishes swim by undulating the trunk and, unlike fishes of more classical shapes, do not use the caudal fin as their main means of propulsion. Pleuronectids and rays use the rather enlarged dorsal-anal or pectoral fins, respectively, for propulsion, but some flatfishes may also undulate the trunk to achieve higher speed or thrust when leaving the seabed or capturing prey. Regulating buoyancy may not be as important to demersal fish as to pelagic organisms. Pleuronectiform fishes, rays, and sharks lack gas bladders, and in many groups the gas bladder serves other purposes, such as sound detection, as well as providing buoyancy. Rays and flatfish are slightly negatively buoyant, while other groups such as squaloid sharks have oily livers that keep them neutrally buoyant. The latter may feed on the bottom, but also in the usually very rich near-bottom zone.

Demersal fishes come in all sizes from small gobies of adult sizes of a few centimeters to the Atlantic halibut (*Hippoglossus hippoglossus*) and Greenland shark (*Somniosus microcephalus*) that may reach 2 and 6.5 m, respectively. The general rule is that abundance declines roughly exponentially with body size. In most demersal shelf assemblages, however, the overall size distribution tends to be log-normal, that is, bell-shaped but skewed toward low sizes. The very smallest fishes may be rare, distributed in other areas, or undersampled, while intermediate sizes are very abundant. Beyond a certain size, there is an exponential decline in numbers with increasing size. The shape of the size frequency distribution seems to be a characteristic feature of a given assemblage.

Many modes of life are possible in the demersal environment. Some fishes are sluggish and have evolved a typical 'lie-and-wait' strategy. Examples are the conger eel (*Conger conger*) which hides in crevices, and the monkfish (*Lophius piscatorius*) which utilizes elaborate camouflage to remain inconspicuous to potential prey animals while swaying its modified first dorsal fin ray as a lure. Other species are very active, and some almost behave as pelagic species, forming aggregations and even schools. Flatfishes and several rays spend part of their time buried in the sand and may only emerge for certain times of the day or tidal cycle to feed. An extreme adaptation is seen for the 5–30-cm sand eels (Ammodytidae) which alternate between a pelagic lifestyle when feeding on zooplankton, to a benthic mode of life during the night and during the winter months when they bury in the sand. Cyclic activity patterns are commonly observed in demersal fish, but may vary seasonally. Most species are more active during the night or at dusk and dawn. Others have rhythms corresponding with tidal cycles. For relatively small demersal fishes these activity patterns tend to reflect a compromise between the need to feed efficiently while not being overly vulnerable to predation from bigger predators.

Camouflage may be attained in many ways, that is, through body shape, skin coloration, and modification of the skin into appendages resembling algae or debris, and may serve two main purposes, either to avoid predators or to remain inconspicuous to the prey passing by. Many species can adjust their body coloration to their environment almost continuously (e.g., many flatfishes), while others have different, persistent color varieties depending on the habitat they grow up and live in. An example of the latter is all the different varieties of gray, brown, golden, and even dark-red Atlantic cod (*Gadus morhua*) found in different parts of the range of this species across the North Atlantic.

Despite sharing some fundamental common features, many morphological adaptations have occurred within fish families enabling closely related species to utilize a variety of resources in a given habitat. Among the gadid fishes of the North Atlantic, the size range is great, for example, from the <20 cm Norway pout (*Trisopterus esmarki*) to Atlantic cod (*Gadus morhua*) and ling (*Molva molva*) that may reach 190–200 cm and weigh 40–50 kg. The Norway pout, saithe (*Pollachius virens*), and blue whiting (*Micromesistius poutassou*) share morphological characters with pelagic fishes, that is, dark or silvery backs, white bellies, big eyes, and almost teethless mouths, and they also form fast-swimming shoals when feeding benthopelagically on plankton or micronekton. Others are particularly well adapted for feeding on epi- and infauna, for example, the haddock (*Melanogrammus aeglefinus*). A typical piscivore with a large gape and well-developed dentition is the whiting (*Merlangius merlangus*) that may become piscivorous even as 10–15-cm juveniles. Other much bigger piscivores are the ling and blue ling (*Molva dipterygia*) that have elongated bodies and probably mostly behave as crepuscular ambush predators. The cod and pollack (*Pollachius pollachius*) appear more as generalists, feeding on benthopelagic and benthic crustaceans, pelagic and demersal fish, and even large pelagic crustaceans, such as hyperid amphipods. Similar variations are found in many other fish families that are predominantly demersal.

Migration

As noted above, diurnal migrations, either vertical or horizontal, are common in demersal fishes. More fascinating, however, are the long-distance seasonal migrations-associated reproduction, feeding, and sometimes overwintering. Many examples were already well described several decades ago based on traditional tagging data. However, the application of modern electronic-tagging techniques, whereby positional information can be derived after retrieval of the tag, has revealed, for example, that plaice may make extensive migrations much more frequently than previously thought. For some species, for example, cod, saithe, halibut, and Greenland halibut (*Reinhardtius hippoglossoides*), traditional tagging experiments have provided evidence of migrations across deep-sea areas. Even trans-Atlantic crossing has been recorded for the cod. Although such basin-wide migrations may not occur regularly, they certainly show that demersal fishes are quite capable of considerable movements over long distances.

An example of regular seasonal migrations is offered by the Northeast Arctic cod stock, which has its main nursery and feeding area in the Barents Sea and along the shelf off Svalbard (**Figure 3**). Within this area, seasonal feeding migrations occur, typically in response to the distribution and migration of one of its principal prey, the small pelagic capelin (*Mallotus mallotus*). However, the several hundred kilometer-spawning migrations of the adult cod are more spectacular. In the middle of the Arctic winter,

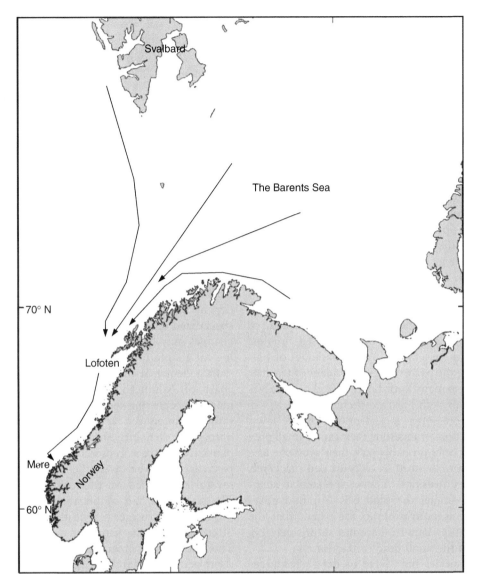

Figure 3 The prespawning migration of the Northeast Arctic cod stock from its Barents Sea feeding areas to the spawning grounds on the coast of Norway.

cod from the northern feeding areas migrate southward along the shelf edge off Norway, mostly against the prevailing currents. In February and March, they aggregate in some primary coastal spawning areas, such as the Lofoten grounds and the banks and fiords of Møre. Spawning in the right place at the right time is apparently essential to cod and many other demersal fishes.

Feeding and Diets

Most demersal fish are carnivores, customarily grouped into three categories: piscivores, benthophages, and zooplanktivores that predominantly eat fish, benthos, and zooplankton, respectively. In many areas, the production of benthic food is low or slow compared with the pelagic production, and in a range of demersal fish communities, there are actually, surprisingly few entirely benthophagous fishes, that is, which feed on epi- or infauna. A lot of demersal species depend on pelagic production by feeding on vertically migrating nekton and zooplankton entering or living in the near-bottom layer.

Piscivores usually have the capacity to deal with live prey that are large compared with their own body size, hence they tend to have big gapes. Some swallow their prey whole, others tear and bite the prey into smaller pieces. However, the techniques vary widely between species within the demersal fish group. Piscivores may either be active hunters, stealthers, or ambush predators. The latter two are often well camouflaged and obtain their prey by slowly approaching the prey or a 'lie-and-wait' strategy. Stealthers rush forward to attack and grip the prey, whereas 'lie-and-wait' strategists may suddenly open their big mouth and ingest the prey by suction (e.g., the monkfish). Piscivores generally have several rows of conical backward-pointing teeth, both on the jaws and within the mouth and pharynx. These primarily prevent the prey from escaping and help in swallowing.

Benthophagous fishes also have several prey capture methods. Some apply suction after having detected prey on the surface of the sediment or within the sediment (e.g., many flatfishes and haddock); others bite off, for example, the siphons of buried bivalves or the arms of brittle stars. Others feed on shelled animals and have very muscular jaws, and both jaw and pharyngeal dentition adapted to crushing bivalve mollusks, coral, and sea urchins. Examples are the wolffishes (Anarichadidae) of boreal waters, and the parrotfishes (Sparidae) of warmer regions, and also many flatfishes that feed on bivalves or other animals with hard exoskeletons.

Zooplanktivores often lack dentition on the jaws, and rely on suction for ingesting their rather small prey whole. Most species probably pick their prey one at a time rather than filter-feed as clupeids and mackerels do at high prey densities. A most spectacular adaptation is seen among the sand eels (Ammodytidae) where the jaw is protrusible to the extent that the mouth size at prey capture becomes surprisingly big compared with the small head of the fish.

Coexisting species may have to divide resources among themselves in a way which reduces competition but enhances growth and survival. Partitioning of food resources among co-occurring species is regarded as a very significant structuring process within demersal fish communities. Many species may occupy the same space at the same time, but their diets tend to differ. The patterns observed today, however, reflect both current processes and evolutionary adaptations (co-evolution).

The resource partitioning among flatfishes is an example of the latter. Careful consideration of species-specific characters involved in the feeding process (e.g., mouth size and morphology, dentition, capability to detect and capture different prey taxa, etc.) reveals that the co-occurring species are actually different in many important ways. It is often found, however, that juveniles have greater diet overlap than the adults. The potential for competition for food would therefore seem greater in areas where juveniles are abundant and co-occur.

In most demersal species there are significant ontogenetic, that is, essentially size-related, changes in diet and feeding ranges. At metamorphosis, when larval characters are lost and most demersal fish take up a demersal lifestyle, pronounced diet changes from feeding on planktonic crustaceans to feeding on benthic or benthopelagic prey may occur. As the fish grows larger, bigger or more energy-rich prey are needed to satisfy its energy requirements. Many species therefore go through a succession of feeding modes during their lifetime, usually starting as planktivores and ending up as piscivores or benthophages. For a given species, pronounced diet shifts tend to happen at rather fixed sizes, but habitat shifts may happen concurrently, and it it sometimes difficult to determine which is more significant. Many species are also quite opportunistic, and their diet varies greatly between habitats and seasons, often depending on the food availability. To other more specialized species the character and extent of diet changes may be more constrained by their morphological adaptations.

In a demersal fish community on the North Sea coast off southern Norway, extensive studies of many co-occurring species were conducted, revealing rather pronounced species-specific diets and also ontogentic diet shifts. Benthophages, piscivores, and planktivores were represented in the area. To the piscivores, sand eels were the most common prey and several species fed on this prey, including the cod. The small cod had significant proportions of other prey, that is, crustaceans, in its diet, but from about 30 cm onward the sand eel dominated (**Figure 4**). This pattern was observed in all seasons except for a short period in March–April when another prey, the herring, became very abundant and seemingly easily available. The herring visits this particular site to deposit its demersal eggs for a 4–6-week period every year. The cod showed a pronounced diet shift in this period, and the response is size-dependent. Only the larger individuals, 50–60 cm and more, fed on adult herring. Smaller cod may not be capable of capturing the 30–35-cm herring, but evidently feed heavily on herring eggs. Only the smallest cod continued to eat crustaceans even though they should be well adapted to locate and ingest herring eggs. Similar responses were observed for many of the demersal fishes in this community, even for planktivores that temporarily became benthophages when turning to egg feeding. In this case, diet overlap appeared to increase significantly both between and within species and this was probably possible because the herring and herring eggs were temporarily extremely abundant compared with the demand from the demersal fishes. The dietary flexibility observed in this area may be

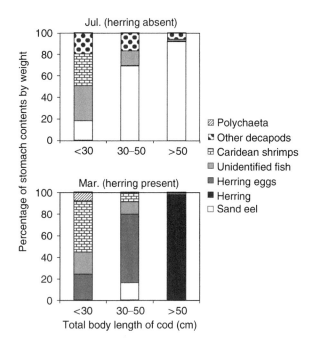

Figure 4 The diet of cod (*Gadus morhua*) on a coastal bank off southwest Norway, illustrating size-related and seasonal variation. Based on data given by Høines Å, Bergstad OA, and Albert OT (1995) The food web of a coastal spawning ground of the herring (*Clupea harengus* L.). In: Skjoldal HR, Hopkins C, Erikstad KE, and Leinaas HP (eds.) *Ecology of Fjords and Coastal Waters*, pp. 385–401. Amsterdam: Elsevier.

very advantageous, especially when living in strongly seasonal environments where the prey abundance varies quickly and perhaps unpredictably.

Sensory Systems

The shelf environment is generally well-lit and most demersal fishes have well-developed vision. In addition, demersal fish may use mechanoreceptory (sense of pressure, motion, and sound), chemosensory (olfaction, gustation, etc.), and electrosensory (i.e., ability to detect electric fields) systems. Different systems may play essential roles during different stages of important processes, such as prey detection, capture and handling, predator avoidance, or reproduction. In general, mechanoreception by the acousticolateralis system (lateral line organ and ear) tends to be most important for the detection of predators, conspecifics, and other physical disturbances in the environment. Olfaction is very important to benthophages that have to search the substrate for food, whereas planktivores and piscivores tend to rely more heavily on vision. Gustation is most important after capturing prey. Benthophages may have elaborate adaptations of their olfactory system. In addition to a well-developed nose, they may have

barbels on the snout (gadids, zoarcids, etc.) and modified and extended first rays of their pectoral fins (e.g., Triglidae) densely packed with olfactory sensory cells.

The electrosensory system is particularly well developed in some sharks and rays, and they may primarily use this system for prey location, and also for orientation and navigation. Some classical studies of the spotted dogfish (*Scyliorhinus canicula*) showed that this species was able to detect weak electric fields produced by an electrode buried in the sand. Indeed, the dogfish attacked the electrode in preference to a readily available piece of fish flesh, suggesting that electric cues were more important than visual and chemical cues in this case. Live prey, such as a buried flatfish favored by dogfish, produce weak electric fields that are modified by muscle activity, for example, during ventilation.

Reproduction and Life History

The majority of demersal teleosts (bony fishes) are oviparous and produce free-floating eggs. Mating and spawning happens at the same time, often in mid-water. Some notable exceptions are the viviparous redfish (*Sebastes*) that releases numerous pelagic larvae, and the wolffish (Anarhichadidae) which deposits a cluster of large eggs that is guarded until hatching. Sand eels (Ammodytidae) also have demersal eggs but do not protect the eggs. Most teleosts have no parental care other than making sure to mate and spawn in an area and habitat that enhances the survival probability of their eggs, larvae, and pelagic juveniles. Most pelagic teleost eggs are small, that is, a few millimeters in diameter, and the fecundity is high, that is, thousands to millions of eggs per female. Batch spawning is common in the highly fecund species, for example, the cod may spawn 10–13 batches of its very small eggs within a spawning season.

Demersal sharks are either ovoviviparous or viviparous, whereas rays and chimaeroids are oviparous, attaching their few large encapsuled eggs to debris or macroalgae. All species produce rather few young in each batch, that is, from a couple to a few tens of eggs/juveniles, and the lifetime production is very low compared with most teleost. Upon hatching or birth, the young resemble the adults, that is, there is no definite larval stage such as in teleosts.

Spawning seasons vary in duration, and the general rule is that the more seasonal the production cycle, the more fixed and limited is the spawning season. High-latitude fishes and those living in monsoonal or upwelling regions have comparatively short spawning seasons, whereas tropical and sub-tropical fishes have protracted seasons or year-round reproduction.

Larvae hatched from small pelagic eggs have a yolk sac that may provide sufficient nutrition for a few days or weeks after which they depend on exogenous feeding, usually on small crustacea, such as copepod nauplii. The mortality in this period is very high indeed, and only a minute fraction of the total number of eggs spawned will eventually result in a surviving demersal juvenile.

The teleost postlarvae pass through a metamorphosis upon reaching a certain size. Characters such as fins and juvenile pigmentation develop, and also sense organs become fully developed. In flatfishes, the eye migration occurs. In general, the larva is transformed into a fish morphologically and behaviorally adapted for demersal life. Associated with this change is the settling on the seabed, at least for parts of the diurnal cycle. Settling areas are often, but not always, separated from the feeding areas of the older fish. This reflects the different habitat and food requirements of juveniles and adults, but may also reduce cannibalism. In the majority of teleost species, the demersal nursery areas are shallower than feeding grounds of older conspecifics. Estuaries and offshore shoals are typical nursery areas, besides rocky shores where the substrate and macroalgae offer protection and a variety of prey. Tidal flats and sandy beaches are typical habitats of juvenile flatfish. A gradual ontogenetic shift in depth distribution happens as the juveniles grow larger.

The expected longevity of demersal species varies greatly. A general pattern is that longevity increases with increasing adult size, but there are many exceptions to this rule. Some species, for example, redfishes (*Sebastes*), can probably live for at least 30–40 years, but this is unusual for shelf species. Small species such as Norway pout (*Trisopterus esmarki*) may at most live for 5–6 years, but in exploited areas this seldom happens. In most shelf waters, the fisheries have influenced age distributions to the extent that life expectancy is significantly reduced.

Somatic growth patterns also vary widely among demersal fishes. In fast-growing species living in strongly seasonal environments, the overall growth trajectory may show seasonality either because food supply and feeding rates vary, or because the somatic growth is influenced by reproductive activity/gonad growth. Growth rate usually declines significantly when the fish becomes mature but never ceases entirely. Attempts have been made to classify species into *r*- and *K*-selected types (*r* and *K* are coefficients expressing rate of reproduction and somatic growth, respectively). The *r*-selected ones are those with short life spans but fast growth and rather small adult size. They mature at a small size and low age, and can rapidly take advantage of short and unpredictable favorable environmental conditions by increasing their numbers. The *K*-species, however, tend to invest more energy in somatic growth by growing slower and maturing later in life, when they are capable of producing many young. The populations of such species comprise many age groups, and they are not, as the *r*-species, so vulnerable to repeated recruitment failures due to low survival of young over a range of years. There are many species that do not readily fit into these classes, and it is unusual for demersal shelf fishes to have the extreme *r*- or *K*-strategies as seen among epipelagic fishes and demersal deep-sea fishes, respectively.

See also

Coral Reef Fishes. Deep-sea Fishes. Fish Feeding and Foraging. Fish: Hearing, Lateral Lines (Mechanisms, Role in Behavior, Adaptations to Life Underwater). Fish Larvae. Fish Locomotion. Fish Migration, Horizontal. Fish Predation and Mortality. Fish Reproduction. Fish Vision. Intertidal Fishes.

Further Reading

Alexander RM (1967) *Functional design in fishes*, 160pp. London: Hutchinson University Library.

do Carmo Gomes M (1993) *Predictions under Uncertainty. Fish Assemblages and Food Webs on the Grand Banks of Newfoundland*, 205pp. St. John's, NL: ISER, Memorial University of Newfoundland.

Evans DH (ed.) (2006) *The Physiology of Fishes*, 2nd edn., 519pp. Boca Raton, FL: CRC Press.

Gerking SD (1994) *Feeding Ecology of Fish*, 416pp. San Diego, CA: Academic Press.

Høines Å, Bergstad OA, and Albert OT (1995) The food web of a coastal spawning ground of the herring (*Clupea harengus* L.). In: Skjoldal HR, Hopkins C, Erikstad KE, and Leinaas HP (eds.) *Ecology of Fjords and Coastal Waters*, pp. 385–401. Amsterdam: Elsevier.

Jobling M (1995) *Environmental Biology of Fishes*, 455pp. London: Chapman and Hall.

Laevastu T (1996) *Exploitable Marine Ecosytems: Their Behaviour and Management*, 321pp. Oxford, UK: Fishing News Books.

Moyle PB and Cech JJ (2004) *Fishes: An Introduction to Ichthyology*, 5th edn., 726pp. New York: Prentice Hall.

Nelson JS (2006) *Fishes of the World*, 4th edn., 624pp. New York: Wiley.

Overholtz WJ and Tyler AV (1985) Long-term responses of the demersal fish assemblages of Georges Bank. *Fishery Bulletin* 83: 507–520.

Pitcher TJ (ed.) (1993) *The Behaviour of Teleost Fishes*, 2nd edn., 436pp. London: Chapman and Hall.

Postma H and Zijlstra JJ (1988) *Ecosystems of the World 27: Continental Shelves*, 421pp. Amsterdam: Elsevier.

Ross ST (1986) Resource partitioning in fish assemblages: A review of field studies. *Copeia* 2: 352–388.

FISH: HEARING, LATERAL LINES (MECHANISMS, ROLE IN BEHAVIOR, ADAPTATIONS TO LIFE UNDERWATER)

A. N. Popper, University of Maryland, College Park, MD, USA

D. M. Higgs, University of Windsor, Windsor, ON, Canada

Introduction

Fishes, like other vertebrates, have a variety of different sensory systems for gathering information from the world around them. Each sensory system provides information about certain types of signals, and all of this information is used to inform the animal about its environment.

While each of the sensory systems may have some overlap in providing information about a particular stimulus (e.g., an animal might see and hear a predator), one or another sensory system may be most appropriate to serve an animal in a particular environment or condition. Thus, for example, visual signals are most useful when a fish is close to the source of the signal, in daylight, and when the water is clear. Chemical signals travel slowly in still water and diffuse in haphazard directions, and so they are generally only effective over short distances in these waters. Where there is a reliable current, chemical signals can dissipate more rapidly and predictably to provide directional information (as in salmon finding their natal stream) but this is limited by current flow. Acoustic signals have a unique advantage in that they travel very rapidly in water and are not interfered with by low light levels or murkiness of the water. Acoustic signals also travel great distances without attenuating (decreasing in level) in open water, and this provides the potential for two animals that are some distance apart to communicate quickly.

Since sound is potentially such a good source of information, fishes have evolved several mechanisms to detect sounds, and many species use sound for communication between conspecifics (members of the same species). Indeed, it is very possible that the vertebrate ability to detect sound arose in fish ancestors in order for these animals to hear nonbiological as well as biological sounds in their environment. Thus, the most primitive vertebrates may have evolved hearing in order to detect sounds that are produced by waves breaking on the shore or by water movement around reefs or the swimming sounds produced by predators. It was probably only later in evolution that fish (and the later-evolving terrestrial vertebrates) evolved the ability to communicate with one another using sounds.

Fishes have two systems for detection of sound and hydrodynamic signals (water motion). The better known of these is the ear, which operates very much like the ears of other vertebrates. The second system is the lateral line, which consists of a series of receptors along the body of the fish both in canals and on the surface. Together, the two systems are known as the octavolateralis system. While it was thought until quite recently that the two systems are related in embryonic origin, nervous innervation, and function, it is now clear that the two systems probably evolved independently of one another, although with a common ancestral structure that included the one feature common to the two systems – the sensory hair cell.

Fish Hearing

What Do Fish Hear?

It is possible to ask fish "what they hear" by using behavioral methods to train fish to respond to the presence of a sound. More recently, investigators have also been using auditory evoked potentials, which are signals recorded directly from the brain, to measure hearing. Using these methods, it is possible to determine the frequencies and intensities of sounds that a fish can detect by changing the signal parameters.

Hearing capabilities have been determined for perhaps 75 species of fishes (of the more than 25 000 extant species). **Figure 1** shows hearing capabilities of several species to illustrate the range and intensities of sound that different species can detect. By way of comparison, a young normal human can generally detect sounds from 20 to almost 20 000 Hz, which means that humans have a much wider hearing range than most fishes.

The goldfish is one of the most sensitive of all fish species and can detect sounds from below 50 Hz to *c.* 3000 Hz. In contrast, other species such as tuna

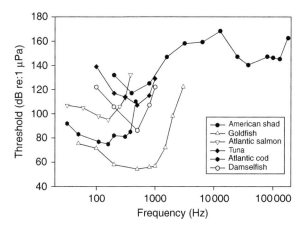

Figure 1 Hearing capabilities of representative teleost fishes measured using behavioral methods. The horizontal axis shows different frequencies while the vertical axis shows best sensitivity at each frequency. The widest hearing bandwidth has been shown for the American shad (*Alosa sapidissima*), a member of the order Clupeiformes. Of the fishes shown, the most sensitive, in terms of detecting the lowest-level sounds, is the goldfish (*Carassius auratus*). This fish can detect sounds from below 50 to c. 3000 Hz and represents the hearing capabilities of many of the fishes that have specialized adaptations for sound detection. The goldfish also has a bandwidth that is typical of most hearing specialists, although the American shad clearly has the widest bandwidth of any fish studied to date. While the Atlantic cod (*Gadus morhua*) also has low sensitivity at lower frequencies, its hearing bandwidth is not as great as that of the goldfish. The Atlantic salmon (*Salmo salar*) can only detect sounds to about 500 Hz, although recent studies have shown that it can also detect infrasound, signals to well below 30 Hz.

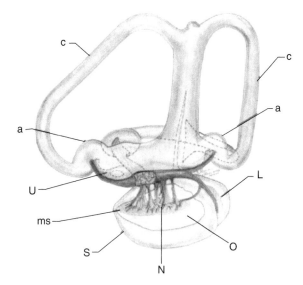

Figure 2 Drawing of the ear of the Atlantic salmon (*Salmo salmo*). Anterior is to the left and dorsal to the top. The figure shows the three semicircular canals (c) and their sensory regions, the ampullae (a). The three otolithic organs, the utricle (U), saccule (S), and lagena (L), each have a single otolith (O) and sensory epithelium (best seen in the saccule – ms). The ear is innervated by the eighth cranial nerve (N). Redrawn from Retzius G (1881) *Das Gehörorgan der Wirbelthiere*, vol. I. Stockholm: Samson and Wallin.

and salmon only hear to several hundred hertz and their sensitivity (lowest sound they can hear) is much poorer than the goldfish. Fish that hear particularly well, such as the goldfish, are called 'hearing specialists' since they have special structures, described below, that enhance their hearing capabilities. Other fishes, such as the salmon, are often called 'hearing generalists' since they have no special adaptation for hearing. While we have data for a small proportion of all of the extant fish species, it appears that most fish fall into the hearing generalist category, and this certainly includes most of the more common food fishes, such as haddock, trout, and salmon.

Of all fishes, those with by far the widest hearing range are some of the herrings and shads (all members of the genus *Alosa*). These fishes have been shown to detect sounds from below 100 Hz to over 180 kHz, a range that is only reached by a few mammals such as some bats and dolphins. While hearing in these fishes is not yet well understood, behavioral studies have shown that *Alosa* react very strongly to dolphin-like sounds, supporting the argument that they use ultrasound detection to avoid being eaten by dolphin predators.

In addition to being able to detect sounds, all vertebrates must be able to discriminate between sounds of different frequencies and intensities, detect the presence of a biological meaningful sound in the presence of biological and nonbiological noises, and determine the direction of a sound source. These capabilities are necessary if an animal is to be able to discriminate between different types of sound sources by their acoustic characteristics and know where a sound is coming from to either escape a predator or find the sound source. Thus, it is not surprising that fish are able to perform all of these auditory tasks. While they do not discriminate or localize sounds as well as mammals, the capabilities of fishes are sufficient to give them a good deal of information about acoustic signals.

How Do Fish Hear?

While fish have no external structures for hearing, as are found in many terrestrial vertebrates, they do have an inner ear which is similar in structure and function to the inner ear of terrestrial vertebrates (**Figure 2**). Unlike terrestrial vertebrates, which require external structures to gather sound waves and change the impedance to match that of the fluid-filled inner ear, sound gets directly to the fish ear since the fish's body is the same density as the water. As a consequence, the fish ear, and the rest of the body, move with the sound field. While this might result in the fish not detecting the sound, the ear also contains very dense structures, the otoliths, which move at a different amplitude and phase from the rest of the body. This provides the mechanism by which fish hear.

The ear of a fish (**Figure 2**) has three semicircular canals that are involved in determining the angular movements of the fish. The ear also has three otolith organs, the saccule, lagena, and utricle, that are involved in both determining the position of the fish relative to gravity and detecting the sound. Each of the otolith organs contains an otolith (a dense calcareous structure) that lies in close proximity to a sensory epithelium.

The sensory epithelium in fish contains mechanoreceptive sensory hair cells that are virtually the same as found in the mechanoreceptive cells of the lateral line (see below) and in the inner ear of terrestrial vertebrates. All parts of the ear have the same kind of cell to detect movement, whether it be movements caused by sound or movements of the head relative to gravity.

The sensory hair cells (**Figure 3**) are not very different from other epithelial cells of the body, except that on their apical (top) ends they have a set of cilia (sometimes called 'hairs', hence the name of the cell) that project into the space above the epithelium and contact the otolith. Each cell has many cilia. Generally these are graded in size, with the longest being at one end of the ciliary bundle. The sensory hair cell responds to bending of the ciliary bundle by a change in its electrical potential. This in turn causes release of chemical signals (neurotransmitters) which excite neurons of the eighth cranial nerve which innervate the hair cells. These neurons then send signals to the brain to indicate detection of a signal.

Bending of the ciliary bundles results from the relative motion between the sensory epithelium (and the fish's body) and the overlying otolith. There is evidence that suggests that the motion of the otolith relative to the body of the fish depends on the direction of the sound source. Since the sensory hair cells are responsive to bending in only certain directions, they can detect the direction of motion of the otolith and provide the fish with information about the direction, relative to the fish, of a sound source.

Why Do Fish Hear?

Fish use information from sound stimuli for many reasons but perhaps the most well studied is to detect sounds produced by other fish, and particularly other fish of the same species. Many species of fish make

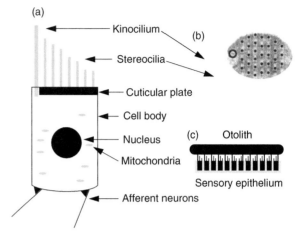

Figure 3 Schematic drawing of a sensory hair cell that is typical of those found in the ear and lateral line of fishes, as well as in the ears of other vertebrates. (a) Side view of a sensory hair cell showing the apically located kinocilium and stereocilia (collectively called the ciliary bundle). These structures penetrate the apical cell membrane. The kinocilium terminates in basal bodies in the cytoplasm (not shown) while the stereocilia terminate in the dense cuticular plate. Other components of the sensory hair cell are represented by the nucleus and mitochondria, but other organelles found in typical epithelial cells are also found in hair cells. The hair cell is typically innervated by afferent neurons of the eighth cranial nerve that carry information from the cell to the brain. Many cells may also have efferent innervation (not shown), which is thought to carry signals from the brain to the hair cell, which modulate their responses to signals. Each cell has a single kinocilium and many stereocilia. Typically, the cilia are graded in height, with the longest lying closest to the kinocilium. (b) Top view of the apical membrane of the sensory hair cell showing the eccentric position of the kinocilium and the rows of stereocilia. The sensory hair cell is stimulated when the ciliary bundle is bent. Bending of the ciliary bundle in the direction directly from the kinocilium to the stereocilia causes maximum response of the sensory cell, while bending in the opposite direction results in least response. Bending in other directions cause a response that is proportional to a cosine function of the direction relative to the axis of maximum stimulation of the sensory cell. (c) Side view of the sensory epithelium with sensory hair cells and the overlying otolith; a thin otolith membrane lies between the otolith and epithelium to keep the two structures next to one another. Relative motion between the otolith and the sensory epithelium in a sound field, resulting from their very different masses, results in bending of the ciliary bundle. Since the hair cells on different epithelial regions are oriented with their kinocilium in different directions (as shown here), motion of the otolith in different directions relative to the sensory cells will give different response levels from different hair cell groups. This information can be used by the fish to tell sound source direction.

and use sounds for communication. These sounds are generally pulses or short bursts of signal that range in frequency from below 30 Hz to c. 800 Hz. The exact frequency range and pattern of the sound varies by species, and sometimes even within species, where different sounds may be used for different functions.

The sounds are produced in a variety of different ways, none of which involves movement of air across a membrane, as happens in terrestrial animals. Sounds in some species are produced by two bony structures hitting or rubbing against one another. In other cases, as in the well-known system of the toadfish (*Opsanus tau*), there are special muscles located on the swim bladder, a bubble of air in the abdominal cavity. These muscles contract at up to several hundred times per second and produce a sound which is amplified by the air in the swim bladder. In still other cases, such as several marine catfish (e.g., *Arius felis*), a muscle from the skull contracts and pulls on another bone which then hits the swim bladder, thereby producing sounds.

Fish sounds are used in a variety of different behavioral situations. The aforementioned toadfish has a repertoire of two different sound types. One sound, which is generally pulsed, is made by both males and females all the year round. A second, long, moaning-like sound, called the boat whistle, is produced only by males during the breeding season. This signal, which sounds very much like a low-frequency boat horn, is used to attract females to the males' nest for breeding purposes. Sounds produced by some species of squirrelfish (Holocentridae) are used for warning of potential enemies in the vicinity, while damselfish (Pomacentridae) males produce sounds to try and scare potential predators away from a nest.

Adaptations for Improvement of Hearing

As shown in **Figure 1**, some fish are hearing specialists. While it is still not clear as to how the American shad and other related species hear very high frequency sounds (ultrasound), it is clear that hearing specialists, such as the goldfish, have special adaptations that improve hearing as compared to other fishes.

All species of fish detect sounds by detecting relative motion between the otoliths and the sensory hair cells. However, other fishes, and most notably the hearing specialists, also detect sounds using the air-filled swim bladder in the abdominal cavity. The swim bladder is used for a variety of different functions in fish. It probably evolved as a mechanism to maintain buoyancy in the water column. In effect, fish can adjust the volume of gas in the swim bladder and make themselves neutrally buoyant at any depth in the water. In this way, they do not have to expend extra energy to maintain their vertical position.

The other two roles of the swim bladder are in sound production and hearing. In sound production, the air in the swim bladder is energized by the sound-producing structures and serves as a radiator of the sound. The swim bladder, because it is filled with air, is also of very different density than the rest of the fish body. Thus, in the presence of sound the gas starts to vibrate. This is capable of re-radiating sound to the ear and is potentially able to stimulate the inner ear by moving the otolith relative to the sensory epithelium. However, in hearing generalists, the swim bladder is quite far from the ear, and any re-radiated sound attenuates a great deal before it reaches the ear. Thus, these species probably do not detect these sounds. Hearing specialists always have some kind of acoustic coupling between the swim bladder and the inner ear to reduce attenuation and ensure that the signal from the swim bladder gets to the ear. In the goldfish and its relatives (e.g., catfish), there is a series of bones, the Weberian ossicles, which connect the swim bladder to the ear. When the walls of the swim bladder vibrate in a sound field, the ossicles move and carry the sound directly to the inner ear. Removal of the swim bladder in these fishes results in a drastic loss of hearing range and sensitivity.

Besides species with Weberian ossicles, other fishes have evolved a number of different strategies to enhance hearing. Most notably, the swim bladder often has anterior projections that actually contact one of the otolith organs. In this way, the motion of the swim bladder walls directly couples to the inner ear of these species. In the herrings, development of these anterior projections coincides with the development of ultrasound detection and also coincides with a dramatic increase in the ability of these fish to detect and avoid their predators, strongly suggesting the natural function of this specialization. Finally, some species have extra air bubbles right next to the ear. For example, in the bubble-nest builders, a bubble of air, analogous to the swim bladder, is trapped in the mouth near the ear. Removal of this air bubble can make the fish temporarily deaf to certain frequencies of sounds. Similarly, elephant-nosed fishes, the Mormyridae, have a small bubble of air that is intimately associated with the sensory region of the inner ear. Behavioral data have shown that this bubble broadens the frequency range of hearing in these species, as well as improves the ability to detect sounds of lower intensity.

Effects of Human-Generated Sound on Fish

There is growing concern that the sounds humans have added to the marine environment may be having negative effects on fish and their survival (just as there are concerns regarding the effect of noise on marine mammals and marine turtles). The types of sounds that humans have added to the environment come from shipping, use of seismic air guns in undersea oil and gas exploration, pile driving in the construction of bridges and other structures, sonar, and similar devices. As a consequence, the ambient sound levels in some (but certainly not all) parts of the oceans (and often smaller bodies of water) have apparently risen over the past decades. Just as increased sound in the environment has an impact on human health and well-being, there is the potential that increased sounds underwater impact marine organisms.

Despite the concerns, there are very few data to indicate if and how the increased ambient sounds, which are, in many cases, within the hearing ranges of fish, affect fish. While there may be no effect whatsoever, there may be effects that involve fishes moving away from a sound source temporarily or even longer-term behavioral effects such as fish permanently leaving a feeding ground or a region in which they reproduce. Other effects may be to increase stress levels, produce temporary hearing loss, or even damage or kill a fish. Of major concern is whether the increased ambient sounds might not harm fish *per se* but yet mask the detection of biologically relevant sounds such as those made by predators or prey, thereby affecting the survival of fish. This masking is similar to what humans encounter when they are in a noisy environment, such as a cocktail party, and have trouble hearing people or other sounds, even when they are nearby.

There are very few actual experimental data on effects of sound on fishes. Whereas there is some evidence that fishes very close to a very intense sound source (such as right next to a large pile-driving operation) may be killed, there are other data that suggest that some fish exposed to other types of sound, such as those produced by seismic air guns used for oil exploration or by very intense low-frequency sonars, are not harmed in any way.

The Lateral Line

The lateral line has been one of the most enigmatic of all vertebrate sensory systems. Over the past 150 years, various investigators have suggested that the lateral line is involved in hearing, temperature reception, chemoreception, touch, and a variety of other functions. However, in the last few years, it has finally become clear that the lateral line is involved as a sensor of water motion, or hydrodynamic stimulation, that arises within a few body lengths of a fish. In other words, the lateral line detects the presence of nearby animals and objects that cause or disrupt water flow.

The lateral line is involved with schooling behavior, where fish swim in a cohesive formation with many other fish. The lateral line tells the fish where the other fish are in the school, and helps the fish maintain a constant distance from its nearest neighbor. In experiments where the lateral line was temporarily disabled, the ability of fish to school was disrupted and fish tended to swim more closely together. The lateral line also is used to detect the presence of nearby moving objects, such as food, and to avoid obstacles, especially in fishes that cannot rely on light for such information, such as the cave fishes that live deep underground. Finally, the lateral line is an important determinant of current speed and direction, providing useful information to fishes that live in streams or where tidal flows dominate.

The lateral line consists of two groups of receptors located on the body surface. One group is in canals, called canal organs (**Figure 4**), while other groups are located on the body surface and are called surface organs. The canal organs are primarily involved in detection of low-frequency (e.g., below 100 Hz) hydrodynamic movements of other fish, whereas the surface receptors appear, at least in some species, to provide fish with information about general water motion and assist the fish in swimming with or against currents.

The lateral line receptors consist of the same sensory hair cells as found in the ear. However, the hair cells are organized into small groups called neuromasts, with perhaps up to 100 cells per neuromasts. The cilia from the neuromasts stick up into a gelatinous sail-like structure called a cupula. Bending of the cupula caused by the movement of water particles results in bending of the cilia on the hair cell and the sending of signals to the neurons

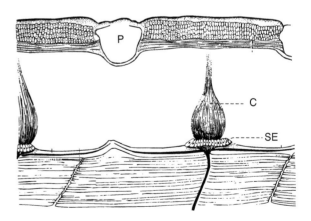

Figure 4 Longitudinal section of a lateral line canal. Each fluid-filled canal is open to the outside via a pore (P). A canal neuromast (SE) with its overlying cupula (C) sits on the floor of the canal, with one neuromast between each pore. The canal neuromasts are innervated by a cranial nerve. From Grassé PP (1958) L'oreille et ses annexes. In: Grassé PP (ed.) *Traité de Zoologie*, vol. 13, pp. 1063–1098. Paris: Masson.

which take signals to the lateral line region of the brain. In essence, the lateral line hair cells are stimulated as a result of the net difference between the motion of the fish and the surrounding water particles.

There is a considerable variation in the exact pattern of the lateral line in different species. Some species have a single canal along the lateral trunk, while other species have multiple canals or even no canals along the trunk. Perhaps the most elaborate canal system, and the most variable, is on the head of fish. The lateral line segments on the head enable surface-feeding fish to detect and locate the source of surface waves produced by prey and may be important for making fine-scale adjustments in position in fish that form particularly tight schools.

Interactions between the Ear and Lateral Line

It is generally thought that the ear and lateral line may be complementary systems. Both detect water motions, but whereas the ear can detect signals that come from great distances, the lateral line only detects signals that are very close to the fish. Significantly, the frequency range over which the two systems appear to work overlaps from about 50 to 150 Hz, although the ear can detect sounds to much higher frequencies and the lateral line can detect hydrodynamic signals to below 1 Hz. Both systems send their signals into nearby regions of the brain. While the initial points of connection in the brain are close to one another, they are not identical. However, after the first connections it appears that the two systems next send inputs directly to the same integrating center (the torus semicircularis), allowing the brain to form an image of the combined mechanosensory information coming from both sensory systems. In addition, the lateral line canals on the head of some species converge in a structure called a lateral recess, a thin membrane situated just above the ear. This may allow the lateral line to be stimulated by vibrations in or around the ear. The functional significance of these connections is not yet known, but their presence supports the idea that fish can use the two systems together to gain a good deal of information about signals in the environment.

Correlations between Structure and Habitat

By examining the structure of the lateral line and ear in fishes occupying a variety of different habitats, some information can be obtained as to the functional significance of specializations in these two systems. In general, species with many superficial neuromasts seem to live in quieter waters while those in running and noisier waters tend to have fewer superficial neuromasts and a better developed canal system. Extensively branched lateral line canals are also commonly found in schooling fishes, possibly allowing for fine spatial discrimination of signals coming from many sources (other fish in the school). These correlations suggest that canals may have evolved as a mechanism to reduce noise in hydrodynamic stimuli. Fewer correlative studies have been done comparing fish hearing to habitat, but it appears that fish with the highest discrimination ability tend to live in shallower, soft bottom habitats while those with less fine-tuned hearing abilities tend to be found in noisier habitats. While there are numerous exceptions to these trends, examination of the lateral line and ears of fish is one way to predict where and how a fish might live.

The habitat in which a fish lives also seems to dictate how important hearing or lateral line inputs can be. This is most evident in fishes such as blind cavefish and blind catfishes (family Ictaluridae) that spend their entire life underground. These animals have no need for vision, as there is no light in their environments, and rely much more heavily on the lateral line and hearing. They have very well developed neuromasts and canal systems and also tend to have a greater proportion of their brain area devoted to receiving hydrodynamic inputs. In deep sea fishes, the lateral line and hearing are highly developed as compared to visual senses, again showing how these senses can take over when others are not available.

See also

Fish Feeding and Foraging. Fish Locomotion. Fish Migration, Horizontal. Fish Migration, Vertical. Fish Predation and Mortality. Fish Reproduction.

Further Reading

Atema J, Fay RR, Popper AN, and Tavolga WN (eds.) (1988) *Sensory Biology of Aquatic Animals*. New York: Springer.

Collin SP and Marshall NJ (eds.) (2003) *Sensory Processing in Aquatic Environments*. New York: Springer.

Coombs S and Montgomery JC (1999) The enigmatic lateral line system. In: Fay RR and Popper AN (eds.) *Comparative Hearing: Fish and Amphibians*, pp. 319–362. New York: Springer.

Fay RR and Popper AN (eds.) (1999) *Comparative Hearing: Fishes and Amphibians*. New York: Springer.

Grassé PP (1958) L'oreille et ses annexes. In: Grassé PP (ed.) *Traité de Zoologie*, vol. 13, pp. 1063–1098. Paris: Masson.

Higgs DM (2004) Neuroethology and sensory ecology of teleost ultrasound detection. In: Mogdans J and von der Emde G (eds.) *The Senses of Fishes: Adaptations for the Reception of Natural Stimuli*, pp. 173–188. New Delhi: Narosa Publishing.

Higgs DM, Mann DA, Souza MJ, Wilkins HR, Presson JC, and Popper AN (2001) Age- and size-related changes in the inner ear and hearing ability of the adult zebrafish (*Danio rerio*). *Journal of the Association for Research in Otolaryngology* 3: 74–184.

Higgs DM, Plachta DTT, Rollo AK, Singheiser M, and Popper AN (2004) Development of ultrasound detection in American shad (*Alosa sapidissima*). *Journal of Experimental Biology* 207: 155–163.

Ladich F and Popper AN (2004) Parallel evolution in fish hearing organs. In: Manley GA, Popper AN, and Fay RR (eds.) *Evolution of the Vertebrate Auditory System*, pp. 95–127. New York: Springer.

Mann DA, Higgs DM, Tavolga WN, Souza MJ, and Popper AN (2001) Ultrasound detection by clupeiform fishes. *Journal of the Acoustical Society of America* 109: 3048–3054.

Montgomery JC, Coombs S, and Halstead M (1995) Biology of the mechanosensory lateral line. *Reviews in Fish Biology and Fisheries* 5: 399–416.

Myrberg AA, Jr. (1981) Sound communication and interception in fishes. In: Tavolga WN, Popper AN, and Fay RR (eds.) *Hearing and Sound Communication in Fishes*, pp. 395–426. New York: Springer.

Plachta DTT and Popper AN (2003) Evasive responses of American shad (*Alosa sapidissima*) to ultrasonic stimuli. *Acoustical Research Letters Online* 4: 25–30 http://scitation.aip.org/getpdf/servlet/GetPDFServlet? filetype = pdf&id = ARLOFJ00000040000020000250000 001&idtype = cvips&prog = normal (accessed Mar. 2008).

Popper AN, Fay RR, Platt C, and Sand O (2003) Sound detection mechanisms and capabilities of teleost fishes. In: Collin SP and Marshall NJ (eds.) *Sensory Processing in Aquatic Environments*, pp. 3–38. New York: Springer.

Popper AN, Fewtrell J, Smith ME, and McCauley RD (2004) Anthropogenic sound: Effects on the behavior and physiology of fishes. *Marine Technology Society Journal* 37(4): 35–40.

Popper AN, Ramcharitar J, and Campana SE (2005) Why otoliths? Insights from inner ear physiology and fisheries biology. *Marine and Freshwater Research* 56: 497–504.

Ramcharitar J, Gannon DP, and Popper AN (2006) Bioacoustics of the family Sciaenidae (croakers and drumfishes). *Transactions of the American Fisheries Society* 135: 1409–1431.

Retzius G (1881) *Das Gehörorgan der Wirbelthiere*, vol. I, Stockholm: Samson and Wallin.

Song J, Matieu A, Soper RF, and Popper AN (2006) Structure of the inner ear of bluefin tuna (*Thunnus thynnus*). *Journal of Fish Biology* 68: 1767–1781.

Zelick R, Mann D, and Popper AN (1999) Acoustic communication in fishes and frogs. In: Fay RR and Popper AN (eds.) *Comparative Hearing: Fish and Amphibians*, pp. 363–411. New York: Springer.

Relevant Websites

http://www.dosits.org
– Discovery of Sound in the Sea (DOSITS).
http://www.life.umd.edu
– Dr. Arthur N. Popper's Laboratory of Aquatic Bioacoustics, University of Maryland College of Chemical & Life Sciences.
http://www.amonline.net.au
– Ichthyology at the Australian Museum.
http://www.uwindsor.ca
– Lab Page of Dennis Higgs, Department of Biology, University of Windsor.
http://www.marine.usf.edu
– Mann Laboratory, College of Marine Science, University of South Florida.
http://www.nbb.cornell.edu
– Page on Andrew H. Bass, Department of Neurobiology and Behavior, Cornell University.
http://oceanexplorer.noaa.gov
– Sound in Sea, Explorations, Ocean Explorer.

MARINE MAMMALS

MARINE MAMMAL OVERVIEW

P. L. Tyack, Woods Hole Oceanographic Institution, Woods Hole, MA, USA

Introduction

The term 'marine mammals' is an ecological grouping that lumps together a phylogenetically diverse set of mammals. The only thing linking them is that marine mammals occupy and rely upon aquatic habitats for all or much of their lives. The cetaceans and the sirenians (see the relevant Encyclopedia articles), or dugongs and manatees, live their entire lives at sea, only coming on land during perilous stranding events. Cetaceans (**Table 1**) evolved from ungulates whose modern members include pigs, while sirenians (**Table 2**) evolved from ungulates related to the modern elephant. One member of the bear family (Ursidae), the polar bear, is categorized as a marine mammal, while two members of the family Mustelidae are considered marine: the sea otter and the marine otter of Chile (**Table 3**). Seals and walruses also evolved from terrestrial carnivores. Most biologists lump the seals and walruses into a suborder called the Pinnepedia, and these are often referred to as pinnipeds, meaning 'finlike feet'. Seals are divided into two families: the Otariidae, or eared seals, and the Phocidae, or true seals (**Table 4**). Walruses are categorized as a separate family, the Odobenidae (**Table 4**). The definition of marine mammals is somewhat arbitrary – river dolphins are considered marine mammals, but river otters are not. Inclusion in this category can have real consequences in the United States, since marine mammals have special protection under the US Marine Mammal Protection Act.

Taxonomy

The basic evolution of marine mammals from carnivores and ungulates is well established, but the details of phylogeny and taxonomy at the species level are in flux, owing in part to recent molecular genetic data. Since there is a relatively well-established nomenclature that has been stable for the past 20 years or so, and which forms the basis for species management, this nomenclature will be used in the following tables, while it is recognized that the developing synthesis of molecular and morphological data will probably change some of the species relations. For example, the US and international agencies responsible for protecting endangered species split right whales of the northern and southern oceans into two species; some taxonomists lump right whales into one species, but the North Pacific and North Atlantic right whales are clearly separated and recent genetic data suggests the division of right whales into three species: Southern, Northern Pacific, and North Atlantic. Where the designation of endangered species focuses on the species level, this lumping of right whales of the North Atlantic and North Pacific, which are highly endangered, with right whales of the South Atlantic, which are doing well, could have profound policy implications.

Adaptations of Marine Mammals

The success of marine mammals is something of a puzzle. How did their terrestrial ancestors, adapted for life on land, compete against all of the life forms that were already so well adapted to the marine environment? Multicellular organisms arose in the oceans of the earth about 700 million years ago (MYa), and for the next 100 million years or so, there was a remarkable burst of evolutionary diversification, as most of the basic body plans of life in the sea evolved. By 350–400 MYa, multicellular animals expanded from the ocean into terrestrial environments, with another evolutionary radiation. Mammals only reentered the sea about 60 MYa, and thus have had less than one-tenth of the time available to the original marine metazoans for adaptation to this challenging environment.

The ocean poses a hostile environment to mammals, yet a remarkable diversity of mammalian groups from carnivores such as bears and otters to ungulates have adapted to the marine environment. Most marine animals maintain their bodies at the temperature of the surrounding sea water and obtain any oxygen required for respiration directly from oxygen dissolved in sea water. Marine mammals need to breathe air and maintain their bodies at temperatures well above the typical temperature of sea water. Many marine animals release thousands to millions of eggs into the hostile marine environment, where the odds are that only a handful will survive. Mammals produce only a handful of offspring at a time, and must rely upon parental care to enhance the survival of their offspring.

Table 1   Scientific and common names of cetaceans along with conservation status[a]

Scientific name	Common name	ESA	MMPA	IUCN
Suborder Mysticeti	Baleen whales			
Family Balaenidae	Right whales			
Balaena mysticetus	Bowhead whale	EN	DEP	
Eubalaena australis	Southern right whale	EN	DEP	
Eubalaena glacialis	Northern right whale	EN	DEP	EN
Family Neobalaenidae				
Caperea marginata	Pygmy right whale			
Family Balaenopteridae	Rorquals			
Balaenoptera acutorostrata	Minke whale			
Balaenoptera borealis	Sei whale	EN	DEP	EN
Balaenoptera edeni	Bryde's whale			
Balaenoptera musculus	Blue whale	EN	DEP	EN
Balaenoptera physalus	Fin whale	EN	DEP	EN
Megaptera novaeangliae	Humpback whale	EN	DEP	VU
Family Eschrichtiidae				
Eschrichtius robustus	Grey whale	EN (NW Pacific)	DEP (NW Pacific)	
Suborder Odontoceti	Toothed whales			
Family Physeteridae	Sperm whales			
Physeter macrocephalus	Sperm whale	EN	DEP	VU
Family Kogiidae				
Kogia breviceps	Pygmy sperm whale			
Kogia simus	Dwarf sperm whale			
Family Ziphiidae	Beaked whales			
Berardius arnuxii	Arnoux's beaked whale			
Berardius bairdii	Baird's beaked whale			
Hyperoodon ampullatus	Northern bottlenose whale			
Hyperoodon planifrons	Southern bottlenose whale			
Mesoplodon bidens	Sowerby's beaked whale			
Mesoplodon bowdoini	Andrew's beaked whale			
Mesoplodon carlhubbsi	Hubb's beaked whale			
Mesoplodon densirostris	Blainville's beaked whale			
Mesoplodon europaeus	Gervais' beaked whale			
Mesoplodon ginkgodens	Ginkgo-toothed beaked whale			
Mesoplodon grayi	Gray's beaked whale			
Mesoplodon hectori	Hector's beaked whale			
Mesoplodon layardii	Strap-toothed beaked whale			
Mesoplodon mirus	True's beaked whale			
Mesoplodon pacificus	Longman's beaked whale			
Mesoplodon peruvianus	Pygmy beaked whale			
Mesoplodon stejnegeri	Stejneger's beaked whale			
Tasmacetus shepardi	Tasman's beaked whale			
Ziphius cavirostris	Cuvier's beaked whale			
Family Monodontidae				
Delphinapterus leucas	Beluga; white whale		DEP (Cook Inlet stock)	VU
Monodon monoceros	Narwhal			
Family Delphinidae	Oceanic dolphins			
Cephalorhynchus commersonii	Commerson's dolphin			
Cephalorhynchus eutropia	Black dolphin			
Cephalorhynchus heavisidii	Heaviside's dolphin			
Cephalorhynchus hectori	Hector's dolphin			EN
Delphinus capensis	Long-beaked common dolphin			
Delphinus delphis	Common dolphin			
Feresa attenuata	Pygmy killer whale			
Globicephala macrorhynchus	Short-finned pilot whale			

Table 1 *Continued*

Scientific name	Common name	ESA	MMPA	IUCN
Globicephala melas	Long-finned pilot whale			
Grampus griseus	Risso's dolphin			
Lagenodelphis hosei	Fraser's dolphin			
Lagenorhynchus acutus	Atlantic white-side dolphin			
Lagenorhynchus albirostris	White-beaked dolphin			
Lagenorhynchus australis	Peale's dolphin			
Lagenorhynchus cruciger	Hourglass dolphin			
Lagenorhynchus obliquidens	Pacific white-sided dolphin			
Lagenorhynchus obscurus	Dusky dolphin			
Lissodelphis borealis	Northern right whale dolphin			
Lissodelphis peronii	Southern right whale dolphin			
Orcaella brevirostris	Irrawaddy dolphin			
Orcinus orca	Killer whale			
Peponocephala electra	Melon-headed whale			
Pseudorca crassidens	False killer whale			
Sotalia fluviatilis	Tucuxi			
Sousa chinensis	Indo-Pacific hump-backed dolphin			
Sousa teuszii	Atlantic hump-backed dolphin			
Stenella attenuata	Pantropical spotted dolphin		DEP (ETP)	
Stenella clymene	Clymene dolphin			
Stenella coeruleoalba	Striped dolphin			
Stenella frontalis	Atlantic spotted dolphin			
Stenella longirostris	Spinner dolphin		DEP (ETP)	
Steno bredanensis	Rough-toothed dolphin			
Tursiops truncatus	Bottlenose dolphin		DEP (mid-Atlantic coastal)	
Family Phocoenidae	Porpoises			
Phocoena dioptrica	Spectacled porpoise			
Neophocaena phocaenoides	Finless porpoise			
Phocoena phocoena	Harbor porpoise			VU
Phocoena sinus	Cochito; Vaquita	EN	DEP	CR
Phocoena spinnipinis	Burmeister's porpoise			
Phocoenoides dalli	Dall's porpoise			
Family Platanistidae	River dolphins			
Platanista gangetica	Ganges river dolphin			
Platanista minor	Indus susu	EN	DEP	
Family Iniidae	River dolphins			
Inia geoffrensis	Boutu; boto			
Lipotes vexillifer	Baiji	EN	DEP	
Pontoporia blainvillei	Franciscana			

[a]CR = Critically endangered; DEP = Depleted; EN = Endangered; ESA = US Endangered Species Act; IUCN = International Union for the Conservation of Nature; MMPA = US Marine Mammal Protection Act; TH = Threatened; VU = Vulnerable. (Sources: *Reynolds and Rommel, 1999*; IUCN red book.)

Some of the mammalian adaptations that are well-suited for terrestrial life appear to create risks and drawbacks of life in the sea, yet these adaptations opened new niches for marine mammals in four of the five trophic levels of marine ecosystems. These adaptations may be particularly important for predators. For example, the mammalian adaptation of homeothermy, or maintaining a constant body temperature, requires a metabolic rate 10–100 times that of an animal whose body stays at ambient temperature. Biochemical reactions occur at different rates at different temperatures, so animals that do not regulate their temperature as precisely as mammals may not have sensory or motor systems operating as rapidly. All a predator needs is a small marginal advantage in ability to detect or locate prey or in locomotor ability and maneuverability in order to succeed. The high metabolic cost of homeothermy in marine mammals may thus be offset by their potential predatory advantage. This high-cost/high-gain strategy is unique in the oceans to marine mammals and marine birds, although some predatory marine fish also warm muscles or the central nervous system for similar advantage. Marine mammals provide

Table 2 Scientific and common names of sirenians along with conservation status[a]

Scientific name	Common name	ESA	MMPA	IUCN
Family Trichechidae				
Trichechus inunguis	Amazonian manatee	EN	DEP	
Trichechus manatus	West Indian manatee	EN	DEP	VU
Trichechus senegalensis	West African manatee	TH	DEP	VU
Family Dugongidae				
Dugong dugon	Dugong	EN	DEP	VU

[a]DEP = Depleted; EN = Endangered; ESA = US Endangered Species Act; IUCN = International Union for the Conservation of Nature; MMPA = US Marine Mammal Protection Act; TH = Threatened; VU = Vulnerable. (Sources: *Reynolds and Rommel, 1999*; IUCN red book.)

Table 3 Scientific and common names of marine mustelids and ursids along with conservation status[a]

Scientific name	Common name	ESA	MMPA	IUCN
Enhydra lutris	Sea otter	TH	DEP	EN
Lutra marina	Marine otter	EN	DEP	EN
Ursus maritimus	Polar bear			

[a]DEP = Depleted; EN = Endangered; ESA = US Endangered Species Act; IUCN = International Union for the Conservation of Nature; MMPA = US Marine Mammal Protection Act; TH = Threatened; VU = Vulnerable. (Sources: *Reynolds and Rommel, 1999*; IUCN red book.)

parental care to their young, and can better afford to put more reproductive effort into fewer offspring than is typical of marine organisms. Marine mammals often feed in areas that are separated from the places where they give birth. This has led to an adaptation unknown among terrestrial mammals (except for bears) – the ability to lactate while fasting.

Mammals evolved an ear specialized to analyze the frequency content of sound. Air-breathing mammals also have a vocal tract well-adapted to producing loud and complex sounds. When mammals took to the sea, these acoustic adaptations took on a special importance. Light does not penetrate more than a few hundred meters in clear open ocean and only a few meters in some coastal areas, so vision, which is a primary distance sense in air, has a limited range under water. Sound, by contrast, propagates extremely well under water. Many marine mammals have evolved specialized abilities of hearing and sound production that are used for communication and exploring the environment. Many of the toothed whales feed on elusive and patchy prey in the dark of the deep sea or at night. Many species have evolved a high-frequency biosonar that allows them to detect and chase prey. Many baleen whales migrate thousands of kilometers

annually, yet are quite social, and they produce low-frequency sounds that can be detected at ranges of hundreds if not thousands of kilometers.

Locomotion

Terrestrial animals have evolved special adaptations to support their bodies and for moving along a solid substrate. Efficient swimming puts very different selection pressures on marine compared to terrestrial locomotion. Water is close to the same density as most animal tissues. This freed aquatic mammals such as the sirenians and cetaceans completely from the need to support their bodies. However, water is also much more viscous than air. This resistance to motion allows marine animals to move by pushing against the medium, but also creates a strong selective pressure for mechanisms to reduce drag. Two different kinds of drag forces act on marine mammals. Viscous drag selects for a smooth skin and a low ratio of surface area to volume. Pressure drag relates to how fluid flows around the body given pressure differences, and this selects for a streamlined hydrodynamic shape. Most marine mammals that swim long distances in the water thus have a slick skin and a low-drag shape that makes them look quite different from their terrestrial ancestors (see **Figure 1**).

How Different Marine Mammals Fall on a Continuum from Aquatic to Terrestrial

While all marine mammals use the marine environment, some species are amphibious and need to function in air and in water. The polar bear, for example, is a strong swimmer but retains a body form quite similar to that of terrestrial bears (see **Figure 1**). Polar bears live on sea ice when they can, and their primary diet is

Table 4 Scientific and common names of pinnipeds along with conservation status[a]

Scientific name	Common name	ESA	MMPA	IUCN
Family Odobenidae	Walrus			
Odobenus rosmarus	Walrus			
Family Otariidae	Eared Seals			
Subfamily Otariinae	Sea lions			
Eumetopias jubatus	Stellar or Northern sea lion	TH	DEP	EN
Neophoca cinarea	Australian sea lion			
Otaria byronia	Southern sea lion			
Phocarctos hookeri	New Zealand sea lion			VU
Zalophus californianus	California sea lion			
Subfamily Arctocephalinae	Fur seals			
Arctocephalus australis	Falkland or South American fur seal			
Arctocephalus forsteri	New Zealand fur seal			
Arctocephalus galapagoensis	Galapagos fur seal			VU
Arctocephalus gazella	Antarctic fur seal			
Arctocephalus philipii	Juan Fernandez fur seal			VU
Arctocephalus pusillus	Australian or S. African fur seal			
Arctocephalus townsendi	Guadelupe fur seal	TH	DEP	VU
Arctocephalus tropicalis	Subantarctic fur seal			
Callorhinus ursinus	Northern fur seal		DEP	VU
Family Phocidae	Earless or True seals			
Subfamily Phocinae	Northern phocids			
Cystophora cristata	Hooded seal			
Erignathus barbatus	Bearded seal			
Halichoerus grypus	Gray seal			EN (NE Atl)
Phoca caspica	Caspian seal			VU
Phoca fasciata	Ribbon seal			
Phoca groenlandica	Harp seal			
Phoca hispida	Ringed seal	EN (Saimaa)	DEP	VU EN (Saimaa)
Phoca larga	Larga seal			
Phoca sibirica	Baikal seal			
Phoca vitulina	Harbor (US) or common (UK) seal			
Subfamily Monachinae	Southern phocids			
Hydrurga leptonyx	Leopard seal			
Leptonychotes weddellii	Weddell seal			
Lobodon carcinophagus	Crabeater seal			
Mirounga angustirostris	Northern elephant seal			
Mirounga leonina	Southern elephant seal			
Monachus monachus	Mediterranean monk seal	EN	DEP	CR
Monachus schauinslandi	Hawaiian monk seal	EN	DEP	EN
Monachus tropicalis	Caribbean monk seal	EN	DEP	EX
Ommatophoca rossii	Ross seal			

[a]CR = Critically endangered; DEP = Depleted; EN = Endangered; ESA = US Endangered Species Act; EX = Extinct; IUCN = International Union for the Conservation of Nature; MMPA = US Marine Mammal Protection Act; TH = Threatened; VU = Vulnerable. (Sources: *Reynolds and Rommel, 1999*; IUCN red book.)

seals, but they often live ashore during four months in the later summer and fall when sea ice has receded. Otariid seals can walk on land, but have a body shape transformed for swimming compared to their terrestrial ancestors (see **Figure 1**). The finlike limbs of seals are designed to push against the water with a large cross-sectional area. Marine mammals with limbs designed for swimming have large muscles attached to

shorter bones, giving more power and a greater mechanical advantage to the swimming motion. The sirenians and cetaceans never come onto land except during a stranding, and their bodies are the most transformed for swimming – neither group has legs or separate hind flippers, but rather a broad tail (see **Figure 1**). Cetaceans and sirenians use their axial musculature running along the entire vertebral column

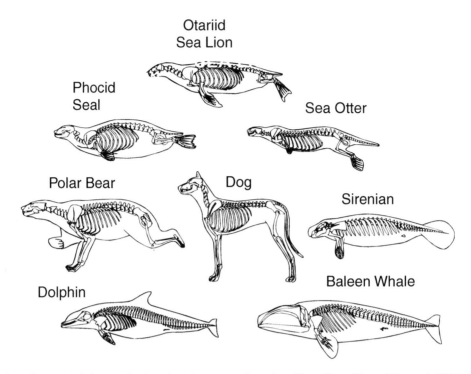

Figure 1 Body outlines and skeletons of selected marine mammal species. (From Reynolds and Rommel (1999), Figures 2–8.)

as they swim. Their tail flukes do not need to function for walking, only for swimming, and they are more efficient at generating thrust than are the seal flippers.

Sensory Systems

Just as these groups differ in their adaptation for locomotion in air and under water, so their sensory systems fall in different places on the continuum from adaptation to terrestrial versus aquatic life. The cetaceans are so fully adapted for marine life that they have lost most olfaction, which senses airborne odors. As mentioned above, sound is a particularly useful modality in the sea, but even for acoustic communication there is variation in the importance of airborne versus underwater sound. The otariid pinnipeds, sea otter, and polar bear communicate primarily in air; some phocid seals communicate both in air and under water; while sirenians, cetaceans, some phocid seals, and the walrus use sound to communicate primarily underwater. There are differences in the relative importance of hearing in air versus water for three different pinniped species whose hearing has been tested in both environments. The Otariid California sea lion is adapted to hear best in air; the phocid harbor seal can hear equally well in air and under water; and the auditory system of the phocid northern elephant seal is adapted for underwater sensitivity at the expense of aerial hearing.

Feeding

Marine mammals feed at a variety of trophic levels from herbivore to top predator. The sirenians are herbivores and eat sea grass that grows in coastal waters of the tropics. Several marine mammals specialize on benthic prey. The walrus feeds primarily on benthic mollusks; sea otters feed on benthic mollusks, echinoderms and decapod crustaceans. Sea otters are a keystone predator; their feeding on echinoderms is thought to structure kelp forest ecosystems. Grey whales feeds on benthic amphipods in the Bering Sea; their feeding turns over 9–27% of the seafloor there each feeding season, which probably enhances the abundance of colonizing benthic species. Baleen whales do not have teeth, and baleen whales other than the grey whale use their baleen to strain prey from sea water. They have been compared to grazers, because they feed on whole patches of prey, but their prey is typically large zooplankton or even schooling fish, which often require a predator-like ability to find and pursue prey. Most seals and toothed whales chase individual prey items, often fish or squid. Very little is known about the feeding behavior of deep diving toothed whales such as sperm whales. Our ignorance of the behavior of sperm whales and their deep squid prey may cause us to underestimate their ecological importance, for it has been calculated that they must take out of the ocean about the same biomass as all human fisheries.

Annual Feeding and Reproductive Cycles of Marine Mammals

Polar Bear and Pinnipeds

Polar bears and seals have a strong annual breeding cycle, with a short birthing and mating season, and delayed implantation that allows them to mate in spring well before they settle into a winter den for a short gestation period. Polar bears feed primarily on seals in the Arctic sea ice. Feeding is particularly intense during the spring birthing season for ringed seals, when up to a quarter of ringed seal breathing holes show predation attempts by polar bears. Up to 44% of the seal pups may be taken by polar bears at this time. Spring is not only the prime feeding season but also the breeding season for polar bears. Pregnant females delay implantation of the fetus until the feeding season is over. By summer, the sea ice habitat breaks up, forcing the polar bears either to summer on pack ice or to come ashore. A female who has come to ashore cannot hunt efficiently, and she may fast. By late September or early October, she will enter a winter den. Once she enters the den, the fetus implants and the pregnancy progresses, with a gestation period of only 3–4 months. The young are small and poorly developed when born. The eyes do not open for 40 days and the infant must nurse several times an hour and rely upon the mother to keep warm.

Cetaceans, sea otters, and sirenians never need to come to shore and do not have any lairs or dens that act as refuges. Seals are like the polar bear in that they must give birth on land or ice. Most seals also have a strong annual breeding cycle, with a short birthing and mating season and delayed implantation. A primary difference between otariid and phocid seals is that otariid mothers will leave their pups on shore as they forage. By contrast, many phocid mothers stay with their pups and fast throughout lactation. Many phocids give birth on the ice. Selection appears to favor a short, intense period of lactation in this setting. For example, the hooded seal suckles her young for an average of 4 days. The milk is >60% fat and the female suckles more than twice an hour. The pup gains more than $7 \, \text{kg} \, \text{d}^{-1}$ during this period.

Annual Cycle of Baleen Whales

Researchers debate whether cetacean swimming is more or less energetically costly than terrestrial locomotion, but living in a buoyant medium has certainly freed cetaceans to grow to larger sizes. The blue whale is the largest animal ever to live on Earth, many times more massive than the largest dinosaurs. Aquatic animals are freed from some of the constraints on size imposed on terrestrial animals, but baleen whales are also larger than any other aquatic animals. This large size is part of a suite of adaptations driven by an annual migratory cycle of most baleen whales, feeding in the summer and mating and giving birth in the winter (see **Figure 2**). Most species of baleen whales have adapted to take advantage of a seasonal burst of productivity in the summer in polar waters. Summer

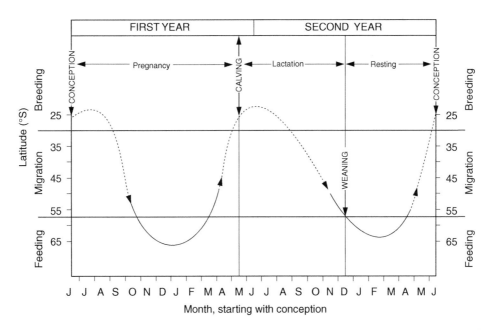

Figure 2 Annual migratory feeding and reproductive cycle of baleen whales. (Adapted from Reynolds and Rommel (1999), Figures 6–19.)

is the feeding season for most migratory baleen whales, and they feed intensively during the summer, building up fat reserves. The primary prey are invertebrate zooplankton such as krill and copepods; some balaenopterid whales also feed on schooling fish. Once the polar pulse of summer productivity is over, there is little reason to stay in these waters. Most baleen whales migrate away from the feeding grounds to separate winter breeding and calving grounds. They seldom feed during these seasons, and live off of their fat reserves from summer feeding.

Most baleen whales migrate to tropical waters for the winter breeding season. It is tempting to assume that they migrate to the tropics to avoid the winter cold and storms in polar seas, but many smaller marine mammals can thermoregulate well in polar areas in winter, so this raises questions about the reasons for migrations to tropical waters in winter. Grey whales migrate roughly 8000 km from feeding grounds in the Bering Sea to breeding grounds in the coastal lagoons of Baja California, but bowhead whales remain in the Bering Sea during winter. There is a need for better modeling of the energetic costs and benefits of migrating thousands of kilometers to warmer water. Many whale species select calm, protected waters with relatively low predation risk for calving, and this may also influence the choice of breeding area.

The annual cycle of baleen whales and the way in which they forage have selected for large size. Baleen whales engulf many prey items within a dense patch of prey, straining prey from sea water using the baleen for which they are named. The habit of feeding on as many prey within a patch as possible selects for large size in the whales. If whales feed in the summer months, and must live off of their energy stores for the rest of the year, then they must be large enough to store enough energy for the whole year. If whales feed in high latitudes and winter in low latitudes, then they must swim thousands of kilometers, and larger animals can swim more efficiently. All of these factors have acted in concert to select for large size in these largest of animals. The growth of a baleen whale calf is truly extraordinary. During the first half-year of life, a blue whale calf will grow on average 3.5 cm d^{-1} and 80 kg/day (**Figure 3**). This is supported entirely by the mother's milk, which comes from fat reserves since the mother is still migrating up to the feeding area. The annual cycle may select for such rapid growth so that the calf is ready to wean during the summer feeding season.

Prolonged Growth and Maturation in Odontocetes

Odontocetes do not have as pronounced an annual feeding cycle as baleen whales. Many odontocetes do have a breeding season, but few reproduce on an annual basis. Most odontocetes have a gestation

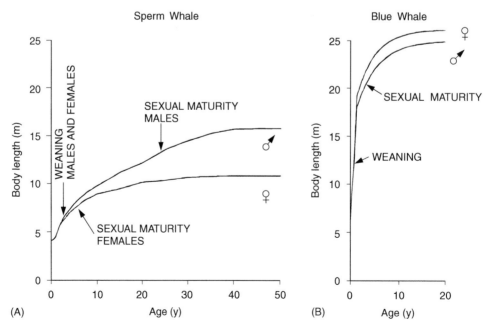

Figure 3 Growth and maturation of (A) sperm whale and (B) blue whale (data from Best, 1979; Lockyer, 1981). Baleen whales live well beyond the 20 years indicated in the figure, which is designed to show ages at which these animals reach full growth. In fact, there is some evidence that bowhead whales may survive for more than a century.

period of about one year, but the larger odontocetes have longer gestation, up to 14–16 months in the sperm and killer whales. Most dolphins and larger toothed whales have a prolonged period of dependency that contrasts particularly strikingly with the record short periods of lactation in some phocid seals. For example, the short-finned pilot whale and sperm whale may suckle young for over 10 years in some cases. The prolonged maturation of toothed whales makes particularly stark contrast with the large baleen whales. **Figure** 3 shows the growth and maturation of the blue whale versus the sperm whale. The blue whale has weaned by 6 months and is sexually mature by 10 years. By contrast, a male sperm whale suckles for many years and may not reach sexual maturity until 20–25 years of age. Even though baleen whales mature more rapidly than sperm whales, they may live longer. Recent evidence suggests that bowhead whales may live more than a century.

Even where lactation may not be this prolonged, there is still often a prolonged period of dependency in odontocetes. For example, bottlenose dolphins in Sarasota, Florida, are sighted with their mothers for 3–4 years, typically until the mother has another calf. There is a paradox in this prolonged dependency. On the one hand, odontocete young are extremely precocial in their senses and motor abilities for vocalizing and swimming. On the other, they appear to need many years of parental care for success. Many biologists believe that this parental care does not just involve the nutritional care of suckling, but may also involve a long period in which the young learn from the mother and other conspecifics. Among the odontocetes, porpoises and river dolphins live life in the fast lane, weaning within a year of age. These species grow and mature more rapidly than most other odontocetes, and do not rely upon as long a period of dependency.

Navigation and Migration

Many marine mammals are highly mobile, and may swim over $100 \, \text{km} \, \text{d}^{-1}$. Some coastal animals such as sea otters, dugongs, or coastal bottlenose dolphins may have home ranges only kilometers to tens of kilometers in scope. All species are mobile enough to require abilities to navigate. Some ice-loving seals range over kilometers, but must find holes in the ice to breathe through. Failure to find these holes could easily be fatal. When Weddell seals are caught at a breathing hole and transported several kilometers by scientists to a new man-made one, they can swim under thick ice to find the original hole. These seals

must be very skilled at timing their dives to be sure they can return to this original hole when they need to breathe. When sunlight is available, antarctic Weddell seals and arctic ringed seals appear to use downwelling light to navigate. It seems that holes and cracks in the ice, and under-ice features, may provide the same kind of cues for landmarks and routes as used by terrestrial animals. If vision is not available seals restrict their diving but appear to be able to use acoustic cues to locate holes. Seals use some sounds to orient, but ringed seals may avoid sounds of stepping or scraping at an airhole. This makes sense, since these sounds may come from a polar bear or Inuit waiting to kill a seal as it surfaces. At close range, even a blindfolded seal can use its vibrissae to center in an airhole. These animals thus rely upon local knowledge and a combination of sensory cues for navigation.

Many marine mammals, such as the baleen whales, migrate thousands of kilometers annually. Scientists have suggested that marine mammals may use visual, acoustic, chemical, and even geomagnetic cues to orient for migration, but little is known about the sensory basis of orientation or migration. Most whale calves will stay with their mother on their initial migration from calving ground to feeding ground, so they may have an opportunity to learn about migration routes. When bowhead whales migrate, they make low-frequency calls, apparently to coordinate movements and to detect ice obstacles in their path. Most pelagic odontocetes live for several years in the group in which they were born, providing opportunities for learning about navigation. Elephant seals also have migrations of thousands of kilometers, but a young seal is left on the beach by its mother. When a weaned seal leaves the beach, it is thought to leave alone and to have to learn diving, foraging, and orientation on its own. The tracks of satellite-tagged migrating seals and whales are often remarkably well-oriented. Coupling such tags with experimental manipulations may illuminate the sensory basis of migration in marine mammals.

Conservation of Marine Mammals

Many populations of marine mammals have been endangered by humans. They have traditionally been tempting targets for commercial hunting because they carry quantities of valuable meat and fat or oil, and because humans can catch them when they surface to breathe. Species that hauled out on land were particularly vulnerable. Human hunters drove the Steller's sea cow (*Hydrodamalis gigas*) to extinction in about 25 years in the mid-eighteenth century. In the twentieth century most of these commercial hunts were regulated to protect populations. The US Marine Mammal Protection Act was passed in large measure to reduce the unintentional killing of dolphins in a tuna fishery in the eastern tropical Pacific. Marine mammals are still killed during fishing activities or by ghost gear. Many of the great whales remain so endangered from commercial hunting that no human take is allowed. In spite of the prohibition on taking severely endangered whales such as the northern right whale, of which fewer than 300 individuals survive, right whales are regularly killed by entanglement in fishing gear or by collision with large ships.

As direct mortality has been increasingly controlled, newer threats to marine mammals have also surfaced. Humans have treated waterways as dumping grounds for toxic wastes, and most contaminants in water ultimately enter the sea. Some marine mammals carry heavy loads of organochlorine compounds such as polychlorinated biphenyls (PCBs) or DDT and toxic elements such as mercury or cadmium. We know little about what pathologies may be linked to these exposures, but some evidence suggests associations between impaired reproduction and exposure to some organochlorine compounds. Whales have died after eating fish contaminated with saxitoxin from a harmful dinoflagellate bloom. Many human seafaring activities also create noise pollution. Humans use sound to explore the oceans with sonar and use intense sounds to explore geological strata below the seafloor. The motorized propulsion of ships has increased ocean noise globally, and underwater explosions have killed endangered whales and river dolphins. All of these threats can be considered forms of habitat degradation.

There has also been growing concern about the effects of climate change upon some marine mammals. Many marine mammals depend upon sea ice. In years with little sea ice, harp seals may have decreased reproduction and increased mortality, and polar bears may not be able to feed as well, leading to lower reproductive rates. If global warming reduces the amount or quality of ice used by marine mammals, this may degrade or eliminate critical habitats. Most current laws to protect marine mammals were designed to prevent humans from killing animals directly. Since marine mammals sample most trophic levels of marine ecosystems and can be counted and observed at the surface by humans, many biologists consider marine mammals to be indicator species, helping us to monitor the health of marine ecosystems. New ways of thinking will be required to protect marine mammal populations from habitat degradation.

See also

Baleen Whales. Copepods. Krill. Marine Mammal Diving Physiology. Marine Mammal Evolution and Taxonomy. Marine Mammal Trophic Levels and Interactions. Primary Production Distribution. Primary Production Processes. Seals. Sea Otters. Sirenians. Sperm Whales and Beaked Whales.

Further Reading

Berta A and Sumich JL (1999) *Marine Mammals: Evolutionary Biology.* San Diego: Academic Press.

Best PB (1979) Social organization in sperm whales, *Physeter macrocephalus.* In: Winn HE and Olla BL (eds.) *Behavior of Marine Mammals: Current Perspectives in Research,* vol. 3: *Cetaceans,* pp. 227–289. New York: Plenum Press.

Castellini MA, Davis RW, and Kooyman GL (1992) *Annual Cycles of Diving Behavior and Ecology of the Weddell Seal. Bulletin of the Scripps Institution of Oceanography.* San Diego: University of California Press.

Le Boeuf BJ and Laws RM (eds.) (1994) *Elephant Seals: Population Ecology, Behavior and Physiology.* Berkeley: University of California Press.

Lockyer C (1981) Growth and energy budgets of large baleen whales from the Southern Hemisphere. *Food and Agriculture Organization of the United Nations Fisheries, Series 5, Mammals in the Seas* 3: 379–487.

Kanwisher J and Ridgway S (1983) The physiological ecology of whales and porpoises. *Scientific American* 248: 110–120.

Reynolds JE III and Rommel SA (eds.) (1999) *Biology of Marine Mammals.* Washington, DC: Smithsonian Press.

Rice DW (1998) *Marine Mammals of the World: Systematics and Distribution.* Lawrence, KS: Society for Marine Mammalogy.

Riedman M (1991) *The Pinnipeds: Seals, Sea Lions and Walruses.* Berkeley: University of California Press.

BALEEN WHALES

J. L. Bannister, The Western Australian Museum, Perth, Western Australia

Diagnostic Characters and Taxonomy

Baleen or whalebone whales (Mysticeti) comprise one of the two recent (nonfossil) cetacean suborders. They differ from the other suborder (toothed whales, Odontoceti), particularly in their lack of functional teeth. Instead they feed on relatively very small marine organisms by means of a highly specialized filter-feeding apparatus made up of baleen plates 'whalebone' attached to the gum of the upper jaw. Other differences from toothed whales include the baleen whales' paired blowhole, symmetrical skull, and absence of ribs articulating with the sternum.

Baleen whales are generally huge. In the blue whale they include the largest known animal, growing to more than 30 m long and weighing more than 170 tonnes. Like all other cetaceans, baleen whales are totally aquatic. Like most of the toothed whales, they are all marine. Many undertake very long migrations, and some are fast swimming. A few species come close to the coast at some part of their life cycle and may be seen from shore; however, much of their lives is spent remote from land in the deep oceans. Baleen whale females grow slightly larger than the males. Animals of the same species tend to be larger in the Southern than in the Northern Hemisphere.

Within the mysticetes are four families: right whales (Balaenidae, balaenids), pygmy right whales (Neobalaenidae, neobalaenids), gray whales (Eschrichtiidae, eschrichtiids); and 'rorquals' (Balaenopteridae, balaenopterids). Within the suborder, 12 species are now generally recognized (**Table 1**).

Right whales are distinguished from the other three families by their long and narrow baleen plates and arched upper jaw. Other balaenid features include, externally, a disproportionately large head (*c.* one-third of the body length), long thin rostrum, and huge bowed lower lips; they lack multiple ventral grooves. Internally, there is no coronoid process on the lower jaw and cervical vertebrae are fused together. Within the family are two distinct groups: the bowhead (*Balaena mysticetus*) of northern polar waters (formerly known as the 'Greenland' right

whale) and the three 'black' right whales (*Eubalaena* spp.) of more temperate seas (so called to distinguish them from the 'Greenland' right whale). (**Figure 1**). All balaenids are robust.

Pygmy right whales (*Capnea marginata*) have some features of both right whales and balaenopterids. The head is short (*c.* one-quarter of the body length), although with an arched upper jaw and bowed lower lips, and there is a dorsal fin. The relatively long and narrow baleen plates are yellowish-white, with a dark outer border, quite different from the all-black balaenid baleen plates. Internally, pygmy right whales have numerous broadened and flattened ribs.

Gray whales (*Eschrichtius robustus*) are also somewhat intermediate in appearance between right whales and balaenopterids. They have short narrow heads, a slightly arched rostrum, and between two and five deep creases on the throat instead of the balaenopterid ventral grooves. There is no dorsal fin, but a series of 6 to 12 small 'knuckles' along the tail stock. The yellowish-white baleen plates are relatively small.

Balaenopterids comprise the five whales of the genus *Balaenoptera* blue, *B. musculus*; fin, *B. physalus*; sei, *B. borealis*; Bryde's, and minke, *B. acutorostrata* & *B. bonaerensis*; whales and the humpback whale (*Megaptera novaeangliae*). All have relatively short heads, less than a quarter of the body length. In comparison with right whales, the baleen plates are short and wide. Numerous ventral grooves are present, and there is a dorsal fin, sometimes rather small. Internally, the upper jaw is relatively long and unarched, the mandibles are bowed outward, and a coronoid process is present; cervical vertebrae are generally free. All six balaenopterids are often known as 'rorquals' (said to come from the Norse 'whale with pleats in its throat'). Strictly speaking, the term should probably be applied to the five *Balaenoptera* species, recognizing the rather different humpback in its separate genus, but many authors now use it for all six balaenopterids.

Baleen whales are sometimes called 'great whales.' Despite their generally huge size, some of the species are relatively small, and it seems preferable to restrict the term to the larger mysticetes (blue, fin, sei, Bryde's, humpback) together with the largest odontocete (the sperm whale, *Physeter macrocephalus*) (**Figure 2**).

In a recent review of the systematics and distribution of the world's marine mammals, Rice (1998)

Table 1 Mysticetes (baleen whales)

Family	Genus	Species	Subspecies	Common name	Maximum length (m)	Generalized distribution
Balaenidae				Right whales		
	Balaena	B. mysticetus		Bowhead whale	19.8	Circumpolar in the Arctic
	Eubalaena	E. glacialis		North Atlantic right whale	17.0	Temperate–Arctic
		E. australis		Southern right whale	17.0	Temperate–N. Atlantic
		E. japonica		North Pacific right whale	17.0	Temperate N. Pacific
Neobalaenidae				Pygmy right whales		
	Neobalaena	Caperea marginata		Pygmy right whale	6.4	Temperate, Southern Hemisphere only
Eschrichtiidae				Gray whales		
	Eschrichtius	E. robustus		Gray whales	14.1	North Pacific–Arctic
Balaenopteridae				'Rorquals'		
	Megaptera	M. novaeangliae		Humpback whale	16.0	Worldwide
	Balaenoptera	B. acutorostrata		Common minke whale		Worldwide
			B. a. acutorostrata	N. Atlantic minke whale	9.2	Temperate–Arctic
			B. a. scammoni	N. Pacific minke whale	?	Temperate–Arctic
			B. a. subsp.	Dwarf minke whale	?	Temperate–subantarctic, Southern Hemisphere only
		B. bonaerensis		Antarctic minke whale	10.7	Temperate–Antarctic
		B. edeni		Bryde's whale	14.0	Circumglobal, tropical–subtropical
		B. borealis		Sei whale	17.7	Worldwide, largely temperate
		B. physalus		Fin whale	26.8	Worldwide
		B. musculus		Blue whale		Worldwide
			B. m. musculus	Blue whale	26.0	N. Atlantic, N. Pacific
			B. m. indica	Great Indian rorqual	?	N. Indian Ocean
			B. m. brevicauda	Pygmy blue whale	24.4	Southern Hemisphere, temperate–subAntarctic
			B. m. intermedia	'True' blue whale	30.5	Southern Hemisphere, temperate–Antarctic

has drawn attention to the derivation of the Latin word Mysticeti and clarified the status of a variant, Mystacoceti. He describes the former as coming from Aristotle's original Greek *mustoketos*, meaning 'the mouse, the whale so-called' or 'the mouse-whale' (said to be an ironic reference to the animals' generally vast size). Mystacoceti means 'moustache whales,' and although used occasionally in the past (and more obviously appropriate for whales with baleen in their mouths), it has been superseded by Mysticeti.

The 12 species in **Table 1** differ somewhat from those listed by Rice. Some authors disagree with his use of the genus *Balaena* for *Eubalaena* and his preference for the single species *glacialis* rather than

the three species, *Eubalaena glacialis*, *E. japonica* and *E. australis* (the North Atlantic, North Pacific and southern right whales). While acknowledging the need for further investigation, they refer to present-day biologists' usage, and genetic information, in preferring a separation between Northern and Southern Hemisphere animals and in recognizing two species in the Northern Hemisphere; *Eubalaena* is, however, the only mysticete genus where one or more separate species is recognized in each hemisphere. Rice also distinguishes between two species of Bryde's whale: *Balaenoptera edeni* and *B. brydei*. The taxonomic status of these 'inshore, smaller' and 'offshore, larger' forms has yet to be determined and here they are subsumed within *B. edeni*. In the case

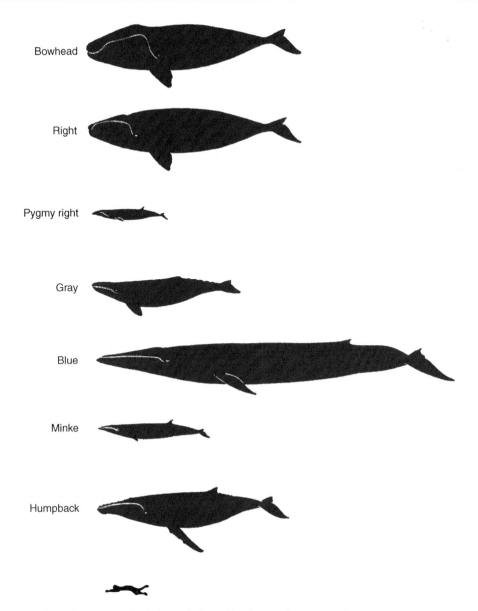

Figure 1 Lateral profiles of representative baleen whales, with a human figure, to scale.

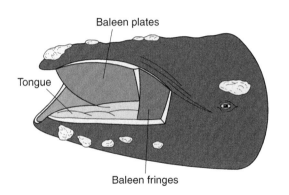

Figure 2 Head of a right whale, showing the arrangement of the filter-feeding apparatus (from Bonner, 1980).

of the blue whale, Rice's inclusion of a northern Indian Ocean form (*B. m. indica,* referred to by Rice as 'the great Indian rorqual') has been followed. Similarly, his listing of three subspecies of minke whale, including the Southern Hemisphere dwarf minke, which has yet to be formally described, has been retained. However, other subspecies, e.g., two sei whales and two fin whales, have not been included.

Distribution and Ecology

Habitat

In addition the subspecies listed in the previous section, many stocks or populations have been recognized, some mainly for management purposes, based on more or less valid biological grounds. Some significant examples include the following.

Bowhead whales In addition to the currently most abundant population (the Bering-Chukchi-Beaufort Seas stock), four others are recognized: Baffin Bay/Davis Strait, Hudson Bay, Spitzbergen, and Okhotsk Sea.

Right whales In the North Atlantic species, two populations are currently recognized, western and eastern, with calving grounds off the southeastern United States and northwestern Africa. The latter may now represent only a relict population(s). In the North Pacific species, the current view is that there well may once have been two or more stocks, based on feeding ground information: at least one now centered in summer on the Sea of Okhotsk and another, although possibly not now a functioning unit, summering in the Gulf of Alaska.

In the southern right whale, there are several populations, defined by currently occupied calving grounds, but these cover only a proportion of the many areas known from historical whaling records to have once been occupied by right whales. Up-to-date information is available on presumed discrete populations off eastern South America, South Africa, southern Australia, and sub-Antarctic New Zealand.

Gray whales A western North Atlantic population may have persisted until the 17th or 18th centuries, but is now extinct. The species now survives only in the North Pacific, where, in addition to a flourishing 'Californian' stock, wintering on the coast of Baja California, and summering in the Bering Sea, animals are now being reported from a remnant western stock, summering in the northern Okhotsk Sea.

Humpback whales In the North Atlantic, two major populations are recognized: one based on animals wintering in the West Indies and the other, now possibly only a relict population, wintering around the Cape Verde Islands. In the North Pacific, three discrete wintering grounds have been recorded: around the Bonin, Mariana, and Marshall Islands in the west; around the Hawaiian Islands in the center; and off Mexico in the east.

In the Southern Hemisphere, seven populations have been postulated. Six are well defined, based on calving (wintering) grounds either side of each continent (one off eastern Australia is closely related to animals wintering off Fiji and Tonga), and a possible seventh in the central Pacific. In the northwest Indian

Ocean, there seems to be a separate population where animals have been reported present throughout the year.

Baleen whales thus occupy a wide variety of habitats, from open oceans to continental shelves and coastal waters, from the coldest waters of the Arctic and Antarctic, through temperate waters of both hemispheres and into the tropics.

Most specialized is the bowhead, *Balaena*, restricted to the harsh cold and shallow seas of the Arctic and sub-Arctic. The black right whales (*Eubalaena*) are more oceanic and prefer generally temperate waters, but come very close to coasts in winter to give birth, particularly in the Southern Hemisphere. Once believed not to penetrate much further south than the Antarctic convergence (c. 50–55°S), there have been recent records in the Antarctic proper, south of 60°S. Whether this is a new phenomenon is unclear: a report by Sir James Clark Ross of many 'common black' (i.e., right) whales in the Ross Sea (eastern Antarctic) at 63°S in December 1840 was discounted when their presence there later that century could not be confirmed. It has been suggested that the currently greatly reduced population of the western North Atlantic right whale, now wintering off the south eastern United States and summering in coastal waters north to the Bay of Fundy (c. 45°N), may represent the peripheral remnant of a more widely distributed stock, formerly summering north to Labrador and southern Greenland, i.e., to at least 60°N (**Figure 3**).

The pygmy right whale (*Caperea*) is restricted to Southern Hemisphere temperate waters, between about 30° and 52°S; it can be found coastally in winter in some areas, and all-year round in others.

Gray whales (*Eschrichtius*) are the most obviously coastal baleen whales. The long coastal migration of the 'Californian' stock, from Mexico to Alaska, supports a major whale-watching industry from December to April. In spring the animals migrate through the Bering Strait into the more open waters of the Bering Sea, but still favoring more shallow waters.

Among the balaenopterids, fin and sei whales are probably the most oceanic, with the former penetrating into colder waters than the latter in summer. Blue whales can be found closer inshore, but are often associated with deep coastal canyons, e.g., off central and southern California. The Southern Hemisphere pygmy blue whale (subspecies *B. m. brevicauda*) has been regarded as restricted to more temperate waters than the 'true' blue whale (*B. m. intermedia*), not often being found much beyond 55°S. The most coastal balaenopterid is the humpback (*Megaptera*), with long migrations between

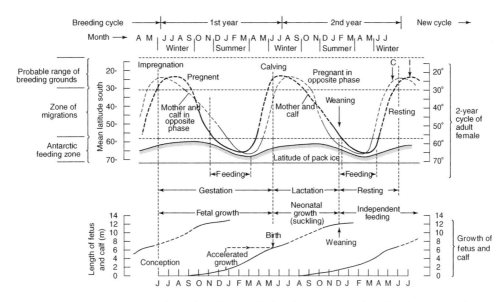

Figure 3 'Typical' life cycle of a southern baleen whale (as modified by Bonner, 1980, from Mackintosh, 1965).

temperate/tropical breeding grounds and cold-water feeding grounds. In the Southern Hemisphere, much of its journey occurs along the east and west coasts of the three continents. In the Northern Hemisphere, humpbacks are rather more oceanic, but still coastal at some stage in their migration: in the North Pacific they can be found wintering off the Hawaiin Islands and summering off Alaska and in the western North Atlantic they winter in the Caribbean and summer between New England, the west coast of Greenland, and Iceland.

Minke whales are wide ranging, from polar to tropical waters in both hemispheres. In the Southern Hemisphere they can, with blue whales, be found closest to the ice edge in summer. Elsewhere they can often occur near shore, in bays and inlets. Their migrations are less well defined and predictable than other migratory balaenopterids; in some regions they are present year-round.

The most localized balaenopterid is Bryde's whale. It is the only species restricted entirely to tropical/warm temperate waters and probably does not undertake long migrations. The two forms – inshore and offshore, in several areas – can differ in their movements. Off South Africa, for example, the inshore form is thought to be present throughout the year, whereas the offshore form appears and disappears seasonally, presumably in association with movements of its food, shoaling fish.

Food and Feeding

Although they include the largest living animals, baleen whales feed mainly on very small organisms and are strictly carnivorous, feeding on zooplankton or small fish. In 'filter-feeding' – sieving the sea – baleen whales are quite different from toothed whales, where the prey is captured individually.

Filter feeding has been described as requiring, in addition to a supply of food in the water, three basic features: a flow of water to bring prey near the mouth; a filter to collect the food but allow water to pass through; and a means of removing the filtered food and conveying it to the stomach for digestion. Baleen whales meet those requirements by (a) seeking out areas where their food concentrates, (b) either swimming open-mouthed through food or gulping it in, (c) possessing a highly efficient filter formed by the baleen plates, and (d) forcing the water containing the food out through the baleen plates and then transfering the trapped food back to the gullet and hence to the stomach. In the latter the tongue is presumed to be involved; in balaenopterids the process is aided by the distensible throat.

While all baleen whales possess a filter based on baleen plates, two rather different systems – essentially 'skimming' and 'gulping' – have evolved to filter a large volume of water containing food. Each relies on a series of triangular baleen plates, borne transversely on each upper jaw. The inner, longer border (hypotenuse) of each plate bears a fringe of fine hairs, forming a kind of filtering 'doormat.' Quite unrelated to teeth (which appear as early rudiments in the gums of fetal baleen whales), baleen is closest in structure to mammalian hair and human fingernails. In the right whales, filtration is achieved with very long and narrow plates in the very large mouth, itself carried in the very large head. The plates, up to 4 m long in bowheads and

2.7 m in the right whales, are accommodated in the mouth by an arched upper jaw and are enclosed in massively enlarged and upwardly bowed lower lips. There is a gap between the rows at the front of the mouth, and the whole arrangement allows the whale to scoop up a great quantity of water while swimming slowly forward. In balaenopterids, with their much smaller heads, the baleen plates are shorter and broader and the rows are continuous at the front. Taking in a large volume of water and food is usually achieved by swimming through a food swarm and gulping, while simultaneously enlarging the capacity of the mouth greatly by extending the ventral grooves and depressing the tongue. The two systems allow, on the one hand, the relatively slow-swimming balaenids to concentrate their rather sparse slow-swimming food over a period, and on the other, the faster-swimming balaenopterids to take in large amounts of their highly concentrated fast-swimming prey over a shorter time.

Typically, baleen whales feed on zooplankton, mainly euphausiids or copepods, swarming in polar or subpolar regions in summer. That is particularly so in the Southern Hemisphere, where the summer distributions of several balaenopterids depend on the presence of *Euphausia superba* known to whalers by the Norwegian word 'krill' (*see* Krill) in huge concentrations in the Antarctic. In the Northern Hemisphere, with a more variable availability of food, balaenopterids are more catholic in their feeding. Humpbacks and fin whales, for example, feeding almost exclusively on krill in the south, commonly take various species of schooling fish in the north.

The variety of organisms taken by the various species in different regions is listed in **Table 2**. While most feeding occurs in colder waters, baleen whales may feed opportunistically elsewhere. All baleen whales but one, the gray whale, feed generally within 100 m of the surface and, consequently, unlike many toothed whales, do not dive very deep or for long periods. Gray whales feed primarily on bottom-living organisms, almost exclusively amphipods, in shallow waters.

Table 2 Baleen Whale Food Items

Species	Subspecies	Common name	Food items	
			Northern Hemisphere	Southern Hemisphere
B. mysticetus		Bowhead whale	Mainly calanoid copepods; euphausiids; occasional mysids, amphipods, isopods, small fish	
E. glacialis		North Atlantic right whale	Calanoid copepods; euphausiids	
E. australis		Southern right whale		Copepods; postlarval *Munida gregaria*; *Euphausia superba*
Caperea marginata		Pygmy right whale		Calanoid copepods
E. robustus		Gray whale	Gammarid amphipods; occasional polychaetes	
M. novaeangliae		Humpback whale	Schooling fish; euphausiids	*E. superba* (Antarctic); euphausiids, postlarval *M. gregaria*, occasional fish (ex-Antarctic)
B. acutorostrata	B. a. acutorostrata	N. Atlantic minke	Schooling fish; euphausiids	
	B. a. scammoni	N. Pacific minke	Euphausiids; copepods; schooling fish	
	B. a. subsp.	Dwarf minke		?Euphausiids, schooling fish
B. bonaerensis		Antarctic minke		*E. superba*
B. edeni		Bryde's whale	Pelagic crustaceans, including euphausiids	Schooling fish; euphausiids
B. borealis		Sei whale	Schooling fish	Copepods, including *Calanus* and *E. superba*
B. physalus		Fin whale	Schooling fish; squid; euphausiids; copepods	*E. superba* (Antarctic); other euphausiids (ex-Antarctic)
B. musculus	B. m. musculus	Blue whale	Euphausiids	
	B. m. indica	Great Indian rorqual	?Euphausiids; copepods	
	B. m. intermedia	'True' blue		*E. superba* (Antarctic); other euphausiids (ex-Antarctic)
	B. m. brevicauda	Pygmy blue		Euphausiids, mainly *E. vallentini*

The baleen plate structure, particularly the inner fringing hairs, to some extent mirrors the food organisms taken or (in the case of *E. superba*) different size classes. Thus there is some correlation between decreasing size of prey and fineness of baleen by species, i.e. gray, blue, fin, humpback, minke, sei, and right whales. Where food stocks are very dense, e.g., around sub-Antarctic South Georgia, fin, blue, and sei whales may all overlap in their feeding on *E. superba*.

Baleen whale food consumption per day has been calculated as some 1.5–2.0% of body weight, averaged over the year. Given that feeding occurs mainly over about 4 months in the summer in the larger species, the food intake during the feeding season has been calculated at some 4% of body weight per day, *c.* 4000 kg per day for a large blue whale. To survive the enormous drain of pregnancy and lactation, it has been calculated that a pregnant female baleen whale needs to increase its body weight by up to 65%. The ability to achieve such an increase in only a few months' feeding indicates the great efficiency of the baleen whales' feeding system.

Predators and Parasites

Apart from humans, the most notable baleen whale predator is the killer whale (*Orcinus orca*). Minke whales have been identified as a major diet item of some killer whales in the Antarctic. Killer whale attacks have been reported on blue, sei, bowhead, and gray whales, although their frequency and success are unknown. Humpbacks often have killer whale tooth marks on their bodies and tail flukes. Humpback and right whale calves in warm coastal waters are susceptible to attack by sharks. There are anecdotal reports of calving ground attacks on humpbacks by false killer whales (*Pseudorca crassidens*).

A form of harassment, only recently described, occurs on right whales on calving grounds off Peninsula Valdes, Argentina. Kelp gulls have developed the habit of feeding on skin and blubber gouged from adult southern right whales' backs as they lie at the surface. Large white lesions can result. The whales react adversely to such gull-induced disturbance and calf development may be affected.

External parasites, particularly 'whale lice' (cyamid crustaceans) and barnacles (both acorn and stalked) are common on the slower-swimming more coastal baleen whales, such as gray, humpback, and right whales. In the latter, aggregations of light-colored cyamids on warty head callosities have facilitated research using callosity-pattern photographs for individual identification. External parasites are much less common on the faster swimming species, although whale lice have been reported on minke whales (in and around the ventral grooves and umbilicus); the highly modified copepod *Penella* occurs particularly on fin and sei whales in warmer waters. The commensal copepod *Balaenophilus unisetus* often infests baleen plates in such waters, especially on sei and pygmy blue whales.

A variety of internal parasites have been recorded, although some baleen whales seem less prone to infection than others. They appear, for example, to be less common in blue whales, but prevalent in sei whales. Records include stomach worms (*Anisakis* spp.), cestodes, kidney nematodes, liver flukes, and acanthocephalans ('thorny-headed' worms) of the small intestine.

The cold water diatom *Cocconeis ceticola* often forms a brownish-yellow film on the skin of blue and other baleen whales in the Antarctic. Because the film takes about a month to develop, its extent can be used to judge the length of time an animal has been there. Its presence led to an early common name for the blue whale 'sulphur bottom.'

For many years the origin of small scoop-shaped bites on baleen whale bodies in warmer waters remained a mystery until they were found to be caused by the small 'cookie-cutter' shark, *Isistius brasiliensis*. Some species are highly prone to such attacks. In Southern Hemisphere sei whales the overlapping healing scars can impart a galvanized-iron sheen to the body.

Life History

Behavior

Sound production Unlike toothed whales, baleen whales are not generally believed to use sound for echo location, although bowheads, for example, are thought to use sound reflected from the undersides of ice floes to navigate through ice fields. However, sound production for communication, for display, establishment of territory, or other behavior, is well developed in the suborder. Blue whales produce the loudest sustained sounds of any living animal. At up to nearly 190 decibels, their long (half-minute or more), very low frequency (<20 Hz) moans may carry for hundreds of kilometers or more in special conditions. Fin whales produce similarly low (20-Hz) pulsed sounds. Minke whales also produce a variety of loud sounds. Right whales produce long low moans; bowhead sounds, recorded on migration past hydrophone arrays in nearshore leads, have been used in conjunction with sightings to estimate population size off northern Alaska.

Southern right whales, at least, seem to use sound to communicate with their calves.

Humpbacks produce the longest, most complex sound sequences in 'songs', described as an array of moans, groans, roars, and sighs to high-pitched squeaks and chirps, lasting 10 or more minutes before repetition, sometimes over hours. It seems that only the adult males sing, generally only in or close to the breeding season. In any one breeding season, all the males sing the same song, changing slightly over successive seasons. Different populations have different songs; so much so, for example, that those off western Australia have a distinctly different song – less complex, less 'chirpy' – than that heard on breeding grounds separated by the Australian continent, off the east coast. 'Songs' may also be heard in migrating humpbacks, but less so on the cold-water feeding grounds, where if they occur at all, they appear generally only as 'snatches' or isolated segments.

Swimming and migration With their stream-lined bodies, rorquals include the fastest-swimming baleen whales. Sei whales have been recorded at around 35 knots (more than 60 km/h) in short bursts; minke and fin whales are also known as fast swimmers, the latter up to 20 knots (37 km/h). Blue whales are among the most powerful swimmers, able to sustain speeds of over 15 knots (28 km/h) for several hours. On migration, humpbacks and gray whales average about 4 knots (8–9 km/h) and bowheads only about 2.7 knots (5 km/h). Migration speeds for southern right whales are not known, but medium range coastal movements off southern Australia indicate 1.5–2.3 knots (2.7–4.2 km/h) over 24 h, for cow/calf pairs.

Baleen whales undertake some of the longest migrations known. Gray whales cover some 10 000 nautical miles (18 000 km) on the round trip between the Baja California breeding grounds and the Alaskan feeding grounds, among the longest migrations of any mammal. Southern Hemisphere humpbacks may cover as much as 50° of latitude either way between breeding and feeding grounds, a round trip of some 6000 nautical miles (11 000 km). Not all baleen whale migrations are so well marked. The biannual movements of Bering Sea bowheads are governed by the seasonal advance and retreat and of the sea ice, which varies from year to year. Although Southern Hemisphere blue and fin whales all feed extensively in the Antarctic in summer, the locations of their calving grounds are not known. Sei whale migrations are relatively diffuse and can vary from year to year in response to changing environmental conditions. By comparison, Bryde's whales hardly migrate at all, presumably being able to satisfy both reproductive and nutritional needs in tropical/warm temperate waters. Even among such migratory animals as humpbacks, it may be that not all animals migrate every year; studies off eastern Australia indicate that a proportion of adult females may not return to the calving grounds each year, and individuals have even been reported in summer further north. However, Southern Hemisphere migrating humpbacks show segregation in the migrating stream: immatures and females accompanied by yearling calves are in the van of the northward migration, followed by adult males and nonpregnant mature females; pregnant females bring up the rear. A similar pattern occurs on the southward journey, with cow/calf pairs traveling last. Very similar segregation is recorded among migrating gray whales.

Baleen whale migrations have generally been regarded as taking place in response to the need to feed in colder waters and reproduce in warmer waters. Explanations for such long-range movements have included direct benefits to the calf (better able to survive in calm, warm waters), evolutionary 'tradition' (a leftover from times when continents were closer together), and the possible ability of some species to supplement their food supply from plankton encountered on migration or on the calving grounds. Corkeron and Connor (1999) have rejected these explanations, suggesting that there may be a major advantage to migrating pregnant female baleen whales in reducing the risk of killer whale predation on newborn calves in low latitudes. It cites in its favor the greater abundance of killer whales in higher latitudes, that their major prey (pinniped seals) is more abundant there, and that killer whales do not seem to follow the migrating animals.

Social activity Large aggregations of baleen whales are generally uncommon. Even on migration, in those species where well-defined migration paths are followed (e.g., gray whales and humpbacks), individual migrating groups are generally small, numbering only a few individuals. It has been stated that predation is a main factor in the formation of large groups of cetaceans, e.g., open-ocean dolphins. Given the large size of most adult baleen whales, predation pressure is low and group size is correspondingly small.

Blue whales are usually solitary or in small groups of two to three. Fin whales can be single or in pairs; on feeding grounds they may form larger groupings, up to 100 or more. Similarly, sei whales can be found in large feeding concentrations, but in groups of up to only about six elsewhere. The same is true for minke whales, found in concentrations on the feeding grounds, but singly or in groups of two or three elsewhere. Social behavior has been studied most intensively in coastal humpbacks, e.g., on calving grounds. Male humpbacks compete for access to females by singing and fighting. The songs seem to act as a kind of courtship display. Males congregate near a single adult female, fighting for position. Such aggression can involve lunging at each other with ventral grooves extended, hitting with the tail flukes, raising the head while swimming, fluke and flipper slapping, and releasing streams of bubbles from the blowhole. As a result of such encounters, individuals can be left with raw and bleeding wounds caused by the sharp barnacles. Among southern right whales, similar 'interactive' groups are often observed on the coastal calving grounds in winter, involving a tight group with up to seven males pursuing an adult female, but not generally resulting in wounded animals. As for humpbacks, it is not yet certain whether such behavior results in successful mating, although, at least in such right whale groups, intromission is often observed.

Feeding balaenopterids have often been reported as circling on their sides through swarms of plankton or fish. It has been suggested that gray whales feed on their right sides, as those baleen plates are more worn down, presumably through contact with the seabed. The most remarkable behavior, however, is reported from humpbacks. In the Southern Hemisphere, on swarms of krill, they may feed in the same 'gulping' way as other balaenopterids. In the Northern Hemisphere, two methods are commonly reported: 'lunging' and 'bubble netting.' In the former, individuals emerge almost vertically at the surface with their mouths partly open, closing them to force the enclosed water out through the baleen. In the latter, an animal circles below the food swarm; as it swims upward, it exhales a series of bubbles, forming a 'net' encircling the prey. It then swims upward through the prey with its mouth open, as in lunging.

Growth and Reproduction

Young baleen whales, particularly the fetus and the calf, grow at an extraordinary rate. In the largest species, the blue whale, fetal weight increases at a rate of some 100 kg/day toward the end of pregnancy. The calf's weight increases at a rate of about 80 kg/day during suckling. During that 7-month period of dependence on the cow's milk, the blue whale calf will have increased its weight by some 17 ton and increased in length from around 7 to 17 meters. Blue whales attain sexual maturity at between 5 and 10 years, at a length of around

22 meters, and live for possibly 80–90 years. Adult female blue whales give birth every 2–3 years, with pregnancy lasting some 10–11 months.

Other balaenopterids follow the same general pattern. Mating takes place in warm waters in winter, with birth following some 11-months later. A 7- to 11-month lactation period may be followed by a year 'resting', or almost immediately by another pregnancy. Most adults are able to reproduce from between 5 and 10 years of age and reach maximum growth after 15 or more years. The smallest balaenopterid, the common minke whale, is born after a pregnancy of some 10 months, at a length of just under 3 meters. Weaning occurs at just under 6 meters, after 3–6 months. The adult female can become pregnant again immediately following birth, but the resulting short calving interval is generally uncommon in baleen whales: 2–3 years is the norm, although female humpbacks can achieve a similar birth rate, enabling their stocks to recover rapidly after depletion (see Further Reading section).

Right whales follow a similar general pattern, but there are some differences. In right whales, gestation lasts about 11 months and weaning for about another year. Females are able to reproduce successfully from about 8 years (there are records of successful first pregnancies from 6 years), but the calving interval is usually a relatively regular 3 years. For bowheads, it has been reported, rather surprisingly, that while growth is very rapid during the first year of life (from c. 4.5 meters), it may be followed by a period of several years with little or no growth. Sexual maturity occurs at 13–14 meters; at the reduced growth rate, that would not be reached until 17–20 years. Similarly, evidence shows considerable longevity in this species: stone harpoon heads found in harvested whales and last known to be used off Alaska early this century suggest that individual animals can be at least 100 years old.

Population Status

For centuries, baleen whales have borne the brunt of human greed, for products and profit. Only the sperm whale, largest of the toothed whales, has rivaled them as a whaling target. Black right whales (*Eubalaena*) were taken in the Bay of Biscay from the 12th century, with the fishery extending across the North Atlantic by the 16th century. Attention then shifted to the Greenland whale (*Balaena*) near Spitzbergen and later off southern and western Greenland. Both species' numbers were reduced to only small remnants, and in several areas (e.g., Spitzbergen and Greenland for *Balaena* and the northeast Atlantic and the North Pacific for *Euba-laena*) the stocks were virtually exterminated. That destruction was undertaken using the 'old' whaling method, with open boats and hand harpoons, on the 'right' species – 'right' because they were relatively easy to approach, floated when dead, and provided huge quantities of products [oil for lighting, lubrication, and soap and baleen ('whalebone') for articles combining flexibility with strength, such as corset stays, umbrella spokes, fishing rods].

Development of the harpoon gun and steam catcher, from 1864, increased the rate of catching greatly, but also allowed attention to turn to the largest baleen whales, the blue and fin whales, whose size, speed, and tendency to sink when dead had prevented capture by the old methods. From its beginning in the North Atlantic, then, by the end of the century, in the North Pacific, 'modern' whaling's next and most intensive phase moved south, first in 1904 at South Georgia in the South Atlantic, just within the Antarctic zone. Initially on humpbacks [up to 12 000 were taken in one year (1912), leading to very rapid stock decline] and then on blue and fin whales, southern whaling based on such land stations – in the Antarctic in summer and the tropics in winter – was overtaken from the late 1920s by the great development of pelagic whaling using floating factory ships. Huge annual Southern Hemisphere catches resulted – a maximum of over 40 000 in 1931 – averaging around 30 000 animals per year in the later 1930s and again after World War II until 1965. Whereas blue whales had been the preferred target in the 1930s, their great reduction in numbers led to a shift in attention progressively over the years to fin whales, to sei whales in the 1940s, and finally to minke. With depletion of stocks and more stringent conservation measures (killing of humpbacks, blue, and fin whales was banned from the mid-1960s, even though some illegal catching continued until the early 1970s or even later), catches fell to between 10 000 and 15 000 per year in the 1970s. The 'old' whaling story had virtually repeated itself – enormous reductions through overfishing of one species or stock leading to exploitation of other species and stocks until, apart from minke whales, only remnants were left. Since 1989, a moratorium on all commercial whaling has eliminated that pressure, with the exception of limited whaling carried out under exemption for scientific research, and since 1993, limited commercial catching of minke whales in the eastern North Atlantic. Some 'aboriginal' whaling has also continued in the Northern Hemisphere, on bowheads off northern Alaska, on gray whales in the Bering Sea, on fin and minke whales off Greenland, and on humpbacks in the Caribbean.

Despite the great scale of the kill in 'old' and 'modern' whaling, no whale species has become extinct through whaling, although a number of individual stocks have been reduced greatly; at least one, the North Atlantic gray whale, has disappeared within the past 200–300 years. In its most recent (1996) 'Red List' of threatened animals, the World Conservation Union (IUCN) (**Table 3**) includes no baleen whale species or stocks as either extinct or critically threatened (the latter within the threatened category). Within the threatened category, seven taxa – three species, one subspecies and three stocks – are listed as endangered (E), four taxa – one species and three stocks – are vulnerable (V). Six taxa – two species, one subspecies, and three stocks – are listed as at lower risk (LR), and two taxa – one species and one subspecies – as data deficient (DD).

Those species under greatest current threat (E) are the Northern Atlantic and North Pacific right, sei, and fin whales, together with the 'true' blue subspecies, two of the five bowhead stocks (Okhotsk Sea, Spitzbergen), and the northwest Pacific gray. Next most threatened (V) are the humpback, two bowhead stocks (Hudson Bay, Baffin Bay/Davis Strait), and the North Atlantic blue. At lower risk (LR) are the southern right and Antarctic minke, one bowhead stock (Bering-Chukchi-Beaufort Seas), the North Atlantic minke, northeast Pacific gray, and North Pacific blue; all but one are further qualified as conservation dependent (cd, not vulnerable because of specific conservation efforts). The exception is the North Atlantic minke, listed as near threatened (nt, not conservation dependent but almost qualifying as vulnerable). The two taxa for which insufficient information is currently available (DD) are Bryde's whale and the pygmy blue.

The International Whaling Commission's Scientific Committee, responsible for the assessments of such stocks' current status, has reported encouraging recent reversals of stock decline for some stocks of some species. One, the northeast Pacific gray whale, has recovered under protection from commercial whaling (but with aboriginal catches up to some 150 per year) to at or near its 'original' (prewhaling) state (*c.* 26 000 animals). Similarly, the northwest Atlantic humpback and several Southern Hemisphere humpback populations have been showing marked increases. The latest estimate of the North Atlantic stock, some 10 600 animals in 1993 (cf. 5500 in 1986), must reflect some population growth in the intervening period, whereas two Southern Hemisphere stocks (off eastern and western Australia) have been increasing steadily, at 10% or more per year since the early 1980s. Indeed, in all areas where surveys have been undertaken recently on Southern Hemisphere humpback populations they

Table 3 IUCN 'Red List' Categories for Baleen Whales (1996)

| Species | Subspecies | Common name | Category | | | | | | | |
| | | | EX | EW | Threatened | | | LR | DD | NE |
					CR	EN	VU			
B. mysticetus		Bowhead whale				*a	*b	* (cd)c		
E. glacialis*		North Atlantic right whale				*d				
E. australis		Southern right whale						* (cd)		
Caperea marginata^e		Pygmy right whale								
E. robustus		Gray whale				*f		* (cd)g		
M. novaeangliae		Humpback whale					*			
B. acutorostrata	B. a. acutorostrata	N. Atlantic minke						* (nt)		
	B. a. scammoni	N. Pacific minke								
	B. a. subsp.	Dwarf minke								
B. bonaerensis		Antarctic minke						(cd)		
B. edeni		Bryde's whale								*
B. borealis		Sei whale				*				
B. physalus		Fin whale				*				
B. musculus	B. m. musculus	Blue whale					*h	* (cd)i		
	B. m. indica	Great Indian rorqual								
	B. m. intermedia	'True' blue				*				
	B. m. brevicauda	Pygmy blue								*

* Includes *E. japonica.*
a Okhotsk Sea bowhead whale, Spitzbergen bowhead whale.
b Hudson Bay bowhead whale, Baffin Bay/Davis Strait bowhead whale.
c Bering-Beaufort-Chuckchi Seas bowhead whale.
d North Atlantic and North Pacific northern right whales.
e Pygmy right whale removed from 1996 Red List.
f Northwest Pacific gray whale.
g Northeast Pacific gray whale.
h North Atlantic blue whale.
i North Pacific blue whale.
Categories: Ex, extract; EW, extinct in the wild; CR, current risk; EN, endangered; VU, vulnerable; LR, lower risk; DD, data deficient; NE, not evaluated; cd, conservation dependent; nt, not threatened.

have been shown to be undergoing some recovery. Three southern right whale stocks (off eastern South America, South Africa, and southern Australia) have been increasing since the late 1970s at around 7–8% per year, although at some 3000, 3000, and 1200 animals, respectively, all are still well below their 'original' stock size. Even the 'true' blue whale, whose future has been of considerable concern, with estimates for the late 1980s at fewer than 500 animals for the whole Antarctic, has shown recent encouraging signs. Based on a series of Antarctic sightings cruises, mainly for minke whales but including other large whales, the most recent calculations (admittedly using only small absolute numbers sighted) show that the population must have been increasing since the last estimate (1991). As yet the analyses do not permit a firm conclusion on the number now present, although it seems likely to be more than 1000.

The one species or stock for which there is now very great concern is the North Atlantic right whale. At very low absolute abundance (only some 300 animals), not recovering despite protection from whaling in the 1930s, now even decreasing through a reduced survival rate and an increase in calving interval, and subject to increasing removals from ship strikes and fishing gear entanglement, the only way to ensure the species survival is to reduce such anthropogenic mortality (ship strike and entanglement) to zero. While research on mortality reduction measures should be pursued, immediate management action is urgently needed.

It has been calculated that the great reduction of baleen whales by whaling, for the Antarctic to around one-third of original numbers and one-sixth in biomass, must have left a large surplus of food – some 150 million tonnes per year – available for other consumers, such as seals, penguins, and fish. (In a different way, earlier whaling in the North Atlantic, particularly on right whales, is believed to have influenced the spread of one sea bird – the fulmar – by providing food in the form of discarded whale carcasses.) In response to an increase in available food, there may well have been increases in growth rates, earlier ages at maturity, and higher rates of pregnancy in some baleen whale species. However, the evidence is equivocal, as it is for competition between individual whale species. For some, e.g., right whales and sei, it has been suggested that an increase in one (right whales) could be inhibited by competition with another (sei whales). In the North Pacific, both sei and right whales can feed on the same prey – copepods – and sei whales can at times be 'skimming' feeders, like right whales. However, evidence that they actually compete on the same prey, in the same area, at the same time, and even on the same prey patch is lacking. Similarly, there has been much debate and speculation on whether the recovery of the Southern Hemisphere 'true' blue whale has been inhibited by an apparent increase in Antarctic minke whales. In that case, there may in fact be very little direct competition for food where the common prey is not limited in abundance (as in the Antarctic) and is available in large patches. The well-authenticated increases in the substantial annual rates for several stocks of Southern Hemisphere humpbacks and right whales and the possibility of at least a limited increase in numbers for the 'true' blue whale suggest that such competition is unlikely, at least where, as in the Antarctic, food supplies are abundant.

Glossary

Baleen Plates of keratin hanging transversely in the roof of the mouth of baleen whales, forming the 'baleen apparatus' for filter feeding on surface plankton; formerly known as 'whalebone' but bearing no resemblance to true bone.

Ventral grooves A series of parallel grooves or plates running longitudinally on the undersurface of the throat and chest region in balaenopterid whales, allowing great expansion of the mouth during feeding.

Krill Planktonic shrimp-like crustaceans of the genus Euphausia, particularly the Antractic Euphautia superba.

See also

Copepods. Krill. Marine Mammal Overview. Marine Mammal Migrations and Movement Patterns. Marine Mammal Social Organization and Communication. Marine Mammal Trophic Levels and Interactions.

Further Reading

Bonner WN (1980) *Whales*. Poole, Dorset; Blandford Press.

Corkeron PJ and Connor RC (1999) Why do baleen whales migrate? *Marine Mammal Science* 15: 1228–1245.

Harrison R and Bryden MM (eds.) (1988) *Whales, Dolphins and Porpoises*. Hong Kong: International Publishing Corporation.

IUCN (1996) *1996 IUCN Red List of Threatened Animals* (plus annexes). Gland, Switzerland: IUCN.

IUCN (1996) *1996 IUCN Red List of Threatened Animals* (plus annexes). Gland, Switzerland: IUCN.

Laws RM (1977) Seals and Whales of the Southern Ocean. *Philosophical Transactions of the Royal Society of London, Series B* 279, 81–96.

Leatherwood S and Reeves RR (1983) *The Sierra Club Handbook of Whales and Dolphins*. San Francisco: Sierra Club Books.

Rice DW (1998) *Marine Mammals of the World:: Systematics and Distribution, Special Publication Number 4*. Lawrence, Kansas: Society for Marine Mammalogy.

SPERM WHALES AND BEAKED WHALES

S. K. Hooker, University of St. Andrews, St. Andrews, UK

Introduction

Sperm whales and beaked whales are among the largest and most enigmatic of the odontocetes (toothed whales). These species tend to live far offshore in regions of deep water, and perform long, deep dives in search of their squid prey. This has generally made the study of these animals much more difficult than that of more accessible, near-shore cetacean species. In addition, the pygmy and dwarf sperm whales, and many species of beaked whale, have superficially similar external morphology, and so are often difficult to identify to species level in the wild. The study of many of these species has therefore been based primarily on examination of stranded and beach-cast animals. As a result, we currently know comparatively little regarding many of these relatively large mammals. For example, one species of beaked whale, Longman's beaked whale, was identified from only two skulls in Australia and Somalia and has only recently been observed in the wild. Another putative species, *Mesoplodon* species 'A', has only ever been observed at sea, and our knowledge of its morphological characteristics remains far from complete. New species of beaked whales are still being discovered. For example, the lesser beaked whale and Bahamonde's beaked whale were only identified in the last decade from specimens collected in Peru and Chile, respectively. Perrin's beaked whale was recently discovered by analyzing DNA sequence data from five archived Californian stranded specimens initially thought to be Hector's beaked whales or Cuvier's beaked whales. Likewise, the dwarf and pygmy sperm whales were only recognized as separate species in the 1960s.

The sperm and beaked whale species about which we know most are the sperm whale, the northern bottlenose whale, and Baird's beaked whale. Much of our information about these species has come from scientific research programs conducted in conjunction with historic whaling operations. Longer-term, nonlethal studies of wild populations only began in the early 1980s. These focused initially on sperm whales, and today include research on populations of northern bottlenose whales, Blainville's beaked whales, and Cuvier's beaked whales. Such studies help provide important behavioral information about these species which has not previously been available from studies of dead animals.

Taxonomy and Phylogeny

There are three superfamilies within the odontocetes: the Physeteroidea (sperm whales), the Ziphioidea (beaked whales), and the Delphinoidea (river dolphins, oceanic dolphins, porpoises, and monodontids). The superfamily Physeteroidea encompasses two families: the Physeteridae which contains the sperm whale, and the Kogiidae which contains the pygmy sperm whale and the dwarf sperm whale. The Ziphioidea encompasses only the family Ziphiidae, which includes at least 21 species of beaked whales (**Table 1**).

Although some genetic studies have challenged the relationship of the sperm whales to other toothed whales, the analytical methods used to determine this have been questioned, and there is general agreement between morphological and other molecular data that the sperm whales and beaked whales are basal odontocetes (**Figure 1**). Physeterids appeared in the fossil record in the early Miocene deposits of Argentina (*c.* 25 Ma). In the past, this family included a diverse array of genera, but today is represented only by the sperm whale. The kogiids are thought to have diverged from the physeterids in the late Miocene and early Pleiocene (*c.* 5–10 Ma). The earliest ziphiids have been found in deposits from the middle Miocene (10–15 Ma). Relationships among the beaked whales are not clear. The six genera in this family have previously been separated into two tribes grouping *Berardius* and *Ziphius*, and grouping *Tasmacetus*, *Indopacetus*, *Hyperoodon*, and *Mesoplodon*. However, it has also been suggested that *Tasmacetus*, with a full set of teeth in upper and lower jaws, may be the sister group to all other living species. Current work investigating the systematics of this group using DNA sequence data in fact suggests that *Tasmacetus* is more closely related to *Ziphius* and *Hyperoodon*.

Anatomy and Morphology

Beaked whales are characterized by the possession of a long and slender rostrum resulting in a prominent beak in most species. An evolutionary trend in ziphiids has led to a reduction in the number of teeth in all genera except *Tasmacetus*. Most species have retained only one or two pairs of teeth, set in varying

Table 1 Sperm and beaked whale species, approximate demographic distribution and size

Species		General location	Adult size (m)	Notes
Family Physeteriidae				
Sperm whale	*Physeter macrocephalus*	Global	12–18	
Family Kogiidae				
Dwarf sperm whale	*Kogia simus*	Tropical and temperate oceanic	2.7–3.4	
Pygmy sperm whale	*Kogia breviceps*	Tropical and temperate, continental shelf and slope	Up to 2.7	
Family Ziphiidae				
Baird's beaked whale	*Berardius bairdii*	North Pacific	11.9–12.8	
Arnoux's beaked whale	*Berardius arnuxii*	Southern Ocean	Up to 9.7	
Cuvier's beaked whale	*Ziphius cavirostris*	Global, common in Eastern Tropical Pacific	7–7.5	
Northern bottlenose whale	*Hyperoodon ampullatus*	North Atlantic	8.7–9.8	
Southern bottlenose whale	*Hyperoodon planifrons*	Southern Ocean	7.2–7.8	
Shepherd's beaked whale	*Tasmacetus shepherdii*	Southern temperate	6.6–7	Few stranded specimens
Longman's beaked whale	*Indopacetus pacificus*	Australia, Somalia	Over 6	Two skulls
Blainville's beaked whale	*Mesoplodon densirostris*	Temperate global	Up to 4.7	
Gray's beaked whale	*Mesoplodon grayi*	Southern temperate circumglobal	Up to 5.6	
Ginkgo-toothed beaked whale	*Mesoplodon ginkgodens*	Temperate/tropical Indian and Pacific Oceans	Up to 4.9	
Hector's beaked whale	*Mesoplodon hectori*	Southern temperate	4.3–4.4	
Hubbs' beaked whale	*Mesoplodon carlhubbsi*	North Pacific	Up to 5.3	
Sowerby's beaked whale	*Mesoplodon bidens*	Northern North Atlantic	5.1–5.5	
Gervais' beaked whale	*Mesoplodon europaeus*	Temperate/tropical Atlantic	4.5–5.2	
True's beaked whale	*Mesoplodon mirus*	Temperate N. Atlantic and temperate Southern Ocean	Up to 5	
Strap-toothed whale	*Mesoplodon layardii*	Southern temperate	5.9–6.2	
Andrew's beaked whale	*Mesoplodon bowdoini*	South Indian and Pacific Oceans	4.6–4.7	
Stejneger's beaked whale	*Mesoplodon stejnegeri*	North Pacific	Up to 5.3	
Lesser beaked whale	*Mesoplodon peruvianus*	Eastern Tropical Pacific	Up to 3.7	Few strandings; tentative sightings
Bahamonde's beaked whale	*Mesoplodon bahamondii*	Peru	Estim. 5–5.5	Partial skull
Perrin's beaked whale	*Mesoplodon perrini*	California (USA)	3.9–4.4	Few strandings
Mesoplodon species 'A'	*Mesoplodon* sp.	Eastern Tropical Pacific	5.5	Sightings only

positions in the lower jaw (**Figure 2**). In most beaked whale species, these teeth only erupt in adult males. From observations of scarring patterns on the animals, these teeth appear to function as weapons in intraspecific combat, and have become much enlarged in some species. Other features which distinguish beaked whales from other groups include the possession of two conspicuous throat grooves or creases which form a forward pointing V-shape, and the lack of a notch in the flukes. The skull morphology of beaked whales is also unique, exhibiting elevated maxillary ridges behind the nasals.

The Physeteroidea are characterized by several features of the skull, including a large supracranial basin. This basin holds the 'spermaceti organ', a fat-filled structure, which lies behind the melon in the forehead, and is unique to these species. This structure was named for the presence of spermaceti, an oily substance thought to resemble semen (after which it was named). It is generally thought that this organ functions in sound transmission. Sperm whales show extreme sexual dimorphism both in their size (up to 18-m length of adult males compared to up to 12-m length of adult females) and in the increased ratio of head to body size such that the adult male head is relatively much larger than the adult female head. Thus whalers valued adult males for their large quantities of spermaceti much more than females. Externally, sperm whales can be differentiated from other species by their narrow lower jaw, and an upper jaw which extends well past the lower one. This group also has reduced dentition. The sperm whale has teeth (18–25 pairs) only in the lower jaw. The dwarf and pygmy sperm whales have reduced numbers of teeth in the lower jaw (generally12–16 pairs in pygmy sperm whales, and 8–11 pairs in the dwarf whale). Some teeth may be present in the upper jaw of both dwarf and pygmy sperm whales, although this is less common in the pygmy sperm whale.

The sperm whales show highly pronounced asymmetry of the skull. Most beaked whales species also have asymmetrical skulls, although *Berardius* and *Tasmacetus* have nearly symmetrical cranial characteristics, suggesting that cranial asymmetry in beaked whales has evolved independently from that in sperm whales. The asymmetry of the sperm whale head extends to the nasal passages in which the right nasal passage has been modified as part of the sound production apparatus while the left nasal passage connects to the blowhole (on the left of the sperm whale's head). The right nasal passage appears to be capable of disconnection from the lungs and the left nasal passage, since sperm whales can breathe and produce clicks at the same time. Thus, unlike other odontocetes which possess two bilaterally placed monkey lips/dorsal bursae complexes, sperm whales have only a single complex.

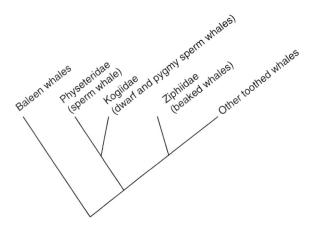

Figure 1 Phylogenetic diagram showing relationship of sperm whales and beaked whales to other cetaceans. Adapted from Heyning JE (1997) Sperm whale phylogeny revisited: Analysis of the morphological evidence. *Marine Mammal Science* 13: 596–613.

Distribution and Abundance

All sperm whales and beaked whales are deep-water oceanic species. However, there is wide variation in species coverage. The sperm whale is found throughout the world's oceans, from the Tropics to the Poles. The pygmy and dwarf sperm whales also

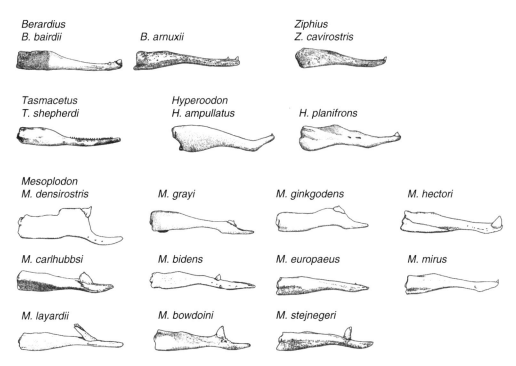

Figure 2 Variation in position, size, and morphology of the lower jaw teeth of adult male beaked whales shown for the majority of recorded beaked whale species. After Jefferson TA, Leatherwood S, and Webber MA (1993) Marine Mammals of the World. Rome: United Nations Environment Program, FAO. http://ip30.eti.uva.nl/bis/marine_mammals.php (accessed Mar. 2008).

have fairly cosmopolitan distributions, and are found in temperate and tropical waters worldwide. In contrast, many beaked whale species have quite limited distributions and only two species, the dense-beaked whale and Cuvier's beaked whale, show similar ranges to the sperm whales (**Table 1**). Many other beaked whale species are limited to a single ocean basin, and several species pairs show an antitropical distribution, for example, *Hyperoodon* (North Atlantic and Southern Oceans) and *Berardius* (North Pacific and Southern Oceans).

There are very few estimates of abundance for sperm and beaked whales. Many of these species are difficult to detect and identify at sea, and so are likely to be more common than sighting records would suggest. The status of all sperm and beaked whales as currently listed by the IUCN (International Union for the Conservation of Nature) Red Book is 'insufficiently known'.

Foraging Ecology

The majority of sperm and beaked whales are thought to feed primarily on squid. The reduced dentition of these species is thought to be due to this dietary specialization. One exception to this is the Shepherd's beaked whale, in which both sexes possess a full set of functional teeth, and the diet appears to consist primarily of fish.

The reduced dentition of the beaked whales, together with their narrow jaws and throat grooves, have been suggested to function in suction feeding. Among the males of species such as the strap-toothed whale, the elaborate growth of the strap-like teeth may limit the aperture of the gape to a few centimeters, and it is difficult to see how prey capture techniques other than suction feeding could be successful. The same mechanism is thought to be used by sperm whales which also have a comparatively small mouth area. Anecdotal evidence suggests that the lower jaw teeth of the sperm whale are not required for feeding, as apparently healthy animals have been seen with broken and badly set lower jaws resulting from past injuries.

Sperm whales and beaked whale species are known to be excellent divers. Dives of up to an hour have been recorded from several beaked whale species, although many of these records have been based on surface observations of diving whales. Similarly, dives of up to 25 min have been recorded from *Kogia* species. Increasingly studies are now using time-depth recorders to monitor dives in more detail from these species. Sperm whales have been recorded to dive repeatedly to depths of 500–800 m for durations of 35–45 min, with a maximum recorded dive depth

of 1860 m. Of beaked whale species, time-depth data loggers have been deployed on northern bottlenose whales, Blainville's beaked whales, and Cuvier's beaked whales (**Figure 3**). All three of these species have also been found to dive regularly to depths of over 800 m, although their dive profiles show distinct species-specific patterns to their diving behavior. Sperm and northern bottlenose whales have a greater frequency of deep dives, but the deepest dive thus far recorded was from a Cuvier's beaked whale at 1950 m, and the longest dive was 85 min. These species are thought to forage at these depths, at times at the seafloor, in search of deep-water squid.

The similarity of ecological niches among beaked and sperm whales might be expected to lead to competition between these species. The relatively discrete distributions of many beaked whale species may have resulted from this. For example, several *Mesoplodon* species coexist in the North Atlantic, but have separate centers of distribution, with little overlap in range: Sowerby's beaked whale has a more northerly distribution than True's beaked whale, which in turn is found to the north of Gervais' beaked whale. On a much smaller spatial scale, there is some suggestion of competitive exclusion between sperm whales and northern bottlenose whales in habitat use of a submarine canyon area off eastern Canada. Prey species (mainly squid) identified from the stomachs of stranded *Kogia* specimens suggest that these species occur primarily along the continental shelf and slope in the epi- and mesopelagic zones. Although the diets of both species overlap, the relative contribution of prey types suggests that the dwarf sperm whale feeds on smaller squid in shallower waters and thus occurs further inshore than the pygmy sperm whale.

Social Organization

The social organization of the majority of sperm and beaked whale species is only poorly known, with the exception of the sperm whale. The social system of the sperm whale appears quite unlike that known for other cetaceans. Groups of females and juveniles are found in temperate and tropical latitudes (**Figure 4**). Males become segregated from these female groups at or before puberty, and migrate to higher latitudes. Younger males are found in 'bachelor schools', which consist of animals of approximately the same age. These schools decrease in size with increasing age of the members, to the point where large mature animals are typically solitary. Sexually mature males return to the tropical waters inhabited by females in order to breed. The sexual dimorphism, scarring

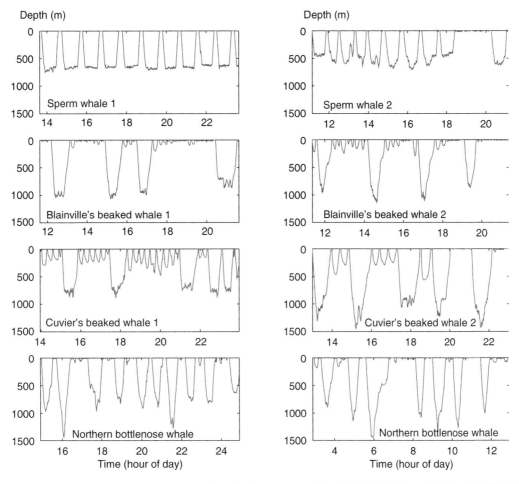

Figure 3 Dive profiles are shown for 10-h segments of each of: two sperm whales (1: Gulf of Mexico, 2003; 2: Gulf of Mexico, 2002), two Blainville's beaked whales (1: Bahamas, 2006; 2: Hawaii, 2004), two Cuvier's beaked whales (1: Liguria, Italy, 2006; 2: Hawaii, 2004), and two records from a single northern bottlenose whale (the Gully, eastern Canada, 1997). Data were kindly provided by Robin Baird, Cascadia Research Collective, USA; Sascha Hooker, Sea Mammal Research Unit, UK; Patrick Miller, Sea Mammal Research Unit, UK; and Mark Johnson, Woods Hole Oceanographic Institute, USA.

patterns, and vocal behavior suggest that adult male sperm whales compete for access to groups of females. Recent mark-recapture data, relative parasite loads, and indications of synchronous estrus suggest that males rove between groups of females, and remain with any one group for only a few hours, although they may revisit groups on consecutive days.

Female sperm whales are found in groups of 20 or so individuals. These groups appear to consist of two or more stable units that associate for periods of *c.* 10 days. Genetic evidence has suggested that these groups are composed of one or more matrilines. However, there are also suggestions of paternal relatedness between grouped matrilines, and recent photoidentification studies suggest that some animals occasionally switch groups, and thus may not be of the same maternal lineages as other group members.

Whalers observed that sperm whale groups often exhibited epimeletic behavior, with individuals supporting and staying with harpooned, injured, and even dead group members. It is thought that this may be a result of the close genetic ties between the individuals in a group. Sperm whales have also been observed to exhibit allomaternal care (babysitting behavior). Calves remain on the surface when a group is feeding, presumably since they are unable to dive to the depths at which adults forage, and adults have been observed to stagger their dives such that there is always an adult at the surface with the calf.

Observations of wild *Kogia* suggest that they typically form small groups of one to four animals, with occasional groups of up to 10 reported. However, almost nothing is known of the composition of these groups or of the behavior of these species at sea.

Many beaked whale species appear to show intraspecific aggression between adult males, presumably for access to females. The prominent and elaborate teeth of many beaked whale species are thought to be used in this male–male conflict (**Figure 2**), resulting in the extensive scarring seen on

Figure 4 Group of female and juvenile sperm whales off the Galapagos Islands. Photograph by Sascha K. Hooker.

adult males. However, in other beaked whale species, males possess only comparatively small lower jaw teeth, and these do not appear to be used for fighting. The northern bottlenose whale is an example of this. Instead, this species shows marked sexual dimorphism in skull structure and associated forehead or melon shape, which is relatively small in females but is enlarged and flattened in adult males. Recent observations have suggested that this melon morphology is also associated with male–male competition, as adult males have been observed to head-butt each other.

Among beaked whales, the composition of social groups is not well known. The two beaked whales for which most data have been collected are the northern bottlenose whale and the Baird's beaked whale. Long-term photoidentification of individual bottlenose whales has in fact suggested stronger associations between males than between females. However, the aggression observed between some associated males makes further interpretation difficult. Anatomical studies of groups of Baird's beaked whales taken in the continuing fishery off Japan are suggestive of a different type of social structure for this species. Among this species, both males and females possess erupted teeth, and females are slightly larger than males. Males appear to reach sexual maturity an average of four years earlier than

females and may live up to 30 years longer. This has led to speculation that males may be providing parental care in this species, although further work is needed to confirm this.

Acoustics, Sound Production, and Sound Reception

The acoustic behavior of sperm whales is relatively well documented. These whales produce broadband clicks with a centroid frequency of 15 kHz. These clicks are thought to function primarily in echolocation, although some repetitive patterned clicks (termed codas) also appear to be used in a social context. Neither *Kogia* species appears to be highly vocal, although echolocation-type signals have been recorded from the pygmy sperm whale. The social whistles characteristic of other odontocete species are absent from the physeterids, and possibly also from some beaked whale species. No whistles were documented in several hours of recordings from northern bottlenose whales, which instead also appear to produce primarily echolocation-type clicks. These were superficially similar to sperm whale clicks, although often at ultrasonic frequencies ($\sim$20–30 kHz). An acoustic recording tag attached to Blainville's and Cuvier's beaked whales has also demonstrated short, directional, ultrasonic echolocation clicks with most of their energy between 35 and 45 kHz. These clicks are produced during dives and are frequency sweeps similar to the echolocation of a bat, with regular clicks produced every 0.2–0.4 s, ending in a rapid increase in slick rate up to 250 clicks per second as the target is approached. Only a few other records of beaked whale acoustic behavior exist and the majority of these were obtained from stranded animals. Other recordings of wild animals have been made of Baird's beaked whale and Arnoux' beaked whale, and these included frequency-modulated whistles, burst-pulse clicks, and discrete clicks in rapid series.

The sound production mechanism used by both sperm and beaked whales for echolocation is homologous with that of other odontocetes, consisting of a sound-producing complex (the 'monkey lips'/dorsal bursae) in the upper nasal passages. Sound is propagated in the water by the melon, a lipid-filled structure which acts as an acoustic lens to focus a directional beam ahead of the animal. The echoes of this sound are then received via the fat body in the lower jaw which connects with the bulla of the middle ear.

The sperm whale head is unique in comparison to other odontocetes in that the blowhole and the sound production mechanism are situated at the front of the head rather than above the eyes. The initial click propagates backward from the front of the head through the spermaceti organ to create an intense forward-directed pulse. Reverberation within the spermaceti organ generates a decaying series of pulses, with the time interval between these pulses related to the size of the head.

Predation

It was previously thought that large size was adequate defense against predators, but female sperm whales (of 10–12 m) have been observed under lethal attack by 'transient' (mammal-eating) killer whales. Additionally, large sharks are thought to be a threat to these species, particularly to juvenile animals. Various methods of defense may be employed. For such a deep-diving species, it is surprising that deep dives are not used as a method of escape from predators. This may be because young calves are unable to dive to the depths or for the same duration as adults. Instead, sperm whales appear to show a behavioral response to the threat of predation, with the adults forming a circle (heads innermost) around the calves.

Pygmy and dwarf sperm whales evacuate reddish-brown intestinal fluid when startled, in a similar manner to inking by squids. The lower intestine is expanded in both species, forming a balloon-like structure filled with up to 12 l (in large specimens) of

this liquid. Additionally, these species possess a crescent-shaped light-colored mark, often called a 'false gill', on the side of the head behind the eye and before the flippers. Along with the underslung mouth, this can lead to the mistaken identification of these animals as sharks. However, whether this patterning functions as camouflage against predation is unknown.

Conservation

The larger species of sperm and beaked whales were all targeted by whaling operations in the past. The sperm whale was the most heavily hunted, primarily due to the prized spermaceti oil that it contained. This fishery spanned the seventeenth to twentieth centuries and at its peak (in the 1960s) average annual catches reached 25 000 animals. Northern bottlenose whales were also quite severely depleted by whaling. The northern bottlenose whale fishery began in the late nineteenth century and between 1880 and 1920 approximately 60 000 bottlenose whales were caught. The other species in this group which has been taken in relatively large numbers is Baird's beaked whale. This species has been hunted in Japan since at least the seventeenth century, but has generally been taken in relatively low numbers (a maximum annual catch of 322 in 1952, but recently averaging 40 whales per year). This fishery continues even today.

Current threats faced by these species range from other factors potentially causing immediate death, such as ship-strikes, to the more insidious threats of ocean plastic, chemical, and acoustic pollution. Since many sperm and beaked whales feed primarily on squid, they are very susceptible to the ingestion of plastics, apparently mistaking it for prey. Stranded animals from several sperm and beaked whale species have been found with plastic in their stomachs and in some cases, this appears to have blocked the normal function of the stomach, causing severe emaciation and likely contributing to their death. The ecological role of odontocetes as long-lived top predators also exposes these animals to increased levels of chemical pollutants. Cetaceans store energy (and pollutants) in their blubber, and have a lower capacity to metabolize some polychlorinated biphenyl (PCB) isomers than many other mammals. Foreign and toxic substances are therefore often biomagnified in odontocete species, and even species living offshore in relatively pristine environments have been found to contain high levels of pollutants. These high pollutant levels can have two major effects: (1) inhibition of immune system capacity to respond to naturally occurring diseases, and (2) potentially causing reproductive failure.

There is also increasing concern about the effect of anthropogenic noise in the marine environment. Sperm whales and beaked whales appear to be particularly susceptible to the effects of such noise. Sperm whales have been observed to react to several types of underwater noise including sonar, seismic activity, and low-frequency sound. Beaked whales also appear to be particularly susceptible to high-intensity underwater sound. Several beaked whale stranding events have coincided with military naval exercises and associated high-intensity underwater sound. However, the mechanism by which such sound is hazardous to these whales is not yet known, thus limiting the likely effectiveness of mitigation efforts. There is currently a high level of concern surrounding this issue.

Glossary

competitive exclusion The principle that two species cannot coexist if they have identical ecological requirements.

cosmopolitan Having a broad, wide-ranging distribution.

echolocation The production of high-frequency sound and reception of its echoes; used to navigate and locate prey.

epimeletic Caregiving behavior.

mandible Lower jaw.

matriline Descendants of a single female.

rostrum Anterior portion or beak region of the skull that is elongated in most cetaceans.

sexual dimorphism Morphological differences between males and females of a species.

ultrasonic High-frequency sounds, beyond the upper range of human hearing.

See also

Dolphins and Porpoises. Marine Mammal Diving Physiology. Marine Mammal Evolution and Taxonomy. Marine Mammal Migrations and Movement Patterns. Marine Mammal Overview.

Marine Mammal Social Organization and Communication. Marine Mammal Trophic Levels and Interactions.

Further Reading

Berta A and Sumich JL (1999) *Marine Mammals: Evolutionary Biology*. San Diego, CA: Academic Press.

Best PB (1979) Social organisation in sperm whales. In: Winn HE and Olla BL (eds.) *Behaviour of Marine Mammals. Vol. 3: Cetaceans*, pp. 227–289. New York: Plenum.

Heyning JE (1997) Sperm whale phylogeny revisited: Analysis of the morphological evidence. *Marine Mammal Science* 13: 596–613.

Jefferson TA, Leatherwood S, and Webber MA (1993) Marine Mammals of the World. Rome: United Nations Environment Program, FAO. http://ip30.eti.uva.nl/bis/marine_mammals.php (accessed Mar. 2008).

Moore JC (1968) Relationships among the living genera of beaked whales with classification, diagnoses and keys. *Fieldiana: Zoology* 53: 209–298.

Perrin WF, Wursig B, and Thewissen JGM (eds.) (2002) *Encyclopedia of Marine Mammals*. San Diego, CA: Academic Press.

Rice DW (1998) *Special Publication No. 4: Marine Mammals of the World – Systematics and Distribution*. Lawrence, KS: The Society for Marine Mammalogy.

Ridgway SH and Harrison R (1989) *Handbook of Marine Mammals, Vol. 4: River Dolphins and the Larger Toothed Whales*. London: Academic Press.

Whitehead H (2003) *Sperm Whales: Social Evolution in the Ocean*. Chicago, IL: Chicago University Press.

Whitehead H and Weilgart L (2000) The sperm whale: Social females and roving males. In: Mann J, Connor RC, Tyack P, and Whitehead H (eds.) *Cetacean Societies: Field Studies of Dolphins and Whales*, pp. 154–173. Chicago, IL: Chicago University Press.

Relevant Websites

http://www.iucnredlist.org
 – IUCN Red List of Threatened Species.
http://www.animalbehaviorarchive.org
 – Macaulay Library Sound and Video Catalog.
http://ip30.eti.uva.nl
 – National Museum of Natural History.

DOLPHINS AND PORPOISES

R. S. Wells, Chicago Zoological Society, Sarasota, FL, USA

Introduction: Physical Descriptions and Systematics

Dolphins and porpoises are members of a diverse group of cetaceans including 44 species belonging to 6 of the 10 families of modern toothed whales, of the suborder Odontoceti (**Table 1**). In general, this group includes the smaller odontocetes, with body lengths ranging across species from about 1.5 m up to about 9 m. The grouping represents two major and very different subgroups. Dolphins include 35 species of the families Delphinidae and Pontoporiidae, primarily inhabiting inshore and/or pelagic marine waters worldwide, and three families of obligate river dolphins (Iniidae, Lipotidae, and Platanistidae), each represented by a single species. The distributions of the three river dolphin species are limited to riverine systems in Asia or South America. Porpoises include the six species of the family Phocoenidae, also primarily inhabiting inshore and pelagic waters.

Dolphins and porpoises exhibit essentially the same basic streamlined spindle or fusiform body shape, which provides for effective movement through the dense water medium (**Figure 1**). They have a single blowhole on top of their head for respiration. In their skulls, they both demonstrate telescoping overlap of the maxillary bones over the frontals in the supraorbital region. Most dolphins and porpoises have numerous teeth (polydonty), and these teeth are homodont (all alike in structure) in all but one species. A few small hairs occur on the beaks of the animals at the time of birth, but these are soon lost.

Postcranially, the bones of the hand and arm have been modified into a pectoral flipper, a solid wing-like structure articulating with the shoulder, and serving as a control surface for maneuvering. Pelvic appendages have been reduced to small, internal pelvic bones (although a bottlenose dolphin captured in Japan in 2006 presented a bilateral pair of fully formed, small fins in the pelvic region). The vertebral column, with its variably fused cervical vertebrae and prominent processes for attachment of the strong musculature used for up and down movement of the tail for propulsion, tapers along the tail stock or peduncle, toward the tail to the flukes, a pair of fibrous horizontal fins. A fibrous dorsal fin located near the middle of the back of most, but not all, dolphins and porpoises serves to stabilize the animals as they swim, as a radiator for cooling the reproductive organs, and as a weapon. This structure can also be useful to researchers, as the dorsal fins of many cetaceans have individually distinctive shapes and natural notch patterns, providing a means of reliable identification for recognizing individuals through time.

Males and females look quite similar and are of similar size in most dolphin and porpoise species, though there may be differences in body and/or appendage size in some, especially among the larger dolphins. Female dolphins typically have a genital and an anal opening in a single ventral groove, with a nipple located in a mammary slit on each side of the genital opening. Males typically have a genital opening in a ventral groove anterior to the separate groove for the anus. Where it occurs, sexual dimorphism is variable across the species, with females of some of the smaller species (*Cephalorhynchus* spp., the tucuxi, *Sotalia fluviatilis*) being slightly larger than the males, whereas the males of the largest species (killer whales, false killer whales (*Pseudorca crassidens*), pilot whales (*Globicephala* spp.)) tend to be much larger than the females. In some cases, features that might be used in competitive displays or affect performance in battle or chasing potential mates, such as dorsal fin height (e.g., killer whales, **Figure 2**), peduncle height, or fluke span, are disproportionately larger for males than for females. Smaller species such as spinner dolphins can demonstrate some degree of dimorphism, with adult males having taller and pointier dorsal fins, and a postanal keel or hump on the ventrum.

Most dolphins and porpoises exhibit countershading, with a darker dorsum and lighter ventrum. Countershading presumably functions as camouflage to facilitate approaching prey and avoiding predators.

Dolphins

There is much variation among the species of dolphins regarding robustness, the presence of a clearly demarcated projecting beak, and the length of the beak, numbers and size of teeth, and the presence of a dorsal fin. The characteristic cone-shaped, or peg-like teeth of dolphins differ across the species in size

Table 1 List of extant species and subspecies of dolphins and porpoises

Sp #	Taxon	Common name	General range	IUCN designation[a]
	Family Delphinidae	Marine Dolphins		
1	Cephalorhynchus commersonii	Commerson's dolphin	Southeastern S. America, Kerguelen and Falkland Islands	DD
	C. commersonii commersonii	Falklands, S. American subspecies	Southeastern S. America, Falkland Islands	NE
	C. commersonii subsp.	Kerguelen subspecies	Kerguelen Island	NE
2	Cephalorhynchus eutropia	Chilean dolphin	Southwestern S. America	DD
3	Cephalorhynchus heavisidii	Heaviside's dolphin	Southwestern Africa	DD
4	Cephalorhynchus hectori	Hector's dolphin	New Zealand	EN
5	Delphinus capensis	Long-beaked common dolphin	Tropical and warm temperate waters of the Pacific, Indian, and S. Atlantic Oceans	LC
6	Delphinus delphis	Short-beaked common dolphin	Tropical to temperate waters of the Pacific and N. Atlantic Oceans	LC
7	Feresa attenuata	Pygmy killer whale	Worldwide tropical and warm temperate waters	DD
8	Globicephala macrorynchus	Short-finned pilot whale	Worldwide tropical to temperate waters	LR(cd)
9	Globicephala melas	Long-finned pilot whale	N. Atlantic and southern S. Pacific, S. Indian, and S. Atlantic Oceans	LC
	G. melas melas	N. Atlantic long-finned pilot whale	N. Atlantic Ocean	NE
	G. melas subsp.	N. Pacific long-finned pilot whale	N. Pacific Ocean	NE (prob. extinct)
	G. melas edwardii	S. Hemisphere long-finned pilot whale	S. hemisphere	NE
10	Grampus griseus	Risso's dolphin, or Grampus	Worldwide, tropical through temperate waters	DD
11	Lagenodelphis hosei	Fraser's dolphin	Tropical and warm temperate waters of all oceans	DD
12	Lagenorhynchus acutus	Atlantic white-sided dolphin	Colder waters of the N. Atlantic	LC
13	Lagenorhynchus albirostris	White-beaked dolphin	Colder waters of the N. Atlantic	LC
14	Lagenorhynchus australis	Peale's dolphin	Southern S. America, S. Atlantic and S. Pacific Oceans	DD
15	Lagenorhynchus cruciger	Hourglass dolphin	Worldwide, Southern Ocean	LC
16	Lagenorhynchus obliquidens	Pacific white-sided dolphin	N. Pacific Ocean	LC
17	Lagenorhynchus obscurus	Dusky dolphin	Southern S. America, Atlantic and Pacific Oceans, S. Africa, New Zealand	DD
	L. obscurus fitzroyi	Falklands, S. American dusky dolphin	Southern S. America, Falkland Islands	NE
	L. obscurus obscurus	S. African, Indian Ocean dusky dolphin	S. Africa	NE
	L. obscurus subsp.	New Zealand dusky dolphin	New Zealand	NE
18	Lissodelphis borealis	N. right whale dolphin	N. Pacific Ocean	LC
19	Lissodelphis peronii	S. right whale dolphin	S. Pacific, S. Atlantic, S. Indian, and Southern Oceans	DD
20	Orcaella brevirostris	Irawaddy dolphin	Coastal Indo-Pacific waters	DD
21	Orcinus orca	Killer whale, or orca	Worldwide	LR(cd)
22	Peponocephala electra	Melon-headed whale	Worldwide, tropical to temperate waters	LC
23	Pseudorca crassidens	False killer whale	Worldwide, tropical through temperate waters	LC
24	Sotalia fluviatilis	Tucuxi	Northeastern S. America	DD
	S. fluviatilis guianensis	Marine tucuxi	Coastal marine waters	NE
	S. fluviatilis fluviatilis	Freshwater tucuxi	S. American rivers	NE
25	Sousa chinensis	Indo-Pacific hump-backed dolphin	Indian Ocean and Indo-Pacific coastal waters	DD

26	*Sousa teuszi*	Atlantic hump-backed dolphin	Northwestern African coastal waters	DD
27	*Stenella attenuata*	Pantropical spotted dolphin	Tropical and warm temperate waters of all oceans	LR(cd)
	S. attenuata subspecies A	E. Pacific offshore spotted dolphin	Eastern tropical Pacific Ocean	NE
	S. attenuata subspecies B	Hawaiian spotted dolphin	Hawaiian Islands	NE
	S. attenuata graffmani	E. Pacific coastal spotted dolphin	E. Pacific Ocean, coastal off Mexico to Colombia	DD
28	*Stenella clymene*	Clymene dolphin	Tropical and warm temperate Atlantic Ocean and Gulf of Mexico	DD
29	*Stenella coeruleoalba*	Striped dolphin	Tropical and warm temperate waters worldwide, including the Mediterranean Sea	LR(cd)
30	*Stenella frontalis*	Atlantic spotted dolphin	Tropical to temperate waters of the Atlantic Ocean and Gulf of Mexico	DD
31	*Stenella longirostris*	Spinner dolphin	Tropical and warm waters of all oceans	LR(cd)
	S. longirstris longirostris	Gray's spinner dolphin	Tropical and warm waters of all oceans	NE
	S. longirostris orientalis	Eastern spinner dolphin	Pelagic waters of eastern Pacific Ocean	NE
	S. longirostris centroamericana	Costa Rican, Central American spinner dolphin	Pacific Ocean, over continental shelf off Central America	NE
	S. longirostris roseiventiris	Dwarf spinner dolphin	Southeast Asian and N. Australian shallow waters	LC
32	*Steno bredanensis*	Rough-toothed dolphin, or Steno	Tropical to temperate waters of all oceans, including the Mediterranean Sea	DD
33	*Tursiops truncatus*	Common bottlenose dolphin	Tropical to temperate waters of all oceans and seas, especially near the coast	DD
34	*Tursiops aduncus*	Indo-Pacific bottlenose dolphin	Tropical to temperate coastal waters of the Indian Ocean	DD
	Family Pontoporiidae	*La Plata Dolphin*		
1	*Pontoporia blainvillei*	Franciscana	Coastal waters of Argentina, Brazil, Uruguay	DD
	Family Iniidae	*South American River Dolphins*		
1	*Inia geoffrensis*	Amazon dolphin, or boto	S. American rivers	VU
	I. geoffrensis humboldtiana	Orinoco dolphin	Orinoco River basin	NR
	I. geoffrensis geoffrensis	Amazon dolphin	Amazon River basin	NE
	I. geoffrensis boliviensis	Bolivian dolphin	Upper Rio Madeira drainage	NE
	Family Lipotidae	*Chinese River Dolphin*		
1	*Lipotes vexillifer*	Baiji or Yangtze dolphin	Yangtze River	CR (likely extinct)
	Family Platanistidae	*S. Asian River Dolphins*		
1	*Platanista gangetica*	Susu, or 'blind' river dolphin	Rivers of India, Pakistan, Nepal, Bangladesh	EN
	P. gangetica gangetica	Ganges dolphin	Ganges–Brahmaputra River system	EN
	P. gangetica minor	Indus dolphin	Indus River	EN
	Family Phocoenidae	*Porpoises*		
1	*Neophocaena phocaenoides*	Finless porpoise	Coastal tropical and warm temperate Indo-pacific waters	DD
	N. phocaenoides phocaenoides	Indian Ocean finless porpoise	Coastal waters of the Indian Ocean to the S. China Sea	NE
	N. phocaenoides sunameri	W. Pacific finless porpoise	E. China Sea to N. Japan	NE

(Continued)

Table 1   Continued

Sp #	Taxon	Common name	General range	IUCN designation[a]
	N. phocaenoides asiaeorientalis	Yangtze River finless porpoise	Yangtze River	EN
2	Phocoena phocoena	Harbor porpoise	N. Atlantic and N. Pacific Oceans, North, Bering, Barents, and Black Seas	VU
	P. phocoena phocoena	N. Atlantic harbor porpoise	N. Atlantic Ocean	NE
	P. phocoena subsp.	W. N. Pacific harbor porpoise	Western N. Pacific Ocean	NE
	P. phocoena vomerina	E. N. Pacific harbor porpoise	Eastern N. Pacific Ocean	NE
3	Phocoena dioptrica	Spectacled porpoise	SW Atlantic and Southern Oceans	DD
4	Phocoena sinus	Vaquita, or Gulf of California harbor porpoise	Gulf of California	CR
5	Phocoena spinipinnis	Burmeister's porpoise	Southern S. America, S. Atlantic and S. Pacific Oceans	DD
6	Phocoenoides dalli	Dall's porpoise	N. Pacific Ocean	LR(cd)
	P. dalli dalli	Dalli-phase Dall's porpoise	N. Pacific Ocean	NE
	P. dalli truei	Truei-phase Dall's porpoise	Western N. Pacific Ocean	NE

[a]Categories include: NE, not evaluated; DD, data deficient; VU, vulnerable; EN, endangered; CR, critically endangered; LR (cd), lower risk, conservation dependent; LC, least concern. Adapted from Reeves RR, Smith BD, Crespo EA, and Notorbartolo di Sciara G (eds.) (2003) *Dolphins, Whales, and Porpoises: 2002–2010 Conservation Action Plan for the World's Cetaceans.* Gland: IUCN/SSC Cetacean Specialist Group.

Figure 1 The basic body form of dolphins is exemplified by the bottlenose. Photo by Randall S. Wells.

Figure 3 Dentition of a bottlenose dolphin, showing rows of homodont teeth, used to grasp prey. Photo by Randall S. Wells.

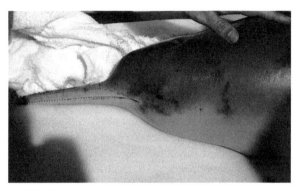

Figure 4 Adult female Franciscana dolphin. Note small body size relative to adult human hand in frame. Photo by Randall S. Wells.

Figure 2 Dorsal fin development in killer whales. Adult males (left) develop much taller and more triangular dorsal fins than subadult males or adult females. Photo by Randall S. Wells.

and number depending on the prey. The pointy teeth are designed for grasping individual prey items, rather than for chewing (**Figure 3**). The number and size of the teeth in turn influence the size of the beak and shape of the mouth of each species. For example, spinner dolphins (*Stenella longirostris*) may have more than 200 small teeth in long, pincers-like jaws for capturing small fish and invertebrates associated with the deep scattering layer. In contrast, killer whales (*Orcinus orca*) have about 50 large teeth for catching large fish and pinnipeds and removing chunks of flesh from a variety of marine mammals. Risso's dolphins (*Grampus griseus*) lack a pronounced beak and have only 10 teeth, all in the lower jaw, for grasping soft-bodied squid prey.

The generally falcate shape of the dorsal fin is a characteristic that distinguishes dolphins from porpoises. Within the dolphins, dorsal fins vary from species to species in height and shape, from the 2-2 m tall triangular fin of adult male killer whales, to the absence of dorsal fins in two species of dolphins (*Lissodelphis* spp.), a rounded fin in Hector's dolphin (*Cephalorhynchus hectori*) and the occurrence of a hump at the base of the fin in others (*Sousa* spp.).

Dolphins range in size from the tiny Franciscana (*Pontoporia blainvillei*) (**Figure 4**), dwarf spinner (*Stenella longirostris roseiventris*), and Hector's dolphins at 1.4–1.5 m and about 50 kg, to adult male killer whales at 9.0 m and 5600 kg.

Color patterns of dolphins vary widely from species to species, including patterns of black, white, gray, brown, orange, and/or pink. Water clarity and its effects on the ability for visual communication may contribute to patterns of coloration, as dolphins living in clear, oceanic water tend to have more complex color patterns on their sides than do dolphins living in the murky waters of estuaries or rivers (**Figure 5**).

Figure 5 Common dolphins in the Gulf of California, showing the complex color pattern exhibited by many oceanic dolphins. Photo by Randall S. Wells.

Figure 6 The Yangtze River dolphin, or baiji, showing the characteristic long beak and reduced eyes. Photo courtesy of Wang Ding, Institute of Hydrobiology, the Chinese Academy of Sciences.

The classification of the numerous members of the family Delphinidae is undergoing much revision with the advent of genetic analysis techniques and the increased efforts by scientists to collect small genetic samples from specimens from around the world. Further confusing taxonomic distinctions for the dolphins, a number of hybrid dolphins (some viable and fertile) have been identified in captive breeding situations and in the wild.

Three families/species of peculiar long-snouted dolphins have been linked under the category 'river dolphin' because their appearances are somewhat similar to one another, very different from other species of dolphins, and because they spend in their entire lives in rivers. These three species have extremely long beaks, flexible necks, relatively large flukes, and flippers, and their eyes are reduced in size (**Figure 6**). The term 'river dolphin' is perhaps a misnomer for these three single-species families, and it is a hold-over from the past. Though they represent the only obligate freshwater dolphin species, other species (Irrawaddy dolphin, *Orcaella brevirostris*, and tucuxi, *Sotalia fluviatilis*) include populations or subspecies that reside permanently in freshwater.

Porpoises

As a group, porpoises tend to be the smallest of the cetaceans, with adult body lengths of less than 2.5 m. They have small, rounded braincases and they lack the pronounced rostrum found in many of the dolphin species, contributing to a condition referred to as paedomorphosis, the retention of juvenile characters in the adult form (this is also seen in dolphins of the genus *Cephalorhynchus*) (**Figure 7**). Raised protuberances occur on the premaxillae. In contrast to the dolphins, porpoises have spatulate, or spade-shaped teeth.

Figure 7 The basic body form of porpoises as exemplified by this vaquita, on the gill net in which it was caught. Photo by Flip Nicklin/Minden Pictures.

Porpoises are stocky and robust, with relatively small appendages. This combination of features may be an adaptation for heat retention for thermoregulation in the cold waters typically inhabited by this family. The dorsal fin, which occurs in all but one (finless porpoise, *Neophocaena phocaenoides*) of the six species, tends to be low and triangular, as compared to typically falcate dolphin fins. All porpoises except Dall's porpoise (*Phocoenoides dalli*) have epidermal tubercles on the leading edge of the dorsal fin. The function of these tubercles is not known. All porpoises have darker patches of pigmentation surrounding the eye, especially the spectacled porpoise (*Phocoena dioptrica*), but the family generally lacks the broad range of colors found in the dolphins.

Earlier confusion regarding the use of the vernacular names dolphin and porpoise seems to have declined over the past few decades, with increased familiarity with the distinctions between the groups. In some parts of the world, the terms were used interchangeably, especially in reference to coastal bottlenose dolphins. In some cases, members of the family Delphinidae have been referred to as porpoises by fishers and management agencies in order to distinguish the mammals from the dolphin fish, the dorado, or mahi mahi (*Coryphaena hippurus*).

Evolution

The earliest whales, the archaeocetes, appear to have evolved from mesonychian condylarths, ungulate ancestors that were primarily terrestrial. Archaeocete fossils from about 52 to 42 Ma have been found in Africa, India, North America, and Pakistan. The recent fossil discovery of a 'walking whale' (*Ambulocetus*) exemplifies the transition from terrestrial to aquatic life. One of the most distinguishing evolutionary developments from archaeocetes to modern cetaceans was the 'telescoping' movement and elongation of the premaxillary and maxillary bones in the skull as the nasal openings migrated to the more effective position on top of the cetacean's head, creating a rostrum or beak. Odontocetes appear to have diverged from the baleen whales (Mysticetes) about 25–35 Ma.

The first modern members of the Delphinidae appeared in the fossil record in the mid–late Miocene, about 10–11 Ma, from kentriodontid-like ancestors, small cetaceans from both the Atlantic and Pacific Oceans that disappeared about 10 Ma. Based on fossil evidence and estimates of divergence between the cytochrome *b* genes, the families Delphinidae and Phocoenidae appeared at about the same time, and both derived from the Kentrodontidae. The family Pontoporiidae appears to have originated in Pliocene and mid-Miocene marine environments of Peru. The adaptation of the different river dolphins to freshwater environments appears to be a convergence. The Iniidae are believed to have entered the Amazon River basin from the Pacific Ocean 15 Ma, or alternatively from the Atlantic about 1.8–5.0 Ma. Platanistid fossils come from mid-Miocene marine deposits in North America and Europe. The Lipotiidae appear to have originated in the Miocene in the North Pacific, with fossils from China and California.

Life History

Little is known about the life history patterns of most dolphins and porpoises. Most of the available information has been derived from examination of carcasses obtained from strandings or fisheries, but some insights have also come from a few long-term studies of wild dolphin populations. Both groups typically produce a single offspring at a time, after a gestation period of about 9–16 months, depending on the species. Larger dolphins, such as pilot whales and killer whales, tend to have the longer gestation periods, with most other species closer to 1 year.

In general, members of the family Delphinidae develop at a slower pace than the porpoises. Dolphins tend to mature later in life than porpoises (typically 6–12 years vs. 3 years for *Phocoena*). At least some porpoises are capable of annual reproduction and can be simultaneously pregnant and lactating, while many dolphins rear calves for multiple years before giving birth to the next calf. Bottlenose dolphins in Sarasota Bay, Florida have been documented to give birth as old as 48 years of age and, with 3–6 -year (successful) calving intervals on average, have been observed with up to eight different calves in a lifetime.

It is believed that few porpoises live longer than 20 years, while the maximum lifespan of most dolphins is likely to be at least 20 years and, in the cases of the larger species, may exceed 60 years. The oldest female bottlenose dolphin in Sarasota Bay is estimated to be 56 years old; the oldest male is 48 years old. Females in some of the large, sexually dimorphic species of dolphins may live 15–20 years longer than males, and become postreproductive. Longevity of river dolphins has yet to be determined, but one baiji (*Lipotes vexillifer*) lived for 22 years in captivity.

Mortality and serious injury of dolphins and porpoises result from a variety of natural sources, including pathogens, biotoxins from Harmful Algal Blooms, predation by sharks and killer whales, and stingray barbs. Anthropogenic sources of mortality that have been clearly identified include directed hunts with harpoons, nets, or drive fisheries, incidental mortality in commercial fishing nets and longlines, ingestion of and entanglement in recreational fishing gear, ingestion of foreign objects, and collisions with boats. Though not yet conclusively demonstrated, the weight of evidence suggests that environmental contaminants such as heavy metals (e.g., mercury), organchlorines (e.g., PCBs and DDT pesticides and their metabolites) along with emerging chemicals such as brominated fire retardants may adversely impact health and/or reproduction, and high-amplitude sounds from some military sonars and underwater explosions may kill cetaceans.

Dolphin and porpoise life history patterns may change in response to changes in abundance such as those resulting from heavy mortality in commercial fisheries. Density-dependent responses have been observed in a variety of cases, such as for dolphins of the genus *Stenella* in the tuna seine net fishery in the Eastern Tropical Pacific Ocean (ETP). As dolphin densities have declined, their age at sexual maturity has also declined, presumably in response to more resources becoming available to remaining individuals. A possible extreme case of this involves the Franciscana dolphin of eastern South America, which reaches sexual maturity at less than 3 years of age, the youngest age known for any odontocete cetacean. It has been suggested that this low age at sexual maturity has come about in response to continuing heavy losses in fishing nets.

Distribution, Ranging Patterns, and Habitat Use

Dolphins and/or porpoises can be found in nearly every marine habitat in the world, as well as in several major river systems (**Table 1**). The most cosmopolitan species, the killer whale (*Orcinus orca*), inhabits waters from the polar ice edge through the tropics. Dolphins reach their highest diversity in tropical and warm temperate waters. Many species are pantropical, while others are limited to one or two ocean basins. There exists only one anti-tropical dolphin species, the long-finned pilot whale (*Globicephala melas*), but there are several anti-tropical species pairs (such as the northern and southern right whale dolphins, *Lissodelphis* spp.).

Pelagic habitats support a number of dolphin species, as well as Dall's porpoises (*Phocoenoides dalli*) and spectacled porpoises (*Phocoena dioptrica*). Offshore deep water areas with a stable mixed layer and a shallow thermocline are home to members of the genus *Stenella* and rough-toothed dolphins (*Steno bredanensis*). More variable offshore areas, where upwelling occurs, is preferred by species such has pilot whales (*Globicephala* spp.), common dolphins (*Delphinus* spp.), striped dolphins (*Stenella coeruleoalba*), and melon-headed whales (*Peponocephala electra*). Coastal waters, including bays, sounds, and estuaries, are preferred by species such as bottlenose dolphins (*Tursiops* spp.), hump-backed dolphins (*Sousa* spp.), Franciscana (*Pontoporia blainvillei*), harbor porpoises (*Phocoena phocoena*), vaquita (*Phocoena sinus*), and Burmeister's porpoises (*Phocoena spinipinnis*).

Dolphin and porpoise habitats are three dimensional, and physical habitat features as well as prey availability influence habitat use. Diving is an adaptation to life in deeper waters, and diving abilities appear to vary greatly both within and across species of dolphins and porpoises. Few data on

diving capabilities are available. Bottlenose dolphins commonly reside in shallow inshore waters, where deep dives are neither necessary nor possible. They utilize resources associated with seagrass meadows, inlets, and other features. However, where *Tursiops* occurs offshore, for example, off the island of Bermuda, it has been documented to dive to depths of 1000 m or more. Rehabilitated rough-toothed dolphins (*Steno bredanensis*) tagged with satellite-linked time-depth recorders and released into pelagic waters rarely dove below 50 m. Based on stomach content data, spinner dolphins (*Stenella longirostris*) are not thought to dive below about 200–300 m. A rehabilitated and tagged Risso's dolphin (*Grampus griseus*) dove occasionally to depths of 400–500 m, but most dives were shallower. None of the porpoises are believed to be deep divers.

The true river dolphins are found only in Asia and South America. Three subspecies of *Inia* spp. inhabit the dynamic Amazon and Orinoco river drainages. These dolphins move from deep channels in time of low water into the rainforest canopy and grasslands during flood season. Two different subspecies of *Platanista* spp., isolated for at least hundreds of years, inhabit the Ganges and Indus river systems. Historically, the baiji (*Lipotes vexillifer*) has inhabited the Yangtze River drainage, including associated large lakes and tributaries, but much of this habitat is no longer accessible to the dolphins due to damming. In addition to the obligate river dolphins, three other small cetaceans also have representatives living in rivers. A subspecies of the finless porpoise (*Neophocaena phocaenoides*) is endemic to the Yangtze River. Populations of the Irrawaddy dolphin (*Orcaella brevirostris*) inhabit rivers in Southeast Asia. A subspecies of tucuxi (*Sotalia fluviatilis*) inhabits rivers in South America, more than 1000 km from the coast.

Ranging patterns are highly variable across the dolphins and porpoises. Some of the pelagic animals, such as spinner dolphins, may range over thousands of square kilometers of open ocean, but where they occur near oceanic islands such as Hawaii, they may be locally resident for decades. Bottlenose dolphins may travel hundreds of kilometers from one seamount or island to the next off Bermuda, or during seasonal migrations along the Atlantic seaboard. However, locally resident populations have been reported from many bays, sounds, and estuaries around the world; one such population has been observed over the past 37 years and across five generations in Sarasota Bay, Florida. Similarly, different groups of killer whales near Vancouver Island, Canada, exhibit patterns of long-term residency versus transience; these ranging patterns are correlated with feeding ecology (see below). Channels and associated habitat limit ranging patterns for river dolphins. One of the smallest documented ranges for a dolphin species was reported for Franciscana dolphins off Argentina, where tagged individuals remained within an area of about 50 km average diameter.

Feeding Ecology

In general, dolphins and porpoises tend to eat fish and/or invertebrates (especially squid), that they can swallow whole or after breaking it into smaller pieces. Their jaws and dentition are designed to capture but not chew prey. The size and number of their teeth relate to the size and kinds of prey eaten. Long-beaked dolphins with numerous tiny teeth (e.g., *Stenella* spp., *Delphinus* spp., *Pontoporia blainvillei*) tend to eat small fish and squid, while dolphins with shorter beaks and fewer and larger teeth will take larger prey. The long beaks of river dolphins facilitate obtaining prey in an obstacle-filled environment. At the other extreme, killer whales, with large, well-anchored teeth, can remove pieces from prey much larger than themselves. Further specialization occurs in Risso's dolphins, where teeth are absent in the upper jaw, and only 10 teeth occur in the lower jaw, for grasping soft-bodied squid. As a departure from all other dolphins and porpoises, the boto (*Inia* spp.) is the only modern dolphin with differentiated dentition. In the front half of the jaws, the teeth are conical, while further back the teeth have a flange on the inside of the crown, reminiscent of molars, presumably for crushing items from their diverse diet of fish, crabs, and turtles.

Different species typically specialize in particular parts of the water column. For example, pantropical spotted dolphins (*Stenella attenuata*) feed on epipelagic prey, right whale dolphins (*Lissodelphis* spp.), spinner dolphins (*Stenellla longirostris*), and Dall's porpoises (*Phocoenoides dalli*) feed on mesopelagic prey, while coastal dolphins and porpoises often take advantage of demersal prey.

Within species, different age, sex, or social groups may specialize on different prey. For example, lactating dolphins may feed on different species and size classes of prey than dolphins in other physiological states, presumably reflecting different energy and hydration requirements. Cultural differences may also occur. Near Vancouver Island, one group of killer whales specializes on fish such as salmon, while another specializes on marine mammals as prey.

Prey distribution influences feeding strategies and behaviors. When prey is relatively evenly distributed, or associated predictably with accessible benthic or shoreline features, then individuals tend to feed alone, as is the case for most porpoises and river dolphins and many coastal dolphins. As prey become more patchy and less predictable, as with schooling fish or squid, then cooperative foraging can be advantageous to dolphins. Schools of dolphins (e.g., *Lagenorhynchus* spp., *Delphinus* spp., and *Stenella* spp.) can spread over relatively large areas to locate rich food patches, and then converge on the prey and work together, circling and concentrating prey schools in order to facilitate prey capture by each individual. Rough-toothed dolphins (*Steno bredanensis*) have been observed to work in pairs to break large fish into smaller pieces. Long-term social relationships likely facilitate cooperative feeding patterns. Killer whales, with their multigenerational matrilineal social groups, exhibit an extreme form of this behavior, when they work together much as a wolf pack to subdue large prey items such as baleen whales. In a variant on this theme, bottlenose dolphins and Irrawaddy dolphins (*Orcaella brevirostris*) have learned to work cooperatively with some net fishermen, driving fish toward the fishermen and presumably increasing their own feeding success through the ensuing confusion of the fish or limitations to their movements by net barriers.

Sensory Systems and Communication

The aquatic medium limits the utility of some mammalian sensory systems, and enhances others. Olfaction has been reduced in dolphins and porpoises, but some level of chemoreception may be possible. Dolphins and porpoises tend to be very tactile animals, especially in terms of physical contact with conspecifics. Water clarity hampers the use of vision by some dolphins and porpoises in their natural environments, but most that have been tested are believed to have reasonably good vision. However, the eyes of the susu (*Platanista gangetica* spp.) lack a crystalline lens, making them essentially blind.

The density of the aquatic medium provides optimal conditions for sound transmission, and both dolphins and porpoises are highly adapted to take advantage of this feature. Both groups produce echolocation clicks over a broad range of frequencies. This sophisticated system of sound production and echo reception presumably facilitates orientation and navigation, and prey finding in murky or dark waters. Echolocation is also believed to be used for group cohesion by some dolphins.

Most dolphins produce pure tone whistles but porpoises do not. A wide variety of whistles are produced in many species. Some dolphins, such as bottlenose (*Tursiops* spp.), produce individually specific whistles referred to as signature whistles. Playback experiments indicate that these whistles function as long-term individual identifiers recognized by kin and close associates. The distinctive information content appears to reside in the whistle structure itself, rather than being associated with any 'voice' features of the individual. Such identifiers may be important for maintaining contact within murky estuarine waters. Dolphins also produce burst-pulse sounds (squawks), primarily in social contexts. It is likely that dolphin social sounds are limited to signaling, emotive, and recognition functions, rather than forming a true language. Killer whales exhibit pod-specific dialects of burst-pulse calls.

Behavior and Social Organization

Dolphins and porpoises spend more than 95% of their time out of sight below the surface of the water. The behaviors and leaps exhibited at or above the surface by some species represent a small fraction of their full repertoire, which varies by species. With the exception of Dall's porpoises, which swim rapidly, creating a distinctive 'rooster tail', and frequently bowride vessels, most porpoises rarely leap or approach boats. Many dolphins will ride in the pressure waves created by vessels, and some exhibit species-specific characteristic leaps, such as the unique multi-revolution spins of spinner dolphins (**Figure 8**), or the high, arcing leaps of pantropical spotted dolphins (**Figure 9**). Murky waters tend to limit observations of the behavior of river dolphins, but susus (*Platanista* spp.) have been reported to swim on their side, with their right flipper near the river bottom, echolocating.

Feeding behaviors can be dramatic. Killer whales and bottlenose dolphins sometimes briefly beach themselves in pursuit of prey on or near the shore. They will also use their flukes to stun prey, a behavior referred to as 'fish-whacking' for bottlenose dolphins. The cooperative efforts of dolphin groups to corral fish schools and drive them to the surface can lead to a feeding frenzy involving large numbers of sea lions and diving birds as well (**Figure 10**). Some dolphins in Australia carry sponges on their rostra, and are believed to use these as tools for obtaining prey. Some of these feeding behaviors are believed to provide evidence for cultural transmission of knowledge as they are passed between generations and spread across social units.

Dolphins and porpoises vary greatly in their degree of sociality. Porpoises and river dolphins tend to live alone or in small groups, with the only persistent groupings being mothers with their most recent

Figure 8 Hawaiian spinner dolphin performing the characteristic spinning leap from which its name is derived. Photo by Randall S. Wells.

Figure 9 Characteristic high leap by a pantropical spotted dolphin. Photo by Randall S. Wells.

Figure 10 'Feeding frenzy' involving common dolphins, sea lions, pelicans, and gulls, in the Gulf of California. Photo by Randall S. Wells.

calves. All dolphins are social to some degree, ranging from a few individuals to thousands in a school. Oceanic dolphins such as spinner and spotted dolphins tend to form large (sometimes hundreds to thousands) and fluid schools, although some associations within these schools may be long term or recurrent. Some spinner dolphins inhabiting waters near isolated atolls maintain very stable groupings over time. Even greater levels of stability are exhibited by the permanent matrilineal pods of killer whales, which represent several generations of related individuals and remain unchanged except by birth, death, or rare separations. Other large dolphins such as pilot whales are also believed to maintain similar stability, based on results of genetic studies. At an intermediate level, bottlenose dolphins such as those in Sarasota Bay, Florida, may form long-term, multigenerational resident communities of about 150 individuals, where group composition is fluid (fission-fusion), but the site fidelity of the animals guarantees repeated contact as they move through the community range. Longer-term associations within the community include mother–calf bonds lasting 3–6 years, and strong male pair bonds that can persist for decades. In Shark Bay, Western Australia, strong bonds among adult male dolphins translate into complex patterns of alliances of males that work together against other alliances to obtain access to females. At the next level, inter-specific associations can be found among the dolphins, with some of the most common being between spinner and pantropical spotted dolphins, bottlenose dolphins and pilot whales, and Pacific white-sided dolphins, Risso's dolphins, and northern right whale dolphins. The reasons for these associations are not entirely clear, but they may relate to foraging or protection from predators.

Conservation Status and Concerns

Dolphin and porpoise populations around the world typically number in the thousands to millions depending on species and stocks. However, of the 44 species listed in **Table 1**, as of 2003 six were considered by the IUCN to be critically endangered ($n=2$), endangered ($n=2$), or vulnerable ($n=2$). Many other species lack sufficient information to make any assessment of their status. In each case where concern has been expressed, human activities have been identified as the primary cause for reductions in abundance.

It is likely that the critically endangered baiji, or Yangtze River dolphin, has been recently driven to extinction, based on a 2006 survey of the remaining habitat of the species during which no baiji were found (**Figure 6**). This species was described by Western scientists in 1918. At that time it was still common from Three Gorges to Shanghai. Beginning in 1958, it was hunted intensively for meat, oil, and leather. The decline was exacerbated by accidental losses in fishing activities, including nets, electrofishing and the use of explosives, and entanglement in illegal rolling hooks. Others died from collisions with vessels in the heavily trafficked river. Pollution from the activities of the billion people served by the river, overfishing, underwater blasting for construction, dredging, and damming of tributaries all likely contributed to the decline of this species.

The other river dolphins are considered to be endangered (*Platanista* spp.) or vulnerable (*Inia* spp.), as a result of their limited habitat, and their close proximity to human activities that directly or indirectly impact the animals, such as habitat fragmentation from damming and fishing. The finless porpoises (*Neophocaena phocaenoides*) that inhabit the Yangtze River have shown recent significant declines as well.

The vaquita, found only in the Gulf of California, is the other of the two critically endangered dolphins or porpoises. Although this porpoise species was only first described in 1958, numbers of remaining vaquita or have been reduced to below 500 individuals due to fishing activities, especially from netting for an endangered sea bass, the totoaba (**Figure 7**).

Found only in the waters of New Zealand, Hector's dolphins (*Cephalorhynchus hectori*) have been dramatically reduced in abundance in recent years from fishing activities, and are now considered to be endangered.

In addition to the vaquita, there are concerns for all of the porpoises except the spectacled porpoise (*Phocoena dioptrica*). The harbor porpoise (*Phocoena phocoena*) is considered vulnerable due to incidental mortalities in fishing nets.

Directed fisheries for some dolphins and porpoises occur in many places around the world. For example, striped dolphins (*Stenella coeruleoalba*) are taken by drive fisheries in Japan, pilot whales (*Globicephala melas*) are obtained similarly in the Faroe Islands, and other species are obtained by netting or harpoons elsewhere, usually for human consumption, or for bait for other fisheries. One of the largest directed fisheries is prosecuted by the Japanese for Dall's porpoises (*Phoceonoides dalli*). The rationale offered for this ongoing fishery is that it provides meat for Japanese markets, partially offsetting the reduction in availability of whale meat from the international whaling moratorium.

As indicated above, incidental mortality in fisheries is one of the most serious threats faced by dolphins and porpoises around the world. In the most dramatic example, millions of dolphins of the several species, but especially *Stenella attenuata* and *Stenella longirostris*, have been killed in the tuna seine net purse seine fishery in the ETP since the late 1950s. Modifications to fishing practices and gear since the 1990s have reduced annual mortalities to a few thousand individuals, but stocks are not recovering at expected rates, suggesting impacts other than direct mortality in the nets. Incidental mortality is experienced in a wide range of commercial and recreational fisheries around the world. It is managed in some countries, but in many there are no effective controls over takes of marine mammals.

Commercial live captures of dolphins for public display, research facilities, and military uses occur in several countries, which supply the rest of the world. Removals of these animals from the wild render them ecologically dead to their native stocks. Typically, these fisheries are prosecuted in the absence of valid scientific assessments of the stocks prior to removals, placing the future for the remaining animals in jeopardy. Institutions in some countries such as the United States have established successful captive breeding programs to avoid these problems, but current breeding programs elsewhere are unable to keep pace with the demands for dolphins for burgeoning programs of interactions with humans around the world, creating pressure for continuing live capture operations.

Environmental contaminants pose a highly insidious and likely underappreciated threat to dolphins and porpoises around the world. Very high concentrations of contaminants that cause health and reproductive problems in other mammals have been measured in dolphin and porpoise tissues. Although direct cetacean mortalities from contaminants have not been frequently identified, the weight of evidence is mounting relating concentrations of toxic chemicals to health problems and reproductive failure.

Because of the odontocetes' acoustic sensitivity, underwater noise is also of increasing concern for marine mammals around the world. Noise from vessel traffic, marine construction and demolition, petroleum exploration and production, and military sonars has been implicated in behavioral changes of dolphins and porpoises, and in some cases is believed to have killed cetaceans.

Glossary

anthropogenic   Of human origin, as in man-made threats to dolphins or porpoises, such as pollution, fishing gear, or noise from industrial or military activities.

anti-tropical   Distributed to the north and to the south of the tropics.

biotoxin   Naturally created toxin produced by organisms such as algae.

by-catch   Organisms caught unintentionally in the course of catching target species.

Cetacean   Whale, dolphin, or porpoise, of the mammalian order Cetacea.

demersal   Near the seafloor, as in prey fish that tend to not move through the water column or to the surface of the water.

drive fishery   Fishery directed at cetaceans in which noise from an array of small boats is used to move the animals onto a beach.

echolocation (sonar)   Sophisticated acoustic orientation system of odontocetes in which clicks produced in air passages are focused and projected forward, and received echoes from the clicks provide information about the cetacean's environment.

epipelagic   Near the water's surface, as in prey fish that do not inhabit the lower portions of the water column or the seafloor.

mesopelagic   Within the water column, as in prey fish that are not commonly found near the surface or the seafloor.

odontocete   A toothed cetacean, of the suborder Odontoceti.

paedomorphosis   The retention of juvenile characteristics in adults.

pantropical   Found in tropical waters throughout the world.

peduncle   The tail stock, or connection between the body and the flukes of a cetacean.

pelagic   Open ocean waters, typically deep and far from shore.

rostrum   The beak or tooth-filled projection of the mouth of a cetacean.

sexual dimorphism   Differences in the body form or size based on gender.

strandings   When live or dead cetaceans become beached.

See also

Marine Mammal Diving Physiology. Marine Mammal Evolution and Taxonomy. Marine Mammal Migrations and Movement Patterns. Marine Mammal Social Organization and Communication. Marine Mammal Trophic Levels and Interactions.

Further Reading

Berta A and Sumich JL (2003) *Marine Mammals: Evolutionary Biology.* San Diego, CA: Academic Press.

Leatherwood S and Reeves RR (eds.) (1990) *The Bottlenose Dolphin.* San Diego, CA: Academic Press.

Mann J, Connor RC, Tyack PL, and Whitehead H (eds.) (2000) *Cetacean Societies: Field Studies of Dolphins and Whales.* Chicago, IL: University of Chicago Press.

Norris KS, Würsig B, Wells RS, and Würsig M (1994) *The Hawaiian Spinner Dolphin.* Los Angeles, CA: University of California Press.

Read AJ, Wiepkema PR, and Nachtigall PE (eds.) (1997) *The Biology of the Harbour Porpoise.* Woerden: De Spil Publishers.

Reeves RR, Smith BD, Crespo EA, and Notarbartolo di Sciara G (eds.) (2003) *Dolphins, Whales, and Porpoises: 2002–2010 Conservation Action Plan for the World's Cetaceans.* Gland: IUCN/SSC Cetacean Specialist Group.

Reeves RR, Smith BD, and Kasuya T (eds.) (2000) *Biology and Conservation of Freshwater Cetaceans in Asia.* Gland: IUCN.

Reynolds JE, III, Perrin WF, Reeves RR, Montgomery S, and Ragen TJ (eds.) (2005) *Marine Mammal Research: Conservation Beyond Crisis.* Baltimore, MD: The Johns Hopkins University Press.

Reynolds JE, III and Rommel SA (eds.) (1999) *Biology of Marine Mammals.* Washington, DC: Smithsonian Institution Press.

Reynolds JE, III, Wells RS, and Eide SD (2000) *The Bottlenose Dolphin: Biology and Conservation.* Gainesville, FL: University Press of Florida.

Rice DW (1998) *Special Publication No. 4: Marine Mammals of the World: Systematics and Distribution.* Lawrence, KS: The Society for Marine Mammalogy.

Ridgway SH and Harrison R (eds.) (1989) *Handbook of Marine Mammals, Vol. 4: River Dolphins and the Larger Toothed Whales.* San Diego, CA: Academic Press.

Ridgway SH and Harrison R (eds.) (1994) *Handbook of Marine Mammals, Vol. 5: The First Book of Dolphins and Porpoises.* San Diego, CA: Academic Press.

Ridgway SH and Harrison R (eds.) (1999) *Handbook of Marine Mammals, Vol. 6: The Second Book of Dolphins and Porpoises.* San Diego, CA: Academic Press.

Twiss JR, Jr. and Reeves RR (eds.) (1999) *Conservation and Management of Marine Mammals.* Washington, DC: Smithsonian Institution Press.

SEALS

I. L. Boyd, Natural Environment Research Council,
Cambridge, UK

Taxonomy

The seals, or Pinnipedia, are the suborder of the
Carnivora that includes the Phocidae (earless or 'true'
seals), Otariidae (eared seals, including fur seals and
sea lions) and the Odobenidae (walrus). They are related to the bears, based on a common ancestry with
terrestrial arctoid carnivores (**Figure 1**). The otariids
retain more of the ancestral characteristics than the
other two groups but all have a more or less aquatic
lifestyle and display highly developed morphological
and physiological adaptations to an aquatic existence.

The Pinnipedia are made up of 34 species and 48
species/subspecies groupings (**Table 1**). However,
with the advent of new methods based on DNA
analysis for examining phylogeny and also because
of new methods used to track animals at sea many of
these groupings are questionable. Several groups that
were thought to have been different species have
overlapping ranges and are likely to interbreed. It
seems most probable that the southern fur seals
(*Arctocephalus* sp., Table 1) are not distinct species.
Conversely, some of the North Atlantic phocid
pinnipeds that are classified as single species are
likely to be better represented as a group of subspecies. The gray seal is a particular example in

which three genetically distinct populations (NW
Atlantic, NE Atlantic and Baltic) are recognized.

Distribution and Abundance

The greatest diversity and absolute abundances of
pinnipeds occurs at temperate and polar latitudes
(**Table 1**). Only three phocid seal species, the monk
seals, are truly tropical species and all of these are
either highly endangered or, in one case, may be
extinct. Among the otariids, fur seals and sea lions
extend their distributions into the tropics but their
absolute abundance in these locations is low compared with the populations at higher latitudes.

Probably >50% of the biomass of pinnipeds in the
world is derived from one species, the crabeater seal.
Some estimates of the abundance of this species have
undoubtedly been exaggerated but it is nevertheless
the dominant species. This is partly because its main
habitat is the vast Antarctic pack ice. The ringed seal
has a comparable distribution in the Arctic and it
also has abundances numbered in the millions. The
relative numbers of the different species are shown in
Table 1.

Although the status of the populations of some
species/subspecies groups is unknown, only seven
(22%) groupings are in decline whereas seventeen
(55%) of these groups are increasing in abundance.
However, twenty-three groups (48%) are classified
as being in need of some form of active conservation
management. Threats to pinnipeds can be represented as: (1) *direct threats* from harvesting and
international trade, incidental catch by commercial
fisheries and direct killing by fishermen; (2) *intermediate threats* from episodic mass mortalities,
habitat degeneration (including environmental pollution, competition for food with humans, disturbance and changes to the physical environment); (3)
longer-term threats from climate change and reduction of genetic diversity.

Morphological and Physiological Adaptations to Aquatic Life

Morphological adaptations include the modification
of limbs to form flippers for swimming, the development of a streamlined fusiform shape, the presence
of insulation in the form of fur and/or a subcutaneous layer of blubber and increased visual acuity
for foraging at extremely low light levels. Unlike

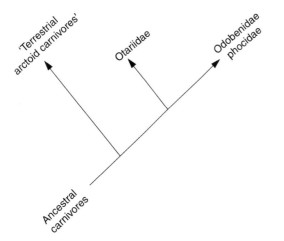

Figure 1 Pinniped phylogeny from a cladogram based on
postcranial morphology. (Reproduced from Berta *et al.*, 1989.)

Table 1 Species and common names, abundances, trends in abundance and conservation status for the Pinnipedia

Taxonomic classification		Common name	Distribution	Abundance[a] ($\log_{10}$ scale)	Trend in abundance	Conservation status
Family Otariidae						
Subfamily Otariinae						
Eumetopias jubatus		Steller sea lion	North Pacific	***	Declining	Threatened
Zalophus californianus subspecies:	californianus	California sea lion	Western USA	****	Increasing	No threat
	wollebaeki	Galapagos sea lion	Galapagos Is.	***	Stable (?)	Rare
	japonicus	Japanese sea lion	Japan	*	Unknown	Possibly extinct
Ontaria byronia		Southern sea lion	South America	***	Declining	Threatened
Neophoca cinerea		Australian sea lion	Southern Australia	**	Stable	Rare
Phocarctos hookeri		Hooker's sea lion	Southern New Zealand	**	Declining	Endangered
Subfamily Arctocephalinae						
Arctocephalus townsendi		Guadalupe fur seal	Guadalupe Is.	**	Increasing	Rare
Arctocephalus galapagoensis		Galapagos fur seal	Galapagos Is.	***	Stable (?)	Rare
Arctocephalus philipii		Juan Fernandez fur seal	Juan Fernandez Is.	**	Increasing	Rare
Arctocephalus australis subspecies:	australis	Falkland fur seal	Falkland Is.	***	Increasing (?)	Locally vulnerable
	gracilis	South American fur seal	South America	****	Increasing	No threat
Arctocephalus tropicalis		Sub-Antarctic fur seal	Sub-Antarctic	****	Increasing	No threat
Arctocephalus gazella		Antarctic fur seal	Antarctic	*****	Increasing	No threat
Arctocephalus pusillus subspecies:	pusillus	South African fur seal	Southern Africa	*****	Increasing	No threat
	doriferus	Australian fur seal	Southern Australia	***	Increasing	No threat
Arctocephalus forsteri		New Zealand fur seal	Australia and New Zealand	***	Increasing	No threat
Callorhinus ursinus		Northern fur seal	North Pacific	*****	Declining (?)	No threat
Family Odobenidae						
Subfamily Odobenidae						
Odobenus rosmarus subspecies:	rosmarus	Atlantic walrus	Arctic	****	Unknown	Commercially threatened
	divergens	Pacific walrus	Arctic	**	Unknown	Commercially threatened
Family Phocidae						
Subfamily Phocinae						
Halichoerus grypus		Grey seal	North Atlantic	****	Increasing	No threat
Phoca vitulina subspecies:	vitulina	Eastern Atlantic harbor seal	NE Atlantic	***	Increasing	No threat
	concolor	Western Atlantic harbor seal	NW Atlantic	***	Increasing	No threat
	stejnegeri	Western Pacific harbor seal	NW Pacific	**	Unknown	Rare
	richardsi	Eastern Pacific harbor seal	NE Pacific	****	Increasing	No threat
	mellonae	Ungava seal	Quebec, Canada	*	Unknown	Rare

(Continued)

Table 1 Continued

Taxonomic classification		Distribution	Abundance[a] (log₁₀ scale)	Trend in abundance	Conservation status
Phoca largha		Spotted seal	Bering Sea	****	Unknown	No threat
Phoca hispida subspecies:		Arctic ringed seal	Arctic	*****	Unknown	No threat
	hispida	Okhotsk sea ringed seal	Sea of Okhtosk	****	Unknown	No threat
	ochotensis	Baltic ringed seal	Baltic Sea	**	Increasing (?)	Threatened
	botnica	Saimaa seal	Lake Saimaa, Finland	*	Unknown	Endangered
	Saimensis	Ladoga seal	Lake Ladoga, Russia	***	Stable – increasing	Vulnerable
	ladogensis					
Phoca caspica		Caspian seal	Caspian Sea	****	Stable	Vulnerable
Phoca sibirica		Baikal seal	Lake Baikal	***	Unknown	Vulnerable
Phoca groenlandica		Harp seal	NW Atlantic & Arctic	******	Increasing	No threat
Phoca fasciata		Ribbon seal	Bering Sea & Arctic	****	Unknown	No threat
Cystophora cristata		Hooded seal	N Atlantic & Arctic	****	Unknown	No threat
Erignathus barbatus subspecies:	*barbatus*	Atlantic bearded seal	Arctic	****	Unknown	No threat
	nauticus	Pacific bearded seal	Arctic	****	Unknown	No threat
Subfamily Monachinae						
Monachus monachus		Mediterranean monk seal	Mediterranean	*	Declining	Endangered
Monachus tropicalis		West Indian monk seal	Caribbean Sea	*	Unknown	Possibly extinct
Monachus schauinslandi		Hawaiian monk seal	Hawaiian Islands	*	Declining	Endangered
Leptonychotes weddellii		Weddell seal	Antarctica	****	Stable	No threat
Ommatophoca rossii		Ross seal	Antarctica	***	Unknown	No threat
Lobodon carcinophagus		Crabeater seal	Antarctica	******	Stable	No threat
Hydrurga leptonyx		Leopard seal	Antarctica	***	Stable	No threat
Mirounga leonina		Southern elephant seal	Sub-Antarctic	****	Declining (?)	No threat
Mirounga angustirostris		Northern elephant seal	Sub-Antarctic	****	Increasing	No threat

[a]The number of asterisks denote the range in the size of the world populations: *, 0–1000; **, 1000–10 000; ***, 10 000–100 000; ****, 100 000–1 000 000; *****, 1 000 000–10 000 000; ******, 10 000 000–100 000 000.

cetaceans, an ability to echolocate has not been confirmed in pinnipeds, although some studies have purported to show that seals are capable of behavior that is consistent with echolocation abilities.

Physiological adaptations include a highly developed dive response. On submergence this involves the rapid reduction of heart rate, reduced peripheral circulation and the sequestration of large amounts of oxygen bound to myoglobin in the muscles. Pinnipeds also have high concentrations of red blood cells in the blood and changed morphology of the red blood cells themselves, which take on a 'cocked-hat' shape. The architecture of the venous system is also modified, especially amongst phocids, to allow a larger volume of blood to be stored. Included within the dive response is the ability, when at the surface between dives, to increase heart rate rapidly to clear and reprocess metabolic waste products and to re-oxygenate the tissue in preparation for the next dive.

Thermal Constraints

Perhaps the greatest single constraint on the evolution of pinnipeds as aquatic animals is the problem associated with thermoregulating in cold water that is 25 times more conductive to heat than air. Since pinnipeds are endothermic homeotherms that normally maintain a body temperature of 36–38°C, they are presented with a significant thermal challenge when they are immersed in water at or close to freezing. The observation that the greatest number and species diversity of pinnipeds is found in temperate and polar regions suggests that they have adapted well to this challenge. However, the cost associated with this seems to be that pinnipeds have retained a non-aquatic phase in their life histories. By virtue of their relatively small body size, newborn pinnipeds cannot survive for long in cold water and so they are born on land or ice and remain there until they have built up sufficient insulation to allow them to go to sea. Unlike cetaceans, pinnipeds do not appear to have solved the problem of giving birth to young with insulation already developed but many cetaceans have a much larger body size than pinnipeds (thus reducing the thermal challenge to the newborn) and many species also migrate to warmer waters to give birth. Cetaceans appear to have developed a different strategy to deal with the cold.

Pinniped Life Histories: The Constraints of Aquatic Life

An important consequence of the necessity for non-aquatic births in pinnipeds is that mothers are more or less separated from their foraging grounds by the need to occupy land or ice during the period of offspring dependency. Food abundance in the marine environment is not evenly distributed and has a degree of unpredictability in space and time. Therefore, pinniped mothers have had to trade-off the necessity to find a location to give birth which is safe from predation, since pinnipeds are vulnerable when on land, with the need to feed herself and her pup throughout the period of offspring dependency. This has apparently led to two different types of maternal behavior.

In small pinnipeds including all the fur seals, most of the sea lions and small phocids, it is not possible, by virtue of body size alone, for mothers to carry sufficient energy reserves at birth to support both her and her offspring until the offspring is able to be independent. This means that mothers must supplement their energy reserves by feeding during lactation. In the case of the smallest pinnipeds, the fur seals, mothers rely almost entirely on the energy from foraging and have very few reserves. Therefore, these small pinnipeds are restricted to breeding at sites which are close enough to food for the mothers to be able to make foraging trips on time scales that are less than the time it would take their pup to starve. Consequently, lactation in these small pinnipeds tends to be extended over several months and, in a few cases, can last over a year.

In contrast, pinnipeds of large body size (the transition in this case between large and small appears to occur at a maternal body mass of about 100 kg) are able to carry sufficient energy reserves to allow mothers to feed both themselves and their offspring while they are ashore. In these species, the tendency is for mothers to make only a single visit ashore and for her not to feed during lactation. As a result, these mothers have a short lactation and, in the case of the hooded seals, this is reduced to only four days, but 15–30 days is more normal.

These types of behaviors, which stem directly from the combined physical restrictions of thermoregulation in newborn pups and maternal body size, have had two further important consequences. The first of these is that larger pinnipeds are better able to exploit food resources at greater distance from the birth site and they have been shown to range over whole ocean basins in search of food. This is a necessary consequence of their larger size because, in contrast to small pinnipeds, they must exploit richer food sources because of their greater absolute food requirement. Since richer food sources are also rarer food sources, the large pinnipeds have fewer options as to where they can forage profitably. Thus, with some exceptions the large pinnipeds occupy much larger ranges than the small pinnipeds.

By mammalian standards all pinnipeds are of large body size. Even though pups are well developed at birth, this means that it takes several years for most pinnipeds to grow to a body size large enough for them to become sexually mature. The minimum duration to reach sexual maturity is about 3 years and the maximum, in species such as the grey seal, is 5–6 years. Thereafter, they only produce a single pup each year and the individuals of most species will fail to produce a pup about one year in four. However, since females may live for 20 to >40 years, they are relatively long-lived animals.

Mating Systems

The second consequence of the physical restriction that thermoregulation in newborn pups places on pinnipeds is the mating system. This feature of pinniped biology has been the subject of intensive investigations, largely because it is much the most dramatic and obvious part of pinniped life-histories. There has been much speculation as to why pinnipeds should mostly have developed mating behavior involving dense aggregations and apparent extreme polygyny but it is likely to be a consequence of the necessity for mothers to give birth out of the water. Restrictions in the availability of appropriate breeding habitat (defined in terms of both its proximity to food and its protection from predation), together with reduced risk of predation that individuals have when they are in groups, probably combined to increase the fitness of those mothers that had a tendency to give birth in groups. It is also considerably more efficient, in energetic terms, to have to return to land only once during each reproductive cycle. Females have made use of an ancestral characteristic involving the existence of a postpartum estrus at which most females are mated and become pregnant. Without this, females would have been required to seek a mate at a time when, for many species, the population would be highly dispersed over a wide area while foraging. This means that appropriate mates would have been more difficult to find than at a time of year when the population is highly aggregated.

Males that are present when there is the greatest chance of mating will be most likely to gain greatest genetical fitness. Thus, in almost all pinnipeds, a competitive mating system has developed around the rookeries of females with their pups. Moreover, this has led to selection for male morphological and behavioral characteristics that confer greater ability to dominate matings. The male hooded seal has developed a deep red septum between his nostrils that can be blown out like a balloon as a display organ; male elephant seals have developed loud vocalizations which, together with their enlarged rostrum, make a formidable display; male harbor and Weddell seals have complex and loud underwater vocalizations that are almost certainly part of a competitive mating system; and in most species there is a marked sexual dimorphism of body size in which males can be six to eight times the mass of females. This sexual dimorphism in mass may serve a double function: increased mass leads to increased muscle power and the ability to fight off rivals for matings and increased mass also confers increased staying power allowing individual males to fast while they are on the breeding grounds and maintaining their presence amongst receptive females for as long as possible. However, this larger body mass also has a cost in that, because of their greater absolute energy expenditure, the larger males must find richer food sources to be able to feed profitably and recover their condition between breeding seasons. A consequence of this is that males have lower survival rates than the smaller females.

An exception to much of this is found in many of the seals that breed on ice. In these cases, mothers often have the option to give birth in close proximity to food and, at least in the Antarctic where there are no polar bears, they are relatively safe from predation. There is no sexual dimorphism in the crabeater seal, a species which gives birth in the Antarctic pack ice without any detectable aggregation of mothers. In the Arctic, the harp seal is the rough ecological equivalent of the crabeater seal and in this species there are large aggregations, known in Canada as whelping patches. It would appear that one of the main contrasts between these species is that harp seals are exposed to polar bear predation whereas crabeater seals are not. In neither case is there marked sexual dimorphism of body size despite evidence for competitive mating. As in the case of the Weddell and harbor seals, which have only small sexual dimorphism of body size, these species mainly mate in the water rather than on land. Therefore, it may be that large body size in male pinnipeds is mainly a characteristic that is an advantage to those that have terrestrial mating.

Diet

Seals are mostly fish-eating although the majority of species have a broad diet that also includes squid, molluscs, crustaceans, polychaete worms and, in certain cases other vertebrates including seabirds and other seals. Even those that prey mainly on fish take a broad range of species although there is a tendency for specialization on oil-bearing species such as herring, capelin, sand eels/lance, sardines and anchovies. This is because these species have a high energy content and they are often in shoals so that they may be an energetically more profitable form of prey than many other species.

Perhaps the most specialized pinniped in terms of diet is the walrus which forages mainly on benthic molluscs, crustaceans and polychaetes. Its dentition is adapted to crushing the shells of molluscs and their tusks are used to stir up the sediment on the sea bed to disturb the prey within. In the Arctic, bearded seals have a similar feeding habit and several other species feed regularly on benthic invertebrates. Among gray seals there is evidence that some individuals specialize in different types of prey. For most species, feeding occurs mainly within the water column and may be associated with particular oceanic features, such as fronts or upwellings of deep water that are likely to contain higher concentrations of prey. Seals may migrate distances of up to several thousand kilometers to find these relatively rich veins of food.

The crabeater seal feeds almost entirely on Antarctic krill (a small shrimp-like crustacean) that it gathers mainly from the underside of ice floes where the krill themselves feed on the single-celled algae that grow within the brine channels within the ice. Antarctic fur seals also feed on krill to the north of the Antarctic pack ice edge and many of the Antarctic seals rely to varying degrees on krill as a source of food. In fact Antarctic krill probably sustain more than half of the world's biomass of seals and also sustain a substantial proportion of the biomass of the world's sea birds and whales. The dentition of crabeater seals is modified to help strain these small shrimps from the water.

Many species of seals will, on occasions, eat sea birds. Male sea lions of several species have been recorded as snacking on sea birds and male Antarctic fur seals regularly feed on penguins. However, the most specialized predator of sea birds and other seals is the leopard seal. This powerful predator is found mainly in the Antarctic pack ice and is credited with being the most significant cause of death amongst juvenile crabeater seals even though few cases of direct predation have been observed. Individual leopard seals may specialize on specific types of prey because the same individuals have been observed preying on Antarctic fur seals at one location in successive years and one of these has been seen at two locations over 1000 km apart where young Antarctic fur seals can be found.

Diving for Food

The development over the past decade of microelectronic instruments for measuring the behavior of pinnipeds has put the adaptations for aquatic life in these animals into a new perspective. Some pinnipeds are capable of very long and very deep dives in search of their prey. The result of this diving ability is that pinnipeds are able to exploit on a regular basis any food that is in the upper 500 m of the water column. In general, larger body size confers greater diving ability mainly because the rate at which animals of large body size use their oxygen store is less than that for small individuals. Ultimately, it is the amount of oxygen carried in the tissues that determines how long a pinniped can stay submerged and time submerged limits the depth to which pinnipeds can dive. Consequently the largest pinnipeds, elephant seals, dive longer and deeper than any others.

On average adult elephant seals dive to what seems to us as a punishing schedule. Average dive durations can exceed 30 minutes with about 2 minutes between dives and elephant seals maintain this pattern of diving for months on end, only stopping every few days to 'rest' at the surface for a slightly longer interval than normal but usually much less than an hour. Technically, elephant seals are more correctly seen as surfacers rather than divers.

Occasionally elephant seals dive to depths of 1500 m and dives can last up to 2 hours with no apparent effect on the time spent at the surface between dives. It is still a mystery to physiologists how elephant seals, and many other species including hooded seals and Weddell seals, manage to have such extended dives. Many physiologists believe that free-ranging seals like elephant seals are able to reduce their metabolic rate while submerged to such an extent that they can conserve precious oxygen stores and they can then rely on aerobic metabolism throughout the dives. This strategy may allow these animals to access food resources that the majority of air-breathing animals cannot reach. As described above, this is likely to be of critical importance to these large-bodied animals because of their need to find rich food sources.

See also

Krill. Marine Mammal Evolution and Taxonomy.

Further Reading

Berta A, Ray CE, and Wyss AR (1989) Skeleton of the oldest known pinniped. *Enaliarctos mealsi. Science* 244: 60–62.

Laws RM (ed.) (1993) *Antarctic Seals*. Cambridge: Cambridge University Press.

Reijnders P, Brasseur S, van der Toorn J, *et al.* (1993) *Seals, Fur Seals, Sea Lions, and Walrus. Status and Conservation Action Plan*. Gland, Switzerland: IUCN.

Reynolds JE and Rommel SA (eds.) (1999) *Biology of Marine Mammals*. Washington and London: Smithsonian Institution Press.

SIRENIANS

T. J. O'Shea, Midcontinent Ecological Science Center, USGS, Fort Collins, Colorado, USA
J. A. Powell, Florida Marine Research Institute, St Petersburg, FL, USA

Introduction

The Sirenia are a small and distinctive Order of mammals. They evolved from ancient terrestrial plant feeders to become the only fully aquatic, large mammalian herbivores. This distinctive mode of life is accompanied by a suite of adaptations that make the Sirenia unique among marine mammals in anatomical and physiological features, distribution, ecology, and behavior. Although the Sirenia have a long history of interaction with humans, some of their biological attributes now render them vulnerable to extinction in the face of growing human populations throughout their coastal and riverine habitats.

Evolution and Classification

Fossil History

The order Sirenia arose in the Paleocene from the Tethytheria, a group of hoofed mammals that also gave rise to modern elephants (Order Proboscidea). The beginnings of the Sirenia were probably in the ancient Tethys Sea, near what is now the area joining Africa and Asia. By the early Eocene, sirenians had reached the New World, as evidenced by fossils from the primitive family Prorastomidae from Jamaica. The prorastomids and the sirenian family Protosirenidae were restricted to the Eocene. Protosirenids and prorastomids were amphibious and could walk on land, but probably spent most of the time in water. The fully aquatic Dugongidae also arose around the Eocene, and persisted to the Recent, as represented by the modern dugong (subfamily Dugonginae) and Steller's sea cow. The dugongines were the most diverse lineage, particularly in the Miocene, with greatest radiation in the western Atlantic and Caribbean but also spreading back into the Old World. Two other subfamilies of dugongids also differentiated but became extinct: the Halitheriinae, which disappeared around the late Pliocene, and the Hydrodamalinae, which was lost with the

extinction of Steller's sea cow in 1768. The hydrodamalines are noteworthy for escaping the typically tropical habitats of sirenians and occupying colder climates of the North Pacific. The family Trichechidae (manatees) arose from dugongids around the Eocene–Oligocene boundary. Two subfamilies of trichechids have been delineated: the Miosireninae, which became extinct in the Miocene, and the Trichechinae, which persists in the three living species of manatees. Early trichechids arose in estuaries and rivers of an isolated South America in the Miocene, where building of the Andes Mountains provided conditions favorable to the flourishing of aquatic vegetation, particularly the true grasses. The abrasiveness of these plants resulted in natural selection for the indeterminate tooth replacement pattern unique to trichechids. When trichechids returned to the sea about a million years ago, this persistent dentition may have allowed them to be more efficient at feeding on seagrasses and outcompete the dugongids that had remained in the Atlantic. Thus dugongids disappeared from the Atlantic, while forms of manatees probably very similar to the modern West Indian manatee spread throughout the tropical western Atlantic and Caribbean, and since the late Pliocene had also dispersed by transoceanic currents to reach West Africa. Beginning in the late Miocene, Amazonian manatees evolved in isolation in what was then a closed interior basin of South America.

Classification, Distribution, and Status of Modern Sirenians

There are four living and one recently extinct species of modern Sirenia. They are classified in two families, the Dugongidae (Steller's sea cow and the dugong) and the Trichechidae (three species of manatees). All four extant species are designated as vulnerable (facing a high risk of extinction in the wild in the medium-term future) by the International Union for the Conservation of Nature and Natural Resources–World Conservation Union.

Steller's sea cow Steller's sea cow, *Hydrodamalis gigas* (**Figure 1**), is placed in the subfamily Hydrodamalinae of the family Dugongidae. These largest of sirenians were found in shallow waters of the Bering Sea around Bering and Copper Islands in the Commander Islands. The first scientist to discover these sea cows was Georg Steller, who first

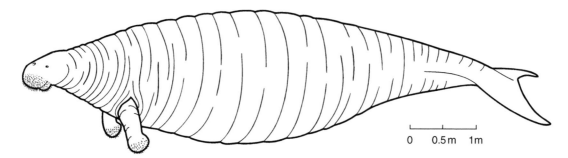

Figure 1 Steller's sea cow (*Hydrodamalis gigas*).

observed them in 1741 during expeditions into the North Pacific led by Vitus Bering. They were hunted to extinction for meat to provision fur hunters about 27 years after their discovery.

Dugongs The dugong (*Dugong dugon*) (**Figure 2**) is the only living member of the family Dugongidae, and is placed in the subfamily Dugonginae. There are no recognized subspecies. Dugongs occur in coastal waters in limited areas of the Indian and western Pacific Oceans. They are currently known from the island of Malagasy and off the east coast of Africa from Mozambique northward to the Red Sea and Persian Gulf; along the Indian subcontinent; off south-east Asia through southern China north to the island of Okinawa in Japan; and through the Phillipines, Malaysia, Indonesia, New Guinea, and most of northern Australia from Shark Bay in Western Australia to Moreton Bay in southern Queensland. Dugongs also occur in very low numbers around the Micronesian islands of Palau. Dugong populations are disjunct and in most areas depleted. Australia provides the major exception, and harbors most of the world's remaining dugongs. One conservative estimate suggests that 85 000 dugongs occur in Australian waters. Dugongs are classified as endangered under the US Endangered Species Act.

West Indian manatees There are two subspecies of West Indian manatees (**Figure 3**): the Florida manatee (*T. manatus latirostris*) (**Figure 4**) of the south-eastern USA, and the Antillean manatee (*T. manatus manatus*) of the Caribbean, Central and South America. West Indian manatees are classified as endangered under the US Endangered Species Act. The Florida subspecies is found year-round in nearshore waters, bays, estuaries, and large rivers of Florida, with summer movements into other states bordering the Gulf of Mexico and Atlantic Ocean. Excursions as far north as Rhode Island have been documented. Winter stragglers occur outside of Florida, sometimes dying from cold exposure. It has not been possible to obtain rigorous population estimates, but it has been estimated that there are between 2500 and 3000 manatees in Florida.

Antillean manatee populations are found in coastal areas and large rivers around the Greater Antilles (Hispaniola, Cuba, Jamaica, and Puerto Rico), the east coast of Mexico, and coastal central America through the Lake Maracaibo region in western Venezuela. There do not appear to be resident manatees along the steep, high-energy Caribbean coastline of Venezuela, but populations are found along the Atlantic coast and far inland in large rivers from eastern Venezuela southward to below the mouth of the Amazon near Recife in Brazil. They have not been documented in the lower Amazon.

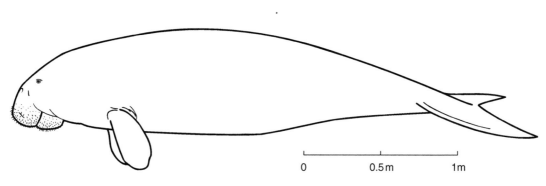

Figure 2 Dugong (*Dugong dugon*).

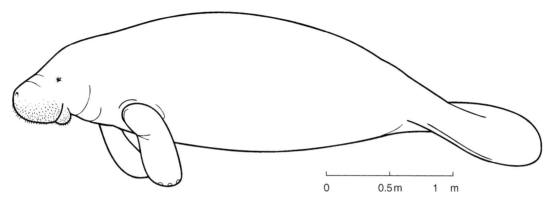

Figure 3 West Indian manatee (*Trichechus manatus*).

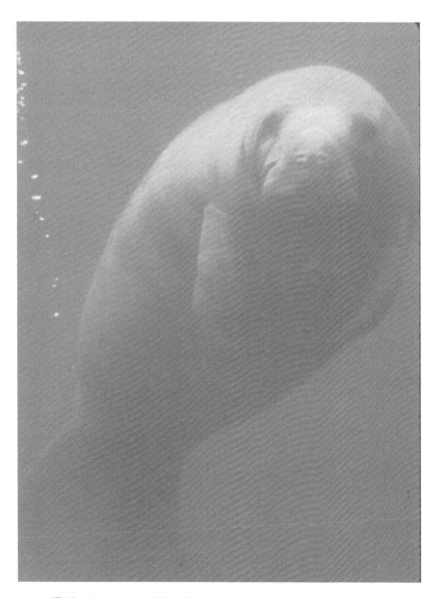

Figure 4 Florida manatee (*Trichechus manatus latirostris*).

Estimates of Antillean manatee population sizes are unavailable, but they are thought to have declined throughout their range.

West African manatees West African manatees (*Trichechus senegalensis*) (**Figure 5**) are found in Atlantic coastal waters of Africa from Angola in the south to Senegal and Mauritania in the north. They extend far inland in large rivers such as the Senegal, Gambia and Niger, into landlocked and desert countries such as Mali and Chad. There are no recognized subspecies. There are no rigorous data on populations, but they have probably suffered widespread declines due to hunting. West African manatees are classified as threatened under the US Endangered Species Act.

Amazonian manatees There are no recognized subspecies of Amazonian manatees (*Trichechus inunguis*) (**Figure 6**). This species is found in the Amazon River system of Brazil, as far inland as upper tributaries in Peru, Ecuador, and Colombia. Areas occupied include seasonally inundated forests. They apparently do not overlap with West Indian manatees near the mouth of the Amazon, and do not occur outside of fresh water. Amazonian manatee populations have declined this century, but there are no firm estimates of numbers. They are classified as endangered under the US Endangered Species Act.

Morphology and Physiology

Adult dugongs range to 3.4 m in length and 420 kg in mass. Florida manatees are larger, ranging to 3.9 m and up to 1655 kg. Morphometric data on West African manatees and Antillean manatees are limited, but they are similar or slightly smaller in size than Florida manatees. Amazonian manatees are the smallest trichechids, ranging up to 2.8 m in length and 480 kg in mass. Steller's sea cow was the largest of all known sirenians, reaching a length of about 9–10 m, and a body mass estimated as high as 10 metric tonnes. The earliest sirenians were about the size of pigs, and had legs, narrow snouts, and the bulky (pachyostotic), dense (osteosclerotic) bones lacking marrow cavities also typical of dugongs and manatees. The heavy skeleton serves as ballast to keep the animals submerged in shallow water. Adaptations for an existence as fully aquatic herbivores also led to rapid loss of hind limbs and development of a broadened down-turned snout. The forelimbs became paddle-like flippers that were positioned relatively close to the head through shortening of the neck, allowing great leverage for steering. Forelimbs are also used to grasp and manipulate

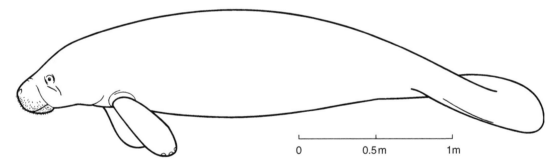

Figure 5 West African manatee (*Trichechus senegalensis*).

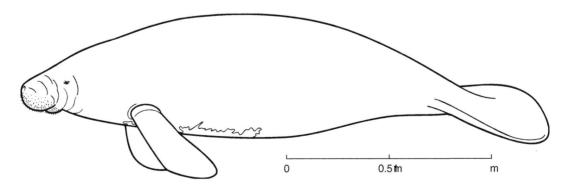

Figure 6 Amazonian manatee (*Trichechus inunguis*).

objects, including food plants, and to 'walk' along the bottom. The flippers of West African and West Indian manatees have nails at their distal ends, but nails are not present in Amazonian manatees or dugongs. Forward propulsion is attained by vertical strokes of the horizontal tail, which is spatulate in manatees, but deeply notched like the flukes of whales in dugongs and Steller's sea cow. Horizontal orientation is important for feeding on plants, and is facilitated by the long, horizontally oriented, and unilobular lungs. The lungs are separated from the other internal organs by a horizontal diaphragm.

Manatees lack notable sexual dimorphism, although female Florida manatees attain larger sizes than males and the vestigial pelvic bones differ in size and shape. Mammary glands of manatees and dugongs consist of a single teat located in each axilla. Incisors erupt in male dugongs (and occasionally in old females) and take the form of small tusks used in mating-related behavior. The dorsal surfaces of dugongs often bear parallel scar marks left from wounds inflicted by tusks of males. The Amazonian manatee has a very dark gray-black appearance and many have white ventral patches. West Indian and West African manatees and dugongs are gray in color, but may also appear brown. Ventral patches are rare and there is little counter-shading in these three species. Hairs are sparsely distributed along the sides and dorsum, and are especially dense and sensitive on the muzzle. These serve a tactile function, which may be particularly important in turbid water and at night. Hairs inside the upper muzzle and lower lip pad are modified into stiff bristles that together with unique facial muscle arrangements form the only prehensile vibrissae known in mammals (**Figure 7**). Steller's sea cow had a small head in relation to the body, and the skin was described as bark-like, dark brown, and in some instances flecked or streaked with white. The forelimbs were reduced to stumps for locomotion along rocky shorelines. They had no teeth and instead relied on horny plates in the mouth to crush kelp. Dugongs have a strongly downwardly deflected snout and are obligate bottom feeders. Dugongs produce a total of six cheek (molariform) teeth in each quadrant of the jaw but also have tough, horny plates in the roof of the mouth. These plates play a significant role in mastication. Dugong teeth are resorbed anteriorly and fall out as they wear, except that the last two molars in each quadrant have persistent pulps and continue to grow throughout life. Manatee cheek teeth, in contrast, continually erupt at the rear of the jaw and move forward as they wear through resorption and reworking of bone. There are no fixed numbers of teeth in manatees, and they erupt throughout life.

Compared with most terrestrial and marine mammals of the same body size, sirenians have unusually low metabolic rates. Metabolic rates of Amazonian manatees are approximately one-third those expected based on measurements of other mammals of similar size. Florida manatee metabolic rates are 15–22% of rates predicted for mammals of their size. They are also poorly insulated, and cannot maintain positive energy balance in cool water, restricting distribution to tropical and subtropical areas. The large size of the Steller's sea cow was in part an adaptation to maintain thermal inertia in the cool North Pacific. Low metabolic rates in the Sirenia can be viewed as adaptations to a diet of aquatic vegetation, which is low in nutritional quality compared with the foods of other marine mammals (such as fish, krill, or squid). Sirenians are hind-gut digesters that may consume up to 10% of body weight per day. The stomach includes a specialized 'cardiac' gland in which special secretory cells are concentrated to avoid abrasion from coarse vegetation. The large intestine is up to 25 m long in dugongs, and passage time of ingesta may take 5–7 days.

Amazonian manatees have an unusual capacity to persist for long periods without food. When food is readily available during the flood season large quantities of fat are deposited, and when individuals are left in isolated pools with receding waters of the dry season they subsist by drawing on this stored energy. Fat deposits coupled with a low metabolic rate may allow Amazonian manatees to go without feeding for up to 7 months. Brains of adult sirenians are very small for mammals of their body size. This may be due to a combination of factors, including low metabolic rates and natural selection for lengthy postnatal growth in body size. Brains also lack marked cerebral convolutions. However, other aspects of neuroanatomy (elaboration of cortical centers and cerebral cytoarchitecture) suggest that sirenians are fully suited for complex behavior. Manatees in captivity can learn a variety of conditioned tasks.

Behavior and Social Organization

Manatees lack strong circadian (24 h) rhythms in activity, although in the more temperate climate of Florida, winter activity can depend on diel changes in water temperature. This lack of circadian activity correlates with the absence of a pineal gland. Seasonal migrations occur within Florida to avoid cool water in winter. Some of these are local movements to constant temperature artesian springs, whereas others span one-way distances of 500 km or more

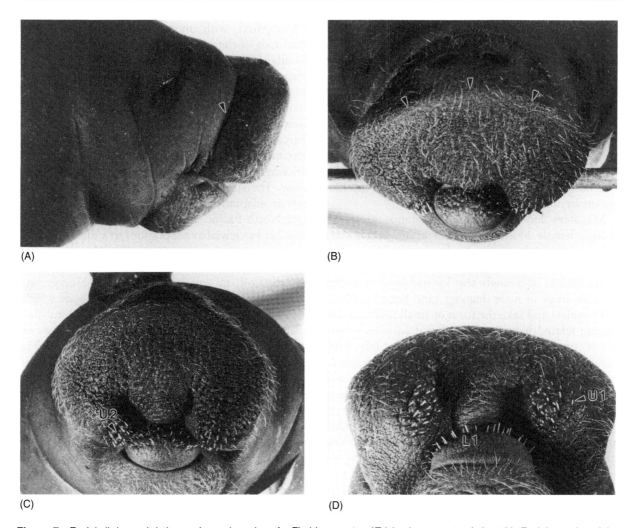

(A)   (B)

(C)   (D)

Figure 7 Facial vibrissae, bristles, and muzzle region of a Florida manatee (*Trichechus manatus latirostris*). Facial muscles of the oral disk and bristles combine to form the only prehensile bristles in mammals. (A) Arrow shows postnasal crease which bounds the supradisk of the prehensile muzzle-bristle apparatus. (B) Arrows show the orofacial ridge of the oral disk. (C) The oral disk is seen face forward, with U2 denoting one of the grasping bristle groups. (D) Bristles are everted at U1 and U2 and on the lower lip pad (L1). (Photographs reproduced with permission of *Marine Mammal Science* and Dr Roger Reep.)

along a north–south gradient. Florida manatees can travel at rates of 50 km per day. Their seasonal destinations can consist of core areas used annually. The methods by which sirenians orient and navigate accurately and directly between destinations through murky waters is unknown. They have small eyes, no external ear, and minute ear openings.

Mating behavior in Florida manatees consists of groups of up to about 20 males actively pursuing females in estrus. The pursuits can last up to 2 weeks, cover distances up to 160 km, and involve vigorous jostling and chasing. Copulations with more than one male have been observed. Dugongs show greater variability in mating systems, and during violent encounters males will inflict cuts with their tusks. In eastern Australia dugongs may exhibit mating behavior similar to that of Florida manatees. In western Australia, however, dugongs also form leks in which single males set up small territories that are visited by individual females. Males advertise their presence on these leks with complex audible underwater vocalizations. Florida manatees also produce audible underwater sounds, but these vocalizations function more as contact calls, may signal simple motivational states, and can serve in individual recognition. Vocal communication is most pronounced between females and calves. West African and Amazonian manatees also produce underwater communication sounds, but these have been less well studied. No stable social organization has been observed in sirenians, other than the long bond between females and calves during lactation. Manatees appear to be primarily solitary but form small transient groups of about 1–20. They may also aggregate in larger numbers around concentrated resources such as freshwater seeps and winter thermal refugia in Florida. Dugongs can aggregate in herds of a few hundred. Herding behavior may have advantages in cultivation grazing, in which feeding activities keep seagrass beds in stages of succession that favor certain food plants.

Ecology and Population Biology

Sirenians are usually found in shallow waters, as aquatic plants do not grow at significant depth where light is restricted. Dugongs feed exclusively on marine angiosperms, the seagrasses of the families Potamogetonaceae and Hydrocharitaceae. The historic distribution of dugongs coincides with the distribution of these food plants. Around tropical Australia, dugongs can be found wherever adequate seagrass beds occur, including distances as far as 60 km offshore and waters to 37 m deep. More delicate, sparsely dispersed deep-water species of seagrasses are fed upon under the latter conditions. Manatees feed on a much wider variety of plants than dugongs, including seagrasses, overhanging mangrove leaves, true grasses along banks and in floating mats, and various rooted, submerged, and floating plants. Predation on sirenians has only rarely been observed. There are uncommon reports of sharks preying on manatees and of the presence of shark bite marks on surviving manatees, but manatees do not typically occur in regions occupied by large sharks. Uncommon reports also exist of predation or possible predation by jaguars on Amazonian manatees, and by sharks, killer whales, and crocodiles on dugongs.

Florida manatees and dugongs have life history characteristics that allow only modest population growth rates, and these characteristics are probably similar in Amazonian and West African manatees. Florida manatees can live 60 years. Litter size is one (with twin births occurring in <2% of pregnancies) after an imprecisely known gestation estimated at about 1 year or slightly longer. Age at first reproduction for females is 3–5 years. Calves suckle for variable periods of 1–2 years, and adult females give birth at about 2.5 year intervals. Survival rates for Florida manatee calves and subadults are not well known. Adult survival, however, has the greatest impact on manatee population growth rates. If adult annual survival is 96%, manatee populations with the healthiest observed reproductive rates can increase at about 7% per year. Growth rates decline by about 1% for every 1% decrease in adult survival. Even when reproductive output is at its maximum, populations cannot remain stable with less than about 90% adult survival. Dugongs have even slower growth rates than manatees. The oldest age attained by adult female dugongs from Australia was 73 years, litter size is one (twins have not been reliably documented) after a gestation period that is probably similar to that of the Florida manatee, and age at first reproduction for females may be 9–10 years or older. The period of lactation is at least 1.5 years, and adult females give birth on a schedule of 3–7 years. Data to support calculations of dugong survival rates are not available, but modeling of life history traits suggest that under best observed reproduction, populations cannot grow by more than about 5–6% annually.

Conservation and Interactions with Humans

Low population growth rates make sirenians vulnerable to modern agents of mortality. Intensifying human activities in coastal areas produce additional sources of death, injury, and habitat change (**Figure 8**). Throughout their recent history, humans have hunted sirenians for meat and fat. People in many indigenous tropical cultures use oil and powders from bones as folk remedies for numerous ailments. Hides have been used for leather, whips, shields for warfare, and even as machine belting during shortages of other materials in World War II. Numerous cultures ascribe magical powers to manatee and dugong bones and body parts. Jewelry and intricate carvings are also made from the dense bones. Ingenious means were employed to hunt sirenians, including the use of box traps in parts of West Africa (**Figure 9**). In most indigenous tropical cultures through the mid-1900s, however, sirenians were hunted principally by hand with harpoons and spear points, and were recognized to be elusive and difficult quarry. Human populations were more sparse, and typically only a few people in any region acquired skills needed to hunt sirenians. Mortality under such conditions may have been sustainable. With the advent of firearms, motors for boats and canoes, and burgeoning numbers of people, overexploitation has occurred. Sirenians have been hunted as a source of bush meat at frontier markets as well as commercially. In Brazil manatee meat was legally sold in processed form, and during peak exploitation in the 1950s as many as 7000 per year were killed for market. In addition, growth in artesanal fisheries has introduced many inexpensive synthetic gill-nets throughout areas used by sirenians. Death due to incidental entanglement has become an additional and significant mortality factor

Figure 8   A Florida manatee showing massive wounds inflicted by a boat propeller. Wounds occur on the dorsum. The nostrils are in the lower center of the photograph, breaking the surface slightly as the animal rises to breathe. (Photograph courtesy of Sara Shapiro, Staff Biologist, Florida Fish and Wildlife Conservation Commission.)

globally. Gill-net commercial fisheries have been excluded from several areas in Australia that are important for dugongs, but few efforts to manage gillnetting for sirenian protection have been instituted elsewhere. Nets deployed to protect bathers from sharks on the coast of Queensland resulted in the deaths of hundreds of dugongs from the 1960s to 1996. Many of these nets have since been replaced with drum lines and deaths have dropped. The vulnerability of sirenians to overexploitation was most markedly illustrated in the case of the Steller's sea cow. Unable to submerge, they were easy quarry as

provisions of meat for fur traders on their long voyages into the North Pacific. This caused extirpation of this sea cow within 27 years of discovery by western science. Today sirenians are legally protected in nearly every country in which they occur, as well as by international treaties and agreements. However, few nations provide active law enforcement or enduring conservation programs for manatees.

Overexploitation of manatees by hunting is no longer an issue in Florida. Instead, extensive coastal development accompanied by technological

Figure 9 A box trap used to capture manatees in West Africa. Walls are made of poles bound by fibrous leaves, with a door at one end that is propped open. Pieces of cassava tubers are left around and in the box as bait. After a manatee swims inside to feed on the starchy cassava, it will jar loose the stick propping the door, which descends and traps the manatee alive.

advances in boating and water diversions have created major new sources of mortality. Overall human-related mortality accounts for about 30% of manatee deaths in Florida each year. Collisions with watercraft account for about 24% of all manatee mortality and have increased more rapidly than overall mortality in recent years (**Figure 10**). About 4% of known manatee deaths are caused by crushing or drowning in flood control structures and canal locks. Other human-related causes of manatee mortality such as entanglement in fishing gear represent approximately 3% of the total.

The east and south-west coasts of Florida have lost between 30 and 60% of former seagrass habitat due to development, dredging, filling, and scarring of seagrass beds caused by motor boats. Some human activities, however, such as dredging of canals and the construction of power-generating plants (that provide warm water for refuge at northern limits of the manatee's winter range), may have opened up previously unavailable habitat. Destruction of quality seagrass habitat through activities that disturb bottoms (*e.g.* dredging and mining) or increase sedimentation and turbidity can also impact dugongs. Loss of seagrasses due to a cyclone

and flooding in Hervey Bay, Queensland, was accompanied by numerous dugong deaths, and a reduction in the estimated population in the bay from about 1500–2000 to <100 dugongs. Direct mortality, reduced immune function, or impaired reproduction of sirenians in relation to environmental contaminants have not been observed. Low positions in marine food webs reduce exposure to many of the persistent organic contaminants that build up in tissues of other marine mammals, but sirenians may be more likely to be exposed to toxic elements that accumulate in sediments and are taken up by plants.

Outlook for the Future

In 1999, at least 268 manatees died in Florida and at least 82 (30%) of these were killed by watercraft. The total number of deaths in 1999 was the highest recorded, except for 1996 when a large red tide event increased manatee deaths in south-west Florida. Watercraft-related mortality is increasing more rapidly than overall mortality. However, aerial survey counts and monitoring of identifiable individual manatees indicate that numbers in particular regions

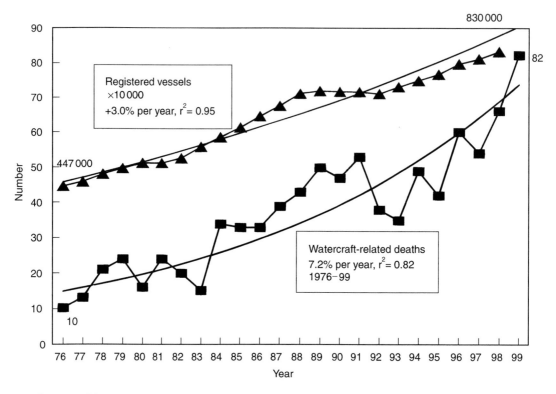

Figure 10 Patterns of increase in the annual numbers of boats registered in Florida and annual totals of boat-killed Florida manatees, 1976–99. Numbers of registered vessels have increased from 447 000 to 830 000. Numbers of boat-killed manatee carcasses recovered have increased from 10 annually to over 80.

of Florida have been slowly increasing. Some argue that there are more manatees now than ever before, so we should expect a higher proportion to die each year as the population grows. Alternatively, population models suggest that the current level of manatee mortality and the low survival of adults, will result or has already resulted in a manatee population decline. Because of inherent variation in counts due to survey conditions, several years of survey data are required to determine conclusively whether a change in population trend has actually occurred. However, the current human population in Florida is about 15 million people, a doubling in the past 25 years. By 2025 the population is expected to reach 20 million. Florida's waterways are a major source of recreation and revenue. Increasing boat traffic and coastal development resulting from Florida's growth will probably accelerate human-related manatee mortality. Other factors that complicate the future of manatees are the loss of warm-water refugia as spring flows decrease from increased ground-water extraction and drought conditions, and the phasing out of coastal power-generating plants because of industry deregulation and competition as more efficient inland plants come on-line. Florida manatees are thus faced with a very uncertain future. Proper planning, good information,

and effective management are critical to the long-term survival of this species in Florida, where societal concern, dedicated financial resources, and formal protection for manatees are among the strongest in the world. West Indian manatees elsewhere in the Caribbean and South America probably have an even less optimistic future. This is because of past suppression or local elimination of populations from hunting and net entanglement, as well as habitat loss and ever-increasing pressure on resources from human populations. This is particularly true for manatees around island countries. On the mainland the outlook is more mixed. In some areas hunting pressure has declined, and conservation efforts have been enhanced. West Indian manatees are likely to persist in very remote and undeveloped areas of South America as long as such conditions prevail.

The outlook for dugongs in Australia is more guardedly optimistic, but recovery of populations and indeed the continued existence of dugongs in most of the rest of their range are much less likely. This is due to severe reductions from hunting and fishing activities in the past, and continued degradation of coastal habitat as human populations burgeon in these areas. Although the distribution of Amazonian manatees has remained similar to historical records, populations have been reduced in many areas of former abundance. Illegal capture for commercial sale of meat and incidental take in fishing nets continue, habitat degradation is increasing, and there is concern about heavy metal contamination of aquatic food plants from mining.

In West Africa the manatee's range has not changed appreciably from historical accounts. However, it is believed that numbers have been reduced due to illegal hunting. Several countries are particularly important for manatees, including Senegal, The Gambia, Guinea-Bissau, Sierra Leone, Ivory Coast, Nigeria, Cameroon, Gabon, and Angola. In these countries manatees are not uncommon, probably because there are extensive areas of optimal habitat located in relatively remote areas where hunting pressure is reduced. Hunting is the primary threat to manatees in Africa. For example, in Guinea-Bissau a single fisherman had sternums from over 40 manatees in his hut. Although manatees are protected throughout their range, the lack of enforcement of hunting laws and the need for supplemental protein in many areas contributes to hunting pressure. Damming of rivers poses an emerging threat, both directly and indirectly. Crushing in dam structures has been reported from Senegal, Ghana and Nigeria (and also in Florida). Dams on many rivers in the manatee's range may cut off needed seasonal migrations. Killing of manatees as they aggregate at freshwater overflows of dams has also been reported. In recent years, there has been increasing trade in West African manatees for commercial display facilities. Accelerating habitat destruction and cutting of mangroves for construction and firewood will have negative effects. There is considerable cause for concern for the manatee's future in several regions of Africa unless hunting is reduced. However, there has been increasing interest in West African manatee conservation, and several countries are moving towards increased protection involving coastal sanctuaries and law enforcement. To prevent future loss of manatees from all but a few remote and protected areas of West Africa, law enforcement and protection must become regional in scale, improvements must be made in economic conditions that contribute to hunting, and modifications will be needed on dams and other structures that kill manatees.

See also

Marine Mammal Evolution and Taxonomy. Marine Mammal Overview.

Further Reading

Bryden M, Marsh H, and Shaughnessy P (1998) *Dugongs, Whales, Dolphins and Seals*. St. Leonards, NSW, Australia: Allen Unwin.

Domning DP (1996) Bibliography and index of the Sirenia and Desmostylia. *Smithsonian Contributions to Paleobiology* 80: 1–611.

Domning DP (1999) Fossils explained 24: sirenians (seacows). *Geology Today* (March–April): 75–79.

Hartman DS (1979) *Ecology and Behavior of the Manatee (Trichechus manatus) in Florida*. American Society of Mammalogists Special Publication 5: 1–153.

O'Shea TJ (1994) Manatees. *Scientific American* 271(1): 66–72.

O'Shea TJ, Ackerman BB and Percival HF (eds) (1995) *Population Biology of the Florida Manatee*. US Department of Interior, National Biological Service Information and Technology Report 1.

Reynolds JE III and Odell DK (1991) *Manatees and Dugongs*. New York: Facts on File, Inc.

Rosas FCW (1994) Biology, conservation and status of the Amazonian manatee. *Trichechus inunguis. Mammal Review* 24: 49–59.

SEA OTTERS

J. L. Bodkin, US Geological Survey, Alaska, USA

Introduction

A century ago sea otters were on the verge of extinction. Reduced from several hundred thousand individuals, the cause of their decline was simply the human harvest of what is arguably the finest fur in the animal kingdom. They persisted only because they became so rare that, despite intensive efforts, they could no longer be found. Probably less than a few dozen individuals remained in each of 13 remote populations scattered between California and Russia. By about 1950 it was clear that several of those isolated populations were recovering. Today more than 100 000 sea otters occur throughout much of their historic range, between Baja California, Mexico, and Japan, although suitable unoccupied habitat remains (**Figure 1**). As previous habitat is reoccupied, either through natural dispersal or translocation, sea otter populations and the coastal marine ecosystem they occupy, can be studied before, during, and following population recovery. Because of concern for the conservation of the species, as well as conflicts between humans and sea otters over coastal marine resources, nearly continuous research programs studying this process have been supported during the past 50 years. Because sea otters occur near coastlines, bring their prey to the surface to consume, and can be easily observed and handled, they may be more amenable to study than most marine mammals. Both the accessibility of the species and the serendipitous 'experimental' situation provided by their widespread removal and subsequent recovery have provided a depth of understanding into the ecological role of sea otters in coastal communities and the response of sea otters to changing ecological conditions that may be unprecedented among the large mammals.

Sea otters are unique, both among the other otters, to which they are most closely related, and the other marine mammals, with which they share the oceans as a common habitat. They are the only fully marine species of Lutrinae, or otter subfamily of mustelids, having evolved relatively recently from their terrestrial and freshwater ancestors. Thus, the natural history of sea otters is a result of both their phylogenetic history and the adaptations required by a life at sea. As a result of these sometimes opposing pressures, sea otters display characteristics of their recent ancestors, the other otters, and also exhibit attributes that result from those adaptations that are common among the other mammals of the sea.

Description

The sea otter is well known as the smallest of the marine mammals but also the largest of the Lutrinae. The sexes are moderately dimorphic with males attaining maximum weights up to 45 kg and 158 cm total lengths. Adult females attain weights up to 36 kg and total lengths to 140 cm. Newborn pups weigh about 2 kg and are about 60 cm in length. Dentition is highly modified with broad, flattened molars for crushing hard-shelled invertebrates, rounded, blunt canines for puncturing and prying prey, and spade-shaped protruding incisors for scraping tissues out of shelled prey (**Figure 2**). The body is elongated and the tail is relatively short (less than one-third of total length) and flattened compared to other otters. The fore legs are short and powerful with sensitive paws used to locate and manipulate prey and in grooming, traits held in common with the clawless otters (*Aonyx* spp. or crab-eating otters) also known to forage principally on invertebrates. The fore legs are not used in aquatic locomotion. The extruding claws present in sea otters are unique among the mustelids and are useful in digging for prey in soft sediment habitats. The hind feet are enlarged and flattened relative to other otters and are the primary source of underwater locomotion. The external ear is small and similar to the ear of the otariid pinniped. Fur color ranges from brown to nearly black and a general lightening from the head downward may occur with aging. The pelage is composed of bundles made up of a single guard hair and numerous underfur hairs, at a ratio of about $1:70$. Hair density ranges to $165\,000\,\text{cm}^{-2}$ and is the densest hair among mammals. Most marine mammals insulate with a blubber layer, but sea otters are the only one to rely exclusively on an air layer trapped in the fur for insulation. Although air is a superior insulator, it requires constant grooming to maintain insulating quality. It is this means of insulation (fur) that made the sea otter so valuable to humans and makes it susceptible to oil spills and other similar contaminants, that can reduce insulation. Additionally, fur and the air it contains, allow the sea otter to be

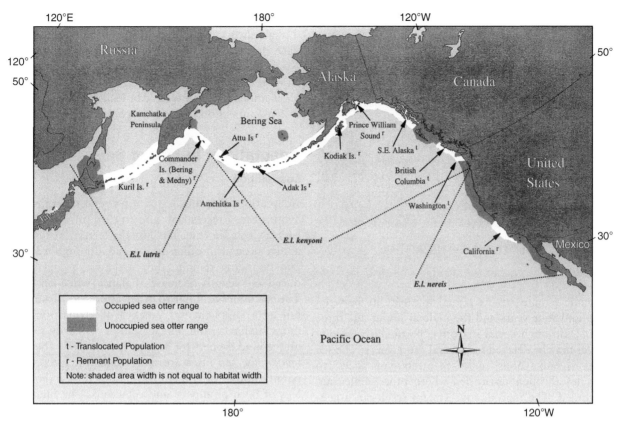

Figure 1 Occupied and unoccupied sea otter habitat in the North Pacific Ocean (2000).

positively buoyant, a trait shared with some marine mammals but with none of the other otters.

Range and Habitat

Primitive sea otters (the extinct *Enhydridon* and *Enhydritherium*) are recognized from Africa, Eurasia, and North America. Specimens of *Enhydra* date to the late pliocene/early Pleistocene about 1–3 million years ago. The modern sea otter occurs only in the north Pacific ocean, from central Baja California, Mexico to the northern islands of Japan (**Figure 1**). The northern distribution is limited by the southern extent of winter sea ice that can limit access to foraging habitat. Southern range limits are less well understood, but are likely to be related to increasing water temperatures and reduced productivity at lower latitudes and constraints imposed by the dense fur the sea otter uses to retain heat.

Although sea otters occupy and utilize all coastal marine habitats, from protected bays and estuaries to exposed outer coasts and offshore islands, their habitat requirements are defined by their ability to dive to the seafloor to forage. Although they may haul out on intertidal or supratidal shores, their habitat is limited landward by the sea/land interface and no aspect of their life history requires leaving the ocean. The seaward limit to their distribution is defined by their diving ability and is approximated by the 100 m depth contour. Although sea otters may be found at the surface in water deeper than 100 m, either resting or swimming, they must maintain relatively frequent access to the seafloor to obtain food. Sea otters forage in diverse bottom types, from fine mud and sand to rocky reefs. In soft sediment communities they prey largely on burrowing infauna, whereas in consolidated rocky habitats epifauna are the primary prey.

Life History

Following a gestation of about six months, the female sea otter gives birth to a single, relatively large pup. In contrast, all other otters have multiple offspring, whereas all other marine mammals have single offspring. Pupping can occur during any month, but appears to become more seasonal with increasing latitude with most pups born at high latitudes arriving in late spring. The average length of pup dependency is about six months, resulting in an average reproductive interval of about one year. Females breed within a few days of weaning a pup, and thus may be either pregnant or with a dependent young throughout most of their adult life. The

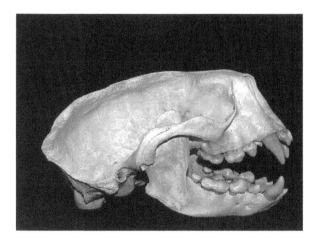

Figure 2   Photo of sea otter skull and dentition.

juvenile female attains sexual maturity at between two and four years and the male at about age three years, although social maturity, particularly among males may be delayed for several more years. The sea otter is polygynous; male sea otters gain access to females through territories where other males are excluded. The reproductive system results in a general segregation of the sexes; most adult females and few territorial males occupy most of the habitat and nonterritorial males reside in dense aggregations. Similar reproductive systems are common among the pinnipeds, whereas most otter species are organized around family units consisting of a mother, her offspring and one or more males.

Survival in sea otters is largely age dependent. Survival of dependent pups is variable ranging from about 0.20 to 0.85, and is likely dependent on maternal experience, food availability, and environmental conditions. Among dependent pups, survival is lowest in the first few weeks after parturition, then increasing and remaining high through dependency. A second period of reduced survival follows weaning and may result in annual mortality rates up to 0.60. Once a sea otter attains adulthood, survival rates are high, around 0.90, but decline later in life.

Maximum longevity is about 20 years in females and 15 years in males. Sources of mortality include a number of predators, most notably the white shark (*Carcharadon charcharias*) and the killer whale (*Orca orcinus*). Bald eagles (*Haliaeetus leucocephalus*) may be a significant cause of very young pup mortality. Terrestrial predators, including wolves (*Canis lupus*), bears (*Ursos arctos*) and wolverine (*Gulo gulo*) may kill sea otters when they come ashore, although such instances are likely rare. Pathological disorders related to enteritis and pneumonia are common among beach-cast carcasses and

may be related to inadequate food resources, although such causes of mortality generally coincide with late winter periods of inclement weather. Nonlethal gastrointestinal parasites are common and lethal infestations are occasionally observed. Among older animals, tooth wear can lead to abscesses and systemic infection, eventually contributing to death.

Adaptations to Life at Sea

Locomotion

Adaptations seen in the sea otter reflect a transition away from a terrestrial, and toward an aquatic existence. The sea otter has lost the terrestrial running ability present in other otters, and although walking and bounding remain, they are awkward. The reduced tail length decreases balance while on land. The sea otter is similar to other otters in utilizing the tail as a supplementary means of propulsion, but more similar to the phocid seal in primarily relying on the hind flippers for aquatic locomotion. The hind feet of the sea otter are more highly adapted compared to other otters, but less modified than the hind legs of both pinnipeds and cetaceans. The hind feet are enlarged through elongation of the digits, flattened and flipper like and provide the primary source of underwater locomotion. While swimming, the extended hind flippers of the sea otter approximate the lunate pattern and undulating movement of the fluke of cetaceans.

The primary method of aquatic locomotion in sea otters while submersed is accomplished by craniocaudal thrusts of the pelvic limbs, including bending of the lumbar, sacral, and caudal regions for increased speed. The sea otter loses swimming efficiency in the resistance and turbulence during the recovery stroke and through spaces between the flippers and tail. Travel velocities over distances less than 3 km are in the range of about $0.5–0.7\,\mathrm{m\,s^{-1}}$, with a maximum of about $2.5\,\mathrm{m\,s^{-1}}$. Estimated sustainable rates of travel over longer distances are in the range of $0.16–1.5\,\mathrm{m\,s^{-1}}$. Rates of travel during foraging dives average about $1.0\,\mathrm{m\,s^{-1}}$.

A unique paddling motion, consisting of alternating vertical thrusts and recovery of the hind flippers is a common means of surface locomotion. The tail is capable of propelling the sea otter slowly, usually during either resting or feeding.

Thermoregulation

Sea otters, similar to all homeotherms, must strike a balance between conserving body heat in cold environments, and dissipating heat when internal production exceeds need. This process is particularly sensitive in the high latitudes where water is cold and heat loss potential is high and when a relatively inflexible insulator such as air is used. The small size of the sea otter magnifies the heat loss problem because of a high surface area to volume ratio. One way to offset high heat loss is through the generation of additional internal heat. The sea otter accomplishes this through a metabolic rate 2.4–3.2 times higher than predicted in a terrestrial mammal of similar size. This elevated metabolic rate requires elevated levels of energy intake. As a result of this increased energy requirement, sea otters consume 20–33% of their body mass per day in food. In northern latitudes, sea otters may haul-out more frequently during the winter as a means to conserve heat.

Although air is a more efficient insulator than blubber (10 mm of sea otter fur is approximately equal to 70 mm of blubber), the fur requires high maintenance costs and does not readily allow heat dissipation that may be required following physical exertion. To maintain the integrity of the fur, up to several hours per day are spent grooming, primarily before and after foraging, but also during foraging and resting activities. To dissipate excess internal heat, and possibly absorb solar radiation the sea otter's hind flippers are highly vascularized and with relatively sparse hair. Heat can be conserved by closing the digits and placing the flippers against the abdomen, or dissipated by expanding the digits and exposing the interdigital webbing to the environment. Because air is compressible, sea otters likely lose insulation while diving, and may undergo unregulated heat loss during deep dives, however the heat loss may be offset to some extent by the elevated metabolic heat produced during diving.

Diet and Diving

Sea otters prey principally on sessile or slow-moving benthic invertebrates in nearshore habitats throughout their range, from protected inshore waters to exposed outer coast habitats. In contrast, most other otters and most pinnipeds and odontocete cetaceans rely largely on a fish-based diet. Although capable of using vision to forage, the primary sensory modality used to locate and acquire prey appears to be tactile, since otters feed in highly turbid water, and at night. Foraging in rocky habitats and kelp forests consists of hunting prey on the substrate or in crevices. Foraging in soft sediment habitats often requires excavating large quantities of sediments to extract infauna such as clams.

Although the number of species preyed on by sea otters exceeds 150, only a few of these predominate in the diet, depending on latitude, habitat type,

season, and length of occupation by sea otters. Generally, otters foraging over rocky substrates and in kelp forests mainly consume decapod crustaceans (primarily species of *Cancer*, *Pugettia*, and *Telmessus*), mollusks (including gastropods, bivalves, and cephalopods) and echinoderms (species of *Strongylocentrotus*). In protected bays with soft sediments, otters mainly consume infaunal bivalves (species of *Saxidomus*, *Protothaca*, *Macoma*, *Mya*, and *Tresus*) whereas along exposed coasts of soft sediments, *Tivela stultorum*, *Siliqua* spp. are common prey. Mussels (species of *Mytilus* and *Modiolis*) are a common prey in most habitats where they occur and may be particularly important in providing nourishment for juvenile sea otters foraging in shallow water. Sea urchins are relatively minor components of the sea otter's diet in Prince William Sound and the Kodiak archipelago, but are a principal component of the diet in the Aleutian Islands. In the Aleutian, Commander and Kuril Islands, a variety of fin fish (including hexagrammids, gaddids, cottids, perciformes, cyclopterids, and scorpaenids) are present in the diet. For unknown reasons, fish are rarely consumed by sea otters in regions to the east of the Aleutian Islands.

Sea otters also exploit normally rare, but episodically abundant prey. Examples include squid (*Loligo* sp.) and pelagic red crabs (*Pleuroncodes planipes*) in California and smooth lumpsuckers (*Aptocyclus ventricosus*) in the Aleutian Islands. The presence of abundant episodic prey may allow temporary release from food limitation and result in increased survival. Sea otters, on occasion, attack and consume sea birds, including teal, scoters, loons, gulls, grebes, and cormorants, although this behavior is apparently rare.

The sea otter is the most proficient diver among the otters, but one of the least proficient among the other marine mammals. Diving occurs during foraging but also while traveling, grooming, and during social interactions. Foraging dives are characterized by rapid ascent and descent rates with relatively long bottom times. Traveling dives are characterized by slow ascent and descent rates and relatively short times at depth (**Figure 3**). The maximum recorded dive depth in the sea otter is 101 m and most diving is to depths less than 60 m. Mean and maximum dive depths appear to vary between sexes and among individuals. Generally, males appear to have average dive depths greater than females, although some females regularly dive to depths greater than some males. Maximum reported dive duration is 260 s. Average dive durations are in the range 60–120 s and differ among areas and are likely influenced by both dive depth and prey availability. Foraging success rates are generally high, from 0.70 to > 0.90

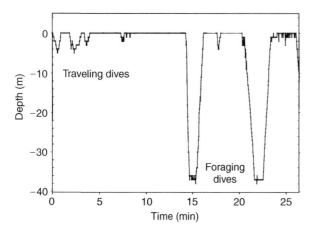

Figure 3 Examples of foraging and traveling sea otter dive profiles. Ascent/descent rates and percentage of dive time at bottom vary between dive types.

although the caloric return per unit effort probably varies relative to prey availability.

Reproduction

In contrast to the pinnipeds and polar bears, as well as the other otters, sea otter reproduction has no obligatory terrestrial component. In this regard, the sea otter is more similar to the cetacea or sirenia. Sea otters throughout their range are capable of reproducing throughout the year. Alternatively, most pinnipeds and cetaceans display strong seasonal, synchronous reproductive cycles. Likely the most conspicuous reproductive adaptation that separates sea otters from the other lutrines and aligns them with the other marine mammals is reduction in litter size. All other lutrines routinely give birth to and successfully raise multiple offspring from a single litter. Although sea otters infrequently conceive twin fetuses, there are no known records of a mother successfully raising more than one pup. This trait of single large offspring appears to be strongly selected for in marine mammals (**Figure 4**) with the exception of polar bears.

The Role of Sea Otters in Structuring Coastal Marine Communities

A keystone predator is one whose effects on community structure and function are disproportionately large, relative to their own abundance. Sea otters provide one of the best documented examples of the keystone predator concept in the ecological literature. Although other factors can influence rocky nearshore marine communities, particularly when sea otters are absent, the generality of the sea otter

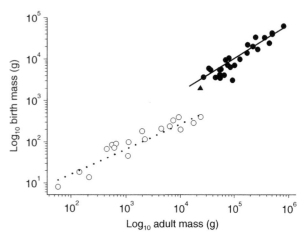

Figure 4 Birth mass versus adult female mass of sea otters relative to the other mustelids and marine mammals. ●, Pinnipeds; ○, Mustelids; ▲, *Enhydra Lutris*; —, Pinniped regression, ..., mustelid regression.

effect is supported with empirical data from many sites throughout the North Pacific rim.

There are several factors that have contributed to our understanding of the role sea otters play in structuring coastal communities. First, sea otters are distributed along a narrow band of relatively shallow habitat where they forage almost exclusively on generally large and conspicuous benthic invertebrates. Because water depths are shallow, scientists, aided with scuba have been able to rigorously characterize and quantify coastal marine communities where sea otters occur. Because sea otters bring their prey to the surface prior to consumption, the type, number, and sizes of prey they consume are straightforward to determine. Furthermore, many of the preferred prey of sea otters are either ecologically (e.g., grazers) or economically (e.g., support fisheries) important, prompting a long-standing interest in the sea otter and its effect on communities. And finally, because sea otters were removed from most of their range prior to 1900, but have recovered most of that range during the last half of the twentieth century, we have been afforded the opportunity to repeatedly observe coastal communities both with and without sea otters present, as well as before and after sea otters recolonize an area. This combination of factors may be nearly impossible to duplicate in other communities that support large carnivores.

Kelp Forests

The majority of studies on sea otter ecosystem effects has taken place in kelp forest communities that occur over rocky reefs and has led to a generalized sea otter paradigm. This scenario describes the rocky reef community in the absence of sea otters as being dominated by the effects of the herbivorous sea urchins (*Strongylocentrotus* spp.) that can effectively eliminate kelp populations, resulting in what have come to be termed 'urchin barrens' (**Figure 5A**). The urchin barren is characterized by low species diversity, low algal biomass, and abundant and large urchin populations. In areas of the Aleutian Islands dominated by urchin barrens the primary source of organic carbon in nearshore food webs results from fixation of carbon by phytoplankton or microalgae.

Sea urchins may be the most preferred of the many species preyed on by sea otters. As a result, when sea otters recolonize urchin-dominated habitat, urchin populations are soon reduced in abundance with few if any large individuals remaining. Following reduced urchin abundance, herbivory on kelps by urchins is reduced. In response to reduced herbivory, kelp, and other algal populations increase in

(A)

(B)

Figure 5 Photos of (A) urchin barrens and (B) kelp forest.

abundance often forming a multiple-layer forest culminating in a surface canopy forming kelp forest (**Figure 5**). The kelp forest in turn supports a high diversity of associated taxa, including gastropods, crustaceans, and fishes. The effects of the kelp forests may extend to birds and other mammals that may benefit from kelp forest-associated prey populations. In areas of the Aleutian Islands dominated by kelp forests, the primary source of organic carbon in nearshore food webs results from fixation of carbon by kelps, or macroalgae.

Soft-Sediment Subtidal

Studies throughout the North Pacific indicate that sea otters can have predictable and measurable effects on invertebrate populations in sedimentary communities. The pattern typically involves a predation-related reduction in prey abundance and a shift in the size-class composition toward smaller individuals, similar to patterns seen in rocky reefs. This pattern has been observed for several species of bivalve clams and Dungeness crabs (*Cancer magister*). The green sea urchin (*Strongylocentrotus droebachiensis*) can be common and abundant in sedimentary habitats in Alaska, particularly where sea otters have been absent for extended periods of time. In such areas, as sea otters recover prior habitat, the green urchin may be a preferred prey. The effect of sea otter foraging on the green urchin in sedimentary habitats is similar to the otter effect observed on urchins in rocky reefs. Urchin abundance is reduced and surviving individuals are small and may achieve refuge by occupying small interstitial spaces created by larger sediment sizes. Although sea urchins may provide an abundant initial resource during sea otter recolonization, clams appear to be the primary dietary item in soft sediment habitats occupied for extended periods.

Cascading trophic effects of sea otter foraging are less well described in sedimentary habitats, when contrasted to rocky reefs. It is likely that the reduction of urchins has a positive effect on algal productivity, especially where the larger sediment sizes (e.g., rocks and boulders) required to support macroalgae are present. In the process of foraging on clams, sea otters discard large numbers of shells after removing the live prey. These shell remains provide additional hard substrate that can result in increased rates of recruitment of some species such as anemones and kelps.

The effects of otter foraging in sedimentary habitats include disturbance to the community in the form of sediment excavation and the creation of foraging pits. These pits are rapidly occupied by the sea star (*Pycnopodia helianthoides*). *Pycnopodia* densities are higher near the otter pits where they may prey on small clams exposed, but not consumed, by sea otters.

Responses to Changing Ecological Conditions

Contrasts between sea otter populations at recently (below equilibrium) and long-occupied sites (equilibrium) as well as contrasts at sites over time provide evidence of how sea otters can modify life history characteristics in response to changing population densities and the resulting changes in ecological conditions. The data summarized below were collected at sites sampled over time, both during and following recolonization, and also at different locations where sea otters were either present for long or short periods of time.

Diet

In populations colonizing unoccupied habitat, sea otters feed largely on the most abundant and energetically profitable prey. Preferred prey species likely differ between areas but include the largest individuals of taxa such as gastropods, bivalves, echinoids, and crustaceans. Over time, as populations approach carrying capacity and the availability of unoccupied habitat diminishes, preferred prey of the largest sizes become scarce and smaller individuals

and less preferred prey are consumed with increasing frequency. Several consequences of this pattern of events are evident, and have been repeatedly observed. One is an increase in dietary diversity over time as otter populations recolonize new habitats and grow toward resource limitation. A relatively few species are replaced by more species, of smaller size, and at least in the Aleutian Islands, a new prey type (e.g., fish) may become prevalent in the diet. This in turn may eventually lead to a new and elevated equilibrium density. Another result of declining prey is an increase in the quantity of time spent foraging as equilibrium density is approached. In below-equilibrium populations, sea otters may spend as little as 5 h each day foraging, whereas in equilibrium populations 12 h per day or more may be required. Finally, declining body conditions and total weights have been seen in sea otters as equilibrium density is attained. Mass/length ratios are consistently greater in below-equilibrium populations compared to those at or near equilibrium At Bering Island as the population reached and exceeded carrying capacity over a 10 year period (**Figure 6**), average weights of adult males declined from 32 kg to 25 kg.

Reproduction

Studies of age specific reproductive rates have produced generally consistent results in populations both below and at equilibrium densities. Although a small proportion of females may attain sexual

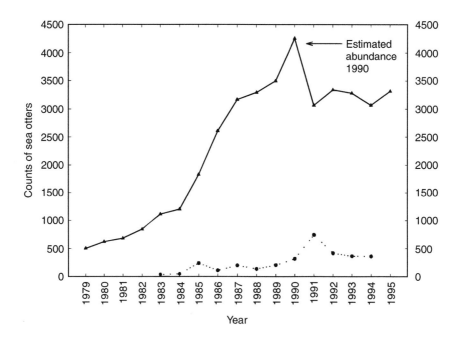

Figure 6 Sea otter abundance (▲, average annual growth 1979–1989 = 0.24) and carcass recovery rates (●) during a period of recolonization and equilibration with prey resources, Bering Island, Russia.

maturity at age two most are mature at age four, regardless of population status. Annual adult female reproductive rates are uniformly high, at around 0.85–0.95, among all populations despite differences in availability of food resources. Under the range of ecological conditions studied, sea otter reproductive output does not appear to play a major role in population regulation, although it seems likely that as ecological or environmental conditions deteriorate, at some point reproductive output should be adversely affected.

Survival

Greater food availability, and the resulting improved body condition can result in significantly higher juvenile sea otter survival rates in below-equilibrium populations. Coupled with high reproductive output, high survival has resulted in sea otter population growth rates in several translocated or recolonizing populations that have approached the theoretical maximum for the species of about 0.24 per year. Declining prey availability does not appear to affect the ages and sexes equally. Survival of dependent pups was nearly twice as high (0.83) in a population below equilibrium, compared to one at equilibrium (0.47). Postweaning survival appears similarly affected, increasing during periods of increasing prey availability. Adult survival appears high and uniform at about 0.90, among both equilibrium and below-equilibrium populations. Survival appears to be greater among females of all age classes, compared to males. At Bering Island in Russia (**Figure 6**) during a year when about 0.28 of the population was recovered as beach cast carcasses, 0.80 of the 742 carcasses were male. Higher rates of male mortality are associated with higher densities of sea otters in male groups and increased competition for food. Thus sex ratios in populations of sea otters are generally skewed toward females. Available data suggest that survival, particularly among juveniles, is the primary life history variable responsible for regulating sea otter populations.

Populations

Trends in sea otter populations today vary widely from rapidly increasing in Canada, Washington and south east Alaska, to stable or changing slightly in Prince William Sound, the Commander Islands, and California, to declining rapidly throughout the entire Aleutian Archipelago. Rapidly increasing population sizes are easily understood by abundant food and space resources and increases should continue to be seen until those resources become limiting. Relatively stable populations can be generally characterized by food limitation and birth rates that approximate death rates. The recent large-scale declines in the Aleutian Archipelago are unprecedented in recent times. Our view of sea otter populations has been largely influenced by events in the past century when food and space were generally unlimited. However, as resources become limiting, it is likely that other mechanisms, such as predation or disease will play increasingly important roles in structuring sea otter populations.

Further Reading

Estes JA (1989) Adaptations for aquatic living in carnivores. In: Gittleman JL (ed.) *Carnivore Behavior Ecology and Evolution*, pp. 242–282. New York: Cornell University Press.

Estes JA and Duggins DO (1995) Sea otters and kelp forests in Alaska: generality and variation in a community ecology paradigm. *Ecological Monographs* 65: 75–100.

Kenyon KW (1969) *The Sea Otter in the Eastern Pacific Ocean. North American Fauna* no. 68. Washington, DC: Bureau of Sport Fisheries and Wildlife, Dept. of Interior.

Kruuk H (1995) *Wild Otters, Predation and Populations*. Oxford: Oxford University Press.

Reynolds JE III and Rommel SE (eds.) (1999) *Biology of Marine Mammals*. Washington and London: Smithsonian Institute Press.

Riedman ML and Estes JA (1990) The sea otter (*Enhydra lutris*): behavior, ecology, and natural history. US Fish and Wildlife Service Biological Report 90(14).

VanBlaricom GR and Estes JA (eds.) (1988) The community ecology of sea otters. *Ecological Studies*, vol. 65. New York: Springer-Verlag.

SEA TURTLES

F. V. Paladino, Indiana-Purdue University at Fort Wayne, Fort Wayne, IN, USA
S. J. Morreale, Cornell University, Ithaca, NY, USA

Introduction

There are seven living species of sea turtles that include six representatives from the family Cheloniidae and one from the family Dermochelyidae. These, along with two other extinct families, Toxochelyidae and Prostegidae, had all evolved by the early Cretaceous, more than 100 Ma (million years ago). Today the cheloniids are represented by the loggerhead turtle (*Caretta caretta*), the green turtle *Chelonia mydas*, the hawksbill turtle (*Eretmochelys imbricata*), the flatback turtle (*Natator depressus*), and two congeneric turtles, the Kemp's ridley turtle (*Lepidochelys kempii*) and the olive ridley turtle *(Lepidochelys olivacea)*. The only remaining member of the dermochelyid family is the largest of all the living turtles, the leatherback turtle (*Dermochelys coriacea*).

Sea turtles evolved from a terrestrial ancestor, and like all reptiles, use lungs to breathe air. Nevertheless, a sea turtle spends virtually its entire life in the water and, not surprisingly, has several extreme adaptations for an aquatic existence. The rear limbs of all sea turtles are relatively short and broadly flattened flippers. In the water, these are used as paddles or rudders to steer the turtle's movements whereas on land they are used to push the turtle forward and to scoop out the cup shaped nesting chamber in the sand (**Figure 1**). The front limbs are highly modified structures that have taken the appearance of a wing. Internally the front limbs have a shortened radius and ulna (forearm bones) and greatly elongated digits that provide the support for the flattened, blade-like wing structure. Functionally the front flippers are nearly identical to bird wings, providing a lift-based propulsion system that is not seen in other turtles.

Other special adaptations help sea turtles live in the marine environment. All marine turtles are well suited for diving, with specialized features in their blood, lungs, and heart that enable them to stay submerged comfortably for periods from 20 min to more than an hour. Since everything they eat and drink comes from the ocean, their kidneys are designed to minimize salt uptake and conserve water and they have highly developed glands in their heads that concentrate and excrete salt in the form of tears.

Despite all their adaptations and highly specialized mechanisms for life in the ocean, sea turtles are inescapably tied to land at some stage of their life. Turtles, like all other reptiles, have amniotic eggs which contain protective membranes that allow for complete embryonic development within the protected environment of the egg. This great advancement in evolution provided many advantages for reptiles over their predecessors: fish and amphibians. However, turtle eggs must develop in a terrestrial environment. Thus, in order to reproduce, female sea turtles need to emerge from the water and come ashore to lay eggs. All seven species lay their eggs on sandy beaches in warmer tropical and subtropical regions of the world.

Eggs are deposited on the beach in nest chambers, which can be as deep as 1 m below the sand surface. The adult female crawls on land, excavates a chamber into which it lays 50–130 eggs, and returns to the water after covering the new nest. Eggs usually take 50–70 days to fully develop and produce hatchlings that clamber to the beach surface at night. The length of the incubation primarily is influenced by the prevailing temperature of the nest during development; warmer temperatures hasten the hatchlings' development rates.

Nest temperature also plays a key role in determining the sex of hatchlings. Sea turtles share with other reptiles a phenomenon known as temperature-dependent sex determination (TSD). In sea turtles, warmer nest temperatures produce females, whereas cooler temperatures generate males. More specifically, it appears that there is a crucial period in the middle trimester of incubation during which temperature acts on the sex-determining mechanism in the developing embryo. Temperatures of $>30°C$ generate female sea turtles, and temperatures cooler than $29°C$ produce males. There is a pivotal range between these temperatures that can produce either sex. During rainy periods it is very common to have nests that are exclusively male, whereas a sunny dry climate tends to produce many nests of all females. Extended periods of extreme weather can produce an extremely skewed sex ratio for an entire beach over an entire nesting season. It is of much concern that global warming trends could have drastic effects on reptiles for which a balanced sex ratio totally depends on temperature.

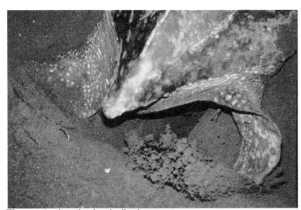

Figure 1 Leatherback digging nest.

During a single nesting season an individual female nests several different times, emerging from the water to lay eggs at roughly two-week intervals. Thus a female's reproductive output can total sometimes more than 1000 eggs by the end of the nesting season. However, once finished, most individuals do not return to nest again for several years. There is much variation in the measured intervals of return, between 1 and 9 years, before subsequent nesting bouts. Alternatively, males may mate at a much more frequent rate. Mating occurs in the water, and has been observed in some areas, usually near the nesting beaches. However, timing and location of mating is not known for many nesting populations, and is virtually unknown for some species.

Sea turtles have evolved a life history strategy that includes long-lived adults, a high reproductive potential, and high mortality as hatchlings or juveniles. As adults they have few predators with the exception of some of the larger sharks and crocodilians. Age to reproductive maturity is estimated at 8–20 years depending on the species and feeding conditions. Their eating habits include herbivores, like the green and black turtle, carnivores like the ridley turtles that eat crustaceans, and even spongivores, like the hawksbill. Almost all sea turtle hatchlings have a pelagic dispersal phase which is poorly documented and understood. Juveniles and adults congregate in feeding areas after the hatchling has attained a certain size. All sea turtles are currently listed in Appendix I of the Convention on International Trade in Endangered Species of Flora or Fauna (CITES Convention) and all species except the flatback (*Natator depressus*) from Australia, are also listed as threatened or endangered by the World Conservation Union (IUCN). Historically populations of sea turtles have declined worldwide over the past 20 years due to loss of nesting beach habitat, harvest of their eggs due to natural predation or human consumption, and the extensive adult mortality brought about by fishing pressure from the expansion of shrimp and drift net or long-line ocean fisheries.

General Sea Turtle Biology

The first fossil turtles appear in the Upper Triassic (210–223 Ma) and the sea turtles of today are descendants of that lineage. The current species of the family Cheloniidae evolved about 2–6 Ma and the sole representative of the Dermochelyidae family is estimated to have evolved about 20 Ma. All sea turtles are characterized by an anapsid skull, in which the region behind the eye socket is completely covered by bone without any openings, and the presence of a bony upper shell (carapace) and a similar bony lower shell (plastron). In all of the cheloniids the carapace and plastron consist of paired bony plates, in contrast to the sole Dermochelyid which has no evident bony plates in the leathery carapace and plastron, but instead interlocking cartilaginous osteoderms.

Like all reptiles, sea turtles have indeterminate growth, i.e., they continue to grow throughout their entire lifetime, which is estimated to be 40–70 years. Growth rates are often rapid in early life stages depending on amount of food available and environmental conditions, but diminish greatly after sexual maturity. As turtles get very old, growth probably becomes negligible. Sea turtles can grow to be quite large; and the leatherback is by far the largest, with most adults ranging between 300 and 500 kg with lengths of over 2 m. The cheloniid turtles are all smaller ranging from the smallest, the Kemp's and olive ridleys, to the flatback, which may weigh as much as 350 kg.

Sea turtles can not retract their necks or limbs into their shell like other turtle species and have evolved highly specialized paddle-like, hind limbs and wing-like front limbs with reduced nails on the forelimbs limited to one or two claw-like growths (**Figure 2**) that are designed to facilitate clasping of the carapace on the female by the male during copulation. Fertilization is internal and sea turtles lay terrestrial nests on sandy beaches in the tropics and subtropics. Adults migrate from feeding areas to nesting beaches at intervals of 1–9 years, distances ranging from hundreds to thousands of nautical kilometers, to lay nests on land. Nest behavior can be solitary or aggregate in phenomena called an 'Arribada' where up to 75 000–350 000 female turtles will emerge to nest on one beach over a period of three to five consecutive evenings (**Figure 3**). Egg clutches of 50–200 eggs are buried in the sand at 20–40 cm below the

Figure 2 Single nail on fore-flipper of a juvenile loggerhead.

surface. The nests are unattended and hatch 45–70 days later into hatchlings that average 15–30 g in weight. The sex of a hatchling sea turtle is determined by the temperature at which the nest is incubated (temperature dependent sex determination) during the middle trimester of development (critical period). For most sea turtles, temperatures above 29.5°C result in the development of a female whereas those eggs incubated at temperatures below 29°C become males. Sea turtles lay cleidoic eggs which are 2–6 cm in diameter and have a typical leathery inorganic shell constructed by the shell gland in the oviduct of the laying females. Nesting is seasonal and controlled by photoperiod cues integrated by the pineal gland and also influenced by the level of nutrition and fat stores available.

Sea turtle physiology is well adapted for deep and shallow diving that can average from 15 min to 2 hours in length. Their tissues contain high levels of respiratory pigments like myoglobin as a reserve oxygen store during the breath hold/dive. Respiratory tidal volumes are quite large (2–6 liters per breath) and allow for rapid washout of carbon dioxide and oxygen uptake during the brief periods spent on the surface breathing (**Figure 4**). Thus in a normal one hour period a sea turtle may spend 50 min under the water and only 10 min at the surface breathing and operating entirely aerobically and accrue no oxygen debt. Cardiovascular adaptations include counter current heat exchangers in the flippers to reduce or enhance heat exchange with the surrounding water. Control of the vascular tree is much higher than the level of arterioles and can permit almost complete restriction of blood flow to all tissues but the heart, brain, central nervous system, and kidney during the deepest dives. They also have evolved temperature and pressure adapted enzymes that will operate well at both the surface and at extreme depths. Leatherbacks are the deepest diving sea turtles; dives of over 750 m in depth have been recorded.

All sea turtles have lacrimal salt glands (**Figure 5**) and reptilian kidneys with short loops of Henle to regulate ion and water levels. Lacrimal salt glands allow sea turtles to drink sea water from birth and

Figure 3 Olive ridley arribada at Nancite Costa Rica.

maintain water balance despite the high levels of inorganic salts in both their marine diet and the salty ocean water they drink. Diets range from: plant materials like sea grass, *thallasia* and algae for herbivores like the green and black sea turtles; crabs and crustaceans for the ridley turtles; sponges for the loggerhead; and jellyfish and other Cniderians are the sole diet of leatherbacks.

Dermochelyidae (1 genera, 1 species)

The genus *Dermochelys*: *Dermochelys coriacea* (Linne') the leatherback turtle Leatherbacks are the largest (up to 600 kg as adults) and most ancient of the sea turtles diverging from the other turtle families in the Cretaceous. About 20 Ma *Dermochelys* evolved to a body form very similar to that seen today. Their shell consists of cartilaginous osteoderms and they do not have the characteristic laminae and plates found in the plastron and carapace of other sea turtles. The leatherback shell is streamlined with seven cartilaginous narrow ridges on the carapace and five ridges on the plastron that direct water flow in a

Figure 4 Leatherback breathing at surface.

Figure 5 Clear salt gland secretion from base of eye.

Figure 6 Leatherback osteoderm with ridges on carapace.

laminar manner over their entire body (**Figure 6**). The appearance is a distinctive black with white spots, a smooth, scaleless carapace skin and a white smooth-skinned plastron. The head has two saber tooth like projections on the upper beak that overlap the front of the lower beak and serve to pierce the air bladder of floating cniderians that are a large component of their diet when available. Their nesting distribution is worldwide with colonies in Africa, Islands across the Caribbean, Florida, Pacific Mexico, the Pacific and Caribbean coastlines of Central America, South America, Malaysia, New Guinea, and Sri Lanka. Remarkably they are found in subpolar oceans at surface water temperatures of 7–10°C while maintaining a core body temperature above 20°C.

Leatherbacks are pelagic and remain in the open ocean throughout their life feeding on soft-bodied invertebrates such as cniderians, ctenophores, and salps. Average adult females weigh about 300 kg and may take only 8–10 years to attain that size after emerging from their nests at about 24 g. Sex is determined by TSD with a pivotal temperature of 29.5°C. They build a simple cup shaped body pit and lay the largest eggs, about the size of a tennis ball, in clutches of 60–110 yolked eggs. Unlike other sea turtles leatherbacks also deposit 30–60 smaller yolkless eggs that are infertile and only contain albumins. These yolkless eggs tend to be laid late in the nesting process and are primarily on the top of the clutch and may serve as a reservoir for air when hatchlings emerge and congregate prior to the frenzied digging to emerge from their nest chamber.

Unlike other sea turtles leatherbacks crawl on their wrists while on land, rotating both front flippers simultaneously and pushing with both rear flippers in unison. This contrasts with the alternating right to left front and rear flipper crawls of the Cheloniidae. In the ocean their enlarged paddle-like front flippers generate enormous thrust providing excellent propulsion and allowing leatherbacks to migrate an average of 70 km per day when leaving the nesting beaches. Leatherback migrations from Central American beaches are along narrow 'corridors' that are about 100 km wide (**Figure 7**). These oceanic migratory corridors provide important insights into the complex reproductive behavior of these animals. Genetic studies have demonstrated that there is excellent gene flow across all the ocean basins with a strong natal homing and distinct genetic haplotypes in different nesting populations.

Cheloniidae (5 genera, 6 species, 1 race)

The genus *Chelonia:Chelonia mydas* (Linne') the green turtle and *Chelonia mydas agassizi* (Bocourt) the black turtle The genus *Chelonia* contains only one living species the green turtle *Chelonia mydas*

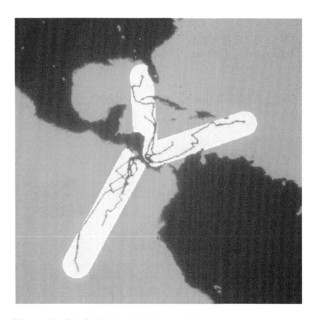

Figure 7 Leatherback migration corridors.

Figure 8   Green turtle.

mydas (Linne') that also has a Pacific–Mexican population that is considered by some researchers as a subspecies or race, the Pacific green turtle or black turtle, *Chelonia mydas agassizi* (Bocourt). The green turtle (**Figure 8**) actually has a brownish colored carapace and scales on the legs, with a yellowish plastron and has one pair of prefrontal head scales with four pairs of lateral laminae. The name 'green turtle' comes from the large greenish fat deposit found under the carapace which is highly desired for the cooking of turtle soup. The black turtle (**Figure 9**) is distinguished by the dark black color of both the carapace and plastron in the adult form. This subspecies or race is found only along the Pacific coastline of Mexico and extends in smaller populations along the Pacific Central American coastline to Panama. *Chelonia* are herbivorous; the green turtle eats marine algae and other marine plants of the genera *Zostera*, *Thallassina*, *Enhaus*, *Posidonia*, and *Halodule* and

Figure 9   Black turtle (with transmitter).

is readily found in the Caribbean in eel-grass (*Zostera*) beds, whereas black turtles rely heavily on red algae and other submerged vegetation. These turtles live in the shallow shoals along the equatorial coastlines as far North as New England in the Atlantic and San Diego in the Pacific and as far south as the Cape of Good Hope in the Atlantic and Chile in the Pacific. The main breeding rookeries include the Coast of Central America, many islands in the West Indies, Ascension Island, Bermuda, the Florida Coast, islands off Sarawak (Malaysia), Vera Cruz coastline in Mexico (center for black turtles), and islands off the coast of Australia. Adult females emerge on sandy beaches at night to construct nests in which they lay 75–200 golf ball-sized leathery eggs. The eggs in these covered nests develop unattended and the young emerge 45–60 days latter. The sex of the hatchlings is determined by the temperature at which the eggs are incubated during the third trimester of development called TSD. Temperatures above 29°C tend to produce all female hatchlings whereas those below 29°C produce males. The hatchlings have a high mortality in the first year and spend a pelagic period before reappearing in juvenile/adult feeding grounds, such as the reefs off Bermuda, the Azores, and Heron Island in the Pacific. As adults they average 150–300 kg with straight line carapace lengths of 65–90 cm. Black turtles tend to be smaller than green turtles. Genetic studies have shown a strong natal homing of the females to the beaches where they were hatched with significant gene flow between rookeries probably due to mating with males from different natal beaches found on common feeding grounds.

The genus *Lepidochelys*: *Lepidochelys kempii* (Garman) the Kemp's ridley, *Lepidochelys olivacea* (Eschsholtz) the olive ridley   The Kemp's or Atlantic Ridley and the olive or Pacific Ridley turtles are the two distinct species of this genus. The Kemp's ridley (**Figure 10**) is the rarest of the sea turtles whereas the olive ridley is the most abundant. Anatomically ridleys are the smallest sea turtles. Adult Kemp's usually have five pairs of grey-colored coastal scutes and olive ridleys have 5–9 olive-colored pairs. The adults have a straight line carapace length of 55–70 cm and weigh on average 100–200 kg. Both species have an interesting reproductive behavior in that they have communal nesting called 'arribadas' (Spanish for arrival). There are a number of arribada beaches where females come out by the thousands on sandy beaches 2–6 km long on three or four successive evenings once a month during the nesting seasons. A

Figure 10 Juvenile Kemp's ridley.

percentage of the total population are solitary nesters that emerge on other nearby beaches or on arribada beaches on nights other than the arribadas. These solitary nesters have a very high hatching success of their individual nests (about 80%). It is unknown what proportion of the total population of females nest in this solitary manner and what role they play in the contribution of new recruits into the reproductive adult numbers. Arribadas in Gahirmatha, India have been described with 100 000 turtles emerging in one night to nest communally only inches apart. Many nests are dug up by successive nesting females in the same or subsequent nights of the arribada. The hatching success of these arribada nests is about 4–8% which is the lowest among all sea turtles. Kemp's ridley turtles have arribadas on only one nesting beach in the world, Rancho Nuevo (Caribbean), Mexico. Historically Kemp's arribadas were estimated at 20 000–40 000 female turtles per night in the 1950s, but recent arribadas average only 300–400 individuals per night. This dramatic decline has been attributed to the numbers of adults and juveniles killed in the shrimping nets of the fisheries in the Gulf of Mexico and the American Atlantic coastline. These genera feed primarily on crustaceans, crabs and shrimp, and as a result have been in direct competition with these well-developed fisheries for many years and have suffered dire consequences. Olive ridleys have arribada beaches in Pacific Mexico, Nicaragua, Costa Rica, and along the Bay of Bengal in India. Very little is known about the ocean life of these sea turtle species and it is believed that after hatching they also spend 1–4 years in a pelagic phase associated with floating *Sargassum* and then reappear in the coastal estuaries and neretic zones worldwide at a straight carapace length of about

20–30 cm. Although Kemp's ridley turtles feed primarily on crabs in inshore areas, olive ridley turtles have a more varied diet that includes salps (*Mettcalfina*), jellyfish, fish, benthic invertebrates, mollusks, crabs, shrimp and bryozoans. This more varied diet may account for the different deep ocean and nearshore habitats in which the olive ridleys are found.

The genus *Eretmochelys*: *Eretmochelys imbricata* (Linne'), the hawksbill This species is the most sought after sea turtle for the beauty of the carapace. Historically eyeglass frames and hair combs were made from the carapace of hawksbills. The head is distinguished by a narrow, elongated snout-like mouth and jaw that resembles the beak of a raptor (**Figure 11**). They have four pairs of thick laminae and 11 peripheral bones in the carapace. They tend to be more solitary in their nesting behavior than the other sea turtles but this may be due to their reduced populations. Other than the kemp's ridley turtles they tend to be the smallest turtles averaging 75–150 kg as adults. They are common residents of coral reefs worldwide and appear as juveniles at about 20–25 cm straight length carapace after a pelagic developmental phase in the floating *Sargassum*. What is impressive about the adults is that their diet is 90% sponges. This is one of the few spongivorous animals and electron micrographs of their gastrointestinal tract has shown the microvilli with millions of silica spicules imbedded in the tissue. Tunicates, sea anemone, bryozoans, coelenterates, mollusks, and marine plants have also been found to be important components of the hawksbill diet.

Hawksbills have the largest clutch size of any of the chelonids averaging 130 eggs per nest. They also lay the smallest eggs with a mean diameter of 37.8 mm and mean average weight of 26.6 g. This results in the smallest hatchling of all sea turtles with an average weight of 14.8 g. Other than possibly the

Figure 11 Raptor-like beak of hawksbill.

olive ridley turtle they have the lowest clutch frequency of only 2.74 clutches per nesting season. Hawksbills also tend to be the quickest to construct their nest when out on a nesting beach; they spend only about 45 min to complete the process whereas other turtles take between one to two hours.

Habitats include shallow costal waters and they are readily found on muddy bottoms or on coral reefs. The genetic structure of Atlantic and Pacific populations indicate that hawksbills like other sea turtles have strong natal homing and form distinct nesting populations. They also have TSD and hatchlings also have a pelagic phase before appearing in coastal waters and feeding grounds at about the size of a dinner plate.

Cuba has requested that these turtles be upgraded to CITES Appendix II status which would allow farming of 'local resident populations.' Genetic studies, however, have conclusively demonstrated that these Cuban populations are not local and include turtles from other regions of the Caribbean. These kinds of controversies will increase as human pressures for turtle products, competition for the same food resources such as shrimp, incidental capture due to longline fishing practices, and beach alteration and use by humans increases.

The genus *Caretta: Caretta caretta* (Linne'): the loggerhead turtle The loggerhead, like olive ridley turtles, are distinguished by two pairs of prefrontal scales, three enlarged poreless inframarginal laminae, and more than four pairs of lateral laminae. They have a beak like snout that is very broad and not narrow like the hawksbill and they have the largest and broadest head and jaw of the chelonids (**Figure 12**). They nest throughout the Atlantic, Caribbean, Central America, South America, Mediterranean, West Africa, South Africa, through the Indian Ocean, Australia, Eastern and Western Pacific Coastlines.

Loggerheads are primarily carnivorous but have a varied diet that includes crabs (including horseshoe crabs), mollusks, tube worms, sea pens, fish, vegetation, sea pansies, whip corals, sea anenomies, and barnacles and shrimp. Their habitats appear to be quite diverse and they will shift between deeper continental shelf areas and up into shallow river estuaries and lagoons. This sea turtle is the only turtle that has large resident and nesting populations across the North American coastline from Virginia to Florida and the nesting rookeries in Georgia and Florida are currently doing well despite severe human pressures.

Loggerheads that nest on islands near Japan have been shown to develop and grow to adult size along

Figure 12 Broad head of juvenile loggerhead.

the Mexican coastline and then migrate 10 000 km across the Pacific to nest. Genetic studies have confirmed that the Baja Mexico haplotypes are the same as the female adults nesting on these Japanese islands confirming strong natal homing in this genus. Loggerheads make a simple nest at night and lay about 100–120 golf ball-sized eggs. The hatchlings spend a pelagic phase and reappear at 25–30 cm straight carapace length in coastal bays, estuaries and lagoons as well as along the continental shelf and open oceans (**Figure 13**). The orientation and homing of loggerhead hatchlings has been extensively studied and it has been demonstrated that these turtles can detect the direction and intensity of magnetic fields and magnetic inclination angles. This ability together with the use of olfactory cues and chemical imprinting on olfactory cues from natal beaches may account for the natal homing ability of all sea turtles.

The genus *Natator: Natator depressus* the flatback This genus has a very limited distribution and is only found in the waters off Australia yet it appears that populations are not endangered at this time. Their nesting beaches are primarily in northern and south-central Queensland. Most of the nesting beaches are quite remote which has protected this species from severe impact by

Figure 13 Juvenile loggerhead turtle.

humans. Apart from the Kemp's ridley this is the only sea turtle to nest in significant numbers during the daytime. It is thought that nighttime nesting has evolved as a behavior to reduce predation and detection but there may also be thermoregulatory considerations. During the daytime in the tropics and subtropics both radiant heat loads from the sun and thermal heat loads from the hot sands may significantly heat the turtles past their critical thermal tolerance. In fact, a number of daytime-nesting flatbacks may die due to overheating if the females do not time their emergence and return to the water to coincide closely with the high tides. If individuals are stranded on the nesting beach when the tides are low and the females must traverse long expanses of beach to emerge or return to the sea during hot and sunny daylight hours, there is a higher potential to overheat and die. On the other hand green turtles in the French Frigate Shoals area of the Pacific are known to emerge on beaches or remain exposed during daylight in shallow tidal lagoons during nonnesting periods and appear to heat up to either aid in digestion or destroy ectoparasites.

Flatback turtles are characterized by a compressed appearance and profile of the carapace with fairly thin and oily scutes. Flatbacks are also distinguished by four pairs of laminae on the carapace and the rim of the shell tends to coil upwards toward the rear. Flatbacks have a head that is very similar to the Kemp's ridley with the exception of a pair of pre-ocular scales between the maxilla and prefrontal scales on the head. These turtles tend to be the largest chelonid, weighing up to 400 kg as adults, and lay the second largest egg with diameters of about 51.5 mm weighing about 51.5 g, smaller only than the leatherback which has a mean egg diameter of 53.4 mm and mean egg weights of 75.9 g. Flatbacks have the smallest clutch size of the chelonids, laying a mean of 53 eggs per clutch, and will lay about three clutches per nesting season. There is only one readily accessible nesting beach for this species at Mon Repos in Queensland, Australia. Other isolated rookeries like Crab Island are found along the Gulf of Carpentaria and Great Barrier Reef.

It is believed that flatbacks may be the only sea turtle that does not have an extended pelagic period in the open ocean. Hatchlings and juveniles spend the early posthatchling stage in shallow, protected coastal waters on the north-eastern Australian continental shelf and Gulf of Carpentaria. Their juvenile and adult diet is poorly known but appears to include snails, soft corals, mollusks, bryozoans, and sea pens.

Further Reading

Bustard R (1973) *Sea Turtles: Natural History and Conservation.* New York: Taplinger Publishing.

Carr A (1986) *The Sea Turtle: So Excellent a Fishe.* Austin: University of Texas Press.

Carr A (1991) *Handbook of Turtles.* Ithaca, NY: Comstock Publishing Associates of Cornell University Press.

Gibbons JW (1987) Why do turtles live so long? *BioScience* 37(4): 262–269.

Lutz PL and Musick J (eds.) (1997) *The Biology of Sea Turtles.* Boca Raton: CRC Press.

Rieppel O (2000) Turtle origins. *Science* 283: 945–946.

MARINE MAMMAL DIVING PHYSIOLOGY

G. L. Kooyman, University of California San Diego, CA, USA

Introduction

Marine mammals are the last major group of vertebrates to adapt widely to the marine environment. Reptiles and birds preceded them by tens if not hundreds of millions of years. The marine reptiles had their greatest success in the Mesozoic. Like the dinosaurs most had disappeared by the end of the Cretaceous, except for sea turtles and crocodiles. Marine diving birds were also present in the Mesozoic, including possibly some penguins. However, as with marine mammals, their greatest diversification occurred during the Tertiary. The lack of competition from such successful and formidable marine reptiles as the mosasaurs, ichthyosaurs, and pleisiosaurs may have enabled this adaptive radiation. Nevertheless, then as now, all three groups had similar physical obstacles to overcome in adapting to marine life. These problems stimulated the evolution of some of the most extreme and unusual physiological and morphological adaptations ever achieved by vertebrates.

The ancestors of whales were the first to begin the invasion of the sea sometime during the Eocene, more than 60 million years ago (Ma). Sea cows, the only herbivorous marine mammal, originated about 50 Ma during the late Eocene, and pinnipeds followed about 30 Ma in the late Oligocene. Pelagic species wander the vast offshore regions of the world's oceans, and dive in waters with depths up to thousands of meters. Because the greatest challenges of the physical environment are preeminent in this region, the pelagic whales and pinnipeds will be discussed in greatest detail.

There are seven major physical obstacles to overcome that require extreme physiological adaptations to life in the oceans.

1. Anoxia: diving into a world that is without oxygen for an air-breathing mammal.
2. Density: just a short distance from the surface the hydrostatic pressure becomes extreme.
3. Breathing: the less time taken for respiration, the more time at depth to search for prey or to avoid being eaten.
4. Vision: even in the best conditions of water clarity in pelagic tropical waters this is a region of twilight to eternal darkness.
5. Acoustics: the limited field of vision underwater increases the importance of hearing over long distances compared to land mammals.
6. Cold: even the warmest tropical sea is 10–15°C cooler than the internal temperature of a hot-blooded marine mammal.
7. Viscosity: there is a reason animals underwater appear to move in slow motion – their movements are slowed by the viscosity of water.

Selection pressure for adaptations to overcome these physical barriers is great and has resulted in some very consistent morphological and physiological adaptations that, in some cases, make it easy to recognize a marine mammal from only a small part of its anatomy. Some of the more salient anatomical features are discussed in relation to their function. Just as there are variations and gradations on the theme of adapting to the marine environment, so too there are extremes that are exemplified by the most pelagic and the deepest divers. **Table 1** shows statistics from each major group regarding the simple assessment of diving ability by the maximum and routine depths and durations. It should be noted that even though the diving ability of some species is impressive, the exploitative ability of marine mammals is superficial considering that the average depth of the world's oceans is 3.5 km and the maximum depth is 11 km. Emphasis will be placed on five of the seven adaptations mentioned.

Adaptations to Anoxia

When marine mammals dive below the surface they enter an anoxic environment even though there is much dissolved oxygen in the surrounding water. Lacking gills or any means of extracting oxygen from the water they are without oxygen, except for that stored within their bodies, until they return to the surface. The brain must not be without oxygen for more than three minutes or irreversible damage occurs, and dysfunction occurs even sooner. With such a short margin of resistance, marine mammals had to develop special adaptations to protect the brain and other organs and tissues sensitive to oxygen deprivation. Some of the broad categories of adaptation are: (1) oxygen stores; (2) redistribution of blood flow by cardiovascular adjustments; (3) reduced

Table 1 Routine and maximum diving characteristics of selected marine mammals from some of the major groups that hunt pelagically

Species	Family	Dive duration (min)		Dive depth (m)	
		Mean	Max.	Mean	Max.
California sea lion (*Zalophus californianus*)	Otariidae	2	10	62	274
Northern elephant seal (*Mirounga angustirostris*)	Phocidae	22	90	520	1581
Sperm whale (*Physeter catodon*)	Physeteridae	35	73	792/466	2035
Bottlenose whale (*Ampullatus hyperoodon*)	Ziphiidae	37	70	1060	1453

metabolism during the dive; (4) behavioral patterns that encourage oxygen conservation; (5) reliance on anaerobic metabolism in organs tolerant to hypoxia.

Oxygen Stores

The oxygen consumed during a breath hold is stored in three compartments, the respiratory system, the blood, and the body musculature (**Figure 1**). The respiratory oxygen store is of marginal value since about 80% of the volume is nitrogen, and because there is little gas exchange between the lung and blood while the animal is at depth. The blood oxygen store is dependent on the blood volume, red cell volume, and the concentration of hemoglobin in the red blood cells. As the cell volume increases, so does viscosity, increasing the resistance to blood flow. Some marine mammals have very high blood oxygen storage capacity, whereas in others the blood oxygen storage capacity is little different from that in terrestrial mammals. In contrast, there is an increased concentration of the oxygen-binding protein myoglobin in muscle in all marine mammals which sets them apart from all other mammals. Myoglobin is 3–15 times more concentrated in the muscle of diving compared to terrestrial mammals, and there may be some relationship between the depth of dives the animal routinely makes and the level of the myoglobin in the muscle (**Tables 1** and **2**).

The deepest divers seem to have the largest oxygen stores. In humans the total store is $20 \, ml \, O_2 kg^{-1}$ body mass, which is about a fifth of that in elephant seals (nearly $100 \, ml \, O_2 \, kg^{-1}$ body mass). Using the human average as a standard of comparison for the typical terrestrial mammal, the elephant seal has a blood volume three times greater, a hemoglobin concentration 1.5 times more, and a myoglobin concentration approximately 10 times more (**Figure 1**).

Cardiovascular Adjustments

Specializations of the cardiovascular system varies among the different families of marine mammals. The least cardiovascular modification appears to be in the sea lion, sea otter and manatee. At the other extreme are the cetaceans with numerous variations or entirely new structures, most of which have unknown function. For example, most toothed whales have an extreme development of the thoracic retia mirabilia. This complex network of arteries is invested in the dorsal aspect of the thorax as well as embedded between the ribs. One of its main functions appears to be to provide the primary blood supply to the brain. This role has been usurped from the internal carotid, which does not reach the brain before it ends as a tapered down and occluded vessel. The reason for this complexity, and other possible functions of the thoracic retia are unknown.

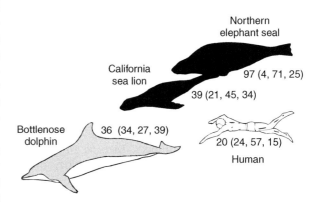

Figure 1 Distribution of oxygen stores within the three major compartments of lung, blood and muscle. Groups represented are the toothed whales (bottlenose dolphin), sea lion and fur seals (California sea lion), and true seals (northern elephant seal). The first number is total body oxygen store; numbers in parentheses are for the lung, blood, and muscle (ml kg⁻¹). (Modified from Kooyman (1989).)

Table 2 Myoglobin concentrations in the major groups of marine mammals

Species	Myoglobin (g 100 g^{-1} wet weight)
Manatee (*Trichechus manatus*)	0.4
Sea otter (*Enhydra lutris*)	3.1
Walrus (*Odobenus rosmarus*)	3.0
California sea lion (*Zalophus californianus*)	2.8
Weddell seal (*Leptonychotes weddellii*)	5.4
Northern elephant seal (*Mirounga angustirostris*)	5.7 (8 months old)
Fraser's dolphin (*Lagenodelphis hosei*)	7.1
Sperm whale (*Physeter catodon*)	5.0

A more universal structure is the aortic bulb, an enlargement of the aortic arch, that functions as a capacious, elastic chamber. This bulb absorbs much of the flow energy developed during systole by the left ventricle. This absorbed pulse is then more evenly spread through the rest of the cardiac cycle. The maintenance of blood flow pressure is especially effective during the bradycardia that occurs during a dive.

There is considerable modification of the venous circulation in seals. The intravertebral extradural vein that lies within the vertebral canal above the spinal cord is responsible for most of the brain return flow. It also drains portions of the back musculature and the pelvic area. Much of the blood volume returns via the intercoastals to the azygous vein and then to the anterior vena cava to the heart. The major blood return of the body is from the inferior vena cava, which drains into a large hepatic sinus. This vessel passes through a narrow restriction, the vena cava sphincter, at the diaphragm before entering the thoracic cavity. The sphincter is a circular muscle capable of reducing flow return to the heart.

Finally, at least in pinnipeds, there is an enlarged spleen that acts as a reservoir for about 50% of the total red blood cell volume while the seal is resting or sleeping. The splenic mass in terrestrial mammals is about 0.5–2% of body mass, whereas it is 4–10% in seals. Once the seal begins to dive these cells are injected into the circulation and raise the hematocrit about 50% above the resting value. Once a diving bout is concluded much of the red blood cells are again stored in the spleen.

The most complete information on cardiovascular function is from seals in which information has been obtained while the animals are making routine foraging dives. Most data from other diving mammals have been collected during trained dives, or during resting submersions. In general, as the dive begins, a rapid onset of bradycardia ensues that may range from 20–90% of the resting heart rate. This is dependent on the duration of the dive, probably the swim speed, and whether the dive is routine or an evasive response that incurs some level of stress. The latter dives are when the most extreme declines in heart rate occur. Concurrent with the reduced heart rate there is the necessary decline in cardiac output, which is greater than would be predicted from the change in heart rate alone. This happens because there is also a reduction of 50% in stroke volume. During the longest dives, there is a reduction in splanchnic blood flow to hypoxic-tolerant organs such as the liver and kidneys, and in somatic blood flow to many of the muscles. The brain, which is not hypoxic tolerant, does not have a reduced blood flow and it becomes a major consumer of the stored oxygen of the circulatory system. What the rate of oxygen consumption is during the dive remains one of the most important unanswered questions in diving physiology.

Metabolic Responses

Even a modest reduction in metabolic rate of the liver, kidney, and gastrointestinal tract will help to conserve the body's oxygen store since they account for about 50% of the total resting oxygen consumption. A small reduction can be easily made up while the animal is ventilating at the surface. The other major consumer is the locomotor muscles. Even under conditions of foraging when the animal may be anxious to swim to the prey patch, the axiom of 'make haste slowly' is applicable. It must conserve the muscle oxygen as much as possible, and one option for doing this is to take advantage of its natural buoyancy. Gliding, the behavior that achieves this result, has now been measured. In four species, two seals, a bottlenose dolphin, and a blue whale, it is now known that during the descent below a depth of 20–40 m the animals glide to greater depths. One elephant seal was observed to glide down to 400 m. In this condition only the brain remains uncompromising in its need of a large oxygen requirement. It has been estimated that gliding to depth can result in an energy saving of over 50% compared to a dive in which there is swimming at all times. These are conservative estimates because the drag caused by the attached camera must have been substantial for all species except the blue whale, and this would incur a cost that in the unfettered animal would be much less. Even at depth there is an additional reduction of metabolic rate for swimming as the streamlined animal strokes intermittently and glides as much as possible. Away from surface effects, the resistance to forward propulsion is at a minimum.

The economy in swim effort results in important savings in oxygen consumption as propulsive muscle does less work. In addition, it is likely that there is a reduced blood flow to muscle as it relies more on the internal store of oxygen rather than that from the circulating blood, which is the main oxygen source for the hypoxic-intolerant brain. The high concentration of myoglobin in all marine mammals indicates that it is a key adaptation for diving. As the muscle oxygen depletes, the need for supplemental energy from anaerobic catalysis of creatine phosphate and glycogen rises. Both of these compounds produce adenosine triphosphate (ATP) that is essential for electron transfer, and the conversion of chemical energy into mechanical work. However, anaerobic glycolysis results in an inefficient use of the glycogen store because of incomplete combustion of glucose to lactate which results in a metabolic acidosis that has a limit of tolerance. Diving mammals have a broad tolerance of acidosis because of their exceptional capacity to buffer the acidity of lactate. Most divers avoid this condition, and reliance on anaerobic metabolism occurs only in exceptional cases when the dive has to be extended beyond the routine foraging durations. This threshold is when a net production of lactate results in a rise in arterial blood lactate after the dive. This has been called the aerobic diving limit (ADL).

When dive durations are plotted against lactate concentration in arterial blood (**Figure 2**) there is a distinct inflection where lactate concentration increases sharply. Beyond this dive duration anaerobic metabolism contributes a greater amount to the energy needs of the animal, mostly in the muscle. There is a cost in addition to the inefficient use of glycogen stores. An imbalance occurs in metabolic end-products, and there is acidification in the cells and the circulatory system which must be restored to normal acid–base balance. To do this there may be an extended surface period for recovery, or if the recovery is continued during the next dive, it is likely to be shorter because of the reduced oxygen and glycogen stores which cannot return to full capacity until the acid–base balance of the body returns to normal levels. In order to avoid the problem inherent in relying on extensive anaerobic metabolism during a dive, most dives of marine mammals are within the

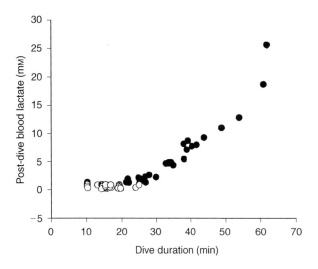

Figure 2 Peak concentration of lactate in arterial blood after dives of different duration in adult Weddell seals. The inflection represents the transitions from completely aerobic dives and is considered the aerobic diving limit (ADL). Open circles reflect blood values of dives in which there is no net production of lactate, and black circles are those in which there was a net production. (Reproduced with permission from Kooyman (2000).)

ADL. Circumstances during which the ADL might be exceeded occur during foraging when a rich source of prey might be at such a depth that the dive has to be longer than the ADL, or during avoidance of a predator when diving unusually deep might aid in a successful escape. For the Weddell seal it might be to travel an exceptionally long distance under ice to a new breathing hole.

Adaptations to Pressure

Once marine mammals developed the capacity to breath hold for a few minutes they were exposed to the second most dominating physical effect of the marine environment – the density of water. Just a short distance from the surface, the hydrostatic pressure becomes overwhelming for any terrestrial animal which lacks the adaptations found in marine mammals. In order to tolerate the effects of pressure, marine mammals have adapted to live with it rather than to resist it as human made submersibles do. Instead of an outer shell that protects the internal organs from the crushing pressure of depth, marine mammals give and absorb the pressure. The chest is almost infinitely compliant allowing the lungs to be compressed to a solid organ. There are no air sinuses in the skull such as the facial sinuses of humans, and the vascular lining of the middle ear can expand and reduce in volume to match the ambient pressure. There are no disparities of pressure between the inside and outside of the body. This has many ramifications, but the most important relates to how diving mammals avoid absorbing the lung gases that comprise 80% nitrogen while diving to great depths. Such a volume is adequate, depending on the distribution of blood, to cause blood and tissue nitrogen tensions to reach a level where gas bubbles could form during the animal's rapid ascent. The result is decompression sickness or the 'bends'.

By yielding to the compressive effects of pressure marine mammals avoid the problem. The lungs of marine mammals are similar in all groups, but distinctive from terrestrial mammals (**Figure 3**). As a result of robust cartilaginous support that is absent from terrestrial mammals the most peripheral airways are less compliant. As the chest wall compresses, it allows the lungs to collapse; this takes place in a graded fashion with the gas from the alveoli being forced into the upper airways where gas exchange does not occur (**Figure 4**). Consequently, the gas is sequestered while the animal is at depth. For some seals this collapse may occur at depths of only 20 m. No matter how much deeper the dive is, the arterial blood nitrogen tension does not rise above 2300 mmHg, within the range where gas bubbles will not form even if the decompression rate is extremely rapid. Curiously, the most robust peripheral airways are not always in the deepest divers.

Furthermore, in some of these species the airways seem to be armored to a far greater degree than is necessary to ensure a graded collapse as they descend to depth.

Adaptations for Ventilation

Because of the compliant chest wall and the robust peripheral airways the lungs of most marine mammals empty to an unusually low volume, about 5–10% of total lung capacity (TLC), compared to the 25% residual volume of terrestrial mammals. This makes possible a vital capacity of about 90% of TLC, which is often equivalent to tidal volume in species that rapidly ventilate such as dolphins. For a fast-swimming dolphin or sea lion to make effective use of such a large tidal volume it needs to be turned over rapidly during the brief time the blow hole or nostrils are near the surface (**Figure 5**). The bottlenose dolphin can turnover about 90% of its TLC within 0.7 s. Indeed, these dolphins and some whales aid this process by exhaling most of the tidal volume just before breaking the surface, so that most of the time at the surface can be used for inhaling. Inhalation is slower because although the chest wall is actively expanded, lung expansion is a passive

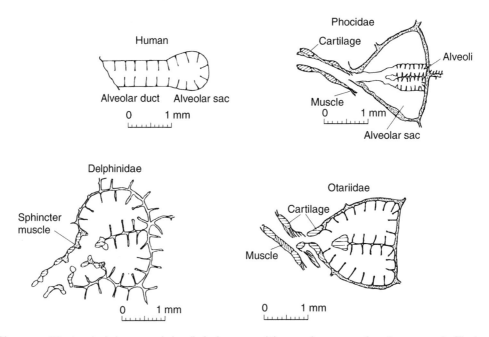

Figure 3 Diagrams of the terminal airways and alveoli of a human and three major groups of marine mammals. The human alveolar duct is thin-walled and exchanges gas with the capillaries. The seal (Phocidae) has a short alveolar duct if present at all and the terminal bronchiole is reinforced with cartilage and muscle. The sea lion (Otariidae) has no gas exchanging surfaces except within the alveolar sac, and the cartilage is robust throughout the terminal bronchiole. The dolphin (Delphinidae) has similar robust cartilage reinforcement within the terminal bronchiole, but in addition there is a series of sphincter muscles. This muscle configuration is unique to the toothed whales. (Modified from Denison DM and Kooyman GL (1973) The structure and function of the small airways in pinniped and sea otter lungs. *Respir. Physiol.* 17: 1–10.)

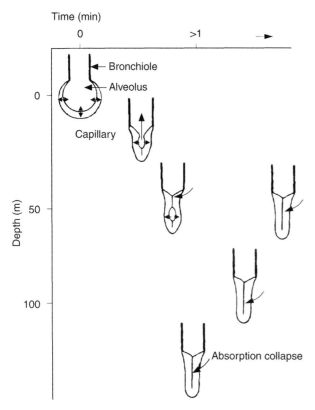

Figure 4 Diagram of the graded collapse of the alveolar gas into the upper airways of a seal as it descends to depth. It is presumed that the same type of collapse occurs in other kinds of marine mammals. The time axis is relative, but emphasizes the short period over which this collapse occurs as the seal descends to depth. The arrows within the alveolus and across the alveolar membrane indicate the direction of gas flow. The curved arrow indicates the further closure of the alveolar space as absorption collapse occurs. The thickness of the arrow's stem indicates the rate. The deeper the seal descends the faster the uptake of gas into the capillaries. Thus, the staircase configuration of the collapsed alveoli showing that the collapse will occur sooner the deeper the animal descends. (Modified from Kooyman GL, Schroeder JP, Denison DM *et al.* (1972) Blood N_2 tensions of seals during simulated deep dives. *American Journal of Physiology* 223: 1016–1020.)

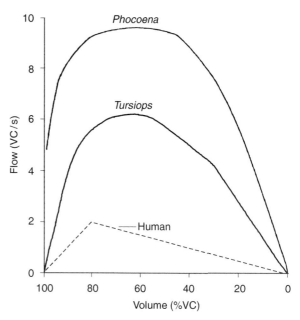

Figure 5 The rapid expiration of gas from the excised lungs of a porpoise (*Phocoena phocoena*) during a forced expiration. The forced expiration curve from a trained bottlenose dolphin (*Tursiops truncatus*). The maximum rate of expiration of a human is shown for comparative purposes. (Modified from Kooyman and Sinnett (1979) Mechanical properties of the harbor porpoise lung. *Respiratory Physiology* 36: 287–300; Kooyman and Cornell (1981) Flow properties of expiration and inspiration in a trained bottlenose porpoise. *Physiological Zoology* 54: 55–61.)

process that relies on the elastic properties of the lung for inflation.

Adaptations for Locomotion

The similarity in shape of different marine mammals, from seals to whales, can be seen in **Figure 1**. There are strict rules for this, which are related to density and viscosity of water (*see* Fish Locomotion). The density of water is about 1000 times greater than air, and at 0°C the kinematic viscosity is about seven times that of air. In order to minimize resistance to movement, marine mammals like many fish had to evolve a special shape of the body. The ideal shape

has a fineness ratio, i.e., body length divided by the maximum body diameter, of about 4.5. This shape greatly reduces form drag, which at the speeds marine mammals swim, is over 90% of the total drag. Also the routine swim speed of all marine mammals is about 2–3 m s^{-1}. At these speeds and through such a dense medium the rates of progress are slower than that of many land mammals, but the burst and glide swimming that are used and the support to the body mass makes their progress seem almost effortless, and accounts for the low metabolic rates discussed earlier.

Adaptations to Light Extremes

Marine mammals are creatures of the night. Even those that may be active during the day search for most of their food at depths where the light level is similar to twilight or less. Hence, they might be thought of as marine bats. Some echolocate to find their prey, whereas others hunt visually. Seals and sea lions are visual hunters and like many nocturnal mammals they have large eyes. The eye of the southern elephant seal has an internal anterior and horizontal diameter of 52 by 61 mm, compared to the human eye of 24 by 24 mm. These eyes have a high concentration of rods, the light-sensitive element within the retina, that is 10–50 times more dense than the human eye. At least some have visual pigments adjusted to the blue light that is the dominant ambient light source at depth. All have a tapetum lucidum layer behind the retinal layer and a large pupillary area. In comparison to humans, a diurnal primate whose pupil area ranges from 3.2 to 50 mm^2, or the domestic cat, whose pupil ranges from 0.9 to 123 mm^2, the northern elephant seal pupil area ranges from 0.9 to 422 mm^2. Less extreme is the California sea lion, which dives to much shallower depths, and whose pupillary dimensions are only 8.4–220 mm^2. At light levels where the sea lion pupil is expanded maximally, which is roughly equivalent to a clear, full-moon night, that of the elephant seal is only at 22% of its maximum diameter. Indeed the maximum pupillary expansion of the elephant seal does not occur until there is total darkness. At this level the eyes are still functional because although no light is produced by the physical environment, biological light is produced by most marine organisms. Hence, when elephant seals descend to depths beyond the limit of surface light there is still much light in the environment. As marine mammals commute with great rapidity to and from the depths it is essential that there is rapid adaptation to the dark. Whereas land mammals may take tens of minutes to adapt to darkness, elephant seals may do so within 4–6 min. The shallower-diving harbor seal takes 18 min and humans about 22 min. Elephant seals may be helped in this adaptation because the contracted pupil is so small that even at the surface the amount of light reaching the retina is not great enough to saturate the rod receptors.

Conclusions

The management of oxygen stores has been a long standing topic of study in the history of marine mammal physiology since the first key works were published in the 1930s and 1940s. Progress in understanding the adaptations of marine mammals to conserve and utilize those stores efficiently has been fitful, depending on the techniques available. The research emphasis has shifted from restrained and forced submersions in the laboratory to experiments on free-ranging animals. New and powerful tools are becoming available that will enhance progress in this field. In other areas such as determining the function of some of the vascular retia of whales, or the function of some of the peculiar lung structure in these animals, progress has ceased despite the fact that some of this poorly understood anatomy appears to play a major role in the adaptation of these animals to the marine environment. The lack of investigations relates to unavailability of the animals, and the intrusive measures that would be required to determine function.

See also

Fish Locomotion. Fish Vision.

Further Reading

Butler PJ and Jones DR (1997) Physiology of diving of birds and mammals. *Physiological Review* 77(3): 837–899.

Kooyman GL (1989) *Diverse Divers: Physiology and Behavior. Zoophysiology Series*, vol. 23. New York: Springer-Verlag.

Kooyman GL (2000) Diving physiology. In: Perrin WF, Wursig B, and Thewissen JGM (eds.) *Encyclopedia of Marine Mammals*. San Diego: Academic Press.

Kooyman GL and Ponganis PJ (1997) The challenges of diving to depth. *American Scientist* 85: 530–539.

LeBoeuf BJ and Laws RM (1994) *Elephant Seals: Population Ecology, Behavior and Physiology*. Berkeley and Los Angeles: University of California Press.

Levenson DH and Schusterman RJ (1999) Dark adaptation and visual sensitivity in shallow and deep-diving pinnipeds. *Marine Mammal Science* 15: 1303–1313.

Shadwick RE (1998) Elasticity in arteries. *American Scientist* 86: 535–541.

Watkins WA, Daher MA, Fristrup KM, and Howald TJ (1993) Sperm whales tagged with transponders and tracked underwater by sonar. *Marine Mammal Science* 9: 55–67.

Williams TM, Davis RW, Fuiman LA, *et al.* (2000) Sink or swim: strategies for cost-efficient diving by marine mammals. *Science* 288: 133–136.

MARINE MAMMAL EVOLUTION AND TAXONOMY

J. E. Heyning, The Natural History Museum of Los Angeles County, Los Angeles, CA, USA

Since Eocene times some 57 million years ago, several groups of land mammals have independently evolved adaptations for a marine existence. With just over 120 modern species (**Table 1**), marine mammals are not taxonomically diverse compared to other groups of marine organisms, yet they are important components of many ecosystems. Marine mammals have evolved from only two terrestrial groups, or clades, of mammals. The first is the Order Carnivora, which includes such familiar creatures as cats, dogs, and bears. From this order arose the pinnipeds (seals, sea lions, walruses), the sea otter, and the polar bear. The other clade is the Ungulata, a group that includes all the orders of modern hoofed animals. Evolving from this rank are the cetaceans (whales, dolphins, and porpoises), the sirenians (dugongs and manatees), and the extinct hippo-like desmostylians.

Table 1 Marine mammals

	Common name	Number of living species
Order Cetacea		
Suborder Mysticeti	Baleen whales	
Family Balaenidae	Right whales	4
Family Neobalaenidae	Pygmy right whale	1
Family Eschrichtiidae	Gray whale	1
Family Balaenopteridae	Rorqual whales	8
Suborder Odontoceti	Toothed whales	
Family Physeteridae	Sperm whale	1
Family Kogiidae	Pygmy sperm whales	2
Family Ziphiidae	Beaked whales	20
Family Platanistidae	Indian river dolphins	1
Family Iniidae	River dolphins	3
Family Monodontidae	Beluga and narwhal	2
Family Phocoenidae	Porpoises	6
Family Delphinidae	Oceanic dolphins	36
Order Carnivora	Carnivores	
Family Phocidae	Seals	19
Family Otariidae	Sea lions and fur seals	16
Family Odobenidae	Walruses	1
Order Sirenia	Sea cows	
Family Dugongidae	Dugongs	1
Family Trichechidae	Manatees	3

Systematics is the science of defining evolutionary relationships among organisms. A phylogeny is a hypothesis of those evolutionary relationships, and is the foundation for any evolutionary study. It is the distribution of characters on a phylogenetic tree that is used to evaluate whether similar features in two organisms are a result of inheritance from a common ancestor or evolved via convergence. No proposed phylogeny can be proven, as proof would require the unattainable knowledge of past evolutionary events. Hence, researchers today can only infer past events from phylogenetic reconstructions of evolutionary relationships.

Most modern systematists use a philosophical approach called cladistics. The basic tenets of cladistics are quite simple: organisms are deemed to be related based on shared derived characters called synapomorphies. Derived characters are defined as having arisen in the common ancestor of the taxa and are subsequently passed on to their descendant taxa. Groups of related taxa and their descendants are called clades regardless of taxonomic rank. Monophyletic groups are those that include the ancestor and all of its descendants. Paraphyletic groups are those which include the ancestor taxa, but not all the descendants. Paraphyletic groups often reflect a certain level of morphological organization or 'grade,' and exclude some or all of the more derived descendants. For example, in classical cetacean taxonomy, the suborder Archaeoceti is unquestionably paraphyletic, as this group of fossil cetaceans includes the most basal species, but not the descendant suborders Mysticeti and Odontoceti.

Order Carnivora

Several groups of carnivores have evolved to a partially aquatic existence, although all must return to land, at least to give birth. The polar bear, *Ursus maritimus*, appears to be a very recent evolutionary divergence of the brown bear, *Ursus arctos*, lineage. Analysis of molecular data suggests that polar bears are most similar to those brown bears from the islands off south-east Alaska. Fossils suggest that the two species diverged in the middle Pleistocene.

There are three species of marine carnivores in the weasel/otter family (Mustelidae). Along the northeast coast of North America, the extinct sea mink, *Mustela macrodon*, has been found primarily from Native American midden sites. The second is the chungungo or marine otter, *Lutra felina*, a poorly known species inhabiting South American waters. Lastly, is the well-known sea otter, *Enhydra lutris*, of the temperate and subarctic North Pacific. These two otters are closely related to the fresh water otters. The sea otter appears to be related to the late Miocene to early Pliocene *Enhydritherium*, found along both coasts of North America and Europe.

Pinnipeds

The pinnipeds (from the Latin meaning 'fin-footed') are a group of the marine mammals, which includes the seals, sea lions, and walrus (**Figure 1**). Pinnipeds arose from the arctoid (bear, dogs, weasels, etc) lineage of carnivores. The pinnipeds consist of three living families, the Phocidae (true seals), the Otariidae (fur seals and sea lions), and the Odobenidae (walruses).

The terrestrial ancestor of the pinnipeds has been the subject of considerable debate. Two schools of thought exist. One, citing primarily biogeographical, and paleontological evidence, supports a diphyletic or dual origin, attributing the sea lions, fur seals, and walruses to an ursid (bear-like) ancestor evolving in the North Pacific and the true seals to a mustelid (weasels, otters, etc.) ancestor evolving in the North Atlantic. The other school, using molecular, karyological, and morphological evidence supports a monophyletic origin for all three pinniped families stemming from an unresolved arctoid ancestor. There is growing consensus that pinnipeds constitute a monophyletic group.

The extant true seals, Phocidae, are distinguished from the sea lions by the lack of external ear flaps, relatively short forelimbs with claws, backward directed hindlimbs that do not permit quadrupedal movement on land, and locomotion in water by sculling the hindlimbs. They are divided into two groups, the Monachinae ('southern' seals) and the Phocinae ('northern' seals). The oldest phocid fossils are from the late Oligocene of the North Atlantic.

The sea lions and fur seals (or 'eared' seals), Otariidae, can be distinguished from the phocids by the presence of external ear flaps, elongate flipperlike forelimbs, rotatable hindlimbs that allow quadrupedal locomotion on land, and locomotion in water by flapping of the fore-flippers. They are divided into two groups, the Arctocephalinae (fur seals) and the Otariinae (sea lions). The diagnostic feature is the presence of abundant underhair in the fur seals. The oldest fossil otariids are from the early Miocene of the North Pacific.

The walrus family includes mostly extinct species consisting of three groups. The most basal are

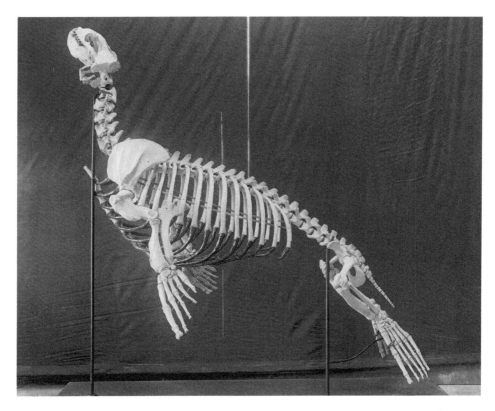

Figure 1   Skeleton of a modern southern sea lion (*Otaria flavescens*). Photo courtesy of the Natural History Museum of Los Angeles County.

archaic nontusked walruses referred to as imagotarines retaining the ancestral fish-eating diet of other pinnipeds whereas the Dusginathinae are an extinct group apparently specializing in squirt/suction feeding behavior for foraging on shellfish and/or crustaceans. Dusignathines have enlarged canines in both jaws providing a four-tusked image to the skull. These two groups are distinguished from the Odobeniinae that only develop tusks in the upper jaw and eventually gave rise to the modern walrus, *Odobenus rosmarus*, which feeds primarily by sucking bivalves out of their shells.

Order Cetacea

Until quite recently it was commonly asserted that whales were fish, and not mammals. Today, the cetaceans (whales, dolphins, and porpoises) are readily recognized as mammals. Their relationship to the ungulates (hoofed terrestrial mammals including horses, cows, pigs, camels, elephants, etc.) is generally accepted, though their closest relation among the living ungulates remains a topic of research. The Order Cetacea consists of three suborders: the Archaeoceti (an extinct group of archaic whales), the Odontoceti (toothed whales), and the Mysticeti (baleen whales).

The hypothesis that the order Cetacea is monophyletic is supported by an overwhelming amount of morphological, cytological, and molecular evidence.

The karyotypes of cetaceans are amazingly conservative when compared to other groups of mammals. The typical chromosome count for most cetaceans is $2N = 44$. The exceptions are sperm whales (Physeteridae and Kogiidae), beaked whales (Ziphiidae), and right whales (Balaenidae). For right whales and beaked whales, the lower counts ($2N = 42$) are a result of the fusion of different chromosome pairs, respectively. The chromosome-banding pattern among cetaceans is also astonishingly conservative.

In modern cetaceans (**Figure 3**), the body shape is fish-like – streamlined with fin-like flippers and flukes. The hind limbs are but vestiges tucked within the body wall, and the nostrils are situated high on the head and termed blowholes. It is because cetaceans differ so strikingly from their terrestrial kin that it is difficult to discern intuitively which, among the other orders of ungulate mammals, are their closest kin.

There is now convincing fossil evidence that land-dwelling extinct mesonychians are closely related to the ancestor of whales; either they are the sister-taxon to the whale ancestor, or whales arose from within the paraphyletic extinct family Mesonychidae (**Figure 2**). There are many similarities between whales and at least some mesonychids in such features as the construction of the cheekteeth, the humerus, and the venous drainage of the skull. An analysis of the postcranial skeleton of the

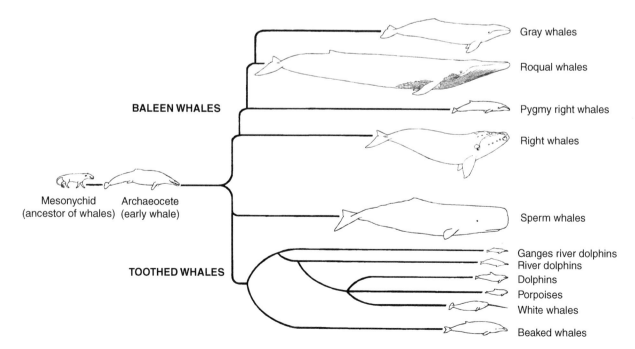

Figure 2 Phylogeny of the modern families of cetaceans based on morphological and molecular data (modified from Heyning, 1995).

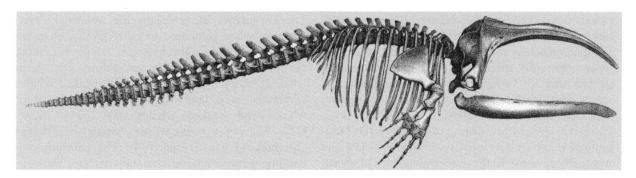

Figure 3   Skeleton of a modern baleen whale (*Balaena mysticetus*). (From Van Beneden and Gervais, 1880.)

mesonychid *Pachyaena* suggests that this animal may have had a body form similar to tapirs, capybaras, or suids (pigs), many of which are excellent swimmers. Some molecular studies indicate that the Cetacea may have actually evolved from within the order Artiodactyla.

Suborder Archaeoceti

Archaeocetes represent a primitive grade-level taxon that includes all cetaceans that lack the derived telescoped pattern of skull bones found in either mysticetes (baleen whales) or odontocetes (toothed whales). Archaeocetes are further characterized by an elongate snout, a narrow braincase, a skull with large temporal fossa and well-defined sagittal and lambdoidal crests, a broad supraorbital process of the frontal bone, and bony nares situated some distance posterior to tip of the snout. The archaeocete grade was 'extinct' by the end of the Eocene. The archaeocetes include five families: Pakicetidae, Ambulocetidae, Protocetidae, Remingtonocetidae, and Basilosaurocetidae.

Most Pakicetids have been unearthed from terrestrial, freshwater, or nearshore deposits of the eastern Tethys, a massive epicontinental sea that divided Eurasia from Africa/India. The oldest and most primitive whale, *Pakicetus inachus* was described from Pakistan. Many of these fossils from the eastern Tethys Sea, some with known hind limbs, are significant in that they serve as morphological transitions between land-dwelling mammals and fully aquatic cetaceans. The Ambulocetidae occupied tidal and estuary habitats. The enigmatic *Ambulocetus natans* differs from all other known archaeocetes in that its eyes are elevated above the overall profile of the skull, and its hind feet are elongate imparting a crocodile-like appearance. These hindlimbs were probably important for aquatic locomotion. The Protocetidae is defined by details of the orbits, teeth, and sacrum. In the eastern Tethys, protocetids are found in nearshore waters, whereas protocetids from the western Tethys died in shallow offshore regions and their North American kin have been found in shallow nearshore deposits. Remingtonocetidae have extremely long and narrow skulls with small eyes and are known from rock units from Pakistan and India. These archaeocetes are found in sediments indicating a coastal environment or nearshore shallow marine deposits. Though they are interpreted to be the most derived of the archaeocete families, the Basilosauridae were among the first archaeocete fossils to be discovered. Basilosaurids are typically divided into two subfamilies: the Durodontinae with unspecialized vertebrae and the Basilosauinae with extremely elongate vertebral bodies. It has been suggested that durodontines were ancestral to both mysticetes and odontocetes. However, further evidence suggests that the durodontines represent the sister taxa to the modern Cetacea clades.

One of the most dramatic morphological changes of early cetaceans was the shift from quadruped locomotion on land to the axial undulation of swimming in the ocean. These include changes of the vertebral column, limb structure, and of the tail flukes. One striking conclusion is how quickly these transitions occurred. At the dusk of the Early Eocene, *Pakicetus* and its contemporaries were quadruped animals that drank fresh water. A few million years later in the Middle Eocene, *Rodhocetus* and its kin were swimming with the aid of tail flukes and drinking sea water. By the Late Eocene, Basilosaurines possessed such exceedingly small hind limbs that it is unlikely that they were of much use in terrestrial locomotion. Hence, one can speculate that by the Late Eocene cetaceans had severed all links to the land.

One hypothesis is that archaeocetes first evolved in fluvial or estuarine environments of the eastern Tethys and subsequently dispersed as more morphologically and physiologically derived forms conquered the oceans. All of the most primitive and chronologically oldest fossil archaeocetes are found along the shores of the eastern Tethys, whereas the more morphologically derived and fully marine protocetids and basilosaurids of the Middle and Late Eocene are found in rocks from Asia, North Africa, North America, New Zealand, and Antarctica.

Suborder Odontoceti

Some cetaceans possess teeth (odontocetes), whereas others have baleen for filtering out prey (mysticetes). There are numerous other synapomorphies that unite the Odontoceti such as the maxilla (upper jaw) which has telescoped back over the frontal bone of the skull. All living and all but the most basal fossil odontocetes have asymmetrical skulls, and all modern species have asymmetrical facial soft anatomy. This asymmetry results from an enlargement of the facial soft anatomy on the right side. All odontocetes have a complex series of air sacs off their nasal passages. All extant odontocetes possess an enlarged fatty melon in front of the nasal passages, which is distinctly different from the diminutive melon-like structure observed in mysticetes. Odontocetes are unique among all tetrapods in that the distal narial passages coalesce to form a single nostril or blowhole. The large melon, complex nasal anatomy, and asymmetry all appear to be correlated with the ability to echolocate.

The sperm whales are represented today by the giant sperm whale (Physeteridae) and the dwarf sperm whales (Kogiidae). This clade is the first to diverge from the lineage of living odontocetes. All sperm whales are recognized by the presence of a spermaceti organ in their head and very asymmetrical skulls. All living species are known or suspected to be deep divers. The beaked whales (Ziphiidae) are a very diverse group with at least 20 known living species. This group is typified by extreme sexual dimorphism in that males of most species have a one or two enlarged pairs of teeth used primarily for fighting other males, presumably for establishing breeding dominance. The river dolphin (genus *Platanista*) of the Indian subcontinent is the sole living representative of the family Platanistidae. The bone pattern of the palate and the elaborate crests of bone on top of the skull define this family. The other living river dolphins and their kin belong to the family Iniidae, with living representatives found in the waters of the Yanzee and Amazon river basins, and coastal waters of eastern South America. This clade is defined by the extreme asymmetry in the nasal sacs. The remaining odontocetes are the closely related families of the Monodontidae (narwhal and beluga), Phocoenidae (porpoises) and the taxonomically diverse oceanic dolphins (Delphinidae).

This clade represents the vast majority of living species, and includes those most familiar to most people.

Suborder Mysticeti

The mysticetes, with their edentulous mouths lined with filtering baleen, are one of the distinct clades among the mammals. The evolutionary transition from capturing single prey to filtering numerous prey items out of mouthfuls of sea water has ramifications not only in the morphology, but also the behavioral ecology of these, the largest of all animals. Mysticetes evolved from cetaceans that possessed teeth. Certain cranial features predate the loss of teeth in the mysticete clade and, therefore, the most basal mysticetes retain teeth.

The oldest known mysticete is the toothed species *Llanocetus denticrenatus* from the Late Eocene of Seymour Island, Antarctica. The next oldest specimens are those of the Oligocene cetotheres whose wide, edentulous palates strongly imply the presence of baleen. As baleen is made of the protein keratin, it typically decomposes with the other soft tissues rarely leaving a fossil trace. These fossil discoveries now represent a moderately good morphological series from the archaeocetes to modern mysticetes.

Three families of extinct mysticetes are recognized. They are the Llancetidae, the Aetiocetidae, and the Cetotheridae. The family Llanocetidae is based on one species, *Llanocetus denticrenatus*, with an estimated total length of perhaps 10 m. Aetiocetids are relatively small toothed mysticetes known only from the shorelines of the North Pacific. *Chonecetus* and some species of *Aetiocetus* retain the primitive eutherian mammal tooth count. *Aetiocetus polydentatus* with its expanded toothcount exhibits an incipient stage of the derived feature of supernumerous teeth as seen in later cetaceans. The Cetotheriidae represent a phylogenetically heterogeneous assemblage that is truly a 'wastebasket' taxon with over 60 named species within 30 or so genera of unknown affinities. The rostrum is typically broad and flat, not dissimilar to that found in the primitive aetiocetids and modern balaenopterids. The oldest cetotheres are *Cetotheriopsis lintianus* from Austria and the relatively complete skull of *Mauicetus lophocephalus* from New Zealand, both from the Late Oligocene.

The modern mysticetes are divided into four families: the Balaenidae, Neobalaenidae, Eschrichtiidae, and Balaenopteridae. The systematics of these families are much less contentious than that for the modern odontocetes, though a few taxa remain elusive with regard to their relationship within and among other modern mysticete groups.

The right whales (Balaenidae) are heavy-bodied mysticetes with large arched heads and cavernous mouths to accommodate the extremely large filtering surface formed by the extraordinarily long baleen plates. Balaenids lack throat grooves. There are two species of modern balaenids, the bowhead and the right whale. Balaenids have skulls that are narrow and highly arched. The baleen is long, narrow, with a fine fringe. The dorsal fin is absent. The oldest fossil of an extant mysticete family is the balaenid *Morenocetus parvus* from the earliest Miocene of Argentina.

The family Neobalaenidae is represented solely by the poorly known Southern Hemisphere pygmy right whale, *Caperea marginata*. Some workers have considered *Caperea* to be a balaenid. However, there is ever-growing consensus, based on morphology and molecular data, that *Caperea* is not an 'aberrant' right whale, but is instead more likely the sister group to the rorqual/gray whale clade. The dorsal fin is small, yet distinctive. The throat grooves of *Caperea* are highly variable in depth, and virtually absent in some individuals. The rostrum is only somewhat arched, intermediate between the conditions seen in gray whales (Eschrichtiidae) and right whales (Balaenidae). The ribs are unique among cetaceans, living or fossil, in that they are broad and overlap each other in profile.

The enigmatic gray whale, *Eschrichtus robustus*, is the sole member of the family Eschrichtiidae. The overall mottled gray color and small dorsal fin followed by a series of dorsal ridges characterize gray whales. The two to five throat grooves are well delineated and are confined to the gular region. The rostrum is attenuate and moderately arched. The yellowish to white baleen is relatively short and moderately wide. Gray whales have lost a digit in their flipper. The only fossil gray whale is a Late Pleistocene individual of superb preservation. However, this animal is indistinguishable from the modern species and its relative young age does not help to resolve the relationship between gray whales and other baleen whales.

The Balaenopteridae, also known as the rorquals, are immediately recognized by their numerous throat grooves. These distinctive throat grooves are numerous, ranging from 14–22 grooves in the humpback whale (*Megaptera novaengliae*) to 56–100 grooves in the fin whale (*Balaenoptera physalus*). Balaenopterids are the 'greyhounds of the sea.' Their bodies are the sleekest among living mysticetes. The dorsal fin is always present and tends to be inversely proportional to body size. The rostrum is extremely broad and flat. Rorqual whales have a very complex interdigitating pattern of bony sutures between the rostral bones and those of the braincase proper. Only four digits are present in the flippers. The humpback is unique among balaenopterids in that it has extremely elongate flippers.

There are three major cranial character suites that have been used to ascertain the phylogenetic position of fossil baleen-bearing whales. These are the position and shape of the occipital shield, the complexity of interdigitation between the bones of the rostrum (nasals, premaxillae, and maxillae) and the braincase proper, and slope of the supraorbital process of the frontal bone. Ancestrally, the occipital shield does not extend very far anteriorly providing dorsal midline exposure of the parietal and frontal bones. The most primitive character state of this feature is seen on the various toothed mysticetes. In the most advanced character state, the occipital is close to the nasals and premaxillae on the vertex. This condition is found in modern balaenopterids, neobalaenid, and also in some cetotheres. The complex interdigitation of the bony sutures is clearly derived. Incipient interdigitation of the rostrum and braincase is seen in cetotheres. Modern balaenopterids uniquely possess a supraorbital process of the frontal that is flat and horizontal until it reaches the braincase and then abruptly turns dorsally to the skull vertex. The result is a large region over the supraorbital process for the greatly enlarged temporalis muscle required to close the mouth after engulfing tons of water. The cetothere *Cetotherium* has distinct crests on the temporal ridge along the contact with the frontals which suggests a condition that foreshadows the state found in modern balaenopterids.

Although differing somewhat in detail, most morphologically based phylogenies of baleen whales as well as molecular-based studies suggest that the balaenids were the first clade to diverge, followed by the Neobalaenidae, then the Eschrichtiidae and Balaenopteridae as sister taxa. Several studies using molecular sequence data have implied that the ancestry of the gray whale (Eschrichtiidae) is located near or within the genus *Balaenoptera* (Balaenopteridae).

Order Sirenia

The sirenians, manatees, dugongs, and their extinct relatives, are a fully aquatic herbivorous group of mammals. The living species are restricted to tropical and subtropical waters; however, fossil species appear to have inhabited temperate waters and the range of the recently extinct Steller's sea cow extended into Arctic waters. Modern dugongs feed primarily on seagrasses, whereas the Steller's sea cow fed on brown algae, and manatees feed on a variety of freshwater aquatic plants. Sirenians are part of the

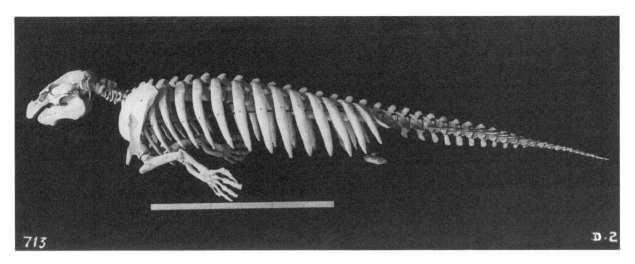

Figure 4 Skeleton of a modern manatee (*Trichechus manatus*). Photo courtesy of the Natural History Museum of Los Angeles County.

ungulate clade called Tethytheria, which also includes the elephants and their extinct relatives, the extinct marine desmostylians, and in some classifications, the hyraxes. All living groups have a unique form of tooth replacement in which new teeth originate at the back of the toothrow, then slowly move forward, and finally the worn down tooth drops out of the front.

The oldest and most primitive sirenians are found in Eocene rocks; *Prorastomus* from the Early and Middle Eocene of Jamaica and *Protosiren* from the Middle Eocene of Pakistan, North Africa, and Europe. Ancient sirenians had four limbs and were amphibious, but as with the cetaceans, modern species have but vestiges of the pelvic girdle and are propelled entirely by tail flukes (**Figure 4**). There are two living groups of sirenians: the manatees (family Trichechidae) and the dugong (Dugongidae). The fossil record is rich with taxa, which is not the case for the living species.

There are but two species of recent dugongids: the dugongs, *Dugong dugon*, of the Indo-Pacific and the recently extinct Steller's sea cow, *Hydrodamalis gigas*. Modern Dugongids have pointed flukes similar to cetaceans and are exclusively marine. The upper incisors are tusk-like in males. The Steller's sea cow was hunted to extinction in its last refugia of Commander Islands in 1768, but it occurred recently (19 000 years ago) as far south as Monterey, California. Dugongids were widespread in Miocene times and apparently spread from the Atlantic into the Pacific prior to the formation of the Isthmus of Panama. Dugongids subsequently became extinct in the Atlantic. There are three closely related species of

manatees all distributed in the Atlantic: the West Indian manatee, *Trichechus manatus*; the African manatee, *T. senegalensis*; and the exclusively freshwater Amazon manatee, *T. inunguis*. Modern manatees have rounded tail flukes and are primarily found in fresh water although the West Indian manatees are not uncommonly found in marine waters. The fossil record for manatees is far sparser than that of dugong.

See also

Baleen Whales. Seals. Sea Otters. Sirenians. Sperm Whales and Beaked Whales.

Further Reading

Berta A and Demérè T (eds.) (1994) *Contributions in Marine Mammal Paleontology Honoring Frank C. Whitmore, Jr.* Proceedings of the San Diego Society of Natural History.

Fordyce RE and Barnes LG (1994) The evolutionary history of whales and dolphins. *Annual Review of Earth and Planetary Sciences* 22: 419–455.

Heyning JE (1997) Sperm whale phylogeny revisited: analysis of the morphological evidence. *Marine Mammal Science* 13: 596–613.

Rice DW (1998) *Marine Mammals of the World: Systematics and Distribution.* The Society for Marine Mammalogy. Special Publication No. 4.

Thewissen JGM (ed.) (1998) *The Emergence of Whales: Evolutionary Patterns in the Origin of Cetacea.* New York: Plenum Press.

MARINE MAMMAL MIGRATIONS AND MOVEMENT PATTERNS

P. J. Corkeron, James Cook University, Townsville, Australia
S. M. Van Parijs, Norwegian Polar Institute, Tromsø, Norway

Introduction

Marine mammals are renowned as great travelers. The migrations of some whales and seals span ocean basins, but other species have home ranges limited to a few square kilometers. Factors that influence marine mammals' movements strongly include how they give birth and mate, and how their food is distributed in space and time. These differ substantially across the marine mammal groups, and so they are dealt with separately in this chapter.

Marine and terrestrial ecosystems differ substantially as environments for air-breathing homeotherms. The density of sea water leads to far lower energetic costs of locomotion for animals living in marine systems. Living in a dense medium also allows mammals to grow to larger sizes than is possible on land, and the energetic cost of travel decreases exponentially with an animal's size. These factors, coupled with the connectivity of oceans, mean that marine mammals generally have far larger ranges, or longer migratory routes, than terrestrial mammals.

However, marine mammals' need for gaseous oxygen means that, unlike most marine animals, they must return regularly to the air–water interface to breathe. Also unlike most marine vertebrates, mammals give birth to live young. These young need to breathe immediately after birth, so they must either be remarkably good swimmers at birth or be born on land. Cetaceans, sea otters, and sirenians give birth at sea, but pinnipeds and polar bears give birth on land. This leads to major differences in their movements.

Locomotion represents an energetic cost to animals, so all individual animals are selected to balance the cost of travel against the benefits gained by travel. Unlike air, which is distributed evenly at the water surface, the food resources of marine mammals are found in patches of differing scales, varying in both space and time. Other marine species are distributed through the oceans' depths, so marine mammals have a third dimension available for their travel. Changes in the density of sea water with depth mean that marine mammals may expend little energy when descending in deep water, but must display physiological adaptations to deal with the extreme pressure associated with these deep dives.

Animals' movements can be considered on several temporal and spatial scales. Migrations, generally on an annual cycle, involve persistent movement between two destinations. An animal's home range is that area within which it carries out most of its normal activities throughout the year. Classically, home ranges are not considered to include migratory movements.

Cetaceans

Two factors appear to be responsible for substantial differences between the movement patterns of mysticetes and odontocetes. Most mysticetes feed in polar or cold temperate waters, highly seasonal environments. Therefore, mysticetes' prey are more heavily based on an annual cycle than the prey of most odontocetes. Also, although all cetaceans give birth aquatically, the location for giving birth appears to be particularly important to mysticetes.

Mysticetes

The annual cycle of most baleen whales is characterized by migrations between polar or cold temperate summering grounds and warm temperate, subtropical or tropical wintering grounds. In general, whales feed on their summering grounds and breed on their wintering grounds. These migrations include movements of nearly 8000 km by some humpback whales, the longest known annual movements of any mammal. Generally, whales' longitudinal (east–west) movements are relatively small when compared with their latitudinal (north–south) movements. Individual animals tend to return to the same summering and wintering areas over several years.

Some species – right (*Eubalaena* spp.), gray (*Eschrichtius robustus*), and humpback whales (*Megaptera novaeangliae*) – migrate relatively close to coastlines. *Balaenoptera* spp. tend to migrate further offshore, and their movements are less well known than those of the coastal migrators. Species also vary in the distance over which they migrate; for example, right whales tend to cover less latitudinal range than humpback or gray whales.

Variation in Migratory Patterns

Not all species of baleen whale demonstrate this annual cycle. Bowhead whales, *Balaena glacialis*, undertake substantial longitudinal movements around the coasts of Alaska, Canada, and Siberia, but the southernmost extent of their movements remains close to the pack ice edge. At the other thermal extreme, some tropical Bryde's whales, *Balaenoptera brydei*, may not migrate at all. A population of humpback whales in the Arabian Sea also appears not to migrate.

There can be great intraspecific variability in the distances traveled by baleen whales. Humpback whales that congregate in Caribbean waters to breed come from discrete feeding groups in Atlantic waters, including western Greenland, off Newfoundland/Labrador, the Gulf of St Lawrence, off the northeastern USA, Iceland, and from north of Norway (**Figure 1**). From Antarctic waters, some humpback whales migrate into equatorial waters in the Northern Hemisphere, but others seem not to migrate north of approximately 15°S.

Individuals of migratory species do not necessarily migrate every year. Some female and juvenile humpback whales, and female southern right whales do not arrive at their wintering grounds every year. Large baleen whales were sighted south of the Antarctic Convergence in winter by early expeditions, and Antarctic minke whales, *Balaenoptera bonarensis*, have been seen inside the pack ice during winter. There are records of blue (*B. musculus*) fin

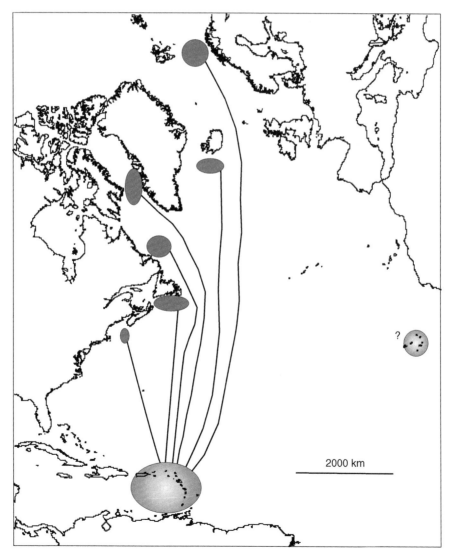

Figure 1 Migratory paths of humpback whales in the North Atlantic.

(*B. physalus*) and humpback whales wintering in high latitude waters of the North Atlantic.

There are records of remarkable longitudinal movements by some baleen whales. Tagging programs during Antarctic whaling demonstrated that individual blue whales could move around the Antarctic continent. Individual humpback whales, apparently males, have been observed overwintering in Hawaiian and Japanese waters in different years. Individual southern right whales have moved between the coast of South America and islands in the central South Atlantic.

As some populations of baleen whales recover from previous overhunting, further variability in migratory behavior is becoming evident. Some gray whales now feed in waters off British Columbia, well south of their primary feeding grounds in Arctic waters. Wintering southern right whales off the east coast of Australia are occurring in more northern waters. Further recoveries of baleen whale populations should reveal more behavioral variability at both wintering and summering grounds.

Why Do Baleen Whales Migrate?

Why baleen whales use high latitude feeding grounds is clear – polar systems produce incredible quantities of baleen whales' prey in the warmer months. The enigmatic aspect of baleen whale migration is why most whales travel to low latitude breeding grounds. Current hypotheses regarding why baleen whales migrate relate to calf survivorship: that calves are born away from polar waters so that in the first few weeks of life they avoid either cold waters, stormy waters, or predation by killer whales, *Orcinus orca*. Information available at present does not allow definitive conclusions on which of these competing (although not mutually exclusive) hypotheses is correct.

Odontocetes

Most odontocetes are smaller than mysticetes, and do not show their annual migratory patterns, nor move over ocean basins. Sperm whales, *Physeter macrocephalus*, are the exception to this. The movements of most odontocete species are not well known.

Sperm Whales

Baleen whales travel long distances latitudinally, but sperm whales are known more for their deep foraging dives. However, male sperm whales also undertake significant latitudinal movements. Females with calves and juveniles live in matrilineal groups in

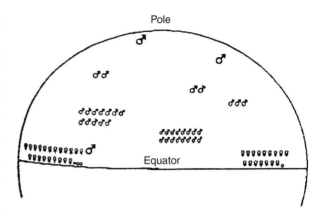

Figure 2 Diagrammatic representation of the latitudinal distribution of sperm whales in the Northern Hemisphere. (Reproduced with permission from Mann *et al.*, 2000.)

tropical and subtropical waters. These groups occupy home ranges with a long axis in the order of 1000 km. Young males leave their natal groups and move into higher latitude waters with conspecifics of similar age. As they mature, male sperm whales become less sociable and move into polar waters. Mature males migrate from their high latitude feeding areas to tropical waters to mate, but whether these are annual migrations is unclear (**Figure 2**).

Despite these long-distance migrations, small areas can be important. On one well-studied (and highly productive) feeding ground of only 20 × 30 km, up to approximately 90 males can be seasonally resident. Females' foraging ranges are larger, and it appears that their large ranges are a strategy to cope with interannual variability in environmental productivity.

Delphinids

Bottlenose dolphins   Bottlenose dolphins, *Tursiops truncatus* and *T. aduncus*, are among the best studied of the delphinids. They are found from temperate to equatorial waters and from shallow waters by the coast to the deep ocean. Throughout their range, two ecotypes occur: inshore and offshore forms, with offshore animals being more robust. The relationship between these two forms and the two species of *Tursiops* is unclear, confounding comparisons. Bottlenose dolphins ranging patterns vary from animals with relatively small home ranges to migratory animals.

Bottlenose dolphins living in sheltered coastal waters, particularly bays, tend to have home ranges of several tens of square kilometers, varying in size with gender, age, and reproductive status. Most individuals appear to remain within these ranges throughout their life. Animals living in shallow waters off open coasts (e.g. California, South Africa) can range over >100 km of coastline. Elsewhere (e.g. eastern Australia) dolphins in shallow waters off open coasts can have small home ranges, similar to those of bay animals elsewhere. In some areas, populations of bottlenose dolphins undertake relatively long migrations. Some inshore dolphins off the US east coast migrate annually over approximately 400 km of coastline, while offshore animals in the same area move even further.

Most information on bottlenose dolphins comes from studies of individually identified animals. Logistically, these studies are likely to concentrate on animals with relatively small ranges. Satellite tracking offshore for bottlenose dolphins has started to reveal the extent over which they can range. Two animals tracked off Florida traveled 2050 km in 43 days, and 4200 km in 47 days. As these animals were rehabilitated after stranding, the extent to which their movements are representative of normal movements is debatable, but they demonstrate the ranging capacities of offshore bottlenose dolphins.

Killer Whales

Killer whales are found in all the world's oceans, from polar to equatorial waters, but their densities in polar waters are substantially higher than in tropical waters. There are two genetically distinct forms of killer whales: residents, feeding mainly on fish, and transients, that are marine mammal predators. Although these forms are best described from the waters off British Columbia, similar forms have been reported from Antarctic waters. Despite the names, the differences in ranging of these two forms of killer whales are not necessarily great. Off British Columbia and Washington, the ranges of transient pods extend to approximately $140\,000\,\mathrm{km}^2$ and those of residents to $90\,000\,\mathrm{km}^2$. As with bottlenose dolphins, logistics limit knowledge of the extent of killer whales' ranges.

Migratory behavior of killer whales remains poorly understood. The longest movements documented to date are of three identified individual transient killer whales that moved 2660 km along the Pacific coast of North America, from 58°41′N to 36°48′N over 3 years. Soviet whaling data from the Antarctic suggest that some killer whales undergo annual migrations to at least temperate latitudes, but sightings of killer whales in the Antarctic pack ice in winter demonstrate that not all animals migrate.

Killer whales' movements seem tied to movements of their prey: north–south movements of some killer whales along the North American west coast appear to be at least partially in response to the presence of migrating gray whales. Killer whales off coastal Norway seem to follow the herring migration. Killer whales appear to move into nearshore waters at several areas in the subantarctic – Peninsula Valdes Argentina – Marion Island, Macquarie Island, the Crozet Archipelago – coincident with seal pupping, although this may reflect the limits of shore-based observation to determine killer whales' real movements.

Other Odontocetes

Some inshore delphinids seem to demonstrate the variability in ranging behavior shown by bottlenose dolphins. Some individual humpback dolphins (*Sousa* spp.) in bay and estuarine environments have small home ranges (tens of square kilometers), others in more coastal waters range along hundreds of kilometers of coastline. Along 2000 km of the open west coast of South Africa there are three distinct matrilineal assemblages of Indian humpback dolphins, *S. plumbea*. Marine Irrawaddy dolphins, *Orcaella brevirostris*, appear to have small ranges (tens of square kilometers), centered around the mouths of rivers.

Movement patterns of delphinids in offshore waters are less well understood, but it is clear that animals range over at least thousands of square kilometers. Surveys off the west coast of the USA, covering over 10 degrees of latitude and extending to 550 km offshore, demonstrated seasonal changes in the distribution of small cetaceans. For example, northern right whale dolphins, *Lissodelphis borealis*, moved into continental shelf waters of the Southern California Bight in the winter, presumably from waters further offshore. Pacific white-sided dolphins, *Lagenorhynchus obliquidens*, and Dall's porpoises, *Phocoenoides dalli*, moved into more southern waters in winter.

Belugas (*Delphinapterus leucas*) and narwhals (*Monodon monoceros*) are ice-associated odontocetes found in Northern Hemisphere waters. Both species' annual movements through Arctic waters are closely tied to the movements of pack ice. One satellite-tagged narwhal traveled >6300 km in 89 days, moving through >6° of latitude. Satellite tagging studies of both species reveal times when they move to the vicinity of glaciers, for reasons that are unclear.

Pinnipeds

Pinnipeds comprise three families: Phocidae, the true or haired seals; Otariidae, the eared or fur seals; and

Odobenidae, the walrus, *Odobenus rosmarus*. Unlike cetaceans or sirenians, pinnipeds have retained some terrestrial traits while adapting to foraging at sea. Although all pinnipeds give birth on land or ice, the importance of terrestrial sites for suckling, rest, mating, molting, predator avoidance, thermoregulation, or saving energy varies considerably across species. Within species, individuals' movement patterns vary depending both upon local environment conditions and individuals' age, sex, or breeding status.

Breeding

Pinniped movement patterns differ from other marine mammals due to the need for females to give birth and (generally) nurse on land. Female pinnipeds exhibit three maternal strategies, 'aquatic nursing,' the 'foraging cycle,' and the 'fasting cycle.' To date, the walrus is the only species to exhibit aquatic nursing. Female walrus give birth, fast for a few days, and then take their calves with them while they forage at sea. Lactation is extended and lasts for up to 2 or 3 years. This strategy bears some resemblance to that of cetaceans and sirenians. Most phocid and several otariid species exhibit the foraging cycle strategy. In these species, females forage at sea during lactation, returning to nurse their pups. Lactation can be relatively prolonged in these cases, most otariids suckle for weeks or months. Lactation is generally less prolonged in phocid species. Harbor seals, *Phoca vitulina*, nurse their pups for 24 days, and forage at sea during the latter stage of nursing (**Figure 3**). Female phocids using the fasting strategy remain on land throughout the whole nursing period. This period is relatively short, around 18 days for gray seals, *Halichoerus grypus*, 23 days for southern elephant seals, *Mirounga leonina*, and only 4 days for hooded seals, *Cystophora cristata*. Therefore, throughout the breeding season, the movements of female seals are constrained to different degrees by the need to suckle.

Although male pinnipeds are not limited by the need to nurse a pup, their movements during the breeding season are constrained by their need to mate. Males either obtain access to females on land or in the water. Approximately half of the pinniped species mate on land. However, walrus, two otariid species, and 15 species of phocid mate aquatically. Land mating pinnipeds remain onshore to defend either territorial access to females, such as in elephant seals, *Mirounga* spp. or access to resources used by females, as in Antarctic fur seals, *Arctocephalus gazella*. Males of aquatic mating pinnipeds, such as harbor seals, were previously thought not to restrict their movements during the mating season. However, recent evidence has shown that males perform vocal and dive displays in small discrete areas and limit their movements to areas frequently used by females throughout the mating season (**Figure 3**). Breeding activities clearly regulate movements of both female and male pinnipeds.

Nonbreeding

There is less variation between movement patterns of males and females outside of the breeding season, although differences do exist. During this period movement patterns are more strongly governed by resource (generally prey) availability.

Pinnipeds exhibit marked variation in the distance they move in order to reach suitable feeding grounds or breeding habitats. Harbor seals remain faithful to a single site or small group of local sites and usually do not forage >50 km from their haul-out sites. Similarly, southern sea lions, *Otaria flavescens*, forage up to 45 km offshore, with occasional longer trips extending to >150 km. In contrast, northern elephant seals, *Mirounga angustirostris*, carry out long-distance migrations of several thousand kilometers, travelling from California to the north-eastern Pacific Ocean twice a year (**Figure 4**).

Pups and juvenile animals are unable to travel as extensively as adults. In harbor seals, mothers restrict their foraging range while accompanied by their pups. Juvenile northern elephant seals do not dive as deeply, move more slowly, and do not migrate as far as adults during the first few years of their lives.

Haul-out Sites

Most pinnipeds remain faithful to a single or small group of haul-out sites. Groups of seals at different sites within a local haul-out area may show consistent differences in sex or age structure. Seasonal changes in haul-out site result from changes in foraging grounds, seasonal availability of prey, and characteristics of haul-out sites. Frequently, females and pups predominate at certain sites during the pupping season. Sheltered isolated areas may be chosen because of the lack of disturbance or terrestrial predators. Other sites may be used predominantly during molting.

Timing of trips to sea in relation to tidal and diel cycles varies considerably both within and between areas. Seals using haul-out sites that are available throughout the tidal cycle have activity patterns dominated by the diel cycle. When seals use haul-out sites that are only available over low tide, the tidal cycle has a more dominant effect. These effects are

Female harbor seal distribution

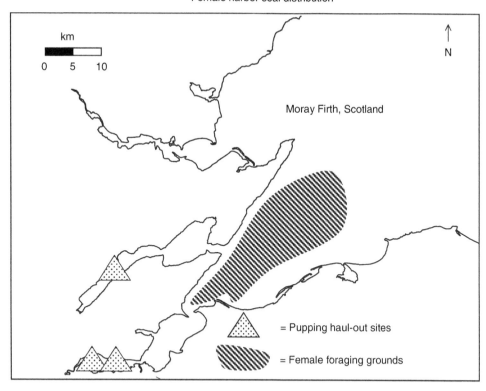

Male harbor seal distribution

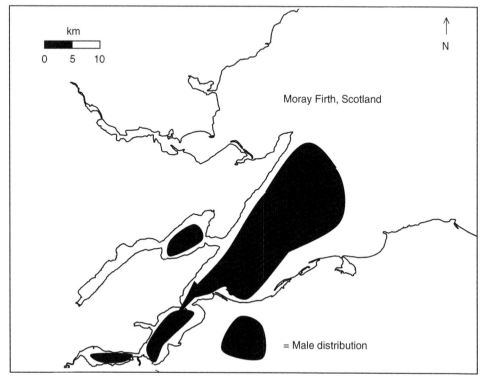

Figure 3 Ranges of harbor seals in the Moray Firth, Scotland. (From Thompson PM, Miller D, Cooper R and Hammond PS (1994) Changes in the distribution and activity of female harbour seals during the breeding season: implications for their lactation strategy and mating patterns. *Journal of Animal Ecology* 63: 24–30. Van Parijs SM, Hastie GD and Thompson PM (1999) Geographic variation in temporal and spatial patterns of aquatic mating male harbour seals. *Animal Behavior* 58: 1231–1239.)

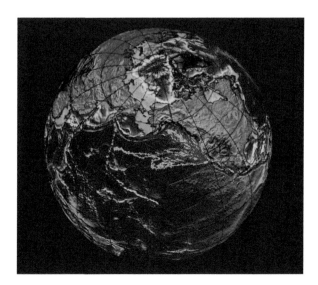

Figure 4 Movements of 22 adult males (red) and 17 adult females (yellow) tracked by satellite during spring and fall migrations from A~o Nuevo, California, during 1995, 1996, and 1997. (Reproduced with permission from Le Boeuf *et al.*, 2000.)

less pronounced when pinnipeds engage in longer foraging trips.

Small-scale movements in the vicinity of haul-out areas can be affected by the risk of predation. For example, northern elephant seals approaching major breeding areas alter their diving behavior in a way that appears to reduce the risk of attack by the white shark, *Carcharodon carcharias*.

Pinnipeds hauling out on sand or rocky shorelines are exposed to different constraints to those hauling out on ice. Ice breeding pinnipeds often show a partiality to hauling out on a particular type of ice. As ice changes often and suddenly, they may be more constrained in the choice of haul-out site at particular periods of the year. Bearded seals, *Erignathus barbatus*, haul out on ice floes, the availability of which alters seasonally, daily, and hourly. Therefore, movements of both female and male bearded seals reflect the availability of a particular type of ice within an area.

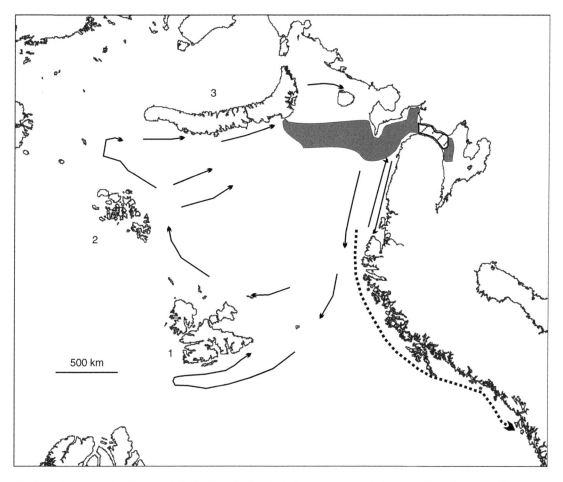

Figure 5 Annual movements of harp seals in the Barents Sea, including movements during recent invasions of the Norwegian coast. (1) Svalbard; (2) Franz Josef Land; (3) Novaja Zemlja. Gray shading shows the molting area, hatching shows the breeding area. The dashed line indicates the extent of movements of harp seals during invasions over recent years. (Data primarily from Haug T, Nilssen KT, Øien N and Potelov V (1994) Seasonal distribution of harp seals (*Phoca groenlandica*) in the Barents Sea. *Polar Research* 13: 163–172.)

The annual movements of harp seals, *Phoca groenlandica*, in the Barents Sea demonstrate their relationship with both ice and food availability. These seals breed in the White Sea in February/March, then molt in April/May in the White Sea and southern Barents Sea. In March and April, adult females seem to move westward on a short feeding migration. Between June and September, they are found in open water or in pack ice from Novaja Zemlja to Svalbard. They tend to be more associated with pack ice edge in September and October, moving east and north east to the vicinity of Franz Josef Land. Through November they seem to migrate ahead of the advancing pack to the southern coast of Novaja Zemlja, from where they move to their breeding grounds (**Figure 5**). Recently, in some years this pattern has been altered, with the westward movement of immature seals along the coast of northern Norway in winter (December–March). As these seals were in poor condition, these movements appear to be related to attempted foraging.

Sirenians

Being generally herbivorous, sirenians' foraging behavior differs substantially from other marine mammals. As animals of tropical and subtropical waters, their movements and food sources also show less seasonal variation than those of most other marine mammals. Dugongs, *Dugong dugon*, are the only extant sirenians that are truly marine, and so they will be the focus of discussion here.

Dugongs occur in shallow (generally $<20 \, m$) waters of the tropical and subtropical Indo-West Pacific. Most dugongs live in relatively small home ranges, in the order of tens to around a hundred square kilometers. Occasionally, satellite-tagged individuals undertake longer movements, up to $600 \, km$ from their home range and then, after a period of up to several weeks, return to their home range. There are no apparent age- or gender-related patterns to this, and reasons for these movements are unclear. Within their home range, dugongs' diel movements are tidally influenced, especially if seagrass beds occur in banks that are $<1 \, m$ deep or exposed at low tide. In at least one area, dugongs graze in large herds at the same site over weeks or months. This ranging and foraging pattern, termed 'cultivation grazing', encourages the growth of the pioneer seagrass species that are dugongs' preferred food. During periods of extreme food shortage in their home range, dugongs are known to travel along several hundred kilometers of coastline in search of new feeding grounds.

Glossary

Migration – Persistent movement between two destinations.

Home range – An animal's home range is that area within which it carries out most of its normal activities throughout the year. Home ranges are not usually considered to include migratory movements.

Haul-out site – Area (land or ice) where seals remove themselves from water.

Foraging – The process by which animals obtain food – includes searching for, capturing, and ingesting food.

Matrilineal (social unit) – Social system where female relatives remain associated, thereby providing the basic unit of the animals' society.

See also

Baleen Whales. Marine Mammal Diving Physiology. Marine Mammal Overview. Marine Mammal Social Organization and Communication. Marine Mammal Trophic Levels and Interactions. Seals. Sea Otters. Sirenians. Sperm Whales and Beaked Whales.

Further Reading

Boness DJ and Bowen WD (1996) The evolution of maternal care in pinnipeds. *Bioscience* 46: 645–654.

Clapham PJ (1996) The social and reproductive biology of humpback whales: an ecological perspective. *Mammal Review* 26: 27–49.

Corkeron PJ and Connor RC (1999) Why do baleen whales migrate? *Marine Mammal Science* 15: 1228–1245.

Gaskin DE (1982) *The Ecology of Whales and Dolphins*. London: Heinemann.

Hammond PS, Mizroch SA and Donovan GP (1990) *Individual Recognition of Cetaceans: Use of Photo-identification and Other Techniques to Estimate Population Parameters*. Reports of the International Whaling Commission, Special Issue 12.

Le Boeuf BJ and Laws RM (eds.) (1994) *Elephant Seals: Population Ecology, Behavior and Physiology*. Berkeley, CA: University of California Press.

Le Boeuf BJ, Crocker DE, Costa DP, *et al.* (2000) Foraging ecology of northern elephant seals. *Ecological Monographs* 70(3): 353–382.

Mann J, Connor RC, Tyack PL, and Whitehead H (eds.) (2000) *Cetacean Societies. Field Studies of Dolphins and Whales*. Chicago: University of Chicago Press.

Marsh H, Eros C, Corkeron PJ, and Breen B (1999) A conservation strategy for dugongs: implications of Australian research. *Marine and Freshwater Research* 50: 979–990.

Moore SE and Reeves RR (1993) Distribution and movement. In: Burns JJ, Montague JJ, and Cowles CJ (eds.) *The Bowhead Whale*, pp. 313–386. Lawrence, KS: Society of Marine Mammalogy.

Preen AR (1995) Impacts of dugong foraging on seagrass habitats: observational and experimental evidence for cultivation grazing. *Marine Ecology Progress Series* 124: 201–213.

Renouf D (ed.) (1991) *Behavior of Pinnipeds*. London: Chapman and Hall.

MARINE MAMMAL SOCIAL ORGANIZATION AND COMMUNICATION

P. L. Tyack, Woods Hole Oceanographic Institution, Woods Hole, MA, USA

Introduction

All animals face the same basic behavioral problems – obtaining food, avoiding predators and parasites, orienting in the environment, finding and selecting a mate, maintaining contact with relatives and group members. When the ungulates and carnivores that ultimately evolved into today's marine mammals entered the sea, the basic problems did not change, but the context in which the animals had to solve them changed radically. Different marine mammal groups have a different balance in the extent to which they use the underwater versus the in-air environments. All sirenians and cetaceans live their entire lives in the sea, and cetaceans show the most elaborate and extreme specializations for life in the sea. All marine mammals other than sirenians, the sea otter, and cetaceans spend critical parts of their lives on land or ice. These other species, including the pinnipeds and polar bear (*Ursus maritimus*), rely upon land refuges for giving birth, and for taking care of the young.

The ocean is a hostile environment for air-breathing mammals. There is little room for error – if an animal misjudges a dive or becomes incapacitated, it may have only minutes to correct the error or there is a risk of drowning. In the days of sail, humans responded to the notion that 'the sea is a harsh mistress' with an apprenticeship system, whereby a young cabin boy spent years learning the ropes before being entrusted to make the decisions required of a captain. Some cetaceans have similarly long periods of dependency when the young can learn how to feed, avoid predators, dive, and orient within their natal group. Pilot whales and sperm whales may even continue to suckle up to 13–15 years of age, and pilot whale females have a post-reproductive period when they switch their reproductive effort fully to parental care. By contrast, some seals that suckle on land have drastically curtailed the period of lactation, so that their young can leave the land-based refuge early. The hooded seal suckles her young twice an hour for an average of just four days before the pup is weaned. Even seals that lactate longer may still leave the young to an early independence. Elephant seals are deep divers on a par with the sperm whale, yet they must learn to navigate the sea alone. When an elephant seal pup is weaned, the mother leaves the pup on the beach. Pups spend about 2.5 months on the rookery, learning to swim and dive. They must fend for themselves as they make their first pelagic trip, lasting about four months. This solo entry into the sea exerts a heavy cost; fewer than half of the pups survive this trip.

Feeding Behavior

Behavioral ecologists divide feeding behavior into a sequence of steps: searching for prey, pursuit, capture, and handling prey. Marine mammals use many different senses to detect their prey. Walruses, sea otters, and gray whales feed on benthic prey. The walrus uses vibrissae in its mustache to sense shells in the mud; experiments with captive walrus show that they can use vibrissae to determine the shape of objects. Most seals are thought to use vision to find their prey – seals that feed at depth have eyes adapted to low light levels. Most toothed whales produce click sounds, and those species tested in captivity have sophisticated systems of echolocation. When a dolphin detects a target and closes in, it usually increases the repetition rate of its clicks into a buzz sound. If other animals intercept this sound, they may learn about prey distribution whether or not the echolocating animal wants to broadcast this information. Captive experiments show that one dolphin can detect an object by listening to echoes from the clicks of another dolphin. Both of these features of echolocation may favor coordinated social feeding. Many questions remain about how some marine mammals search for prey; we have no idea how baleen whales find patches of prey in the water column.

Most marine mammals catch and process prey using their teeth, as most mammals do (**Figure 1A–C**), but the mysticete whales have baleen, a sievelike set of plates that descend from the lower jaw (**Figure 1D**), instead of teeth. Baleen whales are able to engulf many prey items in one gulp and then use this baleen to strain out the sea water. Baleen whales feed on patches of prey, and will often aggregate when they are feeding on large patches; the larger the patch, the

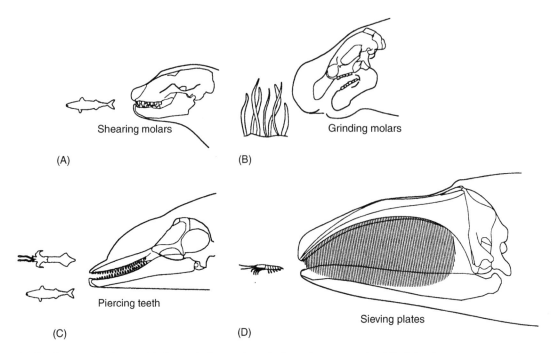

Figure 1 Feeding mechanisms of baleen whales and toothed marine mammals. (A) Polar bears, otters, and most seals have shearing teeth similar to those of terrestrial carnivores. (B) Sirenians have grinding molars that are replaced throughout life. (C) Toothed whales are homodonts; some beaked whales have only one tooth per row, most dolphins have dozens per row. (D) Baleen of a balaenid whale. (Adapted from Figures 2–17 of Reynolds and Rommel (1999).)

more whales in the group. Whales feeding on slow-moving prey, such as copepods, may feed in loose aggregations; those feeding on more mobile prey, such as schooling fish, may feed in a more coordinated fashion. Humpback whales may feed in stable groups in which each individual plays a distinct role.

Toothed whales tend to feed on mobile prey such as fish and squid, and they must catch one prey item at a time in their mouths. When dolphins feed on schooling fish, they usually feed socially in a co-ordinated group. Some dolphins appear to corral the school near the surface, while others swim through capturing a few fish. Marine mammalogists have often described these groups as cooperative, but little is known about the costs and benefits of social feeding, so one must be careful to distinguish between coordinated feeding behavior and behavior following more complex models of cooperation. The normal usage of 'cooperation' in English is just to work together toward a common goal; in ecology, studying cooperation demands measuring the costs and benefits of different behaviors. Different evolutionary models of cooperation involve different kinds of exchanges. One simple model covers the situation in which animals feeding together on a patch may each feed more efficiently than if they were feeding alone – this may apply to some coordinated feeding in whales. One-sided cooperation may be favored when one animal provides benefits to

related animals; kin selection of this sort favors parental care and even cooperation with more distant relatives under some circumstances. More complex models of cooperation between unrelated animals involve separated interactions of reciprocation, in which one animal may provide a benefit to a partner, expecting the partner to reciprocate at some later date. When a lone dolphin discovers a patch, it may direct a feeding call to other animals, which approach to join in. Feeding calls of this sort may evolve through reciprocation. If search costs are high and if a dolphin discovers a patch too large for it to consume, it may benefit from calling to partners in the expectation that the partners may reciprocate.

Sea otters use tools to process food; they dive to catch animals with hard shells, and then bring the prey to the surface, where they break it open on a stone that they carry while foraging. Some marine mammals may also have social mechanisms for processing prey. Some toothed whales catch large oceanic fish such as mahi-mahi (*Coryphaena* spp.), which are almost as big as they are. It may take one animal to hold the prey and another to rip off pieces in order to consume the flesh efficiently.

Defense from Predators

Marine mammals have different strategies for defending themselves from predators. Many of the toothed whales are thought to use their social groups to increase their probability of detecting predators and to protect vulnerable members of the group. Behavioral ecologists studying many animals have found that animals feeding in a group spend less time than lone animals breaking off from feeding to look for predators, and that they also detect the predators at larger ranges, enabling more successful escape strategies. K. Norris has urged that schools of dolphins function in this way to integrate sensory information from each member of the group using social communication to bolster the sensory abilities of each individual.

Some of the large toothed whales use a social defense from predators once the predators are detected. The best observations come from sperm whales. The most dangerous predators of sperm whales are killer whales and human whalers. When killer whales attack a sperm whale group, the adult sperm whales circle around the young with their flukes facing out, and they will attack approaching killer whales with their flukes. If an adult is injured, the sperm whales will circle around that animal to protect it as well. Human whalers knew about this, and they would often harpoon a whale in the group to wound it and then kill each adult in turn as the whales remained nearby to protect the wounded whale.

Other marine mammals appear to have a strategy opposite to grouping for protection from predators. When humpback whale females have a young calf, they do not join with other females but seem to space themselves out in protected clear waters. You might think that these whales are so big they do not need help to defend themselves from predators, but sperm whales are almost as big, yet rely upon social defense. Smaller animals such as seals may also spread out rather than group in response to predator pressure and some seals appear to dive in response to predators, adding a third dimension to this response. The ocean does not have many hiding places, but is so vast that these responses may represent a strategy to spread out to avoid detection.

Finding and Selecting a Mate

Evolutionary biologists assume that selection acts to maximize lifetime reproductive success, including both an animal's own offspring and those of relatives, weighted according to the degree of genetic relationship. Mating behavior is closely related to this goal. The mating behavior of male and female mammals differs in part because of the physiology of mammalian reproduction. Female mammals gestate their young internally and provide nutrition after birth through lactation. Once a female has become pregnant, she must usually wait until her young is born and often even weaned before becoming receptive again. This means that reproduction in females is limited primarily by their ability to gather energy to produce young. While male mammals can provide parental care, it is more common for them not to do so, and paternal care is not known among marine mammals. A male is capable of inseminating another female soon after a previous mating. Males often compete for the opportunity to mate with females, and reproductive success in males is often limited by the number of females they can inseminate. The ratio between the number of receptive males to the number of receptive females is called the operational sex ratio. The more receptive males per female, the higher the selection pressure for competition between males for access to females.

Females have a variety of mating strategies, where 'strategy' is used in an evolutionary rather than purely cognitive sense. A female may have one or more estrous cycles per year, and these may be spontaneous or induced by copulation. A receptive female may search for and select a male either on the basis of specific resources the male may provide, or of judgments of male quality as a mate. A female who is not relying upon resources provided by a male may either elicit competition between the sperm of several males with which she has mated, or incite behavioral competition between males and select the winner.

There are also a variety of mating strategies available to males. A male may defend, from other males, resources used by females. An example of this occurs with fur seals, where males will defend areas on the beach that females use to give birth and for thermoregulation. The goal of the male strategy is to exclude other males from their territory, in hopes that females coming to their territory for the resources will become receptive there. Alternatively, males may defend females directly. Elephant seal females may cluster on a beach. Males establish a dominance hierarchy before the breeding season, and the dominant few males defend receptive females from other males. In bottlenose dolphins, coalitions of 2 or 3 males also have been observed to defend females to prevent them from choosing another mate. Males within a coalition have a strong bond, and are typically sighted together most of the time for many years. The males may chase and herd a female away from the group in which they initially find her, and may escort her for days. Why does this involve coalitions of males? It may be impossible for a lone male to preempt female choice with such maneuverable animals in a three-dimensional medium.

Male strategies often depend upon the distribution of females. In the bottlenose dolphin case, females live in small fluid groups and males search for one receptive female, often guarding her when found. By contrast, sperm whale females live in stable groups of several females with their young. Adult male sperm whales join with these groups for varying amounts of time during the breeding season. Computer models suggest that males should rove between female groups if the duration of estrous is greater than the time it takes to swim between groups. This illustrates a general pattern that the distribution of females is often driven by the distribution of resources, while the distribution of males during the breeding season is often driven by the distribution of receptive females.

Most of the strategies listed above can be viewed as strategies used by males to limit or preempt the ability of a female to select a mate. In situations where males have a limited ability to preempt this choice, males may evolve signals to attract mates and may display to influence female choice. If females select mates on the basis of the display, then the male displays may be under a strong selection pressure to develop whatever features are used by females to select a mate. Darwin distinguished this kind of sexual selection from natural selection. Sexual selection stems from competition between members of the sex with the least parental investment (typically males) for access to mating with the sex that provides the largest parental investment (typically females).

Earlier paragraphs described selection arising from competition between males; this is called intrasexual selection, and often leads to the development of weapons and large body size. Selection arising from competition between males to influence female choice is called intersexual selection.

Intersexual selection often leads to the evolution of elaborate displays, called reproductive advertisement displays. Examples of reproductive advertisement displays include the songs of birds and whales. Songs are usually defined as acoustic displays in which sequences of discrete sounds are repeated in a predictable pattern. Songs are known from a variety of marine mammals. The songs of the humpback whale are well known and sound so musical to our ears that they have been commercial bestsellers. Male humpback whales sing for hours, usually when they are alone during the breeding season. Bowhead whales, *Balaena mysticetus*, produce songs that are simpler than those of humpbacks, consisting of a few sounds that repeat in the same order for many song repetitions (**Figure 2**). As with humpback song, bowhead songs appear to change year after year. Bowhead whales winter in the Bering Sea, and humans have seldom studied them during their winter breeding season, but their songs have been recorded during their spring migrations past Point Barrow, Alaska. The long series of 20 Hz pulses produced by finback whales may also be a reproductive advertisement display. The seasonal distribution of these 20 Hz series has been measured near Bermuda, and it matches the breeding season quite closely.

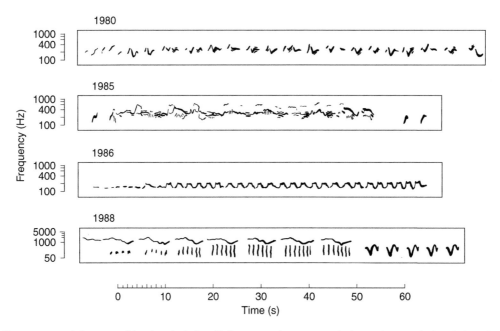

Figure 2 Spectrogram of the song of bowhead whales, *Balaena mysticetus*, recorded over 4 years during their spring migration. (From Figure 4.10 of Tyack and Clark (2000).)

Some pinnipeds also repeat acoustically complex songs during the breeding season. The bearded seal, *Erignatus barbatus*, produces a sirenlike warbling song that includes rapid frequency modulation superimposed upon slower modulation of the carrier frequency (**Figure 3**). The songs of bearded seals are produced by sexually mature adult males during the breeding season. Male walruses, *Odobenus rosmarus*, also perform visual and acoustic advertisement displays near herds of females during the breeding season. Males inflate modified pharyngeal pouches that can produce a metallic bell-like sound. When walruses surface during these displays, they may make loud sounds in air, including knocks, whistles, and loud breaths. They then dive, producing distinctive sounds under water, usually a series of sharp knocks followed by the gonglike or bell-like sounds. Antarctic Weddell seals also have extensive vocal repertoires and males repeat underwater trills during the breeding seasons. Males defend territories on traditional breeding colonies. These trills have been interpreted as territorial advertisement and defense calls.

There is evidence that marine mammal songs play a role both in male–male competition and in female choice. Evidence for intrasexual selection includes observations that aggressive interactions between singers and other males are much more commonly observed than sexual interactions between singers and females and song appears to maintain distance between singers. However, just because the responses of males to song may be seen more frequently than those of females does not mean that the subtler responses of females to singers are not biologically significant. In many species, females will approach and join with a singer, and many acoustic features of these songs are consistent with a role in female choice.

Maintaining Contact with Recognition Calls

Mother–Infant Recognition

When mammalian young are born, they need to suckle for nutrition, and many species depend upon the mother for thermoregulation and protection from parasites and predators. This dependency has created a selection pressure for a vocal recognition system to regain contact when mother and offspring are separated. These problems of recognition between mother and young are acute in colonially breeding otariid seals. A female otariid may leave her young pup on land in a colony of hundreds to thousands of animals, feed at sea for a day or more and, when she returns, must find her pup to feed it. In the Galapagos fur seal, *Arctocephalus galapagoensis*, both mother and pup produce and recognize distinctive contact calls, and the mother often makes a final olfactory check before allowing a pup to suckle.

The young of many dolphin and other odontocete species are born into groups comprising many adult females with their young, and they rely upon a mother–young bond that is more prolonged than that of otariids. As was described in the introduction, many of these species have unusually extended

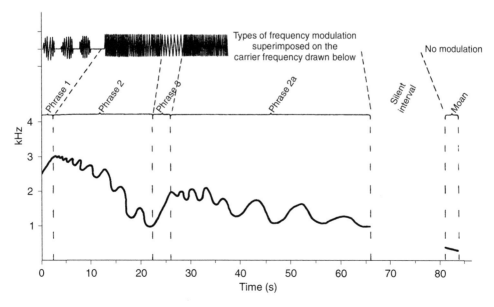

Figure 3 Lower panel: spectrographic portrayal of songs of the bearded seal, *Erignatus barbatus*. Additional frequency modulation is added to this carrier frequency; this warbling modulation is indicated in the upper panel. (From Figure 1 of Ray *et al.* (1969).)

parental care. Bottlenose dolphin calves typically remain with their mothers for 3–6 years. These dolphin calves are precocious in locomotory skills, and swim out of sight of the mother within the first few weeks of life. Young calves often swim with animals other than the mother during these separations. The combination of early calf mobility and prolonged dependence selects for a mother–offspring recognition system in bottlenose dolphins. Dolphin mothers and their young calves use tonal whistles as signals for individual recognition. Observations of captive dolphins suggest that whistles function to maintain contact between mothers and young. When a dolphin mother and her young calf are separated involuntarily in the wild, they whistle at higher rates; during voluntary separations in the wild, the calf often whistles as it returns to the mother. Experimental playbacks to wild dolphins show that mothers and their calves respond preferentially to each others' signature whistles, even after the calves become independent from their mothers.

Individual Recognition in Dolphins and Signature Whistles

Dolphins use whistles not just for mother–infant recognition but also for individual recognition throughout their lives. Calves show no reduction in whistling as they wean and separate from their mother. Adult males are not thought to provide any parental care, but they whistle just as much as adult females. Bottlenose dolphins may take up to two years to develop an individually distinctive signature whistle, but once a signature whistle is developed, it can be stable for decades (**Figure 4**). The signature whistles of dolphins are much more distinctive than similar recognition signals produced by other mammals. These results suggest that signature whistles continue to function for individual recognition in older animals.

Group Recognition in Killer Whales

Many marine mammals live in kin groups, and social interactions within these groups may have a powerful effect on fitness. The different structures of these cetacean societies create different kinds of problems of social living, and there appears to be a close connection between the structure of a cetacean society and the kinds of social communication that predominate in it. For example, stable groups are found in fish-eating or Resident killer whales, *Orcinus orca*, in the coastal waters of the Pacific Northwest of the United States, and these whales also have stable group-specific vocal repertoires. The only way a group of these killer whales, called a pod, changes composition is by birth, death, or rare fissions of very large groups. The vocal repertoire of killer whales includes discrete calls, which are stereotyped with acoustic features that change slowly and gradually over decades. Each pod of Resident killer whales has a group-specific repertoire of discrete calls. Each whale within a pod is thought to produce the entire call repertoire typical of that pod. Different pods may share some discrete calls, but none share the entire call repertoire. Since discrete calls change gradually, pods that diverged more recently produce more similar versions of some calls (**Figure 5**). The entire repertoire of a pod's discrete calls can thus be thought of as a group-specific vocal repertoire. Different pods may have ranges that overlap and pods may associate together for hours or

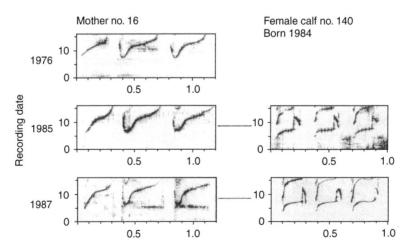

Figure 4 Spectrograms of signature whistles from one wild adult female bottlenose dolphin, *Tursiops truncatus*, recorded over a period of 11 years and of one of her calves at 1 and 3 years of age. Note the stability of both signature whistles. (From Figure 11.11 of Mann *et al.* (2000).)

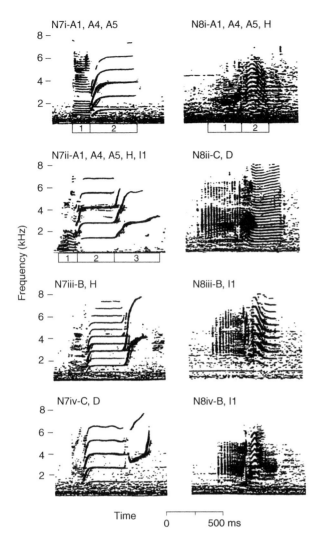

Figure 5 Spectrogram of group-specific subtypes of calls N7 and N8 of resident killer whales, *Orcinus orca*. Not only does each group have a group specific call repertoire, but there are different subtypes for many calls, and different sets of groups produce each subtype. On the upper left of each spectrogram is the subtype of the call and following the dash is a listing of the groups that make this particular call. (From Figure 1.11 of Au *et al.* (2000).)

days before diverging. These group-specific call repertoires in killer whales are thought to indicate pod affiliation, to maintain pod cohesion, and to coordinate activities of pod members.

Correlation of Acoustic Recognition Signals and Social Organization

Most communication signals evolve for the solution of specific problems of social life. In fact, communication and social behavior can be viewed as two different ways of looking at the same thing. There is a clear correlation between the types of social bonds and recognition signals seen in different cetacean groups. Individual-specific signals have been reported for species such as bottlenose dolphins with strong individual social bonds; group-specific vocal repertoires have been reported for species such as killer whales with stable groups, and population-specific advertisement displays have been reported among species such as humpback whales and some seals where adults appear to have neither stable bonds nor stable groups.

See also

Baleen Whales. Marine Mammal Migrations and Movement Patterns. Marine Mammal Overview. Marine Mammal Trophic Levels and Interactions. Sea Otters. Seals. Sirenians. Sperm Whales and Beaked Whales.

Further Reading

Bradbury JW and Vehrencamp SL (1998) *Principles of Animal Communication*. Sunderland, MA: Sinauer Associates.

Norris KS, Würsig B, Wells R, and Würsig M (1994) *The Hawaiian Spinner Dolphin*. Berkeley: University of California Press.

Mann J, Connor R, Tyack PL, and Whitehead H (2000) *Cetacean Societies: Field Studies of Dolphins and Whales*. Chicago: University of Chicago Press.

Ray C, Watkins WA, and Burns JJ (1969) The underwater song of *Erignathus* (Bearded seal). *Zoologica* 54: 79–83.

Reynolds JE III and Rommel SA (eds.) (1999) *Biology of Marine Mammals*. Washington DC: Smithsonian Press.

Trillmich F (1981) Mutual mother–pup recognition in Galápagos fur seals and sea lions: cues used and functional significance. *Behaviour* 78: 21–42.

Tyack PL (1986) Population biology, social behavior, and communication in whales and dolphins. *Trends in Ecology and Evolution* 1: 144–150.

Tyack PL and Sayigh LS (1997) Vocal learning in cetaceans. In: Snowdon CT and Hausberger M (eds.) *Social Influences on Vocal Development*, pp. 208–233. Cambridge: Cambridge University Press.

Tyack PL and Clark CW (2000) Communication and acoustic behavior of dolphins and whales. In: Au WWL, Popper AS, and Fay R (eds.) *Hearing by Whales and Dolphins*. Springer Handbook of Auditory Research Series, pp. 156–224. New York: Springer Verlag.

Whitehead H (1990) Rules for roving males. *Journal of Theoretical Biology* 145: 355–368.

Watkins WA, Tyack P, Moore KE, and Bird JE (1987) The 20-Hz signals of finback whales (*Balaenoptera physalus*). *Journal of the Acoustical Society of America* 82: 1901–1912.

Zahavi A and Zahavi A (1997) *The Handicap Principle*. Oxford: Oxford University Press.

MARINE MAMMAL TROPHIC LEVELS AND INTERACTIONS

A. W. Trites, University of British Columbia, British Columbia, Canada

Introduction

Trophic levels are a hierarchical way of classifying organisms according to their feeding relationships within an ecosystem. By convention, detritus and producers (such as phytoplankton and algae) are assigned a trophic level of 1. The herbivores and detritivores that feed on the plants and detritus make up trophic level 2. Higher order carnivores, such as most marine mammals, are assigned trophic levels ranging from 3 to 5. Knowing what an animal eats is all that is needed to calculate its trophic level.

Marine mammals are commonly thought to be the top predator in marine ecosystems. However, many species of fish occupy trophic levels that are on par or are above those of marine mammals. Some species such as killer whales and polar bears (that feed on other marine mammals) are indeed top carnivores, but others such as manatees and dugongs feed on plants at the bottom of the food web. Thus, marine mammals span four of the five trophic levels.

Marine mammals are a diverse group of species whose behaviors, physiologies, morphologies, and life history characteristics have been evolutionarily shaped by interactions with their predators and prey. It is therefore difficult to generalize about how marine mammals affect the dynamics and structure of their ecosystems. Similarly, it is difficult to generalize about how the interactions between marine mammals and their prey (or between marine mammals and their predators) affect one another, as well as how they affect the dynamics of unrelated species. Nevertheless, some insights into marine mammal trophic interactions can be gleaned from mathematical models and from field observations following the overharvesting of marine mammal populations in the nineteenth and twentieth centuries.

Trophic Levels (Diet Composition)

Trophic levels depend on what a species eats. As an example, a fish consuming 50% herbivorous-zooplankton (trophic level 2) and 50% zooplankton-eating fish (trophic level 3) would have a trophic level of 3.5. Trophic levels (TL) can be calculated from

$$TL = 1 + \frac{\sum_{i=1}^{n}(TL_i \cdot DC_i)}{\sum_{i=1}^{n} DC_i} \quad [1]$$

where n is the number of species or groups of species in the diet, DC_i is the proportion of the diet consisting of species i and TL_i is the trophic level of species i. Thus, the trophic level of the predator is determined by adding 1.0 to the average trophic level of all the organisms that it eats.

Applying eqn [1] to marine mammals shows that sirenians (dugong and manatees) have a trophic level of 2.0, whereas blue whales (which feed on large zooplankton, trophic level 2.2) are at trophic level 3.2 ($= 1.0 + 2.2$). Moving higher up the food chain, Galapagos fur seals have a trophic level of 4.1. Their diet consists of approximately 40% small squids, 20% small pelagic fishes (such as clupeoids and small scombroids), 30% mesopelagic fishes (myctophids and other groups of the deep scattering layer) and 10% miscellaneous fishes (from a diverse group consisting mainly of demersal fish). Substituting these proportions into eqn [1], along with the respective mean trophic levels (TL_i) of these four types of prey (3.2, 2.7, 3.2, and 3.3, respectively), yields a trophic level of 4.11 for Galapagos fur seals. A polar bear that feeds exclusively on ringed seals (3.8) would have a trophic level of 4.8.

Dugongs and manatees occupy the lowest trophic level (2.0) of all marine mammals. They are followed (see **Figure 1**) by baleen whales (3.35: range 3.2–3.7), sea otters (3.45: range 3.4–3.5), pinnipeds (3.97: range 3.3–4.2), and toothed whales (4.23: range 3.8–4.5), with the highest trophic level belonging to the polar bear (4.80).

Trophic interactions between marine mammals and other species can be depicted by flowcharts showing the flow of energy between species in an ecosystem. An example is shown in **Figure 2** for the eastern Bering Sea. Each of the boxes in this flowchart represents a major species or group of species within this system during the 1980s. The boxes are arranged by trophic levels and are proportional in size to their biomass. Lines connecting the boxes show the relative amounts of energy flowing between the groups of species.

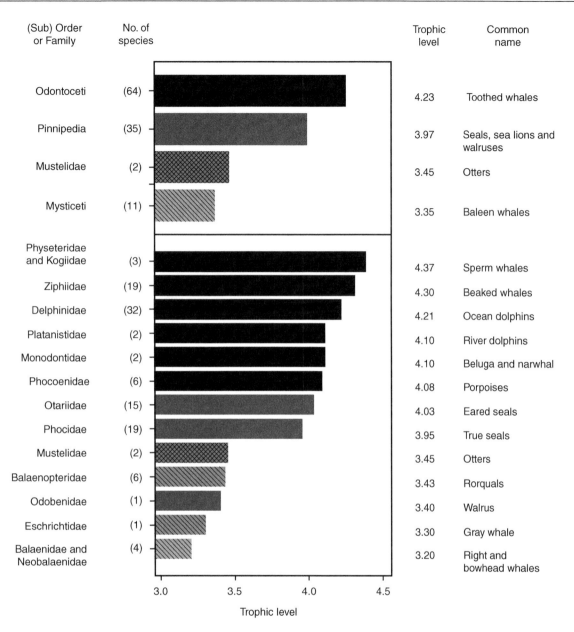

Figure 1 Mean trophic levels for 112 species of marine mammals grouped by families, orders and suborders. Numbers of species averaged within each grouping is shown in brackets. Species not shown are dugong and manatees (Sirenia: trophic level 2.0) and polar bears (Ursidae: trophic level 4.8).

Figure 2 shows a large number of flows in the Bering Sea emanating from three species at trophic level 3 – pollock, small flatfish and pelagic fishes. Major level 4 consumers include large flatfish, deep-water fish, other demersal fishes, marine mammals and birds. Thus, large flatfish and other species of fish share the pedestal with marine mammals as top predators of marine ecosystems. These fish are also major competitors of marine mammals.

Trophic levels depicted in **Figures 1** and **2** are approximate, and are based on generalized diets and the mean trophic levels of prey types. In actual fact,

trophic levels of most marine mammals probably vary from season to season, or from year to year, because diet is unlikely to remain constant. How much they might vary is not known, but is probably within ±0.2 trophic levels.

Trophic Levels (Stable Isotopes)

Diets have traditionally been described from stomach contents of shot or stranded animals. This has been augmented by the identification of prey from

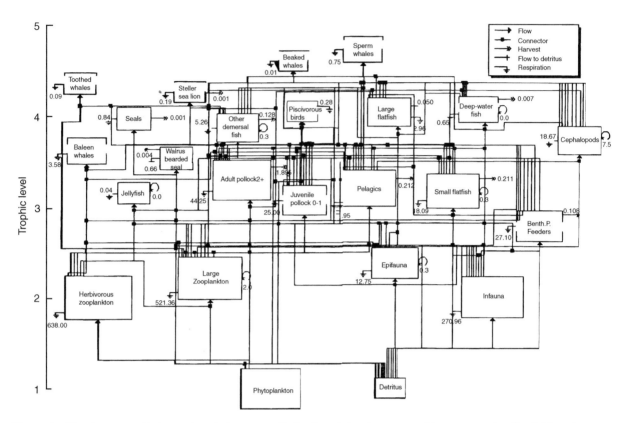

Figure 2 Flowchart of trophic interactions in the eastern Bering Sea during the 1980s. All flows are in t km^{-2} y^{-1}. Minor flows are omitted as are all backflows to the detritus. Note that size of each box is roughly proportional to the biomass therein, and that each box is placed according to its trophic level in the ecosystem.

the bony remains found in feces (primarily from pinnipeds), and from fatty acid signatures of prey species that have been laid down in the blubber of marine mammals. Unfortunately, the diets of most species of marine mammals are poorly understood due to incomplete sampling across time and space.

There is another way to estimate trophic levels without stomach contents or other dietary information. It is referred to as stable isotope analysis and relies on the relative concentration of two isotopes (nitrogen-14 and nitrogen-15). Marine mammals and other organisms tend to accumulate the heavier isotope (nitrogen-15) in their tissues. Thus, as matter moves from one trophic level to the next, the ratio of the two isotopes shifts by a roughly constant amount. Trophic levels can be calculated by dividing the difference between the isotopic ratio in the marine mammal tissue and the isotopic ratio of the organism at the bottom of the local food chain, by this constant difference between trophic levels.

A comparison of the isotopic estimates of trophic levels for species in Prince William Sound, Alaska with estimates derived from dietary analysis (eqn [1]) suggests that the two techniques produce comparable results. One of the strengths of the isotopic analysis is that it can be conducted from biopsy samples and does not require killing the animal to examine stomach contents. This is particularly useful for assessing the trophic levels of cetaceans. Stable isotope analysis can also be used to probe the past to learn about the trophic levels that marine mammal populations once occupied. As predators, marine mammals are better samplers of the marine environment than biologists. Thus, analyzing seasonal and annual changes in the nitrogen concentrations contained along growing whiskers and baleen can provide a time series of dietary information. Similarly, trophic levels can be calculated from nitrogen concentrations in bones and teeth archived in museums or recovered from archaeological digs.

Another useful stable isotope ratio is the relative concentration of carbon-13 and carbon-12. Studies have shown that there is a very slight enrichment of carbon from one trophic level to another (0.1–0.2%). In the marine environment, slight enrichment occurs at low trophic levels, but not among vertebrate consumers. Thus, isotopic carbon ratios are not useful for assessing trophic level, but they are useful for tracking carbon sources through a food chain and for assessing long-term changes in ocean productivity.

Isotopic analyses of marine mammal tissues have shown that species inhabiting the northern oceans have higher nitrogen-isotope ratios than those from southern oceans. This indicates that southern species feed at lower trophic levels, and presumably consume larger amounts of invertebrates. Measuring the isotopic carbon ratio of baleen plates from bowhead whales further shows that primary productivity declined in the Bering Sea through the 1970s–1990s. This drop in primary productivity may reflect an overall lowering of carrying capacity and may have a bearing on the observed decline of Steller sea lions, harbor seals, and northern fur seals during this period. Thus, isotopic analysis is a useful tool for estimating trophic levels of marine mammals, and for detecting shifts in ocean productivity and diets of marine mammals.

Trophic Interactions

Changes at one level of a food web can have cascading effects on others. One of the best ways to explore the direct and indirect impacts of competition and predation by marine mammals on other species is with mathematical descriptions of ecosystems (i.e., ecosystem models). Ecosystem models, such as the one developed for the Bering Sea (**Figure 2**), allow changes in abundance to be tracked over time, and predictions to be made about the strength and significance of predator–prey interactions on each other, and on other components of their ecosystem.

Major changes have occurred in the abundance of a number of species in the Bering Sea since the mid-1970s. Most notable has been the decline of Steller sea lions, harbor seals, crabs, shrimp and forage fishes (such as herring and capelin). In contrast, populations of walleye pollock and large flatfish (mostly arrowtooth flounder) increased through the 1970s and 1980s. Some have felt that commercial whaling prompted these changes by removing a major competitor of pollock – the baleen whales. Mathematically, removing whales can be shown to positively affect pollock by reducing competition for food. However, whaling alone is insufficient to explain the 400% increase in pollock that is believed to have occurred. Overall, the models developed to date suggest that changes in the biomass of marine mammals have little or no effect on changes in the biomass of other groups in the Bering Sea. Most impacts on this northern marine ecosystem appear to

be associated with changing the biomass of lower trophic levels (such as primary production).

The conclusions drawn from the eastern Bering Sea model may be indicative of marine mammals in long-chained food webs, and may not reflect the role of marine mammals in shorter-chained food webs such as in the Antarctic. A case in point is the increase in abundance of krill-eating Antarctic fur seals, crabeater seals, leopard seals, and penguins that followed the cessation of commercial whaling. Commercial whaling removed over 84% of the baleen whales from the Antarctic and 'freed up' millions of tons of krill for other species to consume. Some believe that the increase in these other krill-eating species is now impeding the recovery of Antarctic whales.

Sea otters and sea urchins form another short-chained food web with strong trophic interactions. By the turn of the twentieth century, sea otters had been hunted to near extinction. Without predation by otters, sea urchin populations grew unchecked and overgrazed the fleshy algae along the Pacific coast of North America. The once productive kelp forests became underwater barrens. With the reintroduction of sea otters however, productivity increased three-fold as urchins were removed and kelp and other fleshy algae began to regenerate. Kelp provides habitat for fish and invertebrates, changes water motion, and can affect onshore erosion and the recruitment of fish and invertebrates. Thus, sea otters can change the state of near-shore ecosystems and the way they function.

Other examples of marine mammals affecting their prey include harbor seals in freshwater lakes, and killer whales preying on sea otters in Alaska. A number of lakes in Quebec, Canada, are home to land-locked harbor seals that feed on trout. Studies have shown that the trout in these lakes are younger and spawn at younger ages than adjacent lakes without harbor seals. The trout also grow faster and attain smaller sizes in the lakes inhabited by harbor seals.

Marine mammals may also significantly affect prey abundance, as in the case of killer whales eating sea otters, Steller sea lions, and other warm-blooded species. Killer whales were observed eating sea otters along the Aleutian Islands in the 1990s and may be responsible for reported declines in sea otter population abundance. Killer whales have also been implicated as a contributing factor in the decline of Steller sea lions and may be impeding their recovery.

Despite the apparent effects of some species of marine mammals on their prey, there are a number of cases where mass removals of marine mammals did not appear to have a major effect on other components of their ecosystems. Examples are the overhunting of elephant seals and California sea lions along the coast of California, the overhunting of northern fur seals in the Bering Sea, and the culling of harbor seals in British Columbia. One explanation for the lack of tractable impacts in these cases is that their food webs are more complex relative to other systems (i.e., predators consuming many different species of prey, may have no noticeable impact on any single prey type). Another reason might be related to the type of marine ecosystems that these species inhabit (i.e., whether they inhabit shelf or deep-water systems, or whether they are primarily benthic or mid-water feeders). Further insights might be gained by developing ecosystem models for these systems.

Quantifying the feeding relationships between marine mammals and other species provides a means for assessing competition between species at similar trophic levels. Some species may significantly compete with more than one species. In the Bering Sea, for example (**Figure 2**), baleen whales and pollock have high overlaps in their diets (73–86%). There is also a significant amount of competition between seals and adult pollock for prey. Toothed whales, for example, compete primarily with beaked whales and seals, whereas the largest competitors of sea lions appear to be seals, toothed whales, and large flatfish. Fish, it turns out, can be major competitors of marine mammals.

Competition can affect body growth, reproduction and survival of marine mammals. In the Bering Sea and Gulf of Alaska, for example, the growth of Steller sea lions and northern fur seals (as measured by length) appears to have been stunted during the 1980s compared to the 1970s. Eastern Pacific populations of gray whales also appear to be in poorer condition (as measured by the ratio of girth to body length) in the 1990s compared to earlier decades. These changes in body size may be density-dependent responses to reduced prey availability or may be indicative of populations that have approached or attained their carrying capacities.

Reductions in prey abundance have been recorded in the Antarctic (i.e., krill), and along the coasts of California and South America during El Niño events. Pinniped pups born during these periods of reduced prey abundance incur high rates of mortality (typically 2–3 times normal levels) and are weaned at lower weights than normal (typically 15–20% lighter). Lactating females must also spend longer periods of time away from their pups to search for prey. Such temporal changes in prey abundance may result in the loss of an entire year class, and may be

one of the evolutionary forces that shaped the life history of marine mammals (i.e., they are long-lived, have low reproductive rates and can endure short-term reductions in prey abundance).

Although it has not yet been demonstrated for marine mammals, reductions in prey availability can theoretically delay the onset of sexual maturity, and reduce fertility (by causing a female to not ovulate, or by causing a fetus to be reabsorbed or aborted). Reduced nutrition may also compromise an organism's resistance to disease, and may increase vulnerability to predation. Food deprivation may mean, for example, that a seal must spend increased amounts of time searching for prey and less time hauled out on shore away from predators such as killer whales and sharks.

Conclusions

Calculating trophic levels is a necessary first step to quantifying and understanding trophic interactions between marine mammals and other species in marine ecosystems. This can be achieved using dietary information collected from stomachs and scats, or by measuring isotopic ratios contained in marine mammal tissues. These data indicate that marine mammals occupy a wide range of trophic levels beginning with dugong and manatees (trophic level 2.0), and followed by baleen whales (3.35), sea otters (3.45), seals (3.95), sea lions and fur seals (4.03), toothed whales (4.23), and polar bears (4.80).

With the aid of ecosystem models and other quantitative analyses, the degree of competition can be quantified, and the consequences of changing predator–prey numbers can be predicted. These analyses show that many species of fish are major competitors of marine mammals. A number of field studies have also shown negative effects of reduced prey abundance on body size and survival of marine mammals. However, there are fewer examples of marine mammal populations affecting their prey due perhaps to the difficulty of monitoring such interactions, or to the complexity of most marine mammal food webs.

See also

Baleen Whales. Marine Mammal Diving Physiology. Marine Mammal Evolution and Taxonomy. Marine Mammal Migrations and Movement Patterns. Marine Mammal Overview. Marine Mammal Social Organization and Communication. Seals. Sea Otters. Sirenians. Sperm Whales and Beaked Whales.

Further Reading

Bowen WD (1997) Role of marine mammals in aquatic ecosystems. *Marine Ecology Progress Series* 158: 267–274.

Christenson V and Pauly D (eds) (1993) *Trophic Models of Aquatic Ecosystems*. ICLARM Conference Proceedings 26.

Estes JA and Duggins DO (1995) Sea otters and kelp forests in Alaska: generality and variation in a community ecological paradigm. *Ecological Monographs* 65: 75–100.

Greenstreet SPR and Tasker ML (eds.) (1996) *Aquatic Predators and Their Prey*. Oxford: Fishing News Books.

Kelly JF (2000) Stable isotopes of carbon and nitrogen in the study of avian and mammalian trophic ecology. *Canadian Journal of Zoology* 78: 1–27.

Knox GA (1994) *The Biology of the Southern Ocean*. Cambridge: Cambridge University Press.

Pauly D, Trites AW, Capuli E, and Christensen V (1998) Diet composition and trophic levels of marine mammals. *Journal of Marine Science* 55: 467–481.

Trillmich F and Ono K (1991) *Pinnipeds and El Niño: Responses to Environmental Stress*. Berlin: Springer-Verlag.

Trites AW (1997) The role of pinnipeds in the ecosystem. In: Stone G, Goebel J, and Webster S (eds.) *Pinniped Populations, Eastern North Pacific: Status, Trends and Issues*, pp. 31–39. Boston: New England Aquarium, Conservation Department.

Trites AW, Livingston PA, Mackintosh S *et al.* (1999) *Ecosystem Change and the Decline of Marine Mammals in the Eastern Bering Sea: Testing the Ecosystem Shift and Commercial Whaling Hypotheses*. Fisheries Centre Research Reports 1999, Vol. 7(1).

BIRDS

SEABIRDS: AN OVERVIEW

G. L. Hunt, Jr., University of Washington, Seattle, WA, USA

Seabirds, also known as marine birds, are species that make their living from the ocean. Of the approximately 9700 species of birds in the world, about 300–350 are considered seabirds. The definition of what constitutes a seabird differs among authors, but generally includes the penguins (Sphenisciformes), petrels and albatrosses (Procellariiformes), pelicans, boobies and cormorants (Pelicaniformes), and the gulls, terns, and auks (Charadriiformes) (**Table 1**). Sometimes included are loons (Gaviiformes), grebes (Podicipediformes), and those ducks that forage at sea throughout the year or during the winter (Anseriformes). Bird species that are restricted to obtaining their prey by wading along the margins of the sea, such as herons or sandpipers, are not included in this definition.

The distribution of types of seabirds shows striking differences between the Northern and Southern Hemispheres, particularly at high latitudes. Best known are the restrictions of the auks (Alcidae) to the Northern Hemisphere, and the penguins (Spheniscidae), diving petrels (Pelecanoididae), and most species of albatrosses (Diomedeidae) to the southern oceans. There are also many more species of shearwaters and petrels (Procellariidae) that nest in the Southern Hemisphere than in the north.

Patterns of seabird species distributions differ between the Antarctic and the Arctic due to the underlying oceanic currents. In the Antarctic, distributions are annular or latitudinal, with strong similarities in species composition of seabird communities in all ocean basins at a given latitude. In the Arctic, communities are arranged meridionally, and show strong differences between ocean basins and, at a given latitude, between sides of ocean basins. These differences between the seabird communities in the Northern and Southern Hemispheres reflect differences in the patterns of flow of major ocean current systems. At smaller spatial scales, in both hemispheres, the species composition of seabird communities is sensitive to changes in water mass characteristics.

Some species of seabirds are among the world's record holders for distance covered during migration. Arctic terns nest along the coasts of the northern North Atlantic and North Pacific Oceans as well as around the coasts of the Arctic Ocean. These birds annually migrate from these prey-rich northern waters to the pack ice of the Southern Ocean. Several species of Southern Hemisphere Procellariiformes conduct transequatorial migrations from their breeding grounds in the Antarctic and subantarctic to spend the austral winter in summer conditions in north temperate to subarctic waters. One of the longest of these migrations is performed by the short-tailed shearwater, which migrates from its breeding grounds in southeastern Australia to forage in June through October in the Bering and Chukchi Seas, and even in the Arctic Ocean. When not breeding, many Southern Ocean albatrosses circumnavigate Antarctica during year-long migrations.

The cost of flight varies greatly among seabird species. Some groups, such as diving petrels and auks, have heavy wing loading (ratio of body mass to wing surface area) and require flapping flight that is relatively expensive. Others have low wing-loading and are able to make use of wind energy and soar extensively, flapping only occasionally. These species, including albatrosses and many of the petrels, can travel relatively long distances with minimal energy expenditure. Between these extremes are birds that alternate flapping flight with soaring or gliding. These species-specific differences in cost of flight likely had profound effects on the types of birds that were able to survive in different climate domains of the oceans. Seabirds in the Southern Hemisphere tend to be efficient fliers that cover large areas in search of food by using the wind to enhance their soaring flight. These birds may range up to 5000 km from their colonies in search of prey. In contrast, in the Northern Hemisphere, most species of seabirds depend on relatively expensive flapping flight, and forage within 100 km or less of their colonies. Costs of flight have also clearly had a strong effect on the types of foraging techniques and chick-provisioning routines used by different groups of seabirds. These energetic constraints have also been shown to affect what water masses a seabird species might use. Those species with high costs of flight, particularly those that pursuit dive for prey, require productive coastal waters and frontal systems with high concentrations of prey, whereas species with inexpensive flight costs range over vast expanses of oceanic waters in search of widely dispersed prey.

During the breeding season, most seabirds nest in dense colonies on predator-free islands or on inaccessible cliffs. A few nest at low densities or as scattered individuals. All seabirds provision their

Table 1 Distribution and species richness of families of seabirds, based primarily on Harrison (1983)

Common name of family	Family	Total number of species	Number of species nesting south of 30°S	Number of species nesting between 30°S and 30°N	Number of species nesting north of 30°N	Period of sea use	Flight type (flapping, flap-gliding or soaring)	Foraging region (primarily neritic or oceanic)
Penguins	Spheniscidae	18	16	2	0	Year-round	Wing-propelled swimming	Mostly neritic
Loons or divers	Gaviidae	5	0	0	5	Migration and winter	Flapping	Neritic
Grebes	Podicipedidae	21	12	13	8	Migration and winter	Flapping	Neritic
Albatrosses	Diomedeidae	17	14	3	3	Year-round	Soaring	Oceanic
Petrels and shearwaters	Procellariidae	66	44	20	15	Year-round	Flap-gliding	Mostly oceanic
Storm petrels	Hydrobatidae	20	9	11	10	Year-round	Flap-gliding	Mostly oceanic
Diving petrels	Pelecanoididae	4	4	1	0	Year-round	Flapping	Mostly neritic
Frigate birds	Fregatidae	5	2	5	2	Year-round	Soaring	Mostly oceanic
Tropic birds	Phaethontidae	3	2	3	3	Year-round	Flapping, with some soaring	Oceanic
Gannets and boobies	Sulidae	9	6	7	5	Year-round	Flapping, with some gliding	Both neritic and oceanic
Pelicans	Pelecanidae	8	4	7	5	Most species fresh water	Flapping, and some gliding	Neritic
Cormorants	Phalacrocoracidae	28	20	17	10	Most species year-round	Flapping	Neritic
Sea ducks	Anatidae (part)	19	5	0	11	Migration and winter	Flapping	Neritic
Phalaropes	Scolopacidae	3	0	0	3	Migration and winter	Flapping	Neritic
Skuas, gulls, terns, noddies, and skimmers	Laridae	100	36	49	59	Most species year-round	Flapping	Mostly neritic; some tropical terns and noddies oceanic
Auks	Alcidae	23	0	2	22	Year-round	Flapping	Neritic

Based on Harrison P (1983) *Seabirds: An Identification Guide*. Boston: Houghton Mifflin.

young at the nest, and feed them at intervals from once every few hours to up to 15 days, depending upon the species. Because of their need to periodically return to their nests with food, seabirds are good examples of central place foragers.

The life history patterns of seabirds include long life spans, delayed reproduction, and small numbers of young produced in any one year. Interannual survival for adult birds, where estimated, is generally thought to be on the order of 90–95% or better. In contrast, survival through the first winter after fledging may be below 50%. The age of first reproduction varies between species as a function of expected life span. Most seabirds do not commence breeding until they are at least 3 or 4 years old, and some of the albatrosses delay the year of first breeding until they are 15 years of age. Clutch size is small. Most oceanic foragers lay only one egg per year. Some species of neritic foragers lay clutches of two or three eggs, with cormorants laying clutches of six or more eggs. Clutch sizes for species nesting in productive upwelling zones may be larger than is typical for their taxonomic relatives in less productive areas.

For most species of seabirds, reproductive rates are low and, except for those species adapted to upwelling regions, there is little opportunity to increase reproductive output in years of high prey abundance. In some ocean regions, years with successful reproduction are the exception. In those cases, it is likely that a few good years and a small percentage of particularly able parents may account for the majority of the young produced. Reproductive success improves with experience, and for many species mate retention from one year to the next is high, with diminished reproductive success associated with the changing of mates. Thus, high survival rates for adult birds are critical to maintain stable populations.

Seabirds obtain their prey by a wide variety of methods including pursuing fish and plankton at depths as great as 180 m or more, seizing prey on the wing from the surface of the ocean, and stealing prey from others (**Figure 1**). Prey most commonly taken by seabirds includes small fish and squid and large

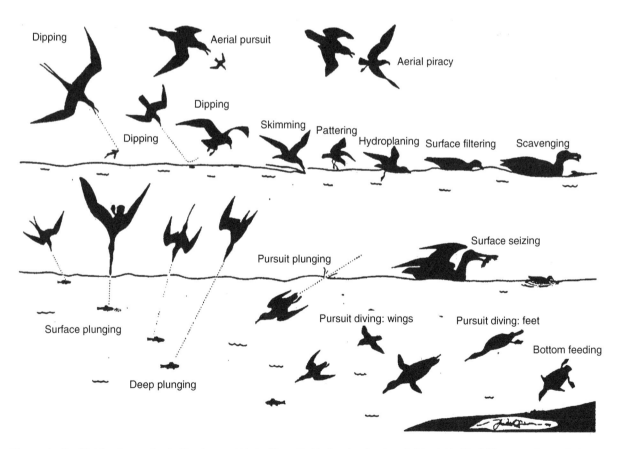

Figure 1   Seabird feeding methods. The types of birds silhouetted in the drawing from left to right. (Top) Skua pursuing a phalarope; skua pursuing a gull; (second row) frigate bird, noddy, gull, skimmer, storm petrel, prion, Cape petrel, giant petrel; (third row) tern, pelican, tropic bird, gannet, albatross, phalarope; (under water) shearwater, murre, diving petrel, penguin, cormorant, a sea duck. Reproduced from Ashmole NP (1971) Seabird ecology and the marine environment. In: Farner DS and King JR (eds.) *Avian Biology*, vol. 1, pp. 224–286. New York: Academic Press, with permission from Elsevier.

zooplankton such as euphausiids (krill) and amphipods. A few seabird species specialize on copepods. The use of gelatinous zooplankton is known for albatrosses, petrels, and auks. Gelatinous zooplankton may be more commonly consumed by other species than is appreciated, as remains of gelatinous species are difficult to detect in stomach contents or preserved food samples.

Foraging methods used by seabirds vary with the density of prey and may also vary with water clarity. Plunge-diving for prey and surface-seizing of prey are particularly common in clear tropical waters. In more turbid and prey-rich polar and subpolar waters, many seabirds pursue their prey beneath the sea. In the Southern Ocean, where continental shelf regions are deep and limited in area, most seabirds other than penguins are efficient fliers that can cover vast areas in search of patchy prey. In the Northern Hemisphere, most species of seabirds forage over broad, shallow continental shelves where interactions between currents and bathymetry result in predictable concentrations of prey. These prime foraging locations are often close to shore where tidal currents create convergence fronts and other physical features that force aggregations of zooplankton upon which small fish forage. Important in both hemispheres are frontal systems associated with shelf edges. Recent studies using satellite telemetry have shown that some species of seabirds forage along the edges of mesoscale eddies.

Marine birds are assumed to play only an insignificant role in oceanic carbon cycling. However, where their consumption has been examined, they have been found to take considerable amounts of prey. In the southeastern Bering Sea, seabirds have been estimated to consume between 0.12 and $0.29 \, \mathrm{g} \, \mathrm{C} \, \mathrm{m}^{-2} \, \mathrm{yr}^{-1}$, depending on the region of the shelf and the year. Since the seabirds are foraging between the second and third trophic levels, this consumption accounts for between 12 and $29 \, \mathrm{g} \, \mathrm{C} \, \mathrm{m}^{-2} \, \mathrm{yr}^{-1}$ of the $200–400 \, \mathrm{g} \, \mathrm{C} \, \mathrm{m}^{-2} \, \mathrm{yr}^{-1}$ of primary production over the shelf. Where estimates have been made of the proportion of secondary production consumed, seabirds have been found to take between 5% and 30% of local secondary production (**Table 2**). In the North Sea, seabird consumption of sand eels is on the order of 197 000 t. Over half of this consumption is concentrated near the Shetland Islands. However, seabird consumption of sand eels in the North Sea as a whole is small compared to the overall production of sand eels or the take of these fish by the industrial fishery which focuses on offshore banks. Recent calculations of the fraction of total exploitable stocks in the eastern Bering Sea that are consumed by seabirds suggest that about 3% of walleye pollock and 1% of herring are taken by birds. Recent shifts in the diets of shearwaters in the eastern Bering Sea could mean that the estimate for pollock is low by a factor of 2 or more. In the North Pacific, depending on the region of concern, estimates of prey consumption by marine birds vary between 0.01 and $1.72 \, \mathrm{mt} \, \mathrm{km}^{-2}$ for the summer months of June, July, and August.

The effects of fisheries on seabirds are almost always greater than the effect of birds on fisheries. Seabirds have been used by fishers as indicators of the presence of large predatory fish. Commercial fishing activity frequently provides offal and discards

Table 2 Community energetics models of fish consumption by seabirds

Location	Estimated % pelagic fish production consumed	Major consumers	Major prey species
Oregon coast	22	Shearwaters, storm petrels, cormorants, murre	Northern anchovy, juvenile hake
Foula, Shetland Islands	29	Fulmar, murre, shag, puffin	Sand eels
North Sea	5–8	Fulmar, gulls, terns, murre, puffin	Sand eels
North Sea	5–10	Fulmar, gannet, shag, gulls, kittiwake, terns, razorbill, murre, puffin	Sand eels
Saldanha Bay, South Africa	29	Penguin, gannet, cormorant	Pilchard
Benguela Region	6	Gannet, cormorant	Pilchard
Vancouver Island, British Columbia[a]	11, 17, and 21	Shearwater, murre	Juvenile herring

[a]Logerwell EA and Hargraves NB (1997) Seabird impacts on forage fish: Population and behavioral interactions. In: *Proceedings of Forage Fishes in Marine Ecosystems*. Alaska Sea Grant College Program AK-SG-97-01.
Modified from Hunt GL, Jr., Barrett RT, Joiris C, and Montevecchi W (1996) Seabird/fish interactions: An introduction. *ICES Cooperative Research Reports* 216: 2–5.

to seabirds, and a number of seabird populations have benefited from this supplemental source of food. Indeed, juvenile albatrosses in the North Pacific may be experiencing increased winter survival rates thanks to the availability of discards and offal from commercial fisheries. However, seabirds attracted to fishing vessels do not distinguish between discarded fish and baited hooks. Between 1990 and 1994, over 30 000 Laysan albatrosses and 20 000 black-footed albatrosses were killed on long-lines off Hawaii. In the Southern Hemisphere, up to 20 000 shy albatrosses and 10 000 wandering albatrosses are entangled annually in the bluefin tuna long-line fishery. Approximately 10% of the world population of wandering albatrosses is killed annually. A loss of this magnitude cannot be sustained in a species that has a life history strategy that depends on extraordinarily high annual rates of adult survival. Not surprisingly, populations of 6 of 17 species of albatross are now declining rapidly.

Seabirds have proven to be useful indicators of changes in marine ecosystems. Because of their position at the top of marine food webs, they tend to accumulate a number of pollutants. For example, dichlorodiphenyldichloroethylene (DDE), which concentrates in lipids, can result in eggshell thinning in some species. Eggshell thinning in various pelican and cormorant populations was one of the first indicators that DDE was present in certain coastal waters. Likewise, recent work has shown that seabirds bioaccumulate mercury and other heavy metals. Seabirds also provide information about the status of populations of prey on which they depend. Changes in prey abundance are reflected by changes in annual rates of production of young, or in extreme cases, changes in seabird population size. Climate-driven changes in prey abundance, at scales from years to hundreds of years, are reflected by changes in seabird distribution, abundance, and reproductive output. Although a goal of a number of studies, the ability to estimate the standing stocks of prey populations based on indices derived from seabirds has yet to be accomplished. Calibration of the responses of the seabirds to shifts in prey abundance has been difficult to achieve.

The most serious threats to the conservation of seabirds are those that result in the deaths of adult birds, particularly those individuals that have been successful breeders. Seabird populations can usually recover from a single instance of mortality, but chronic elevated rates of mortality are devastating. Seabirds are extremely vulnerable to predation in their colonies. Thus, the presence of foxes, cats, rats, and other introduced predators on islands where seabirds nest is of great concern. Past introductions

of predators have resulted in the extirpation of nesting birds, and the removal of predators has resulted in their return. Also of great concern is the annual drowning of high numbers of seabirds in gill nets and on long-lines. The situation with respect to albatrosses and other procellariiform birds is of particular concern, as these birds require long adult life spans to insure the production of sufficient young to maintain stable populations. Oil spills can have devastating local effects, but if the resulting pollution is not long-lasting, populations are likely to recover. Chronic pollution by oil or other chemicals may have both lethal and sublethal effects, and can damage populations of seabirds over time. Competition for resources, such as nesting space or food, is occasionally of concern. Development of islands and beaches affects a few populations of seabirds. There is growing concern about potential impacts on migrating seabirds from the development of wind farms at-sea. Likewise, there are a few instances where fisheries may be competing with seabirds for particular size classes or species of prey. However, since most fisheries target large predatory fish and discard offal and small fishes, many of which would otherwise have not been available to seabirds, it is unclear how widespread competitive interactions with fisheries may be.

See also

Alcidae. Laridae, Sternidae and Rynchopidae. Pelecaniformes. Phalaropes. Procellariiformes. Seabird Conservation. Seabird Migration. Sphenisciformes.

Further Reading

Ashmole NP (1971) Seabird ecology and the marine environment. In: Farner DS and King JR (eds.) *Avian Biology*, vol. 1, pp. 224–286. New York: Academic Press.

Croxall JP (1987) *Seabirds: Feeding Ecology and Role in Marine Ecosystems*. Cambridge, UK: Cambridge University Press.

Enticott J and Tipling D (1997) *Seabirds of the World*, 234pp. Mechanicsburg, PA: Stackpole Books.

Furness RW and Monaghan P (1987) *Seabird Ecology*. London: Blackie.

Garthe S and Hüppop O (2004) Scaling possible adverse effects of marine wind farms on seabirds: Developing and applying a vulnerability index. *Journal of Applied Ecology* 41(4): 724–734.

Harrison P (1983) *Seabirds: An Identification Guide*, 488pp. Boston: Houghton Mifflin.

Hunt GL (1990) Marine ecology of seabirds in polar oceans. *American Zoologist* 31(1): 131–142.

Hunt GL, Jr., Barrett RT, Joiris C, and Montevecchi W (1996) Seabird/fish interactions: An introduction. *ICES Cooperative Research Reports* 216: 2–5.

Hunt GL, Jr., Mehlum F, Russell RW, Irons D, Decker MB, and Becker PH (1999) Physical processes, prey abundance, and the foraging ecology of seabirds. In: Adams NJ and Slotow R (eds.) *Proceedings of the 22nd International Ornithological Congress, Durban*, pp. 2040–2056. Johannesburg: BirdLife South Africa.

Hunt GL, Jr. and Nettleship DN (1988) Seabirds of high-latitude northern and southern environments. In: Oulette H (ed.) *Proceedings of the XIX International Ornithological Congress, 1986*, pp. 1143–1155. Ottawa: University of Ottawa Press.

Hunt GL and Schneider D (1987) Scale dependent processes in the physical and biological environment of marine birds. In: Croxall JP (ed.) *Seabirds: Feeding Biology and Role in Marine Ecosystems*, pp. 7–41. Cambridge, UK: Cambridge University Press.

Livingston PA (1993) Importance of predation by groundfish, marine mammals and birds on walleye pollock and Pacific herring in the eastern Bering Sea. *Marine Ecology Progress Series* 102: 205–215.

Logerwell EA and Hargraves NB (1997) Seabird impacts on forage fish: Population and behavioral interactions. In: *Proceedings of Forage Fishes in Marine Ecosystem*. Alaska Sea Grant College Program AK-SG-97-01.

Murphy RC (1936) *Oceanic Birds of South America*, vols. 1 and 2. New York: Macmillan and American Museum of Natural History.

Nelson JB (1978) *The Sulidae: Gannets and Boobies*. Oxford, UK: Oxford University Press.

Pennycuick CJ (1989) *Bird Flight Performance. A Practical Calculation Manual*. Oxford, UK: Oxford University Press.

Phillips RA, Silk JRD, Croxall JP, Afanasyev V, and Bennett VJ (2005) Summer distribution and migration of nonbreeding albatrosses: Individual consistencies and implications for conservation. *Ecology* 86: 2386–2396.

PICES Working Group 11 Final Report (1999) *Consumption of Marine Resources by Seabirds and Marine Mammals*. Sidney, BC: PICES.

Schneider DC, Hunt GL, Jr., and Harrison NM (1986) Mass and energy transfer to seabirds in the southeastern Bering Sea. *Continental Shelf Research* 5: 241–257.

Shaffer SA, Tremblay Y, Weimerskirch H, et al. (2006) Migratory shearwaters integrate oceanic resources across the Pacific Ocean in an endless summer. *Proceedings of the National Academy of Sciences of the United States of America* 103: 12799–12802.

Tasker ML and Furness RW (1996) Estimation of food consumption by seabirds in the North Sea. In: Hunt GL, Jr. and Furness RW (eds.) *ICES Cooperative Research Report, 216: Seabird/Fish Interactions, with Particular Reference to Seabirds in the North Sea*, pp. 6–42. Copenhagen: ICES.

Warham J (1990) *The Petrels: Their Ecology and Breeding Systems*. London: Academic Press.

Warham J (1996) *The Behaviour, Population Biology and Physiology of the Petrels*. London: Academic Press.

Whittow GC and Rahn H (1984) *Seabird Energetics*. New York: Plenum.

ALCIDAE

T. Gaston, National Wildlife Research Centre, Quebec, Canada

Introduction

The auks are seabirds of the northern hemisphere. They form a family (Alcidae) or subfamily (Alcinae) of the order Charadriiformes, which includes the gulls, terns, jaegers, sandpipers, and plovers. They are often considered the northern hemisphere equivalents of the penguins, being well-adapted to underwater swimming. However, all extant species retain the power of flight. They are almost entirely confined to Arctic, Subarctic and Boreal waters and are marine throughout the year, being most common in waters of the continental shelf and slope. Twenty-three species are currently recognized and one other became extinct in historic times (great auk *Pinguinus impennis*). Six species of five genera occur in the North Atlantic and 20, of 10 genera in the North Pacific, south of the Chukchi Sea. Two murres (*Uria* spp.), the dovekie (*Alle alle*) and the black guillemot (*Cepphus grylle*) are circumpolar in distribution.

Many species of auks are extrememly abundant, with common and thick-billed murres (*Uria aalge, U. lomvia*), least and crested auklets (*Aethia pusilla, A. cristatella*), dovekies and Atlantic puffins (*Fratercula arctica*) all having world populations of > 10 million individuals. They form the dominant avian biomass over large areas of Arctic and Subarctic waters throughout the year and may be significant predators on large zooplankton and small fishes in areas around their breeding colonies.

Evolution and Systematics

Recent work on mitochondrial DNA and allozymes divides the family Alcidae into 5 tribes and 12 genera (**Table 1**), with the puffins (Fraterculini) and auklets (Aethiini) being sister tribes, more closely related to one another than to the other auks. Otherwise, the molecular data suggest that the tribes originated from an initial rapid divergence among early members of the family, possibly about 10–12 million years ago.

There is extensive fossil material on the family. The first definite records of auks are from the middle Miocene, about 15 million years ago (Ma), although putative proto-auks have been described from as far back as the Eocene. A subfamily of flightless auks, the Mancallinae, was present in the Pacific during the late Miocene (7–4 Ma). Deposits from California have yielded at least five species of the flightless *Mancalla*, ranging in size from about 1 to 4 kg.

Auks probably originated in the Pacific, where all but two (*Alca, Alle*) of the extant genera are widely distributed, colonizing the Atlantic soon after. Since then, Pacific and Atlantic auks evolved largely in isolation. The Pliocene auk fauna from California was similar in diversity to the recent fauna, but the diversity of auks represented in the Pliocene sediments of eastern North America suggests that the current, relatively small, community of auks in the Atlantic (six species, including the great auk) is not a historical consequence of the lack of connections with the Pacific. Instead, it is a consequence of extinctions during the Pleistocene, presumably as a result, directly or indirectly, of the ice ages.

Characteristics and Adaptations

Auks are well-adapted to underwater swimming. They have compact, streamlined bodies, short wings and very short tails. The feet are placed far back on the body, webbed, with no hind toe; claws are narrow and the tarsus is laterally compressed. The bill is variable in shape and may be highly ornamented in the breeding season. There are 11 primary wing feathers and 16–21 secondaries, the outermost primary being very small and the longest being, usually, the tenth. The feather tracts of the back and belly are

Table 1 Tribes and genera of the Alcidae (after Gaston and Jones 1998)

Tribe	Genus
Fraterculini	Fratercula
	Cerorhinca
Aethiini	Aethia
	Cyclorhynchus
	Ptychoramphus
Brachyramphini	Brachyramphus
Cepphini	Cepphus
	Synthliboramphus
Alcini	Alle
	Pinguinus
	Alca
	Uria

continuous and, beneath the contour feathers, down feathers are present and dense all over the body. There are 6–8, occasionally 9, pairs of tail feathers.

The fusiform body, with the feet set far back, is common to all underwater swimming birds. Other characters that are common to underwater swimmers, but absent in Charadriiformes that do not dive, are the presence of strongly developed vertebral hypophyses on the last cervical vertebra, and the enlarged number of thoracic vertebrae (8–10 compared to 5–7 in other Charadriiformes). An increase in the number of vertebrae allows for a longer body, while retaining flexibility of movement.

Auks fly with rapid wing-beats and without gliding or soaring, using the spread feet for steering and braking. They generally take off from land with difficulty and use the feet to taxi when taking off from water. Underwater, they swim using the wings as paddles, like penguins. Maneuvering underwater is achieved mainly by asymmetrical strokes of the wings. Air is released from plumage before diving by forcing it from the breast feathers using subcutaneous muscles. On land, the Alcini and Brachyramphini rest on the belly, or on the tarsi, rising to their feet only in walking, whereas Aethiini and Fraterculini are more agile and normally stand with the tarsus erect, rather than horizontal.

The anatomy of the auks is similar to that of other Charadriiformes, except for their specializations towards underwater swimming. They have shorter wings and legs than gulls and shore birds, and the relative length of the humerus is reduced. The covert feathers are stiffer and more extensive, both above and below the wing, an arrangement that reinforces the trailing edge of the wing and closes gaps between adjacent flight feathers. Primaries 6–10 form a closely knit unit with little independent movement, making the wing more effective as an underwater paddle. The rigidity is caused by a greater development of connective tissue, compared to other Charadriiformes. Despite the fact that the primary feathers are very stiff, the leading edge of the wing is very curved on the downstroke, presumably because of water resistance.

The articulating surface of the humerus is much larger in auks than in other Charadriiformes, allowing greater force to be transmitted in swimming. However, like other flighted birds, but unlike penguins, the auks retain considerable flexibility of movement at the articulation of the humerus. The supracoracoideus muscle, which raises the wing during the recovery stroke (2–3% of body mass) is larger than in shore birds and gulls (1.5%) and the pectoralis muscles, the main source of power on the downstroke, although similar in size to those of gulls and shorebirds, are more elongated, with the supracoracoideus lying directly below them. This arrangement improves streamlining, and also improves insulation by distributing stored fat in a thin subcutaneous layer over most of the body, rather than depositing it in discrete pockets, as occurs in many birds.

Underwater, auks swim very jerkily, with a rapid acceleration at each downstroke. Hence, forward propulsion on the upstroke, if any, is relatively small. The wings are held sharply bent at the wrist, reducing the functional surface area to slightly less than half that when spread in flight. The fact that the whole wing area is unnecessary for effective underwater propulsion is emphasized by the fact that all genera except *Aethia* and *Cyclorhynchus* lose their flight feathers simultaneously during the annual postbreeding moult, becoming flightless for a month or more.

The auks have developed many physiological adaptations for prolonged diving, including high blood volume and high levels of myoglobin (>1% body mass), both of which enhance the bird's ability to store oxygen. Despite these enhancements, many long dives undertaken by thick-billed murres, and probably other auks, exceed the estimated limit for aerobic respiration, requiring the birds to respire anaerobically for a portion of the time underwater. Heart rate and peripheral blood flow also may be restricted.

The digestive system includes a well-developed proventriculus, a muscular stomach and a relatively short intestine. A functional crop is present only in auklets, although there is some croplike development of the lower proventriculus in the puffins. Auklets and dovekies develop diverticulae in the throat while breeding, in which they transport food for their young. Nasal glands, which are used to excrete salt and maintain ion balance, are very well developed.

The dominant plumage color of auks is black above and white below, but *Cepphus* species are mainly black in summer, mainly white in winter, and *Brachyramphus* species are mottled brown in summer (**Figure 1**). Temporary nuptial ornaments, including ornamental plumes on the head and deciduous, brightly colored, horny plates on the bill, are present in the Aethiini and Fraterculini. Sexual dimorphism is very small: males of most species are slightly larger in some dimensions, especially bill depth. There is no plumage dimorphism and no distinctive immature plumage. First-winter birds generally appear similar to winter-plumage adults, although young *Cepphus* are readily distinguished by their greater mottling.

Figure 1 Examples of the Alcidae. (1) Razorbill (*Alca torda*). Adult breeding plumage. Length: 42 cm; approximate body mass: 710 g. Range: North Atlantic Ocean. (2) Brunnich's guillemot (*Uria lomvia*). Other name: thick-billed murre. (a) Adult breeding plumage; (b) adult non-breeding plumage; (c) transitional plumage. Length: 40 cm; approximate body mass: 950 g. Range: North Pacific, North Atlantic and Arctic Oceans. (3) Least auklet (*Aethia pusilla*). Length: 15 cm; approximate body mass: 85 g. Range: North Pacific Ocean. (4) Marbled murrelet (*Brachyramphus marmoratus*). Adult breeding plumage. Length: 25 cm; approximate body mass: 235 g. Range: eastern North Pacific Ocean. (5) Crested auklet (*Aethia cristatella*). (a) adult breeding plumage; (b) adult, non-breeding plumage. Length: 25 cm; approximate body mass: 260 g. Range: North Pacific Ocean. (6) Atlantic puffin (*Fratercula arctica*). Other name: puffin. Adult, breeding plumage. Length: 35 cm; approximate body mass: 470 g. Range: North Atlantic Ocean. (7) Little auk (*Alle alle*). Other name: dovekie. Adult breeding plumage. Length: 20 cm; approximate body mass: 165 g. Range: North Pacific and North Atlantic Oceans, Arctic Ocean. Illustrations from Harrison P (1985) *Seabirds, an identification guide. Revised edition*. Boston, Massachusetts: Houghton Mifflin.

Diet and Foraging Ecology

The diet of the smaller auks consists largely of zooplankton, especially copepods and euphausiids and larval fishes, although the parakeet auklet *Cyclorhynchus psittacula* takes substantial quantities of jellyfish medusae. The larger auks take a mixture of zooplankton and small fishes and squid, with common murres and *Cepphus* spp., and possibly *Brachyramphus* spp., being mainly piscivorous. Nestling *Uria*, *Alca*, *Brachyramphus*, *Cepphus*, *Cerorhinca*, and *Fratercula* species are all fed predominantly on fish.

Auks that feed mainly on fish tend to have narrower, more pointed bills (low width/gape ratio), and narrower, more cornified, tongues than those that feed on plankton. Planktivorous auks, apart from having broader, shorter bills, have large numbers of fleshy denticles on the palate and the upper surface of the tongue. The convergence of these characters for auks feeding exclusively on plankton is dramatically demonstrated by the parallels between the auklets and the dovekie. These adaptations are also shown within the genus *Uria*, with the predominantly fish-eating common murre having a narrower tongue and fewer palatal denticles than the thick-billed murre, which feeds on a greater variety of prey. Bill depth seems to be unrelated to diet, suggesting that the very deep bills of puffins and auklets may have evolved for secondary sexual purposes, rather than as feeding adaptations.

Auks feed predominantly in continental shelf waters, with species of *Brachyramphus* and *Cepphus* feeding entirely in inshore waters and most other genera occurring mainly within 50–100 km of land. However, *Fratercula* species are found far offshore in winter, including waters beyond the continental shelf. All species feed entirely underwater, with the smaller species diving to maximum depths of 30–40 m and the larger species to over 100 m. The deepest dives are achieved by *Uria* species, which may reach 200 m on occasions. Normal foraging depths for smaller auks are 10–30 m and for larger auks 30–60 m.

Many species take advantage of prey aggregations caused by oceanographic fronts and tidally induced upwellings. Striking examples occur in the passes among Aleutian Islands and other complex archipelagos, where feeding auklets may reach great densities. The prey involved are usually slow-swimming zooplankton such as pteropod mollusks, copepods, euphausiids, and amphipods.

Reproduction

Breeding sites are coastal, except in the marbled and Kittlitz's murrelets, which are also the only solitary nesters, and in the dovekie. Most auks breed on islands, and some species breed exclusively on remote islands well offshore, because islands offer refuge from terrestrial predators and at the same time may be close to foraging areas. Hence, the distribution of breeding auks is much influenced by the distribution of suitable breeding islands.

Most species are highly social while breeding, with a prolonged prebreeding period and extensive courtship activity in which both vocal and visual displays are prominent. All species are socially monogamous and the sexes play equal roles in incubation and in rearing chicks to the age when they leave the colony. Most show a strong tendency to return to the colony where they were reared.

Auks are highly variable in the type of nesting site that they use. None builds a nest: the *Aethia* auklets and some *Synthliboramphus* murrelets use crevices under boulders or among scree, and murres lay their eggs on open cliff ledges or on flat rocky islets (common murre), while puffins, ancient murrelet, and Cassin's auklet often dig extensive burrows in soil. Most of the smaller auks make use of sites that protect them from surface predators, such as gulls and crows: only the largest, the murres, nest in the open in colonies, where their densely packed ranks form a defense against nest predators. Cliffs are frequently used, especially in the Arctic. The other open nesters, the *Brachyramphus* murrelets, are solitary nesters, either on the horizontal limbs of mature trees (marbled murrelet *Brachyramphus marmoratus*, commonly) or on the ground on remote mountain tops (Kittlitz's murrelet *B. brevirostris*).

Like many other small seabirds, some of the auks are nocturnal in their coming and goings to their breeding sites. The *Synthliboramphus* murrelets and whiskered and Cassin's auklets (*Aethia pygmaea*, *Ptychoramphus aleuticus*) are invariably nocturnal, and the rhinoceros auklet (*Cerorhinca monocerata*) is largely nocturnal, but diurnal or crepuscular in some parts of its range. *Brachyramphus* species are normally crepuscular. Everything about the breeding strategies of the auks suggest the dominant influences of predation and kleptoparasitism.

Egg formation is a lengthy process and females spend most of their time away from the colony during the last 10 days before they lay. Clutches consist of one or two eggs and only one brood is reared annually (California populations of Cassin's auklet sometimes rear two). Eggs are thick-shelled and either white, buff, tan, or bluish in ground colour, sometimes with prominent black or brown markings. After clutch completion, the sexes alternate incubation duty, taking equal shares. Laying, in practically all populations, is confined to a span of about 6 weeks, with most eggs being laid within 2–3 weeks. Incubation periods range from 29 to 45 days. The time from the start of incubation to chick departure varies from a minimum of 35 days in the precocial *Synthliboramphus* species to a maximum of more than 80 days in the *Cerorhinca*.

Most species, including the puffins, auklets and razorbill (*Alca torda*) that lay only a single egg, have two lateral brood patches. The murres and the dovekie have a single, central patch, set far back on the belly. Brood patches are a potential site of heat loss. They defeather rapidly at the start of incubation, refeather as soon as hatching occurs, and are small relative to the size of the eggs, so that during incubation only part of the egg's surface is in contact with the patch. Nonbreeders either do not form brood patches or develop only partial patches.

Notwithstanding these adaptations, the insulation of the auks appears to be relatively poor, considering the climates in which they live. To maintain normal avian body temperature, they rely principally on a very high rate of metabolism: basal metabolic rates (the metabolic rate of resting, nondigesting birds) among auks are exceptionally high, compared to those of other birds of similar size.

Hatchlings are covered with a dense, woolly, down plumage. They are active within 1–2 days after hatching and capable of thermoregulation either immediately (*Synthliboramphus*) or within a few days (the rest). All genera except *Synthliboramphus* are semiprecocial and chicks are reared at the nest site for a minimum of 2 weeks. There is no postfledging parental care except in the dovekie, although in *Synthliboramphus*, *Uria*, and *Alca* the young are cared for until some time after departing the colony, which they leave when only partly grown and incapable of sustained flight. Most breeders leave the colony either with their chicks (Alcini and *Synthliboramphus* spp., where there is parental care after departure), or within 1–2 weeks following chick departure. Juveniles, family parties, and postbreeding adults disperse rapidly from the colony area (except young whiskered auklets).

Postbreeding and Wintering

A complete prebasic molt (except in the puffins, where it involves the body plumage only) follows rapidly on the termination of breeding and involves the shedding of nuptial ornaments and the adoption of a generally distinctive nonbreeding (winter) plumage (no change in Craveri's and Xantus' murrelets). In the auklets, primary molt is sequential and begins during chick-rearing, but in other species the primaries are dropped simultaneously, making flight

impossible for a period. Most species remain scattered offshore while molting.

Following the annual prebasic molt, most species disperse or shift toward distant wintering areas. However, some populations of common murres and black guillemots return to the area of the breeding colony and commence periodic attendance at the breeding site. This also applies to some populations of marbled murrelets. These birds feed in the same areas practially year-round. However, among most species breeding in Arctic and Subarctic waters, the bulk of the population moves substantially farther south. Young birds tend to disperse farthest and there is usually a disproportionately high representation of first-year birds in 'wrecks'; – the periodic casting ashore of large numbers of weakened birds, often during prolonged storms. Wintering areas and the behavior of auks on the wintering grounds have been less studied than their activities during the breeding season. Detailed studies of feeding ecology in winter have mostly been carried out on species and populations that occur in inshore waters and the winter ecology of many is essentially unknown. Movement toward the breeding colonies begins in February–April, depending on latitude.

Population Dynamics

Annual survival of adult auks is generally greater than 85%, and greater than 95% in some populations of common murres and Atlantic puffins (*Fratercula arctica*), making them among the longest-lived birds. In the longer-lived species, average age at first breeding is 5 years or more, whereas Cassin's auklets, ancient murrelets, and probably some *Aethia* spp. may begin to breed at 2, many at 3 years. Populations contain substantial numbers of nonbreeders. Those in their second summer or older often attend the breeding colony to select mates and breeding sites. Reproductive success increases with age for the first few years of breeding. Comparisons of reproductive success are complicated by the fact that different species depart from the colony at different stages of development. The maximum productivity is about 1.5 young/pair per year and most average <1.

Auks and People

Auks and their eggs have been harvested by people from the earliest times and their bones are frequent constituents of middens throughout the coastal areas of the northern hemisphere. The remains of the great auk have been discovered in 40 middens in Norway alone, as well as some in Britain, Iceland, Greenland, and the United States. Excavations in Newfoundland dating to 4000 BP contain many Great Auk bones and they are also found in middens in Florida dating to 3000 BP. All major colonies of thick-billed murres in the eastern Canadian Arctic show traces of Eskimo occupation nearby, usually with associated remains of thick-billed murres. In areas surrounding the Straits of Georgia, British Columbia, common murre remains are widespread in Indian middens dating from the pre-European period, while middens in the Aleutians, some dating to 4000 BP, contain the remains of auklets and puffins, as well as murres.

A variety of techniques were developed for catching auks. One of the most widespread, used by Bering Sea Inuit, Icelanders, and Faeroese to catch puffins, and by the Inuit of north-west Greenland to catch dovekies, was a net at the end of a long pole. The hunter sheltered behind a stone wall, or depression in the ground, and suddenly raised the net in the path of low-flying birds, which were unable to turn in time to avoid it. In Iceland, snares placed on rafts floating offshore were also used to good effect, trapping murres, puffins, and razorbills.

The technique of netting birds flying over the colony is a very efficient way to harvest auks, as the prebreeding component of the population often circles constantly, making them much more vulnerable than the breeders. Similarly, snares placed on boulders or floating rafts used by displaying birds, a method used on St. Lawrence Island to capture least and crested auklets, and in Thule, Greenland, to capture dovekies, select mainly prebreeders. The removal of these birds has much less effect on the population than the killing of breeders and may partly explain how early societies managed to coexist successfully with their prey. It has been estimated that 150 000–200 000 Atlantic puffins were taken annually in Iceland, without any apparent effect on population levels. In contrast, the shooting of breeding birds at colonies that became widespread in Greenland in this century has been the main cause of the drastic decline suffered by thick-billed murre populations there over the past 50 years.

The harvesting of auk eggs has been very common throughout their range and may well have been a factor controlling their distribution on inshore islands accessible to permanent human settlements. On St. Kilda, harvesting of puffin and murre eggs was a regular activity, while in the Queen Charlotte Islands, and throughout the Alaskan islands, the excavation of ancient murrelet and auklet burrows for eggs was a routine spring harvest. Thick-billed murre eggs are harvested in Greenland, Canada, the Pribilof Islands, and Russia.

In addition to their use as food, the skins and ornaments of certain auks were valued for clothing and decoration. Inuit on St. Lawrence Island and Aleuts in the Aleutian chain sewed parkas out of auk skins, especially crested auklets and horned puffins. Elsewhere, puffin and dovekie skins were sewn into inner garments, to be worn under furs. In north-west Greenland, dovekie skins were made into undershirts, while farther south, in Upernavik District, murre skins were made into capes. Dovekie skins had to be softened by chewing; only elderly women did this, as their teeth were worn smooth enough not to damage the delicate skins. The spectacular beaks of puffins and auklets were also used as ornaments by the Aleuts and Inuit of the Bering Sea region, hundreds sometimes being sewn on the outside of a garment, along with the golden crests of tufted puffins.

Commercial exploitation of auk colonies by postindustrial societies resulted in substantial declines. In the Gulf of St. Lawrence, the huge auk colonies visited by Audubon in 1827 were reduced to a mere remnant by the late nineteenth century, while at Funk Island, Newfoundland, the great auk was exterminated largely for its feathers. The common murre population at the Farallon Islands, California, and several large thick-billed murre populations in the Russian Far East and Novaya Zemlya were decimated by egg harvesting.

Substantial harvesting of auks still occurs in several areas. Traditional harvests of murres and puffins continue in Iceland and the Faroes, although reduced from former levels. Relatively small numbers of thick-billed murres are taken by Inuit in the Canadian Arctic, although they form important components of the summer diet at a few settlements. Much larger numbers are taken in West Greenland, where regulations prohibiting the shooting of birds at their colonies were introduced only in 1978 and were still more or less unenforced in 1987. Shooting away from the colonies is still permitted throughout the year in some districts, although subject to seasonal limits in the more populated areas. The same populations are hunted more heavily off Newfoundland and Labrador in winter, with the annual kill estimated at about 200 000 since 1993. Although it is only legal to kill murres, some razorbills and dovekies, and a few puffins, are also shot.

Although direct harvests have affected several auk species and were responsible for the extermination of the great auk, the effects of mammalian predators, introduced either deliberately, or accidentally, have probably had a much greater impact on auk populations worldwide. The main agents of destruction were foxes, introduced throughout the Alaskan islands for fur farming. Rats have also caused many declines and extirpations. Raccoons and mink have an important impact in some areas, and rabbits, through their effects on vegetation and soil, may also have caused problems for some burrowing species.

Japanese, Craveri's, Xantus' and marbled murrelets are all considered endangered or threatened in one way or another. It is certain that the majority of auk populations are smaller, in many cases much smaller, than they would have been a few centuries ago. Probably, we will see little change in that situation, although programs to eliminate introduced predators from certain important Pacific islands may improve the situation for some species. All auks are very susceptible to contamination by oil and they have formed the majority of seabirds killed in oil spills off Europe and North America. Unlike gulls, they have not profited at all from fisheries wastes. Protection from egging has led to increases of some species in the twentieth century. However, overall, the auks remain precariously dependent on human goodwill for their future survival.

See also

Fish Predation and Mortality. Laridae, Sternidae and Rynchopidae. Seabird Conservation. Seabird Foraging Ecology. Seabird Migration. Seabird Population Dynamics.

Further Reading

Gaston AJ and Jones IL (1998) *The Auks:: Family Alcidae*. Oxford: Oxford University Press.

Birkhead TR and Nettleship DN (1985) *The Atlantic Acidae*. London: Academic Press.

Harris MP (1989) *The Puffin*. London: T & AD Poyser.

Gaston AJ and Elliot RD (eds.) (1991) *Conservation Biology of the Thick-billed Murre in the Northwest Atlantic*. Ottawa: Canadian Wildlife Service.

Sealy SG (ed.) (1990) *Auks at Sea*. Studies in Avian Biology 14, Cooper Ornithological Society.

Friesen VL, Baker AJ, and Piatt JF (1996) Phylogenetic relationships within the Alcidae (Aves: Charadriiformes) inferred from total molecular evidence. *Molecular Biology and Evolution* 13: 359–367.

LARIDAE, STERNIDAE AND RYNCHOPIDAE

J. Burger, Rutgers University, Piscataway, NJ, USA
M. Gochfeld, Environmental and Community
Medicine, Piscataway, NJ, USA

Introduction

Gulls belong to the family Laridae and terns to the family Sternidae, although many authorities treat them as subfamilies (Larinae, Sterninae) of a single family, the Laridae. Skimmers belong to the Rynchopidae. Gulls, terns, and skimmers, all members of the order Charadriiformes, are similar in many respects, but they differ significantly in many morphological, behavioral, and ecological ways.

Gulls and terns are generally diurnal species that perform most of their breeding, foraging, and migrating activities during the day, while skimmers are largely nocturnal, and forage and court mostly at night. The only gulls that are primarily nocturnal during the breeding season are the swallow-tailed gull of the Galapagos Islands and the gray gull that breeds in the deserts of northern Chile.

Of all seabirds, gulls are among the least specialized, and occupy a wide variety of habitats from the high Arctic and subAntarctic islands, to tropical sea coasts, and even to interior marshes and deserts. In both breeding and feeding, gulls are generalists, and their overall body shape reflects their lack of specialization to any one foraging method, food type, or nesting habitat. Gulls are highly gregarious birds that breed, roost, feed, and migrate in large colonies or flocks.

Terns and skimmers are more specialized than gulls, both in their breeding habitat and in their foraging behavior, and skimmers have a highly specialized morphology and feeding behavior. While individual species of gulls, such as herring gull, may breed in many different habitats, ranging from dry land to cliffs, species of terns and skimmers breed in fewer habitats, and some are quite stereotypic in their habitat selection. Gulls feed in more different habitats on many different foods, while terns feed mainly over water by plunge-diving or dipping. Skimmers have one of the most unique feeding methods, skimming the water surface.

Taxonomy

The gulls are a worldwide group of about 51 currently recognized species with the main diversity occurring in both north and south temperate latitudes. Terns are also a worldwide group of about 44 species, with the main diversity occurring in tropical as well as temperate latitudes, while each of the three species of skimmers has a more limited distribution, one each in the Americas, Africa, and Asia (scientific names given in **Tables 1** and **2**). There is a tendency for taxonomists working at higher categories to lump genera and families together, where specialists on particular groups are more likely to emphasize differences within the group, by generic splitting – the approach followed here.

Gulls

There are several natural subgroups among the gulls, most of which can be assigned either to the large white-headed or the small dark-hooded tribes. On behavioral grounds, emphasizing the commonality of display patterns, Moynihan treated all gulls in the genus *Larus*. Most taxonomists, however, separate some relatively unique gulls into their own genera, including the swallow-tailed gull (*Creagrus*), of the Galapagos, and several Arctic species, including Ross's gull (*Rhodostethia*), ivory gull (*Pagophila*), kittiwakes (*Rissa*) and Sabine's gull (*Xema*). Less often two south temperate species, the dolphin gull (*Leucophaeus*) and occasionally the Pacific gull (*Gabianus*) are separated as well.

Terns

The main groups of terns include the black-capped terns (mostly in the genus *Sterna*), marsh terns (*Chlidonias*), noddies (*Anous, Procelsterna, Gygis*), and the Inca tern (*Larosterna*). The capped terns include small and medium sized birds in the genus *Sterna* and the distinctive, large, crested terns (appropriately considered a separate genus *Thalasseus*, although often placed into *Sterna*). Some unique capped terns are the gull-billed (*Gelochelidon*), Caspian (*Hydroprogne*), and large-billed (*Phaetusa*) terns. *Sterna* is a relatively homogenous assemblage. Noddies are uniformly colored, either all dark (black, brown, blue and gray noddies) or all white (white tern), while the unique Inca tern is all dark with dramatic yellow wattles at the gape.

Skimmers

The three species of skimmers are closely related and form a superspecies. The Indian skimmer breeds

Table 1 Gulls of the world with general breeding range

Gull	Scientific name	General breeding range (winter range)
Dolphin gull	*Leucophaeus scoresbii*	Southern South America
Pacific gull	*Larus[a] pacificus*	Southern Australia
Band-tailed gull	*Larus belcheri*	Pacific Coast of South America
Olrog's gull	*Larus atlanticus*	Uruguay to Argentina
Black-tailed gull	*Larus crassirostris*	Siberia to Japan (east China)
Gray gull	*Larus modestus*	Western South America (coastal)
Heermann's gull	*Larus heermanni*	West Mexico (California)
White-eyed gull	*Larus leucophthalmus*	Red Sea
Sooty gull	*Larus hemprichii*	Middle East (East Africa and Pakistan)
Mew gull	*Larus canus*	Eurasia and western North America (Africa, east Asia)
Audouin's gull	*Larus audouinii*	Western Mediterranean and West Africa
Ring-billed gull	*Larus delawarensis*	Central and eastern North America (to northern South America)
California gull	*Larus californicus*	Central North America (Pacific Coast to west Mexico)
Great black-backed gull	*Larus marinus*	Holarctic (south-eastern US, western Europe)
Kelp gull	*Larus dominicanus*	Southern South America, Australia, Africa
Glaucous-winged gull	*Larus glaucescens*	Eastern Asia to western North America
Western gull	*Larus occidentalis*	Pacific Coast of North America
Yellow-footed gull	*Larus livens*	Gulf of California
Glaucous gull	*Larus hyperboreus*	Holarctic
Iceland gull	*Larus glaucoides*	Eastern Canadian Arctic and Greenland (North America and Europe)
Thayer's gull	*Larus thayer[b]*	Arctic Canada (Pacific Coast of North America)
Herring gull	*Larus argentatus*	Holarctic (to northern South America and South-east Asia)
Yellow-legged gull	*Larus cachinnans*	Eurasia (North Africa, Middle East to Western India)
Armenian gull	*Larus armenicus[b]*	Armenia, Turkey (Middle East)
Slaty-backed gull	*Larus schistisagus*	Siberia to Japan (south to Taiwan, few in Alaska)
Lesser black-backed gull	*Larus fuscus*	Eurasia (Africa, Arabia, south-western Asia)
Greater black-headed gull	*Larus icthyaetus*	Central Asia (southern and south-east Asia)
Brown-headed gull	*Larus brunnicephalus*	Central Asia (southern and south-east Asia)
Gray-headed gull	*Larus cirrocephalus*	Southern South America, Africa
Hartlaub's gull	*Larus hartlaubii*	South Africa
Silver gull	*Larus novaehollandiae*	Australia, New Caledonia
Red-billed gull	*Larus scopulinus*	New Zealand
Black-billed gull	*Larus bulleri*	New Zealand
Brown-hooded gull	*Larus maculipennis*	Southern South America (northward along coasts)
Black-headed gull	*Larus ridibundus*	Eurasia (Africa, southern Asia) few in North America
Slender-billed gull	*Larus genei*	Southern Europe to western Asia (Middle East)
Bonaparte's gull	*Larus philadelphia*	Alaska–Canada (North American coasts to Caribbean)
Saunder's gull	*Larus saundersi*	North-eastern China (China to Japan)
Andean gull	*Larus serranus*	Andes (Pacific Coast of South America)
Mediterranean gull	*Larus melanocephalus*	Southern Europe (north Africa)
Relict gull	*Larus relictus*	Central Asia (unknown)
Lava gull	*Larus fuliginosus*	Galapagos endemic
Laughing gull	*Larus atricilla*	Eastern US, Caribbean to South America
Franklin's gull	*Larus pipixcan*	Central North America (South America)
Little gull	*Larus minutas*	Eurasia (Europe, North Africa)
Ivory gull	*Pagophila eburnea*	Holarctic
Ross's gull	*Rhodostethia rosea*	Siberia to Alaska
Sabine's gull	*Xema sabini*	Holarctic
Swallow-tailed gull	*Creagrus furcatus*	Galapagos and Malpelo I (Colombia) (South America)
Black-legged kittiwake	*Rissa tridactyla*	Holarctic
Red-legged kittiwake	*Rissa brevirostris*	Bering Sea endemic

[a] Often placed in the monotypic genus *Gabianus*.
[b] The status of this as a species is in question.

in southern Asia from Pakistan to Cambodia; the African skimmer breeds along rivers in Africa, and the black skimmers breeds in North and South America.

Physical Appearance

In all species of these families, males and females are indistinguishable on the basis of plumage. Moreover in gulls and terns there is very little sexual size

Table 2 Terns and skimmers of the world with general breeding range

Tern/skimmer	Scientific name	General breeding range (winter range)
Gull-billed tern	Gelochelidon[a] nilotica	Old and New World
Caspian tern	Hydroprogne[a] caspia	Old and New World
Elegant tern	Thalasseus[a] elegans	Pacific Coast of North America (to South America)
Lesser crested tern	Thalasseus bengalensis	Old World tropics
Sandwich tern	Thalasseus sandvicensis	American tropics and Europe (Africa)
Chinese crested tern	Thalasseus bernsteini	China: Taiwan; nearly extinct (Indonesia)
Royal tern	Thalasseus maximus	American tropics and West Africa
Greater crested tern	Thalasseus bergii	Old World tropics
Large-billed tern	Phaetusa simplex	South American rivers (coastal South America)
River tern	Sterna aurantia	India and south-east Asia
Roseate tern	Sterna dougallii	Pan-tropical
Black-naped tern	Sterna sumatrana	Old World tropics
White-fronted tern	Sterna striata	New Zealand (south-east Australia)
South American tern	Sterna hirundinacea	South American coasts
Arctic tern	Sterna paradisaea	Holarctic (subAntarctic)
Antarctic tern	Sterna vittata	Subantarctic (southern Africa)
Kerguelen tern	Sterna virgta	Crozets, Prince Edward, Marion, Kerguelen
Forster's tern	Sterna forsteri	North America (Central America)
Trudeau's tern	Sterna trudeaui	Southern South America
Little tern	Sterna albifrons	Eurasia, Africa, Australasia
Saunder's tern	Sterna saundersi	Indian Ocean (Africa)
Least tern	Sterna antillarum	North America (northern South America)
Yellow-billed tern	Sterna superciliaris	South American rivers (coastal South America)
Peruvian tern	Sterna lorata	Pacific Coast of South America
Fairy tern	Sterna nereis	Western and southern Australia
Damara tern	Sterna balaenarum	South Africa (to West Africa)
White-cheeked tern	Sterna repressa	Indian Ocean and Middle East
Black-bellied tern	Sterna acuticauda	India and south-east Asia
Aleutian tern	Sterna aleutica	Siberia and Alaska
Gray-backed tern	Sterna lunata	Central and western Pacific
Bridled tern	Sterna anaethetus	Pan-tropical
Sooty tern	Sterna fuscata	Pan-tropical
Black-fronted tern	Sterna albistriata	New Zealand rivers (New Zealand coast)
Whiskered tern	Chlidonias hybridus	Eurasia, Africa, Australasia
White-winged tern	Chlidonias leucopterus	Eurasia (Africa, Australasia)
Black tern	Chlidonias niger	North America and Europe (South America, Africa)
Brown noddy	Anous stolidus	Pan-tropical
Black noddy	Anous minutus	West Indies, Australia, south-west Pacific
Lesser noddy	Anous tenuirostris	Indian Ocean to western Australia
Blue noddy	Procelsterna caerulea[b]	Tropical western Pacific
Gray noddy	Procelsterna albivitta[b]	Temperate Pacific
White tern	Gygis alba	Pan-tropical
Inca tern	Larosterna inca	Pacific coast of South America
Black skimmer	Rynchops niger	North American coasts and South America inland
Africa skimmer	Rynchops flavirostris	West, central, and East Africa (Egypt)
Indian skimmer	Rynchops albicollis	North Indian subcontinent to Cambodia

[a]These genera are often placed in the genus Sterna.
[b]These two species have until recently been considered a single species.

dimorphism, while skimmers are among the most sexually dimorphic in size of any bird. The skimmers are heavy-bodied, with very long narrow wings, and large, narrowly compressed or knife-like bills for skimming the water. Adult gulls, terns, and skimmers of most species have the coloration of plunge-diving sea birds. They are generally white below, which is believed to serve as camouflage against the pale sky, reducing their conspicuousness to their underwater prey. Young birds are generally spotted or blotched or streaked, affording camouflage on the various substrates they occupy, particularly during the critical pre-fledging period when they rely on cryptic coloration to avoid predation (**Figure 1**).

Figure 1 Examples of gulls, terns and skimmers. (1) common tern (*Sterna hirundo*) Adult, breeding plumage. Length: 36 cm; wingspan: 80 cm; approximate body mass 120 g. Range: Widely distributed in Northern Hemisphere during breeding season, and coastal oceans of the world during the non-breeding season. (2) Sooty tern (*Sterna fuscata*). Other names: Wideawake. (a) Adult; (b) juvenile. Length: 44 cm; wingspan: 90 cm; approximate body mass: 195 g. Range: Tropical Oceans. (3) Black skimmer (*Rynchops niger*). Adult, breeding. Length: 45 cm; wingspan: 117 cm; Range: Coastal waters and some rivers of North and South America. (4) Herring gull (*Larus argentatus*). Adult, breeding plumage. Length: 61 cm; wingspan: 139 cm; approximate body mass: 1160 g. Range: Widespread in costal waters of North Pacific and North Atlantic Oceans. Also along coasts of Mediterranean Sea and Indian Ocean. (5) Great black-backed gull (*Larus marinus*). Adult breeding plumage. Length: 75 cm; wingspan: 160 cm; approximate body mass: 1680 g. Range: North Atlantic Ocean. (6) Kittiwake (*Rissa tridactyla*). Other name: black-legged kittiwake. Adult breeding plumage. Length: 41 cm; wingspan: 91 cm; approximate body mass: 385 g. Range: North Pacific, North Atlantic and Arctic Oceans. Illustrations from Harrison P (1985) *Seabirds, an identification guide. Revised edition.* (Boston, Massachusetts: Houghton Mifflin).

Gulls

As a group gulls are generally heavy-bodied, long-winged birds with intermediate length necks and tarsi, webbed feet, and heavy, slightly-hooked bills. Gulls range in weight from as little as 100 g for a little gull to 2000 g for some great black-backed gulls. Gulls exhibit no sexual dimorphism in plumage patterns, but females are slightly smaller than males and in some species show more slender necks and beaks. All species have 12 rectrices and the tails are rounded in all but a few species (Sabine's gull, swallow-tailed gull).

Gulls are generally white-bodied, with a darker mantle varying from pale silvery-gray to black (except for the all-white ivory gull). Several of the smaller species exhibit a pale pink or cream-colored bloom on their breast early in the breeding season. In the Ross's gull (not roseate tern) this is very pronounced and persistent. Gulls have either a dark hood (nearly obscure in some species), dark mask, or all white head during the breeding season. Generally the larger gulls are white-headed, although the great black-headed gull, one of the largest species, is an exception. The smaller gulls are generally either dark-masked or dark-headed. In almost all species the wingtips are black, the melanin pigment offering resistance to wear. Most species have a complex pattern of white 'windows' on the black outer primaries which may differ among closely related species. Several Arctic species (glaucous, Iceland, ivory gulls), have white wing tips.

Some species of gulls reach adult plumage in 2 years (the smaller hooded and masked gulls), while the larger species have 3–4 distinct plumage year classes. Some do not reach fully adult plumage until their fifth year. Most gulls molt their flight feathers twice a year, and their body feathers once a year. Franklin's gull, which migrates farther than any other gull from its breeding range in the prairies of North America to South America, undergoes two complete molts each year. However, Sabine's gull migrates almost as far, but has only one complete molt each year.

Terns

Terns have narrower, more elongated bodies than gulls and proportionately longer, more slender and pointed wings. Their beaks are generally slender and sharply pointed. Most species of terns forage at least occasionally, by plunge-diving for fish, and accordingly their body is stream-lined. Most terns are white below and gray above, with a black crown in nuptial plumage, although a few species are all dark or all white. The roseate tern has a pale pinkish 'bloom' on the breast early in the breeding season. This is quickly lost by wear. There are no sexual differences in plumage patterns.

The plumage patterns of the gull-billed tern, crested terns, typical black-capped, and large-billed tern are similar: mainly white below, gray above, with a black cap during the breeding season. During the nonbreeding season most species lose part or all of the black crown. Terns have one complete molt and one body molt each year.

Skimmers

Skimmers are the size of the large terns, and are black above and pure white below, although there is a pale cream-colored tinge particularly on the flanks, early in the nesting season. The Indian skimmer has a broad white collar and the other two species gain such a collar in their post-breeding molt. The beak is bright reddish-orange with a yellow tip in the African and Indian skimmers and is red at the base with the distal half black in the black skimmer. The knife-like bill is adapted for slicing through the water surface when the birds are skimming to catch fish. A most unusual feature of skimmers is that the upper mandible is shorter than the lower. Skimmers have two molts a year, including a complete molt and a body molt.

Habitat

Gulls and terns are quite variable in their habitat preference, although almost all species feed and nest in association with water. Gulls forage in a wider range of terrestrial and aquatic habitats, while terns and skimmers forage almost exclusively over water. Exceptions include the gull-billed tern which feeds extensively on insects obtained by plunging to the surface or hover-dipping over grasslands.

Gulls

Gulls use a wide variety of habitats; for any single species the range may be narrow or broad. In the breeding season, colonies of nesting gulls can be found in coastal and estuarine habitats, as well as inland. Most species of gulls nest along the continental coasts or on large inland lakes, showing a strong preference for islands or inaccessible sites. A few (brown-hooded, Franklin's, relict) nest mainly on inland lakes or marshes, while two, the lava and swallow-tailed gulls nest on remote oceanic islands, the Galapagos. Franklin's gulls breed on inland freshwater marshes but winter on South American sea coasts.

Gulls occupy a wide variety of nesting habitats, including sandy or rocky islets or beaches, with or without vegetation, marshes, riverine or lake sand bars, wind-swept sand dunes and cliffs, trees, and even buildings. The unique gray gull breeds in the barren, montane deserts of Chile, flying each day over the Andes to the Pacific Coast to obtain food for their young. Gull colonies are generally located in habitats removed from mammalian predators, but some gulls will breed on terrestrial habitats where both avian and mammalian predators are threats.

During migration gulls fly to coastal and estuarine habitats, and in winter, they generally remain along coasts or on large lakes. Although they can be found out to the continental shelf, most are not truly pelagic.

During the winter, their distribution is a function of food availability. Since they eat a wide variety of foods, they can be found in nearly all aquatic habitats. Outside the breeding season, gulls are found at virtually all latitudes where open water is available.

Terns

Terns occur throughout the world, and breed on all continents, including Antarctica (Antarctic tern). Although individual tern species are less variable in their breeding habitats than gulls, overall, terns occupy a wide range of breeding habitats, including inland and coastal marshes, islands in rivers, lakes and estuaries, sandy or rocky beaches, cliffs, and oceanic islands. Since most terns nest on the ground, they seek remote or inaccessible places as an antipredator strategy against mammalian predators. Several of the noddy species nest on trees.

During the nonbreeding season, most species of terns migrate to coastal estuaries and the open ocean, although some never leave their inland marshes and rivers. Some species are truly pelagic during most of the nonbreeding season. Sooty tern, the most oceanic tern species, maintains a pelagic existence from the time of fledging until returning to land to breed for the first time at several years of age. Arctic terns wintering in Antarctica often roam through the loose pack ice along the edge of the ice pack.

Skimmers

The skimmers in South America, Africa, and Asia are largely restricted to nesting on riverine sand bars and islands in lakes. The black skimmers of North America are mainly coastal, nesting on beaches and islands, although small numbers breed inland. Except in North America, the nesting of skimmers is influenced by rains, for they must wait until water levels fall and their islands are exposed. Skimmers are the least pelagic members of this family, and rarely are observed far from land.

Breeding Ecology and Behavior

Gulls, terns, and skimmers normally breed in colonies, either in monospecific colonies, or in monospecific groups within colonies that include other species. Terns are the most gregarious and they generally breed, forage, and migrate in flocks that can range from a few individuals to many thousands or even millions (sooty tern).

Most gulls and terns in temperate zones breed at the same time of year, breed once a year, and breed every year. While North American skimmers do likewise, skimmers that are dependent on the formation of sand bars and sandy islands in rivers must wait until such sand bars are exposed. Some tropical terns have shorter breeding seasons, and can breed every 8 months.

Territory size generally increases with body size for gulls, and decreases with body size for terns (**Figures 2** and **3**). Based on an examination of 49 species of gulls and terns nesting in aquatic and terrestrial habitats, variability in density was accounted for by colony size, body length, and wingspan, and variability in nearest neighbor distance was accounted for by body size.

Gulls and terns of several species, such as common tern, can be very aggressive at mobbing potential predators, including human intruders into their colony. Such mobbing behavior can deter avian predators, but is less effective in discouraging mammalian predators. Small species often nest in colonies with larger species to take advantage of the larger size and antipredator behavior of these species, and skimmers often nest in colonies with other species to take advantage of their mobbing behavior, because they do not normally do so.

Both members of the pair engage in territory defense, incubating, and protection and feeding of the young. The eggs of most gulls and terns are brown with dark splotches, while the base color of skimmer eggs is white. Clutch size in most gulls and terns is two or three, while in skimmers it is more variable, ranging from two to four or occasionally five. Since incubation begins after laying the second egg in skimmers, there is asynchronous hatching, and the young can be very variable in size. The first two chicks may hatch only a day apart, while the fourth may not hatch until 5 days later. Pairs that lose eggs or chicks may initiate a repeat nesting attempt. Following the breeding season, young skimmers and gulls may remain with their parents for a few days or weeks while they improve foraging skills, however,

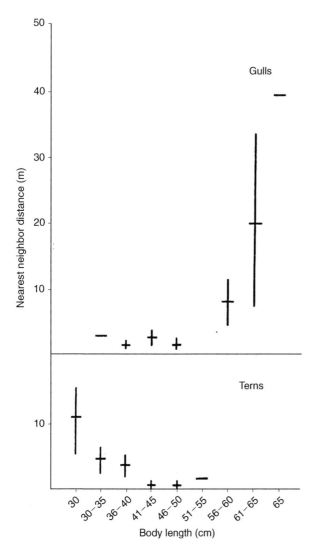

Figure 2 Relationship between body size and territory size (nearest neighbor distance) showing that smaller gulls and larger terns tend to nest more densely.

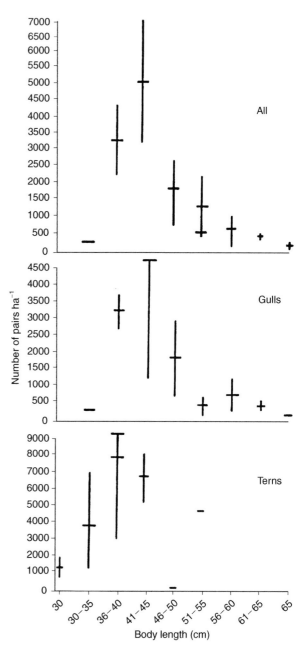

Figure 3 Relationship between body size and colony size showing that smaller gulls tend to nest in larger colonies while smaller terns tend to nest in smaller colonies.

young terns remain with their parents for weeks or months, perfecting the difficult task of plunge-diving. Some terns migrate with their parents and remain with them much of the winter.

In a normal breeding season the phenology includes (1) arrival in the colony vicinity a few weeks before occupation, (2) colony occupation for a few weeks before initiation of nest-building and egg-laying, (3) territorial defense and courtship during this period, (4) courtship feeding of the female by the male and selection of a nest site and nest construction, (5) a copulation period of a few days to 2 weeks, (6) an egg-laying period of 1–3 weeks, (7) an incubation period of 20–30 days resulting in incubation activities in the colony over about 6 weeks, and (8) a brooding or pre-fledging phase of 4–6 weeks. These time periods are slightly less for the

small gulls, and slightly more for the large gulls. Environmental constraints such as low food supply or bad weather impose synchrony on the breeding cycle.

Gulls

Most species of gulls nest in a wide range of habitats, although some species are specialists, nesting only in one habitat (Franklin's and brown-hooded gulls nest only in marshes, kittiwakes nest on cliff ledges or buildings, and Bonaparte's gulls normally nest in trees).

Most gulls breed in colonies of few to several hundred pairs, although some species will occasionally breed solitarily (herring gull, Bonaparte's gull), and others will breed in colonies of several thousand pairs (**Figure 3**). The small gulls generally nest in more dense colonies than do the larger gulls. The large gulls nest in more open colonies, which is partly the result of cannibalism.

Species nesting in or adjacent to gull colonies include: grebes, ducks, cormorants, herons and egrets, gannets, alcids, and even penguins. Some of these apparently do so because they derive early warning and antipredator advantages from the gulls, although they also risk predation by the gulls.

The nesting period usually lasts from 3 months for the smaller gulls, and up to 5 months for the larger species. A prominent exception to the annual cycle is the swallow-tailed gull of the Galapagos which breeds in all months of the year, and even within a colony there is breeding asynchrony among pairs.

The age of first breeding in gulls varies from 2 (for some smaller species) to 5 years of age (for some large species). Females are likely to start breeding at a younger age than males. Factors that influence age of breeding within a species include availability of mates, nesting sites, food, and weather conditions. Reproductive success usually increases over the first few years as birds gain breeding experience; success remains relatively high for a number of years, and decreases only in very old gulls.

A typical cycle is exemplified by the widespread herring gull. Early in the breeding cycle, gulls begin to gather on 'clubs' on their nesting colonies, where they solicit for mates or re-establish pair bonds. Gradually the gulls spread out over the nesting colony to delineate and defend nesting territories. Unmated males defend territories while courting potential mates; mated pairs defend the territory together and engage in pair-bond maintenance. Once pairs have formed, display frequency increases, and pairs may spend hours every day engaging in 'mew calls', mutual 'long-calling', and 'head-tossing'. Within a few days, the pair begins to select a nest site, and each member shows the other its choice for a nest site with a 'choking' display. Nest building is an integral part of pair formation and pair maintenance for some species, and is less important for others, depending upon physical constraints. All gulls construct a nest, and use vegetation to form a nest cup. In most species males courtship-feed their mates in the pre-egg-laying period, although the level of courtship feeding is less than it is for terns. Copulation is always accompanied by a staccato, repetitive 'copulatory call' that is accompanied by 'wing-flagging' which is visible far across the colony. This leads to contagious copulation by neighboring pairs, and is believed to lead to greater breeding synchrony.

Most gulls have a modal clutch size of three eggs, although some species such as Hartlaub's, black-billed, silver gulls, and kittiwakes usually lay only two eggs, and swallow-tailed gulls only one. Both members of the pair engage in incubation, which lasts an average of 24–26 days. Incubation bouts usually range from 1 to 4 h during daylight, with one member of the pair remaining on the nest at night. The first two chicks usually hatch within a few hours of one another, but the third chick may hatch a day or two later, giving it a distinct disadvantage when competing with siblings for food. The cryptic downy chicks are usually a pale gray or pale tan with dark splotches. The chicks are brooded until they are 1–2 weeks old, and usually are guarded until they fledge.

Both parents feed chicks, although during the first week the male performs more provisioning, while the female remains on the territory and broods the chicks. After the first week or two, both members of the pair take turns with foraging trips and guarding the chicks. Unless disturbed by storms, floods or humans, they use the nest throughout the brooding period. The pre-fledging period in gulls ranges from 4 to 7 weeks, depending on the size of the species. The full extent of post-fledging parental care has not been established for gulls, but for some it clearly extends for a few weeks post-fledging.

In all species of gulls studied, pair-bonds are monogamous, with a relatively high degree of mate fidelity from year to year. Pairing for life is characteristic of gulls that have been studied. However, divorce occurs when pairs are unable to work out incubation and brooding activities, or when they are unsuccessful at raising offspring. Pairs that remain together lay earlier, and raise more young than pairs that have found new mates.

Study of the mating systems of gulls has been stimulated because of the relatively recent reports of female–female pairs in ring-billed, herring, western, red-billed, and California Gulls. Female–female pairs constitute at least 10% of the pairs of western gulls in some colonies and 6% of the pairs in red-billed gulls. Female–female pairs apparently result from a shortage of breeding males, which may be a consequence of differential survival rates. Laboratory experimentation with western gull eggs indicated DDT-induced feminization of genetically male gull embryos. The levels of DDT that caused feminization were of the same magnitude as those found in seabird eggs in Southern California in the early 1970s.

Developmental feminization of males may be associated with inability to breed as adults, and may explain the highly skewed sex ratios and reduced number of male gulls in some populations. Some promiscuity occurs in gulls, with both sexes copulating with birds other than their mates.

Males and females invest approximately equivalent effort, averaged over the entire season, although males often provide more of the initial territory defense, perform all of the courtship feeding, and may perform the bulk of territorial defense, as occurs in western gulls and ring-billed gulls. Females sometimes perform more of the incubation duties and chick defense than males. Moreover, in territorial clashes, males tend to defend against males, and females defend against females when both are present on the territory.

Terns

Although terns have a worldwide distribution, some individual species have a very restricted range. Terns breed in colonies that may be used for several years or decades, and are abandoned when the breeding site becomes unsuitable because of increased predation, human disturbance, habitat changes, or weather stresses such as severe storms or floods. Philopatry varies as a function of habitat stability, and species that nest on remote oceanic islands or on cliffs occupy the same colony site for years. Changing habitats, such as freshwater marshes, riverine sand bars, and coastal sandspits, may be used for only a few years or even only 1 year. In some species individual terns are known to return to their previous site, assess its suitability, and then either occupy the site or move elsewhere.

The breeding season may last only 2 months for Arctic species, 3–4 months for temperate species, and 3–5 months for tropical species. Terns usually spend only 2–3 weeks in the vicinity of the colony before they begin to settle on the colony for another breeding attempt. Temperate species of terns typically leave the breeding grounds with their chicks, and disperse before migrating to wintering grounds. The chicks may remain with their parents for many weeks or months, slowly decreasing their dependence on their parents for food. Post-fledging parental care has been documented only in those species that migrate and overwinter near land where parents can land to feed their chicks.

Nesting colonies range in size from a few widely dispersed pairs (e.g. Damara and Inca terns) to colonies of a million or more (sooty terns). Intermediate-sized terns, such as the common tern, usually nest in colonies of tens to hundreds of pairs (with a few

colonies exceeding a thousand). The large terns usually nest in colonies on the order hundreds of pairs (**Figure 3**).

All species of terns are territorial, although the size of the territory varies from 40 cm to a few meters. Unlike gulls, the inter-nest distance decreases as the size of the tern increases (**Figure 2**). That is, the larger species (royal, Caspian, crested, and sandwich terns) nest in very dense colonies where an incubating bird can almost reach its neighbor, the intermediate-sized terns (common, roseate, Arctic, Forster's, noddies) nest in fairly dense colonies where inter-nest distances may range from 1 to 3 m, and the small terns often nest in fairly dispersed colonies where inter-nest distances may be 5 m or more. The larger species rely on their dense nesting pattern to prevent any ground or aerial predators from getting within the colony, while the smaller species rely on dispersion and cryptic coloration to reduce predation, and they rely on antipredator behaviors such as mobbing and aerial attacks to repel predators.

Although some species of terns nest in monospecific colonies, many species nest with other species of terns, gulls, skimmers, boobies, alcids, and ducks, and with sea turtles. Some terns (Forster's black) choose to nest with other species, and they select the colony site after the other species is already nesting. These terns are smaller than the gulls, and derive antipredator protection from the larger gulls that mob and attack aerial predators such as hawks. Terns also nest in colonies that are adjacent to colonies of other species, such as albatrosses, boobies, and cormorants.

The normal breeding phenology is to select territories, defend territories, solicit mates or reestablish pair-bonds with former mates, engage in courtship and courtship feeding for a few days or weeks, to have an egg-laying period of a few days to 2 or 3 weeks, to have an incubation period of 22–26 days, and to have a brooding and pre-fledging phase of about 4 weeks. After fledging birds may linger at the colony or depart, resulting in a colony breeding phase of up to 6–10 weeks. Smaller terns require somewhat shorter and larger terns longer, breeding times. Within a colony breeding is generally much more synchronous among Arctic species and can be much less synchronous among tropical species. The former are constrained by short breeding season and the latter by less abundant food supplies.

During courtship both mated and unmated birds engage in 'fish flights', which begin when a mated male brings back a fish to a female. After he lands on the territory, both take off and fly high in the air, often joined by one or two other birds. They glide and circle on stiffly bowed wings, uttering unique flight calls. Once pairs have formed, courtship displays continue, with the addition of courtship feeding. During this period the female often remains on the territory, defending it against intruders, while the male spends all day bringing back fish to courtship feed her. This is said to offer the female the opportunity to assess the prospective paternal qualities of her mate.

Nest site selection occurs when both members of the pair agree on a given location. Most terns do not construct a nest, but merely make a scrape in the sand or roll a few pebbles or shells around a scrape, or find a suitable cup-shaped place in the coral or rock for their eggs. Terns nesting in marshes (for example Forster's terns and black terns), construct a nest of vegetation, on which eggs can float up if flooding occurs. Tree-nesting species such as some noddies must select a branch or shelf where they can build a nest of twigs and feathers cemented by their excreta. Cliff or ledge-nesting species such as brown noddies often move the small coral rubble to the edge of the cliff, providing a smooth surface for the eggs and chicks, and a slight lip to prevent the egg from rolling off. The white tern makes no nest at all, but places its egg on a branch, or ledge, or artificial object, where it is hardly protected from falling.

Clutch size in terns varies from one egg (royal, elegant, white-fronted, black-naped, sooty, noddies, white tern), to two eggs (Caspian, roseate, Arctic, sandwich, least), to three eggs (gull-billed, common). For species that normally lay two or three eggs, clutch size is dependent on food supply. In low food years, terns can reduce the average clutch size, delay breeding, or forego breeding. Members of pairs share incubation duties and chick care, although early in the brooding phase males may bring back more food and females may incubate them more often.

The birds leave their overnight roost or breeding colony to search for food at dawn. These feeding flights may involve large flocks (particularly of non-breeding birds), or prolonged streams of individuals and small groups. During the breeding season, terns that are not incubating often rest on their territory with their mates. Terns are often more active in the early morning and late afternoon. The daily activity patterns of coastal-nesting species of terns are often influenced by tidal cycles.

Skimmers

The breeding behavior of skimmers is similar, except for the following: (1) they shift colony sites more often because of the changing nature of their nesting habitat, (2) colonies may consist of only a few pairs, (3) males defend the territory more than females, (4) a male will usually bring back a fish prior to copulation, pass it to his mate, and she may hold it while they copulate, (5) clutch size is more variable, and may range from two to five, depending upon food supply, (6) since clutch size is larger than in gulls and terns, and the parents may start incubating almost immediately, there can be greater size disparity between the offspring, and a pair rarely raises more than two young, even when food is not especially scarce, (7) males provision the chicks more often during the first 2 weeks of their lives, and (8) if disturbed, they move the chicks soon after hatching. Once chicks are more than a week old, the parents may not brood them, and they continue to be vulnerable to prolonged chilling rains, which can produce high mortality among chicks in the 1–2 week age range. At the same time the parents may be brooding the much smaller, last-hatched chicks, which may, under these circumstances, survive.

Foraging

Gulls have the most diverse foraging behavior and feed on the greatest variety of foods, while terns primarily plunge-dive or hover-dip for fish, and skimmers skim the surface of the water for fish. Both terns and skimmers are limited to feeding over water, while gulls feed on land, along the shore, and over water. Both gulls and terns engage in piracy, although terns generally pirate from conspecifics or other terns, while gulls pirate from a range of species. Foraging efficiency increases with age in gulls, terns, and skimmers.

Gulls forage in a variety of natural habitats including the open ocean, the surf zone, intertidal mudflats, rivers and rivermouths, rocks and jetties, estuaries, bays, lakes, reservoirs and rivers, wet meadows and farm fields, sewage outfalls, refuse dumps, and even in the air. Gulls are an integral part of coastal and estuarine habitats, and many species feed along the shore on a variety of fish and invertebrates. Gulls are particularly characteristic of the intertidal zones. They also feed in a variety of man-influenced situations, including on landfills, behind ploughs or boats, and by pan-handling from people at fast-food places or along the shore. They forage using a wide range of techniques, including walking on the ground, swimming in the water and dipping for food, and plunge-diving. They also drop invertebrates from some distance to crack open their shells, a behavior shared only with Corvids. In some species, individuals have specialized diets or foraging techniques when compared with their populations as a whole.

Table 3 Conservation status of selected gulls, terns and skimmers[a]

Species	Status
Pacific gull	Near-threatened. Nowhere common; is declining probably due to kelp gull pressure.
Olrog's gull	**Vulnerable**, and probably endangered. Only 1200 pairs known in six colonies, none secure. Competition from increased fishing; vulnerable to oil exploration.
Heermann's gull	At least vulnerable with only eight colonies, and most of its large population in one colony.
White-eyed gull	**Vulnerable**. Poorly known, with fewer than 7500 pairs. Vulnerable to egging, oiling, exploitation, tourism.
Audouin's gull	Formerly **endangered**, but increasing under intense protection. Still fewer than 20 000 pairs, but now in about 30 colonies. Fishing competition potentially severe.
Yellow-footed gull	Vulnerable due to very low numbers and small range. Probably <5000 pairs in less than a dozen colonies.
Armenian gull	Vulnerable. Low numbers and very small range in Arnebua to north-western Iran. Formerly considered a race of herring gull.
Hartlaub's gull	Probably vulnerable. About 30 known breeding colonies, and over 25% of population in one colony (albeit protected).
Saunder's gull	**Endangered**. Fewer than 2500 pairs (probably <1500 pairs). All seven known colony sites are being developed or slated for development. Intense management and protection urgently required.
Relict gull	Probably endangered. Small range, few colonies, small population. Fewer than 2000 pairs and declining.
Lava gull	**Vulnerable** and probably threatened. The world's rarest gull. Although there are fewer than 400 pairs, most of the habitat is currently well-protected. Increased tourism and fishing jeopardize its survival.
Red-legged kittiwake	**Vulnerable**. A Bering Sea endemic that is declining. Only seven known colony sites. Although population is still large, it has declined precipitously through the 1990s.
Chinese crested tern	Critically **endangered**, breeding area unknown, often listed as extinct, but a few recent sight records on wintering grounds.
River tern	Status poorly known, vulnerable
Roseate tern	Not globally threatened (c. 50 000 pairs), but North American and European populations are **endangered**, Caribbean population is **threatened**.
Kerguelen tern	**Vulnerable** and probably threatened. No truly secure colony. Total population <2500 pairs, mostly in Kerguelen where cat predation is a continuing problem.
Saunder's tern	Status unknown, not known to be common anywhere. No secure colonies documented.
Least tern	Not globally threatened, but all races vulnerable, and California race is **endangered**.
Peruvian tern	Status uncertain, a vulnerable species. World population probably around 5000 pairs.
Fairy tern	**Vulnerable**. About 5000 pairs in Australia. New Zealand population is **endangered** with fewer than 10 pairs.
Damara tern	Near-threatened with fewer than 10 000 pairs. Breeding areas are neither protected nor heavily disturbed at this time.
Black-bellied tern	**Vulnerable**. Fewer than 10 000 pairs and continually declining.
Black-fronted tern	**Vulnerable**. Fewer than 10 000 pairs, mostly in small colonies on braided rivers being invaded by lupine.
Lesser noddy	Not globally threatened, but status poorly known. Only a few colonies known.
Gray noddy	Not globally threatened, but eastern colonies on Desadverturas and Easter Island are at least threatened if not endangered.
African skimmer	Not globally threatened, but fewer than 10 000 pairs, breeding in many very small colonies.
Indian skimmer	**Vulnerable**. Declining, but poorly documented over most of its range.

[a]Official global or regional status is given in boldface. The list provides both official and unofficial status of tern and gull species that are in trouble. For many of the remaining species that are globally secure, there may be many colonies, populations, or regions where it is declining or in danger of disappearance. The hallmark of security is having many colonies, including those in protected areas. The consequence of having many colonies is that not all can be protected.

Conservation

Of the 99 species of gulls, terns and skimmers, world population estimates range from a few hundred pairs (lava gull) to several million pairs (herring gull) and several tens of millions (sooty tern). Saunder's gull and Chinese crested tern are endangered. Ten species are considered 'threatened', and one skimmer is considered 'vulnerable' (see **Table 3**).

Although egging, hunting, and exploitation for the millinery trade resulted in sharp declines for many species in the last two centuries, human persecution has ceased in many (but not all) parts of the world.

'Egging', the collection of bird eggs for food, continues to be a major threat in tropical regions, (e.g. roseate terns in the Caribbean). Current threats to species in this family include habitat loss, habitat degradation, increased predation (often caused by predators, such as cats, introduced and maintained by humans), and overfishing by humans that reduces food supplies. Populations can also be threatened by pollution, particularly oil spills that occur near nesting colonies or in favorite foraging grounds. In the 1960s and 1970s eggshell thinning due to DDT was a problem, and in the 1980s in the Great Lakes of North America, organochlorides contributed to decreased hatching rates, lowered parental attendance, and lower reproductive success.

Gulls, terns, and skimmers suffer heavy and increased predation from predators introduced by man, including cats, dogs, and by predators that have profited by man's activities, such as foxes. In addition, many species of terns and smaller gulls have been negatively impacted by competition with larger gulls. In many temperate regions, the large white-headed gulls (e.g. herring gulls, ring-billed gulls) expanded their numbers and ranges dramatically during the twentieth century, abetted by the availability of human refuse in uncontrolled garbage dumps. This new food source greatly increased the survival of juvenile gulls. The large gulls displaced smaller gulls and terns from their traditional nesting sites. They also preyed on the eggs and chicks of these species. New landfill practices and alternative refuse management have reduced this food source in many urban areas; the populations of some gulls have begun to decline.

Conservation measures include protecting colonies from direct exploitation (hunting of adults, egging), creation of suitable nesting space, stabilization of ephemeral nesting habitats, construction of artificial nesting islands or platforms, removal of predators (feral cats, large gulls, foxes), protection from other predators, and reduction of human disturbance at colony sites through sign-posting, fencing, or even wardening. More difficult, but equally important, is the protection of foraging sites and the prey base, which may involve fisheries management.

See also

Seabird Conservation. Seabird Foraging Ecology. Seabird Migration.

Further Reading

Burger J and Gochfeld M (1996) Laridae (gulls). In: delHoyo J, Elliot A, and Sargatal J (eds.) *Handbook of the Birds of the World*, pp. 572–623. Barcelona, Spain: Lynx Ed.

Gochfeld M and Burger J (1996) Sternidae (terns). In: del Hoyo J, Elliott A, and Sargatal J (eds.) *Handbook of the Birds of the World*, pp. 624–667. Barcelona, Spain: Lynx Ed.

Grant PJ (1997) *Gulls: A Guide to Identification.* Vermillion, South Dakota: Buteo Books.

Olsen KM and Larsson H (1995) *Terns of Europe and North America.* Princeton, NJ: Princeton University Press.

Tinbergen N (1960) *The Herring Gulls World.* New York: Basic Books.

Zusi RL (1996) Rynchopidae (skimmers). In: del Hoyo J, Elliott A, and Sargatal J (eds.) *Handbook of the Birds of the World*, pp. 668–677. Barcelona, Spain: Lynx Ed.

PELECANIFORMES

D. Siegel-Causey, Harvard University, Cambridge MA, USA

Introduction

Who does not know what a pelican looks like? Pelicans, and the other members of the Pelecaniformes, are common inhabitants of the marine littoral and are found almost everywhere along the margins of the world ocean. They are medium-sized to very large aquatic birds found commonly near coastal marine and inland waters, feeding mainly on fish and less commonly on small arthropods and mollusks. Pelicans typify two of the common features of this order of birds, including a large distensible pouch under the bill and a totipalmate foot (four toes connected by three webs). Other features common to Pelecaniformes are less exclusive, but altogether are diagnostic of the group. The feet are set far back on the body for efficient swimming, but on land most of this group are clumsy and almost immobile. The eggs and chicks are incubated on the feet, the adults lacking brood patches. In most, chicks are born blind and naked, and down feathers appear after a few days. Chicks feed on regurgitated food taken by inserting their head into the parent's throat. Most species breed in large, dense colonies, and in places the birds are a valuable source of guano – a source of nitrates and a valuable fertilizer. Colonies are susceptible to disturbance by humans and small mammals when eggs and chicks are lost through temperature imbalances after the adults leave the nest, or by predation. Because of their proximity to humans, and dense aggregations in feeding and breeding activities, most pelecaniforms – cormorants and pelicans in particular – are persecuted by fishermen under the mistaken belief that they deplete fishing stocks.

Systematics

General

On the surface, Pelecaniformes are an order of birds easily identified by a few key characters, the most noticeable being a totipalmate foot, that is, one with all four toes joined by a web. No other assemblages of birds possess this trait: gulls and waterfowl have only three toes joined by a web, and some birds have only a single web between two toes. Other features have been associated with this group, including a gular pouch, or the unfeathered, loose folds of skin underneath the bill, and a lack of brood patches on the abdomen. Six families are now considered to be placed in the Order: tropicbirds (Phaethontidae), frigatebirds (Fregatidae), pelicans (Pelecanidae), gannets (Sulidae), cormorants (Phalacrocoracidae), and darters (Anhingidae). Recent studies on the morphology and molecular variation in the traditional species of pelecaniform birds suggest, however, that these groups may not be as closely related as formerly thought. For example, there are doubts that tropicbirds and frigatebirds are pelecaniform, and instead ought to be placed in other groups, such as the tubenoses (Procellariformes), herons and storks (Ciconiiformes), or their own orders. It also appears that, except for the lack of toe webbing, the shoebill (*Balaeniceps rex*) and Hammerhead (*Scopus umbretta*) herons are most closely related to pelicans, and may represent aberrant members of the Pelecaniformes.

The pelecaniform birds and other large waterbird orders (i.e., Sphenisciformes, Gaviiformes, Podicipediformes, Procellariiformes, Ciconiiformes) are considered by most systematists to comprise the early branching lineages of birds with flight. The earliest known fossil pelecaniform is thought to be *Elopteryx* from the Cretaceous (70 million years ago), but as with all early bird fossils there is considerable debate about its identity – even whether or not it is avian! Cormorants are among the earliest known pelecaniforms, with fossil elements recovered from as early as the Eocene–Oligocene boundary (approximately 55 million years ago). There are approximately an equal number of extinct and extant species described, but the fossils are often from single bones and of dubious affinities.

The species-level taxonomy has been very neglected, with some families having many species and subspecific forms described (e.g., cormorants with 39 species and 57 taxa), while others have had little attention. For example, the darters are variously described as comprising one, two, or four species, but with few data to back up any of the competing schemes. As a whole, the group is drab or monochrome or both, and sexes are dimorphic only in the darters and the frigatebirds.

Order Pelecaniformes

The groups now considered to be members of the Order Pelecaniformes comprise 6 families, and include 6 genera, 67 species, and about 120 described subspecies, races, or types (**Figure 1**). Cormorants are the most diverse group, with half of the extant species; tropicbirds and anhingas are the least diverse, with 3–4 species known at present. Pelecaniform birds are found on every continent, along most marine coasts and major fresh water drainages, and are absent only from extreme deserts and the ice-covered regions. The nest is built by the female with material usually brought by the male, and is situated on the ground or in trees, selected to be well protected from predators. Reproductive behavior is complex and involves primarily visual displays near the nest site; vocalizations are rare and often described as being grunts, croaks, or strident.

Phaethontidae A single genus, *Phaethon*, and three species are found throughout the tropical waters of the world ocean. Tropicbirds differ from the typical pelecaniform birds and really only share a single feature – the totipalmate foot – with the other species. The chicks hatch fully covered with down; they are fed by the adults from the bill rather than the throat as in other species; courtship displays are aerial, noisy, and synchronized; the bill lacks a terminal hook; the gular pouch is feathered and obsolete; and there are many other distinctions.

All are medium-sized – about the size of a small gull – white with black eye and wing barring, with long tapering wings and gull-shaped bills, and with very elongated tail feathers forming streamers often longer than the rest of the body. The white parts of the plumage are often tinted pink or gold; the sexes are alike in appearance. Species range widely over the warm tropical and subtropical waters, and represent the most oceanic of all pelecaniform birds. They only come to land for breeding, on remote oceanic islands, and select nest sites that are inaccessible to terrestrial predators and have some shade, such as might be provided by overhanging rocks on shrubs, or within cavities.

Little is known about the ecology and behavior of the family; the birds are rarely seen in groups or flocks, often only singly, and usually in flight. Feeding seems to be in low light (i.e., early morning or late evening) and by plunge-diving into local concentrations of cephalopods and flying-fish. Hovering over schools and ship-following have been noted, as well as gathering with multispecies feeding flocks. More commonly, though, tropicbirds tend to forage solitarily, dive abruptly, capture food close to surface, and then take to the air soon after swallowing their prey.

Fregatidae Frigatebirds are found throughout tropical and subtropical oceans of the world, commonly associated with the regions of trade winds. They are large – roughly the size of large gulls – and predominantly black, with some species having white breasts or bellies. Five species are described, but there is substantial confusion on the status of certain island forms, plumage variants, and size differences. The wings are large, elongated, and pointed, with a distinctive open W-shaped flight silhouette, and all species are quite deft in the air. During breeding, males inflate their scarlet gular pouch to nearly half the size of the body, and do so in groups and alone, on ground and in the air, and at times will vibrate it with the bill, as though drumming.

Frigatebirds will leave the roosts at dawn and spend the rest of the day foraging alone or in small groups. They are known for feeding by theft ('kleptoparasitism'; hence their common names of frigate birds or man-of-war birds) but commonly feed on shoals of flying fish, cephalopods, and rarely krill. They rarely dive in the water, preferring instead to feed by surface dipping or even aerial captures of emergent flying fish. Every tropical seabird colony seems to be attended by several frigatebirds, and unattended eggs and chicks are taken by swooping after an aerial dive. Fish are snatched from the bills of other seabirds, whether at sea or on the nest.

The breeding season seems to be determined by local conditions, such as seasonal onset of oceanographic or weather patterns, abundance of food, and appearance (or nonappearance) of neighboring breeding species. The reproductive cycle is longer than for most of the other Pelecaniformes, with incubation lasting as long as 60 days, fledging as long as 7 months, postfledging dependence on adults lasting on average between 9 and 12 months, and age of first breeding from 6 to 11 years. This seems to be related to the unpredictable nature of the food supply, low chick productivity, and low breeding success. While the birds are not threatened at present except in a few localities, these aspects of natural history make them particularly vulnerable to human disturbance and habitat destruction.

Pelecanidae Pelicans are the largest members of the order, and are distinguished by their large body, long, heavy bill, and huge distensible gular pouch. Seven species are recognized, and all are found not far from water and rarely out of sight of land. Plumages are generally light colored – only the brown pelican is dark – and in the breeding season the bare facial and gular skin, and often bills, become brightly colored in both sexes. Because of size and large broods, pelicans are voracious feeders. They predominately eat fish, and feed by

Figure 1 Examples of Pelecaniform Seabirds. (1) Northern gannet (*Sula bassana*). Other names: North Atlantic gannet, gannet. Length: 93 cm; wingspan: 172 cm; approximate body mass: 3090 g. Range: North Atlantic Ocean. (2) Brown booby (*Sula leucogaster*). Length: 69 cm; wingspan: 141 cm; approximate body mass: 1150 g. Range: Tropical Pacific, Atlantic and Indian Ocean. (3) Brown pelican (*Pelecanus occidentalis*). Length: 114 cm; wingspan: 203 cm; approximate body mass: 3500 g. Range: west coast of the Americas from British Columbia to Ecuador; east coast of the Americas from New Jersey to the Amazon River mouth, and Caribbean waters. (4) Red-tailed tropicbird (*Phaethon rubicauda*). Other names: Bosunbird. Length: 46 cm; wingspan: 104 cm; approximate body mass: 715 g. Range: Tropical Pacific and Indian Oceans. (5) Great frigatebird (*Fregata minor*). Other names: Man o'War Bird. Adult female. Length: 93 cm; wingspan: 218 cm; approximate body mass: 1375 g. Range: Tropical Pacific, Atlantic and Indian Oceans. (6) Great cormorant (*Phalacrocorax carbo*). Other names: common cormorant. (a) Adult; (b) immature. Length: 90 cm; wingspan: 140 cm; approximate body mass: 2280 g. Range: Widely distributed in coastal waters of world except not found along west coast of the Americas or east coast of Central and South America. (7) Imperial shag (*Phalacrocorax atriceps*). (a) Adult breeding; (b) non-breeding plumage. Length: 72 cm; wingspan: 124 cm; approximate body mass: 3120 g. Range: Southern Ocean. Illustrations from Harrison P (1985) *Seabirds, an identification guide. Revised edition*. Boston, Massachusetts; Houghton Mifflin.

plunge-diving into surface shoals of fish with the bill opened. Most pelicans are found near lakes and large water bodies, and only the brown pelican is primarily marine.

Pelicans are gregarious, and are rarely encountered alone. They breed almost exclusively in large colonies; the prefledged young form large crèches within the breeding colony; and roosting and flight are done in large groups. Pelicans have the best-documented communal feeding, in which large groups will coordinate fishing in fresh or brackish waters as a group. From 10 to 50 birds will alight on the water in a row, and drive surface-schooling fish toward shallow water by paddling, flapping wings, and plunging their bills in the water, sometimes feeding and sometimes not. Other birds plunge-dive into the water after fish, or swim from beneath; when captured, the fish are retained in the gular pouch as the water drains away. They are then eaten, often after being flipped up into the air to be swallowed head-first.

Pelicans, along with cormorants, are among the birds most affected by human disturbance and environmental pollution, particularly by chlorinated pesticides such as DDT and dieldrin. Pelicans and humans have long been associated, and freshwater species are vulnerable to habitat loss and pollution. Fishermen have persecuted pelicans and other fish-eating birds in times of low fish catch, but most studies indicate that the take by pelicans is primarily of noncommercial fish. In every case, low fishing yields are due to poor management or poor environmental stewardship, and not due to pelican–human competition. Pesticide contamination of the marine foodweb led to drastic reductions in brown pelicans throughout their range midway through the twentieth century, but pollution abatement and environmental protection efforts have allowed populations to recover. Pelicans now serve as an icon for conservation success, but many populations are still under threat.

Sulidae The sulids, gannets and boobies (genera *Morus* and *Sula*) include 9 species found throughout the tropical and temperate regions of the marine environment. Generally, they are duck-sized, with a long conical bill finely serrated along the edges. Gannets are found in high-latitude waters, while boobies tend to be more tropical; both are exclusively marine and piscivorous. Sulids feed by plunge-diving, then often shallow pursuit (usually to no greater than 15 m depth) after fish in the upper water levels. The plumage is thought to reflect this behavior, as the light chest and abdomen will provide little contrast when seen from below, and prove indistinct when seen from above. Whether related to camouflage or swimming ability, sulids are very efficient in capturing small

schooling fish such as anchovy (*Engralis*) and sardines (*Sardinops*).

The timing of breeding seems mostly associated with local conditions of food and nest site availability, and occurs annually in most species. In very good conditions, the tropical boobies will breed at intervals less than 12 months, and the red-footed booby in the Galapagos often breeds once in 15–18 months. These nonannual periods have been thought to be a response to bad (or good) oceanographic conditions related to the El Niño Southern Oscillation, or conversely La Niña, or as a means to desynchronize breeding with potential nest predators such as frigatebirds, or both. In both gannets and boobies, the reproductive cycle is similar to other seabirds, in that chicks fledge, leave the nest, and are independent before the end of the first year. Sulids usually do not breed for the first time until they are 3–6 years old, and live 10–20 years.

The largest clutches are found in the species breeding in upwelling regions with high fish production (e.g., Peruvian and blue-footed boobies), and average 2–3 eggs. Other tropical boobies such as brown and masked boobies will lay two eggs, laid 4–6 days apart; the larger and stronger chick – not always the first one hatched – will often push the weaker chick from the nest to its ultimate demise. All other sulids lay a single egg. The most aberrant and presumed the most primitive sulid is Abbott's booby, which is now restricted to Christmas Island, Indian Ocean. Chicks take 20–24 weeks to fledge, and are fed by adults for an additional 20–40 weeks. This slow development is attributed as an evolutionary response to the unpredictable and distant feeding grounds of the adults, but the evidence is not clear. Abbott's booby is also the most threatened as a result of habitat loss, constriction of breeding range, and low productivity in raising chicks to breeding adults.

Phalacrocoracidae There are nearly as many species of cormorants and shags as all of the remaining members of the Pelecaniformes. This family has about 60 described taxa; these are sometimes lumped into as few as 26 species or as many as 40 or more. The discrepancy lies in the many related island forms found throughout the Southern Ocean, including Antarctic, Subantarctic, and Southern Subtemperate waters. There are two distinct subfamilies, the cormorants (Phalacrocoracinae) and the shags (Leucocarboninae), although the common names often do not correspond so neatly: the pied shag of the Antipodes is in fact a cormorant, and the pelagic cormorant of the North Pacific is a shag. The two groups are distinguished by clear morphological and genetic differences, and represent an early divergence as far back as the Miocene, perhaps 15–20 million years ago. Cormorants are generally heavy-bodied birds with labored flight, commonly found on fresh water systems, often far into the interior. Shags are smaller and slimmer, are more powerful fliers, and are common inhabitants of the marine littoral, neritic, and pelagic waters.

In many species, particularly those of the Northern Hemisphere, the plumage is basically black or very dark, often with metallic or iridescent sheens to the feathers. Most shags and cormorants of the Southern Hemisphere have in addition a white abdomen; in a few species such as the red-legged shag and spotted shag (*Stictocarbo gaimardi* and *S. punctatus*), the plumage is predominantly gray. In breeding, adults will often undergo dramatic changes in color and appearance that affect both bare parts and the plumage. The naked skin of the face, gular pouch, bill lining, and rarely feet, become more intensely colored, change color, or even assume color, and these changes are often accompanied by enlarged or increased carunculations near the eyes and face. Many species produce long white filoplumes and breeding plumes on the head and neck (great cormorant, *Phalacrocorax carbo sinensis*), and in a few species (red-faced, pelagic cormorants *Stictocarbo urile, S. pelagicus*) a conspicuous white patch appears on the thigh.

Unlike many other Pelecaniformes, cormorants and shags have very flexible requirements for nesting and nest construction. They are known to breed in diverse habitats including on level ground, on cliffs and embankments, in trees, on bridge supports, and on wharfs: generally, on objects large and sturdy enough to support the weight of the nest and occupants while affording protection from ground predators. Proximity to water is an absolute requirement because cormorants are nearly exclusively piscivorous, and the size of breeding colonies in continental interiors correlates with how close they are to feeding areas (**Figure 2**). By virtue of wing morphology and aerodynamics, cormorants are indifferent fliers and do not range far from roosting or breeding areas. Colony and perch sites as a consequence are located near foraging areas, tend to be patchily distributed throughout the landscape, and concentrate large numbers of birds. There is ongoing debate whether cormorants in colonies share information in some way about feeding areas, but the phenomenon of social feeding in cormorants – as in pelicans – is well observed. Two patterns are known: line hunting, in which cormorants move through the water in a straight line in a rolling flock, and zigzag hunting, in which individuals search and change directions. Line hunting is usually associated with

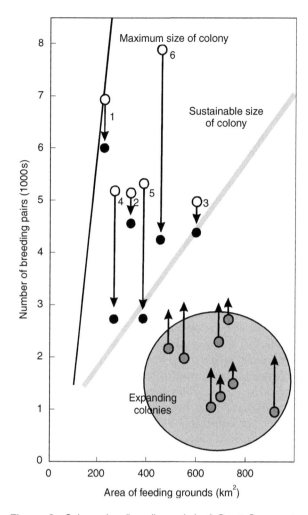

Figure 2 Colony size (breeding pairs) of Great Cormorants *Phalacrocorax carbo* in relation to available feeding habitat (0–20 m water depth) without overlap with their colonies with a range of 20 km from the colony. Maximum colony size (open circles) and, if different, most recent colony size (dark circles) indicate effects of oversaturation of feeding resources. The recent colonies in Denmark (shaded circles) are undersaturated with respect to available feeding resources and are fast expanding in size. Both effects suggest the existence of 'sustainable' and a 'maximum' density. The oversaturated effect is particularly noticeable in the Brændegård colony (1) in Denmark. Numbers refer to monitored colonies in Denmark (1–3) and The Netherlands (4–6). (Redrawn with permission from van Eerden and Gregersen (1995).)

smaller fish like smelt (*Osmerus*) and ruffe (*Gymnocephalus*), whereas zigzag fishing is seen when cormorants pursue larger fish like roach (*Rutilus*) and Perch (*Perca*).

Anhingidae Darters are fairly large birds ('goose-sized') with a very long, slender neck with a pronounced 'S-shaped' kink, a long spearlike bill, and a large fanlike tail resembling that of a turkey. The taxonomy is contentious: variously one, two, four or more species have been described for the family. Darters are primarily aquatic and spend their life in water or on branches overhanging protected, usually fresh water, streams and ponds, especially swamps and marshy areas. Darters are widely distributed throughout tropical and subtropical zones, and are sometimes found in warm temperate habitats. They are the least marine of the Pelecaniformes, but can be observed in brackish coastal waters, mangrove and cypress swamps, and coastal lagoons.

They are similar in appearance to cormorants, but are easily distinguished in flight, on roosts, and on the water. Darters are good fliers and often alternate powered flight with glides; cormorants use sustained flapping and seem always at risk of not quite keeping airborne. Darters will use thermals for soaring and will soar for extended periods, perhaps while surveying new areas for feeding or roosting. Their long tail and tapered wings allow a much greater agility in air than shown by shags or cormorants, and they are deft at flying through enclosed swamps and thickets. While it is swimming on the surface, the body of a darter is usually submerged, with only the head and snakelike neck visible, making it obvious why the term 'snake bird' is often used for them. The wettable plumage of darters results in considerable loss of body heat under water, and they therefore spend large amounts of time sunning and drying feathers.

Darters have several unique anatomical features apparently related to feeding. The distal portions of the upper and lower bill have fine, backward-pointing serrations for holding fish. Modifications to the eighth and ninth cervical vertebrae allow a right-angled kink in the neck, and the extensor ligaments of the neck pass through a 'pulleylike' loop that allows a fulcrum for the straight-line stabbing motion with which this species spears its prey. Darters have an unusual bony articulation at the base of skull, allowing an even greater forward acceleration of the head and neck, and a hairlike pyloric valve in the stomach. Unlike all other pelecaniform birds, darters will stab fish with the bill, rather than grasping them between the bill, and then flip the fish into the throat head first. This type of fishing technique is well-suited for the shallow, murky waters they inhabit, and deadly to the slow moving, laterally compressed fish the birds pursue. In North America, centrarchid and cyprinodontid fish are most preferred, but freshwater decapods often are eaten when abundant.

Darters are monogamous, as are most Pelecaniformes, with the pair bond lasting many years and breeding pairs returning to the same nest each season. Nesting is usually solitary, but small aggregations are known, and darters will commonly breed in mixed colonies with other waterbirds such as cormorants, herons, egrets, ibises, and storks. Darters are sedentary, rarely dispersing after breeding, but some of the populations breeding in temperate regions will disperse to warmer areas in winter. Owing to their wariness, solitary nature, and preference for impenetrable swampy areas, comparatively little is known about their behavior or natural history.

Ecology

Color has been implicated as an important factor in pelecaniform feeding; it is commonly stated that light-colored plumage affords a minimal contrast against daylight or sky, while dark plumage is especially cryptic in dim light. Pelecaniform birds range in color as adults from mostly white (e.g., gannets, tropicbirds) to mostly black (e.g., cormorants, frigatebirds), and with many intermediate shades, colors, and contrasts (e.g., grey plumage in red-footed cormorants, mottled plumage in boobies and brown pelicans). However, juvenile and immature frigatebirds, sulids, and pelicans are generally darker than adults, while in cormorants and darters, and in some frigatebirds and pelicans, the juveniles are lighter. Shags, boobies, and some other pelecaniforms, have white and black contrasting patterns on their abdomens and chests, and in species such as brown booby (*Sula variegata*) and rock shag (*Stictocarbo magellanicus*), there are morphological variants with the same patterns. Many have speculated that these plumage patterns relate to greater efficiency in capturing fish or avoiding detection, but other factors seem as important. Melanistic plumage is often more durable than light-colored plumage, and thermoregulation is affected strongly by overall coloration. Much more study is required to understand what may be important in plumage color.

Primarily fish-eating birds, the Pelecaniformes nonetheless utilize a diversity of feeding methods, marine and aquatic habitats, and food items (**Table 1**). Gregariousness is prominent in the group, with some being groups well-known for social feeding (pelicans, cormorants), and with most breeding in large, dense colonies. Nearly every type of fresh water and marine habitat is exploited, but, except for tropicbirds and frigatebirds, the pelecaniform birds are limited to littoral and neritic waters, rarely encountered far from land. Their preference for small schooling fish, and for breeding in large, dense colonies near the feeding grounds, makes their guano production a desirable attribute from the human viewpoint, but interactions with humans are for the most part neutral or negative. Cormorants and

Table 1 Summary of Ecological, and behavioral characteristics of Pelecaniforms

	Phaethontidae	Fregatidae	Pelecanidae	Sulidae	Phalacrocoracidae	Anhingidae
Sociality						
Colonial	+	+	+	+	+	+
Flocking	?	+	+	+	+	
Solitary	+	+	+	+	+	+
Feeding method						
Pursuit						
Deep plunge	+			+	+	+
Shallow plunging	?		+	+		
Surface feeding	?			?	+	
Skimming		+		+		
Kleptoparasitism		+				
Feeding period						
Day	+	+	+	+	+	+
Crepuscular	+	+	+	+	+	+
Night	?	?	+	?	?	?
Feeding depth						
Bottom					+	
Midwater				+	+	+
Surface	+	+	+	+		+
Feeding habitat						
Littoral					+	+
Neritic (marine)			+	+	+	
Pelagic (marine)	+	+		+		
Diet						
Marine v. freshwater	M	M	M/F	M	M/F	F
Fish	+	+	+	+	+	+
Cephalopood	+	+		+		
Crustacean	?	?	+		+	+
Scavenging		+	+			

pelicans are often – wrongly – implicated in the depredation of fingerling and fry of economically important fish such as trout and salmon, and consequently suffer culling and killing campaigns to reduce their numbers.

Behavior

Mostly nonvocal and colonial, pelecaniform birds utilize ground-based visual displays much more prominently than other waterbirds (see **Figure 3**). When birds live and feed together in close proximity in colonies, the interactions between individuals is much greater than found in solitary nesting or dispersed breeders. Further, because space is limited at most sites suitable for nesting, colonial birds breed in extremely dense conditions, with territories squeezed to a minimal area around the nest, or even nonexistent. Most of the pelecaniform birds have a basic courtship sequence that serves to establish the pair bond and maintain it, often for many years. In nearly every case, the male selects the nest site and advertises for a mate; the females choose among males through assessment of no doubt many factors, including the nature of the displays, the excellence of the nest site, and prior history. Copulation occurs on the nest site, usually between bouts of nest building by the male, sometimes by both partners. After the pair bond has been established, both birds take turns guarding the nest, incubating eggs and chicks, and feeding young. With exceptions noted above, parental care stops soon after fledging, and young birds will forage independently soon after leaving the nest. Some pelecaniform birds will begin breeding as soon as the second year (e.g., cormorants), although three years of age seems to be the norm, while in other species (e.g., Abbott's booby) many years may pass before individuals breed.

Figure 3 Behavioral displays in the rock shag (*Stictocarbo magellanicus*). Appeasement displays: (A) landing display approaching a nesting cliff; (B) postlanding or hopping; (C) nest-touching after a threat display. Courtship displays: (D) beginning and (E) ending phase of darting by the male; (F) beginning and (G) ending of throat-clicking by a pair. Pairing displays that terminated in copulation: (H) initial phase of wing-waving by the male (bottom) after approach by the female (top); (I) beginning phase of hop by female, male is wing-waving; (J) conclusion of hop by female followed by nest-indicating, male is wing-waving; (K) full neck extension by male during wing-waving, female is gaping; (L) bill-biting by female and bowing by male. (Redrawn with permission from Siegal-Causey (1986).)

Status and Conservation

Few of the Pelecaniformes are globally threatened, although some island and inland populations of pelecaniform birds have become extinct in the face of habitat loss and contamination. In some species (e.g., double-crested cormorant, white pelican), numbers are increasing with near-exponential growth, while in others (e.g., Dalmatian pelican) population abundance is rapidly decreasing. Island species and populations have been especially hard-hit by the introduction of terrestrial predators such as cats, rats, and pigs; loss of traditional breeding habitat has played a greater role in aquatic and inland-nesting species. Some islands such as Christmas Island (Indian Ocean) and Ascension Island (Atlantic Ocean) are breeding grounds for several pelecaniform species, and their recent protected status has promised a much better future than might have been possible only a few years earlier.

See also

Seabird Conservation.

Further Reading

Ashmole NP (1971) Sea bird ecology and the marine environment. In: Farner DS, King JR, and Parkes KC (eds.) *Avian Biology*, Vol. 1. New York: Academic Press.

Cairns DK (1986) Plumage colour in pursuit diving seabirds: why do penguins wear tuxedos? *Bird Behaviour* 6: 58–65.

Diamond AW (1975) The biology of tropicbirds at Aldabra Atoll, Indian Ocean. *Auk* 92: 16–39.

Frederick PC and Siegel-Causey D (2000) Anhinga (*Anhinga anhinga*). In: Poole A and Gill F (eds.) *The Birds of North America, No. 522*. Philadelphia, PA: The Birds of North America, Inc.

Johnsgard PA (1993) *Cormorants, Darters, and Pelicans of the World*. Washington, DC: Smithsonian Institution Press.

Nelson JB (1978) *The Sulidae: Gannets and Boobies*. Aberdeen: Oxford University Press.

Nelson JB (1985) Frigatebirds, aggression, and the colonial habit. *Noticias Galapagos* 41: 16–19.

Siegel-Causey D (1986) Behaviour and affinities of the Magellanic Cormorant. *Notornis* 33: 249–257.

Siegel-Causey D (1988) Phylogeny of the Phalacrocoracidae. *Condor* 90: 885–905.

Siegel-Causey D (1996) The problem of the Pelecaniformes: molecular systematics of a privative group. In: Mindell DL (ed.) *Avian Molecular Evolution and Systematics*. New York: Academic Press.

Van Eerden MR and Voslamber B (1995) Mass fishing by cormorants *Phalacrocorax carbo sinensis* at Lake IJsselmeer, The Netherlands: a recent successful adaptation to a turbid environment. *Ardea* 83: 199–212.

Van Tets GF (1965) *A comparative study of some social communication patterns in the Pelecaniformes*. *Ornithological Monographs, No. 2*. Washington, DC: American Ornithological Union.

PHALAROPES

M. Rubega, University of Connecticut, Storrs, CT, USA

Introduction

Phalaropes are shore birds that have largely abandoned the shore. Rather than wading at the boundary of water and land, as most shore birds do, red (*Phalaropus fulicaria*) and red-necked (*P. lobatus*) phalaropes spend up to 9 months of the year swimming on the open ocean. (Red phalaropes are commonly called gray phalaropes in Europe.) A third species, Wilson's phalarope (*P. tricolor*), does not have a marine life history (and thus will not be much considered here) but swims most of the year on saline lakes in the interior of North and South America, where its biology is similar to that of the marine phalaropes. During the breeding season, all three species frequent fresh-water wetlands. At around 20 cm long, phalaropes are among the smallest of oceanic birds. Phalaropes display un-waderlike adaptations for swimming, including lobed toes, legs that are flattened to reduce drag when paddling, and dense, waterproof feathers.

Those air-trapping feathers, combined with small body size, result in the corklike buoyancy of phalaropes, as a result of which it is almost comically difficult for the birds to dive. Restricted to surface waters, phalaropes are famous among birdwatchers and biologists for frenetic spinning in small circles. This behavior generates a miniature upwelling that brings prey from deep in the water column within reach. Once grasped, prey are rapidly moved up the beak into the mouth by a process that uses the surface tension of water.

The sex roles are reversed in phalaropes: larger, brightly colored females lay eggs for their mates, while smaller males with dull feathers perform all care of eggs and chicks. This unusual breeding system is further distinguished by polyandry: female phalaropes sometimes have multiple mates in a single season. Committed to an aquatic lifestyle, phalaropes do virtually everything but lay eggs on the water, nesting on land near pools or ponds. The marine phalaropes breed circumpolarly in the Arctic and sub-Arctic and migrate to pelagic wintering areas by flying out to sea. Some red-necked phalaropes also migrate overland by a series of short flights visiting every imaginable body of water, but especially saline lakes.

The population status of the two marine phalaropes is uncertain at best. The pelagic biology of these birds is poorly known, compared to other seabirds, and this hampers efforts to monitor populations and obtain data about their numbers. Breeding biology is better understood, but few breeding populations are being monitored. Large flocks at sea and on saline lakes imply that phalaropes are abundant, but human disturbance has reduced the breeding range of phalaropes. More disturbing, massive flocks have simply disappeared from former migratory staging sites, for example, the western Bay of Fundy on the north-eastern coast of Canada. We do not know whether this represents a real reduction in the population, or a shift to a new staging location caused by a change in the availability of plankton.

Phalaropes at Sea

Appearance

Phalaropes are small sandpipers, with the small heads, needle-shaped beaks, and elongate necks typical of the family Scolopacidae (**Figure 1**). As they paddle on the ocean surface, their long wader's legs are generally hidden from an observer. Their small size and shape make them unmistakable for any other kind of bird at sea, although they are easily mistaken for each other. Red and red-necked phalaropes are more similar to each other, and more closely related (see Systematics below), than either is to Wilson's phalarope. These degrees of relatedness show in appearance, structure, and behavior. When seen at sea in their white and gray nonbreeding plumage (**Figure 1B**) red and red-necked phalaropes are difficult to distinguish. Red-necked phalaropes are slightly smaller, and more lightly built, with an exceptionally fine needlelike black bill; red phalaropes are more heavily built, with a broader, deeper, more yellow bill. Either species is easily distinguished from Wilson's phalarope (which is the largest of the three species and lacks a dark eye patch and wing bars), which is only rarely seen in the marine environment, and never pelagically.

Female phalaropes are slightly larger than males and have brighter feathers during the breeding season (**Figure 1a**) a condition known as reverse sexual dimorphism (in most birds males are the larger and

Figure 1 Examples of Phalaropes. Red phalarope (*Phalaropus fulicarius*). Other names: Grey Phalarope. (a) Adult female, breeding; (b) adult, non-breeding, both sexes; (c) adult, late summer/autumn. Length: 20 cm; wingspan: 37 cm; approximate body mass: 50 g. Range: Breeds throughout the Arctic and into the subarctic; migrates and winters throughout the North and South Atlantic Oceans, the North Pacific Ocean and the Eastern South Pacific Ocean. Illustrations from Harrison P (1985) *Seabirds, an identification guide. Revised edition.* Boston, Massachusetts: Houghton Mifflin.

more colorful sex). These physical differences are accompanied in phalaropes by reversed sex roles (see 'Phalaropes on land' below). During most of the time they are at sea, however, males and females do not differ in their plumage and cannot be distinguished (**Figure 1b** and c). Some researchers have suggested that measures of body size can be used to discriminate between male and female phalaropes; however, the discriminant functions developed for this purpose will only reliably identify large females. Small females overlap males in body size and cannot be distinguished from them with a high degree of certainty on the basis of body size alone.

Aquatic Adaptations

All phalaropes have similar adaptations for an aquatic lifestyle that are not shared by the other sandpipers. The form of the adaptations tends to be less pronounced in Wilson's phalarope, which is less aquatic in its habits.

Phalaropes swim by paddling, and their toes are bordered by flaps or lobes of flesh (**Figure 2**). These toe lobes are flexible: when the foot is drawn forward through the water, the lobes fold flat behind the toe, reducing drag; on the backstroke the lobes flare out, boosting thrust. Phalarope legs are laterally flattened, another modification that reduces drag as they swing their legs through the water.

As befits birds that spend their time swimming, phalaropes have waterproof feathers, with particularly

Figure 2 The lobed toes of phalaropes. These flaps of skin flare out to provide thrust when the bird is paddling through the water; they fold back around the toe on the upstroke to reduce resistance. Photo by author.

dense belly plumage. Immersion in salt water may present a constant challenge for them: red-necked phalaropes at saline lakes avidly seek freshwater sources in which to bathe and drink. It is unknown whether they seek out microhabitats with less saline water at sea for the same purpose.

Retaining the hollow bones of more terrestrial birds, rather than the denser bones common in diving seabirds, phalaropes are very buoyant and ride lightly on the surface of the water. This buoyancy, combined with the air trapped in their dense belly feathers, makes diving very difficult for them. They dive shallowly and very seldom, generally only when pursued persistently by a predator at close range. When forced to dive, they quickly pop back to the surface.

Distribution

Phalaropes are pelagic during all but the short breeding season. Both red and red-necked phalaropes winter in or near the Humboldt Current, off the western coast of South America. Significant numbers of red phalaropes also winter off western Africa, and are found in the Pacific off the western United States from June through March. Red-necked phalaropes mix with reds off the western United States during migration, from July through early November. The Eurasian population of Red-necked phalaropes winters in the Arabian Sea, and from central Indonesia to western Melanesia.

Food and Feeding

Diet Phalaropes are planktivores, specializing on copepods, euphausiids, and amphipods. The marine phalaropes are size-selective; copepods taken apparently do not exceed 6 mm long by 3 mm wide. They also take almost anything else that is small and floats, including other crustaceans, insects, invertebrates (including hydrozoans, molluscs, polychaetes, and gastropods), small fish, and fish eggs. In addition, phalaropes regularly ingest nonnutritious materials, including small quantities of seeds, sand, feathers, and plastic particles.

Feeding behavior and mechanics Phalaropes are visual hunters. They typically swim in meandering, sinusoidal tracks, leaning forward and peering into the water. They peck at prey on, or just beneath, the water's surface, and where prey densities are high their peck rates may climb to 180 pecks per minute. They will occasionally seize a flying insect from the air or, rarely, catch aquatic organisms by rapidly swiping the bill sideways through the water in a motion known as scything. Unlike other planktivores, phalaropes are not filter feeders; they capture zooplankton one at a time, tweezering them out of the water between the tips of their beak.

When prey are successfully seized, red-necked phalaropes use the surface tension of water to move their tiny catch from the tip of their beaks to their mouths (**Figure 3**). They accomplish this by suspending a drop of water containing the prey between their upper and lower jaws. Since water molecules are attracted to one another, drops of water tend to assume shapes with the tightest packing of molecules, i.e., shapes with the least possible surface area. Work is required to pull enough molecules out of the center of the drop to make new surface area; the amount of work required in any particular instance of water temperature and salinity is a measure of the surface tension of water. Feeding red-necked phalaropes open their jaws; this action stretches the prey-containing drop, increasing its free surface area. Once the drop is stretched, the surface tension of the water drives the drop, and the prey it contains, to the back of the jaws, where the free surface area is minimized and the prey can be swallowed. This method of transporting prey, called 'surface tension feeding', can be completed in as little as 0.02 s. Wilson's phalaropes also use this method of prey transport; the feeding mechanics of red phalaropes have not been studied in detail.

When prey are below the surface, a bird may submerge its head and neck or (even more rarely) up-end, but a more typical response to prey deep in the water column is a conspicuous toplike spinning. All the phalaropes engage in this behavior; indeed, it is probably their best-known characteristic. Old accounts of their behavior attributed spinning variously to courtship behavior or stimulation of prey in cold water, but most authors suspected that spinning functioned to 'stir up' prey from the bottom of ponds and pools. This explanation was based on observations of birds on small ponds near breeding areas, and did not account for phalaropes spinning while at sea, when the bottom was hundreds of meters below.

Spinning does produce subsurface water flow that concentrates and lifts prey nearer the bird, but not by creating flow against the bottom. The whirling motion is produced by kicking harder and with higher frequency with the outer leg than with the inner leg. (Observations of captive phalaropes indicate that individuals are 'handed' — birds spin both clockwise and counterclockwise, but each individual only spins in one direction.) Birds spin around at about one complete rotation per second. This rapid cycling kicks water at the surface away from the axis of the bird's rotation. This deflection of surface water generates an upward-momentum jet of subsurface water; in other words, phalaropes make their own small upwellings in the center of the area they are circling by pushing surface water away so quickly that subsurface water must flow upward to replace it.

Birds watch this area of upwelling for rising prey in essentially the same way that a spinning ballerina keeps from falling over. They fix their gaze on one spot, holding their heads immobile until their rotating bodies force head movement, then they snap their heads one-quarter of a turn while still looking at the same spot. Birds can generate flows to as deep

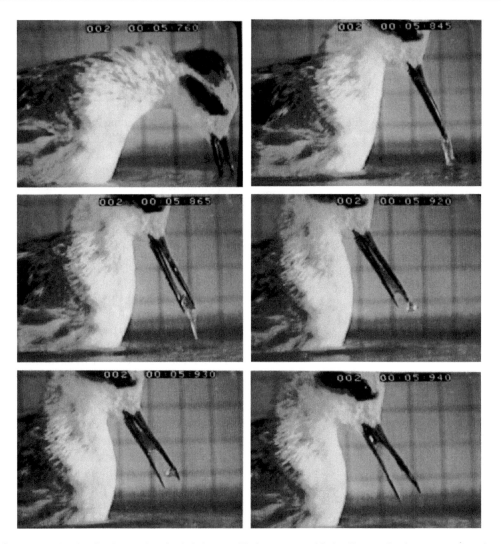

Figure 3 Surface tension feeding in a red-necked phalarope. Phalaropes use this feeding mechanism to transport tiny prey to the mouth, where it can be swallowed. Jaw-spreading (also called mandibular spreading) stretches the drop; the surface tension of the water drives the drop to the back of the beak, where it will have the smallest possible free surface area. (Reproduced with permission from Rubega MA and Obst BS (1993) Surface tension feeding in phalaropes: discovery of a novel feeding mechanism. *Auk* 110: 169–178.)

as 0.5 m, and each cycle of the spin is slightly off to the side of the previous one, so that birds slowly progress, and process water, along a track about one body length wide. As can be imagined by anyone who has ever carried water in a bucket, lifting water is an energetically expensive proposition for a phalarope: about twice as expensive, per unit of water inspected, as simply swimming in a straight line to feed. Hence, they should only spin when absolutely necessary, and they generally select pelagic habitats in which spinning is unnecessary.

Habitat

The pelagic biology of phalaropes appears to revolve around food. Although the marine phalaropes spend

the majority of their life cycle at sea, they do not breed in oceanic environments. Thus, unlike other marine birds such as petrels or penguins, their use of the ocean is not influenced by factors such as access to islands with nesting sites. This freedom to concentrate on prey shows in their relationship to physical structure in the ocean. Although they occasionally take small fish, phalaropes are fundamentally planktivores, and their habitats at sea are characterized by features of the ocean that concentrate and bring to the surface dense concentrations of zooplankton. They are commonly found at fronts, thermal gradients, convergences, upwellings, and slicks. Up to two million migratory birds have been estimated at a single upwelling near Mount Desert Rock off the Maine coast. Red-necked phalaropes in migration use a wide variety of aquatic habitats and are consistently associated within them with small-scale hydrographic features. At Mono Lake, in eastern California, red-necked and Wilson's phalaropes concentrate their feeding activities in areas where currents help to raise prey to the surface, and at drift lines where prey are concentrated.

When physical oceanography fails them, phalaropes make use of other marine organisms to locate and gain access to food. The few phalaropes found in the outer continental and slope domains of the South Atlantic Bight off the eastern coast of the United States are associated with mats of *Sargassum* seaweed, which are themselves the product of convergences. Red phalaropes associate with feeding whales and schools of fish that force plankton to the surface incidental to their feeding activities. The relationship of red phalaropes to whales is sufficiently dependable that whalers formerly used them as an indicator of the presence of whales. There are even reports of phalaropes picking parasites off the backs of whales. They are sometimes parasites themselves: red-necked phalaropes commonly dash in to seize prey that have been spun to the surface by the effort of another phalarope.

Phalaropes on Land

All oceanic birds must make landfall to breed and reproduce, and phalaropes are no exception. The marine phalaropes are essentially on land only during the brief breeding season; red-necked phalaropes also migrate over land to some extent, and are especially numerous on saline lakes (see 'Movements' below). In contrast, Wilson's phalaropes are almost entirely continental in their distribution, albeit more aquatic in their habits than most shore birds. What follows is a brief summary of the biology of phalaropes on land, and its bearing on their life at sea. Readers with an interest in the breeding biology of phalaropes will find more detail in works listed in the bibliography.

Appearance

Each species of phalarope has an extremely colorful, distinctive appearance during the breeding season, with striking patterns of black, gray, and red or rusty markings on the neck and face (**Figure 1**a). They are easily distinguished from one another and from other shore birds. All phalaropes exhibit reverse sexual dimorphism; females are more sharply and brightly colored than males when breeding. On average females are also larger.

Breeding Behavior

The breeding behavior of phalaropes is notable for the degree to which it is conducted on water. The nest may be on land, but most significant behavioral components of breeding are carried out on the waters of small pools and ponds. Breeding displays and fights over mates usually occur on the water, as do the resulting copulations. This is in contrast to most oceanic birds, which return to land in order to engage in these behaviors.

Phalaropes also differ from other oceanic birds in their unusual breeding system. First, in an arrangement that is rare for any bird, the roles of the sexes are reversed; females compete vigorously with one another for males, and provide no care for eggs or chicks, while males build the nest, brood the eggs, and care for the young when they hatch. Second, phalaropes, like most seabirds, are usually monogamous, but they differ in being polyandrous when the opportunity arises. When more males than females are present, females will breed with more than one male in a single season; female red phalaropes have laid second clutches of eggs within a few hundred meters of their first mate's nest. Whether monogamous or polyandrous, females normally leave a male after having laid the last of the 3–4 eggs in the clutch. Chicks hatch after about 20 days of incubation, can walk and swim a few hours later, and fledge within about 20 days. Although the male tends them, they feed themselves, and may become completely independent before they are able to fly.

Unlike many oceanic birds, phalaropes do not breed colonially, although pairs may breed in relatively close proximity when habitat is limited; the density of nests at any one site is highly variable from year to year. They show little tendency to return to any particular breeding site, or to the site where they were hatched. Red phalaropes will exploit the aggressive nature of colonial birds to ward off predators, by nesting in the colonies of other seabirds such as Arctic terns (*Sterna paradisaea*).

Distribution

The marine phalaropes breed in tundra habitats circumpolarly in Arctic and sub-Arctic, along coasts of the Arctic Ocean. The red-necked phalarope breeds farther south, and further into the interior of Eurasia, Alaska, and northern Canada than does the red phalarope. Red-necked phalaropes also breed in small numbers in the Aleutian Islands, Scotland, and Ireland. The nonpelagic Wilson's phalarope, in contrast, breeds only in the Nearctic, primarily in the interior of western North America.

Food and Feeding

Diet During the breeding season phalaropes essentially remain planktivores, eating small aquatic prey, especially the larvae of dipteran flies. However, on the breeding grounds their diet expands to contain adult dipteran flies (which they snap out of the air), mosquito larvae, dragonfly nymphs, water boatmen, backswimmers, caddisflies, beetles, bugs, ants, spiders, mites, snails, crustaceans, molluscs, and annelid worms. When food is limited they may eat seeds and other plant materials.

The diets of red-necked phalaropes at saline lakes consist almost entirely of brine flies, with third-instar larvae predominating, plus adult and larval dipterans. Brine shrimp are very abundant at saline lakes but are rarely taken by red-necked phalaropes. This lack of interest is the product of a nutrient limitation; captive birds are reluctant to eat brine shrimp, and those restricted to a brine shrimp diet lost about 5% of their body weight while eating three times their body weight over a 12-hour period. In contrast, Wilson's phalaropes do eat significant amounts of brine shrimp at saline lakes; their ability to extract nutrition from them has not been investigated in detail.

Movements

All phalaropes are migratory; they differ chiefly in the degree to which they move over land versus sea. Red phalaropes virtually always migrate pelagically; red-necked phalaropes migrate over both land and ocean; Wilson's phalaropes are thought to migrate southward over the Pacific after leaving North America, apparently without ever landing on the water, while the north-bound migration occurs almost entirely over land. In all three species the timing of movements is similar: nonbreeding birds of both sexes leave the breeding grounds first, followed consecutively by females, males, and juveniles.

The migration of red phalaropes is perhaps least well understood, since it is least easily observed. What is known about their migratory routes is inferred as much from information about where they are not seen as from sightings of them on the move. Nearctic breeders winter off western and southwestern Africa. Until recently large flocks occurred in the western Bay of Fundy during migration (see Conservation and Threats), but few have been seen farther south near the western Atlantic coasts. Thus, they presumably fly directly across the Atlantic to their destinations off the African coast after leaving the Bay of Fundy. Many red phalaropes winter in the Humboldt Current, off western South America, and

large flocks are present during migration off the Pacific coast of North America. Sightings of red phalaropes in the central Pacific Ocean during the migratory period indicate that those breeding in the Siberian Arctic cross the Pacific to winter in the Humboldt current.

Red-necked phalaropes mix travel over land and sea to a much greater extent. Those breeding in Europe and western Siberia move through the Caspian Sea and overland via lakes across the former Soviet Union and Iran to arrive at their Arabian Sea wintering grounds through the Gulf of Oman. Birds that have bred in Fenno-Scandinavia move southeast through the gulfs of Bothnia and Finland. Breeding populations from eastern Siberia migrate overland and offshore of Japan to winter in the East Indies. Nearctic populations move south across Canada and the western United States, where tens of thousands stop at hypersaline lakes, along with smaller numbers at every conceivable body of water in their southward path; from these lakes they move out to sea and south to their wintering grounds at the northern edge of the Humboldt Current. Huge flocks estimated at 2 000 000 individuals occurred in the Bay of Fundy until recently (see Conservation and Threats), and these may include birds from Greenland, Iceland, and the Nearctic. Where these birds winter is a mystery; they do not winter off western Africa with the red phalaropes with which they mingle in the Bay of Fundy, and only small flocks of red-necked phalaropes have been seen farther south in the western Atlantic in the winter. It has been suggested that they may fly to the Humboldt Current via routes crossing the Caribbean and Central America, but no more than a few red-necked phalaropes have been seen in these areas.

Systematics

The names and scientific classifications of phalaropes have histories nearly as colorful as the birds themselves (in South America, red-necked phalaropes are called 'pollito del mar,' roughly 'little chicken of the sea'). Both common and scientific names have changed repeatedly (**Table 1**). Each species was once considered to form a distinct genus, and was given a separate Latin first name (*Phalaropus*, *Lobipes*, and *Steganopus* for red, red-necked, and Wilson's phalarope, respectively). Subsequently, and for many years, the phalaropes have been placed in a single group in the family Scolopacidae (sandpipers), subfamily Phalaropodinae, genus *Phalaropus*.

The idea that all the phalaropes descended from a single common ancestor is sometimes disputed.

Table 1 The diversity of names for phalaropes

	Red phalarope	Red-necked phalarope	Wilson's phalarope
Other common names	Grey phalarope	Northern phalarope	
	Red coot-footed tringa	Hyperborean phalarope Cock coot-footed tringa	
Present scientific name	*Phalaropus fulicaria*	*Phalaropus lobatus*	*Phalaropus tricolor*
Previous scientific names	*Tringa fulicaria*	*Tringa tobata/lobata*	*Steganopus tricolor*
		Lobipes lobatus	some authorities have resurrected this name for Wilson's phalarope
Synonymous scientific names	*Phalaropus glacialis, rufus, platyrhynchus, rufescens, griseus, cinereus*	*Lobipes hyperboreus, anguirostris, antarcticus, fuscus, ruficollis, tropicus, vulgaris*	*Steganopus wilsoni, incanus, frenatus, stenodactylus*

Analyses of both their genetic material and skeletons showed that red and red-necked phalaropes are each other's closest relatives. Analyses of skeletal materials united the more distantly related Wilson's phalarope to the other two in a single group, but genetic analyses do not unambiguously support the idea that all three phalaropes belong to a single group. Both kinds of evidence indicate that phalaropes are either closely related to the Tringine or Scolopacine sandpipers. Thus, their scientific classification has recently depended on the authority consulted. For example, Sibley and Monroe's 1993 taxonomy of birds, based on DNA–DNA hybridization studies, put the phalaropes in the subfamily Tringinae, with red and red-necked phalaropes in the genus *Phalaropus*, and Wilson's phalarope returned to the genus *Steganopus*. The American Ornithologists' Union continues to classify them in a single genus *Phalaropus*, in the subfamily Phalaropidinae.

Conservation and Threats

Population Status

Phalarope populations are not thought to be threatened on a global scale, but their status is poorly known. Thus, significant population declines would be difficult to document with any degree of certainty. This lack of information arises from the peculiar life history of phalaropes; most seabirds are counted at their breeding colonies, while most shore birds are counted at migratory staging areas or wintering grounds because they do not nest in colonies. Phalaropes do not nest colonially, and in the nonbreeding season gather far out at sea, where it is difficult to find and count them.

Local declines of breeding birds have been documented; for instance, red-necked phalaropes no longer breed anywhere in Britain apart from in Scotland and Ireland (and there only irregularly) because of egg collecting in the nineteenth century. Breeding populations elsewhere are largely unstudied and, where they are monitored, information about population trends is equivocal. The population of male red-necked phalaropes at LaPerouse Bay, in Churchill, Manitoba declined by 94% between 1980 and 1993, but nesting densities have increased since 1981 near Prudhoe Bay, Alaska.

At sea, apparent declines are even more difficult to understand. The number of phalaropes staging for fall migration in the western Bay of Fundy declined from estimates of two million to almost nothing in the mid-1990s. This disappearance may represent a true population decline, or simply a shift of currents and prey, and thus of birds, to some as-yet undiscovered area of the Bay or the western Atlantic. Similar declines have been reported in the number of phalaropes seen off coastal Japan in spring. Limited evidence suggests that the numbers of birds passing through the Bay of Fundy during spring migration is unchanged.

Threats

Compared to many oceanic birds, phalaropes probably face relatively few threats. Their breeding populations are widely distributed and thus, unlike those of many colonially nesting seabirds breeding on islands, are resistant to depredations of introduced predators. Their predators on the breeding grounds include raptorial birds such as pomarine and parasitic jaegers, mammals such as arctic and red foxes and short-tailed weasels, and chick and egg predators such as glaucous gulls, sandhill cranes, and arctic ground squirrels. They are safe from most of these when at sea. However, they are not invulnerable even at sea: four red-necked phalaropes were once found in the stomach of a common dolphin taken off Baja California, Mexico. With the exception of minor subsistence hunting by indigenous northerners, phalaropes are not hunted by humans and as surface-swimming planktivores, are not incidentally taken in fishing nets, as so many seabirds are. They are potentially vulnerable to spilled oil, particularly since oil and food particles may be concentrated at the same convergence zone.

As for all oceanic organisms, human-caused disruption and destruction of marine environments is likely the most serious threat facing phalarope populations.

See also

Baleen Whales. Copepods. Seabird Foraging Ecology. Seabird Migration. Seabird Population Dynamics.

Further Reading

Colwell MA and Jehl JR, Jr (1994) Wilson's phalarope (Phalaropus tricolor). In: Poole A and Gill F (eds.) *The Birds of North America, No. 83. Philadelphia: The Academy of Natural Sciences*; Washington, D.C: The American Ornithologists' Union.

Cramp S and Simmons KEL (eds.) (1983) *Handbook of the Birds of Europe, the Middle East and North Africa: The Birds of the Western Palearctic, vol. 3*. Oxford: Oxford University Press.

Del Hoyo J, Elliott A, and Sargatal J (eds.) (1992) *Handbook of the Birds of the World, vol. 1*. Barcelona: Lynx Edicions.

Höhn EO (1971) Observations on the breeding behaviour of grey and red-necked phalaropes. *Ibis* 113: 335–348.

Murphy RC (ed.) (1936) *Oceanic Birds of South America, vol. II*. New York: American Museum of Natural History.

Rubega MA, Schamel DS, and Tracy D (2000) Red-necked phalarope (Phalaropus lobatus). In: Poole A and Gill F (eds.) *The Birds of North America, No. 538. Philadelphia: The Academy of Natural Sciences*; Washington, D.C.: The American Ornithologists' Union.

Tinbergen N (1935) Field observations of East Greenland birds. I. The behavior of the red-necked phalarope (*Phalaropus lobatus* L.) in spring. *Ardea* 26: 1–42.

PROCELLARIIFORMES

K. C. Hamer, University of Durham, Durham, UK

Introduction

The procellariiformes (commonly referred to as petrels) are a monophyletic group of seabirds containing about 100 species in four families: the albatrosses (Diomedeidae, 13 species), the shearwaters, fulmars, prions, and gadfly petrels (Procellariidae, 65 species), the storm-petrels (Hydrobatidae, 21 species), and the diving-petrels (Pelecanoididae, 4 species). It has recently been suggested that the Diomedeidae and Procellariidae may in fact contain up to 21 species and 79 species, respectively (bringing the total to 125 species), but this remains open to debate and so the more conservative taxonomy has been retained here. Petrels range in body size from the least storm-petrel *Halocyptena microsoma* (wing span 32 cm, body mass 20 g) to the royal albatross *Diomedea epomophora* (wing span 300 cm, body mass 8700 g), but they can all be identified by a single diagnostic feature: the external nares open at the end of a prominent horny tube on the upper mandible (**Figure 1**). They are usually considered a separate and ancient order of birds, probably derived from a late Cretaceous ancestor, although DNA evidence has suggested a more recent origin, with the petrels joining the divers (loons), frigate-birds, and penguins within the Order Ciconiiformes.

Petrels occur from tropical to polar regions in all oceans, but most species breed at cool temperate latitudes, with the highest species richness at 45–60°S for albatrosses and diving-petrels, and 30–45°S for the Procellariidae. However, storm-petrels have highest species richness in warm temperate waters at 15–30°N (**Table 1**). Overall, about 70% of petrel species and probably more than 80% of individuals breed in the southern hemisphere. They generally nest on islands, headlands, or mountains isolated from mammalian predators. About 20% of petrel species are surface-nesters while the rest breed below ground in burrows or beneath boulders and scree. Surface-nesters are either too large to burrow or are smaller, mainly tropical, species without predators at the colony. Burrow-nesters usually return to land only by night, probably as a defense against avian predators. Petrels nest colonially, often in high numbers, with colonies of several species exceeding 2 million breeding pairs. Many species undertake regular annual migrations between breeding colonies and separate locations used outside the breeding season. These often involve transequatorial movements, although some species change latitude without crossing the equator and others move east or west without much change in latitude.

Population Ecology

Demography

Petrels are very long-lived for their size, with some storm-petrels reaching 20 years of age and some albatrosses reaching 80 years. Sexual maturation occurs at 2–13 years of age (later in larger species; **Figure 2**) and all species have only a single-egg clutch. Breeding success is typically about 0.4–0.7 chicks fledged per nest, increasing with age and experience. Most breeding failures occur during incubation and early chick-rearing (especially at hatching), with a secondary peak in losses occurring around fledging in some species, mainly due to naive birds being captured by predators. Most petrels are annual breeders but some albatrosses, which have very long breeding seasons, are biennial, while female Audubon's shearwaters *Puffinus lherminieri* at the Galapagos Islands lay at roughly 9-month intervals. In many species a substantial proportion of breeders (up to 30% in some cases) misses a year occasionally, probably to replenish body reserves depleted during the previous breeding season. Because petrels are long-lived and have low annual reproductive rates, their populations are more sensitive to changes in adult survival than to changes in juvenile survival or breeding success. For instance, in wandering albatrosses *Diomedea exulans*, a reduction of 1% in adult survival would require an increase of 6% in juvenile survival to prevent a decline in population size.

Regulation of Population Size

Under normal circumstances, some petrel populations may be regulated in a density-dependent manner by intraspecific competition for food or breeding space. Although there is little direct evidence of prey depletion, the large size of many petrel colonies implies a strong potential for foraging

Figure 1 Examples of Procellariiform Seabirds. (1) Wandering albatross (*Diomedea exulans*). Adult. Length: 115 cm; wingspan: 300 cm; approximate body mass: 8750 g. Range: Most ocean areas south of 30 S. (2) Mottled petrel (*Pterodroma inexpectata*). Other names: scaled petrel, Peale's petrel. Length: 34 cm; wingspan: 74 cm; approximate body mass: 330 g. Range: Pacific Ocean. (3) Sooty shearwater (*Puffinus griseus*). Length: 44 cm; wingspan: 99 cm; approximate body mass: 780 g. Range: North and South Pacific Ocean, North and South Atlantic Ocean, Southern Ocean. (4) Common diving-petrel (*Pelicanoides urinatrix*). Other names: sub-antarctic diving-petrel. Length: 22 cm; wingspan: 35 cm; approximate body mass: 135 g. Range: Southern oceans. (5) Cape petrel (*Daption capense*). Other names: Cape pigeon, pintado petrel. Length: 39 cm; wingspan: 86 cm; approximate body mass: 450 g. Range: most ocean areas south of 30 S. (6) Wilson's storm-petrel (*Oceanites oceanicus*). Length: 17 cm; wingspan: 40 cm; approximate body mass: 35 g. Range: North and South Atlantic Ocean, Southern Ocean.

success to be reduced by competition for food, and the notion of competitive exclusion is supported by segregation of feeding zones between sexes, age classes, and populations of some species, particularly southern albatrosses. However, the large foraging ranges of petrels and their wide dispersal at sea reduce the likely importance of prey depletion in most cases, and direct interference competition is unlikely to have a major influence on foraging success except under particular circumstances such as when

Table 1 Species richness of breeding petrels by latitude

| Latitude (deg) | Number of species | | | | |
	Diom	Procel	Hydro	Pelec	Total
75.1–90N	0	1	0	1	1
60.1–75N	0	2	2	0	4
45.1–60N	0	2	3	0	5
30.1–45N	2	10	7	0	20
15.1–30N	2	15	9	0	27
0–15N	0	3	1	0	4
0–15S	1	9	4	1	15
15.1–30S	0	17	5	1	23
30.1–45S	6	32	4	2	44
45.1–60S	9	21	4	3	37
60.1–75S	0	6	2	0	8
75.1–90S	0	2	1	0	3

Taxonomy follows *Warham (1990)* except that Levantine shearwater Puffinus yelkouan (previously regarded as a Mediterranean subspecies of P. puffinus) is given full specific status, raising the number of Procellariidae breeding at 30.1–45°N from 9 to 10. DiomDiomedeidae, ProcelProcellariidae, HydroHydrobatidae, PelecPelecanoididae.

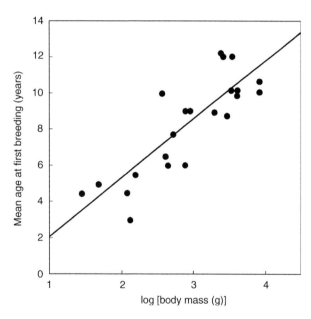

Figure 2 The relationship between body mass and mean age at first breeding in 22 species of petrel. (Data from Table 15.4 in Warham (1990) and Table 1.3 in Warham (1996).)

scavenging behind fishing vessels. Shortage of space may be an important regulatory factor for some species, especially burrow-nesters, when competition for nests may delay the age of first breeding and hence reduce recruitment. Predation at the colony seems to have little impact on petrel populations under natural conditions, although introduced alien predators can have large impacts (see Conservation below).

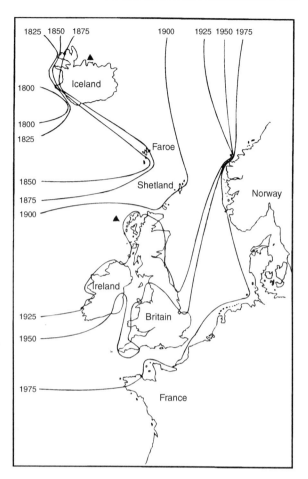

Figure 3 The spread of northern fulmars in the eastern Atlantic Ocean. Contours are at 25-year intervals. The triangles show the position of the two preexpansion colonies at Grimsey, north of Iceland and St. Kilda, west of Scotland. (Redrawn from Williamson (1996) with permission.)

Changes in Distribution

Most petrel species have fairly stable geographical distributions. However, one major exception to this is the northern fulmar *Fulmarus glacialis*, which has undergone a massive range expansion in the northeast Atlantic Ocean during the past 150 years (**Figure 3**). Historically, its breeding distribution was restricted to Arctic regions, with only two fulmar colonies in the temperate North Atlantic, one on the island of Grimsey about 40 km north of Iceland, the other at St Kilda, about 50 km west of NW Scotland. Fulmars spread to the Westman Islands off the south coast of Iceland sometime between 1713 and 1757, then to Faroe and Britain, colonizing the island of Foula, Shetland, in 1878. By 1975 their range included France, Germany (Helgoland), and Norway, and there are now colonies in Greenland, Labrador, and Newfoundland. The most likely cause of this spread was the appearance of a novel food supply in the form of offal and discards, first from Arctic whalers then from industrial and commercial trawlers. In keeping with this view, fulmars at high Arctic colonies appear less reliant on fishery discards than individuals at more southerly colonies. However, the latter also consume many other types of prey which they catch for themselves. These include fast-growing planktonivorous fish such as sand eels *Ammodytidae* and sprat *Sprattus sprattus*, which increased during the second half of the twentieth century following overfishing of predatory fish such as herring *Clupea harengus* and mackerel *Scomber scombrus*. Commercial fishing may thus have benefited fulmars in recent years almost as much by causing these dramatic changes in marine ecosystem structure as by the direct provision of discards.

Another species currently expanding its geographical range is the Laysan albatross *Diomedea immutabilis*. Following increases in populations on small islands in Hawaii, this species has established (or possibly reestablished) colonies on the main Hawaiian Islands since 1970 and has recently founded a number of new colonies in western Mexico. Two species undergoing lesser range expansions are the Manx shearwater *Puffinus puffinus*, which spread from NW Europe to establish a breeding colony in Newfoundland during the 1970s, and the shy albatross *D. cauta* which recently colonized the Crozet Islands, far from its nearest colony in the Bass Strait, Australia.

Food and Foraging

With the exception of diving-petrels, which forage close inshore, all petrels are pelagic feeders. During the breeding season, some albatrosses may travel up to 3000 km or more on a single foraging trip lasting several days, while trips by northern fulmars may be as short as 6 hours. Most species exploit a variety of epipelagic fish, squid, and crustacea, with a greater reliance on fish and squid by most albatrosses, shearwaters, fulmars, and gadfly petrels (most of the latter relying mainly on squid), and a greater reliance on crustacea by storm petrels, diving-petrels, and prions. In the Southern Ocean, a number of petrels prey extensively on the swarming crustacea *Euphausia superba* and *E. chrystallorophias* (krill). Most petrels exploit carrion when available, especially discards from industrial and commercial fishing vessels. Some species, notably northern fulmars, also include other prey such as polychaete worms and medusae in their diet at some colonies. Diets can differ markedly between breeding and nonbreeding seasons, especially in migratory species. For instance,

short-tailed shearwaters *Puffinus tenuirostris* consume a mixture of krill, fish, and squid during the breeding season in the Southern Ocean, but during the nonbreeding season in Alaska they prey almost exclusively on the euphausiid crustacean *Thysanoessa raschii*. In most petrels the diets of males and females are similar. However, in giant petrels, *Macronectes giganteus* and *M. halli*, males primarily scavenge for carrion from penguin and seal colonies while females feed primarily at sea.

About 80% of petrel species obtain food by seizing it from the surface of the water, usually while in flight but sometimes while swimming. Other species dive below the surface and some pursue prey underwater, sometimes at great depth (up to 71 m in short-tailed shearwaters). Feeding often occurs in association with dolphins, tuna (*Thunnus* spp.), and other marine predators that force prey toward the surface. Some species also forage extensively at tidal fronts, or at polynyas in polar regions, where prey tend to aggregate.

Stomach Oil

With the exception of diving-petrels, most petrels alter the chemical composition of captured prey during transport to the nest by differential retention of the aqueous and lipid fractions of the digesta within the proventriculus: following liquefaction of the food, the denser aqueous fraction passes into the duodenum first, leaving a lipid-rich liquid termed stomach oil in the proventriculus. Stomach oil has a much higher caloric density than the prey (up to 30 times higher in some cases) and may be essential in some species to allow adults to carry enough energy back to the chick. The extent to which adults form stomach oil increases with foraging range and trip duration.

Breeding Ecology

Petrels are socially monogamous and form long-lasting pair bonds, although they do switch partners on occasion, particularly after an unsuccessful breeding attempt. Pairing usually takes place at the colony and involves elaborate visual displays in diurnal species or auditory displays in nocturnal species. Young prebreeding birds return to the colony progressively earlier each year to display, pair off, and establish a breeding territory. This process takes several years and initial breeding attempts mostly fail.

In established breeders, the period between arrival at the colony each year and egg-laying is generally 30–40% of the whole reproductive cycle, except in albatrosses where it comprises 7–14% of the total. Sedentary species, which occur in all families except the Diomedeidae, may also visit the colony outside the breeding period. Precopulatory behavior may be complex (as in albatrosses) or simple (as in diving petrels) but generally involves mutual preening of the head, nape, cheek, and throat. Copulation takes place on the ground, usually at the nest. Individuals may copulate many times and females sometimes solicit or accept copulations from males other than their partner, although these seldom result in extra-pair paternity. For instance, in a study of northern fulmars, females copulated up to 54 times with their male partner and up to 17 times with an extra-pair male. The final copulation was always with the pair male and there was no evidence of any extra-pair paternity. In all petrels, after copulating there is a period of several days to several weeks, termed the prelaying exodus, when some or all of the established breeders return to sea during egg formation. This may allow birds access to highly productive waters distant from the colony. Females stay away longer than males and in some species only the females leave.

After returning to the colony, the female almost immediately lays a single egg weighing 5–30% of her own body mass (heavier than in other birds that lay a single-egg clutch) with a relatively large yolk. The embryo within the egg grows slowly, resulting in a long incubation period (about 39–79 days depending largely on body mass). The male is responsible for the first long incubation shift and thereafter the two parents alternate between incubating the egg and foraging at sea, at average intervals of about 1–20 days (generally longer in larger species but shortest in diving-petrels and longest in some gadfly petrels and albatrosses of intermediate body mass; **Figure 4**). The two sexes usually spend about the same amount of time incubating, although in some species the male spends longer. In both cases, incubation shifts by both sexes shorten toward the end of incubation, which increases the probability that the incubating bird has food for the chick when it hatches.

If for whatever reason the foraging bird does not return sufficiently quickly, the incubating bird may head to sea before its partner returns, leaving the egg unattended. Such eggs are vulnerable to predators, especially in surface-nesters, but they can remain viable for many days without further incubation (up to 23 days in Madeiran storm-petrel *Oceanodroma castro* at the Galapagos Islands). This resistance to adverse effects of chilling may be a major adaptation of petrels to pelagic foraging, permitting them to nest far from their main food sources. However, cooling of the embryo does retard growth, resulting in a delay in hatching.

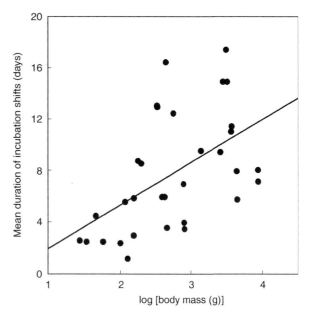

Figure 4 The relationship between body mass and incubation shift length in 32 species of petrel. (Data from Table 14.1 and 15.4 in Warham (1990).)

Like embryos before hatching, petrel chicks grow slowly for their size. This results in petrels having very long breeding seasons, with two extremes for the period from egg-laying to chick-fledging being 125 days in snow petrels *Pagodroma nivea* and 380 days in wandering albatrosses. Petrel chicks also accumulate very large quantities of nonstructural body fat during posthatching development (up to 30% of body mass in northern fulmars). This results in chicks attaining peak body masses far in excess of adult body mass (up to 170% of adult mass in yellow-nosed albatross *Diomedea chlororhynchos*) before losing mass prior to fledging.

Newly hatched petrel chicks are unable to regulate their own body temperatures and they are brooded more or less continuously for the first few days after hatching. Small chicks also have limited gut capacity and so, during this period, they are fed small meals several times a day by the attending parent. Once chicks gain thermal independence, they are left unattended for most of the time while both parents forage simultaneously at sea. This change in parental attendance is accompanied by an increase in meal size and a decrease in feeding frequency. For most of the nestling period, parents deliver meals weighing 5–35% of adult body mass (proportionally smaller in heavier species) to the chick. Overall feeding frequency, resulting from provisioning by both parents, is generally between one meal every two days and two meals per day (higher in fulmars and diving petrels than in other species). Feeding frequency then declines at the end of the nestling period, so that in most cases total food delivery is insufficient to meet the chick's nutritional requirements (**Figure 5**). Chicks thus enter negative energy balance and lose mass as they presumably deplete their fat stores. However, northern fulmar chicks continue to accumulate fat until fledging, and their prefledging drop in body mass is due entirely to the loss of water from embryonic tissues as they attain full size and functional maturity. Grey-headed albatrosses *Diomedea chrysostoma* show a similar pattern, and further data are required to determine how prefledging declines in body mass relate to changes in body fat in other species.

Large fat stores were until recently thought to provide chicks with an energy reserve to tide them over long intervals between feeds. However, in many species, the intervals between feeds are too short to account for the quantities of fat stored, and the combined cumulative effects of variability in both feeding frequency and meal size are probably more important than long intervals *per se*. Fat stores may also be important to chicks after fledging, while they

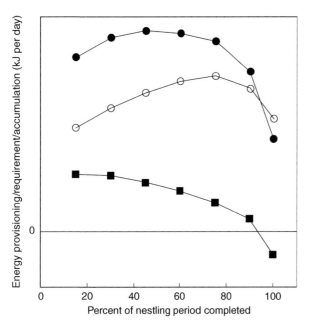

Figure 5 A generalized scheme for the relationship between food provisioning and growth of petrel chicks. Solid circles indicate mean energy provisioning rate (more or less independent of chick age except at the end of the nestling period): open circles indicate mean daily energy requirement (determined by body mass and structural (lipid-free) growth rate, and reaching a maximum after about 75% of the nestling period): solid squares indicate net rate of accumulation of energy within the body (about 70% of the difference between provisioning rate and requirement when growth is positive; >100% of the difference when growth is negative). N.b. In some species, despite a drop in energy provisioning rate at the end of the nestling period, supply does not fall below energy requirement, and so chicks continue to accumulate energy, principally as body fat, until fledging.

learn to forage for themselves, and this view is supported by the fact that northern fulmars have larger fat stores at fledging than at any previous time. In addition to body fat, petrel chicks also store large quantities of stomach oil, which acts as a convenient energy store and is metabolically more efficient than converting the oil into body fat.

Reproductive Effort

Petrels are long-lived and so they need to balance the requirements of current and future reproduction in order to maximize their lifetime reproductive output. During chick-rearing, they need to judge how much energy to invest in the chick without impairing their own ability to breed again. In species that make long foraging trips and feed their chick comparatively infrequently, this seems to be achieved by adults delivering food according to an intrinsic rhythm that is sensitive to changes in the parent's body condition but not to changes in the chick's nutritional requirements. Other species that make shorter trips on average appear to monitor the chick's nutritional status and will reduce subsequent food delivery to a well-fed chick. However, experimental evidence indicates that they either cannot or do not increase food delivery to a chick in poor condition. Moreover, experimental manipulations that increase the work that adults need to perform to obtain food generally result in a reduction in the rate of food provisioning of the chick rather than a decline in the parent's nutritional reserves. In some species, adults also adopt a dual foraging strategy, in which comparatively short trips undertaken to obtain food for the chick are interspersed with less frequent long trips on which adults obtain food for themselves. The switch between short and long trips seems to be determined by a threshold body condition, below which adults forage only for themselves. In some cases the areas of ocean visited on short and long trips are far apart. For instance, short-tailed shearwaters breeding in SE Australia forage comparatively close to the colony to obtain food for their chick but obtain food for themselves in the highly productive waters of the Polar Frontal Zone at least 1000 km away.

Conservation

Petrels are adversely affected by a wide range of factors related to human activities, including ingestion of plastic particles and exposure to a wide variety of marine pollutants, but, historically, introduction of alien mammals to remote breeding colonies has been perhaps the greatest threat to petrel populations, with the greatest impacts being due to introductions of cats *Felis catus* and rats *Rattus* spp. Small species of petrels have been particularly affected, but even large species such as albatrosses have not been immune. Eradication programs have been developed to remove introduced mammals from a number of islands where petrels breed, especially in the southern hemisphere, and these have met with some success. For instance Marion Island, in the sub-Antarctic Indian ocean, was cat-free prior to 1949, when five cats were introduced as pets. By 1975 the cat population had risen to about 2000 individuals consuming 450 000 burrowing petrels annually. An eradication program was started in 1977, when the cat population was over 3000 individuals, and by 1993 there were no signs of live cats on the island. The elimination of cats resulted in a marked increase in the breeding success of several species of petrel, although a combination of small populations and delayed maturity has resulted in only slow population recoveries. It should also be noted that on islands that have been colonized by more than one species of predator, eradication of one species could lead to a disproportionate increase in other species, to the detriment of petrels. This is likely to be a particular problem with eradication of cats from islands that have also been colonized by rats.

One of the major current threats to petrel populations, especially in the southern hemisphere, is longline fishing using baited hooks to catch demersal and pelagic fish, particularly tuna and Patagonian toothfish *Dissostichus eleginoides*. Petrels can be drowned after striking at baited hooks and, while the rate of by-catch is generally very low, the large number of hooks set means that total mortality can be sufficient to reduce population sizes of some species. For instance, the Japanese pelagic longline fishery for southern bluefin tuna *T. maccoyi* deployed up to 100 million hooks or more annually during the 1990s. In Australian waters, the average rate of sea bird by-catch from this fishery was 1.5 birds per 10 000 hooks set, but this resulted in an annual mortality of up to 3500 birds per year. Longline fishing operations have been implicated in serious population declines of a number of albatrosses, including wandering albatross, grey-headed albatross, black-browed albatross *Diomedea melanophrys*, yellow-nosed albatross, and light-mantled sooty albatross *Phoebetria palpebrata*. Some populations have declined by as much as 90% and are facing imminent extinction if current catch rates are not reduced.

See also

Krill. Seabird Foraging Ecology. Seabird Migration. Seabird Population Dynamics.

Further Reading

Brooke M (1990) *The Manx Shearwater*. London: Academic Press.

Croxall JP and Rothery P (1991) Population regulation of seabirds: implications of their demography for conservation. In: Perrins CM, Lebreton J-D, and Hirons GJM (eds.) *Bird Population Studies. Relevance to Conservation and Management*, pp. 272–296. Oxford: Oxford University Press.

Fisher J (1952) *The Fulmar*. London: Collins.

Robertson G and Gales R (eds.) (1998) *Albatross Biology and Conservation*. Chipping Norton, Australia: Surrey, Beatty Sons.

Tickell WLN (2000) *Albatrosses*. Yale: Yale University Press.

Warham J (1990) *The Petrels. Their Ecology and Breeding Systems*. London: Academic Press.

Warham J (1996) *The Behaviour, Population Biology and Physiology of the Petrels*. London: Academic Press.

Weimerskirch H and Cherel Y (1998) Feeding ecology of short-tailed shearwaters: breeding in Tasmania and foraging in the Antarctic? *Marine Ecology Progress Series* 167: 261–274.

Williamson M (1996) *Biological Invasions*. London: Chapman and Hall.

SPHENISCIFORMES

L. S. Davis, University of Otago, Dunedin,
New Zealand

What is a Penguin?

Penguins, with their upright stance and dinner-jacket plumage, constitute a distinct and unmistakable order of birds (Sphenisciformes). Granted there are a few embellishments here and there – the odd crest, a black line or two on the chest – but otherwise, penguins conform to a very conservative body plan. The design of penguins is largely constrained by their commitment to an aquatic lifestyle. Penguins have essentially returned to the sea from which their ancestors, and those of all tetrapods, came. In that sense, they share more in common with seals and sea turtles than they do with other birds. Their spindle-shaped bodies and virtually everything about them have evolved in response to the demands of living in water (**Table 1**).

The loss of flight associated with their aquatic makeover is the penguins' most telling modification. While isolated examples of flightlessness can be found in virtually all other groups of waterbirds, penguins are the only group in which all members cannot fly. Among birds generally, they share that distinction only with the ratites (the kiwis, ostriches, emus, and their ilk), where flight has been sacrificed for large size and running speed.

Despite earlier claims to the contrary, it is clear that penguins have evolved from flying birds. The evidence from morphological and molecular studies suggests that penguins are closely related to loons (Gaviiformes), petrels, and albatrosses (Procellariiformes), and at least some families of the Pelicaniformes, most notably frigate-birds. Despite this, the exact nature of the relationship between penguins and these groups remains unresolved: at the moment it would seem to be a dead heat between loons and petrels as to which group is the sister taxon of penguins (**Figure 1**). (On the surface, loons may seem strange candidates to be so closely allied to penguins – penguins are found in the Southern Hemisphere, loons in the Northern Hemisphere; penguins are wing-propelled divers, loons are foot-propelled divers. However, it seems that loons, or their ancestors, were wing-propelled divers in their past.)

If the relationship of penguins to other birds seems confusing and controversial, the relationships among penguins themselves are no less so. Penguins are confined to the Southern Hemisphere and the distribution of fossilized penguin bones discovered to date mirrors their present-day distribution. Fossils have been found in New Zealand, Australia, South America, South Africa, and islands off the Antarctic Peninsula. The oldest confirmed fossil penguins have been described from late Eocene deposits in New Zealand and Australia, dating back some 40 million years. However, fossils from Waipara, New Zealand, unearthed from late Paleocene/early Eocene deposits that are about 50–60 million years old, represent possibly the earliest penguin remains. These have still to be described fully, but they show a mixture of attributes from flying birds and those of penguins (the bones are heavy and nonpneumatic; the wing bones are flattened in a way that is consistent with being a wing-propelled diver). The Waipara fossils may very well be near the base of the penguin radiation, when the transition was being made from flyer to swimmer. In any case, by 40 million years ago, penguins were already very specialized, in much the same way as modern penguins, for underwater swimming.

Diving versus Flying

There is a trade-off between diving performance and flying. Even so, it would be wrong to conclude that flying birds, as a matter of course, cannot dive well; that somehow the riches that the sea has to offer are denied to them much beyond the surface. The truth is that some flying birds – for example, the diving-petrels and, especially, the alcids (auks, auklets, and murres) – can literally fly underwater as well as above. Although they are not closely related to penguins, auks are often considered to be the Northern Hemisphere's ecological equivalent of penguins, and in proportion to their size, auks can dive as deeply as many penguins. Nevertheless, the requirements for efficient flight – light bodies and flexible wings with a large surface area/small wing loadings – are not the same as those needed for diving efficiency – large, heavy bodies and stiff, powerful wings. Wingloadings are a measure of the body mass of the animal relative to the surface area of the wing. They give an indication of the lift that can be provided by the wing. If the linear measurements of a bird were simply scaled up, because body

Table 1 Some adaptations of penguins for an aquatic lifestyle

Attribute	How modified	Purpose
Wings	Shorter, more rigid. Flipper acts as paddle/propeller. Feathers much reduced to decrease resistance	Increase diving capacity. By eliminating flight, the birds no longer need to keep body light.
Body shape	Spindle shape. Very low coefficient of drag	Reduce drag and increase efficiency of swimming. Water is more dense and offers more resistance than does air.
Bones	Nonpneumatic	Unlike flying birds, which have spaces in their bones to make them light (pneumatic), penguins have solid bones to increase strength and density.
Feathers	Short, rigid and interlocking	Feathers trap air beneath them, creating a feather survival suit that provides most of the insulation necessary for a warm-blooded animal to exist in water.
Fat	Subdermal layer of fat	Whereas flying birds cannot afford to carry too much fat, the subdermal layer of fat in penguins contributes a little to their insulation and also enables them to endure long periods of fasting on land necessary for incubation and molting.
Coloration	Dark back, light belly	To aid concealment in the open ocean, like many pelagic predators, penguins have a dark back, so that they merge with the bottom when viewed from above, and a light belly, so that they merge with the surface when viewed from below.
Legs	Placed farther back. Very short tarsometatarsus. Upright stance on land	To reduce drag in water. Feet act as rudder.
Eyesight	Variable	The eye is able to be altered to accommodate refractive differences when moving between water and land.
Circulation	Countercurrent blood system	Allows penguins to reduce heat loss in the water or very cold environments, while aiding heat dissipation when hot.

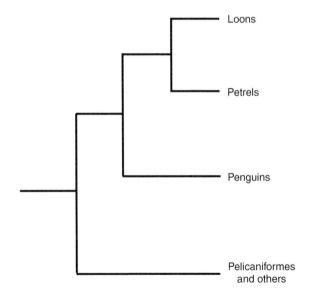

Figure 1 Nearest living relatives of penguins (adapted from Davis and Renner (in prep.)).

mass is a function of volume, the loading on the wing would become greater. This means that, to generate the same lift, larger birds must have wings with a disproportionately larger surface area.

Birds with relatively light bodies expend energy simply counteracting the buoyancy of their bodies in water. The time a bird can stay underwater and the depth to which it can travel are physiologically related to the size of the bird, providing another advantage to large birds. But as body mass increases, the surface area of wings needs to increase to a much greater extent just to maintain the same lift. There appears to be a cut-off point at about 1 kg beyond which it is not possible to both dive and fly efficiently. Indeed, auks that exceeded this size threshold, such as the extinct great auk, were flightless like penguins.

Logically, then, it seems that the transition from flighted to flightless would have taken place in birds around the 1 kg threshold, which is roughly the size of today's little penguins (*Eudyptula minor*) and consistent with the size of the smallest fossil penguins. However, once penguins were freed of the need to balance the requirements of flight with the requirements for underwater diving, they radiated rapidly and were able to attain much greater sizes. Many fossil penguins were larger than living penguins, with the tallest being up to 1.7 m and well over 100 kg.

Living Penguins

The species living today represent but a small remnant of the penguin diversity from times past. Exactly how modern penguins relate to the fossil

Table 2 Species of living penguins[a]

Species	Scientific name	Principal location	Latitude (°S)	Body mass (kg)	Foraging type	Migratory	Nest type	Diet
Galapagos	Spheniscus mendiculus	Galapagos	0	2.1 (M), 1.7 (F)	☐	☐	☐	☐
Humboldt	Spheniscus humboldti	Peru, Chile	5–42	4.9 (M), 4.5 (F)	☐	☐	☐	☐
African	Spheniscus demersus	South Africa	24–35	3.3 (M), 3.0 (F)	☐	☐	☐	☐
Magellanic	Spheniscus magellanicus	Argentina, Chile, Falklands Is.	29–54	4.9 (M), 4.6 (F)	■	■	☐	☐
Little	Eudyptula minor	Australia, New Zealand	32–47	1.2 (M), 1.0 (F)	□	□	☐	□
Yellow-eyed	Megadyptes antipodes	New Zealand, Auckland Is.	46–53	5.5 (M), 5.1 (F)	☐	☐	□	□
Fiordland	Eudyptes pachyrhynchus	New Zealand	44–47	4.1 (M), 3.7 (F)	■	■	□	□
Snares	Eudyptes robustus	Snares Islands	48	3.3 (M), 2.8 (F)	■	■	■	□
Erect-crested	Eudyptes sclateri	Antipodes Is., Bounty Is.	47–49	6.4 (M), 5.4 (F)	■	■	■	□
Rockhopper	Eudyptes chrysocome	Subantarctic	37–53	2.5 (M), 2.3 (F)	■	■	■	□
Macaroni/ Royal	Eudyptes chrysolophus	Subantarctic	46–65	5.2 (M), 5.3 (F)	■	■	■	□
Gentoo	Pygoscelis papua	Subantarctic, Antarctic	46–65	5.6 (M), 5.1 (F)	☐	□	■	■
Chinstrap	Pygoscelis antarctica	Antarctic	54–69	5.0 (M), 4.8 (F)	■	■	■	■
Adelie	Pygoscelis adeliae	Antarctic	54–77	5.4 (M), 4.8 (F)	■	■	■	■
King	Aptenodytes patagonicus	Subantarctic	45–55	16.0 (M), 14.3 (F)	■	■	■	☐
Emperor	Aptenodytes forsteri	Antarctic	66–78	36.7 (M), 28.4 (F)	■	■	■	□

Key

	Foraging	Migratory	Nest	Diet
☐	inshore	resident	burrow	fish
□	either	either	forest	fish, cephalopods, crustaceans
■	offshore	migratory	open	crustaceans

[a]M, Male; F, Female.

penguins is not at all clear, as most of the extant (living) genera show up in the fossil record only relatively recently (within the last 3 million years or so). While there is argument about the precise number of species living today (most authorities list between 16 and 18 species), there is agreement that the extant penguins fall into six distinct genera (Table 2).

Sphensicus

In contrast to the popular misconception that penguins are creatures of the snow and ice, representatives of the *Spheniscus* penguins breed in tropical to temperate waters, with one species, the Galapagos penguin (*S. mendiculus*), breeding right on the equator.

There are four species: the African (*S. demersus*), Humboldt (*S. humboldti*) and Magellanic penguins (*S. magellanicus*), in addition to the Galapagos penguin. They were the first penguins to be discovered by Europeans. After Vasco de Gama's ships sailed around the Cape of Good Hope in 1497, they encountered African penguins in Mossel Bay, South Africa. However, the account of this discovery was not published until 1838 and the first announcement to the world at large concerning penguins was to come from another famous voyage, that of Magellan's circumnavigation of the globe (1519–1522). A passenger, Pigafetta, described in his diary a great number of flightless 'geese' seen on two islands near Punta Tombo, Argentina, home to a large concentration of Magellanic penguins.

Figure 2 Magellanic penguins. (Photograph: L.S. Davis).

The *Spheniscus* penguins are characterized by black lines on their chests and distinct black and white bands on their faces (as a consequence, they are sometimes referred to as the 'ringed' or 'banded' penguins) (**Figure 2**). Their faces are also distinguished by having patches of bare pink skin, which help to radiate heat. Warm-blooded animals like penguins and seals must be well insulated if they are to maintain a constant body temperature when in the water, because the sea acts as a huge heat sink; but this creates problems of overheating for penguins when ashore, especially in the hot climates that the *Spheniscus* penguins inhabit. For this reason, all these species nest in burrows or, where available, in caves, in clefts in rocks, or under vegetation.

Penguins, with their limited foraging range due to their being flightless, require a consistent and good food supply near to their breeding areas. Tropical waters are typically not very productive and are unable to support the dense swarms of fish or krill on which penguins depend. This has probably acted as a barrier, restricting penguins to the Southern Hemisphere. The *Spheniscus* penguins are able to breed as far north as they do because of wind-driven upwelling of nutrient-rich water and because the cold-water Benguela and Humboldt Currents, which run up the sides of southern Africa and South America, bring nutrient-rich waters from farther south. *Spheniscus* penguins feed mainly on small pelagic schooling fish such as sprats and anchovies. The top mandible is hooked at its far end, which helps the penguin to catch and hold fish.

With the exception of the Magellanic penguin, which is at the southern extreme of their range, the *Spheniscus* penguins are inshore foragers making foraging trips of relatively short duration (1–2 days) throughout the breeding period. They lay two similar-sized eggs. Breeding can occur in all months of the year (African and Galapagos), at two peak times

(Humboldt), or just once (Magellanic): a pattern that roughly corresponds with the more pronounced seasonality the farther south they breed.

Eudyptula

Little penguins (*Eudyptula minor*) are the smallest living penguins and are found only in Australia (often called fairy penguins) and New Zealand (often called little blue or blue penguins). While most authorities recognize only the single species, whether this really is a monotypic genus is a matter of debate. Six subspecies have been described and it has been suggested in some quarters that at least one of these, the white-flippered penguin (*E. m. albosignata*), is deserving of separate species status in its own right. However, the latter freely hybridize with other subspecies and analyses of isozymes and DNA do not support such a conclusion at this stage. More extensive genetic studies of this genus are warranted.

Apart from their small size, little penguins are characterized by being nocturnal (coming ashore after dark) and nesting in burrows, although they will breed in caves or under suitable vegetation in some locations. Morphologically they are quite similar to *Spheniscus* penguins and also feed on small schooling fish. The plumage of all penguins conforms to the dark top and white undersides common to pelagic marine predators, but in little penguins the plumage on the backs has more of a blue hue compared to the blackish coloration of other penguins (**Figure 3**).

Patterns of breeding are very variable throughout their range in both the duration of the breeding period and the number of clutches per year (either one or two). Little penguins lay clutches of two similarly-sized eggs.

Typically they are described as inshore foragers (with foraging trips lasting 1–2 days), but there is evidence of plasticity in their feeding strategies. In some locations or in years of poor food supply, they may feed considerably farther offshore, with a concomitant increase in the duration of their foraging trips (up to 7 days or more).

Megadyptes

The representative of another monotypic genus, the yellow-eyed penguin (*Megadyptes antipodes*), is often touted as the world's rarest penguin. (The Galapagos penguin may well be rarer, and the Fiordland penguin is not much better off.) It breeds on the south-east coast of New Zealand and the sub-Antarctic Auckland Islands. The yellow-eyed is a medium-sized penguin, distinguished by having pale yellow eyes and a yellow band of plumage that runs

Figure 3 Little penguin chick, almost ready to fledge, begging for a meal from its parent. (Photograph: M. Renner.)

Figure 4 Yellow-eyed penguin on nest. (Photograph: J. Darby.)

where opportunities for mate switching and extra-pair copulations are rife.

Another factor contributing to the apparent faithfulness of yellow-eyed penguins is that adults remain more-or-less resident at the breeding site throughout the year, even though breeding occurs only between September to February. They are inshore foragers, feeding mainly on fish and cephalopods, and typically foraging trips are from 1 to 2 days during both incubation and chick rearing. The clutch consists of two equal-sized eggs and there appears to be little in the way of evolved mechanisms for brood reduction: if both chicks hatch, they have an equally good chance of fledging.

Eudyptes

In contrast, the members of the genus *Eudytptes*, which is closely related to *Megadyptes* and may have been derived from ancestors that were very like the yellow-eyed penguin, are famous for their obligate brood reduction (i.e., they lay two eggs but only ever fledge one chick). Collectively, the six Eudyptid species are known as the crested penguins. They are distinguished by plumes of yellow or orange feathers, of varying length, that arise above their eyes like out-of-control eyebrows (**Figure 5**).

For the most part, they are penguins of the sub-Antarctic, although Fiordland penguins (*Eudyptes pachyrhychus*) breed on the south-west corner of New Zealand (though some would argue that conditions there are not too dissimilar from the sub-Antarctic, especially as the Fiordland penguin breeds during winter). Other crested penguins also have quite restricted distributions: the Snares penguin (*E. robustus*) breeds only on the Snares Islands south of New Zealand; the erect-crested penguin (*E. sclateri*) breeds on the Bounty and Antipodes Islands, also near New Zealand; and the royal penguin (*E. schlegeli*) breeds only on Macquarie Island.

around the back of its head from its eyes (**Figure 4**). But perhaps its most unique characteristic is a behavioral one: it nests under dense vegetation and typically nests are visually isolated from each other. Whether this visual isolation is an absolute requirement or simply a consequence of their need for dense cover to escape the sun's heat, there is little doubt that yellow-eyed penguins are among the least overtly social of the penguins. Although they nest in loose colonies, individual nests can be from a few meters to several hundred meters apart. Perhaps partly as a consequence of this, mate fidelity is higher than in the cheek-by-bill colonies of other species

Figure 5 Erect-crested penguins. (Photograph: L.S. Davis.)

However, there is a considerable body of evidence and opinion that argues that the royal penguin is really only a pale-faced subspecies of the more ubiquitously distributed macaroni penguin (*E. chrysolophus*). The rockhopper penguin (*E. chrysocome*), the smallest of the crested species, has a circumpolar distribution, but there are three recognized subspecies and preliminary analysis of DNA would suggest that they are at least as distinct as royal penguins are from macaroni penguins.

All the crested penguins are migratory, spending two-thirds of the year at sea (their whereabouts during this period are largely unknown) and, with the exception of the Fiordland penguin, returning to their breeding grounds in spring. Despite laying a single clutch of two eggs, they only ever fledge one chick. Moreover, the eggs are of dramatically different sizes, with the egg laid second being substantially bigger than that laid first (this form of egg-size dimorphism is unique among birds). In those species with the most extreme egg-size dimorphism (erect-crested, royal/macaroni), the second egg can be double the size of the first egg (**Figure 6**). Furthermore, although the laying interval between the first and second eggs is the longest recorded for penguins (4–7 days), the chick from the large second-laid egg usually hatches first.

The evolutionary reasons for this bizarre breeding pattern have long been matters of debate and, it is fair to say, have yet to be determined satisfactorily. Crested penguins are probably unable adequately to feed two chicks. They are all offshore foragers (feeding on a mixture of fish, cephalopods, and swarming crustaceans). The logistics of provisioning chicks from a distant food source probably make it too difficult to bring back enough food to fledge two chicks; but this idea has not been tested properly yet. While in some species it can be argued that the small

first egg provides a measure of insurance for the parents in case the second egg is lost (or the chick from the second-laid egg dies), this cannot be a universal explanation. Studies of erect-crested and royal penguins have found that first-laid eggs are usually lost before or on the day the second egg is laid and it has even been suggested that the parents may actively eject the first eggs. While the evidence for egg ejection must be viewed as equivocal, first eggs cannot provide much of an insurance policy in these species. In any case, none of this explains why it is the first egg that is the smallest and that usually does not produce a chick.

For penguins breeding in higher latitudes, overheating when on land is less of a problem, permitting them to breed in colonies in the open. Some, like macaroni penguins, form vast colonies, while Fiordland penguins nest in small colonies in forest or caves.

Pygoscelis

The three members of this genus are the classic dinner-jacketed penguins of cartoons. They all breed in the Antarctic to varying degrees. The gentoo penguin (*P. papua*) consists of two subspecies, one of which breeds on the Antarctic Peninsula and associated islands (*P. p. ellsworthii*) and the other (*P. p. papua*) on several sub-Antarctic islands. Chinstrap penguins (*P. antarctica*) breed below the Antarctic Convergence, mainly on the Antarctic Peninsula, but also on a few sub-Antarctic islands. The Adélie penguin (*P. adeliae*), with very few exceptions (Bouvetoya Island), breeds only on the Antarctic continent and its offshore islands.

The three species are medium-sized penguins with white undersides and black backs. They differ most obviously in the markings on their faces. Adélie penguins have black faces and throats, and prominent white eye-rings. The proximal ends of the mandibles are covered by feathers, giving the impression that the bill is short (**Figure 7**). Gentoo penguins also have black faces and white eye-rings, but have white patches above the eyes and a bright orange-red bill. Chinstrap penguins have white faces with a black crown, and a black line running under the chin, from ear-to-ear, producing the effect from which they derive their name.

All three species breed in colonies in the open and nest just once per year during the austral summer. A clutch of two eggs is laid, with the first slightly larger than the second. The northern subspecies of gentoo penguins (*P. p. papua*) lay their eggs in nests made from vegetation. They are inshore foragers and remain resident at the breeding site throughout most

Figure 6 Extreme egg-size dimorphism in erect-crested penguin eggs. (Photoraph: L.S. Davis.)

Figure 7   Adélie penguin with chick. (Photograph: L.S. Davis.)

Figure 8   King penguin chicks on the Falkland Islands. (Photograph: L.S. Davis.)

of the year. They nest in colonies, but the location of these colonies can shift from season to season as the area becomes soiled. Despite this, mate fidelity is high. In contrast, the other Pygoscelid penguins are migratory, make their nests out of stones, and the nest sites are more permanent. Chinstrap and Adélies are offshore foragers, with foraging trips commonly lasting two or more weeks during incubation. All the Pygoscelid penguins feed on a type of swarming crustacean known as krill (*Euphausia* spp.).

Aptenodytes

This genus is comprised of the two largest of the living penguins: emperor (*A. forsteri*) and king (*A. patagonicus*). Apart from their large size, these species are distinguished by their bright yellow (emperor) or orange (king) auricular patches and the fact that they lay a single large egg. Both are offshore foragers, and it seems unlikely that they could possibly bring back enough food to successfully rear two of their very large chicks (**Figure 8**). Each of them has a truly remarkable pattern of breeding.

Emperor penguins are notable for breeding during the heart of the Antarctic winter. They breed on ice in dense colonies: there are no nests, the single egg (or small chick) is carried upon the parent's feet. The male incubates the egg entirely by himself, while the female goes off to feed for about 2 months. Together with the period they are at the colony during courtship, that means the males go without food for up to 3.5 months. Emperor penguins feed on fish, cephalopods, or crustaceans, with the relative importance of each varying according to the location or time. To help reduce heat loss during the Antarctic winter, the penguins stand in a tight huddle.

In some ways the breeding of king penguins is even more remarkable. It takes them over a year (14–16 months) to provide enough food for their chicks to reach a sufficient size to fledge. As a consequence, king penguins can breed only twice every three years. During the winter months the parents must forage at great distances and the chicks are left to fast, sometimes for as long as 5 months! In anyone's book, that is a long time between dinners. The king penguin breeds on sub-Antarctic islands. It does not build a nest, carrying the large single egg on its feet like an emperor penguin, but it does defend a territory within the colony. Some colonies can be enormous, with 100 000 or more birds. Mate fidelity is low in both king and emperor species. During incubation, foraging trips can last 2 weeks or more. They feed largely on Myctophid fish.

Life in the Sea

Tests using life-size plastic models of penguin bodies have shown that they have lower coefficients of drag than those for any cars or planes that scientists and engineers have been able to manufacture. When swimming, penguins move their flippers to generate vortex-based lift forces, in essence creating a form of underwater jet propulsion. The usual underwater traveling speed of most penguins is around 2 m s^{-1}. However, penguins are capable of high-speed movements known as 'porpoising', whereby they clear the water surface to breathe. Porpoising speeds can be 3 m s^{-1} or higher. The speed at which a penguin must be traveling to porpoise is related to its body size, and it is likely that for large penguins, like emperor and king penguins, porpoising requires just too much effort. All smaller penguins use porpoising to escape from predators. Porpoising is energetically expensive compared to underwater swimming and, consequently, inshore foraging penguins tend not to use porpoising simply for traveling to and from their foraging areas. For offshore foragers, however, the time savings gained from porpoising may outweigh the energetic costs, and it seems likely that porpoising may sometimes, at least, be used in these species as an expedient means of travel and not just an escape reaction.

Maximum dive depth and duration are correlated with body mass. Emperor penguins have been recorded diving to over 500 m and all but little penguins are readily capable of diving to depths of 100 m or more. Even so, most foraging takes place at much shallower depths: penguins are visual predators and so light levels in the water column will limit their effectiveness as hunters (albeit bioluminescent prey, such as squid, enable some penguins to hunt at very low light levels, either at night or at great depth). The prey of penguins tends to vary with latitude and phylogenetic grouping (**Table 2**).

A Life in Two Worlds

An important determinant of the breeding success of penguins is how they manage to balance the time at sea needed for foraging against the time required ashore for breeding and molting. Nesting duties are shared, and the male and female of a pair take alternating turns being in attendance at the nest or at sea foraging. The foraging penguin needs to find enough food to sustain itself, to cover the costs of getting the food and, once chicks hatch, to ensure the growth of its chicks. However, if it spends too long at sea either the partner on the nest might desert or, if the chicks have hatched, the chicks might starve to death.

For inshore foragers, which make short and frequent trips to sea, the risks of desertion and starvation are not as acute as for those that need to forage farther afield. All adult penguins are capable of enduring relatively long periods of fasting, yet there are finite limits to how long an incubating bird can live off its fat reserves (e.g., male Adelie penguins arriving at the colony at the start of the breeding season will endure just over a month without food). For many offshore foragers, foraging trips during the incubation period tend to be from 10 to 20 days (albeit female emperor penguins are away for about 2 months). The situation changes dramatically once the chicks hatch, as chicks of all species need to be fed frequently, requiring that foraging trips be relatively short during at least the early stages of chick rearing. For offshore foragers that take long foraging trips during incubation, this creates a dilemma: they must switch over to short foraging trips by the time the chicks hatch or risk losing them to starvation. There is a growing body of evidence that indicates that penguins have an internal mechanism that measures the duration of the incubation period, precipitating the early return of a foraging bird if its chicks are about to hatch. The remnants of the yolk sac provide a little insurance against a tardy parent by permitting penguin chicks to survive about 3–6 days without being fed from the time they first hatch. In the case of the emperor penguin – in which the male undertakes all the incubation – if the female has not returned by the time the chick hatches, the male is able to feed it an oesophageal 'milk' made from breaking down his own tissues even though he has not eaten himself for over 3 months.

In contrast to flying birds, which cannot afford to store much fat because of the weight, the ability of penguins to store enough fat to withstand long fasts is important not only when breeding but also when molting. Although a small amount of their insulation is provided by the subdermal layer of fat, most insulation comes from the feathers. But feathers wear and, to provide an effective barrier against heat-stealing waters, they must be renewed to maintain their integrity. This is an energetically expensive process, whereby a new suit of feathers is grown beneath the old one. During this process, which can take 2 or 3 weeks, the birds typically remain on land and are unable to eat.

Threats and Conservation

Desertions and starvations can be major causes of breeding failure for some species of penguins and the likelihood of these occurring will be exacerbated by anything that increases the time penguins must spend foraging. This makes penguins especially vulnerable to perturbations in the marine environment. Threats to penguins come from anything that reduces their food supply (causing a concomitant increase foraging times), such as commercial fisheries or pollution.

The major threat facing penguins may be from global warming. Warmer water associated with ENSO (El Niño Southern Oscillation) events can reduce the amount of upwelling of nutrients and seriously affect the availability of prey in various areas. ENSO events can be especially critical for nonmigratory species like Galapagos and yellow-eyed penguins, which rely on a persistent and steady food supply in a localized area. Breeding dates of little penguins throughout Australia and New Zealand are correlated with sea surface temperatures, with laying commencing later and fewer birds attempting to breed or being successful when water temperatures are warmer.

The commitment penguins made to the sea by becoming flightless has made them doubly vulnerable. Not only are they at risk from factors affecting their food supply, but their lack of agility on land makes them potentially easy targets for predators. It is for this reason that flightlessness in water birds has usually evolved on offshore islands or isolated places relatively free of predators. However, humans have managed to undo much of that by introducing exotic predators to many areas where penguins breed. Mustelids, rats, and feral cats have had serious impacts upon penguins when they have been introduced to places like New Zealand, South Africa, and the sub-Antarctic islands. During the nineteenth and early twentieth centuries, humans were often significant predators themselves, killing adult penguins for their oil and their skins and taking eggs for food. It has been estimated that in one area alone, over 13 million eggs were harvested from African penguins during a 30-year period. Humans also have an impact on penguins by reducing the availability of suitable habitat through harvesting guano (this is especially so of African and Humboldt Penguins in South Africa and South America, respectively) or deforestation. The latter dramatically reduced the numbers of yellow-eyed penguins breeding on mainland New Zealand: this trend has been reversed in recent years by extensive replantings.

Until fairly recently, the species inhabiting the Subantarctic and Antarctic have been able to rely on their isolation for protection. However, as humans reach ever more into these nether regions of the planet, they are no longer immune to the threats from overfishing, pollution, disturbance from tourists, introduced predators, and introduced diseases. On a positive note, however, the design of penguins has enabled them to survive, if not flourish, for over 40 million years. They have been through periods of vast climatic change in the earth's history. While they have had to conform to a design shaped by the requirements of living in water, paradoxically, this has given them a great deal of versatility with respect to the environments they can exploit on land. They are the only 100-degree birds: the only birds capable of breeding at temperatures from $-60°C$ (midwinter in the Antarctic) to $+40°C$ (midsummer in Peru). It is to be hoped that these qualities will serve them well for the next 40 million years.

See also

Seabird Conservation. Seabird Foraging Ecology. Seabird Population Dynamics.

Further Reading

Ainley DG, LeResche RE, and Sladen WJL (1983) *Breeding Biology of the Adelie Penguin*. Berkeley: University of California Press.

Dann P, Norman I, and Reilly P (eds.) (1995) *The Penguins: Ecology and Management*. Chipping Norton, NSW, Australia: Surrey Beatty Sons.

Davis LS (1993) *Penguin: A Season in the Life of the Adélie Penguin*. London: Pavilion.

Davis LS and Darby JT (eds.) (1990) *Penguin Biology*. San Diego: Academic Press.

Stonehouse B (ed.) (1975) *The Biology of Penguins*. London: Macmillan.

Williams TD (1995) *The Penguins*. Oxford: Oxford University Press.

SEABIRD CONSERVATION

J. Burger, Rutgers University, Piscataway, USA

Introduction

Conservation is the preservation and protection of plants, animals, communities, or ecosystems; for marine birds, this means preserving and protecting them in all of their diverse habitats, at all times of the year. Conservation implies some form of management, even if the management is limited to leaving the system alone, or monitoring it, without intervention or human disturbance. The appropriate degree of management is often controversial. Some argue that we should merely protect seabirds and their nesting habitats from further human influences, leaving them alone to survive or to perish. For many species, however, this solution is not practical because they do not live on remote islands, in inaccessible sites, or places that could be totally ignored by people. For other species, their nesting and foraging habitats have been so invaded by human activities that they must adapt to new, less suitable conditions. For some species, their declines have been so severe that only aggressive intervention will save them. Even species that appear to be unaffected by people have suffered from exotic feral animals and diseases that have come ashore, brought by early seafarers in dugout canoes or later by mariners in larger boats with more places for invading species to hide.

For marine birds there is compelling evidence that the activities of man over centuries have changed their habitats, their nesting biology, and their foraging ecology. Thus, we have a responsibility to conserve the world's marine birds. For conservation to be effective, the breeding biology, natural history, foraging ecology, and interactions with humans must be well understood, and the factors that contribute to their overall reproductive success and survival known.

Marine Bird Biology and Conservation

In this article, seabirds, or marine birds, include both the traditional seabird species (penguins, petrels and shearwaters, albatrosses, tropicbirds, gannets and boobies, frigate-birds, auks, gulls and terns, and pelicans) and closely related species that spend less time at sea (cormorants, skimmers), and also other species that spend a great deal of their life cycle along coasts or at sea, but may nest inland, such as shore birds. Seabirds are distributed worldwide, and nest in a variety of habitats, from remote oceanic islands that are little more than coral atolls, to massive rocky cliffs, saltmarsh islands, sandy beaches or grassy meadows, and even rooftops. While most species of seabirds nest colonially, some breed in loose colonies of scattered nests, and still others nest solitarily. Understanding the nesting pattern and habitat preferences of marine birds is essential to understanding the options for conservation of these birds on their nesting colonies. Without such information, appropriate habitats might not be preserved.

The attention of conservationists is normally directed to protecting seabirds while they are breeding, but marine birds spend most of their lives away from the breeding colonies. Although some terns and gulls first breed at 2 or 3 years of age, other large gulls and other seabirds breed when they are much older. Some albatrosses do not breed until they are 10 years old. Nonbreeders often wander the oceans, bays, and estuaries, and do not return to the nesting colonies until they are ready to breed. They face a wide range of threats during this period, and these often pose more difficult conservation issues because the birds are dispersed, and are not easy to protect. In some cases, such as roseate terns (*Sterna dougallii*), we do not even know where the vast majority of overwintering adults spend their time. Since many species forage over the vast oceans during the nonbreeding season, or before they reach adulthood, conditions at sea are critical for their long-term survival. While landscape ecology has dominated thought for terrestrial systems, few of its tenets have been applied to oceanic ecosystems, or to the conservation needs of marine birds.

A brief description of the factors that affect the success of marine bird populations will be enumerated before discussion of conservation strategies and management options for the protection and preservation of marine bird populations.

Threats to Marine Birds

Marine bird conservation can be thought of as the relationship between hazards or threats, marine bird vulnerabilities, and management. The schematic in **Figure 1** illustrates the major kinds of hazards faced by marine birds, and indeed all birds, and the different kinds of vulnerabilities they face. The outcomes shown in **Figure 1** are the major ones; however, there are many others that contribute to the overall decline in population levels (**Figure 2**). Conservation involves some form of intervention or management for each of these hazards, to preserve and conserve the species.

Marine Bird Vulnerabilities

Factors that affect marine bird vulnerability include the stage in the life cycle, their activity patterns, and their ecosystem (**Figure 3**). Marine birds are differentially vulnerable during different life stages. Many of the life cycle vulnerabilities are reduced by nesting in remote oceanic islands (albatrosses, many petrels, many penguins) or in inaccessible locations, such as cliffs (many alcids, kittiwakes, *Rissa tridactyla*) or tall trees. However, not all marine birds nest in such inaccessible sites, and some sites that were inaccessible for centuries are now inhabited by people and their commensal animals.

For many species, the egg stage is the most vulnerable to predators, since eggs are sufficiently small that a wide range of predators can eat them. Eggs are placed in one location, and are entirely dependent upon parents for protection from inclement weather, accidents, predators, and people. In many cultures, bird eggs, particularly seabird eggs, play a key role, either as a source of protein or as part of cultural traditions. In some cultures, the eggs of particular species are considered aphrodisiacs and are highly prized and sought after.

Egging is still practiced by humans in many places in the world, usually without any legal restrictions. Even where egging is illegal, either the authorities overlook the practice or it is impossible to enforce, or it is sufficiently clandestine to be difficult to apprehend the eggers.

Chicks are nearly as vulnerable as eggs, although many seabirds are semiprecocial at birth and are able to move about somewhat within a few days of hatching. The more precocial, the more likely the chick can move about to hide from predators or people, or seek protection from inclement weather. Nonetheless, chicks are unable to fly, and thus cannot avoid most ground predators and many aerial

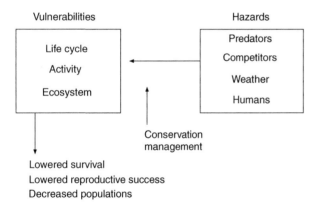

Figure 1 Marine avian conservation is the relationship between the hazards marine birds must face, along with their vulnerabilities.

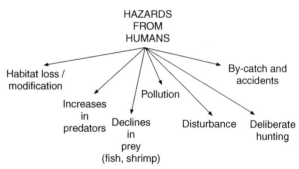

Figure 3 Humans provide a wide range of hazards, including direct and indirect effects.

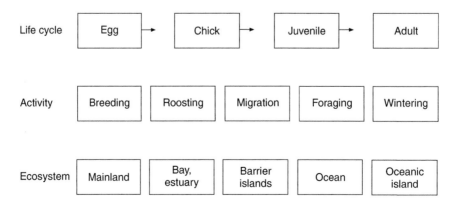

Figure 2 The primary vulnerabilities marine birds face deal with aspects of their life cycles, activity patterns and the ecosystems they inhabit.

predators if they cannot hide sufficiently. The pre-fledging period can last for weeks (small species such as terns) to 6 months for albatrosses.

The relative vulnerability of juveniles and adults is usually the same, at least with respect to body size. Most juveniles are as large as adults and can fly, and are thus able to avoid predators. Juveniles, however, are less experienced with predators and with foraging, and so are less adept at avoiding predators and at foraging efficiently. The relative vulnerability of juveniles and adults depends on their size, habitat, type of predator, and antipredator behavior. For example, species that nest high in trees (Bonaparte's gull, *Larus philadelphia*) or on cliffs (e.g., kittiwakes, some murres, some alcids) are not exposed to ground predators that cannot reach them. Species that nest on islands far removed from the mainland are less vulnerable to ground predators (e.g., rats, foxes), unless these have been introduced or have unintentionally reached the islands. Marine birds that are especially aggressive in their defense of their nests (such as most terns) can sometimes successfully defend their eggs and young from small predators by mobbing or attacks.

Activity patterns also influence their vulnerability. When marine birds are breeding, they are tied to a particular nest site, and must either abandon their eggs or chicks or stay to protect them from inclement weather, predators, or people. At other times of the year, seabirds are not as tied to one location, and can move to avoid these threats. Foraging birds are vulnerable not only to predators and humans, but to accidents from being caught in fishing gill nets, drift nets, or longlines, from which mortality can be massive, especially to petrels and albatrosses.

The choice of habitats or ecosystems also determines their relative vulnerability to different types of hazards. Marine birds nesting on oceanic islands are generally removed from ground predators and most aerial predators but face devastation when such predators reach these islands. Cats and rats have proven to be the most serious threat to seabirds nesting on oceanic and barrier islands. The threat from predators increases the closer nesting islands are to the mainland, a usual source of predators and people. Similarly, the threat from storm and hurricane tides is greater near-shore, particularly for ground-nesting seabirds.

Marine Bird Hazards

The major hazards and challenges to survival of marine birds are from competitors (for mates, food, nesting sites), predators, inclement weather, and humans (**Figure 1**). Of these, humans are the greatest problem for the conservation of marine birds and strongly influence the other three types of hazards. Humans affect marine birds in a wide range of ways, by changing the environment around seabirds (**Figure 4**), ultimately causing population declines. While other hazards, such as competition for food and inclement weather are widespread, marine birds have always faced these challenges.

Habitat loss and modification are the greatest threats to marine birds that nest in coastal regions, and for birds nesting on near-shore islands. Direct loss of habitat is often less severe on remote oceanic islands, although recent losses of habitat on the Galapagos and other islands are causes for concern. Habitat loss can also include a decrease in available foraging habitat, either directly through its loss or through increased activities that decrease prey abundance or their ability to forage within that habitat.

Humans cause a wide range of other problems:

- Introducing predators to remote islands, and increasing the number of predators on islands and coastal habitats. For example, rats and cats have been introduced to many remote islands, either deliberately or accidentally. Further, because of the construction of bridges and the presence of human foods (garbage), foxes, raccoons and other predators have reached many coastal islands.
- Decreasing available prey through overfishing, habitat loss, or pollution. Coastal habitat loss can decrease fish production because of loss of nursery areas, and pollution can further decrease reproduction of prey fish used by seabirds.
- Decreasing reproductive success, causing behavioral deficits, or direct mortality because of pollution. Contaminants, such as lead and mercury, can reduce locomotion, feeding behavior, and parental recognition in young, leading to decreased reproductive success.
- Decreasing survival or causing injuries because birds are inadvertently caught in fishing lines, gillnets, or ropes attached to longlines.
- Decreasing reproductive success or foraging success because of deliberate or accidental disturbance of nesting, foraging, roosting, or migrating marine birds. For some marine birds the presence of recreational fishing boats, personal watercraft, and commercial fishing boats reduces the area in which they can forage.
- Deliberate collection of eggs, and killing of chicks or adults for food, medicine, or other purposes. On many seabird nesting islands in the Caribbean, and elsewhere, the eggs of terns and other species

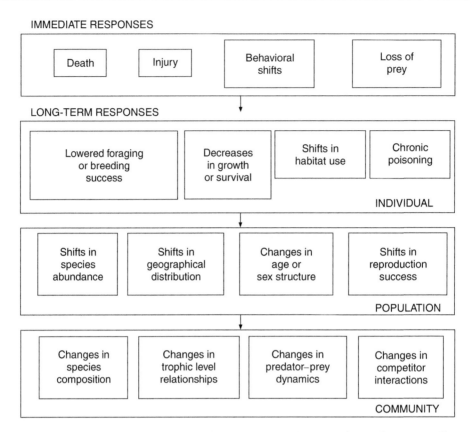

Figure 4 Human activities can affect marine birds in a variety of ways, including immediate and long-term effects.

are still collected for food. Egging of murres and other species is also practiced by some native peoples in the Arctic.

Conservation and Management of Marine Birds

Conservation of marine birds is a global problem, and global solutions are needed. This is particularly true for problems that occur at sea, where the birds roost, migrate, and forage. No one governmental jurisdiction controls the world's population of most species. Education, active protection and management, international treaties and agreements, and international enforcement may be required to solve some of the major threats to marine birds. However, the conservation and management of marine birds also involves intervention in each of the above hazards, and this can often be accomplished locally or regionally with positive results.

Habitat loss can be partly mitigated by providing other suitable habitats or nesting spaces nearby; predators can be eliminated or controlled; fishing can be managed so that stocks are not depleted to a point where there are no longer sufficient resources for marine birds; contamination can be reduced by legal

enforcement; human disturbance can be reduced by laws, wardening, and voluntary action; by-catch can be reduced by redesigning fishing gear or changing the spatial or temporal patterns of fishing; and deliberate or illegal hunting can be reduced by education and legal enforcement. Each will be discussed below.

Habitat creation and modification is one of the most useful conservation tools because it can be practiced locally to protect a particular marine bird nesting or foraging area. In many coastal regions, nesting habitat has been created for beach-nesting terns, shore birds, and other species by placing sand on islands that are otherwise unsuitable, extending sandy spits to create suitable habitat, and removing vegetation to keep the appropriate successional stage. In some places grassy meadows are preserved for nesting birds, while in others sand cliffs have been modified, and concrete slabs have been provided for cormorants, gannets, and boobies (albeit to make it easier to collect the guano for fertilizer). Habitat modification can also include creation of nest sites. Burrows have been constructed for petrels; chick shelters have been created for terns; and platforms have been built for cormorants, anhingas, and terns.

The increase in the diversity and number of introduced and exotic predators on oceanic and other islands is a major problem for many marine birds. One of the largest problems marine birds face worldwide is the introduction of cats and rats to remote nesting islands. Since most marine birds on remote islands nest on the ground, their eggs and chicks are vulnerable to cats and rats. Some governments, such as that of New Zealand, have devoted considerable time and resources to removal of these two predators from remote nesting islands, but the effort is enormous.

Many marine birds evolved on remote islands where there were no mammalian predators. These species lack antipredator behaviors that allow them to defend themselves, as seen in albatrosses, which do not leave their nests while rats gnaw them. Most sea birds on remote islands nest on, or under the ground, where they are vulnerable to ground predators, and they do not leave their nests when approached. Rats and cats have proven to be the most significant threat to sea birds worldwide, and their eradication is essential if some marine birds are to survive. New Zealand has invested heavily in eradicating invasive species on some of its offshore islands, allowing sea birds and other endemic species to survive. Cats, however, are extremely difficult to remove, even from small offshore islands, and up to three years were required to remove them completely from some New Zealand islands. Such a program involves a major commitment of time, money, and personnel by local or federal governments.

Simply removing predators, however, does not always result in immediate increases in seabird populations. Sometimes unusual management practices are required, such as use of decoys and playback of vocalizations to attract birds to former colony sites. Steve Kress of National Audubon reestablished Atlantic puffins (*Fratercula arctica*) on nesting colonies in Maine by a long-term program of predator removal, decoys, and the playback of puffin vocalizations.

Increasing observations of chick mortality from starvation or other breeding failures have focused attention on food availability, and the declines in fish stocks. Declines in prey can be caused by sea level changes, water temperature changes, increases in predators and competitors, and other natural factors. However, they can also be caused by overfishing that depletes the breeding stocks, and reducing the production of small fish that serve as prey for seabirds. There are two mechanisms at work: in some cases fishermen take the larger fish, thereby removing the breeding stock, with a resultant decline in small fish for prey. This may have happened in the northern

Atlantic. In other cases, fishermen take small fish, thereby competing directly with the seabirds, as partially happened off the coast of Peru.

Overfishing is a complicated problem that often requires not only local fisheries management but national and international treaties and laws. Even then, fisheries biologists, politicians, importers/exporters, and lawyers see no reasons to maintain the levels of fish stocks necessary to provide for the foraging needs of sea birds. Nonetheless, the involvement of conservationists interested in preserving marine bird populations must extend to fisheries issues, for this is one of the major conservation challenges that seabirds face.

By-catch in gill nets, drift nets, and longlines is also a fisheries problem. With the advent of longlines, millions of seabirds of other species are caught annually in the miles of baited lines behind fishing vessels. The control and reduction in the number of such fishing gear is critical to reducing seabird mortality. Longlines are major problems for seabirds in the oceans of the Southern Hemisphere, although Australia and New Zealand are requiring bird-deterrents on longline boats.

Pollution is another threat to sea birds: Pollutants include heavy metals, organics, pesticides, plastics, and oil, among others. Oil spills have often received the most attention because there are often massive and conspicuous die-offs of sea birds following major oil spills. Usually, however, the carcass counts underestimate the actual mortality because the spills happen at sea or in bad conditions where the carcasses are never found or do not reach land before they are scavenged or decay. Although direct mortality is severe from oil spills, one of the greatest problems following an oil spill is the decline of local breeding populations, as happened following the *Exxon Valdez* in Alaska. Ten years after the spill some seabird species had still not recovered to prespill levels. Partially this resulted from a lack of excess reproduction on nearby islands, where predators such as foxes kept reproduction low.

While major oil spills have the potential to cause massive die-offs of birds that are foraging and breeding nearby, or migrating through the area, chronic oil pollution is also a serious threat. Many coastal areas, particularly near major ports, experience chronic oil spillage that accounts for far more oil than the massive oil spills that receive national attention. Chronic pollution can cause more subtle effects such as changes in foraging behavior, deficits in begging, weight loss, and internal lesions.

When there are highly localized population declines as a result of pollution, such as oil spills, or of inclement weather, predators, or other causes, the management options are limited. However, one method to encourage rapid recovery is to manage the breeding colonies outside of the affected area, allowing them to serve as sources for the depleted colonies. In the case of the *Exxon Valdez*, for example, there were numerous active colonies immediately outside of the spill impact zone. However, reproduction on many of these islands was suboptimal owing to the presence of predators (foxes). Fox removal would no doubt increase reproductive success on those islands, providing surplus birds that could colonize the depleted colonies within the spill zone itself.

Management for reductions in marine pollutants, including oil, can be accomplished by education, negotiations with companies, laws and treaties, and sanctions. For example, following the *Exxon Valdez*, the U.S. government passed the Oil Pollution Act that ensured that by 2020 all ships entering U.S. waters would have double hulls and many other safety measures to reduce the possibility of large oil spills.

Another threat to marine birds is through atmospheric deposition of mercury, cadmium, lead, and other contaminants. At present, mercury and other contaminants have been found in the tissues of birds throughout the world, including the Arctic and Antarctic. While atmospheric deposition is greatest in the Northern Hemisphere, contaminants from the north are reaching the Southern Hemisphere. The problem of atmospheric deposition of mercury, and oxides of nitrogen and sulfur, can be managed only by regional, national, and international laws that control emissions from industrial and other sources, although some regional negotiations can be successful.

Marine birds have been very useful as indicators of coastal and marine pollution because they integrate over time and space. While monitoring of sediment and water is costly and time-consuming, monitoring of the tissues of birds (especially feathers) can be used to indicate where there may be a problem. Declines in marine bird populations such as occurred with DDT, were instrumental in regulating contaminants. Marine birds have been especially useful as bioindicators in the Great Lakes for polychlorinated biphenyls (PCBs), on the East Coast of North America and in northern Europe for mercury, and in the Everglades for mercury.

Human disturbance is a major threat to seabirds, both in coastal habitats and on oceanic islands. While the level and kinds of human disturbance to marine birds in coastal regions is much higher than for oceanic islands, the birds that nest on oceanic islands did not evolve with human disturbance and are far less equipped to deal with it. Disturbance to

breeding and feeding assemblages can be deliberate or accidental, when people come close without even realizing they are doing so, and fail to notice or be concerned. Sometimes colonial birds mob people who enter their colony, but the people do not see any eggs or chicks (because they are cryptic), and so are unaware they are causing any damage. Chicks and eggs, however, can be exposed to heat or cold stress during these disturbances.

Human disturbance can be managed by education, monitoring (by volunteers, paid wardens, or law enforcement officers), physical barriers (signs, strings, fences, barricades), laws, and treaties. In most cases, however, it is worth meeting with affected parties to figure out how to reduce the disturbance to the birds while still allowing for the human activities. This can be done by limiting access temporally and keeping people away during the breeding season, or by posting the sensitive location but allowing human activities in other regions. Compliance will be far higher if the interested parties are included in the development of the conservation strategy, rather than merely being informed at a later point. Moreover, such people often have creative solutions that are successful.

The deliberate collecting of eggs and marine birds themselves can be managed by education, negotiations, laws and treaties, and enforcement. In places where the collection of eggs or adult birds is needed as a source of protein or for cultural reasons, mutual education by the affected people and managers will be far more successful. In many cases, indigenous peoples have maintained a sustainable harvest of seabird eggs and adults for centuries without ill effect to the seabird populations. However, if the populations of these people increase, the pressure on seabird populations may exceed their reproductive capacity. People normally took only the first eggs, and allowed the birds to re-lay and raise young. Conservation was often accomplished because individuals 'owned' a particular section of the colony, and their 'section' was passed down from generation to generation. There was thus strong incentive to preserve the population and not to overuse the resource, particularly since most seabirds show nest site tenacity and will return to the same place to nest year after year. When 'governments' took over the protection of seabird colonies, no one owned them any longer, and they suffered the fate of many 'commons' resources: they were exploited to the full with devastating results. Whereas subsistence hunting of seabirds and their eggs was successfully managed for centuries, the populations suffered overnight with the advent of government control and the availability of new technologies (snowmobiles, guns). More recently, the use of personal watercraft has increased in some coastal areas, destroying nurseries for fish and shellfish, disturbing foraging activities, and disrupting the nesting activities of terns and other species.

Hunting by nontraditional hunters can also be managed by education, persuasion, laws, and treaties. However, both types of hunting can be managed only when there are sufficient data to provide understanding of the breeding biology, population dynamics, and population levels. Without such information on each species of marine bird, it is impossible to determine the level of hunting that the populations can withstand. Extensive egging and hunting of marine birds by native peoples still occurs in some regions, such as that of the murres in Newfoundland and Greenland.

On a few islands, some seabird populations have both suffered and benefitted at the hands of the military. Some species nested on islands that were used as bombing ranges (Culebra, Puerto Rico) or were cleared for air transport (Midway) or were directly bombed (Midway, during the Second World War). In these cases, conservation could only involve governmental agreements to stop these activities, and of course, in the case of war, it is no doubt out of the hands of conservationists. However, military occupancy may protect colonies by excluding those who would exploit the birds.

Conclusions

Conservation of marine birds is a function of understanding the hazards that a given species or group of species face, understanding the species vulnerabilities and possible outcomes, and devising methods to reduce or eliminate these threats so that the species can flourish. Methods range from preserving habitat and preventing any form of disturbance (including egging and hunting), to more complicated and costly procedures such as wardening, and attracting birds back to former nesting colonies.

The conservation methods that are generally available include education, creation of nesting habitat and nest sites, the elimination of predators, the cessation of overfishing, building of barriers, use of wardens and guards, use of decoys and vocalization, creation of laws and treaties, and the enforcement of these laws. In most cases, the creation of coalitions of people with differing interests in seabirds, to reach mutually agreeable solutions, will be the most effective and long-lasting. Although ecotourism may pose the threat of increased disturbance or beach development, it can be managed as a source of revenue to sustain conservation efforts.

It is necessary to bear in mind that conservation of seabirds is not merely a matter of protecting and preserving nesting assemblages, but of protecting their migratory and wintering habitat and assuring an adequate food supply. Assuring a sufficient food supply can place marine birds in direct conflict with commercial and recreational fishermen, and with other marine activities, such as transportation of oil and other industrial products, use of personal watercraft and boats, and development of shoreline industries and communities. Conservation of marine birds, like many other conservation problems, is a matter of involving all interested parties in solving a 'commons' issue.

See also

Alcidae. Laridae, Sternidae and Rynchopidae. Pelecaniformes. Procellariiformes. Seabird Conservation. Seabird Foraging Ecology. Seabird Migration. Seabird Population Dynamics. Sphenisciformes.

Further Reading

Burger J (2001) *Tourism and ecosystems. Encyclopedia of Global Environmental Change.* Chichester: Wiley.

Burger J and Gochfeld M (1994) Predation and effects of humans on island-nesting seabirds. In: Nettleship DN, Burger J, and Gochfeld M (eds.) *Threats to Seabirds on Islands*, pp. 39–67. Cambridge: International Council for Bird Preservation Technical Publication.

Croxall JP, Evans PGH, and Schreiber RW (1984) *Status and Conservation of the World's Seabirds.* Cambridge: International Council for Bird Preservation Technical Publication No. 2.

Kress S (1982) The return of the Atlantic Puffin to Eastern Egg Rock, Maine. *Living Bird Quarterly* 1: 11–14.

Moors PJ (1985) *Conservation of Island Birds.* Cambridge: International Council for Bird Preservation Technical Publication.

Nettleship DN, Burger J, and Gochfeld M (eds.) (1994) *Threats to Seabirds on Islands.* Cambridge: International Council for Bird Preservation Technical Publication.

Vermeer K, Briggs KT, Morgan KH, and Siegel-Causey D (1993) *The Status, Ecology, and Conservation of Marine Birds of the North Pacific.* Ottawa: Canadian Wildlife Service Special Publication.

SEABIRD FORAGING ECOLOGY

L. T. Balance, NOAA-NMFS, La Jolla, CA, USA
D. G. Ainley, H.T. Harvey Associates, San Jose CA, USA
G. L. Hunt Jr., University of California, Irvine, CA, USA

Introduction

Though bound to the land for reproduction, most seabirds spend 90% of their life at sea where they forage over hundreds to thousands of kilometers in a matter of days, or dive to depths from the surface to several hundred meters. Although many details of seabird reproductive biology have been successfully elucidated, much of their life at sea remains a mystery owing to logistical constraints placed on research at sea. Even so, we now know a considerable amount about seabird foraging ecology in terms of foraging habitat, behavior, and strategy, as well as the ways in which seabirds associate with or partition prey resources.

Foraging Habitat

Seabirds predictably associate with a wide spectrum of physical marine features. Most studies implicitly assume that these features serve to increase prey abundance or availability. In some cases, a physical feature is found to correlate directly with an increase in prey; in others, the causal mechanisms are postulated. To date, the general conclusion with respect to seabird distribution as related to oceanographic features is that seabirds associate with large-scale currents and regimes that affect physiological temperature limits and/or the general level of prey abundance (through primary production), and with small-scale oceanographic features that affect prey dispersion and availability.

Water Masses

In practically every ocean, a strong relation between sea bird distribution and water masses has been reported, mostly identified through temperature and/or salinity profiles (**Figure 1**). These correlations occur at macroscales (1000–3000 km, e.g. associations with currents or ocean regimes), as well as mesoscales (100–1000 km, e.g. associations with warm- or cold-core rings within current systems). The question of why species associate with different water types has not been adequately resolved. At issue are questions of whether a seabird responds directly to habitat features that differ with water mass (and may affect, for instance, thermoregulation), or directly to prey, assumed to change with water mass or current system.

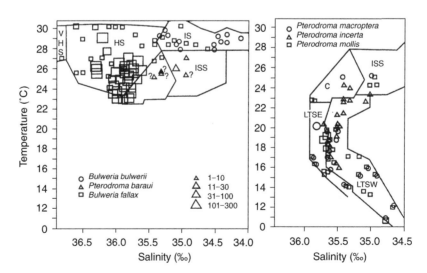

Figure 1 Distribution of gadfly petrels in the Indian Ocean corresponding to various regimes of surface temperature and salinity: (A) warm-water species, and (B) cool-water species. The relative size of symbols is proportional to the number of sightings. Symbols for water masses as follows: VHS, very high salinity; HS, high salinity; IS, intermediate salinity; ISS, intermediate salinity south; C, common water; LTSE, low temperature southeast; LTSW, low temperature southwest. (Redrawn from Pocklington R (1979) *Marine Biology* 51: 9–21.)

Environmental Gradients

Physical gradients, including boundaries between currents, eddies, and water masses, in both the horizontal and vertical plane, are often sites of elevated seabird abundance. Seabirds respond to the strength of gradients more than the presence of them. In shelf ecosystems, e.g. the eastern Bering Sea shelf and off the California coast, cross-shelf gradients are stronger than along-shelf gradients, and sea-bird distribution and abundance shows a corresponding strong gradient across, as opposed to along the shelf. At larger scales the same pattern is evident, e.g. crossing as opposed to moving parallel with boundary currents.

Physical gradients can affect prey abundance and availability to seabirds in several ways. First, they can affect nutrient levels and therefore primary production, as in eastern boundary currents. Second, they can passively concentrate prey by carrying planktonic organisms through upwelling, downwelling, and convergence. Finally, they can maintain property gradients (fronts, see below) to which prey actively respond. In the open ocean, where currents and dynamic processes are less active, prey behavior should be the principal mechanism responsible for seabird aggregation. In these cases, locations of aggregations are unpredictable, and this has important consequences for the adaptations necessary for seabirds to locate and exploit them. In contrast, in continental shelf systems, currents impinge upon topographically fixed features, such as reefs, creating physical gradients predictable in space and time to which seabirds can go directly. Thus, the first and second mechanisms are the more important in shelf systems, and aggregations are so predictable that seabirds learn where and when to be in order to eat.

Fronts

Much effort has been devoted successfully to identifying correlations between seabird abundance and fronts, or those gradients exhibiting dramatic change in temperature, density, or current velocity (**Figure 2**). Results indicate a considerable range of variation in the strength of seabird responses to fronts. This may be due to the fact that fronts influence seabird distribution only on a small scale. The factors behind the range of response is of interest in itself.

Nevertheless, fronts are important determinants of prey capture. Two hypotheses have been proposed to account for this: (1) that frontal zones enhance primary production, which in turn increases prey supply, e.g. boundaries of cold- or warm-core rings in the Gulf Stream; and (2) that frontal zones serve to

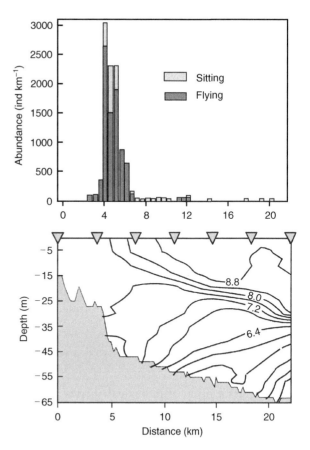

Figure 2 The aggregation of foraging shearwaters at fronts, in this case the area where an elevation in bottom topography forces a transition between stratified and well-mixed water. The bottom panel shows isotherms. (Redrawn from Hunt GL *et al., Marine Ecology Progress Series* 141: 1–11.)

concentrate prey directly into exploitable patches, e.g. current rips among the Aleutian Island passages.

Topographic Features

Topographic features serve to deflect currents, and can be sites of strong horizontal and vertical changes in current velocity, thus, concentrating prey through a variety of mechanisms (**Figure 2**). For example, seamounts are often sites of seabird aggregation, likely related to the fact that they are also sites of increased density and heightened migratory activity for organisms comprising the deep scattering layer. A second example are topographic features in relatively shallow water, including depressions in the tops of reefs and ridges across the slope of marine escarpments, which may physically trap euphausiids as they attempt to migrate downward in the morning. A third example is the downstream, eddy effect of islands that occur in strong current systems.

Depth gradient itself is sometimes correlated with increased abundance of seabirds, and water depth in general has long been related to seabird abundance and species composition. Depth-related differences in primary productivity explain large-scale patterns between shelf and oceanic waters. Within shelf systems themselves, several hypotheses explain changes in species composition and abundance with depth. First, primary production may be diverted into one of two food webs, benthic or pelagic, and this may explain differences in organisms of upper trophic levels in inner versus outer shelf systems, e.g. the eastern Bering Sea. Alternatively, the fact that interactions between flow patterns in the upper water column and bottom topography will be strong in inner shelf areas but decoupled in outer shelf areas, may result in differences in predictability of prey and consequently, differences in the species that exploit them, e.g. most coastal shelf systems. Finally, depending on diving ability and depth, certain seabirds may be able to exploit bottom substrate, whereas others may not.

Sea Ice

A strong association of individual species and characteristic assemblages exists with sea ice features. On one hand, certain species are obligate associates with sea ice; on the other, sea ice can limit access to the water column, and in some cases, aggregation of sea birds near the ice margin is a simple response to this barrier (**Figure 3**).

Sea ice often enhances foraging opportunities. First, there is an increased abundance of mesopelagic organisms beneath the ice, believed to be a phototactic response of these organisms to reduced light levels. Second, primary production is enhanced beneath the ice or at its edge, and in turn leads to increased abundance of primary and secondary consumers. The abundance and degree of concentration of this sympagic fauna varies with ice age, ice structure, and depth to the bottom. As a result, Arctic ice, often multi-year in nature, has a speciose sympagic fauna as compared to Antarctic ice, which is often annual. The ice zone can be divided into at least three habitats, a region of leads within the ice itself, the ice edge, and the zone seaward of the ice. Each zone is exploited by different seabird species, and the relative importance of zones appears to differ between Arctic and Antarctic oceans, with the seaward zone being particularly important in Antarctic systems.

Foraging Behavior

Most seabird species take prey within a half meter of the sea surface. This they accomplish by capturing prey that either are airborne themselves (as a result of escaping subsurface predators, see below), or are shallow enough that, to grasp prey, the bird dips its beak below the surface (dipping) or crashes into the surface and extends its head, neck and upper body

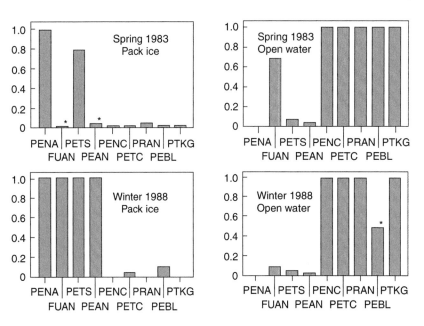

Figure 3 The correspondence of various seabird species either to pack ice or open water immediately offshore of the ice; Scotia-Weddell Confluence, Southern Ocean. Abbreviations: PENA, Adélie penguin; FUAN, Antarctic fulmar; PETS, snow petrel; PEAN, Antarctic petrel; PENC, chinstrap penguin; PETC, pintado petrel; PRAN, Antarctic prion; PEBL, blue petrel; PTKG, Kerguelen petrel. (Redrawn from Ainley et al. (1994) Journal of Animal Ecology 63: 347–364.)

downward (surface plunging, contact dipping). Other species feed on dead prey floating at the surface. Another group takes prey within about 20 m of the surface either by flying into the water to continue flight-like wing movements below the surface (pursuit plunging) or by using the momentum of an aerial dive (plunging).

Finally, a number of seabirds can dive using either their wings or feet for propulsion. Dive depth is related to body size, which in turn relates to physiological capabilities of diving. The deepest dives recorded by a bird (penguin) reach 535 m, but most species are confined to the upper 100 m. A highly specialized group of species capture prey by stealing them from other birds.

Foraging Strategies

The issue of how seabirds locate their prey is far from completely understood. Below is a summary of known strategies, most of which depend on sea-birds locating some feature which itself serves to reliably concentrate prey.

Physical Features

Physical features are important to foraging sea birds, because they serve to concentrate prey in space and time, and because they often occur under predictable circumstances (**Figure 2**). This issue was addressed above.

Subsurface Predators

Subsurface predators commonly drive prey to the surface because the air–water interface acts as a boundary beyond which most prey cannot escape. Under these circumstances, seabirds can access these same prey from the air. A wide array of subsurface predators are important to seabirds: large predatory fishes (e.g. tuna), particularly in relatively barren waters of the tropics; marine mammals, both cetaceans and pinnipeds; and marine birds themselves.

Subsurface predators increase the prey available to birds in at least three ways. First, they drive prey to the surface. Second, they injure or disorient prey, which then drift to the surface and are accessible to surface foragers. Third, they leave scraps on which seabirds forage, particularly when the prey themselves are large.

Feeding subsurface predator schools can be highly visible and the degree of association of birds with these prey patches is often great. One investigation found 79% of the variability in seabird density could be explained by the number of gray whale mud plumes; this visibility may have been responsible for

the higher correlation than typically reported for other studies attempting to relate seabird and prey abundance (see below).

Feeding Flocks

Seabirds in most of the world's oceans exploit clumped prey by feeding in multispecies flocks. Studies in all latitudes report that sea birds in flocks, often in a very few disproportionately large aggregations, account for the majority of all individuals seen feeding. Although flocks can result from passive aggregation at a shared resource, evidence indicates that seabirds benefit in some way from the presence of other individuals. First, as noted above, some seabird species act as subsurface predators, making prey available to surface-feeding sea birds, e.g. alcids driving prey to the surface for larids in coastal Alaska. Second, a small number of seabird species are kleptoparasitic, obtaining their prey from other sea birds, e.g. jaegers, skuas and a few other species. Third, vulnerability of individual fish in a school may increase with the number of birds feeding in the flock. Finally, flocks are highly visible signals of the location of a prey patch, e.g. species keying on frigate birds circling high over tuna schools.

Certain species are disproportionately responsible for these signals, simply through their highly visual flight characteristics. Such species are termed 'catalysts'. There is strong evidence that seabirds follow these visual signals, in some cases distinguishing between searching and feeding catalyst species, and between those feeding on a single prey item and those feeding on a clumped patch.

Nocturnal Feeding

Many fishes and invertebrates remain at depth during the day and migrate to the surface after dark. During crepuscular or dark periods, surface densities of prey can be 1000 times greater than during the day. This migration is more significant in low than high latitudes, and in oceanic than neritic waters. Many studies indirectly infer nocturnal feeding from the presence of vertically migrating species in seabird diets or from circadian activity patterns. Little direct evidence exists.

Among the indirect data on how seabirds might locate prey at night is a considerable body of information on olfaction. In particular, members of the avain order Procellariiformes possess olfactory lobes disproportionately large compared to the total brain size and compared to most vertebrates. Experiments show a marked ability of these birds to differentiate among odors and, especially, to find sources of odors that are trophically meaningful.

Maximization of Search Area

Feeding opportunities in the open ocean are often patchily distributed requiring seabirds to travel over large areas in search for prey. This ability to search large areas is tightly linked to adaptations for flight proficiency. Seabirds capable of wide-area search exhibit morphological adaptations of the wing and tail that enhance energy-efficient flight. They also modify their flight behavior to take advantage of wind as an energy source. Penguins, loons, grebes, cormorants and alcids, the pre-eminent sea bird divers (see above), sacrifice wide-area search capabilities for what is needed for diving: high density and small wings. Therefore, divers are limited to areas of high prey availability.

An investigation into the relationship between wind and foraging behavior has revealed taxon-specific preferences in flight direction dependent on wind direction, and in turn, to wing morphology and presumed prey distribution. Procellariiformes, primarily oceanic foragers, preferentially fly across the wind, whereas Pelecaniformes and Charadriiformes, the majority of which forage over shelf and slope waters, preferentially fly into and across headwinds. Because prey on a global scale are more patchy in oceanic than shelf and slope waters, across-wind flight may allow Procellariiformes to cover more area at lower energetic cost, whereas headwind flight allows slower ground speeds, possibly increasing the probability of detecting prey and decreasing response time once a prey item is detected. Flying up- or across-wind among procellariids also maximizes probabilities of finding prey using olfaction.

Associations With Prey

Positive or significant correlations have been identified between seabird and prey abundance, in the Bering Sea, eastern North Atlantic, Barents Sea, and in locations throughout the Antarctic, although rarely at scales smaller than about 2–3 km. In some studies, however, the correlation is weak to nonexistent, or a negative correlation is reported. From these results, several general principles have arisen. First, the strength of the correlation increases with the spatial scale at which measurements are made. Second, correlations between planktonic-feeding seabirds and their prey are lower than correlations between piscivorous seabirds and their prey. The reasons for this pattern may relate to differences in patch characteristics dependent on prey species, or to differences in search mode of various predators. Third, correlations are not as strong as expected and in many cases, a correlation is found only with repeated surveys. Many hypotheses have been proposed to explain the latter, including: (1) seabirds are unable consistently to locate large prey patches; (2) prey are sufficiently abundant that seabirds do not need to locate largest prey patches; (3) prey are actively avoiding sea birds; (4) prey patches are continuously moving so that a time lag exists between patch formation, discovery by the sea bird, and measurement by the researcher; (5) extremely large prey patches are disproportionately important to seabirds so that they spend much of their time searching for or in transit to and from such patches; and (6) our means to measure prey patches (usually hydroacoustically) is a mismatch to the biology and attributes of the predators. Finally, different seabird species may respond on the basis of threshold levels of prey abundance, and these thresholds vary seasonally as well as with reproductive status of the bird (breeders require more food than nonbreeders, which comprise at least half the typical sea-bird population).

Resource Partitioning

Food is considered to be an important resource regulating seabird populations. Accordingly, much research has focused on identifying differences in the way coexisting species exploit prey. At sea, the fact that different oceanic regimes or currents support different prey communities as well as different seabird communities has been used to support the idea that the geographic range of seabird species is a response to the presence of specific prey. Contrasting these patterns, however, several colony-based studies report high diet overlap between species, leading to speculation that dietary differences may reflect differences in foraging habitat as opposed to prey selection. Evidence from at-sea research indicates broad overlap in the species and/or sizes of prey taken by coexisting seabird species, often despite species-specific feeding methods, body size or habitat segregation.

Prey Selection

Under certain circumstances seabirds do make choices as to what prey they will attempt to capture (**Figure 4**). Among breeding species of the western North Atlantic and North Sea, in cases where a prey stock, such as capelin or sandlance, are being exploited by a large array of species, birds key in on fish that provide the highest energy package, in this case fish in reproductive condition. In the Bering Sea, breeding auklets have been observed to ignore smaller zooplankton to take the most energy-dense

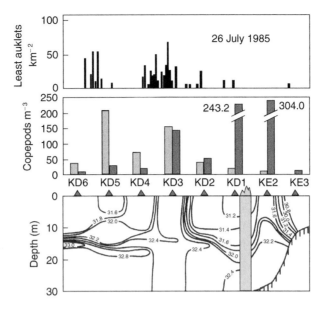

Figure 4 Aggregation of least auklets over concentrations of the copepods *Eucalanus bungii* and *Neocalanus* spp. (light bars), and *Calanus marshallae* (dark bars). All three are confirmed prey of least auklets. Birds virtually ignored huge concentrations of *C. marshallae*, which were much closer to the breeding colony on King Island to preferentially feed on larger and presumably more energy-rich *Neocalanus* spp. (Redrawn from Hunt GL Jr and Harrison NM (1990) *Marine Ecology Progress Series* 65: 141–150.)

copepod species available. Finally, in the Antarctic, during winter with almost constant darkness or near-darkness when the mesopelagic community is near the surface most of the time, seabirds have been documented to avoid smaller prey (euphausiids) to take larger and more energy-rich prey (myctophids; **Figure 5**). However, although this is evidence for active prey selection, in these cases there is broad dietary overlap among seabird predators.

Prey/Predator Size

Body-size differences among coexisting sea bird species have been used to imply diet segregation by prey size (**Figure 5**). In general, discounting penguins but realizing there are a number of exceptions, the larger seabirds (i.e. those > 1500 g) tend to take fish and squid; the smaller species tend to take juvenile or larval fish and squid, along with zooplanktonic invertebrates. It comes down to energetic cost-efficiency of foraging and the morphology of the seabird foraging apparatus: the bill, which picks one prey item at a time (the lone exception, perhaps, being one or two species groups, e.g. prions, that may filter-feed). Some degree of dietary size segregation is apparent among the few studies that have investigated all species breeding at single sites, e.g. a tropical oceanic island. Other studies at sea report little, if any, dietary separation, even though a 1000-fold difference in seabird size can exist. The implication is that seabirds often forage opportunistically depending on the availability of prey in their preferred

habitat, and that differences in habitat are more important than differences in prey selection in facilitating predator coexistence.

Habitat Type/Time

Species or assemblages often segregate according to habitat with little evidence of interactions among sea birds that significantly influence their pelagic distributions. Instead, the implication is that species respond to physical and biological characteristics of environments according to their individual needs and flight or diving capabilities. Spatial segregation can occur with respect to simple habitat features. For example, the Antarctic avifauna is divided into one assemblage associated with pack-ice covered waters and the other with ice-free waters (**Figure 3**). The species composition of these assemblages changes little over time; assemblage distribution tracks the distribution of ice features in the absence of differences in prey communities between the two habitat types, and there is little spatial overlap between the two assemblages. Spatial segregation, with respect to species or assemblages, can also occur along environmental gradients with respect to physical, chemical, and biological features of a sea bird's habitat, in both the horizontal and vertical dimension. This has been well documented especially in shelf waters, where differences in foraging habitat, particularly as determined by depth, lead to differences in diet; it is also evident in the pelagic tropical waters along productivity gradients.

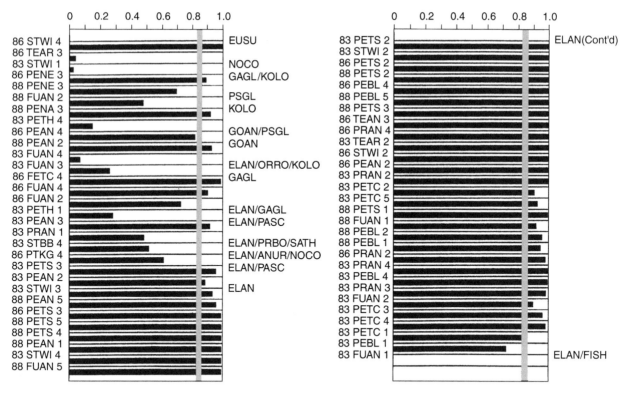

Figure 5 Diet overlap among Antarctic seabirds. Even though a 1000-fold difference in predator size existed, there was no appreciable separation of diet for many sea-bird species. The vertical stippled line indicates the level at which diet is considered to be similar on the basis of prey species overlap. Bird species to the left of bars, prey species to the right. Each bird species name is preceded by the year in which collection was made, and followed by a number that denotes habitat (1, open water; 2, sparse ice; 3, heavy ice). Bird species: PENE, emperor penguin; PENA, Adélie penguin; FUAN, Antarctic fulmar; PETS, snow petrel; PEAN, Antarctic petrel; PENC, chinstrap penguin; PETC, pintado petrel; PRAN Antarctic prion; PEBL, blue petrel; PTKG, kerguelen petrel; STWI, Wilson's storm-petrel; STBB, black-bellied storm-petrel; TEAR, Arctic tern; TEAN, Antarctic tern. Prey names: ANUR, *Anuropis* spp.; EUSU *Euphausia superba*; ELAN, *Electrona antarctica*; GAGL, *Galiteuthis glacialis*; GOAN, *Gonatus antarcticus*; KOLO, *Kondakovia longimana*; NOCO, *Notolepis coatsi*; ORRO, *Orchomene rossi*; PASC, *Pasiphaea scotia*; PRBO, *Protomyctophum bolini*; PSGL, *Psychroteuthis glacialis*; SATH, *Salpa thompsoni*. (Redrawn from Ainley DG *et al.* (1992) *Marine Ecology Progress Series* 90: 207–221.)

A recurrent theme is that seabird species sort out along prey density gradients, regardless of prey identity. Such segregation has been recorded even within the same prey patch, with certain species exploiting the center and others the periphery. The idea that oligotrophic waters, having reduced prey availability, can only be exploited by highly aerial species with efficient locomotion, whereas productive waters are necessary for diving species is one that occurs in a wide variety of studies conducted in tropical, temperate, and polar systems.

Finally, segregation can occur with respect to time, specifically with respect to those species that feed at night versus during the day. A few seabird species are adapted to feed only at night.

Mutualism and Kleptoparasitism

The seabird flocking community in the North Pacific comprises species having complementary foraging behaviors, thus, indicating a degree of integration within the community. In particular, the feeding behavior of catalyst and diving species could be interpreted as mutualistic, catalysts signaling the location of a prey patch, and divers increasing or maintaining prey concentration. Certain authors have speculated that this relationship could have resulted from co-evolution of behavior designed to increase the mutualistic benefit of the association. Kleptoparasitism was also proposed to stabilize these feeding flocks by forcing alcids to forage at the edges of a prey patch where they are less vulnerable to piracy, thus, maintaining patch density and ultimately, increasing prey availability to all flock members.

Morphological or Physiological Factors

Differential resource use is sometimes ascribed to species-specific morphological or physiological factors affecting flight or diving capabilities. Several examples exist. First, terns differ in their ability to feed successfully in dense flocks over predatory fishes as a function of a given species' ability to hover for prolonged periods of time. Second, differential metabolic demands may be responsible for species-specific differences in the threshold prey density to which alcids respond. Finally, differential flight costs correlate with species-specific patterns in resource use, e.g. along productivity gradients in the tropics, or the amount of foraging habitat that can be exploited (near-shore vs. offshore).

Ultimately, many of these morphological and physiological adaptations are driven by body size. Body size influences depth of dive capabilities, cost of transport, and basal metabolic rate. Additionally, body size can frequently be used to predict the outcome of interference competition (below).

Competition

Interference competition apparently does occur between seabirds at sea. It is referred to most often in the context of feeding flocks, taking the form of aggressive encounters, and collisions between feeding birds. The proximate limiting resource identified in many of these cases is access to prey, i.e. space over the prey patch. In another situation, shearwaters in the North Pacific feed by pursuit plunging in large groups, by which they disperse, decimate, or drive prey deeper into the water column thereby reducing the availability of prey to surface-feeding species. This same mechanism has been proposed for tropical boobies, which by plunge diving may also drive prey beyond the reach of surface feeders.

Despite widespread discussion of trophic competition, supporting data are sparse and some evidence indicates it to be not important in structuring some seabird communities. For example, one study in the Antarctic found no habitat expansion of the pack-ice assemblage into adjacent open waters seasonally vacated by another community (**Figure 3**), a shift that might be expected if competition affected community structure and habitat selection. In that study, sufficient epipelagic prey were available in the ice-free waters to be exploited successfully by seabirds (**Figure 5**).

Competition with Fisheries

Many of the forage species sought by seabirds are the same sought by industrial fisheries. The result is conflict, particularly in eastern boundary currents, where clupeid fishes are dominant and are of ideal size and shape to be consumed by seabirds. The tracking of bird populations with fish stocks has been especially well documented in the Benguela and Peru

currents, where not only have fish stocks been heavily exploited but so have guano deposits accumulated by the seabirds. The bird populations have responded closely to geographic, temporal and numerical variation in the fish stocks. Well documented, also, have been fish stocks and avian predator populations in the North Sea. There, commercial depletion of predatory fish benefited sea-bird populations by reducing competition for forage fish; when fisheries turned to the forage fish themselves, sea-bird populations declined. In some areas, it has been proposed to use statistical models of predator populations as an indicator of fish-stock status independent of fishery data, for instance, the Convention for the Conservation of Antarctic Living Marine Resources. Much information is needed to calibrate seabird responses to prey populations before sea birds can be used reliably to estimate prey stocks.

See also

Seabird Migration.

Further Reading

Ainley DG, O'Connor EF, and Boekelheide RJ (1984) *The Marine Ecology of Birds in the Ross Sea, Antarctica.* American Ornithologists' Union, Monograph 32. Washington, DC.

Ainley DG and Boekelheide RJ (1990) *Seabirds of the Farallon Islands: Ecology, Dynamics and Structure of an Upwelling-system Community.* Stanford, CA: Stanford University Press.

Ashmole NP (1971) Seabird ecology and the marine environment. In: Farner DS, King JR, and Parkes KC (eds.) *Avian Biology*, vol. 1, pp. 223–286. New York: Academic Press.

Briggs KT, Tyler WB, Lewis DB, and Carlson DR (1987) *Bird Communities at Sea off California: 1975–1983.* Studies in Avian Biology, No. 11. Berkeley, CA: Cooper Ornithological Society.

Burger J, Olla BL, and Winn WE (eds.) (1980) *Behavior of Marine Animals, vol. 4: Birds.* New York: Plenum Press.

Cooper J (ed.) (1981) *Proceedings of the Symposium on Birds of Sea and Shore.* African Seabird Group, Cape Town, Republic of South Africa.

Croxall JP (ed.) (1987) *Seabirds: Feeding Ecology and Role in Marine Ecosystems.* London: Cambridge University Press.

Furness RW and Greenwood JJD (1983) *Birds as monitors of environmental change.* London and New York: Chapman Hall.

Furness RW and Monaghan P (1987) *Seabird Ecology.* London: Blackie.

Montevecchi WA and Gaston AJ (eds.) (1991) *Studies of high latitude seabirds 1: Behavioural, Energetic and*

Oceanographic Aspects of Seabird Feeding Ecology. Canadian Wildlife Service, Occasional Papers 68. Ottawa, Canada.

Nettleship DN, Sanger GA, and Springer PF (1982) Marine Birds: Their Feeding Ecology and Commercial Fisheries Relationships. Canadian Wildlife Service, Special Publication. Ottawa, Canada.

Vermeer K, Briggs KT, and Siegel Causey D (eds.) (1992) Ecology and conservation of marine birds of the temperate North Pacific. Canadian Wildlife Service, Special Publication. Ottawa, Canada

Whittow GC and Rahn H (eds.) (1984) *Seabird Energetics.* New York: Plenum Press.

SEABIRD MIGRATION

L. B. Spear, H.T. Harvey Associates,
San Jose, CA, USA

Introduction

Bird migration is one of the most fascinating phenomena in our living environment, and accordingly has been studied since ancient times, particularly among nonmarine species. Studies of migration and navigation of nonmarine species have become quite sophisticated, examining in detail subjects including orientation and navigation, and physiological and morphological adaptations. In contrast, studies of migration among marine birds have been fewer and more simplistic. Indeed, until recently, much of the information on migration of seabirds had come from the recovery of individuals ringed (i.e., metal rings are attached to the legs, each stamped with a unique set of numbers) at their breeding sites. Although ringing has revealed considerable information about the migrations of species that stay close to coasts (thus, facilitating recoveries), it has provided little insight into the movements of species that stay far at sea during the nonbreeding period. In addition, studies at sea have been few, owing to the immense size of the world's oceans and the inherent logistical difficulties. Fortunately, however, an upsurge in pelagic investigations has occurred in the past 20 years, owing to the advent of ground-position satellite (GPS) telemeters that can be attached to larger avian species (e.g., albatrosses), and ship-board studies of the flight direction and flight behavior of birds on the high seas.

Background

Migration among seabirds ultimately is a response to the seasonally changing altitude of the sun's position, causing changes in environmental conditions to which the birds must adapt to survive and reproduce. Individual seabirds have the ability to go to a precise wintering location and then return to a precise breeding location. The duration between trips to and from wintering and breeding sites can be annual (as in adults), or last several years in the case of subadults of some species (notably the Procellariiformes, see below). In the latter case, fledgelings go to sea

and do not return to land until reaching sexual maturity at ages of up to 10 years or more, such as in the wandering and royal albatrosses (*Diomedea exulans* and *D. epomophora*). On reaching breeding age, many seabirds return to the same colony from which they originated, in fact, they often nest on or adjacent to the exact location where they were hatched. After first breeding, a large proportion also return each year to the same nest site to breed. Furthermore, individual adults have the ability to return with pinpoint precision to the same wintering location each year following breeding. These locations are usually those where these individuals foraged during their subadult years.

Seabirds can home in on their breeding sites during all types of weather, during darkness (e.g., some species return to nesting burrows under dense vegetation only during the night, even during dense fog), and can navigate distances at sea approaching a global scale. The latter was demonstrated in two experiments in which the Manx shearwater (*Puffinus puffinus*) and 18 Laysan albatrosses (*Phoebastria immutibilis*) were taken from their breeding sites (where they were attending eggs or young) and airfreighted to locations thousands of kilometers away. Many were released at locations where they surely had not been before, and in environments for which they are not adapted. The shearwater returned to its nest site in Wales in 12.5 days, covering the 5200 km (shortest) distance from its release site at Boston, Massachusetts, at a rate of 415 km/day. Fourteen of the 18 albatrosses returned to their nest sites on Sand Island, Hawaii, with median trip duration and distance flown of 12 days, and 275 km/day (straight-line), respectively. One bird, released in Washington, took 10.1 days to cover the 5200 km distance back to Sand Island, although it probably flew a longer, tacking course because of the headwinds that it would have encountered if it flew directly to the island.

Sensory Mechanisms Used for Orientation and Navigation

As noted above, studies among terrestrial species have provided many insights into the sensory mechanisms by which birds navigate over long distances. Given the ability of seabirds to navigate long distances across the open ocean, they most likely use one or a combination of the sensory mechanisms indicated for terrestrial species. These mechanisms include endogenous (genetically transmitted) vector navigation, olfactory and time-compensated sun orientation; star, magnetic, UV, and polarized light orientation.

Endogenous vector navigation, in which birds are genetically programmed to follow the correct course and to start their migrations at the correct time, is thought to explain how young birds that have never migrated before, and that frequently do not accompany experienced individuals (as in the case for many seabirds), find their wintering areas. In this regard, spatial and temporal orientation appear to be coded to both celestial rotation and the geomagnetic field, this information being contained within a heritable, endogenous program.

Yet, migrants encounter many uncertainties due to unpredictable weather which can disrupt vector navigation. Through a series of experimental studies examining hypotheses addressing this problem, it has become the general consensus that birds have a compass sense as well as a very extensive grid/mosaic map sense. Birds can integrate combinations of time-compensated sun inclination (particularly at sunset), star, magnetic, UV, and polarized light cues, although the possibility that these capabilities vary among species remain open. Nevertheless, the evidence indicates that for short-term orientation, magnetic cues take precedence over celestial, that visual cues at sunset over-ride both the latter, and that polarized sky light is used during dusk orientation. The basis for the map aspect employed for navigation is not well studied and consists of two hypotheses: perceptions of a magnetic grid and/or an olfactory mosaic/gradient.

Physiological and Behavioral Adaptations

Many species of terrestrial birds have major shifts in their physiology just preceding the migration period, including dramatic increases in food intake and body fat, hypertrophy of breast muscles, increased hematocrit levels, and increases in body protein. Although few physiological studies of seabirds exist, most indicate a lack of, or smaller, physiological changes then occurs in land birds. During the post-breeding phase of migration, seabirds are generally lighter, with lower fat reserves, than when returning to their colonies at the beginning of the next breeding season. For example, adult sooty shearwaters (*P. griseus*) weigh 20–25% less during the post-breeding migration than when returning during the prenuptial period. Similar differences in the pre- and post-breeding body mass also occur in several Charadriformes (e.g., gulls, terns, auks).

Several factors could be responsible for the lack of more obvious physiological adaptations for migration among seabirds. First, most of the terrestrial species in which major physiological changes occur have small body masses ($< 75\,g$) and, thus, lower flight efficiency than larger species such as most seabirds. Second, and probably most importantly, many of the terrestrial species perform long-distance nonstop flights, often at high elevations, over obstacles where they cannot feed (e.g., large bodies of water, or deserts). In contrast, such migrations are rare among seabirds, because seabirds nearly always migrate at low elevations over the ocean, facilitating frequent feeding along the migration route. Hence, seabirds are seldom far from a habitat offering feeding opportunities, even during transequatorial migrations by species that feed mostly in temperate or boreal latitudes.

The post-breeding migration of seabirds is generally more leisurely than the pre-breeding migration. One reason is that seabirds, especially those moving longer distances, feed more during the post-breeding than the pre-breeding movement. This behavior is probably related to the poorer body condition of seabirds just after breeding. Two examples include the sooty shearwater and the Arctic tern. Both perform transequatorial migrations, although the shearwater is a southern hemisphere breeder, with its post-breeding migration during the boreal spring, whereas the tern is a northern hemisphere breeder that leaves its breeding grounds in the boreal autumn. Thus, the post-breeding movements by the two species are 6 months out of phase, indicating that seasonal differences in ocean productivity in equatorial waters (highest in the boreal autumn) are unrelated to the low feeding rate during the pre-breeding migration. Faster prenuptial migration probably occurs because ample fat reserves have been obtained on the winter grounds (i.e., feeding is not required). In addition, higher wing loading (from higher fat reserves) facilitates faster flight. Thus, the seasonal differences in fat reserves among species of seabirds are not likely adaptations for migration *per se*, but instead, are important in the life histories of many because early arrival on the breeding grounds facilitates the acquisition of a higher quality nesting territory favorable for successful breeding and because the amount of time available for foraging after arrival at the colony is greatly reduced.

Morphological Adaptations

The distance that seabirds migrate is strongly related to morphology. The most important morphological feature is the shape of the wings, measured as the aspect ratio, a dimensionless value calculated as the wing span^2 divided by total wing area. Hence, birds with high aspect ratios have narrower wings. They also have less profile drag (i.e., less friction with the air), lower air turbulence, and generally migrate longer distances compared with birds with lower aspect ratios. For instance, in the Laridae many species of terns, and smaller gulls and skuas with long narrow wings are transequatorial migrants, whereas larger gulls and skuas with lower aspect ratios usually move much shorter distances or are even sedentary.

Wing loading (wing area divided by body mass) is also related to migration patterns in seabirds, although this relationship differs with flight styles used by different seabird taxa, and is also confounded with aspect ratio. Higher wing loading requires swifter flight for the birds to remain airborne. Within taxa of seabirds that typically use gliding or flap-gliding flight (e.g., albatrosses, shearwaters, and petrels) those with higher wing loading tend to have longer migrations. The gliding species with higher wing loading also tend to have higher aspect ratios (**Figure 1**), which increases their flight efficiency. Swift, energy-efficient flight equates to longer distances travelled; however, the gliding species are heavily dependent on wind energy for flight because of their flight mode and high wing loading. As a result, those with higher wing loadings are confined to higher latitudes where the wind is usually stronger, although some species with moderate to high wing loading do make transequatorial crossings.

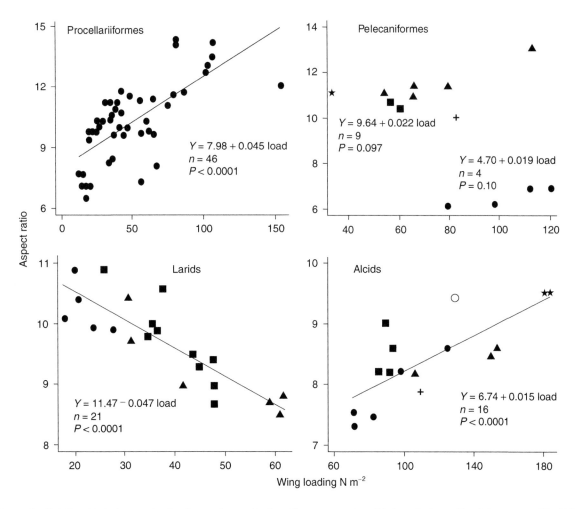

Figure 1 Relationship between wing loading and aspect ratio of four groups of seabirds, as indicated by data presented in 1997 by Spear and Ainley. Each point denotes the average for a given species. Lines indicate the best regression fit. Values of *n* are number of species, and *P* values indicate the level of significance by which slopes deviated from zero. Among the Pelecaniformes, the cormorants (●) were considered to be distinct from six species representing pelicans (+), tropic birds (■), boobies (▲), and frigate birds (★). Larids included terns (●), skuas (▲), and gulls (■). Alcids included auklets (•) murrelets (■), puffins (▲), pigeon guillemot (+), razorbill (○), and murres (★).

In contrast to gliders, seabirds that use flapping flight, such as Larids, have an inverse relationship between wing loading and migration distance (**Figure 1**). This is undoubtedly related to the lower flight efficiency of flapping species with higher wing loading compared with those with lower wing loading, all other factors being equal. Another factor is the inverse relationship between wing loading and aspect ratio when comparing gliding versus flapping species (with the exception of Alcids – see below). That is, aspect ratio decreases (and drag increases) with increases in wing loading among most seabird taxa that typically use flapping flight.

As noted above, Alcids (auks, auklets, murres, murrelets, and guillemots) are an exception to the wing loading versus aspect ratio relationship for species using flapping flight. Aspect ratios of Alcids increase with wing loading (**Figure 1**), i.e., a relationship similar to that of Procellariiformes (primarily gliders). However, unlike the Procellariiformes and Larids, there does not appear to be a relationship between wing loading, or aspect ratio, and migration tendency among Alcids. This may be because Alcid wings are highly adapted for underwater propulsion (i.e., similar to penguins) resulting in small wing sizes and the highest wing loading among avian species. Thus, their foraging ranges and nonbreeding movements are short. Indeed, many Alcid species are flightless during part of the dispersal period due to primary molt, indicating that a significant part of the dispersal is traversed by swimming.

Types of Migration

Several types of 'migration' are recognized. These include 'true migration' in which all members of a population move from a breeding area to a wintering area disjunct from the former; and 'partial migration', in which some members of a population migrate and others do not. Yet another type of movement, 'dispersal', is found in populations in which individuals move various distances after the breeding season, such that they occur at all distances within a given radius of the breeding site. Although a large proportion of the terrestrial avifauna exhibits true annual migration, this is rare among seabirds, probably because the marine environment is more stable in that it does not have the extreme seasonal warming and cooling of land masses. In addition, seabirds usually expand their at-sea ranges without having to cross as many barriers as are encountered during movements by terrestrial species. Even so, seabirds have the notoriety of having the longest

distance migrants among the animal kingdom. The following is a review of the movement patterns of seabirds.

Sphenisciformes

Penguins (Spheniscidae) The pelagic ranges of the penguins are confined to the southern hemisphere. Relatively little is known of the post-breeding movements of these 17 flightless species. The four species of the genus *Spheniscus* have the lowest latitude distributions among penguins and probably move the shortest distances. Two, the Galapagos and Humboldt penguins (*S. mendiculus* and *S. humboldti*), are relatively sedentary along the coasts of the Galapagos Islands, and Peru to northern Chile, respectively. These species rarely are found more than 10 km from shore. The two other *Spheniscus* species usually stay within 50 km of the coasts of Africa (African penguin, *S. demerus*) and southern South America (Magellanic penguin, *S. magallenicus*), although there have been sightings of the latter to 250 km offshore. Satellite telemetry has shown that the Magellanic penguins breeding on Punta Tombo, Argentina, disperse for up to 4000 km northeastward to wintering areas off Brazil, although other population members winter along the coast of Argentina, closer to the colony. Three of the four telemetered Magellanic penguins which moved to coastal Brazil traveled 20 km/day.

Dispersal by some of the species breeding in higher, temperature to subpolar, latitudes (king, *Aptenodytes patagonicus*; rockhopper, *Eudyptes chrysocome*; Snares Island, *E. robustus*; Fiordland crested, *E. pachyrhynchus*; erect-crested, *E. sclateri*; royal, *E. schlegeli*; macaroni, *E. chytsolophus*; Gentoo, *Pygoscelis papua*; yellow-eyed, *Megadyptes antipodes*; and little penguins, *Eudyptula minor*) is probably more extensive than is true for *Spheniscus*, as they are often seen hundreds of kilometers from shore. Finally, two of the three penguin species that breed only on the Antarctic continent (Adelie and chinstrap penguins, *Pygoscelis adeliae* and *P. antarctica*) migrate north to areas near or inside of the ice pack (Adelies; e.g., polynyas, leads) or open water (chinstraps) after waters near the continent become covered with solid ice. Movements of the third species, the Emperor penguin (*Aptenodytes forsteri*), are not well know; however, there are sighting records for this species from the coasts of Argentina and New Zealand.

Procellariiformes

Albatrosses (Diomedeidae) The albatrosses are under extensive taxonomic reclassification, and number between 14 and 24 species. Post-breeding movements of albatrosses are mostly longitudinal; no species undertakes transequatorial movements. These birds range long distances from their colonies during the nonbreeding period as well as during foraging trips undertaken while they are breeding. For example, six wandering albatrosses equipped with satellite transmitters flew 3660–15 200 km during single foraging trips after being relieved by their mates from incubation duties at the nest. Even breeding waved albatrosses (*Phoebastria irrorata*), with the smallest pelagic range among the group, have a round-trip commute of no less than 2000 km between the breeding colony on the Galapagos Islands and the nearest edge of their foraging area along the coast of Ecuador and Peru.

Most albatross species range farthest from their colonies during the nonbreeding period. In fact, many species are partially migratory (as opposed to being dispersers). As explained above, these species are considered as partially migratory because individuals are found in both the wintering and breeding areas during winter, but do not winter (or winter in small numbers) between the two locations. For example, Buller's (*Thalassarche bulleri*), Chatham (*T. eremita*), Salvin's (*T. salvini*), and shy (*T. cauta*) albatrosses breed on islands near New Zealand (the first three) and Australia (shy). Although some birds stay near the colonies throughout the year, large proportions of the Buller's, Chatham, and Salvin's albatrosses migrate at least 8500 km eastward across the South Pacific to the coast of South America, and many shy albatrosses migrate westward across the Indian Ocean to the coast of South Africa.

Three other species, the wandering, black-browed (*T. melanophris*), and gray-headed (*T. crysostoma*) albatrosses, have breeding colonies located circumpolarly across southern latitudes near 50°S. The South Georgia populations may be partially migratory, although the distinction from dispersive is not clear. South Georgian wandering albatross fly north to waters off Argentina, and then eastward to important wintering areas off South Africa. Some continue to Australian waters and may even circumnavigate the Southern Ocean. One of several of these birds equipped with a satellite transmitter averaged 690 km/day. A large proportion (*c.* 85%) of South Georgian black-browed albatrosses also winter off South Africa, and many South Georgian gray-headed albatrosses are thought to fly westward to waters off the Pacific coast of Chile, and then to New Zealand.

The waved albatross is unique among the albatross group. Besides having the smallest pelagic range, it breeds near the Equator, i.e., at a latitude

>25° lower than any of the other species. Furthermore, the foraging area used while breeding is the same, or nearly the same, as that used post-breeding. Thus, the classification of this species as a 'disperser', or even as 'partially migratory', is appropriate only in that it leaves the colony post-breeding (and does not occupy the 900 km stretch between the Galapagos and the mainland), although the size of the foraging area changes little.

The post-breeding movements of yellow-nosed (*T. chlororhynchos*), sooty (*Phoebetria fusca*), light-mantled sooty (*P. palpebrata*), royal, black-footed (*Phoebastria nigripes*), short-tailed (*P. albatrus*), and Laysan albatrosses are less clear, although each apparently disperses post-breeding to seas adjacent to their colonies.

Fulmars, Shearwaters, Petrels, Prions, and Diving Petrels (Procellariidae)   The migration tendencies of this family of 78 species is not well known, although those that have been studied are either partially migratory or dispersers. The pelagic range of 38 species (49%) is confined to the southern hemisphere, including the 18 species of fulmarine petrels and prions, three of the four species of diving petrels, seven (33%) of the shearwaters, and 10 (29%) of the gadfly petrels (**Table 1**). Another 19 (25%) of the 78 species perform extensive transequatorial migrations. These include nine species (43%) of the shearwaters and 10 species (29%) of gadfly petrels. Twenty (26%) others are primarily tropical, including one species of diving petrel, five species (24%) of shearwaters, and 14 (40%) gadfly petrels. Many of the 'tropical' species disperse across the Equator, but like species having non-transequatorial movements, movements of these species usually have a greater longitudinal than latitudinal component, and are usually of shorter distances than those of transequatorial migrants.

The detailed migration/dispersal routes of most Procellariids are poorly known. Two exceptions are the sooty shearwater and Juan Fernandez petrels, which are very abundant and appear to be partial migrants. Two populations of sooty shearwaters exist, one breeding in New Zealand and the other in Chile. Many individuals from each population migrate to and from the North Pacific each year, and none winter in the equatorial Pacific. Observations of flight directions in the equatorial Pacific indicate that many complete a figure of eight route (*c.* 40 500 km). The route apparently involves easterly flight from New Zealand to the Peru Current in winter, northwesterly flight to the western North Pacific in spring, eastward movement to the eastern North Pacific during summer, and

Table 1 Migration tendencies (transequatorial, nontransequatorial, and tropical) of the 78 species of Procellariids and 19 species of Oceanitids[a]

PROCELLARIIDAE
FULMARINE PETRELS Nontransequatorial (12)
Northern giant petrel *Macronectes halli*
Southern giant petrel *M. giganteus*
Northern fulmar *Fulmarus glacialis*
Southern fulmar *F. glacialoides*
Antarctic petrel *Thalassoica antarctica*
Cape petrel *Daption capnse*

Snow petrel *Pagodroma nivea*
White-chinned petrel *Procellaria aequinoctialis*
Parkinson's petrel *P. parkinsoni*
Westland petrel *P. westlandica*
Grey petrel *P. cinerea*

GADFLY PETRELS
Nontransequatorial (10)
Great-winged petrel *Pterodroma macoptera*
Atlantic petrel *P. incerta*
Kerguelen petrel *P. brevirostris*
Magenta petrel *P. magentae*
Soft-plumaged petrel *P. mollis*
Barau's petrel *P. baraui*
White-headed petrel *P. lessoni*
Bonin petrel *P. hypoleuca*
Chatham petrel *P. axillaris*
Defilippe's petrel *P. defilippiana*

Transequatorial (10)
Mottled petrel *P. inexpectata*
Murphy's petrel *P. ultima*
Solander's petrel *P. solandri*
Kermadec petrel *P. neglecta*
Juan Fernandez petrel *P. externa*
White-necked petrel *P. cervicalis*
Cook's petrel *P. cooki*
Black-winged petrel *P. nigripennis*
Pycroft's petrel *P. pycrofti*
Stejneger's petrel *P. longirostris*

Tropical (14)
Phoenix petrel *P. alba*
Trinidad petrel *P. arminjoniana*
Herald petrel *P. heraldica*
Hendersons's petrel *P. atrata*
Tahiti petrel *P. rostrata*
Mascarene petrel *P. aterrima*
Bermuda petrel *P. cahow*
Black-capped petrel *P. hasitata*
Hawiian petrel *P. sandwichensis*
Galapagos petrel *P. phaeopygia*
White-winged petrel *P. leucoptera*
Collared petrel *P. brevipes*
Bulwer's petrel *Bulweria bulwerii*
Jouanin's petrel *B. fallax*

Unknown (1)
Macgillivray's petrel *P. macgillivrayi*

PRIONS Nontransequatorial (7)
Blue petrel *Halobaena caerulea*
Broad-billed prion *Pachyptila vittata*
Antarctic prion *P. desolata*
Salvin's prion *P. salvini*
Fairy prion *P. turtur*
Fulmar prion *P. crassirostris*
Slender-billed prion *P. belcheri*

DIVING PETRELS
Nontransequatorial (3)
Georgian diving petrel *Pelecanoides georgicus*
Common diving petrel *P. urinatrix*
Magellan diving petrel *P. magellani*

Tropical (1)
Peruvian diving petrel *P. garnoti*

SHEARWATERS (21 species)
Nontransequatorial (7)
Little shearwater *Puffinus assimilis*
Black-vented shearwater *P. opisthomelas*
Fluttering shearwater *P. gavia*
Hutton's shearwater *P. huttoni*
Heinroth's shearwater *P. heinrothi*
Balearic shearwater *P. mauretanicus*
Levantine shearwater *P. yelkouan*

Transequatorial (9)
Streaked shearwater *Calonectris leucomelas*
Cory's shearwater *C. diomedea*
Pink-footed shearwater *Puffinus creatopus*
Flesh-footed shearwater *P. carneipes*
Great shearwater *P. gravis*
Buller's shearwater *P. bulleri*
Sooty shearwater *P. griseus*
Short-tailed shearwater *P. tenuirostris*
Manx shearwater *P. puffinus*

Tropical (5)
Wedge-tailed shearwater *P. pacificus*
Christmas shearwater *P. nativitatis*
Newell's shearwater *P. newelli*
Townsend's shearwater *P. auricularis*
Audubon's shearwater *P. lherminieri*

Table 1 *Continued*

OCEANITIDAE
STORM PETRELS

Nontransequatorial (8)	**Transequatorial (5)**
Gray-backed storm petrel *Garrodia nereis*	Wilson's storm petrel *Oceanites oceanicus*
White-faced storm petrel *Pelagodroma marina*	British storm petrel *Hydrobates pelagicus*
Black-bellied storm petrel *Fregetta tropica*	Leach's storm petrel *Oceanodroma leucorhoa*
White-bellied storm petrel *F. grallaria*	Black storm petrel *O. melania*
Tristram's storm petrel *O. tristrami*	Matsudaira's storm petrel *O. matsudairae*
Swinhoe's storm petrel *O. monorhis*	
Ashy storm petrel *O. homochroa*	**Tropical (6)**
Least storm petrel *O. microsoma*	Elliot's storm petrel *Oceanites gracilis*
	White-throated storm petrol *Nesofregetta fuliginosa*
	Wedge-rumped storm petrel *Oceanodroma tethys*
	Harcourt's storm petrel *O. castro*
	Markham's storm petrel *O. markhami*
	Hornby's storm petrel *O. hornbyi*

[a]Tropical species are those in which most individuals stay between the tropic of Cancer/Capricorn.

southwest flight to New Zealand during autumn (**Figure 2**). Most are probably nonbreeders, possibly from both the New Zealand and Chilean populations. Many, probably breeders from both populations, likely use shorter routes to and from the North Pacific (*c.* 28 000–29 000 km). Other (nonmigratory) individuals from both populations apparently stay in the southern hemisphere. Migration routes are coordinated with wind regimes in the Pacific, such that the usual flight direction utilizes quartering tail winds (**Figure 2**).

Juan Fernandez petrels breed in the Juan Fernandez archipelago off Chile. Many migrate into the North Pacific where they winter mostly between 5°N and 20°N. Another large component of the population stays in the South Pacific, mostly between 12°S and 35°S. Collections of specimens indicate that the great majority found in the North Pacific are subadults, whereas those in the South Pacific are predominantly adults.

It is likely that many other Procellariids also perform partial migration (e.g., Cook's, white-winged, black-winged, and mottled petrels), or even complete migrations (e.g., Cook's petrel, Hutton's shearwater, Magenta and Bonin petrels.

Storm Petrels (Oceanitidae) The 19 species of storm petrels include eight with nontransequatorial movements, five transequatorial, and six that stay primarily in tropical latitudes. Storm petrels are mostly dispersive. Leach's and Wilson's storm petrels disperse the farthest and also are the most abundant, with circumpolar distributions. The former breeds mostly between 25°N and 50°N, and winters as far south as waters near New Zealand (about 35°S). Similarly, the Wilson's storm petrel

breeds from about 45°S to 60°S, and winters north to about 40°N.

Two species of storm petrels that may be migratory, or partially migratory, include the white-faced and white-bellied storm petrels. Many white-faced storm petrels (race *maoriana*) that breed adjacent to New Zealand apparently migrate east to waters off Chile and Peru. In warm-water (El Niño) years, some birds continue westward from the Peru Current, out along the Equator, in association with waters of the South Equatorial Current. The *grallaria* race of white-bellied storm petrel is represented by a very small population breeding on islands north of New Zealand and Australia. This population is particularly interesting in that new information from the equatorial Pacific indicates that many or all of these birds migrate about 2500 km to a relatively small (1 million km^2) section of waters between 135°W and 145°W and between 5°S and about 12°S, adjacent to the Marquesas.

Pelecaniformes

Pelicans (Pelecanidae, 4 marine species), boobies and gannets (Sulidae, 9 species), cormorants (Phalacrocoracidae; 29 marine species), frigate birds (Fregatidae; 5 species), and tropic birds (Phaethontidae, 3 species) Distributions of the marine species of pelicans and cormorants are coastal, whereas those of boobies, frigate birds, and tropic birds are pelagic. Movements of most Pelecaniformes are dispersive, and none of the nontropical species performs transequatorial migrations. Many species, including all boobies, frigate birds, and tropic birds, are tropical; most of the cormorants, pelicans, and gannets prefer temperate to polar latitudes. Movements of the tropical species are primarily

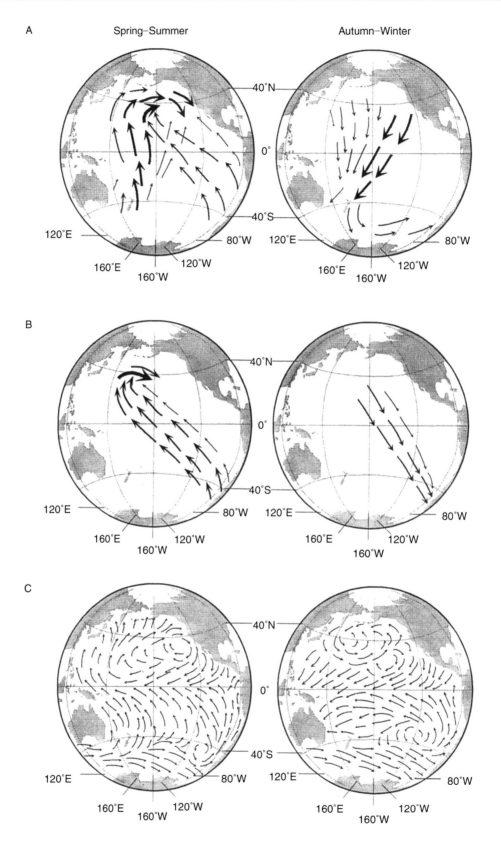

Figure 2 Suggested migration routes of sooty shearwaters from colonies near (A) New Zealand and Australia, and (B) Chile. Sizes of vectors reflect differences in the number of shearwaters, as suggested in 1974 by Shuntov from observations in the South and North Pacific, and in 1999 by Spear and Ainley, from observations in the equatorial Pacific. (C) Wind regimes of the Pacific Ocean during two seasonal periods as depicted in 1966 by Gentilli.

nondirectional (i.e., direction can include combinations of north, south, east, and west orientation), whereas those of temperate to polar species usually have a stronger latitudinal component.

Charadriiformes

Phalaropes (Scolopacidae) Two of the three phalarope species breed in the northern hemisphere, either in the continental interiors (red-necked phalarope, *Phalaropus lobatus*) or Arctic slopes (red phalarope, *P. fulicarius*), and perform extensive migrations to marine habitat. Movements of a large proportion of both populations, particularly those of the red phalarope, are transequatorial, with many individuals wintering along the west coasts of South America and Africa. A major concentration also winters in the Panama Bight.

Skuas (Stercoracidae) Like the two phalaropes, the three smaller skuas (pomarine, *Stercorarius pomarinus*; Arctic, *S. parasiticus*; and long-tailed, *S. longicaudus*) breed on the Arctic slope and winter in oceanic waters. A large proportion of these birds also perform extensive transequatorial migrations, with large numbers wintering off the west coasts of South America and South Africa. Individuals are also occasionally seen off Australia, New Zealand, and in the Indian Ocean. Another large percentage winters off Mexico, Central America, and northern Africa. Finally, a minority, primarily adults, stay in temperate to subpolar latitudes of the northern hemisphere during winter.

The four larger skuas (great, *Catharacta skua*; brown, *C. lonnbergi*; South Polar, *C. maccormicki*; and Chilean, *C. chilensis*) generally perform only short, dispersive post-breeding movements. An exception is the South Polar skua, many of which disperse widely from their Antarctic breeding sites, even into the more northern latitudes (e.g., Alaska) of the northern hemisphere.

Gulls and Terns (Laridae) The 48 gull species (subfamily, Larinae) are mostly dispersive, although for some species the post-breeding dispersal of some birds can extend for thousands of kilometres. As noted above, the larger species, with higher wing loading, tend to move shorter distances post-breeding than do smaller species. The five migratory species include the Franklin's, Sabine's, Thayer's, lesser black-backed, and California gulls. Of these, the movements of three (6% of the 48 species) are transequatorial (**Table 2**). In fact, these three species are the only ones with regular transequatorial

movements among the gulls that have nontropical breeding grounds. Of the remaining species, the range of 40 (83%) is confined to one hemisphere or the other, and that of five others (10%) is confined within tropical latitudes.

Of the five migrants, only the Sabine's and Thayer's gulls breed in Arctic latitudes. Interestingly, other species that also breed in the Arctic, including the Ivory, Ross's, glaucous, and Iceland gulls remain there, or disperse relatively short distances to sub-Arctic latitudes, during winter. It is likely that different foraging habitats or prey requirements are responsible for these differences in movement patterns.

The post-breeding, dispersive movements of larger gull species have been studied in detail. These studies indicate that breeding adults leave the colony and fly quickly to specific locations, such as a particular bay or fishing port. The locations, 'vacation spots', are those with which they have become familiar during their subadult years (first 4 years of life), such that the birds know the foraging logistics and availability of a predictable food supply. This is important. After reaching the vacation location the adults must molt and replace the primary wing feathers (making them less mobile for about a month) and replenish body reserves in preparation for the next breeding season.

The 44 species of terns (subfamily, Sterninae) are mostly smaller than gulls and have higher aspect ratios (i.e., longer, narrower wings). Not surprisingly, this group is represented by a higher proportion of species that migrate, including eight species (18%) whose migrations are transequatorial, compared with 14 species (32%) whose movements are mostly confined to one hemisphere, and 21 species (50%) that remain primarily within tropical latitudes.

Arctic terns are unique because they have the longest migrations known among the world's animal species. These terns breed in the Arctic to 80°N, and winter in the Southern Ocean to 75°S. Based on band returns and observations at sea, some Arctic terns from Scandinavia and eastern Canada fly to and from waters off Australia, New Zealand, and the Pacific (**Figure 3**). The shortest, round-trip flight to the Pacific exceeds 50 000 km. These birds can live up to 25 years, indicating that the lifetime migration distance could exceed 1 million km.

Murres, murrelets, auks, auklets, puffins, and guillemots (Alcidae) The 22 species of Alcids, confined to the northern hemisphere, have dispersive post-breeding movements. Compared with other seabirds, their dispersal distances are short (see Morphological Adaptations and Flight Behavior). The primary reasons are: (1) their very

Table 2 Migration tendencies of the Larids (transequatorial, nontransequatorial, and tropical), including 48 species of gulls and 44 species of terns[a]

GULLS

Nontransequatorial (40)

Great black-backed gull *Larus marinus*
Western gull *L. occidentalis*
Yellow-footed gull *L. livens*
Herring gull *L. argentatus*
Kelp gull *L. dominicanus*
Glaucous-winged gull *L. glaucescens*
Glaucous gull *L. hyperboreus*
Slaty-backed gull *L. shistisagus*
Great black-headed gull *L. ichthyaetus*
Indian black-headed gull *L. brunnicephalus*
Chinese black-headed gull *L. saundersi*
Pacific gull *L. pacificus*
Band-tailed gull *L. belcheri*
Iceland gull *L. glaucoides*
Kumlien's gull *L. kumlieni*
Thayer's gull *L. thayeri*
California gull *L. californicus*
Black-tailed gull *L. crassirostris*
Sooty gull *L. hemprichii*
White-eyed gull *L. leucophthalmus*
Dolphin gull *L. scoresbii*
Common gull *L. canus*
Ring-billed gull *L. delewarensis*
Black-headed gull *L. ridibundus*
Laughing gull *L. atricilla*
Bonaparte's gull *L. philadelphia*

Relict gull *L. relictus*
Hartlaub's gull *L. hartlaubii*
Heermann's gull *L. heermanni*
Brown-hooded gull *L. maculipennis*
Silver gull *L. novaehollandiae*
Black-billed gull *L. bulleri*
Little gull *L. minutus*
Audouin's gull *L. audouinii*
Mediterranean gull *L. melanocephalus*
Slender-billed gull *L. genei*
Black-legged kittiwake *Rissa tridactyla*
Red-legged kittiwake *R. brevirostris*
Ross's gull *Rhodostethia rosea*
Ivory gull *Pagophila eburnean*

Transequatorial (3)

Lesser black-backed gull *L. fuscus*
Franklin's gull *L. pipixcan*
Sabine's gull *L. sabini*

Tropical (5)

Lava gull *L. fuliginosis*
Gray gull *L. modestus*
Gray-headed gull *L. cirrocephalus*
Andean gull *L. serranus*
Swallow-tailed gull *L. furcatus*

TERNS

Nontransequatorial (14)

Caspian tern *Sterna caspia*
South American tern *S. hirundinacea*
Antarctic tern *S. vittata*
Kerguelen tern *S. virgata*
Forster's tern *S. forsteri*
Trudeau's tern *S. trudeaui*
Roseate tern *S. dougalii*
White-fronted tern *S. striata*
Aleutian tern *S. aleutica*
Fairy tern *S. nereis*
Black-fronted tern *S. albostriata*
Damara tern *S. balaenarum*
Little tern *S. albifrons*
Least tern *S. antillarum*

Transequatorial (8)

Sandwich tern *S. sandvicensis*
Common tern *S. hirundo*
Arctic tern *S. paradisaea*
Royal tern *S. maxima*
Elegant tern *S. elegans*
Black tern *Childonias niger*
Whiskered tern *C. hybridus*
White-winged tern *C. leucopterus*

Tropical (22)

Large-billed tern *S. simplex*
Gull-billed tern *S. nilotica*
Indian River tern *S. aurantia*
White-cheeked tern *S. repressa*
Black-napped tern *S. sumatrana*
Black-billed tern *S. meganogastra*
Gray-backed tern *S. lunata*
Bridled tern *S. anaethetus*
Sooty tern *S. furcata*
Amazon tern *S. superciliaris*
Peruvian tern *S. lorata*
Crested tern *S. bergii*
Lesser-crested tern *S. bengalensis*
Chinese crested tern *S. bernsteini*
Cayenne tern *S. eurygnatha*
Saunder's little tern *S. saundersi*
Blue-gray noddy *procelsterna cerulea*
Brown noddy *Anos stolidus*
Black noddy *A. minutus*
Lesser noddy *A. tenuirostris*
Inca tern *Larosterna inca*
White tern *Gygis alba*

[a]Tropical species are those in which most individuals stay between tropic of Cancer/Capricorn.

high wing loading and, thus, inefficient flight; (2) they are highly adapted pursuit divers that can exploit a range of subsurface habitats; and (3) they occur in waters of the Arctic and boundary currents where prey are abundant. In summary, long distance flights by Alcids are impractical, and are not required. This life history trait is like that of penguins, another group highly adapted for pursuit

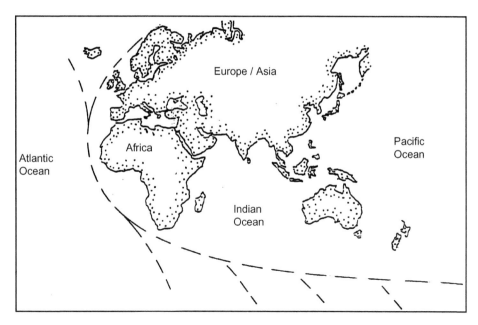

Figure 3 Suggested migration route (dashed lines) to-and-from wintering areas by Arctic Terns breeding in Scandanavia and eastern Canada, from band returns summarized in 1983 by Mead, and from at-sea observations in the Indian Ocean reported in 1996 by Stahl *et al.*

diving, but is in marked contrast to the movements of other seabirds with poorer diving abilities. Alcids with the longest distance dispersals are the little auk (*Alle alle*), tufted puffin (*Fratercula cirrhata*), horned puffin (*F. corniculata*), Atlantic puffin (*F. arctica*), and parakeet auklet (*Cyclorrhynchus psittacula*). Some individuals representing these species disperse up to 1000 km or more into the pelagic waters of the North Atlantic and North Pacific.

See also

Alcidae. Laridae, Sternidae and Rynchopidae. Pelecaniformes. Phalaropes. Procellariiformes. Seabird Conservation. Seabird Foraging Ecology. Seabird Population Dynamics. Sphenisciformes.

Further Reading

Able KP (1995) Orientation and navigation: A perspective of fifty years of research. *Condor* 97: 592–604.

Berthold P (1993) *Bird Migration: A General Survery.* Oxford: Oxford University Press.

Gentilli J (1966) Wind principles. In: Fairbridge RW (ed.) *The Encyclopedia of Oceanography*, pp. 989–993. New York: Reinhold.

Harrison P (1983) *Seabirds, an Identification Guide.* Boston, MA: Houghton-Mifflin.

Mead C (1983) *Bird Migration.* New York: Facts On File, Inc.

Pennycuick CJ (1989) *Bird Flight Performance.* New York: Oxford University Press.

Prince PA, Croxall JP, Trathan PN, and Wood AG (1998) The pelagic distribution of albatrosses and their relationships with fisheries. In: Robertson G and Gales R (eds.) *Albatross Biology and Conservation*, pp. 137–167. Chipping Norton, Australia: Surrey Beatty Sons.

Shuntov VP (1974) *Seabirds and the Biological Structure of the Ocean.* Springfield, VA: US Department of Commerce. [Translated from Russian.]

Spear LB and Ainley DG (1997) Flight behaviour of seabirds in relation to wind direction and wing morphology. *Ibis* 139: 221–233.

Spear LB and Ainley DG (1999) Migration routes of Sooty Shearwaters in the Pacific Ocean. *Condor* 101: 205–218.

Stahl JC, Bartle JA, Jouventin P, Roux JP, and Weimerskirch H (1996) *Atlas of Seabird Distribution in the South-west Indian Ocean.* Villiers en Bois, France: Centre National de la Recherche Scientifique.

SEABIRD POPULATION DYNAMICS

G. L. Hunt, University of California, Irvine, CA, USA

The population biology of seabirds is characterized by delayed breeding, low reproductive rates, and long life spans. During the breeding season, the distribution of seabirds is clumped around breeding colonies, whereas when not breeding, birds are more dispersed. These population traits have important consequences for interactions between people and seabirds. The aggregation of large portions of the adult population of a species in colonies means that a single catastrophic event, such as an oil spill, can kill a large segment of the local breeding population. Although seabird populations can withstand the failure to produce young in one or even a few years without suffering severe population-level consequences, the loss of adults has an immediate and long-lasting impact on population dynamics. Even a small decrease in adult survival rates may cause population decline.

Seabirds breed in colonies on islands, cliffs, and other places where they are protected from attacks by terrestrial predators. Species that forage at large distances from their colonies usually choose locations for their colonies that are less vulnerable to incursions by predators than are the colonies of species that forage in the immediate vicinity of the colony. For the offshore species, the cost of increased travel to a more protected site may be minor compared with the benefit of freedom from unwanted visitors. In contrast, for species that need to forage close to their colonies, even a short increase in the distance traveled between colony and foraging site may mean that it is uneconomical to occupy a particular breeding site.

Colony size tends to vary with the distance that a species travels in search of food. Inshore-foraging seabirds may nest singly or in small groups, whereas species that forage far at sea may have colonies that are comprised of hundreds of thousands of pairs. Two hypotheses have been offered to explain this trend. One hypothesis focuses on the issue of food availability. If birds forage far from their colonies, there is a much greater area in which food may be encountered than if foraging is restricted to a small radius around the colony, and thus a larger size colony can be supported. This hypothesis assumes that seabird colony size is limited by food availability. For species that forage near their colonies,

there is evidence that reproductive parameters sensitive to prey availability, such as chick growth rates and fledging success, vary negatively with colony size. Likewise, there is evidence that colony size and location may be sensitive to the size and location of neighboring colonies. Evidence that seabirds depress prey populations near their colonies is limited. The second hypothesis focuses on the role that colonies may play in the process of information acquisition by birds seeking prey. When birds forage far from their colony, there may be a need for large numbers of birds so that those flying out from the colony are able to observe successful returning foragers and thereby work their way to productive foraging areas using the stream of birds returning to the colony for guidance. The longer the distance from the colony, the greater the number of birds that are required to provide an unbroken stream of birds to guide the out-bound individuals. The evolution of a system of this sort is possible because each individual will benefit from information on food resources gained by being part of a large colony. Selection for large colony size will continue so long as the colony is not so large that food supplies are severely depressed.

Seabirds show considerable philopatry, with individuals often returning to the same colony, or even the same part of the colony from which they fledged. Once a nest site or territory is established, individuals and pairs may use the same site in subsequent years. Pairing tends to be for multiple seasons ('for life'), particularly when pairs are successful at raising young. Divorce does occur, and may be most frequent after failure to raise young. However, experience is important, including experience with a particular partner, so changing partners may result in a period of adjustment to a new partner and consequent lower reproductive success.

The philopatry of seabirds and their tendency to remain in the same part of a colony once they are established may have important implications for the genetic structure of seabird populations. Few data are presently available, but there is some evidence for closer genetic ties for individuals nesting in close proximity. If this is proven to be generally true, there would be important ramifications for understanding a genetic basis for the evolution of coloniality based on inclusive fitness and the reciprocal aid of relatives.

Thus far, there is scant information regarding the genetic distance between individuals in different colonies. The issue of whether colonies are discrete populations (stocks in fisheries parlance) or whether there is considerable exchange between colonies has important implications not only for the evolution of local variation, but also for the conservation and management of seabirds. Little information is available as to whether some colonies are net exporters of recruits (sources) and others net importers (sinks), or if when a colony is decreased in size by a catastrophe unrelated to food resources, it will be quickly replenished by recruits from other colonies.

Delay in the age at which breeding commences is a striking characteristic of seabird population biology. Most species do not begin breeding until they are in their third or fourth year, and some groups, such as the albatrosses, delay breeding until they are 10 or more years of age. Birds that commence breeding at a younger age than is usual for their species have reduced reproductive success in their first year of breeding when compared with individuals that wait to breed at an older age. Additionally, birds that commence breeding at an early age tend to have an elevated mortality rate in the first year of breeding. The delay in the commencement of breeding may be a reflection of the long period needed for young birds to acquire the skills for efficient foraging. Several studies have shown that sub-adult birds are less efficient foragers, and have a lower success rate in capturing prey than do adult birds foraging in the same area. Additionally, sub-adult birds may visit the colony where they will breed for one or more years before they make their first attempt at breeding. This time may be necessary for learning where prey is to be found in the vicinity of the colony. The delay of breeding is particularly great in the Procellariiform birds, and these birds forage over vast areas of the ocean. It may be particularly challenging for them to learn where prey may be most predictably located within the potentially huge foraging arena available to them. The delay in the commencement of breeding may also provide time for birds in newly forming pairs to learn each others behavioral rhythms so that they will be more effective parents, although evidence to test this hypothesis is lacking. Certainly, coordination between the members of a pair is an essential ingredient of successful reproduction.

Most species of marine birds lay small clutches of eggs, with typical clutch size being one or two eggs. There is a tendency for species that forage far from their colonies to have smaller clutches (one egg) than those which forage inshore near the colony (two or three eggs). Likewise, species that live and forage in areas of strong upwelling tend to have considerably larger clutch sizes (three or four eggs) than closely related species that forage in mid-ocean regions. Two factors may come into play here. First, birds that forage at great distances from their colonies may be

unable to transport sufficient prey sufficiently quickly to raise more than one offspring. In many of these species, individual parents may visit the colony to provision their young at intervals from 1 to 10 days or more. In some species of Procellariiformes, adults alternate long and short foraging trips while provisioning their young. After short trips, chicks gain mass, but adults lose body mass, whereas, after long trips, adults have gained mass, although the chicks will have not benefited as greatly as after short trips. These results point to the constraints on the ability of these birds to raise larger broods.

Many species that forage close to their colonies (inshore-foraging species) raise larger broods than offshore-foraging species. Parent birds of the inshore-foraging species usually make multiple provisioning visits to the colony during a day. Although they may not be able to carry more food per trip than similar-sized species that forage offshore, the possibility of multiple trips allows sufficiently high rates of food delivery to permit supplying the needs of more than one offspring. Multiple-chick broods are common in pelicans, cormorants, gulls, and some species of terns, and are also found in the most-inshore-feeding alcids.

Species breeding in upwelling systems and subject to the boom or bust economy of an ecosystem with extreme interannual variation in productivity may be a special case. In these species, periodic declines in ecosystem production, as may occur in El Niño years off Peru or California, can result not only in reproductive failures, but also considerable mortality of adults. In these cases, the potential for high reproductive output in the years when prey is plentiful may be necessary to offset adult mortality rates that may be higher than is typical for most seabird species.

For the offshore-foraging species, annual fluctuations in reproductive output may be small, particularly when compared with inshore-foraging species. This low variability may be a reflection of the wide expanse of ocean over which they can search for food. For the inshore species, there may be many years in which no young are fledged and only a few years in which they are successful at fledging young. This interannual variation may reflect the dependence of inshore-foraging species on localized upwelling and other forms of physical forcing of prey patches, which may show considerable interannual variation in the amount of prey present. Thus, the 'good' years become disproportionately important for the success of the local population. Additionally, there is limited evidence that suggests that a few individuals within a seabird population may account for a large proportion of the young produced. The implications of these findings, if further work shows them to be generally true, will be considerable for management efforts toward the conservation of seabirds. Loss of the most productive individuals or disruption of breeding in one of the rare years in which food resources are sufficient to lead to strong production of young will have a disproportionate impact on population stability. However, identifying the most productive individuals, or predicting which years will be critical for good reproductive output will be a challenge.

The third component of seabird life histories that must accompany delayed commencement of reproduction and low rates of reproduction is an extended period of survival during which reproduction is possible. The life spans of seabirds are among the longest of any birds. Survival to the age of 20 years or more is probably common in most species, and the larger species of Procellariids are known to live in excess of 40 years. Knowledge of the actual life spans of seabirds is difficult to obtain, as most of the marking devices used until recently have had much shorter life spans than the birds whose life spans were being measured. The result is a considerable underestimate of the expected life spans of seabirds, and thus of their possible life time reproductive potential. Despite these problems, estimates of adult survival rates of 92–95% are the norm. Thus, even a small decrease in the rate of adult survival will have a proportionally large impact on mortality rates.

Survival of recently fledged young and juveniles is considerably lower than that of adults. The highest rates of mortality are most likely to occur in the first few months of independence, when young are learning to fend for themselves. Mortality is also high during the first winter, again probably due to experience. Possibly as many as 50% of fledglings fail to survive to their first spring. Survival rates for juveniles are higher, but there may be an increase in mortality as birds begin to enter the breeding population and encounter the aggression of neighbors and the increased energetic demands of caring for young.

There are conflicting hypotheses about what limits the size of seabird populations, although in most cases it is believed that food rather than predation or disease is the most likely limiting factor. Phillip Ashmole has argued that seabirds are likely to be most stressed for food during the breeding season, when large numbers of individuals are concentrated in the vicinity of breeding colonies. In contrast, during winter, seabirds are spread over vast reaches of ocean and are not tied to specific colonies and their associated foraging areas. During winter, seabirds should be free to move about and take advantage of prey, wherever it may be found. As discussed above, there is a modest suite of data that suggests that prey availability may limit colony size

and reproductive output of breeding seabirds. There are even fewer data to test the hypothesis favored by David Lack, that seabird populations are limited by wintertime food supplies and survival rates. Few studies of seabirds have focused on winter ecology during winter conditions, although this is the time when 'wrecks' (masses of dead birds driven ashore) are most common. Marine birds are sensitive to ocean conditions, with foraging success being reduced during periods of high winds. It is likely that in winter, even if there is no change in the amount of prey present, prey may be harder to obtain. Additional information on winter food stress comes from species that perform transequatorial migrations, thus allowing them to winter in the summer of the opposite hemisphere. Even under these presumably benign conditions, Southern Hemisphere shearwaters wintering in the North Pacific Ocean and Bering Sea experience occasional episodes of mass mortality, apparently from starvation. There is increasing evidence that decreases in the productivity of the California Current system are causing declines in the numbers of both migrant shearwaters and locally breeding species. The question as to which is the most stressful season for seabirds remains to be resolved.

Because of the sensitivity of seabird population dynamics to changes in adult survival rates, any factor that increases adult mortality is potentially detrimental to the conservation of seabirds. Thus, the loss of adult birds in fishery bycatch is of great concern. Breeding adults caught in gill nets and/or on long lines result in the loss not only of the adult, but also the chick for which it was caring. The loss of a breeding adult may also result in lower subsequent reproductive output by the surviving parent because it will likely have lower reproductive success during the first year with a new partner. The group most vulnerable to bycatch on long lines appears to be the Procellariiformes, which make shallow dives to grab baited hooks as they enter or leave the water. These are amongst the longest-lived of seabirds, and the curtailment of their breeding lives has a severe impact on their populations. Indeed, the populations of many species of albatross are declining at an alarming rate. A second major threat to seabirds is the presence of introduced predators on the islands where the birds breed. Rats, cats, foxes, and even snakes kill both chicks and attending adults. Again, loss of adult breeding birds has the most potentially serious impact on the future stability of the population. Reduction of anthropogenic sources of adult mortality in seabirds must be one of the most urgent imperatives for conservation biologists and managers.

See also

Alcidae. Procellariiformes. Seabird Conservation.

Further Reading

Ainley DG and Boekelheide RJ (1990) *Seabirds of the Farallon Islands: Ecology, Dynamics, and Structure of an Upwelling-system Community.* Stanford, California: Stanford University Press.

Ashmole NP (1963) The regulation of numbers of tropical oceanic birds. *Ibis* 103b: 458–473.

Ashmole NP (1971) Seabird ecology and the marine environment. In: Farner DS and King JR (eds.) *Avian Biology*, Vol. 1, pp. 224–286. New York: Academic Press.

Furness RW and Monaghan P (1987) *Seabird Ecology.* London: Blackie Books.

Gaston AJ and Jones IL (1998) *The Auks: Family Alcidae.* Oxford: Oxford University Press.

Hunt GL Jr (1980) Mate selection and mating systems in seabirds. In: Burger J, Olla BL, and Winn HE (eds.) *Behavior of Marine Animals*, Vol. 4, pp. 113–151. New York: Plenum Press.

Lack D (1966) *Population Studies of Birds.* Oxford: Clarendon Press.

Lack D (1968) *Ecological Adaptations for Breeding in Birds.* London: Chapman and Hall.

Nelson JB (1978) *The Sulidae: Gannets and Boobies.* Oxford: Oxford University Press.

Warham J (1990) *The Petrels: Their Ecology and Breeding Systems.* London: Academic Press.

Warham J (1996) *The Behaviour, Population Biology and Physiology of the Petrels.* London: Academic Press.

Williams TD (1995) *The Penguins.* Oxford: Oxford University Press.

Wittenberger JF and Hunt GL (1985) The adaptive significance of coloniality in birds. In: Farner DS and King JF (eds.) *Parkes KC Avian Biology*, Vol. VIII. Orlando, FL: Academic Press.

Wooller RD, Bradley JS, Skira IJ, and Serventy DL (1989) Short-tailed shearwater. In: Newton I (ed.) *Lifetime Reproduction in Birds.* London: Academic Press.

APPENDIX

APPENDIX 9. TAXONOMIC OUTLINE OF MARINE ORGANISMS

L. P. Madin, Woods Hole Oceanographic Institution, Woods Hole, MA 02543, USA

Introduction

This appendix is intended as a brief outline of the taxonomic categories of marine organisms, providing a classification and description of groups that are mentioned elsewhere in the Encyclopedia. It is presented at the level of Phylum (or equivalent) in most cases, but to the level of Class in large groups with many marine species. The few categories that contain no known marine species are not included in this list.

The outline is intended as a list rather than any kind of phylogenetic tree. The three Domains listed are now considered, on genetic evidence, to be the three primary categories of living organisms on Earth. The definitions and relationships of many higher taxonomic groups are currently in flux owing to the advent of new molecular genetic data. This is particularly true for the Archaea, Bacteria, and Protista (also called Protozoa or Protoctista), in which numerous phylogenetic lineages have been identified in recent years that are neither consistent with classical categories based on morphological characters nor arranged in a comparable hierarchy. The accepted systematics of the Protista will probably change in the near future, reflecting this new genetic information. However, as these new categories are still being defined, and have not yet come into common usage in ocean sciences, this outline retains a more classical and widely known system for protists (Margulis and Schwartz, 1988), with names that will more likely be familiar to marine scientists. Where possible, the approximate number of known (named) species is given, with an indication of how many of these are marine. For many groups the described species may be only a small fraction of the probable true number of species.

It should be apparent from perusal of the list that the ocean environment is home to the vast majority of the biological diversity on Earth. Notwithstanding the large numbers of insect species and flowering plants on land, the remarkable diversity of different body plans defining the higher taxonomic levels occurs almost entirely in the sea.

Readers interested in recent work on the phylogeny and classification of any of these organisms are directed to the Further Reading list. Often the most up-to-date information will be found on the Web sites listed.

DOMAIN ARCHAEA

Prokaryotic cells having particular cell lipids and genetic sequences that distinguish them from all other organisms. Many are extremophiles living at high temperatures or under unusual chemical conditions. Few have been cultured. "Species" are defined and classification is based mainly on molecular evidence. Increasing numbers of Archaea are being detected in marine environments.

Korarchaeota

A poorly-known group of hyperthermophilic organisms thought to be near the evolutionary base of the Archaea, and perhaps the most primitive organisms known.

Crenarchaeota

Primarily hyperthermophilic forms, including many sulfur-reducing species, but also species living at very low temperatures in the ocean.

Euryarachaeota

A group containing methane-producing forms and others that live in extremely saline conditions.

DOMAIN BACTERIA

A tremendously diverse domain of prokaryotic cells with particular genetic sequences and cell constituents. All photosynthetic and pathogenic forms are in this domain. A dozen major phylogenetic groups can be identified on genetic evidence, of which all but one include marine forms. Numbers of species are almost impossible to specify.

Aquificales

Primitive hyperthermophiles with a chemoautotrophic metabolism. Known from hydrothermal environments.

Green non-sulfur bacteria

A group including some photosynthetic and multicellular filamentous forms. Related forms may have produced ancient stromatolite structures in shallow seas.

Proteobacteria

Purple bacteria; a large, metabolically and morphologically diverse group including some photosynthetic, chemoautotrophic, and nitrogen-fixing forms that occur in the ocean. There are also many pathogenic species.

Gram-positive bacteria

Common soil-dwelling forms, but some also marine. Many form resting stages called endospores enabling survival under harsh conditions.

Cyanobacteria

An ancient and diverse lineage of photosynthetic bacteria that include some of the most abundant and important primary producers in the ocean.

Bacteroides, Flavobacterium and related forms

Gram-negative microbes occurring in freshwater and marine environments, as well as soils and deep-sea sediments.

Green sulfur bacteria

Anaerobic forms that oxidize sulfur compounds to elemental sulfur and occupy a variety of marine environments.

Deinococci

Highly radiation-resistant cells, which include forms found in terrestrial and marine thermal springs.

Planctomyces

One of only two groups having cell walls that lack the component peptoglycan; found in freshwater and marine environments.

Thermotoga

A group of anaerobic microbes found at shallow and deep-sea hydrothermal vents, and which includes some of the most thermophilic bacteria known.

Spirochaetes

Flagellated cells with a unique helical morphology; most are parasites and some inhabit marine animals.

DOMAIN EUKARYA

Single-celled or multicelled organisms with membrane-bounded nuclei containing the genetic material, and other specialized cellular organelles. Cell division is by some form of mitosis, and metabolism is usually aerobic. All protists, plants and animals are eukaryotes.

PROTISTA (also Protoctista, Protozoa)

Single-celled (and some multicellular) organisms that are neither animals nor true plants. It is a diverse and polyphyletic group that includes protozoans, algae, seaweeds and slime molds. These categories are defined mostly by morphology.

Phylum Dinoflagellata

Marine protozoans having unusual organization of their DNA, two locomotory flagella, and frequently encased in a rigid test. Dinoflagellates can be photosynthetic or heterotrophic; some live symbiotically in the tissues of other organisms. Approximately 3000 species, mostly marine.

Phylum Rhizopoda (also Amoebozoa)

Amoebas; single-celled organisms lacking flagella or cilia and moving by pseudopodia. Some are encased in a test. There are several thousand species in terrestrial, freshwater and marine environments, as well as parasitic and pathogenic forms.

Phylum Chrysophyta

A large group of unicellular algae that lack sexual stages and reproduce only by asexual division and formation of "swarmer" cells for dispersal. Most live in fresh water; the silicoflagellates are the only marine representatives.

Phylum Haptophyta

Marine photosynthetic cells that alternate between a flagellated free-swimming stage and a resting cocco-lithophorid stage covered with calcareous plates. These stages were previously thought to be distinct species. Several hundred species.

Phylum Euglenophyta

Flagellated protozoa that may be photosynthetic or heterotrophic, solitary, or colonial. They have flexible cell walls made of protein, and complex internal organelles. About 800 species including some marine.

Phylum Cryptophyta

Also called cryptomonads, these cells are widely distributed in fresh and salt water, and as internal parasites. They lack sexual reproduction and swim with two flagella.

Phylum Zoomastigina

Nonphotosynthetic cells with one to many flagella, including free-living, symbiotic, and parasitic species, with both sexual and asexual reproduction. The group is probably polyphyletic and includes hundreds of described species, with probably thousands more unknown; many are marine.

Phylum Bacillariophyta

The diatoms; photosynthetic cells enclosed in elaborate siliceous tests consisting of two halves or valves. Diatoms are important components of the food chain in marine and fresh water, with about 12 000 species.

Phylum Phaeophyta

The brown algae and seaweeds; macroscopic, photosynthetic, multicellular plantlike forms inhabiting intertidal, subtidal, or pelagic marine environments. About 1500 species, all marine.

Phylum Rhodophyta

Red algae; complex multicellular seaweeds inhabiting intertidal and subtidal environments worldwide. All contain particular photosynthetic pigments that give them a reddish color. About 4000 species, mainly marine.

Phylum Chlorophyta

Green algae, including single-celled forms and some multicellular seaweeds. They are important primary producers in all aquatic environments. There are about 7000 species, including many marine forms.

Phylum Actinopoda

Heterotrophic protozoa having long filamentous cytoplasmic extensions called axopods, supported by silica-based or strontium-based skeletal elements. Important marine forms are the radiolarians and acantharians. Some harbor algal symbionts or form macroscopic colonies. About 4000 species, mostly marine.

Phylum Foraminifera

Amoeboid protozoans with internally chambered, calcified, or agglutinated shells or tests. All are marine, living in benthic and pelagic habitats. About 4000, mainly benthic, extant species, but 30 000 fossil species.

Phylum Ciliophora

The ciliates; single-celled organisms covered with short cilia that are used for locomotion and/or food gathering. Ciliate cells have two nuclei and reproduce by fission. The 8000 species live in freshwater and marine environments.

Phylum Cnidosporidia

A diverse, polyphyletic group of heterotrophic microbes that are parasites and pathogens of animals, including many marine invertebrates and fish. About 850 species.

Phylum Labyrinthulomycota

Slime nets; colonial protozoans that construct networks of slime pathways on the surface of various substrates. The osmotrophic cells move along the slimeways toward food sources. Only about 10 described species, including marine forms.

PLANTAE

Plants are multicellular, photosynthetic organisms that develop from an embryo that is produced by sexual fusion. Plant cells have rigid cell walls and contain chloroplasts where photosynthesis occurs. Most of the 235 000 described species are terrestrial, with a few secondarily adapted to shallow marine environments.

Phylum Angiospermophyta

Flowering plants; virtually all the familiar grasses, flowers, vegetables, shrubs, and trees on Earth belong in this group, comprising about 230 000 species. A few grasses live in salt marsh and shallow subtidal marine environments.

FUNGI

Multicellular organisms that are neither motile nor photosynthetic, and form spores for reproduction. The basic structural elements of fungi are threadlike hyphae, which are partially divided into separate cells. Fungi range from microscopic yeasts and molds to large mushrooms and shelf fungi, and include some pathogenic forms. The vast majority of the 100 000 species are terrestrial.

Phylum Ascomycota

A diverse group including yeasts, molds, and truffles. In the marine environment filamentous ascomycotes grow and feed on decomposing plant material. A few of the 30 000 species are found in marine environments.

Phylum Basidiomycota

Complex, mainly terrestrial fungi including rusts, smuts, and mushrooms. Of some 25 000 species, only a handful are known from the marine environment, where they grow on marine grasses.

ANIMALIA

Animals are motile, heterotrophic, multicellular organisms, all of which develop from a ball of cells called a blastula, which originates by fusion of gametes. Most animals have complex tissues, organs, and organ systems, and higher animals have well-developed nervous and sensory capabilities.

Phylum Placozoa   A simple, tiny multicellular marine organism resembling a large amoeba, lacking tissues or organs. Only one species known.

Phylum Porifera   There are about 10 000 species of sponges, animals with skeletons composed of spicules, but which lack tissues, organs, or definite symmetry. Sponges have free-swimming larvae and sessile adults that filter-feed. All but a few hundred species are marine.

Class Calcarea.   Sponges with skeletons made up of calcareous spicules. About 500 species, all marine.

Class Demospongia.   Sponges with skeletons of spongy protein and/or silica, mainly marine. About 9500 species.

Class Hexactinellida.   Glass sponges, with skeletons of six-rayed silica spicules. About 50 deep-sea species.

Phylum Cnidaria   Radially symmetric animals with distinct tissues, including the jellyfishes, corals, anemones, and hydroids. All cnidarians are predators, using cnidocysts (nematocysts) to sting prey. Body forms include the polyp and medusa. Over 10 000 described species, nearly all marine. The group Myxozoa, previously considered protozoans or degenerate metazoans, are now thought to belong with the Cnidaria.

Class Anthozoa.   Corals and sea anemones, having the polyp form only. About 6200 benthic marine species.

Class Hydrozoa.   Most hydrozoans have a life cycle that alternates between an asexual polyp (hydroid) stage and a free-swimming, sexual medusa. Hydroid stages are usually colonial. Some coastal hydroid species lack the medusa and some oceanic species lack the hydroid. About 3000 species, nearly all marine.

Class Scyphozoa.   More complex, larger jellyfish with simpler or absent polyp stages. About 200 marine species.

Class Cubozoa.   Medusae with cuboidal body shape, well-developed nervous system and eyes, and highly toxic nematocysts. About 30 mainly tropical species.

Phylum Ctenophora   Comb jellies; transparent gelatinous animals that use fused plates of cilia (comb plates) for locomotion and sticky tentacles to capture prey. They have biradial symmetry and a more complex digestive system than Cnidarians. All 100 species are marine, mainly planktonic.

Phylum Rhombozoa   Simple, microscopic organisms that live as internal parasites in the kidneys of cephalopods. They have complex life cycles, and the group is sometimes considered a class of the phylum Mesozoa, along with the Orthonectida. About 65 species.

Phylum Orthonectida   About 20 species of simple, small organisms that are internal parasites of various marine worms, mollusks, and echinoderms.

Phylum Platyhelminthes   The flatworms, bilaterally symmetrical worms with three cell layers and distinct tissues, but no body cavity (coelom) and guts with only one opening. Most of the approximately 18 000 species are parasites in a wide range of hosts, but there are many free-living forms in all environments.

Class Turbellaria.   Free-living flatworms that are mainly predators or scavengers of other small organisms. Most are hermaphroditic. About 4500 species, including many marine forms.

Class Monogenea.   Ectoparasitic flatworms, mainly on skin or gills of marine fishes. Although previously included in the Trematoda, this group now appears to be evolutionarily distinct. About 1100 described species, but possibly many more.

Class Trematoda.   Parasitic flatworms or flukes, having digestive systems and complex life cycles, often among alternating hosts. There are about 8000 species of flukes, which infect both invertebrate and vertebrate hosts and cause some human diseases.

Class Cestoda.   Tapeworms; parasitic, segmented flatworms that lack digestive systems and live in the alimentary tracts of vertebrate hosts, including humans. About 5000 species, some in marine fishes or turtles.

Phylum Nemertea   Ribbon worms; long unsegmented worms with a complete digestive tract and a large cavity containing a proboscis that can be extended to sample the environment or capture prey. There are about 900 species, mainly benthic marine forms, but some freshwater or terrestrial.

Phylum Gnathostomulida   Minute, wormlike animals that live interstitially in marine sands and sediments. They feed on bacteria and protozoa using a specialized jaw, and are hermaphroditic. About 100 described species, but probably many more undiscovered.

Phylum Gastrotricha   Small wormlike organisms in freshwater and marine environments, living in sediments or on plants or animals. They feed on bacteria, protozoa, and detritus, using cilia to collect particles. About 500 species, half of them marine.

Phylum Rotifera   Small aquatic organisms with ciliated structures and complex jaws at the head. They have internal organs and complete guts, and feed either on particles or on small animals. Reproduction is sexual, but males are rare or unknown in many species. Of the 2000 species only about 50 are marine.

Phylum Kinorhyncha   Small, segmented animals with external spines that live interstitially in marine sediments or on the surfaces of seaweeds or sponges. There are about 150 species known.

Phylum Loricifera   Microscopic marine animals encased in a covering of spiny plates called a lorica, into which the head and neck can retract. Described only in 1983, the 10 known species of loriciferans live between and clinging to sand grains.

Phylum Acanthocephala   Parasitic worms in the guts of vertebrates, where they anchor to the intestine wall by spines on their head. About 1100 species., some living in marine fishes, turtles, and mammals.

Phylum Cycliophora   Described in 1995, this phylum comprises one known species, a microscopic animal that lives attached to the mouthparts of lobsters and collects particulate food. The life cycle is unusual and complex, with sexual and asexual stages.

Phylum Entoprocta   Small filter-feeding animals on stalks that live attached to various substrates either as single organisms or as colonies. A ring of ciliated tentacles surrounds the mouth and anus and creates water currents to collect food. About 150 species, all but one marine.

Phylum Nematoda   Roundworms; unsegmented worms with a layered cuticle, which molts during growth. Nematodes are among the most ubiquitous and numerous animals on Earth. They live in all environments and as parasites of most plants and animals. Of the 16 000 described species, a few thousand are marine. It is likely that many times more species exist.

Phylum Nematomorpha   Long, wiry, unsegmented worms, sometimes called horsehair worms. The gut is reduced or absent. Larval stages are internal parasites in arthropods and adults do not feed at all. A few of the 325 species are marine.

Phylum Bryozoa   Also called the Ectoprocta, a group of small colonial organisms that filter-feed using a tentaculate structure called the lophophore. Individual bryozoans are encased in tubular or boxlike housings and reproduce asexually to produce encrusting or plumose colonies attached to hard substrates. About 5000 species, all but 50 are marine.

Phylum Phoronida   Phoronids are tube-dwelling marine worms that also use a lophophore to collect particulate food. They are common in mud or sand, or attached to rocks or pilings. About a dozen widely distributed species are known.

Phylum Brachiopoda   Brachiopods are lophophorate, filter-feeding animals whose bodies are enclosed in bivalve shells. Most live secured by a stalk to hard substrates or in sediments, at depths from intertidal to 4000 m. Only about 335 living species, but over 30 000 fossil ones known. The living genus *Lingula* dates back over 400 million years.

Phylum Mollusca   A large and diverse phylum containing the familiar clams, snails, squid, and octopus. Mollusks possess mantle tissue that secretes a carbonate shell around the body, a belt of teeth called the radula for feeding, and a muscular foot variously modified for digging, crawling, or swimming. A diverse, widespread, and economically important group, mollusks have a long and complex taxonomic history, with between 50 000 and 100 000 described species. Most mollusks are marine but there are many freshwater and terrestrial snails

Class Monoplacophora.   Small, single-shelled animals living on hard surfaces, usually in the deep sea. Primitive in structure and thought to be similar to ancestral forms. Only 11 known species.

Class Aplacophora.   Small wormlike animals with calcareous spicules but no true shell. They lack the typical molluscan foot, but creep with cilia. About 250 species are known from various benthic marine environments.

Class Caudofoveata.   Shell-less wormlike animals that live in burrows in deep-sea sediments. Little is known of the ecology of the 70 known species.

Class Polyplacophora.   The chitons; mollusks having a shell of eight overlapping, articulating plates. All are marine and most live on intertidal or subtidal rocks, where they feed by scraping algae with their radulas. About 600 species.

Class Gastropoda.   The largest and most diverse class of mollusks, gastropods include aquatic and terrestrial snails, slugs, limpets, and nudibranchs. In most, the body sits on a muscular foot used for locomotion, and is enclosed in a conical or coiled shell. Gastropods may be filter feeders, grazers, or predators. By various counts there are 40 000 to 80 000 species, about half of them marine.

Class Bivalvia. Bivalves or Pelecypods, mollusks with the body enclosed between two valves or shells hinged together, and closed by an adductor muscle. Most are filter feeders, drawing water into the shell cavity and filtering particles from it. Some bivalves attach to surfaces; others burrow into sediments. Most of the 8000 species are marine.

Class Scaphopoda. Mollusks with a conical, tusk-shaped shell that is open at both ends. Scaphopods burrow into marine sediments and collect small food organisms with specialized tentacles. About 350 species.

Class Cephalopoda. Squid, octopus, and *Nautilus*; in cephalopods the molluscan foot is modified into tentacles surrounding the mouth. Cephalopods are actively swimming predators with highly developed nervous and sensory systems. *Nautilus* and most extinct cephalopods have external chambered shells, while squid have reduced internal skeletons and octopus have none. All 650 species are marine.

Phylum Priapulida Marine worms that burrow into sediments with only their mouths exposed at the surface. They are predatory on other small worms. The 10 known species live from estuarine to abyssal environments.

Phylum Sipuncula About 320 species of unsegmented marine worms, with a retractable proboscis called the introvert. They are benthic, often living in sediments or among other animals. Tentacles around the mouth collect detritus and other particulate food.

Phylum Echiura Unsegmented, benthic marine worms having an extensible proboscis that is used to collect detrital food. They live mainly within burrows in sediments. Considered by some to be a class of the Annelida. About 140 species.

Phylum Annelida Segmented worms, a large group of diverse species, most having bodies divided into segments by internal septa, and with chitinous setae on the exterior body. There are about 12 000 species of annelids, in all aquatic and terrestrial environments.

Class Polychaeta. Worms usually with distinct head region, numerous setae and paddle- or leg-like parapodia for locomotion. The group includes mobile, burrowing, attached, and symbiotic forms, feeding as predators, scavengers, filter or deposit feeders. About 8000 species, almost all marine.

Class Oligochaeta. Worms lacking parapodia and with few setae; terrestrial forms include earthworms. Most of the 3100 species are freshwater or terrestrial.

Class Hirudinea. Leeches; the body is not segmented internally and lacks setae on the exterior. Most are ectoparasites, feeding on blood of other animals, but some are predators. About 500 species, many marine.

Phylum Tardigrada Water bears; minute animals with eight short legs that live in aquatic or moist terrestrial environments and suck juices from plants or animals. They are able to remain in a dried state for long periods, returning to active metabolism on rehydration. About 550 species, a few of them marine.

Phylum Arthropoda The arthropods, or jointed-leg animals, are one of the most successful and widespread metazoan groups. All possess segmented bodies and articulated exoskeletons, which are molted during growth. Insects and arachnids are the dominant arthropods on land, but almost entirely absent from the sea, where crustaceans predominate. About 1 million described species, mostly insects, but many more probably exist.

Class Merostomata. An ancient group of chelicerates now containing only 4 species of horseshoe crabs. They live in subtidal environments, and feed as predators and scavengers.

Class Pycnogonida. Sea spiders; lacking the well-developed head, thorax, and abdomen of other arthropods. About 1000 species, entirely marine, which feed on body fluids of other animals and plants.

Class Crustacea. Largely aquatic arthropods including copepods, amphipods, shrimp, barnacles, crabs, and lobster. Often with a calcified carapace covering the segmented body. All forms have a nauplius larva as the first of many molt stages. About 35 000 species, almost entirely marine, in all habitats.

Phylum Pogonophora   Thin, wormlike animals that live in tubes in sediment or attached to benthic surfaces. Pogonophorans lack digestive systems and obtain nutrition by absorption of dissolved organic nutrients. Pogonophorans are thought by some to be aberrant annelids. About 100 species, mostly in deep water.

Phylum Vestimentifera   Closely related to pogonophorans, these larger marine worms rely on symbiotic bacteria in their tissues to generate nutrition from the metabolism of inorganic chemical compounds. They are best known from deep-sea hydrothermal vents, where they can be over 2 m long. About a dozen species have been described.

Phylum Echinodermata   An entirely marine phylum including the sea stars, urchins, and brittle stars. All have a five-part radial symmetry, a water-vascular system, tube feet used for locomotion, respiration and feeding, and a skeleton made of minute calcareous ossicles or spicules. Over 6000 species.

Class Crinoidea. Most ancient of the living echinoderms, crinoids have multiple pinnate arms used for filtering food particles from the water. Some are attached to the bottom by a stalk, others swim by movement of the arms. About 600 subtidal and deep-sea species.

Class Asteroidea. The seastars, most with five radial arms, are slow-moving predators in intertidal and subtidal environments. About 1500 species.

Class Ophiuroidea. Brittle stars, having five slender and flexible arms radiating from a central disk. Some are deposit or filter feeders, others predatory. About 2000 species, including many deep-sea forms.

Class Concentricycloidea. Small discoidal organisms known only from submerged wood in the deep sea. They lack five-part symmetry and arms, and the body is covered by overlapping calcareous plates. Two known species.

Class Echinoidea. Sea urchins; with a rigid, globular or flattened test made of calcareous ossicles and a complex mouth structure for grazing and chewing. Most are free-living in subtidal environments, but some burrow in sediments or rock. Approximately 950 species.

Class Holothuroidea. Sea cucumbers; with an elongate, flexible body, bilateral symmetry and no arms. Most of the 1500 species are benthic deposit or filter feeders.

Phylum Chaetognatha   Arrow worms; planktonic marine organisms that are predatory on small zooplankton, using chitinous spines around the mouth to catch prey. The 100 species are mainly planktonic with a few benthic forms.

Phylum Hemichordata   Wormlike marine organisms that burrow in sediments or form colonies on hard substrates. Most are deposit or suspension feeders. About 100 species from shallow tropics to deep sea.

Phylum Chordata   A large and diverse phylum including the familiar vertebrates. All have a dorsal nerve cord that can form a brain, a notochord that becomes the vertebral column in vertebrates, and gill slits in the throat at some stage of development. Chordates live in all environments and are one of the most successful and widespread groups. Perhaps 45 000 species, about half marine forms (mainly fish).

Subphylum Urochordata The tunicates; sessile or motile animals with the body enclosed in a tough, flexible tunic. Most are filter feeders. The notochord and dorsal nerve are seen only in larval stages, and sessile adults may be asymmetrical in form. About 3000 species, all marine.

Class Ascidiacea. Sea squirts; sessile, filter-feeding tunicates that live mainly on hard benthic substrates. About 2700 species.

Class Sorberacea. A small group of solitary, deep-sea tunicates that appear to prey on live organisms instead of filter-feeding.

Class Larvacea. Minute planktonic tunicates (also called appendicularians) with small bodies and long tails. They filter feed using an external mucous structure that concentrates small particulate material for ingestion by the larvacean. About 200 species.

Class Thaliacea. Pelagic tunicates with gelatinous, transparent bodies. They filter feed by creating a water current through their bodies. All have complex life-cycles with sexual and asexual, solitary, and colonial stages. About 100 species.

Subphylum Cephalochordata Lancelets or "Amphioxus"; small fish-shaped animals with notochords extending the length of the body. They burrow into substrates with the head end exposed and filter particulate food. About 20 species.

Subphylum Vertebrata Chordates with a backbone replacing the notochord, and a distinct head region with brain. Approximately 42 000 species in all environments.

Class Agnatha. Lampreys and hagfish; eel-like jawless fishes without scales, bones or fins. Most are scavengers or parasites on other fish. About 60 marine and freshwater species.

Class Chondrichthyes. Sharks and rays; fish with cartilaginous bones and small denticle scales embedded in the skin. The 850 species are virtually all marine.

Class Osteichthyes. The bony fishes; having bone skeletons, scales and often air bladders for buoyancy. Highly diverse and widely distributed in all marine and freshwater habitats, with about 25 000 species.

Class Reptilia. Turtles, snakes and lizards. Most of the 6000 species are terrestrial except for a few marine turtles, crocodiles, and snakes.

Class Aves. Birds; about 9000 species in all terrestrial habitats and many marine forms including penguins, albatrosses, gulls, etc.

Class Mammalia. Four legged, endothermic animals usually with fur or hair, which mainly give live birth and suckle the young. Most of the 4500 species are terrestrial; marine forms include whales, dolphins, seals, and otters.

Further Reading

Atlas RM (1997) *Principles of Microbiology.* Dubuque, IA: William C. Brown.

Brusca RC and Brusca GJ (1990) *Invertebrates.* Sunderland, MA: Sinauer Associates.

Cavalier-Smith T (1998) A revised six-kingdom system of life. *Biological Reviews of the Cambridge Philosophical Society* 73: 203–266.

Margulis L, Corliss JO, Melkonian M, and Chapman DJ (1990) *Handbook of Protoctista.* Boston: Jones and Bartlett.

Margulis L and Schwartz KV (1988) *Five Kingdoms: An Illustrated Guide to the Phyla of Life on Earth.* NewYork: WH Freeman.

Nielsen C (1995) *Animal Evolution: Interrelationships of the Living Phyla.* Oxford: Oxford University Press.

Patterson DJ (1999) The diversity of eukaryotes. *American Naturalist* 154(supplement): S96–S124.

Pechenik JA (2000) *Biology of the Invertebrates*. Boston: McGraw-Hill.

Williams DD (2000) *Invertebrate Phylogeny*. Scarborough: CITD Press, University of Toronto. CD ROM.

Websites

"Microscope": http://www.mbl.edu/baypaul/microscope/general/page_01.htm

"Tree of Life" http://phylogeny.arizona.edu/tree/phylogeny.html

INDEX

Notes

Cross-reference terms in italics are general cross-references, or refer to subentry terms within the main entry (the main entry is not repeated to save space). Readers are also advised to refer to the end of each article for additional cross-references – not all of these cross-references have been included in the index cross-references.

The index is arranged in set-out style with a maximum of three levels of heading. Major discussion of a subject is indicated by bold page numbers. Page numbers suffixed by *T* and *F* refer to Tables and Figures respectively. vs. indicates a comparison.

This index is in letter-by-letter order, whereby hyphens and spaces within index headings are ignored in the alphabetization. For example, 'oceanography' is alphabetized before 'ocean optics.' Prefixes and terms in parentheses are excluded from the initial alphabetization.

Where index subentries and sub-subentries pertaining to a subject have the same page number, they have been listed to indicate the comprehensiveness of the text.

Abbreviations used in subentries

ENSO - El Niño Southern Oscillation

Additional abbreviations are to be found within the index.

A

Abbott's booby, 525
  see also Sulidae (gannets/boobies)
Abyssal gigantism, 130
Abyssal pelagic zone, 270
Abyssal zone, 126*T*, 129–130, 130
Acartia spp. copepods, 7
Acidification, threat to cold-water coral
  reefs, 197
Acorn barnacle (*Semibalanus balanoides*),
  203
Actinologica Britannica (Gosse), 188,
  188*F*
Actinopterygii (ray-finned fishes), 210
Adaptations
  Antarctic fishes, 218–219
  benthic boundary organisms, 203–204
  benthic organisms, 123–124
  bioluminescence, 75
  demersal fishes, 365, 366
  eels, 265
  fish, 215
  fish hearing, 375
  fish locomotion, 299, 300–301*F*
  fish vision, 350
  gelatinous zooplankton, 51
  intertidal fishes, 233–234
  mesopelagic fishes, 242–245
  micronekton, 75–76
Adélie penguin (*Pygoscelis adeliae*), 548*T*,
  551, 552*F*

Aerobic diving limit (ADL), marine
  mammals, 461, 461*F*
Aethia, 503*T*
  reproduction, 506
  see also Alcidae (auks)
Aethia cristatella (crested auklet), 503,
  505*F*
Aethia pusilla (least auklet), 503, 505*F*
African penguin, 548, 548*T*, 549,
  553–554
Aggregation(s)
  foraging seabirds, 565
    see also Seabird(s)
  krill *see* Krill (Euphausiacea)
Alabaminella weddellensis foraminifer,
  157*F*, 158
Alaska pollock *see* *Theragra*
  chalcogramma (Alaska/walleye
  pollock)
Albacore tuna (*Thunnus alalunga*),
  309–310, 309*F*
Albatrosses (Diomedeidae), 497,
  500–501, 539
  importance of krill, 69
  migration, 574–575
  population declines, longline fishing and,
  545
  see also Procellariiformes (petrels);
    Seabird foraging ecology
Alca, 503*T*
  diet, 506
  see also Alcidae (auks)

Alca torda (razorbill), 505*F*, 507
  see also Alcidae (auks)
Alcidae (auks), 505*F*, **503–509**
  abundance, 503
  characteristics and adaptations,
    503–504
    digestive system, 504
    plumage, 504, 505*F*
    prolonged diving, 504
    underwater swimming, 504
  diet and foraging ecology, 504–506
  distribution, 503
  evolution, 503
  fossils, 503
  harvesting by humans, 507–509
    commercial, 508
    eggs, 508
    skins, 508
    techniques, 508
  migration, 579–581
  population dynamics, 507
  postbreeding, 507
    prebasic molt, 507
  reproduction, 506–507
    breeding sites, 506
    chick-rearing, 507
    egg incubation, 506–507
  species, 503
  systematics, 503, 503*T*
    genera, 503, 503*T*
    tribes, 503, 503*T*
  thermoregulation, 507

Alcidae (auks) (*continued*)
  hatchlings, 507
  threats
    human exploitation, 508
    predation, 508–509
  wintering, 507
  see also Seabird(s)
Aleutian Archipelago, sea otter
    population, 449
Algae
  definition, 140
    comparison to 'true' plants, 140
    cyanobacteria, 140
    primary production, 140
  diversity *see* Algae, diversity
  general definition, 28–29
  symbiotic relationships
    gelatinous zooplankton, 52
    radiolarians, 28–29, 32
      see also Algal blooms; Cyano-
        bacteria; Microphytobenthos;
        Phytobenthos; Phytoplankton
        blooms; Seaweed(s)
Algae, diversity (phytobenthos), 140–142
  habitats, 142
    epiphytes, 142
    microphytobenthos, 142
    seaweeds, 142
  size diversity, 141
  structural diversity, 141–142
    advanced forms, 141
    calcium carbonate secretion, 141
    example of kelp, 141
    heterotrophic production, 141–142
    simple forms, 141
  taxonomic classification, 142
    comparison to higher plants, 142
    distinguishing features, 142
    seaweed divisions, 142
Algal blooms
  control by viruses, 18
  see also Phytoplankton blooms
Alginate, 145
Alle, 503T
  see also Alcidae (auks)
Alle alle (little auk), 503, 505F
Alosa spp. (shad), hearing range, 373
Amazonian manatee (*Trichechus
    inunguis*), 434–435
  see also Manatees
Ambulocetidae, 468
  see also Cetaceans
Ammodytidae (sand eels), 284, 300–301F,
    366
Ammonium (NH$_4^+$)
  phytoplankton growth reaction, 94
  requirement by microphytobenthos, 138
  species selectivity, 138
  see also Nitrogen cycle
Amphipods
  aggregation, 205T
Ampullatus hyperoodon (bottlenose
    whale), diving characteristics,
    459T
Anableps anableps (four-eyed fish), 352F
Anchovies (*Engraulis* spp.), 250

Anemonefishes (Pomacentridae), 223, 224
Angelfish (Pomacanthidae), 300–301F
Anglerfishes (Lophiiformes), 214, 282
Anguilla (eels), **262–271**
  see also Eels
Anguilla anguilla (common eel), 268
Anguilliformes (eels), 300–301F
Anhingidae (darters), 522, 526–527
  breeding patterns, 527
  characteristics, 526–527
  cormorants *vs.*, 527
  distribution, 526–527
  feeding patterns/adaptations, 527, 528T
  species, 526–527
  see also Pelecaniformes
Antarctic fish(es), **218–221**
  adaptations, 218–219
    antifreeze, 219
    antifreeze glycopeptides, 219, 220F
    freezing points, 219
    kidney adaptations, 219
    cardiovascular adaptations, 219–221
    high oxygen tension areas, 221
    increased volume of blood, 220–221
    lack of hemoglobin, 220
    oxygen availability in seawater,
      219–220
    theories of oxygen uptake, 220
    cold adaptation, 218–219
    metabolic rates, 218–219
  changes in taxonomic methods, 218
  fish fauna, 218, 218T
  low diversity, 218
Antarctic ice, zones, seabird associations,
    564
Antarctic krill *see Euphausia superba*
    (Antarctic krill)
Antarctic neritic krill *see Euphausia
    crystallarophias* (ice krill)
Anthomedusae medusas, 52, 53F
Anthropogenic impacts, 422
  beaked whales, 409
  cold-water coral reefs, 195–197
  coral reef fishes, 222
  demersal fishes, 370
  dolphins and porpoises, 417, 421
  of fisheries (effects)
    benthos, 174–175
    fish hearing, 375–376
    fish populations, 209
    of overfishing, 284
  seabirds, 501
  sperm whales, 409
  threats to deep-sea fauna, 185–186
    see also Human activities, adverse
      effects; Large marine ecosystems
      (LMEs); Pollution
Anticyclonic circulation, 270
Anti-tropical distribution, definition, 422
Appendicularia tunicates, 58, 59F, 60
Aptenodytes (penguins), 552
  breeding patterns, 552
  characteristics, 548T, 552, 552F
  distribution, 548T, 552
  feeding patterns, 548T, 552
  migration, 574

  nests, 548T, 552
  species, 548T, 552
  see also Sphenisciformes (penguins)
Aptenodytes forsteri (emperor penguin),
    548T, 552, 553
Aptenodytes patagonicus (king penguin),
    548T, 552, 552F
Aquaculture
  copepod pests, 49
  eels, 269
Archaea, 21–22
Archaeocetes, 467, 468–469
  see also Cetaceans
Arctic charr (*Salvelinus alpinus*), 255
Arctic ice
  zones, seabird associations, 564
Arctic Ocean
  benthic foraminifera, 150T
  krill, 64, 65–66
  see also Polar ecosystems
Arctocephalinae (fur seals), 385T, 425T,
    466
  see also Otariidae (eared seals)
Aristostomias spp. (barbeled
    dragonfishes), 358, 360F
Arnoux' beaked whale, 408–409
Arribada, 451–452, 452F
Arrow worms (Chaetognatha),
    114–115
Ashmole, Phillip, 584
Asparagopsis armata algae, 143
Aspridonotus taeniatus (saber-toothed
    blenny), 282
Astrocopus spp. (stargazers), 216
Astyanax spp. fish, 377
Atlantic *see* Atlantic Ocean
Atlantic bluefin tuna *see Thunnus
    thynnus* (Atlantic bluefin tuna)
Atlantic cod (*Gadus morhua*), 283
  camouflage, 366
  diet, 368–369, 369F
  migration, 311, 311F
  ontogenetic feeding shifts, 284
  vertical migration patterns, 317
Atlantic herring (*Clupea harengus*), 246,
    280, 368–369
  migration, 310, 310F
  ontogenetic feeding shifts, 281, 281F
  schooling decisions, 345–346, 346F
  vertical migration, 347F
Atlantic mackerel (*Scomber scombrus*),
    309–310, 310F
Atlantic Ocean
  benthic foraminifera, 150T
  geohistorical studies, 158
  krill, 63–64, 63F, 64F, 66
  low macrobenthic diversity, 171
    variations, 171
  North *see* North Atlantic
  North Atlantic Oscillation Index, 270
  pelagic fisheries, 250
  planktonic foraminifera, 37F, 38F
    see also North Atlantic; North Atlan-
      tic Oscillation
Atlantic puffin, 503, 505F
  see also Alcidae (auks)

Atlantic ridley turtle, 455–456, 455F
 see also Sea turtles
Atlantic salmon (Salmo salar), 282
 diet, 256
 distribution, 253
 ears, 373F
 food, 256
 life cycle, 254F
 migration, 257F, 258F, 259F
 movements, 255–256
 predation, 257–258
 taxonomy, 252
Atmosphere
 measure/unit of pressure, 130
Atmospheric pressure, 130
Audubon's shearwaters (Puffinus
  lherminieri), 539
 see also Procellariiformes (petrels)
Auklet(s)
 crested, 503, 505F
 least, 503, 505F
 migration, 579–581
 see also Alcidae (auks)
Auks see Alcidae (auks)
Aulostomus maculatus (trumpet fish), 282

B

Bacteria
 benefits to benthos, 126–127, 165
 bioluminescence, 81, 111, 112T, 115,
  118–119
 chemoautotrophic, 130, 178–179
 controlling agents, 13
 denitrification, 138–139
 estuaries, 9
 mats see Bacterial mats
 microbial loops see Microbial loops
 mortality, viruses contributing to, 13,
  16–17
 photosynthesis, 140
 removal by viruses, 13
 symbiotic relationships
  deep-sea animals, 130
  fish, 115–116
  squid, 81, 115
  vent animals, 178–179
   see also Bacterioplankton; Cyano-
    bacteria; Microbial loops
Bacterial mats, 184
Bacterial production, 23F, 24, 25–26
Bacterial respiration, 24, 25–26
Bacteriophage, 15
Bacterioplankton, 21–27
 abundance
  compared with phytoplankton, 24–25
  euphotic zones, 25F
  gradient across zones, 24–25
 biomass, 24–26
  compared with phytoplankton, 25,
   26T
  environmental assessments, 25
  estimates, 25
 food webs and biogeochemical cycles,
  26–27
  bacteriovores, 26

coastal and estuarine food webs, 27
nutrient cycling, 27
nutrient regeneration, 27
viruses, 27
 see also Primary production measurement methods; Primary production processes
functions, 21
growth and production, 24–26
identification methods, 22F
identity and taxonomy, 21–23
 domain Archaea, 21–22
 domain Bacteria, 22–23
 identification methods, 21
nutrition and physiology, 23–24
 bacterial growth efficiency, 24
 averages, 24
 use of substrates, 24
 dissolved organic matter, 21, 23
 growth and primary production,
  23–24, 23F
 nutrient limitation, 23
 oligotrophs vs. copiotrophs, 24
 sea water culture experiments, 23
production
 compared with phytoplankton, 26T
 measurement methods, 25–26
 research improvements, 21
 total bacterial carbon utilization, 26
 see also Microbial loops
Bahamonde's beaked whale, 403
Baird's beaked whale (Berardius bairdii),
 403, 408, 409
Balaenids see Right whales (balaenids)
Balaenoptera musculus see Blue whale
Balaenopterids (rorquals), 391, 392T, 470
 growth and reproduction, 399
 trophic level, 489F
 see also Baleen whales
Baleen, definition, 402
Baleen whales (Mysticeti), 382–383T,
 468F, 391–402
 annual feeding and reproductive cycle,
  387–388, 387F
 behavior, 397–398
  growth and reproduction, 399–402
  migration, 398
  social activity, 399
  sound production, 398
  swimming, 398–399
 characteristics, 386F, 391–393, 393F,
  467
 competition between species, 401–402
 distribution, 393–394
 ecology, 393–394
 evolution, 469
 extinct families, 469
 food/feeding, 396T, 397
  daily consumption, 397
  filter-feeding apparatus, 393F,
   395–396, 481–482, 482F
 habitat, 393–394
 importance of krill, 68
 life history, 397–398
 migration, 472–474
  reasons for, 474

surveillance, 474
 variation in patterns, 473–474, 473F
  annual, 473–474
  intraspecific, 473
parasites, 397
population status, 399
 assessment, International Whaling
  Commission's Scientific Committee,
  400–401
 recovery, 474
 World Conservation Union 'Red List'
  categories, 400, 401T
predators, 397–398
 killer whales, 397
skeleton, 386F, 467F
taxonomy, 391, 391–393, 392–393,
 392T, 469
 family Balaenopteridae see
  Balaenopterids
 family Neobalaenidae see Pygmy right
  whales
threatened species, 400, 401T
threats, 399–400
toothed whales vs., 391
trophic level, 488, 489F
 see also Cetaceans
Barbeled dragonfishes (Aristostomias
 spp.), 360F
Barnacles (Cirripedia), wave resistance,
 203
Barracuda (Sphyraena spp.), 300–301F
Barreleyes (Opisthoproctus spp.), 352F
Basilosauridae, 468
 see also Cetaceans
Basking shark (Cetorhinus maximus),
 280–281
Bathyal zone, 126T, 129–130, 131
Bathygobius soporator (frillfin goby), 235
Bathymetry
 three dimensional, 190F
Bathyphotometers, 117
 factors affecting measurements,
  117–118
 types, 116F, 117–118
Bathysiphon filiformis foraminifer, 151F
Batoidea (rays), 216, 300–301F
BBL see Benthic boundary layer (BBL)
Beaked whales (Ziphiidae), 382–383T,
 404T, 403–410
 acoustics and sound, 408–409
  acoustic behavior, 408–409
  sound production mechanism, 409
 anatomy and morphology, 403–405
  distinguishing features, 403–404
  lower jaw morphology, 405F
  rostrums, 410
  skull asymmetry, 405
 conservation, 409–410
  current threats, 409
  anthropogenic noise, 410
  whaling industry, 409
 distribution and abundance, 405–406
  lack of information, 406
  limited distributions, 405–406
 foraging ecology, 406
  competition between species, 406

Beaked whales (Ziphiidae) (*continued*)
  diving ability, 406, 407F
    squid diet, 406
    suction feeding, 406
  predation, 409
    predators and defence mechanisms,
      409
  relationship to other cetaceans, 405F
  research history, 403
  social organization
    composition of groups, 408
    intraspecific aggression, 407–408
  taxonomy and phylogeny, 403
    fossil records and evolution, 403
    ongoing classification, 403
    species diversity, 403
  trophic level, 489F
Bearded seal (*Erignatus barbatus*)
  song, 485, 485F
    see also Phocidae (earless/'true' seals)
Behavior
  acoustic, beaked whales (Ziphiidae),
    408–409
  baleen whales see Baleen whales
    (Mysticeti)
  copepods see Copepod(s)
  coral reef fish see Coral reef fish
  demersal fishes, 365–366
    see also Demersal fish
  fish schooling, 338F
  foraging, seabirds, 564–565
  harvesting see Fisheries
  intertidal fishes, 234–236
  marine mammals, 481–482
  mesopelagic fishes, 241–245
  Odobenidae (walruses), 481
  Pelecaniformes, 528, 528T, 529F
  Phaethontidae (tropic birds), 523, 528T
  predation see Predation behaviors
  sirenians, 437
Benthal environment, definition, 131
Benthic, definition, 123
Benthic boundary layer (BBL)
  adaptations to resist shear stress,
    203–204
    behavioral adaptations, 203–204
    skimming flow, 204
    structural adaptations, 203
    tube adaptations, 204, 204F
    wave resistance, 203
  aggregation of organisms, 204, 205T
    achievement, 204
    behavioral reasons, 204
    bivalve reefs, 204
  definition, 199
  effects on organisms, **199–206**
  flow adaptations, 201
    'evolutionary adaptation' defined, 201
  hydrodynamic measures, 199
  juvenile/adult stage adaptations, 205
    feeding adaptations, 202, 205
    shear stress resistance, 203, 205
    tube building, 204, 205
  larval stage adaptations, 201–202, 205
    adaptive strategies, 202T
    dispersal, 201

dispersal distances, 202
  epifaunal life cycle, 201, 201F
  settlement process, 202
  size, 201–202
organisms, 199–200
  epifauna, 199–200, 200T
  epifauna types, 199–200
  near-bottom swimmers, 199
  suprabenthos (hyperbenthos),
    200–201, 200T
research needs, 205
structure and depth, 199
suspension-feeding adaptations,
  202–203, 203T
  active suspension feeders, 202–203
  deposit/suspension feeders, 203
  facultative passive/active feeders, 203
  passive suspension feeders, 202
  requirement of rapid growth, 202
    see also Benthic organisms; Ekman
      layer; Macrobenthos
Benthic fish(es), 227
  see also Deep-sea fish
Benthic foraminifera, **147–158**
  ecology, 150–151
    abundance and diversity, 151–152
    hard-substrate habitats, 151
    soft sediment habitats, 150
    well-oxygenated sites, 150, 150T
  environmental distribution controls,
    155–156
    $CaCO_3$ dissolution, 153
    currents, 155
    depth, 153–154
    lateral advection of water masses,
      153–154
    organic matter fluxes, 154
    species' optimal habitats, 154–155,
      155F
  low-oxygen environments, 156–158
    foraminifera tolerances, 155–156, 156
    related to organic flux, 155
  microhabitats and temporal
    variability, 153–155
    deep-sea diversity, 151F
    factors influencing distribution, 152,
      152–153
    food and oxygen variability, 153,
      154F
    species distributions, 152, 152F, 153F
  role in benthic communities, 152–153
    biostabilization, 151–152
    bioturbation, 151–152
    organic carbon cycling, 151
    place in food webs, 151
  examples, 148F, 149F
  general characteristics, 147
    cell body, 147
    test, 147
  morphological/taxonomic diversity,
    147–150
    range of morphologies, 147
    sizes, 147
    taxonomic test characteristics, 147
  research history, 147
    multidisciplinary research, 147

research methodology, 150–151
  collection methods, 148
  distinguishing live and dead
    individuals, 149–150
  influence of mesh size, 150
  use in geological research, 147
  use in paleo-oceanography, 156T, 158
    example, 158
    factors making foraminifera useful,
      156–157
    limitations of accuracy, 157–158
    paleoenvironmental attributes
      studied, 157
    see also Planktonic foraminifera
Benthic gigantism, 168F
Benthic infauna
  classification, 124–125
    alternative groupings, 126
    by size groups, 125, 125F
    communities, 125
    percentage of benthos, 125
Benthic organisms, 270, **123–131**
  'benthic' defined, 123
  bioluminescence, 113–114
  boundary layer see Benthic boundary
    layer (BBL)
  classification, 124–126
    epifauna, 124–125
    percentage of benthos, 125
    infauna, 124–125
    alternative groupings, 126
    communities, 125, 126T, 127T
    percentage of benthos, 125
    size groups, 125, 125F
  classification of zones, 126T
  deep-sea environment, 129–130
    abyssal gigantism, 130
    biodiversity theories, 130
    cyclical events, 130
    depth and food availability, 130
    dominant species, 130
    energy sources, 130
    zones, 129–130
      see also Deep-sea fauna; Deep-sea
        fish
  depth divisions, 123
  distribution
    sediment influence on, 127
  feeding habits, 126–127
    bacterial breakdown of food,
      126–127
    dependence on detritus, 126–127
    detrivores, 127
    grazing and browsing, 127
    predatory behavior, 127
    sediment influence on distribution,
      127, 127T
  foraminifera see Benthic foraminifera
  larvae, pelagic vs. nonpelagic, 128, 128F,
    129, 129F
  physical conditions, 123–124, 124F
    exposure to air, 124F
    level-bottom sediment, 124
    light, 123, 124F
    salinity, 124, 124F
    substratum material, 124, 124F

temperature, 123–124, 124F
turbulence, 124F
water level, 123
water pressure, 123
primary productivity, 100
reproduction, 128–129
fecundity and mortality, 129
oviparity, 128
settlement process, 129
sexual/asexual reproduction, 128–129
strategies for dispersal/nondispersal, 128
viviparity, 128
spatial distribution, 127–128
competition for space, 127
horizontal, 127–128
vertical, 128
see also Benthic boundary layer (BBL); Benthic foraminifera; Demersal fish(es); Macrobenthos; Meiobenthos; Microphytobenthos; Phytobenthos
Benthopelagic fishes, 227
Benthos, 131
see also Benthic organisms
Berardius bairdii (Baird's beaked whale), 403, 408, 409
Bering Sea
trophic interactions, 488, 490F, 491
Beroida ctenophores, 55F, 56
Bichir (Polypterus spp.), 210
Billfishes (Istiophotidae), 300–301F
Biological pump
micronekton diurnal vertical migrations, 73
Bioluminescence, **111–119**
applications, 118–119
oceanographic, fisheries and medical, 118–119
biochemistry, 111
luciferin/luciferase system, 111
cephalopods, 81
colors produced
blue, 113, 117
blue-green, 113
green, 113
red, 117
yellow, 113–114
copepods, 48–49
definition, 111
fishes, 115–117
forms, 112T
flashes, 111, 113
glow, 111, 114–115, 115
secretions, 111–113, 115
waves of light, 113
measuring, 117–118
bathyphotometers, 116F, 117
bioluminescent signatures, 118
identifying organisms, 118
seasonal and diel variability, 117–118
seasonal/diurnal variability, 117–118
stimulable bioluminescent potential, 117
see also Bathyphotometers
micronekton see Micronekton

microorganisms, 111
bacteria, 111, 115
dinoflagellates, 111
radiolarians, 111
types, 111
occurrence, 111, 112T
phenomena, 118
photocytes, 113, 113F, 114–115
photophores, 114–115
accessory optical structures, 114F
fishes, 115–116
occlusion, 115, 115F
pigments and reflectors, 113F
shrimps and krill, 115
squids, 115
plankton, 111–115
cnidarians, 113
copepods, 111–113
ctenophores, 113
echinoderms, 113–114
ostracods, 111–113
tunicates, 114–115
worms, 113–114
squid and octopods, 115
see also Fish vision
Biomineralization, 30
Biostabilization, 136
Biotoxins, 422
Bioturbation, 177–178, 180
benthic foraminifera ecology, 151–152
Black-browed albatross, 545, 575
see also Albatrosses
Black guillemot (Cepphus grylle), 503
see also Alcidae (auks)
Black-legged kittiwake see Kittiwake
Black Sea
introduction of Mnemiopsis leidyi, 60–61
Black skimmer (Rynchops niger), 513F
see also Rynchopidae (skimmers)
Black surf perch (Embiotica jacksoni), 281–282
Black turtle, 454–455, 454F
see also Sea turtles
Blainville's beaked whale (Mesoplodon densirostris), 406, 407F, 408–409
Blennies
Aspridonotus taeniatus (saber-toothed blenny), 282
Hemiemblemaria simulus (wrasse blenny), 328
Blind cavefish (Astyanax spp.), 377
Bluefin tuna
foraging strategies, behavior optimization, 282–283
southern (Thunnus maccoyii) see Thunnus maccoyii (southern bluefin tuna)
Blue mussel (Mytilus edulis), 201F
Blue penguin, 549
see also Little penguin
Blue tuskfish (Choerodon albigena), 358F
Blue whale
growth and maturation, 388F, 389, 399
lateral profile, 393F

sound production, 397–398
see also Baleen whales
Bluntsnout smooth-head (Xenodermichthys copie), 357F
Boobies see Sulidae (gannets/boobies)
Bosunbird, 524F
see also Phaethontidae (tropic birds)
Boto (Inia geoffrensis), 419
Bottlenose dolphins (Tursiops truncatus), 415F
calving, 417
feeding behaviors, 419, 420
habitats, 474
home ranges, 474–475
inshore forms, 474
mating strategies, 483–484
migration, 474–475
offshore forms, 474, 475
oxygen stores, 459F
signature whistles, 419–420
social interactions, 420–421
see also Oceanic dolphins
Bottlenose whale
diving characteristics, 459T
see also Beaked whales (Ziphiidae)
Bowhead whale (Balaena mysticetus)
growth and reproduction, 399
habitat, 394
lateral profile, 393F
song, 484, 484F
see also Baleen whales
BP (bacterial production), 23F, 24, 25–26
BR (bacterial respiration), 24, 25–26
Brachyramphus, 503T
diet, 506
feeding patterns, 506
plumage, 504, 505F
reproduction, 506
see also Alcidae (auks)
Brachyramphus marmoratus (marbled murrelet), 505F
Bristlemouths (Gonostomatidae), 71–73
Brittle stars (Ophiuroidea), 130, 180F
Brown booby, 524F, 525
see also Sulidae (gannets/boobies)
Brown pelican, 523–525, 524F
see also Pelecanidae (pelicans)
Brown seaweeds (Phaeophyta spp.), 142
Brunnich's guillemot (Uria lomvia), 503, 505F
see also Alcidae (auks)
Buoyancy
deep-sea fishes, 231
Burrow(s)
benthic infauna, 177F
Butterfly fish (Chaetodontidae), 300–301F
By-catch, 422
seabirds see Seabird(s)

C

Calanoida copepods, 40
body forms, 42F
Calanoides carinatus, 49
Calanoides carinatus, 49

Calanus finmarchicus (zooplankton), 10, 40, 42F, 48, 49F
Calcium (Ca, and Ca²⁺), 102
Calcium carbonate (CaCO₃)
    foraminiferal distributions, 153
    secreted by seaweeds, 141
California sea lion (*Zalophus californianus*)
    adaptation to light extremes, 463–464
    diving characteristics, 459T
    myoglobin concentration, 460T
    oxygen store, 459F
    see also Otariinae (sea lions)
Calycophorae siphonophores, 54, 55F
Camouflage
    cephalopods, 80–81
    demersal fishes, 366
    mesopelagic fishes, 242
Cape petrel (Cape pigeon), 540F
    see also Procellariiformes (petrels)
Carbon (C)
    C:N:P ratios, 102
    cycle *see* Carbon cycle
    fixation, 93
    see also Carbon dioxide (CO₂);
        Nutrient(s)
Carbon cycle
    importance of bacterioplankton, 26, 27
    role of seabirds, 500
    see also Primary production measurement methods; Primary production processes
Carbon dioxide (CO₂)
    fixing, 103
    reduction, 100
    sequestration by phytoplankton, 85, 90–91
    see also Carbon cycle
Caretta caretta (loggerhead turtle), 452F, 456, 456F
    see also Sea turtles
Caridea (decapod shrimps), bioluminescence, 115
Carnivores, 465, 465–467, 465T
    family Mustelidae *see* Mustelids
    pinnipeds *see* Pinnipeds
    see also Marine mammals
Carp (*Cyprinus carpio*), 213F
Catadromy, 270
Catfish (Ictaluridae), 377
Cephalopods, **78–84**
    biology, 78–80
        brain and senses, 80
        buoyancy and jet propulsion, 78–80
        buoyancy, 80
        jet propulsion, 80
    central nervous system, 80
    color and pattern, 80
        ability to change, 80
        chromatophore muscles, 80
        chromatophores, 80
        function, 80–81
        iridophores, 80
        leucophores, 80
        escape and luminescence, 80–81
        visual capabilities, 80–81

bioluminescence, 115
Cranchiidae family, 57F, 58
description of class, 78
diagnosis(classification/features) of Cephalopoda, 78, 79T
    compared to other Mollusca, 78
    compared to teleost fishes, 78–80
distribution, 78
ecology, 82
feeding and growth, 81
    feeding methods, 81
    food, 81
    rates, 81
fisheries, 78, 83
    commercial harvest, 83
    historical and contemporary, 83
    global biomass, 83
        estimates, 83
        uncertainty of data, 83
life cycle, 81
    common features, 78
population biology, 82
    single generation life cycle, 82F, 83
reproduction, 81
    maturation and mating, 81
spawning and death, 81–82
    breeding seasons, 82
    death, 82
    fecundity, eggs and hatchlings, 81–82
    locational spawning differences, 81
trophic relations, 82–83
    consumption rates of predators, 83
    importance as food, 83
    prey and predators, 78
Cephalorhynchus dolphins, 411
Cephalorhynchus hectori (Hector's dolphin), 415, 421
Cephaloscyllium ventriosum (swell shark), 353F
Cepphus, 503T
    diet, 506
    plumage, 504
    see also Alcidae (auks)
Cepphus grylle (Black guillemot), 503
Cerorhinca, 503T
    diet, 506
    reproduction, 506–507
    see also Alcidae (auks)
Cestida ctenophores, 56
Cetaceans, 422
    conservation status, 382–383T
    evolution, 467
    migration and movement patterns, 472
        odontocetes *vs.*, 472
    modern, 467, 467F
    phylogeny, 467–468, 468F
    suborders, 467
        Archaeoceti, 467, 468–469
        Mysticeti *see* Baleen whales
        Odontoceti *see* Odontocetes
    thermoregulation, 427
        see also Beaked whales (Ziphiidae);
            Dolphins and porpoises; Marine
            mammals; Sperm whales (Physeteriidae and Kogiidae)

Cetorhinus maximus (basking shark), 280–281
Chaetognatha (arrow worms), 114–115
Channichthyidae (icefishes), 220
Charadriiformes
    Alcidae *see* Alcidae (auks)
    Laridae *see* Laridae (gulls)
    migration, 579
    Rynchopidae *see* Rynchopidae (skimmers)
    Sternidae *see* Sternidae (terns)
    see also Seabird(s)
Charr (*Salvelinus*), 252
    Salvelinus alpinus (arctic charr), 255
Chauliodus macouni (viperfish), 76F
Chelonia mydas (green turtle), 313–314, 454–455, 454F
    see also Sea turtles
Chelonia mydas agassizi (black turtle), 454–455, 454F
    see also Sea turtles
Cheloniids, 450, 454–455
    genera, 450
        Caretta, 456–457
        Chelonia, 454–455
        Eretmochelys, 456
        Lepidochelys, 455–456
        Natator, 456
        see also Sea turtles
Chemolithotrophs, 100
Chemolithotrophy, 100
Chimaera monstrosa (Chimaeriformes), 357F
Chinook salmon *see* *Oncorhynchus tshawytscha* (chinook salmon)
Chinstrap penguin (*Pygoscelis antarctica*), 548T, 551
Chlorophyta (green seaweeds), 142
Choerodon albigena (blue tuskfish), 358F
Chum salmon *see* *Oncorhynchus keta* (chum salmon)
Cichlid (*Nannacara anomala*), 361F
Circatidal vertical migration, 270
Cirripedia (barnacles), wave resistance, 203
Clades, 270
Clams, 179F
Classification/taxonomy (organisms)
    bacterioplankton, 21–23
    beaked whales, 403
    benthic organisms *see* Benthic organisms
    intertidal fishes, 233
    meiobenthos, 159
    phytobenthos *see* Phytobenthos
    radiolarians, 30
    salmonids, 252
    sirenians, 384T, 431–432, 465, 465T
Cleaner wrasse
    Thalassoma bifasciatum, 328
Climate change
    effects on food webs, 284
    influence of primary production, 106–107
    krill as indicator species, 69–70
    see also El Niño Southern Oscillation (ENSO); Plankton and climate

Climate research, 28–29
  see also Plankton and climate
Climatic warming *see* Global warming
Clines, 270
Clupea pallasii (Pacific herring), 246
Clupea sprattus (sprats), 249
Cnidarians
  bioluminescence, 112T, 113
  medusae, 52
Coho salmon *see Oncorhynchus kisutch*
      (coho salmon)
Cold-water coral reefs, **188–198**
  cold-water corals, 188–189
    species and distribution, 188–189
  description, 188
  Enallopsammia profunda, 188–189
  feeding, growth and reproduction,
      191–192
    factors required for growth, 191
    reproductive ecology, 191–192
    understanding of cold-water coral
        growth, 191
  future research, 197
  genetic diversity, 192–193
    molecular technologies, 192–193
  Goniocorella dumosa, 188–189
  habitats and biodiversity, 193–194
    animal communities, 193, 193F
    fish habitats, 194
    functional relationships between
        communities, 193–194
    parasitism of corals, 194
    sampling methodologies, 194
    structural complexity, 193
  historical background, 188
    knowledge and research, 188
    submersible research, 189F
  Lophelia pertusa, 188–189, 188F, 189F,
      190F, 191
    associated with *Eunice norvegica*
        worm, 193–194
    growth on oil platforms, 196
    molecular research, 192–193
    parasites, 194
    reproductive ecology, 191–192, 192F
  Madrepora oculata, 188–189, 191,
      192–193, 193–194
  mapping/photographing, 190F
  new research technologies
    benthic landers, 194, 194F
    molecular technologies, 192–193
  Oculina varicosa, 188–189
  reef distribution and development,
      189–191
    complexity, 189, 191F
    distribution factors, 189–190
    global distribution, 192F
    hydraulic theory, 190
    limited understanding, 190–191
    longevity, 189
  Solenosmilia variabilis, 188–189
  threats, 195–197
    acidification of the oceans, 197
    deep seabed mining, 197
    fisheries, 195, 197F
    oil exploration, 195

  sediment loads and, 196
  timescales and archives, 194–195
    environmental archives, 195, 196F
    global circulation patterns and, 195
    reef ages, 195
Comb jellies *see* Ctenophores
Commercial fisheries *see* Fisheries
Commercial whaling
  baleen whales, 399–400
  history, 399–400
Common cormorant, 524F, 526, 526F
  see also Cormorants
Common diving petrel, 540F
  see also Procellariiformes (petrels)
Common dolphins (*Delphinus* spp.), 416F
Common tern, 513F
  see also Sternidae (terns)
Competition
  baleen whales (Mysticeti) species,
      401–402
  beaked whales (Ziphiidae) species, 406
  fish feeding *see* Fish feeding and foraging
  seabird foraging *see* Seabird(s)
Competitive exclusion, 410
Conger eel (*Conger conger*), 366
Conservation
  definition, 558
  marine mammals, 389–390
  seabird, **555–561**
    biology and, 555
    methods, 558–561, 561
    education, 560
    elimination of predators, 559
    fishing management, 558, 559
    habitat creation/modification, 558
    management of human disturbance,
        560
    management of hunting, 561
    pollution reduction, 559
  sperm and beaked whales, 409–410
Continental shelf, 131
  distribution and characteristics, 9–10
  plankton communities *see* Plankton
      communities
Contranatant migration, 270
Copepod(s), **40–50**
  Acartia spp., 7
  Alaskan Gyre, 10
  behavior, 47–48
    diurnal vertical migration, 47–48, 48F
    factors influencing, 48
    sensory mechanisms, 48
    swimming, 48
    nauplii, 48
    speeds, 48
    swimming-feeding interdependency,
        48
  biogeochemical role, 49–50
    production of fecal pellets, 50
  bioluminescence, 48–49, 111–113
  Calanoides carinatus, 49
  Calanus finmarchicus, 40, 42F, 48, 49F
    see also Calanus finmarchicus
        (zooplankton)
  continental shelves, 9
  distribution and habitats, 42–44

  hyperbenthos, 43–44
  marine caves, 43–44
  marine sediments, 43
  most abundant areas, 43
  Pacific Ocean abundances, 45F
  parasites on other animals, 43
  under polar ice, 45
  sea surface, 45
    see also Upwelling ecosystems
  diversity, 40, 163
  estuaries, 7
    life stages concentrations, 7
    persistence, 7
  feeding, 44–45
    appendages, 45
    behavior, 45
    feeding rate, 45
    food, 45
    rate, 46F
  food source for krill, 68
  growth and development, 45–46
    egg development, 46
    growth rates, 46–47, 46F
    molting, 45–46
    nauplii, 44F
    nauplii and copepodites, 46
  importance, 40
  Lepeophtheirus salmonis (salmon louse),
      50
  Lernaeocera spp., 50
  life histories, 48–49
    bathypelagic species, 49
    diapause, 49, 49F
    variations by habitat, 49
  metabolism, 46–47
    enzymatic activity, 47
    excretion, 47
    respiration, 47
  morphology, 40–42
    appendages, 44F
    Calanoida, 41F
    Cyclopoida, 42F
    body morphology, 40–42
    body sizes, 40
    early life stages, 42–43
    external anatomy, 43F
    internal anatomy, 43F
    maxilla, 45F
  Myticola intestinalis, 50
  Neocalanus plumchrus, 10
  Oithona spp., 10, 11
  as pests, 49
    aquaculture and fisheries, 50
  predation by higher trophic animals, 163
  reproduction, 47
    diapause eggs, 47
    fecundity and spawning, 47
    mating behavior, 47, 47F
  roles in ecosystem, 50
    control of phytoplankton, 50
    fisheries, 50
    food for other species, 50
    nutrient recycling, 50
  subtropical gyres, 10–11
  taxonomy, 40
    orders included, 40

Copiotrophs, 24
Coral(s)
  cold-water, 188–189
    species and distribution, 188–189
  parasitism of, 194
Coral reef fish(es), **222–226**
  behavior, 223
    interactions, 223
    mutualism, 223
    piscivory and defense, 223–224
    modes of piscivory, 223–224
    predation-avoidance, 223–224
    territoriality, 223
  distribution and diversity, 222
    anthropogenic threats, 222
    diversity, 222
    latitudinal distribution, 222
  ecology, 225
    advantages as research subjects, 225
    community structure, 225
    feeding guilds, 225
    resource partitioning, 225
    population dynamics, 225
    metapopulations, 225
    regulating mechanisms, 225
  fisheries see Coral reef fisheries
  life cycle, 224–225
    pelagic larval stage, 224
    settlement, 224–225
    spawning, 224
  maintenance of diversity, 225–226
    diversity hypotheses, 225–226
  morphology, 222–223
    typical traits, 222–223
    vision and coloration, 223
  reproduction, 224
  sex reversal, 224
Coral reef fisheries, 222
  sustainability and overfishing, 222
Cormorants
  breeding patterns, 526, 526F
  darters vs., 527
  human exploitation/disturbance, 525
  line hunting, 526
  plumage, 526, 527
  shags vs., 525–526
  zigzag hunting, 526
  see also Phalacrocoracidae
Cornet fishes (Fistulariidae), 300–301F
Coronatae medusas, 52, 53F
Cosmopolitan distribution, 410
Countershading, 270
Crabs
  fisheries
    methods, 456
    see also Crustacean(s)
Cranchiidae cephalopods, 57F, 58
Crested auklet, 503, 505F
  see also Alcidae (auks)
Crested penguins see Eudyptes (crested
      penguins)
Crustacean(s)
  amphipodal aggregation, 205T
  bioluminescence, 112T
  Cystisoma spp., 58
  deep-sea communities, 130

Phronima spp., 58
  tanaids, 180F
    see also Benthic boundary layer (BBL);
        Copepod(s)
Ctenophores, 54
  Beroida, 55F, 56
  bioluminescence, 112T, 113
  Cestida, 56
  Cydippida, 54, 55F
  Lobata, 54–56, 55F, 56
  Platyctenida, 54
  Thalassocalycida, 54, 54–56
Cubomedusae medusas, 52–54, 53F
Cumaceans (Crustacea), 180F
Current(s)
  effects on migration, 312–313
  geostrophic see Geostrophic currents
Cuttlefish (Sepia spp.)
  buoyancy, 80
  spawning, 81
Cuvier's beaked whale (Ziphius
      cavirostris), 406, 407F, 408–409
Cyanobacteria, 22–23, 101–102
  in epipelic biofilms, 132
  iron requirement, 103
  viruses, 15
    see also Microphytobenthos;
        Phytobenthos
Cyclopoida copepods, 40, 42F
Cyclorhynchus, 503T
  see also Alcidae (auks)
Cyclothone spp. fish, 209, 214
Cydippida ctenophores, 54, 55F
Cyprinus carpio (carp), 213F
Cystonectae siphonophores, 54, 55F

D

Dactylopteridae (flying gurnads),
      300–301F
Daily vertical migration, 270
  see also Diurnal vertical migration
Dall's porpoise (Phocoenoides dalli), 416,
      420, 421
Damsel fish (Eupomacentrus planifrons),
      283
Damselfishes (Pomocentridae), 223
Daption capense (Cape petrel), 540F
Darters see Anhingidae (darters)
Dasyatis sayi (stingray), 351F
DDT (dichloro-diphenyl-trichloroethane)
  seabirds as indicators of pollution, 560
Decapod shrimps (Caridea),
      bioluminescence, 115
Deep sea, definition, 176–177
Deep seabed mining
  threat to cold-water coral reefs, 197
Deep-sea environment, 129–130
Deep-sea fauna, **176–187**
  anthropogenic threats, 185–186
    fishing, trawling and dredging,
        185–186
    mining, 186
    pollutants, 185–186
    precautionary approach, 186
    reasons for concern, 186

  vulnerability of fauna, 186
  waste disposal, 186
  changing perceptions
    about biodiversity, 176
    about stability, 176
  comparison of benthic communities by
      depth, 183F
  defining the organisms, 179–180
    feeding modes, 180
    size groupings, 179–180, 180F
  depth, biomass and density relations,
      180, 181F
  diversity, 176
  diversity, theories about, 185
    areal relationship theory, 185
    current research, 185
    intermediate disturbance theory, 185
    patch mosaic model, 185
    predator theory, 185
    stability-time hypothesis, 185
  diversity patterns, 182–183
    depth patterns, 182–183
    discovery, 182
    latitudinal patterns, 183
  environmental characteristics, 176
  habitats, 176–178, 177F
    'deep-sea' defined, 176–177
    environmental characteristics,
        177–178
    seasonality, 178
    small-scale heterogeneity, 178
  hydrothermal vents, 178–179
    biodiversity, 178, 179F
    causes and characteristics, 178
    chemoautotrophic bacteria, 178–179
    see also Hydrothermal vent(s)
  low diversity environments, 184–185
    areas of regular disturbance, 184
    deep-sea trenches, 184
    hydrothermal vents, 184
    hypoxic areas, 184
    young areas, 184–185
  new questions, 176
  numbers of species, 183–184
    estimates, 183, 184F
    number of marine animal phyla, 183
    small size of sampled area, 183–184
    studies required, 184
  sampling, 180–182
    ALVIN corers, 181, 182F
    box corers, 181, 182F
    epibenthic sled, 180–181, 182F
    gear for specialist research, 181–182
    multicorer, 181
    quantitative samplers, 180–181
    submersibles, 181, 182F
    trawls, 180–181
    visual surveys, 181
  seamounts, 178
    habitat characteristics, 178
  typical landscape, 177F
    see also Benthic foraminifera; De-
        mersal fish(es); Macrobenthos;
        Meiobenthos; Pelagic biogeography
Deep-sea fish(es), **227–232**
  benthic/benthopelagic species, 227

buoyancy, 231
  methods employed, 231
categories, 227
definition, 227
depth-related abundances, 227–230
  faunal provinces, 227–229
  food sources, 229–230, 229F
  relation to food supply, 227
diet, 230
  feeding strategies, 230
  research methods, 230
evolution, 227
life histories, 232
  'bigger-deeper' phenomenon, 232
  incomplete information, 232
longevity, 231–232
  aging methods, 231, 231F
morphologies, 228F
regional differences/similarities, 227
reproduction, 232
  early life stages, 231–232
  hermaphroditism, 231
  spawning, 231
sampling techniques, 227
sensory systems, 230
  olfaction, 230
  sight, 230
  sound production, 230
  touch, 230
    see also Bioluminescence; Demersal
      fish(es)
Deep-sea trenches, 184
Deep-water species see Demersal fish
Delphinidae see Oceanic dolphins
Delphinus spp. (common dolphins), 416F
Demersal fish(es), **363–371**
  appearance and behavior, 365–366
    activity patterns, 366
    adaptations to benthic habitat, 365
    adaptations to specific habitats, 366
    camouflage, 366
    modes of life, 366
    sizes, 365–366
  body shapes, 364F
  continental shelves, 363
    high productivity, 363
    patterns, 363
  definition, 363
  distribution and diversity, 363–365
    distribution, 363, 365F
    environments/habitats, 363–365
    influence of shelf width, 363–365
    latitudinal patterns, 363
    taxonomic diversity, 363
  feeding/diet, 367–369
    carnivorous categories, 367
    benthophages, 368
    piscivores, 367–368
    zooplanktivores, 368
  co-occurring species, 368–369
    dietary flexibility, 368–369
    ontogenetic changes, 368
    partitioning of resources, 368
    flatfish, 368
  migration, 366–367
    diurnal vertical migrations, 366

horizontal migrations, 366
  Northeast Atlantic cod, 366–367,
    367F
  reproduction and life history, 369–370
    habitats for hatchlings and young, 370
    larvae, 370
    longevity, 370
    sharks, 369
    somatic growth patterns, 370
    spawning seasons, 369–370
    teleost postlarvae, 370
    teleosts, 369
  sensory systems, 369
    roles of different systems, 369
    sharks and rays, 369
      see also Deep-sea fauna; Large
        marine ecosystems (LMEs)
Demersal organisms, 422
Denatant migration, 270
Denitrification
  bacteria, 138–139
  role of microphytobenthos, 138–139
    see also Nitrogen cycle
Dentition
  dolphins and porpoises, 411–415, 415F
  killer whale (Orcinus orca), 411–415
  polar bear, 482F
  sea otter, 442–443, 444F
Deposit feeders, 166
Depth
  effect on fish, 273–274
  gradient, seabird abundance and,
    563–564
Dermochelyids, 450, 452–453
Detritus, 131, 166, 177–178, 270
Detrivores, 127, 127–128, 131
Diapause, copepods, 49, 49F
  eggs, 47
Diatom(s)
  epipelic biofilms, 132
  epipsammic assemblages, 132–133
  extracellular polysaccharide production,
    136
  influence of NH_4^+ concentrations, 138
    see also Microphytobenthos;
      Phytoplankton
Diazotrophy, 101
Diet
  benthic organisms, 126–127
  copepods, 45
  demersal fishes, 367–369
  dolphins and porpoises, 419
  fish larvae, 288–289
  intertidal fishes, 237
  macrobenthos, 166–167
  micronekton, 75–76
  radiolarians, 29–30
  salmonids, 256
  seabirds, 499–500, 500, 500T
  sperm and beaked whales, 406
  suspension feeders, 202–203, 203T
    see also Fish feeding and foraging
Dinoflagellates
  bioluminescence, 111, 112T
  symbiotic relationship with radiolarians,
    51–52

    see also Microphytobenthos
Diomedea cauta (shy albatross), 542, 575
Diomedea chlororhynchos (yellow-nosed
  albatross), 543, 545
Diomedea chrysostoma (gray-headed
  albatrosses), 543–544, 545, 575
Diomedea epomophora (royal albatross),
  539
Diomedea exulans (wandering albatross),
  539, 540F, 543, 545, 575
Diomedea immutabilis (Laysan albatross),
  expansion of geographical range,
  542
Diomedea melanophrys, 545
Diomedeidae see Albatrosses
Direct deposit feeders, 127, 131
Dissolved organic matter (DOM)
  transient accumulations in upper ocean,
    23–24
  use by bacterioplankton, 21, 23
Distribution
  Anguilla eels, 262, 263T
  beaked whales, 405–406
  cephalopods, 78
  cold-water corals, 188–189, 189–191
  copepods, 42–44
  coral reef fishes, 222
  dolphins and porpoises, 418
  fish, 214–215
  fish tolerances/limits, 273
  gelatinous zooplankton, 60
  herring, 246
  intertidal fishes, 233, 234T
  mackerels, 250
  macrobenthos, 165
  meiobenthos, 162
  mesopelagic fishes, 239–240
  micronekton, 71
  microphytobenthos, 136–137
  plankton, 4
  planktonic foraminifera, 33
  primary production, **105–110**
  radiolarians, 28–29
  salmonids, 253
  sardines, 249
  seabirds, 497
  sperm whales, 405–406
  sprats, 249
Diurnal vertical migration, 270
  copepods, 47–48, 48F
  deep-sea fishes, 230
  fish larvae, 292
  mesopelagic fishes, 241–242
  micronekton, 73
  plankton, 3
Diving
  Alcidae (auks), 504
  beaked whales (Ziphiidae), 406, 407F
  bottlenose whale, 459T
  elephant seals see Elephant seals
  mammals (marine) see Marine mammals
  seabirds, 565
    flying vs, 546–547
  sea otter, 445–446, 446F
  sperm whales (Physeteriidae and
    Kogiidae), 406, 459T

Diving petrels, 539
  migration, 575–577, 576T
  see also Procellariiformes (petrels)
Dolphins (Delphinidae), 411–416
  beak and teeth, 411–415, 415F
  body shape, 415F
  classification, 416
  color patterns, 415, 416F
  dorsal fin, 415, 415F
  oceanic *see* Oceanic dolphins
  river, 382–383T
  river dolphins, 416, 416F
    distribution, 418
    trophic level, 489F
  size ranges, 415, 415F
    see also Dolphins and porpoises;
      Odontocetes (toothed whales)
Dolphins and porpoises, **411–423**
  behavior and social organization,
      420–421
    feeding, 420
    leaping and bowriding, 420, 420F
    multi-species feeding frenzy, 420F
    socializing, 420–421
  conservation status/concerns, 421–422
    contaminants, 422
    directed fisheries, 421
    incidental mortality, 421–422
    live captures, 422
    noise pollution, 422
    statistics, 421
  distribution, ranging patterns and
      habitats, 418–419
    diving ability, 418
    global distribution, 418
    habitats, 418
    ranging patterns, 418–419
  dolphins *see* Dolphins (Delphinidae)
  evolution, 417
    archaeocetes, 417
    rise of modern families, 417
  feeding ecology, 419
    influence of prey distribution, 419
    specialization within groups, 419
    specialization within water column,
        419
    tooth size and prey, 419
  life history, 417–418
    density-dependent responses, 417–418
    development pace, 417
    gestation period and calf numbers,
        417
    life spans, 417
    threats, 417
  physical descriptions and systematics,
      411–416
    appendages, 411
    body shape, 411
    countershading, 411
    cranial features, 411
    genital and anal openings, 411
    peduncles, 422
    rostrums, 422
    sexual dimorphism, 411
    species diversity, 411
    vertebral column, 411

porpoises *see* Porpoises
sensory systems and communication,
    419–420
  echolocation, 419
  smell, vision and touch, 419
  whistles, 419–420
species diversity, 412–414T
vernacular names confusion, 416–417
DOM *see* Dissolved organic matter
    (DOM)
Dovekie (*Alle alle*), 503, 505F
Dragonfishes (*Pachystomias* spp.), 358
Drive fisheries, 422
Dugong (*Dugong dugon*), 432, 432F,
    470–471
  conservation status, 384T
  cultivation grazing, 479
  ecology, 437
  exploitation, 437–438
  future outlook, 440–441
  home ranges, 479
  mating behavior, 436–437
  morphology, 434–435
  movement patterns, 479
  population biology, 437
  trophic level, 488
    see also Sirenians
Dugongidae, 384T, 471
  classification, 431
  extinct species, 471
    Steller's sea cow, 431–432, 432F,
        470–471, 471
    see also Sirenians
Dwarf spinner dolphin (*Stenella
    longirostris roseiventris*), 415

E

Echinoderms
  bioluminescence, 112T, 113–114
  sea urchins, 146
Echolocation, 410, 422
  dolphin *see* Oceanic dolphins
Ecology
  baleen whales (Mysticeti), 393–394
  benthic foraminifera *see* Benthic
      foraminifera
  cephalopods, 82
  cold-water coral reefs, 191–192, 192F
  coral reef fish *see* Coral reef fish(es)
  dolphins and porpoises, 419
    see also Dolphins and porpoises
  fish horizontal migration, 308
  gelatinous zooplankton *see* Gelatinous
      zooplankton
  manatees, 437
  Phaethontidae (tropic birds), 523, 528T
  phytobenthos *see* Phytobenthos
  planktonic foraminifera *see* Planktonic
      foraminifera
  seabirds *see* Seabird(s)
  sirenians, 437–439
Ecophenotypes, 270
Eels (*Anguilla*), 300–301F, **262–271**
  catadromous species, 262
  diversity/distribution, 262, 263T

reproduction requirements, 262
evolution/paleoceanography, 268–269
  evolution, 269
  separation, 269
fisheries/aquaculture, 269
  culture operations, 269
  glass eel fisheries, 269
  world catches, 269
genetics/panmixia, 268
  American, Japanese, Australian and
      New Zealandic eels, 268
  European, African and Icelandic eels,
      268
  relationships debates, 268
  Tucker's hypothesis, 268
  failure on genetic grounds, 268
  failure on oceanographic grounds,
      268
growth rate, 265
  age determination, 265
  influencing factors, 265
  variations, 265
life cycle, 262–264, 264F
  eggs, 264
  elvers and yellow eels, 264–265
  pigmentation development, 264–265
  glass eels, 264
  metamorphosis, 264, 268
  knowledge base, 262
  leptocephali, 264, 270
  larval duration, 264
  mode of nutrition, 264
  morphology, 264
  overview, 262
  silver eels, 266
  fecundity, 266, 266T
  spawning areas/times, 262–264
  American/European eels, 263
  Japanese/Australian/New Zealandic/
      tropical eels, 264
  lack of observation in nature,
      262–263
migration, 308–309
morphology of life stages, 262
oceanic migration, 266–267, 308–309
  glass eels, 268
  metamorphosis, 268
  selective tidal transport, 268
  leptocephali, 267–268
  daily vertical migration, 267
  European, 268
  horizontal migration, 267–268
  vertical distribution, 267
  silver eels, 266, 266–267
  migration lengths, 267, 267T
  research technology, 267
  selective tidal transport, 266
  swimming ability, 266–267
  travel routes/rates, 267
orders, 262
plasticity/adaptation, 265
  diet, 265
  habitat selection, 265
  metamorphosis, 264, 265–266, 268
  size at, 265
  yellow eels to silver eels, 265–266

sex determination/differentiation, 265
sexual dimorphism, 265
size and age at maturity, 265, 266T
status of populations, 269–270
downward trend, 269
recruitment, 269
Eggs
Alcidae (auks), 508
cephalopods, 81–82
copepods, 47
eels (*Anguilla*), 264
Ekman layer, 270
Elasmobranchs, 213–214, 213–214, 214F
body fluids and osmoregulation, 216
buoyancy, 215
differences from teleosts, 213–214
fins, 298
main features, 214F
osmoregulation, 216
paucity of habitats, 214
skeleton, 214
Elephant seals
adaptation to light extremes, 463–464
diving, 481
ability, 459T
adaptations, 459, 459F, 460T
mating strategies, 483–484
myoglobin concentration, 460T
see also Phocidae (earless/'true' seals)
El Niño Southern Oscillation (ENSO)
impact on seabirds
penguins, 553
models *see* El Niño Southern Oscillation (ENSO) models
see also Climate change
Embiotica jacksoni (black surf perch), 281–282
Emperor penguin, 548T, 552, 553
see also Aptenodytes
Enallopsammia profunda coral, 188–189
Engraulis (anchovies), 250
Engraulis ringens (Peruvian anchoveta), 250, 280
Enhydra lutris see Sea otter
ENSO *see* El Niño Southern Oscillation (ENSO)
Environmental gradients, seabird abundance and, 562–563
Environmental protection, from pollution *see* Pollution
Epifauna, 124–125, 126F, 131, 165, 169
see also Benthic boundary layer (BBL)
Epifluorescence microscopy
use in bacterioplankton studies, 21, 24–25
Epipelagic zone, 270, 422
see also Pelagic biogeography
Epistominella exigua foraminifera, 157F, 158
Erect-crested penguin *see Eudyptes sclateri* (erect-crested penguin)
Eretmochelys imbricata (hawksbill turtle), 455F, 456
see also Sea turtles
Erignatus barbatus (bearded seal)

song, 485, 485F
see also Phocidae (earless/'true' seals)
Esox spp. (pike), 300–301F
Estuaries
plankton communities, 7
Eubalaena glacialis see North Atlantic right whale
Eudyptes (crested penguins), 550–551
breeding patterns, 551
characteristics, 548T, 550, 550F
distribution, 548T, 550–551
egg-size dimorphism, 551, 551F
feeding patterns, 548T
migration, 574
species, 548T, 550
see also Sphenisciformes (penguins)
Eudyptes chrysolophus (marconi penguin), 548T, 550–551
Eudyptes robustus (Snares penguin), 548T, 550–551
Eudyptes sclateri (erect-crested penguin), 548T, 550–551, 550F, 551
egg-size dimorphism, 551, 551F
see also Eudyptes (crested penguins)
Eudyptula, 549
migration, 574
species, 549
see also Sphenisciformes (penguins)
Eudyptula albosignata albosignata, 549
Eunice norvegica worm, 193–194
Euphausiacea (krill) *see* Krill (Euphausiacea)
Euphausia crystallarophias (ice krill), 66
Euphausia pacifica (Pacific krill), 66, 69
Euphausia superba (Antarctic krill), 62, 63F
aggregations, 67F, 68
see also Krill
diurnal vertical migration, 64, 68
fisheries, 69
food sources, 66
life span, 65
winter survival, 66
see also Krill (Euphausiacea)
Euphausiids, 71, 73F
see also Krill (Euphausiacea)
Euphotic zone, 93
Eupomacentrus planifrons (damsel fish), 283
European eel (*Anguilla anguilla*), 268
Euthecosomes, 57F
Evolution
beaked whales, 403
deep-sea fishes, 227
dolphins and porpoises, 417
eels, 269
fish, 209–210, 272
intertidal fishes, 233–234
krill, 62
planktonic foraminifera, 33
salmonids, 252
Exotic species, 60–61
Exploitation *see* Human exploitation
Extracellular polysaccharide production, 132

F

Fairy penguin, 549
False killer whale (*Pseudorca crassidens*), 411
FAO *see* Food and Agriculture Organization (FAO)
Feeding
baleen whales (Mysticeti), 387–388, 387F, 396T, 397
cephalopods *see* Cephalopods
cold-water coral reefs *see* Cold-water coral reefs
copepods *see* Copepod(s)
deep-sea fauna, 180
demersal fish *see* Demersal fish(es)
filter, right whales (balaenids), 393F
fish *see* Fish feeding and foraging
gray whale (*Eschrichtius robustus*), 397
krill (Euphausiacea), 68
macrobenthos *see* Macrobenthos
marine mammals *see* Marine mammals
odontocetes (toothed whales), 482, 482F
pinnipeds (seals), 387, 476
planktonic foraminifera, 36
polar bear, 387
radiolarians, 29–30
seabirds *see* Seabird(s)
suction
beaked whales (Ziphiidae), 406
sperm whales (Physeteriidae and Kogiidae), 406
surface-tension
red-necked phalarope, 533, 534F
Wilson's phalarope, 533
Feeding behavior
bottlenose dolphins (*Tursiops truncatus*), 419, 420
killer whale (*Orcinus orca*), 419, 420
Odobenidae (walruses), 481
Phalaropes *see* Phalaropes
Feeding guilds, coral reef fishes, 225
Feeding patterns
Anhingidae (darters), 527, 528T
Aptenodytes (penguins), 548T, 552
Brachyramphus, 506
Eudyptes (crested penguins), 548T
Fregatidae (frigatebirds), 523, 528T
little penguin (*Eudyptula minor*), 548T, 549
Pelecanidae (pelicans), 523–525, 528T
Pelecaniformes, 527–528, 528T
Phalacrocoracidae, 528T
Pygoscelis, 548T, 551–552
Spheniscus, 548T, 549
Sulidae (gannets/boobies), 525, 528T, 569
yellow-eyed penguin, 548T, 550, 553
Feeding shifts
Atlantic cod (*Gadus morhua*), 284
Atlantic herring (*Clupea harengus*), 281, 281F
herring (*Clupea harengus*), 281, 281F
Finless porpoise (*Neophocaena phocaenoides*), 416, 418, 421

Fiordland penguin, 548T, 550–551
  see also Eudyptes (crested penguins)
Fish(es), **209–217**
  abundance, 209
    diversity and numbers, 209
    species, 209
  adaptations, 215
    body fluids and osmoregulation,
      215–216
    buoyancy, 215
    elasmobranchs vs. teleosts, 215
    electroreception, magnetic fields and
      navigation, 216–217
    electroreceptors, 216–217
    generation of pulses, 216
    navigation via electroreceptors, 217
    navigation via magnetoreceptors, 217
    locomotion see Fish locomotion
    vision see Fish vision
    warm blood, 216
    body areas warmed, 216
    purposes for warming, 216
  benthic see Benthic fish(es)
  bioluminescence, 112T, 115–116
    counterillumination camouflage,
      116–117
    luciferin/luciferase system, 116
    photophores, 116–117
    symbiotic bacteria, 115–116
  coral reef see Coral reef fish(es)
  deep-sea see Deep-sea fish(es)
  demersal see Demersal fish(es)
  distribution, 214–215
    bathypelagic zone, 214
    benthopelagic zone, 214
    coral reefs, 214–215
    euphotic/epipelagic zone, 214
    mesopelagic zone, 214
  diversity and origins, 209–210
    agnathans and gnathostomes, 210
    common ancestor, 210
    earliest ray-finned fishes, 210
    fossil record, 209–210
    species numbers, 209
  farming see Aquaculture
  feeding/foraging see Fish feeding and
    foraging
  food sources
    copepods, 50
    krill, 68–69
  freshwater diversity, 209
  habitats and adaptations, 209
  hearing see Fish hearing and lateral lines
  human value, 217
    advances in endocrinology, 217
    contributions to/from other
      disciplines, 217
  intertidal see Intertidal fish(es)
  larvae see Fish larvae
  lateral lines see Fish hearing and lateral
    lines
  locomotion see Fish locomotion
  mesopelagic see Mesopelagic fish(es)
  migration
    horizontal see Fish horizontal
      migration

    vertical see Fish vertical migration
  out of water, 209
  pelagic see Pelagic fish(es)
  reproduction see Fish reproduction
  schooling see Fish schooling
  species numbers
    fossil species, 209–210
    living species, 209
  threats, 209
  types, 210–213
    comparisons, 210–212
    elasmobranchs, 213–214
    teleosts, 212–213
  types and relationships, 210
    ancient relationships, 211F
    cladistic models of taxonomy, 210
    example, 210, 211F
    classification by morphology, 210
    classification difficulties, 210
  vision see Fish vision
  see also Antarctic fish(es); Flatfish;
    Salmonids
Fish ecophysiology, **272–278**
  basic features, 272, 273F
  biotic/abiotic distribution factors,
    272–273
    examples, 272–273
  distribution tolerances/limits, 273
    oxygen tension, 276–277
    avoidance of hypoxic areas,
      276–277
    pH, 277
    control of acid-base balance, 277
    presence of water, 273
    necessity of water for most fish, 273
    problems of immersion, 273
    salinity, 275–276
    acclimation, 276
    elasmobranch fish, 275
    fresh/saltwater species differences,
      275
    marine teleost fish, 275–276, 276F
    osmorespiratory compromise, 276
    temperature, 274–275
    adaptations to avoid freezing, 275
    adaptations to different temperatures,
      275
    ectothermic nature of fish, 274
    factors determining thermal niche,
      274
    susceptibility to freezing, 275
    warm-water specialist example,
      274–275
    water depth, 273–274
    depth-related constraints, 273–274
  diversity of fish, 272, 272T
  optimal abiotic conditions, 277
    aquaculture perspective, 277
    biogeographical perspective, 277,
      277F
  origins/evolution of fish, 272
  plasma and hemolymph osmolarity,
    273F
    see also Antarctic fish(es); Deep-sea
      fish(es); Eels; Intertidal fish(es);
      Salmonids

Fisheries
  cephalopods, 78, 83
  cold-water coral reefs, 195, 197F
    as threat to, 195, 197F
  copepod pests, 49
  copepods, 50
  coral reef fishes, 222
  dolphins and porpoises, 421
  eels, 269
  fish schooling, 342–348, 343F
    see also Fish schooling
  food webs and see Food webs
  herring, 247
  importance of pelagic fishes, 251
  krill, 62, 69, 69–70
  mackerels, 250
  sardines, 249
  threat to cold-water coral reefs, 195,
    197F
  tunas, 250
    see also Large marine ecosystems
      (LMEs); Whaling industry
Fish eyes see Fish vision
Fish feeding and foraging, **279–285**
  complex interactions, 284–285
    effects of climate change, 284–285
    ontogenetic feeding shifts, 284
  food chains, 284
    critical links, 284
    prey species, 284
  foraging strategies, 282–284
    optimization of behavior, 282
    bluefin tuna example, 282–283
    game theory, 283–284
    sharing of resources, 283
  modes of feeding, 279, 279–282, 279T
    adaptations, 279
    ambushers, 282
    carnivores, 281
    benthic feeders, 281
    tactics used to catch prey, 282
    carrion feeders, 280
    competition, 283
    defence of territory, 283
    hierarchy development, 283
    scramble competition, 283
    detrivores, 280
    feeding guilds, 279
    fluidity of tactics, 282
    food type classifications, 279
    herbivores, 280
    non-exclusive categories, 280
    optimization of behavior, 282
    planktivores, 280
    basking sharks, 280–281
    herring, 281
    migrations, 280
    proportions of feeding modes, 280T
    specialist exploitation, 282
    stalkers, 282
    use of camouflage, 282
    use of lures, 282
    use of speed, 282
Fish hearing and lateral lines, **372–378**
  acoustic signals, 372
  ear-lateral line interactions, 377

complementary systems, 377
ears, 374
  mechanics, 374
  mechanoreceptive hair cells, 374,
    374F
  hearing, 372–373
    mechanics, 373–374
    inner ear, 373–374, 373F
    purposes, 374–375
    behavioral situations, 375
    communication, 374
    sound production mechanisms,
      374–375
    sounds heard, 372–373
    discrimination/understanding, 373
    hearing ranges, 372, 373F
    specialist/generalist hearing, 372–373
    testing fish hearing, 372
  hearing adaptations, 375
    air bubbles, 375
    hearing specialists, 375
    sound production and hearing, 375
    swim bladder, 375
  human-generated sound, 372, 375–376
    conflicting evidence, 376
    few data, 376
    growing concern, 375–376
  lateral line, 376–377, 376F
    description, 376
    hydrodynamic stimulation, 376
    pattern variations, 377
    purposes, 376
    sensory hair cells, 376–377
  octavolateralis system, 372
  structure and habitat, 377
    evolution of specialized structures,
      377
    habitat and hearing importance, 377
Fish horizontal migration, **307–315**
  ecology, 308
    regional production cycles, 308
    spawning and homing, 308
  life histories, 308
    amphidromous species, 309
    Japanese ayu, 309
    anadromous species, 308
    families included, 308
    salmon, 308
    catadromous species, 308–309
    anguillid eels, 308–309
    European/American eels, 309
    families included, 308
    oceanodromous species, 309–312
    Arcto-Norwegian cod, 311
    cod, 311
    herring, 310, 310F
    plaice, 311–312
    scombrids, 309–310, 310F
    tuna, 309–310, 309F
    walleye pollock, 310–311
  migration circuits, 307
  migration mechanisms, 312
    movement and navigation, 312
    ocean currents, 312–313
    oceanic gyres, 312
    orientation and navigation, 314

  tidal streams, 313–314, 313F, 314F
  migration speed, 312–313, 313F
  vertical movements, 312–313, 313F
  migration occurrence, 307
    importance to world fisheries, 307,
      307T
    numbers of migrating species, 307
  tag data, 314
  terminology, 307–308
    categories of migrating fish, 307–308
    see also Eels; Fish locomotion; Fish
      vertical migration; Pelagic
      fish(es); Salmonids; Tide(s)
Fish larvae, 330–331, 331F, **286–296**
  behavior, 292
    diurnal vertical migration, 292
    selective tidal stream transport, 292
    swimming behaviors, 292
  development see Fish reproduction
  examples, 287–288F
  factors controlling survival, 286
  foods/feeding, 288–289
    common prey, 288–289
    evaluation of nutritional condition,
      289
    feeding behaviors, 289, 290F
    growth, 289
    prey size, 288–289
  general description, 286
  integration, 292–294
    age determination, 293–294
    cohort biomass, 292, 293F
    density-dependent processes, 293
    density-independent processes, 293
    larval stage dynamics models, 294
    recruitment, 293F
  interactions with other plankton, 286
    distributions within nurseries, 286
    ichthyoplankton, 286
    sampling methods, 286
  predation, 289–290, 291F
    effect of larval growth rate, 289
    predator types, 289
    vulnerability, 289–290
      see also Fish predation and
        mortality
  properties, 295T
  qualities of survivors, 295
  recruitment, 295
  research needs, 294–295
    better abundance estimates, 294
    better knowledge of trophic
      relationships, 294
    new technologies, 294–295
    time series of abundances, 294
  size and survival, 295
  survival and recruitment, 286, 295
  survival hypotheses, 286–288
    critical period, 286–287
    lottery, 288
    match-mismatch, 287–288
    retention, 288
    stable ocean, 287–288
  temperature and salinity, 290–292
    effects of salinity levels, 292
    importance, 290–291

  optimum salinity-temperature levels,
    292
  temperature effects on bioenergetics,
    292
  temperature effects on growth rates,
    291–292
  see also Fish feeding and foraging;
    Fish predation and mortality;
    Fish reproduction; Large marine
    ecosystems (LMEs); Plankton
Fish locomotion, **297–306**
  adaptations, 215, 299, 300–301F
    generalists vs specialists, 299
    stream-lining, 215
    swim muscles, 215
  apparatus, 297–299
    body curvature waves, 297
    body shape, 298
    fins, 298
    mucus, 299
    muscles/myotomes/myosepts,
      297–298, 297F
    scales, 298
    skin, 298
    vertebrae, 297
  basic principles, 297
  energy cost, 303–304
    cost of swimming, 303–304, 303F
    optimum speed and mass, 304, 304F
    overcoming drag, 303–304
  energy-saving behaviors, 304–305
    burst-and-coast swimming, 304, 304F
    schooling behavior, 304–305
  fish wakes, 299–303
    creation of vortex rings, 299–302,
      302F
    quantitative flow visualization
      techniques, 302–303, 302F, 303F
  methods, 299
    body curvature waves, 299
    dorsal/anal fin propulsion, 299
    pectoral/median fin propulsion, 299
  speed and endurance, 305–306
    maximum burst speeds, 305
    maximum sustained speeds, 305
    muscle groups, 305–306
    oxygen limitations, 305
    relationships, 305, 305F
  styles, 299
Fish predation and mortality, **322–329**
  community structure, 327
    influencing factors, 327
  diversity of predators, 322–323, 324T
    fish, 323
    invertebrates, 323
    types and sizes, 322–323
  evolution, 327–328
    avoidance tactics, 327–328, 328F
    feeding-predator avoidance trade-off,
      328
  life stages and predation, 323F, 325,
    326F
    predators of adults, 325
    predators of eggs, 325
    predators of juveniles, 325
    predators of larvae, 325

Fish predation and mortality (*continued*)
  life transitions and predation, 325–326
    highest predation rates, 325–326
    metamorphosis, 326
  mortality
    fishing, 322
    natural, 322, 323F
  predation equation, 324–325, 325F
    changes in behavior, 325
    escape, 325
    factors affecting predation rates,
        324–325
    sequence of predation, 324
  recruitment, 326–327
    density-dependent predation, 327
    early life stage predation, 327
    effects on population levels, 326–327
    time in life stage, 327
  removal of fish by predator type, 322F
  studying predation, 323–324
    assessing impact, 323, 324T
    modelling, 323–324
    see also Fish feeding and foraging;
        Fish schooling; Seabird foraging
        ecology
Fish reproduction, **330–336**
  behavior, 335–336
    antipredator mechanisms, 335
    food acquisition, 335
    high mortality rates, 336
    light, 335
  characteristics of most fish, 330
  deep-sea fish *see* Deep-sea fish(es)
  demersal fish *see* Demersal fish(es)
  egg and larval development, 334–335
    description/development of eggs, 334,
        334F
    flexion, 334
    larval feeding, 334
    larval physiology, 334
    larval stages and metamorphosis,
        335F
    metamorphosis, 334–335
    sensory systems, 334
    see also Fish larvae
  eggs, 330–331, 330F
    effect of temperature on development,
        331F
    intertidal fishes, 237
    parental care, 238
  factors affecting reproductive ability,
      330
  fecundity and egg size, 332, 332F
    relationship, 332
    reproductive strategies, 332
  intertidal fish *see* Intertidal fish(es)
  larvae, 330–331, 331F
    see also Fish larvae
  life histories (general), 330–331, 335F
    exceptions, 331
  spawning behavior and parental care,
      332–334
    egg cases, 333F
    oviparity, 333
    parental care, 333–334
    seasonal migrations, 333

  viviparity, 333
  spawning season, 331–332
    high latitudes, 331–332
    low latitudes, 332
    see also Fish larvae; Life histories
        (and reproduction)
Fish schooling, **337–349**
  behavior, 338F
  decisions, 346F
  definitions, 337
    social group types, 337
  dynamics of decisions
    mature fish, 346
    young fish, 345–346
  fisheries applications, 342–348, 343F
    adjustment models, 343, 345F
    catchability/catch rate, 343
    distribution dynamics, 347T
    fidelity to shoal, 344, 348
    fishing technologies, 342
    genetic relatedness, 344–345, 348
    human response to population
        collapse, 342
    metapopulation model, 343
    models, 344F
    patterns, 347
    population collapse/range reduction,
        342
    predation minimizing, 346–347
    retention zones, 343–344
    schooling decisions, 345–346
    shoaling behavior, 347–348
    stock collapse/range collapse,
        342–343, 344F
  food and predators, 339–341
    anti-predator tactics, 340
    competition strategies, 339
    feeding methods, 339
    food location, 339
    grouping advantages, 339–340
    predator avoidance, 339
    herding, 341
    pack hunting, 341
    splitting off individuals, 341
    'twilight hypothesis', 341
    predator inspection behavior, 340
    'tit-for-tat' altruistic behavior, 341
    shoal detection by predators, 339
    shoaling behavior key factors, 339
    shoal size and success, 340
  genetic basis for behavior, 341–342
    British minnows experiment, 341–342
  'how' and 'why' questions, 337, 338T
  inter-fish/inter-school distances, 348F
  learned behavior vs genetic basis, 341
  other functions, 342
    direction correction, 342
    energy efficiency, 342
    parasite reduction, 342
  school rules and size, 337–338
    effects of individual size, 337–338
    influence of food/predation regime,
        337
    join, leave or stay (JLS) decisions, 337
    spatial relationships, 337
  sensory system, 338–339

  lateral line system, 338–339
  senses used, 338
  vision, 338
    see also Fish feeding and foraging;
        Fish locomotion; Fish predation
        and mortality
Fish swimming *see* Fish locomotion
Fish vertical migration, **316–321**
  multiple controls model, 320–321, 321F
  pattern examples, 316
    type I, 316, 319
    type II, 316, 316F, 317F, 319
  pattern variations, 316–317
    changes between types I and II, 317
    genus/species, 317, 318F
    individual patterns, 316–317
    influence of environmental conditions,
        317
    influence of size, 317
    population/site, 317
    seasonal patterns, 317
  periodicities, 316
  potential factors influencing, 317–318
    commensal species, 319
    endogenous circadian rhythm, 317
    estuarine fish, 318–319
    food, 319
    intertidal fish, 318
    jellyfish associations, 319
    light, 317–318
    migration of prey, 319
    sensor mechanisms, 319
    tides, 318–319
  reasons for study, 316
  survey difficulties, 321
  theoretical explanations, 319–320
    bioenergetics, 319–320
    feeding-predator avoidance balance,
        320
    optimization models, 320
    predation, 320
    predator avoidance, 320, 320F
    see also Demersal fish(es); Fish
        feeding and foraging; Fish hori-
        zontal migration; Fish
        locomotion
Fish vision, **350–362**
  adaptations, 216
    optimal visual pigment, 216
    use of light as camouflage, 216
    use of photophores, 216
    varied visual environments, 216
  diversity of environments, 351F
  environmental adaptations, 350
  image formation, 350–351
    asymmetrical/odd-shaped eyes, 350,
        351F, 352F
    focusing, 350–351, 352F
    lens, 352F
    resting eye, 350
  light/dark adaptation, 355
    structural changes, 355, 359F
  ocular filters, 351–353
    deep-sea fishes, 353
    overcoming drawbacks, 352–353,
        354F

reflective corneal layers, 351–352
  short-wave absorbing, 351–352
pupils, 351
  constriction/dilation, 351, 353F
retinal structure, 353–355
  basic structure, 353, 355F
  regional variations, 354–355, 358F
  rods/cones, 356F, 357F
  species variation, 353–355
  cone arrangement, 354, 358F
  multiple cones, 354
  reflective tapeta, 354, 357F
  rods/cones, 353–354
typical teleost eye, 351F
visual abilities, 359
  absolute sensitivity, 359
  contrast, 359–360
  measuring, 360
  detection, 359
  distance perception, 361–362
  visual angle, 361–362
  importance, 359
  polarization, 362
  polarization sensitivity, 362
  spatial resolution, 360–361
  acuity, 360–361
  minimum resolving angle, 360–361
  neural processing, 361
  spectral responses, 361
  limits, 361
  proving color vision, 361
visual pigments, 355–356
  spectral absorption, 356–359, 359F
  bioluminescence detection, 358, 360F
  cone visual pigment, 358–359, 361F
  fresh/saltwater species differences,
    357–358
  optical environment, 358, 360F
  optimization, 359
  range, 356–357
  structure, 355–356
  chromophore and opsin, 355–356
  factors affecting absorption spectrum,
    356
  see also Bioluminescence
Fistulariidae (cornet fishes), 300–301F
Flatback turtle (Natator depressus), 456
  see also Sea turtles
Flatfish (Pleuronectiformes), 300–301F
Florida manatee (Trichechus manatus
    latirostris), 432, 433F, 436F, 438F
  see also Manatees
Flow cytometry (FCM), 24–25
Flying fish (Exocoetidae), 274–275,
    300–301F
Flying gurnads (Dactylopteridae),
    300–301F
Food and Agriculture Organization
    (FAO), 239
Food webs, 284
  bacterioplankton, 26, 26–27
    see also Bacterioplankton
  cephalopods, 82–83
  copepods, 50
  krill, 68
  meiobenthos, 159, 163

micronekton, 72F
plankton viruses, 17–18
  see also Upwelling ecosystems
Foraging
  definition, 479
  fish see Fish feeding and foraging
  marine mammals see Marine mammals
  seabirds see Seabird(s)
Foraging ecology, beaked whales
    (Ziphiidae), 406
  see also Beaked whales (Ziphiidae)
Fossil(s)
  Alcidae, 503
  pelecaniform birds, 522
  penguins
    modern penguins and, 547–548
    size, 547
    Waipara (New Zealand), 546
  sea turtles, 451
  sirenians, 431
Four-eyed fish (Anableps anableps), 352F
Franciscana dolphin (Pontoporia
    blainvillei), 415, 415F, 417–418
Fraser's dolphin
  myoglobin concentration, 460T
  see also Oceanic dolphins
Fratercula, 503T
  diet, 506
  see also Alcidae (auks)
Fratercula arctica (Atlantic puffin), 503,
    505F
Fregata minor (great frigatebird), 524F
Fregatidae (frigatebirds), 522, 523
  breeding patterns, 523
  characteristics, 523, 528T
  distribution, 523
  feeding patterns, 523, 528T
  migration, 577–579
  species, 523, 524F
  see also Pelecaniformes
Frillfin goby (Bathygobius soporator),
    235
Frontal zone, 270
Fronts
  seabird abundance and, 563, 563F
Fulmar(s), 539
  migration, 575–577, 576T
  northern see Northern fulmar (Fulmaris
    glacialis)
  see also Procellariiformes (petrels)
Fundulus heteroclitus (killifish), 276
Fur seal see Arctocephalinae (fur seals)

G

Gadfly petrels, 539
  migration, 575, 576T
  see also Procellariiformes (petrels)
Gadiformes, 300–301F
Galapagos penguin (Spheniscus
    mendiculus), 548, 548T, 549, 553
Gannets see Sulidae (gannets/boobies)
Gasteropelecidae (hatchet fishes),
    300–301F
Gasterosteus aculeatus (three-spined
    stickleback), 283

Gelatinous zooplankton, **51–61**
  body form evolution
    adaptations, 51
    common form, 51
    environmental impact, 51
  ecology, 60–61
    fragility, 60
    geographical and depth distribution,
      60
    outcompeting other animals, 60–61
    research method advances, 60
    seasonal blooms, 60
    trophic niche distribution, 60
  taxonomic groups, 51
    cephalopods, 56–58
    crustaceans, 58
    ctenophores, 54
    heteropods, 56
    holothurians, 58
    medusae, 51–52
    pelagic tunicates, 58
    Doliolida, 58–59
    Larvacea/Appendicularia, 60
    Pyrodomida, 58
    Salpida, 59–60
    polychaete worms, 58
    pteropods, 56
    Radiolaria, 51
    siphonophores, 52–54
  vertical migration, 319
Genomic techniques, 21
Gentoo penguin (Pygoscelis papua), 548T,
    551
Georges Bank, North Atlantic, 363–365,
    365F
Geostrophic currents, 270
Giant petrels, 542
  see also Procellariiformes (petrels)
Glass catfish (Kryptopterus bicirrhis),
    356F, 361F
Glass eels see Eels
Global carbon cycle see Carbon cycle
Global warming
  effect on ocean pH, 197
  threat to penguins, 553
  see also Climate change
Globicephala (pilot whales), 411,
    420–421, 421
Globicephala melas (long-finned pilot
    whale), 418
Globigerina bulloides foraminifera,
    158
Gobies (Gobiidae), 223
Gobiidae (gobies), 223
Goldfish (Carassius auratus), 361F
  hearing range, 372–373
Goniocorella dumosa coral, 188–189
Gonostomatidae (bristlemouths),
    71–73
Gorgonacea (sea whips), 177F
Gosse, Philip Henry, 188, 188F
Gradients, physical, seabird abundance
    and, 562–563
Grampus griseus (Risso's dolphin),
    411–415
Gravitational tides see Tide(s)

Gray-headed albatross (*Diomedea chrysostoma*), 543–544, 545, 575
see also Albatrosses
Gray whale (*Eschrichtius robustus*), 391, 392T, 470
  feeding, 397
  habitat, 394
  lateral profile, 393F
  migration, 398
  trophic level, 489F
  see also Baleen whales
Grazing
  benthic organisms, 127
  phytoplankton blooms see Phytoplankton blooms
Great auk (*Pinguinis impennis*), 503
  overexploitation, 503, 508–509
  remains, 507–508
  see also Alcidae (auks)
Great black-backed gull (*Larus marinus*), 513F
Great cormorant (*Phalacrocorax carbo*), 524F, 526, 526F
  see also Cormorants
Great frigatebird (*Fregata minor*), 524F
  see also Fregatidae (frigatebirds)
Green seaweeds (*Chlorophyta* spp.), 142
Green turtle (*Chelonia mydas*), 313–314, 454–455, 454F
  see also Sea turtles
Grey phalarope see Red phalarope
Guanine (C₅H₅N₅O), 270
Guillemot(s)
  black, 503
  Brunnich's, 503, 505F
  migration, 579–581
  see also Alcidae (auks)
Gulls see Laridae (gulls)
Gymnoptera pteropods, 56–58
Gymnosomata pteropods, 56, 57F

H

Hadal zone, 126T, 129–130, 131
Haeckelia rubra, 54
Hagfish (Myxinidae), 280
Hake (Merluccidae), 363–365
Halocyptena microsoma (least storm petrel), 539
  see also Procellariiformes (petrels)
Harbor porpoise (*Phocoena phocoena*), 421
  see also Dolphins and porpoises; Porpoises
Harbor seals (*Phoca vitulina*)
  lactation, 476
  movement patterns during breeding season, 476, 477F
  see also Phocidae (earless/'true' seals)
Hardy, Sir Alister, 281
Harpacticoida copepods, 40
Harp seal (*Phoca groenlandica*)
  migration and movement patterns, 478F, 479
  see also Phocidae (earless/'true' seals)

Hatchet fishes (Gasteropelecidae), 300–301F
Haul-out site(s)
  definition, 479
  pinnipeds (seals), 476–479, 477F
Hawksbill turtle (*Eretmochelys imbricata*), 455F, 456
  see also Sea turtles
Hearing
  fish see Fish hearing and lateral lines
  marine mammals, 386
Hector's dolphin (*Cephalorhynchus hectori*), 415, 421
Hemiemblemaria simulus (wrasse blenny), 328
Herbivores, fish feeding/foraging, 280
Herring (*Clupea*), 246, 246–249
  description and life histories, 246–247
  distribution, 246
  fisheries history, 247
  fishery locations, 247–249
  migration, 246–247, 247F, 248F
  Northeast Atlantic spawning stocks, 247
  Northwest Atlantic spawning stocks, 247
  overexploitation, 249
  Pacific spawning stocks, 247, 248F
  relationship between fecundity and egg size, 332, 333F
Herring gull (*Larus argentatus*), 513F
  see also Laridae (gulls)
Hessler, Robert, 182
Heteropods, 56
High-nutrient, low-chlorophyll (HNLC) regimes, 102–103, 103, 106–107, 108
Hippocampus (sea horse), 300–301F
HNLC regimes, 102–103, 103, 106–107, 108
Holothuroidea (sea cucumbers), 58, 130, 203
Home range
  definition, 472, 479
Horse mackerel (*Trachurus trachurus*), 250, 251
Huchen (*Hucho* spp.), 252
Hucho spp. (huchen), 252
Human activities, adverse effects
  management see Conservation
  on seabirds, 556F, 557, 558F, 560
    immediate, 558F
    long-term, 558F
    see also Anthropogenic impacts; Climate change
Human exploitation
  management of see Conservation
  seabirds, 556F, 557
  management see Conservation
Humboldt penguin (*Spheniscus humboldti*), 548, 548T, 549, 553–554
Humpback dolphins (*Sousa* spp.), 415
Humpback whale (*Megaptera novaeangliae*)
  defense from predators, 483
  habitat, 394–395

lateral profile, 393F
migration, 398, 473, 473F
sound production, 398
  see also Baleen whales
Hydrobatidae, 539
  see also Procellariiformes (petrels)
Hydrodamalis gigas (Steller's sea cow), 431–432, 432F, 438–439
  see also Dugongidae
Hydrothermal vent(s)
  biodiversity, 178, 179F, 184
  causes and characteristics, 178, 179F
  chemoautotrophic bacteria, 178–179
  deep-sea fauna see Deep-sea fauna
  discovery, 176, 178
  energy in benthic environment, 130
  transient nature, 178–179
Hyperbenthos (suprabenthos), 166, 176, 199
  see also Benthic boundary layer (BBL)
Hyperoodon ampullatus (northern bottlenose whale), 403, 406, 407–408, 407F, 409

I

Ice
  pinniped (seal) haul-out sites, 478
  sea see Sea ice
  see also Sea ice
Icefishes (Channichthyidae), 220
Ice krill (*Euphausia crystallarophias*), 66
Ictaluridae (catfish), 377
Imperial shag, 524F
  see also Shag(s)
Indian mackerel (*Rasterlliger kanagurta*), 250
Indian Ocean
  benthic foraminifera, 150T
  bioluminescent phenomena, 118
  krill species, 63
Indicator organisms
  climate change
    krill as, 69–70
Infauna, 124–125, 131, 165, 169
  burrows, 177F
  communities, 125
    latitudinal differences, 125–126, 126F
    Petersen's, 125, 126T
    Thorson's, 127T, 169
  size groups, 125F
Inia (South American river dolphins), 418, 421
Inia geoffrensis (boto), 419
International Union for Conservation of Nature (IUCN), 'Red List' categories for baleen whales, 400, 401T
International Whaling Commission (IWC)
  Scientific Committee, baleen whale population status, 400–401
Intertidal fish(es), **233–238**
  classification, 233
  feeding ecology and predation impact, 237

diet, 237
impact on prey abundance, 237
habitats, abundance and systematics, 233
abundances by habitat, 233
diversity/distribution, 233, 234T
life histories and reproduction, 237–238
eggs, 237
larval development, 238
distribution of larvae, 238F
life spans, 237
mating behavior, 237
parental care of eggs, 238
spawning, 237, 237–238
traits and adaptations, 233–234
behavior, 234–236
circatidal rhythms, 235–236, 236F
homing, 235
intertidal movements, 235
locomotion, 234–235
movement, 235
resident/visitor differences, 235
thigmotaxis, 234–235
tidal synchronization, 235–236
combating intertidal stresses, 233
evolution, 233–234
morphology, 234, 234F
physiology, 236–237
gas exchange, 236
respiration, 236–237
water loss, 236, 236F
Ionic equilibrium, 270
Iron (Fe)
effect on N:P ratio, 103
effect on primary production, 106–107
requirement for primary productivity, 101, 103
see also Trace element(s)
IRONEX experiments, 103
definition, 103
Irrawaddy dolphin (Orcaella brevirostris), 418, 419
Isosmotic conditions, 270
Istiophoridae see Billfishes (Istiophotidae)
Istiophorous albicans (larger sailfish), 282
IUCN (International Union for Conservation of Nature), 'Red List' categories for baleen whales, 400

J

Japanese ayu (Plecoglossus altivelis), 309

K

Kelp forests, 447F
sea otters and, 446–447, 491
Kemp's ridley turtle (Lepidochelys kempii), 455–456, 455F
see also Sea turtles
Killer whale (Orcinus orca), 415F
distribution, 418, 475
dorsal fin, 415
feeding behaviors, 419, 420
group recognition calls, 486–487, 487F

home ranges, 475
migration and movement patterns, 475
prey movements and, 475
predation by sperm whales, 483
resident form, 475
sexual dimorphism, 411
social interactions, 420–421
teeth, 411–415
transient form, 475
see also Odontocetes (toothed whales)
Killifish (Fundulus heteroclitus), 276
Kilopascal, 131
King penguin (Aptenodytes patagonicus), 548T, 552, 552F
see also Aptenodytes
King salmon (Oncorhynchus tshawytscha), 297F
Kittiwake (Rissa tridactyla), 513F
see also Laridae (gulls)
Kleptoparasitism, in seabird foraging, 565, 568
Kogiidae (dwarf and pygmy sperm whales)
trophic level, 489F
see also Odontocetes (toothed whales); Sperm whales (Physeteriidae and Kogiidae)
Krill (Euphausiacea), 63F, 71, 73F, **62–70**
age determination techniques, 64–65
aggregations, 67, 67–68, 67F
biological influences, 67
densities, 67–68
percentage of regional biomass, 68
physical influences, 67–68
in baleen whale diet, 396
bioluminescence, 115
definition, 402
diversity, 62
evolutionary development, 62
fisheries, 62, 69–70
management, 69
recent catch levels, 69
general characteristics, 62
growth, development and physiology, 67–68
Euphausia crystallorophias, 66
Euphausia pacifica, 66
Euphausia superba, 65
food sources, 66
habitat-dependent strategies, 64–65
latitude-dependent strategies, 67
life spans, 65
lipid utilization, 66
Meganyctiphanes norvegica, 66
potential for rapid growth, 66
sex-related differences, 66
Thysanoessa inermis, 65–66, 66
Thysanoessa macrura, 66
Thysanoessa raschii, 65–66
winter survival, 66
influence of ocean currents, 62
place in food webs, 62
role in food web, 68
food consumption, 68–69
feeding strategies, 68
food sources, 68

nutrient cycling, 68
predators, 69
baleen whales, 68
fishes, 68–69
impact of krill distribution patterns, 69
seabirds, 69
seals, 68
in seal diet, 429
spatial distribution, 68
diurnal vertical migration, 67, 68
dynamism, 68
species separation and distribution, 62–64, 64F
aggregations, 68
Arctic Ocean, 64
Atlantic and Pacific Oceans, 63–64
northern limits, 64
distribution in water column, 63
diurnal vertical migration, 64, 68
effect of ocean currents, 63
evolutionary separation, 62–63
Indian Ocean, 63
latitudinal distribution, 63
Southern Ocean, 63, 65F
variability, 69
competition with salps, 69
impact of ocean currents, 69
impact of physical environment, 69
importance of sea ice, 69
indicator species for climate change, 69–70
Kryptopterus bicirrhis (glass catfish), 356F, 361F

L

Labridae (wrasses), 223, 224, 300–301F
Labroides (wrasse), 282
Lack, David, 584
Lagenodelphis hose (Fraser's dolphin)
myoglobin concentration, 460T
see also Oceanic dolphins
Laminaria hyperborea algae, 145–146
Lamna ditropis (salmon shark), 216
Lamniformes (mackerel sharks), 298
Lampris (opah), 300–301F
Lanice conchilega polychaete worm, 204, 204F
Lanternfishes (Myctophidae), 71–73, 317, 318F
bioluminescence, 116–117
Large marine ecosystems (LMEs)
planktonic foraminifera, 36
role of copepods, 50
see also Demersal fish(es); Pelagic fish(es); Plankton
Larger sailfish (Istiophorous albicans), 282
Laridae (gulls), 511T, **510–521**
body size
colony size and, 515, 516F
territory size and, 515, 516F
breeding ecology/behavior, 510, 515, 516–518
age of first breeding, 517

Laridae (gulls) (continued)
   chick-rearing, 517
   clutch size, 515–516, 517
   egg incubation, 517
   male vs. female investment, 518
   mobbing, 515
   nesting, 516–517
   pairing, 517
   phenology, 516
   range, 511T
   studies, 517–518
  conservation, 519–521
   measures, 521
   status, 520, 520T
  distribution, 510, 511T
  foraging, 519
  habitat, 510, 514, 514–515, 516–517
   foraging, 519
   nesting, 514, 516
   winter, 515
  migration, 515, 579, 580T
  physical appearance, 511–512,
    512–514, 513F
   plumage, 514
   Sternidae (terns) vs., 514
  species, 510, 511T
  taxonomy, 510
  threats, 520–521
  see also Seabird(s)
Larus argentatus (herring gull), 513F
Larus marinus (great black-backed gull),
   513F
Larvacean(s), 58, 59F, 60
  'houses', 270
Larvae
  adaptive strategies, 202T
  fish see Fish larvae
  lecithotrophic, 128, 131, 202
  pelagic, 131
   benthic organisms, 128
   settlement process, 129, 173–174
  planktotrophic, 131, 166, 202
   benthic organisms, 128, 129
   settlement process, 129, 173–174, 202
  time in plankton, 202
Latimeria spp., 210
Laysan albatross (Diomedea immutabilis)
  expansion of geographical range, 542
  see also Albatrosses
Least auklet, 503, 505F
  see also Alcidae (auks)
Least storm petrel, 539
  see also Procellariiformes (petrels)
Leatherback turtle (Dermochelys
   coriacea), 450
  characteristics, 451F, 453–454, 453F
  distribution, 452–453
  migration corridors, 453, 453F
  physiology, 452, 452F
  reproduction, 453
  see also Sea turtles
Lemon sole (Microstomus kitt), 281
Lepeophtheirus salmonis (salmon louse),
   50
Lepidochelys kempii (Kemp's ridley
   turtle), 455–456, 455F

see also Sea turtles
Lepidochelys olivacea (olive ridley turtle),
   452F, 455–456
see also Sea turtles
Lepidorhombus wiffiagonis (megrim), 282
Leptocephali, 264, 270
  see also Eels
Leptomedusae medusas, 52, 53F
Lernaeocera spp. copepods, 50
Leucocarboninae see Shag(s)
Life histories (and reproduction)
  cephalopods, 81
  cold-water corals, 191–192
  copepods, 47, 48–49
  coral reef fishes, 224
  deep-sea fishes, 232
  demersal fishes, 369–370
  dolphins and porpoises, 417–418
  eels, 262–264, 264F
  herring, 246–247
  intertidal fishes, 237–238
  krill, 67–68
  mackerels, 250
  mesopelagic fishes, 240–241
  micronekton, 75
  planktonic foraminifera, 33, 35
  radiolarians, 31
  salmonids, 253T, 254T
  sardines, 249
  seabirds, 499
  sprats, 249
  tunas, 250
  viruses, 13–14, 14F
   see also Fish larvae; Fish reproduction;
   Reproduction
Light
  effects of water depth, 123
  effects on fish behavior, 335
  impact on vertical migration, 317–318
  influence on primary production, 106
  penetration into sediments
   cohesive sediments, 132
   impact on photosynthesis, 135
   noncohesive sediments, 132–133
  penetration through water column, 107
Light-mantled sooty albatross, 545
  see also Albatrosses
Limnomedusae medusas, 52
Lipophrys pholis (shanny), 235–236,
   236F
Lipotes vexillifer (Yangtze river dolphin),
   416F, 417, 418, 421
Lissodelphis spp. (right whale dolphins),
   415, 418
Little auk, 503, 505F
  see also Alcidae (auks)
Little penguin (Eudyptula minor), 548T,
   549
  breeding patterns, 549, 553
  characteristics, 548T, 549, 550F
  feeding patterns, 548T, 549
  nests, 548T, 549
Littoral zone, 126T
Lobata ctenophores, 55F, 56
Locomotion
  fish see Fish locomotion

marine mammals, 384, 472
  sea otter, 444
Loggerhead turtle (Caretta caretta), 452F,
   456, 456F
see also Sea turtles
Long-finned pilot whale (Globicephala
   melas), 418
Longline fishing
  seabird by-catches, 545, 584
Longman's beaked whale, 403
Lophelia pertusa coral, 188–189, 188F,
   189F, 190F, 191
  associated with Eunice norvegica worm,
   193–194
  growth on oil platforms, 196
  molecular research, 192–193
  parasites, 194
  reproductive ecology, 191–192, 192F
Lophiiformes (angler fishes), 214, 282
Lophius piscatorius (monkfish), 366
Louvar (Luvarus imperialis), 300–301F
Luciferin/luciferase system, 111, 111–113
  dietary requirement for luciferin, 116
  synthesis of luciferin, 116–117
Lutra marina (marine otter)
  conservation status, 384T
  see also Sea otter
Lutrinae see Sea otter
Luvarus imperialis (louvar), 300–301F

M

Macaroni penguin, 548T, 550–551
  see also Eudyptes (crested penguins)
Mackerel (Scomber scombrus), 250,
   250–251, 280
  description and life histories, 250
  distribution and catches, 250
Mackerels (Scombridae), 250, 250–251
Mackerel sharks (Lamniformes), 298
Macrobenthos, **165–175**
  composition and succession, 168–169,
   170F
   global similarity of patterns, 169
   rich diversity, 168–169
  description, 165
  epifauna, 165
  food sources and feeding methods, 166,
   166–167
   environmental factors, 166–167, 167F
   influence of body size, 167
  functional importance, 172–174
   large-scale disturbances, 173
   larval settlement process, 173–174
   macrofauna actions on sediments,
    173, 174T
  global biomass pattern, 165
   food availability and depth, 165, 166F
   latitudinal differences, 165
   trenches, 165
  importance in environmental
   assessment, 174–175
   establishing baseline, 174–175
   impact of trawling, 174–175
   monitoring changes, 174
  infauna, 165

large-scale diversity patterns, 169–171
  communities, 169
  deep-sea *vs.* shallow water, 171
  depth-related patterns, 171
  epifauna *vs.* infauna, 169
macrofauna *sensu lato*, 168
macrofauna *sensu stricto*, 168
relationship between diversity and
    depth, 165
research history and size limits, 165–166
  benthos categories, 166
  size limits, 166
size spectra, 167–168
  peaks pattern, 167–168, 168F
  size *vs.* function differentiation, 168
small-scale diversity patterns, 171–172
  disruption problems, 172
  three-dimensional spatial patterns,
    172, 172F
species numbers, 169
  discoveries of new species, 169
  research difficulties, 169
  size of unexplored area, 169
substrate differences, 165
  see also Benthic boundary layer (BBL);
    Benthic foraminifera; Meiobenthos;
    Microphytobenthos; Phytobenthos
Macrocystis (giant kelps), 145–146
Macrofauna, 125, 125F, 131, 179–180,
    180F
  see also Macrobenthos
Macronectes giganteus, 542
  see also Procellariiformes (petrels)
Macronectes halli, 542
  see also Procellariiformes (petrels)
Macrouridae (rattail fish), 180F
Madeiran storm petrel, 543
  see also Procellariiformes (petrels)
Madrepora oculata coral, 188–189, 191,
    192–193, 193–194
Magellanic penguin, 548, 548T, 549,
    549F
Malacosteus spp. (stoplight loosejaws),
    358, 360F
Manatees, 384T, 471
  behavior, 435–436
    mating, 436–437
  conservation status, 384T
  ecology, 437
  exploitation, 437–438, 439F
  future outlook, 439–440, 441
  morphology, 434–435, 435–437, 436F
  physiology, 435
    myoglobin concentrations, 460T
  population biology, 437
  skeleton, 471F
  threats, 437–438
    destruction of seagrass habitat, 439
    hunting, 437–438, 439F, 441
    watercraft collisions, 438–439, 438F,
      439–440, 440F
  trophic level, 488
  see also Sirenians
Mandibles, 410
Manganese nodules, 186
Manx shearwater (*Puffinus puffinus*)

expansion of geographical range, 542
  see also Shearwater(s)
Marbled murrelet, 505F
  see also Alcidae (auks)
Mare, Molly, 166
Marine birds *see* Seabird(s)
Marine mammals, **381–390**
  adaptations, 381–384, 481
    amphibious, 384–386
  bioacoustics
    recognition calls, 485–486
    group recognition, 486–487
    individual recognition, 486
    mother-infant, 485–486
    social organization and, 487
    songs, 484, 484F, 485F
  conservation, 389–390
  defense from predators, 482–483
  diving physiology, **458–464**
    adaptations for locomotion, 463
    adaptations for ventilation, 462–463,
      463F
    adaptations to anoxia, 458–459
    cardiovascular adjustments, 459–460
    metabolic responses, 460–461, 461F
    myoglobin concentrations, 459, 460T
    oxygen stores, 459, 459F, 464
    adaptations to light extremes,
      463–464
    adaptations to pressure, 461–462,
      462F, 463F
    aerobic diving limit, 461, 461F
    gliding, 460–461
    obstacles to overcome, 458, 458–459,
      461–462, 462–463, 463–464
    research, 464
  evolution, 458, **465–471**
  exploitation *see* Human exploitation
  feeding, 386
    annual cycles, 386–387
    behavior, 481–482
    mechanisms, 481–482, 482F
  locomotion, 384, 472
  lungs, 462, 462F, 463F
  migration and movement patterns, 389,
    472–480
  navigation, 389
  reproduction
    annual breeding cycles, 386–387
    mating strategies, 483–485
    advertisement displays, 484, 484F,
      485F
    female, 483
    male, 483, 483–484
  sensory systems, 386
    hearing, 386
    vision, 463–464
  social organization and communication,
    481–487
  taxonomy, 381, **465–471**
    Order Carnivora *see* Carnivores
    Order Cetacea *see* Cetaceans
    Order Sirenia *see* Sirenians
  trophic level(s), **488–494**
    estimation, 492
    dietary analysis, 488–489, 489F, 492

equation, 488
  stable isotope analysis, 489–491, 492
  interactions, 490F, 491–492
  prey-predator, 491, 492
  variation, 489
Marine otter *see* Sea otter
Marine silica cycle, 30
Marine snow
  sedimentation of calcite, 37–38
Masked booby, 525
  see also Sulidae (gannets/boobies)
Masu salmon (*Oncorhynchus masou*),
    255, 256
Matrilineal (social unit), definition, 479
Matrilines, 410
Maurolicus muelleri (pearlsides), 320
Medusae, 52
  Anthomedusae, 52, 53F
  Coronatae, 52, 53F
  Cubomedusae, 52, 52–54, 53F
  Leptomedusae, 52, 53F
  Limnomedusae, 52
  Narcomedusae, 52, 53F
  Rhizostomae, 52
  Semaeostomae, 52, 53F
  Trachymedusae, 52, 53F
Megadyptes, 549–550
  see also Sphenisciformes (penguins);
    Yellow-eyed penguin
Megafauna, 179–180, 180F
Megrim (*Lepidorhombus wiffiagonis*),
    282
Meiobenthos, **159–164**
  abundance and diversity, 163
  collection and extraction, 161–162
    qualitative samples, 162
    quantitative samples, 161–162
    sample types and methods, 161
  definitions and taxa, 159, 161F
    taxa diversity, 159, 160T
  distribution, 162
    geographic distribution, 162
    large-scale spatial distribution, 162
    small-scale spatial distribution,
      162–163
  functional roles, 163
    food for higher trophic levels, 163
    general roles, 163
    importance in food web, 163
    mineralization and nutrient
      regeneration, 163–164
    pollution monitoring, 163
    see also Microbial loops
  habitats, 159–160
    epibenthic and interstitial meiofauna,
      159–160
    non-sediment habitats, 160–161
    sediment habitats, 159–160
    types of meiofauna, 160T
    ubiquitous nature, 159
  identification, 159
  meiofauna and pollution, 164
    assessment approaches, 164
    see also Pollution
  role in food web, 159
    see also Fish feeding and foraging

Meiobenthos (*continued*)
  use in pollution monitoring, 159
    see also Benthic foraminifera; Co-
      pepod(s); Deep-sea fauna; Macro-
      benthos; Microphytobenthos
Meiofauna, 125, 125*F*, 131, 179–180,
  180*F*
Mercury (Hg)
  seabirds as indicators of pollution, 560
Meridional distribution, 270
Merluccidae (hakes), 363–365
Mesopelagic fish(es), **239–245**
  adaptations, 242–245
    camouflage, 242
    diurnal migration, 244
    eye morphologies, 242
    influence of light stimuli, 242
    metabolic rates, 244–245
    mouth morphologies, 242
    muscles, 243–244, 244*F*
    ventral light organs, 243
  behavior, 241–245
    diurnal migration, 241–242
    daytime depths, 242
    influence of light, 242
    orientation in water column, 241
    research difficulties, 241
  definition, 239
  distribution, 239–240
  diversity, 239, 239*T*, 240*F*
  general morphologies, 239–240, 240*F*
  history of scientific interest, 239
  life histories, 240–241
    early life stages, 240–241
    fecundity and mortality, 240
    geographical variations, 241
    growth and age composition, 241
    growth patterns, 241
    protandry, 241
    sexual dimorphism, 241
    size, 240
    spawning seasons, 240–241
  major groups, 242, 243*T*
    see also Pelagic fish(es)
Mesopelagic zone, 270, 422
Mesoplodon densirostris (Blainville's
  beaked whale), 406, 407*F*,
  408–409
Microbial loops
  effect on plankton communities, 5, 11
  role of meiobenthos, 163–164
    see also Bacterioplankton; Primary
      production distribution; Primary
      production measurement methods;
      Primary production processes
Microfauna, 125, 125*F*, 131
Micronekton, **71–77**
  bioluminescence, 73–75
    adaptations, 75
    methods of production, 74–75
    uses, 75
  distributions, 71
    biology and morphology, 71
    species variations, 71
  diurnal vertical migrations, 73
    adaptive significance, 73

biological pump, 73
    migration theories, 73
    biomass extent, 73
    role of sunlight, 73
    sound-scattering layers, 73, 75*F*
    species variations, 73, 75*F*
  food habits, 75–76
    deep-water food paucity, 75–76
    midwater *vs.* deep-water nekton,
      75–76
    predator avoidance, 76
  life history, 75
    finding a mate, 75
    life cycle, 75
    reproductive patterns, 75
    mesopelagic examples, 74*F*
  midwater micronekton, 71–73
    distribution, 71–73
    diurnal vertical migration, 73
    position in food webs, 72*F*
    taxonomic groups included, 71
Microorganisms, 179–180
Microphytobenthos, **132–139**
  common genera, 132*T*
  depth, oxygen and production, 135*F*
  description, 132
  distribution and biomass, 136–137
    large-scale heterogeneity, 137
    influence of sediment properties, 137
    intertidal mudflats, 137
    subtidal habitats, 137
    seasonal variations, 137
    small-scale heterogeneity, 136–137
    temporal variation, 137
  EPS production and sediment
    biostabilization, 136
    extracellular polysaccharide
      production, 136
    sediment biostabilization, 136
  functions, 132
  light penetration and photosynthesis,
    134–136
    diurnal vertical migration, 135–136
    irradiance gradients, 135
  light sensitivity, 135
  nutrient cycling, 138–139, 138*F*
    denitrification, 138–139
    effect of biofilms, 138
      see also Microbial loops; Nitrogen
      cycle
  nutrient limitation, 137–138
    effect on photosynthesis, 137–138
    effect on species composition, 138
    influence of sediment properties, 137
  photosynthesis, 133–134
    diurnal vertical migration, 135–136
    impact of light penetration, 135
    production rates, 133, 133*T*
    photosynthesis measurement techniques,
      133, 133*F*
    bicarbonate uptake, 133–134
    chlorophyll *a* fluorescence, 134, 134*F*
    oxygen exchange, 133–134
    oxygen production, 134
  primary production, 133–134
    effects of temperature, 136

tidal differences in temperature, 136
    variability by biomass and irradiance,
      135
      see also Primary production meas-
      urement methods
  response to nutrients, 137–138
  types, 132
    epipelic biofilms, 132
    epipsammic assemblages, 132–133
    influence of sediment properties, 132
      see also Benthic boundary layer
      (BBL); Phytobenthos
Microstomus kitt (lemon sole), 281
Migration, 333
  cod, 311, 311*F*
  definition, 472, 479
  demersal fishes, 366–367
  eels, 266–267
  green turtles, 313–314
  herring, 246–247, 247*F*, 248*F*, 310,
    310*F*
  Japanese ayu, 309
  marine mammals, 389, **472–480**
  plaice, 311–312, 313*F*, 313*F*, 314*F*
  salmon, 308
  salmonids, 255, 256*F*, 257*F*, 258*F*, 259*F*
  seabird *see* Seabird migration
  seabirds, 497
  tunas, 309–310
  walleye pollock, 310–311
    see also Circatidal vertical migration;
      Contranatant migration; Daily ver-
      tical migration; Denatant migra-
      tion; Diurnal vertical migration;
      Fish horizontal migration; Fish ver-
      tical migration
Minke whales
  habitat, 395
  lateral profile, 393*F*
    see also Baleen whales
Minnows (*Phoxinus* spp.), 341–342
Mitochondrial DNA, 270
Mnemiopsis leidyi (a comb jelly), 60–61
Mola mola (sunfish), 300–301*F*
Molecular phylogeny, 270
Molluscs *see* Mollusks
Molluskan fisheries
  cephalopods, 83
    see also Phytoplankton blooms
Mollusks
  aggregation, 205*T*
  bioluminescence, 112*T*, 115
  heteropods, 56, 57*F*
  pteropods, 56
    euthecosomes, 57*F*
    Gymnoptera, 56–58
    Gymnosomata, 56, 57*F*
    Pseudothecosomata, 56, 57*F*
    Thecosomata, 56
Molybdenum (Mo)
  requirement for primary productivity,
    101
    see also Trace element(s)
Monachinae (southern phocids), 385*T*,
  425*T*
    see also Phocidae (earless/'true' seals)

Monkfish (*Lophius piscatorius*), 366
Monodontidae, 382–383T
 trophic level, 489F
  see also Odontocetes (toothed whales)
Monostroma algae, 143
Mottled petrel (*Pterodroma inexpectata*),
  540F
 see also Procellariiformes (petrels)
Mugilidae, 280
Mullets (Mugilidae), 280
Murre(s)
 common, 503
 migration, 579–581
 thick-billed, 503, 505F
  see also Alcidae (auks)
Murrelets
 migration, 579–581
  see also Alcidae (auks)
Mussels (*Mytilus*)
 wave resistance, 203
Mustelids, 381, 384T, 465
 conservation status, 384T
 trophic level, 489F
Mutualism, in seabird foraging, 565, 568
Myctophidae (lanternfishes), 71–73
 bioluminescence, 116–117
Myctophids (lanternfishes), 317, 318F
Myoglobin concentrations, marine
  mammals, 459, 460T
Myomeres, 270
Mysticeti
 derivation of word, 391–392
  see also Baleen whales (Mysticeti)
Myticola intestinalis copepod, 50
Mytilus see Mussels (*Mytilus*)
Mytilus edulis (blue mussel), 201F
Myxinidae, 280

N

Nannacara anomala (cichlid), 361F
Narcomedusae medusas, 52, 53F
Narwhal, 382–383T
 movement patterns, 475
 trophic level, 489F
  see also Odontocetes (toothed whales)
Natator depressus (flatback turtle), 456
 see also Sea turtles
Nauplii, copepods, 44F, 46
Nautilus (Nautilidae), 78
Nekton, **71–77**
 definition, 71
 habitats, 71
 problems, 71
 taxonomic groups included, 71
  see also Micronekton
Nematodes, 180F
Neocalanus plumchrus copepods, 10
Neogloboquadrina pachyderma
  foraminifer, 158
Neophocaena phocaenoides (finless
  porpoise), 416, 418, 421
Nitrate (NO$_3^-$)
 phytoplankton growth reaction, 94
  see also Denitrification; Nitrogen;
   Nitrogen cycle

Nitrogen (N)
 C:N:P ratios, 102
 cycle *see* Nitrogen cycle
 denitrification, 138–139
  see also Denitrification
 limiting factor to photosynthesis, 101,
  103
  see also Nitrogen cycle
Nitrogen cycle
 marine
  denitrification *see* Denitrification
 role of copepods, 50
  see also Primary production processes
Nitrogen oxides, pollution, 560
North Atlantic gannet *see* Northern
  gannet
North Atlantic krill (*Meganyctiphanes
  norvegica*), 63F
 diurnal vertical migration, 64, 68
 growth and development, 66
North Atlantic Oscillation Index, 270
North Atlantic right whale, 392T
 life cycle, typical, 395F
 as threatened species, 394, 401, 401T
  see also Baleen whales
Northern bottlenose whale (*Hyperoodon
  ampullatus*), 403, 406, 407–408,
  407F, 409
Northern fulmar (*Fulmaris glacialis*)
 changes in distribution, 541–542
 food/foraging, 542
  see also Procellariiformes (petrels)
Northern gannet (*Sula bassana*), 524F
 see also Sulidae (gannets/boobies)
Northern phocids *see* Phocinae (northern
  phocids)
North Pacific
 sedimentary communities, sea otter
  effects on invertebrate populations,
  447
Notacanthus chemnitzii (spiny eel), 357F
Notothenioidei, 218, 219F
Nuclear DNA, 270
Nucleotide, 270–271
Nutrient(s)
 effect on primary production, 106
 requirements for primary productivity,
  101, 102
Nutrient cycling
 bacterioplankton, 27
 benthic foraminifera, 151, 158
 copepods, 47, 50
 meiobenthos, 163–164
 microphytobenthos, 138
 plankton, 3
 plankton communities, 5
  see also Microbial loops; Phytoplankton blooms; Primary production distribution; Primary
   production measurement methods;
   Primary production processes

O

Ocean currents *see* Current(s)
Ocean gyre(s)

Alaskan Gyre, 10
Labrador-Irminger Sea Gyre, 10
North Atlantic Subtropical Gyre
  (NASG), 10, 10–11
North Pacific Central Gyre (NPCG), 10,
  10–11
Norwegian Sea Gyre, 10
Oceanic dolphins, 382–383T
 body outline and skeleton, 386F
 echolocation, 481
 home ranges, 474–475, 475
 lungs, 462F
 migration and movement patterns,
  474–475, 475
 recognition calls
  individual recognition, 486
  mother-infant, 485–486
  signature whistles, 485–486, 486,
   486F
 trophic level, 489F
  see also Odontocetes
   (toothed whales)
Oceanites oceanicus (Wilson's storm
  petrel), 540F
Oceanitidae *see* Storm petrels
Oceanodroma castro (Madeiran storm
  petrel), 543
Ocean tides *see* Tide(s)
Octopodidae (octopuses)
 bioluminescence, 115
 brain functions, 80
 spawning, 81
Oculina varicosa coral, 188–189
Odobenidae (walruses), 381, 385T, 425T,
  466–467
 abundance, 425T
  trends, 425T
 acoustic advertisement displays, 485
 conservation status, 425T
 diet, 429
 distribution, 425T
 feeding behavior, 481
 species, 385T, 425T
 trophic level, 489F
  see also Pinnipeds
Odontocetes (toothed whales), 382–383T,
  468F
 baleen whales *vs.*, 391
 defense from predators, 482–483
 definition, 422
 evolution, 469
 extant, 469
 feeding, 482, 482F
 feeding mechanisms, 482, 482F
 growth and maturation, prolonged,
  388–389, 388F
 migration and movement patterns, 474,
  475
 cetaceans *vs.*, 472
 taxonomy, 382–383T, 469
  family Delphinidae *see* Oceanic
   dolphins
  family Kogiidae *see* Kogiidae
  family Phocenidae *see* Porpoises
  family Physeteridae, 382–383T
  family Platanistidae, 382–383T

Odontocetes (toothed whales) (*continued*)
  family Ziphiidae *see* Beaked whales
    (Ziphiidae)
  trophic level, 488, 489*F*
  see also Marine mammals
Oil exploration, threat to cold-water
    coral reefs, 195
Oil platforms, *Lophelia pertusa* (coral)
    growth, 196
Oil pollution
  seabirds and, 559
Oithona spp. copepods, 10, 11
Olfaction
  deep-sea fishes, 230
  dolphins and porpoises, 419
Oligocottus maculosus (tidepool sculpin),
    235
Oligotrophs, 24
Olive ridley turtle (*Lepidochelys
    olivacea*), 452*F*, 455–456
  see also Sea turtles
Oncorhynchus gorbuscha (pink salmon),
    254–255, 256, 256*F*
Oncorhynchus keta (chum salmon), 255,
    256, 256*F*
Oncorhynchus kisutch (coho salmon),
    255, 256
Oncorhynchus masou (masu, cherry
    salmon), 255, 256
Oncorhynchus mykiss (rainbow trout),
    358*F*
Oncorhynchus nerka (sockeye salmon),
    253–254, 256, 256*F*, 259, 305*F*
Oncorhynchus tshawytscha (chinook
    salmon), 255, 256
Oncorhynchus tshawytscha (king
    salmon), 297*F*
Opah (*Lampris* spp.), 300–301*F*
Ophiuroidea (brittle stars), 128, 130
Opisthoproctus (barreleyes), 352*F*, 354*F*
Opsanus tau (toadfish), 374–375
Orcaella brevirostris (Irrawaddy dolphin),
    418, 419
Ostracods, 111–113, 180*F*
Otariidae (eared seals), 381, 385*T*, 386,
    386*F*, 425*T*, 466–467
  abundance, 425*T*
    trends, 425*T*
  conservation status, 385*T*, 425*T*
  distribution, 424, 425*T*
  lungs, 462*F*
  mother-infant recognition, 485
  species, 385*T*, 425*T*
  subfamily Otariinae *see* Otariinae (sea
    lions)
  trophic level, 489*F*
  see also Pinnipeds
Otariinae (sea lions), 385*T*, 425*T*, 466
  skeleton, 386*F*, 424*F*
  see also Otariidae (eared seals)
Otoliths, 231*F*, 271
Overfishing
  effects on seabirds, 559
    detrimental, 557
  management, 559
Oviparity, 131, 333, 333*F*

Ovoviviparity, 333
Oxygen (O$_2$)
  availability in deep-sea, 177–178
  influence on fish distribution, 276–277
  stores, marine mammals, 459, 459*F*, 464
  use by Antarctic fishes, 219–221

P

Pachystomias spp. (dragonfishes), 358
Pacific *see* Pacific Ocean
Pacific herring (*Clupea pallasii*), 246
Pacific krill (*Euphausia pacifica*), 66, 69
Pacific Ocean
  benthic foraminifera, 150*T*
  krill species, 63–64
  North *see* North Pacific
  pelagic fisheries, 250
  see also North Pacific
Pacific ridley turtle, 452*F*
  see also Sea turtles
Pacific salmon (*Oncorhynchus*), 252, 256,
    257
  diet, 256
  predation, 257
  taxonomy, 252
  vertical movement, 256
Paedomorphosis, 422
Pagodroma nivea (snow petrels), 543
  see also Procellariiformes (petrels)
Pakicetidae, 468
  see also Cetaceans
Panmixia, 271
Pantropical organisms, 422
Pantropical spotted dolphin (*Stenella
    attenuata*), 420*F*, 421–422
Parasites, associated with baleen whales,
    397
Parental care
  of eggs, intertidal fish, 238
  fish, 333–334
Parrotfishes (Scaridae), 224
Peale's petrel, 540*F*
  see also Procellariiformes (petrels)
Pearlsides (*Maurolicus muelleri*), 320
Pearly nautilus (*Nautilus* spp.), 78
Pectinidae (scallops), adaptations to resist
    shear stress, 203–204
Pelagic biogeography
  planktonic foraminifera, 33
Pelagic fish(es), **246–251**
  benthopelagic, 227
  clupeoids, 246–249
    herring, 246–249
    sardines, 249–250
    sprats *see* Sprat
  groups
    clupeoids, 246, 246–249
    mackerels, 246, 250–251
    tunas, 246, 250
    see also Clupeoids; Tuna
  importance to fisheries, 251
  sprats, 249
  see also Fish horizontal migration;
    Mesopelagic fish(es)
Pelagic habitat, 422

Pelagic zones, 271
Pelecanidae (pelicans), 522, 523–525
  breeding patterns, 525
  characteristics, 523–525, 528*T*
  feeding patterns, 523–525, 528*T*
  human disturbance, 525
  migration, 577–579
  species, 523–525, 524*F*
  threats, 525
  see also Pelecaniformes
Pelecaniformes, 524*F*, **522–530**
  behavior, 528, 528*T*, 529*F*
  breeding patterns, 528
  characteristics, 522
  conservation, 528–530
  distribution, 522–523
  ecology, 527–528
  feeding patterns, 527–528, 528*T*
  fossils, 522
  migration, 577–579
  plumage, 527
  population status, 530
  taxonomy, 522, 522–523
    family Fregatidae *see* Fregatidae
      (frigatebirds)
    family Phaethontidae *see*
      Phaethontidae (tropic birds)
    family Sulidae *see* Sulidae (gannets/
      boobies)
  see also Seabird(s)
Pelecanoididae *see* Diving petrels
Pelecanus occidentalis see Brown pelican
Pelicanoides urinatrix (common diving
    petrel), 540*F*
Pelicans *see* Pelecanidae (pelicans)
Penguins *see* Sphenisciformes (penguins)
Perca fluviatilis (perch), 359*F*
Perch (*Perca fluviatilis*), 359*F*
Perrin's beaked whale, 403
Peruvian anchoveta *see Engraulis ringens*
    (Peruvian anchoveta)
Petersen, C.G.J.
  classification of the benthos, 124–125,
    125, 126*T*
  macrobenthos studies, 173
Petrels *see* Procellariiformes (petrels)
pH
  influence on fish distribution, 277
Phaeophyta (brown seaweeds), 142
Phaethon rubicauda, 524*F*
Phaethontidae (tropic birds), 522, 523
  characteristics, 523, 528*T*
  distribution, 523
  ecology and behavior, 523, 528*T*
  migration, 577–579
  species, 523, 524*F*
  see also Pelecaniformes
Phalacrocoracidae, 522, 525–526
  breeding patterns, 526, 526*F*
  characteristics, 525–526, 528*T*
  feeding patterns, 528*T*
  migration, 577–579
  species, 524*F*, 525–526
  see also Pelecaniformes
Phalacrocorax atriceps (imperial shag),
    524*F*

Phalacrocorax carbo (great cormorant), 524F, 526, 526F
  see also Cormorants
Phalaropes, **531–538**
  conservation, 537
  definition, 531
  diversity of names, 536, 537T
  females *vs.* males, 531, 531–532, 532F, 535
  on land, 535
    appearance, 532F, 535
    breeding behavior, 535
    male-female role reversal, 531, 535
    diet, 536
    distribution, 535
    migration and movement patterns, 536, 579
  population status, 531, 537
    declines, 537
  at sea, 531–532
    appearance, 531–532, 532F
    aquatic adaptations, 532–533
    lobed toes, 532, 532F
    diet, 533
    distribution, 533
    feeding behavior/mechanics, 533–534
    spinning, 533
    surface tension feeding, 533, 534F
    habitat, 534–535
  systematics, 536–537, 537T
  threats, 537
  see also Seabird(s)
Phalaropus lobatus see Red-necked phalarope
Phoca groenlandica (harp seal)
  migration and movement patterns, 478F, 479
  see also Phocidae (earless/'true' seals)
Phocidae (earless/'true' seals), 381, 385T, 425T, 466
  abundance, 425T
    trends, 425T
  body outline and skeleton, 386F
  conservation status, 385T, 425T
  distribution, 424, 425T
  lungs, 462F
  species, 385T, 425T
  subfamily Phocinae *see* Phocinae (northern phocids)
  trophic level, 489F
  see also Pinnipeds
Phocinae (northern phocids), 385T, 425T
  see also Phocidae (earless/'true' seals)
Phocoena dioptrica (spectacled porpoise), 416, 421
Phocoena phocoena (harbor porpoise), 421
  see also Porpoises
Phocoena sinus (vaquita), 416F, 421
Phocoenidae *see* Porpoises
Phocoenoides dalli (Dall's porpoise), 416, 420, 421
Phoebastria irrorata (waved albatross), 574–575, 575
  see also Albatrosses

Phoebetria palpebrata (light-mantled sooty albatross), 545
  see also Albatrosses
Phosphorescence, 111
Phosphorus (P)
  C:N:P ratios, 102
  limiting factor to photosynthesis, 101, 103
Photolithotrophs, 100
Photolithotrophy, 100
Photons
  flux density, 103
Photophores, 114–115
  effects of accessory optical structures, 114F
  effects of pigment and reflectors on light emission, 113F
  fishes, 115–116
  occlusion, 115, 115F
  shrimps and krill, 115
  squids, 115
Photosynthesis
  algae, 140
  cyanobacteria, 140
  measuring, 93–94
  microphytobenthos *see* Microphytobenthos
  phytoplankton *see* Phytoplankton
  see also Primary production; Primary production distribution; Primary production measurement methods; Primary production processes
Phoxinus spp. (minnows), 341–342
Physalia physalis (Portuguese man-of-war), 54
Physeteridae, 382–383T
  see also Odontocetes (toothed whales)
Physeter macrocephalus (sperm whale), 403, 408F
Physical gradients, seabird abundance and, 562–563
Physonectae siphonophores, 54, 55F
Phytobenthos, **140–146**
  'algae' defined, 140
  cyanobacteria, 140
  definitions, 140
    descriptions, 140
  ensuring commercial supplies, 145–146
    biotic considerations, 146
    manipulation of stocks, 146
    sustainable harvesting, 145–146
    value-dependent approaches, 145
  human uses, 144–145
    alginate, 145
    fertilizers, 145
    high value chemicals, 145
    human consumption, 145
    principle uses, 145
  primary production, 140
  seaweed ecology, 143–144
    see also Seaweed ecology
  seaweed life cycles, 142–143
  validity of laboratory cultures, 143
    Asparagopsis armata and *Falkenbergia rufanolosa*, 143
    Monostroma spp., 143

possibility of false results, 143
  see also Algae; Primary production distribution; Primary production measurement methods; Primary production processes
Phytoplankton, 3–4
  bacterioplankton comparison, 24–25
  blooms *see* Phytoplankton blooms
  CO_2 sequestration, 85
  growth reactions, 94
  microphytoplankton, 88–89
  nanophytoplankton, 88–89
  photosynthesis, 85
  picophytoplankton, 88–89
  primary production, 93
  Prochlorococcus, 89–90
  size structure *see* Phytoplankton size structure
  see also Diatom(s); Primary production distribution; Primary production processes
Phytoplankton blooms, 9, 109F
  control by copepods, 50
  influence of viruses, 18
  planktonic foraminifera, 36–37
  see also Algal blooms; Nutrient cycling; Primary production
Phytoplankton size structure, **85–92**
  controlling factors, 90
    factors favoring large-celled organisms, 90
  ecological/biogeochemical implications, 90–91, 91T
    CO_2 sequestration, 90–91
    ecosystem differences, 91
    food web interactions, 90–91
  patterns, 88–89
    size-abundance spectra, 89–90
    ecosystem differences, 89–90, 90F
    size-fractionated chlorophyll *a*, 88–89, 88–89, 89F
  range of cell sizes, 85
  size, metabolism and growth, 85–87
  size and growth rates, 88
    influence of nutrients, 88
    variable values, 88
  size and loss processes, 87–88
    large size, advantages/disadvantages, 88
    respiration and exudation, 87
    sinking velocity, 88
  size and resource acquisition, 85–87
    deviation from general allometric theory, 87
    light absorption, 86
    metabolic rates, 85–86
    nutrient limitation, 85–86
    nutrient uptake, 85, 86F
    optical absorption cross-section, 86
    package effect, 86–87, 86F
    photosynthesis, 87, 87F
Pike (*Esox* spp.), 300–301F
Pilchard (*Sardina pilchardus*), 280
Pilot whales (*Globicephala* spp.), 411, 420–421, 421

Pinguinus, 503T
  see also Alcidae (auks)
Pink salmon (*Oncorhynchus gorbuscha*),
    254–255, 256, 256F
Pinnipeds (seals), 381, 385T, **424–430**
  abundance, 424, 425T
    trends, 424, 425T
  adaptations to aquatic life, 424–427
    morphological, 424–427
    physiological, 427, 460
  annual feeding and reproductive cycle,
    387, 476
  conservation status, 385T, 425T
  constraints of aquatic life, 427–428
    maternal body size, 427
  diet, 428–429
    fish, 428–429
    krill, 429
    seabirds, 429
  distribution, 424, 425T
  diving, 427
    cardiovascular adaptations, 460
    for food, 429, 481
  evolution, 466–467
  importance of krill, 68
  lactation, 427
  mating systems, 428
    competitive, 428
    postpartum estrus, 428
    sexual dimorphism, 428
    terrestrial, 428
  migration and movement patterns,
    475–476
    breeding season, 476, 477F
    haul-out sites, 476–479, 477F, 479
    nonbreeding season, 476, 478F
  phylogeny, 424, 424F, 466
  taxonomy, 391–393, 424, 425T, 466
    family Otariidae *see* Otariidae (eared
      seals)
  thermoregulation constraints, 427
    newborn pups, 427, 427–428, 428
  threats, 424
  trophic level, 488, 489F
  see also Marine mammals
Pintado petrel, 540F
  see also Procellariiformes (petrels)
Pipefishes (Syngnathidae), 300–301F
Plaice (*Pleuronectes platessa*)
  larval stages and metamorphosis, 335F
  life history, 335F
  migration, 311–312, 313F, 313F,
    314F
Plainfin midshipman (*Porichthys notatus*),
    353F, 355F
Plankton, 271, **3–4**
  abundance and productivity, 3–4
  assemblages and communities, 4
    factors limiting distribution, 4
    transition zones, 4
  bioluminescence, 111–115
  climate and *see* Plankton and climate
  communities *see* Plankton communities
  definition, 3
  diurnal vertical migration, 3
  foraminifera *see* Planktonic foraminifera

holoplanktonic *vs.* meroplanktonic
    species, 3
Hutchinson's 'paradox of plankton', 18
krill *see* Krill (Euphausiacea)
nutrient cycling, 3–4
phytoplankton, 3–4
  see also Phytoplankton
roles in marine environment, 3
size, 3
subdivisions, 3
uneven distributions, 4
viruses *see* Plankton viruses
zooplankton, 4
  see also Gelatinous zooplankton;
    Zooplankton
  see also Bacterioplankton; Copepod(s);
    Demersal fish(es); Fish larvae; Pri-
    mary production processes
Plankton and climate, 34, 38
  see also Upwelling ecosystems
Plankton communities, **5–12**
  characteristics, 8T
  current research methodologies, 11
  definition of 'community', 5
  effects of microbial loops, 11
  future research methodologies, 11
  general features, 5–7
    continuous operation/functioning, 5
    microbial loop, 5
    nutrient recycling, 5
    place in water column, 5
    quantitative assessments, 6–7
    size groupings, 5
    sizes and relationships, 5, 6F
    sun energy, 5
  specific communities, 7
    continental shelves, 7–10, 9
    environmental characteristics, 9
    episodic nature of nutrient
      availability, 9
    interactions across thermoclines, 9
    short-lived plankton communities, 10
    estuaries, 7
    copepod species, 7
    environmental characteristics, 7
    open ocean, 10–11
    gyres, 10
    regional and seasonal variation, 10
    subpolar gyres, 10
    subtropical gyres, 10–11
    temperate gyres, 10
    types of environments, 7
Planktonic foraminifera, **33–39**
  applications of research, 38
    climate change studies, 38
  cellular structure, 34–35, 34F
    chamber formation, 34
    cytoplasm and organelles, 34, 35F
    interaction with surrounding water,
      34–35
    inter-chamber connections, 34–35
    spines, 34–35, 36F
  description and distribution, 33
  ecology and distribution, 36–37
    biomass distributions, 36–37
    depth habitat, 36

faunal provinces, 36, 37F
  north Atlantic distribution, 37F
evolutionary history, 33
general life history, 33
molecular biology research, 35–36
  diversity studies, 35–36
reproduction and ontogeny, 35
  life cycle, 35
  reproductive method, 35
research history, 33
  uses in other studies, 33
research methods, 33–34
  molecular biology, 34
  reconstruction information from
    shells, 34
  reconstruction studies, 34
  sampling and observation, 33–34
sedimentation of calcite, 37–38
  global calcite production, 37–38
  preservation/remineralization of
    shells, 37–38
  seasonality, 38F
  see also Calcium carbonate
    ($CaCO_3$)
symbionts, commensals and parasites,
    35, 36F
  spinose *vs.* nonspinose species, 35
trophic demands, 36
  food and feeding, 36
Plankton viruses, **13–20**
  benefits of infection
    increase in production, 19
    nutritional gains, 19
    resistance to superinfection, 19
  comparison to mortality from protists,
    16–17
  effects on host species, 18
    control of algal blooms, 18
    species composition, 18
    vulnerability of dominant hosts, 18
  food web and geochemical cycles, 17–18
    enabling host to produce toxins,
      17–18
    prokaryote-viral loop, 17, 18F
    release of cell contents, 17
    relevant features of viruses, 17
    viral lysis, 17
  general properties, 13–14
  genetic transfer, 19
    mechanisms, 19
  history of knowledge, 13
  host resistance, 18–19
    laboratory *vs.* field systems, 18–19
  importance to planktonic communities,
    20
  observation, 14–15
    electron microscopy, 14–15
    epifluorescence microscopy, 14–15,
      15F
    preparation methods, 14–15
  types of viruses, 15
    bacteriophages, 15
    molecular techniques, 15
    viruses of cyanobacteria, 15
  viral activities, 15–16
    lysogeny and chronic infection, 16

lytic infection, 16
viral roles in system functions, 13
virus abundance, 15
    environmental variations, 15
    virus:prokaryote ratios, 15, 16F
Platanista spp. (susus), 418, 419, 420, 421
Platanistidae, 382–383T
    see also Odontocetes (toothed whales)
Platyctenida ctenophores, 54
Plecoglossus altivelis (Japanese ayu), 309
Pleuronectes platessa (plaice), 311–312, 313F, 313F, 314F
Pleuronectiformes (flatfish), 300–301F
Polar bear (*Ursus maritimus*)
    annual feeding and reproductive cycle, 387
    body outline and skeleton, 386F
    shearing teeth, 482F
    trophic level, 488, 489F
    see also Marine mammals
Polar ecosystems
    copepod habitat, 45
Pollack (*Pollachius pollachius*), 361F
Pollution
    benthic studies, 169, 171F
    oil spill *see* Oil pollution
    threat to deep-sea fauna, 185–186
Polychaetes/polychaete worms, 57F, 58, 179F, 180F
    aggregation, 205T
    bioluminescence, 112T, 113–114
    deep-sea communities, 130
Polychlorinated biphenyls (PCBs)
    seabirds as indicators of pollution, 560
Polypterus spp. (bichir), 210, 211F
Pomacanthidae (angelfish), 300–301F
Pomacentridae (anemonefishes), 223, 224
Pontoporia blainvillei (Franciscana dolphin), 415, 415F, 417–418
Porichthys notatus (plainfin midshipman), 353F, 355F
Porphyra seaweeds, 143, 145, 146
Porpoises, 382–383T, 416–417
    appendages, 416
    body shape, 416F
    color patterns, 416
    cranial features, 416
    trophic level, 489F
    see also Dolphins and porpoises; Odontocetes (toothed whales)
Portuguese man-of-war (*Physalia physalis*), 54
Predation
    Alcidae (auks), 508–509
    Atlantic salmon (*Salmo salar*), 257–258
    beaked whales (Ziphiidae), 409
    copepods, by higher trophic animals, 163
    fish *see* Fish predation and mortality
    fish larvae *see* Fish larvae
    killer whale (*Orcinus orca*) by sperm whales, 483
    Pacific salmon (*Oncorhynchus*), 257
    Procellariiformes (petrels), 544–545
    salmonids, 256–258

seabirds, 557
    sperm whales *see* Sperm whales (Physeteriidae and Kogiidae)
    Sphenisciformes (penguins), 553–554
Predation behaviors
    herding, 341
    pack hunting, 341
    splitting off individuals, 341
    'twilight hypothesis', 341
    see also Fish predation and mortality
Predator avoidance
    coral reef fishes, 223–224
    fish larvae, 335
    juvenile fish, 335–336
    micronekton, 76
    schooling behavior, 339
    vertical migration, 320
    see also Fish predation and mortality
Predator-prey interactions
    sperm and beaked whales, 409
Pressure
    marine mammals adaptations, 461–462, 462F, 463F
Prey
    cephalopods, 78
    fish larvae, 288–289
    foraging seabirds and, 566
    killer whale migration and, 475
    migration, fish vertical migration, 319
    seabirds, 499–500
Primary production
    definition, 93, 106
    distribution *see* Primary production distribution
    estimation/measurement *see* Primary production measurement methods
    global, marine contribution to, 103
    microphytobenthos, 133–134
    processes *see* Primary production processes
    see also Photosynthesis; Phytoplankton; Phytoplankton size structure; Primary production processes
Primary production distribution, **105–110**
    different quantities measured, 105
    estimation methods, 105, 105T
    factors influencing variations, 106–107
        light, 106
        micronutrient availability, 106–107
        nutrient availability, 106
        phytoplankton biomass, 106
        species composition/succession, 107
        temperature, 106
    horizontal distribution, 108
        areas of high productivity, 108
        HNLC regimes, 108
        regional/seasonal variations, 108
    influencing factors, 106
    mathematical modeling techniques, 107
    measurement compilations, 108
    measuring by remote sensing, 108–110, 109F
        advantages, 109
        improvements expected, 109–110
        methodology, 108
        problems, 109–110

successful models, 108–109
    requirement for extrapolation schemes, 106
    research difficulties, 105
    space-timescale links, 105
    vertical distribution, 107–108
        critical depth, 107–108
        euphotic zone, 107
        mixed *vs.* stratified layers, 108
        see also Primary production measurement methods; Primary production processes
Primary production measurement methods, **93–99**
    approaches, 94
        ^{13}C method, 95T, 96
        ^{14}C method, 95–96, 95T
        ^{18}O method, 95T, 96
        change in O_2 concentration, 94, 95T
        change in total CO_2, 94–95, 95T
        technical objectives, 94
    definitions, 93
    methodological considerations, 96
        containers, 97, 98T
        filtration or acidification, 98
        incubation, 96–97
            artificial incubators, 95T, 97
            in situ, 95T, 96–97
            simulated *in situ*, 95T, 97
            types, 96
        incubation duration, 97–98, 99T
        interpretation of carbon uptake, 98–99
    photosynthesis and phytoplankton growth, 93–94
        photosynthesis reaction, 93–94
        photosynthetic quotient, 94
        phytoplankton growth reaction, 94
        quantifying photosynthesis, 93–94
    planktonic primary production, 93
    sampling, 96
        contamination avoidance, 96
        light exposure avoidance, 96
        trace metal-clean procedures, 96
        turbulence avoidance, 96
    significance of uncertainties, 99
    uncertainty of interpretations, 99
    uncertainty of measurements, 99
Primary production processes, **100–104**
    benthic primary productivity, 100
    cell processes, 101–102
        cell size, 102
        diversity of photosynthesizers, 101–102
        phylogenetic differences, 102
        pigmentation and photon absorption, 102
    characteristics of primary producers, 100
        mineral dependencies, 102
        movement, 102
        pigmentation, 102
    chemolithotrophy, 100
    determinants of productivity, 102–103
        C:N:P ratios, 102

Primary production processes (*continued*)
  geophysical and ecological factors, 102–103
    limiting nutrients, 103
  marine contribution to global primary production, 103
  net primary productivity by habitat, 101*T*
  ocean habitat, 100–101
    absorption of solar radiation, 100
    nutrients, 101
    stratification, 101
    see also Carbon cycle; Nitrogen cycle
  photolithotrophy, 100
    bacteriochlorophyll-based, 100
    O₂-evolvers, 100
    rhodopsin-based, 100
    see also Primary production distribution; Primary production measurement methods
Prions, 539
  migration, 575–577, 576*T*
  see also Procellariiformes (petrels)
Procellariidae, 539
  migration, 575–577, 576*T*
  see also Procellariiformes (petrels)
Procellariiformes (petrels), **539–545**
  breeding ecology, 539, 542–544, 583
    chick-rearing, 543–544, 544, 544*F*, 583
    copulation, 542–543
    displays, 542
    egg incubation, 543, 543*F*
  characteristics, 539, 540*F*
  conservation, 544–545
  food/foraging, 542
    nocturnal, 565
    stomach oil, 542, 544
    see also Seabird foraging ecology
  life spans, 539, 584
  migration, 574–575
  population ecology, 539
    demography, 539, 541*F*
    distribution, 539, 541–542
      changes in, 541–542, 541*F*
    regulation of population size, 539–541
  reproductive effort, 544
  species, 539, 540*F*
    richness by latitude, 539, 541*T*
  threats, 544–545
    human activities, 544–545
    longline fishing, 545, 584
    predation, 544–545
  see also Seabird(s)
Protista, 28–29
Protocetidae, 468
  see also Cetaceans
Pseudorca crassidens (false killer whale), 411
Pseudothecosomata pteropods, 56, 57*F*
Pteriomorpha (mussels), 203
Pterodroma inexpectata (mottled petrel), 540*F*
  see also Procellariiformes (petrels)

Pteropods, 56
  Gymnosomata, 56
  Thecosomata, 56
Ptychoramphus, 503*T*
  see also Alcidae (auks)
Puffins, 503, 505*F*
  migration, 579–581
  see also Alcidae
Puffinus lherminieri (Audubon's shearwaters), 539
Puffinus puffinus (Manx shearwater), expansion of geographical range, 542
Puffinus tenuirostris (short-tailed shearwater), 542
Pycnogonids (sea spiders), 112*T*, 180*F*
Pygmy right whales (neobalaenids), 391, 392*T*, 470
  habitat, 394
  lateral profile, 393*F*
  trophic level, 489*F*
  see also Baleen whales
Pygoscelis, 551–552
  breeding patterns, 551, 551–552
  characteristics, 548*T*, 551, 552*F*
  distribution, 548*T*, 551
  feeding patterns, 548*T*, 551–552
  migration, 574
  nests, 548*T*, 551–552
  species, 548*T*, 551
  see also Sphenisciformes (penguins)
Pygoscelis antarctica see Chinstrap penguin
Pygoscelis papua (gentoo penguin), 548*T*, 551
Pygoscelis pygoscelis ellsworthii, 551
Pygoscelis pygoscelis papua, 551, 551–552
Pyrosoma spp. tunicates, 114–115

R

Radioisotope tracers
  techniques, 21
Radiolarians, **28–32**
  algal symbionts, 28–29
    location within cell body, 29–30
    role, 32
  bioluminescence, 111, 112*T*
  biomineralization, 30–31
    definition and process, 30
    deposition rate, 31
    growth forms, 30–31
    sedimentation rate, 31
  cellular morphology, 29–30
    cell parts and organization, 29–30
    parts and organization, 29*F*
  distribution, 28–29
  feeding, 29–30
  gelatinous colonies, 51–52
  morphology, 28–29, 28*F*
  reproduction, 31
    asexual/sexual, 31
  siliceous skeleton, 29–30
  symbiotic relationship with dinoflagellates, 51–52

taxonomy, 30
  diversity, 30
  groups and subgroups, 30
  zoogeography, 31–32
    effect of temperature, 31–32
    factors influencing abundance, 31–32
    food, 32
    growth requirements, 32
    role of algal symbionts, 32
    shallow-water species zones, 31–32
Rainbow trout (*Oncorhynchus mykiss*), 358*F*
Rasterlliger kanagurta (Indian mackerel), 250
Rattail fish (Macrouridae), 180*F*
Ray-finned fishes (Actinopterygii), 210
Rays (Batoidea), 216, 300–301*F*
Razorbill (*Alca torda*), 505*F*, 507
  see also Alcidae (auks)
Recruitment, 271
Red algae (*Rhodophyta* spp.), 142
Red-necked phalarope, 531
  appearance, 531
  diet, 536
  distribution, 533, 535
  habitat, 534–535
  migration, 536
  names, 536, 537*T*
  population declines, 537
  surface-tension feeding, 533, 534*F*
  see also Phalaropes
Red phalarope, 531
  appearance, 531, 532*F*
  distribution, 533
  habitat, 535
  migration, 536
  names, 536, 537*T*
  see also Phalaropes
Red-tailed tropic bird (*Phaeton rubricauda*), 524*F*
  see also Phaethontidae (tropic birds)
Regime shifts, 271
Remingtonocetidae, 468
  see also Cetaceans
Reproduction
  balaenopterids (rorquals), 399
  baleen whales (Mysticeti), 399–402
  benthic organisms *see* Benthic organisms
  bowhead whale (*Balaena mysticetus*), 399
  Brachyramphus, 506
  cephalopods, 81
  Cerorhinca, 506–507
  cold-water coral reefs *see* Cold-water coral reefs
  copepod(s) *see* Copepod(s)
  deep-sea fish *see* Deep-sea fish(es)
  demersal fish *see* Demersal fish(es)
  eels (*Anguilla*), 262
  fish *see* Fish reproduction
  leatherback turtle (*Dermochelys coriacea*), 453
  marine mammals *see* Marine mammals

planktonic foraminifera *see* Planktonic foraminifera
radiolarians, 31
sea otter (Lutrinae), 443–444, 446, 446F, 448–449
sea turtles *see* Sea turtles
see also Life histories (and reproduction)
Resource partitioning
coral reef fishes, 225
seabirds *see* Seabird(s)
Reynolds number, 199
Rhizostomae medusas, 52
Rhodophyta spp. (red algae), 142
Right whale dolphins (*Lissodelphis* spp.), 415, 418
Right whales (balaenids), 391, 392T, 470
filter-feeding apparatus, 393F
growth and reproduction, 399
habitat, 394
lateral profile, 393F
sound production, 397–398
trophic level, 489F
see also Baleen whales
Risso's dolphin (*Grampus griseus*), 411–415
River dolphins, 418, 421
trophic level, 489F
see also Odontocetes (toothed whales)
Rockhopper penguin, 548T, 550–551
see also Eudyptes (crested penguins)
Rock shag
behavioral displays, 529F
see also Shag(s)
Rorquals *see* Balaenopterids (rorquals)
Rough-toothed dolphin (*Steno bredanensis*), 419
Royal albatross (*Diomedea epomophora*), 539
see also Albatrosses
Royal penguin, 548T, 550–551, 551
see also Eudyptes (crested penguins)
Rynchopidae (skimmers), 510, **510–521**
breeding ecology and behavior, 510, 515, 519
chick-rearing, 519
clutch-size, 515–516, 519
egg incubation, 515–516, 519
phenology, 516
range, 512T
conservation, 519–521
measures, 521
status, 520, 520T
distribution, 510, 510–511, 512T, 515
exploitation, 519
foraging, 519
habitats, 514, 515
nesting, 515
physical appearance, 511–512, 513F, 514
plumage, 514
sexual dimorphism, 511–512
species, 510–511, 512T
taxonomy, 510, 510–511
threats, 520–521

of world, 512T
see also Seabird(s)
Rynchops niger (black skimmer), 513F

S

Saber-toothed blenny (*Aspridonotus taeniatus*), 282
Saccopharynx spp., 76F
Salinity, 124
effects on fish larvae, 292
influence on fish distribution, 275
Salmonids, **252–261**
anadromous/nonanadromous forms, 252
diet, 256
Atlantic/Pacific salmon, 256
Atlantic salmon postmolts, 256
masu salmon, 256
pink, chinook and coho salmon, 256
sockeye, pink and chum salmon, 256
distribution, 253
arctic charr, 255
Atlantic salmon, 253
chinook salmon, 255
chum salmon, 255
coho salmon, 255
masu salmon, 255
pink salmon, 254–255
sea trout, 253
sockeye salmon, 253–254
environmental factors, 258
sea surface temperature, 258–260
homing, 261
directed navigation, 261
life histories, 253T, 254T
Atlantic salmon, 254F
similarities/differences, 252–253
migration, 255, 255–256, 256F, 257F, 258F, 259F, 308
Atlantic salmon, 255–256
food-dependency, 255
movement patterns, 255
offshore movement, 255
Pacific salmon, 256
variation in postmolt patterns, 255
ocean climate affecting, 260–261
Atlantic Ocean, 260
effects of variations, 260
factors affecting abundance, 260–261
Pacific Ocean, 260
origins, 252
debates, 252
evolution, 252
predation, 256–258
Atlantic salmon predators, 257–258
Pacific salmon predators, 257
rates of oceanic travel, 259T
sea surface temperature
Atlantic Ocean, 259–260, 260F
Pacific Ocean, 259
river and ocean, 260
sockeye salmon, 259
surface salinity, 258
effects, 258
taxonomy, 252

identifying features, 252
see also Fish feeding and foraging; Fish horizontal migration; Fish larvae; Salmon
Salmon louse (*Lepeophtheirus salmonis*), 50
Salmon shark (*Lamna ditropis*), 216
Salpidae (salps), competition with krill, 69
Salvelinus (charr), 252
Salvelinus alpinus (arctic charr), 255
Sampling
deep-sea fauna, 180–181
macrobenthos, 169, 172, 173
meiobenthos, 161
Sand eels (Ammodytidae), 284, 300–301F, 366
Sanders, Howard, 182
Sardina (sardines), 249
Sardina pilchardus (pilchard), 280
Sardines (*Sardina, Sardinops, Sardinella*), 249, 249–250
anchoveta fisheries, 249–250
anchovy fisheries, 249–250
description and life histories, 249
distribution, 249
fisheries history, 249
pilchard fisheries, 249–250
Scaled petrel, 540F
see also Procellariiformes (petrels)
Scallops (Pectinidae)
adaptations to resist shear stress, 203–204
Scaridae (parrotfishes), 224
Schmidt, J., 262–263
Scolopacidae, Phalaropes *see* Phalaropes
Scomber scombrus (Atlantic mackerel), 280, 309–310, 310F
see also Mackerel
Scombridae, 300–301F
Indian mackerel, 250
mackerels, 250
migration, 309–310
tunas, 250
Scopelarchus analis (short fin pearleye), 354F
Scyliorhinus canicula (spotted dogfish), 369
Seabird(s), **497–502**
by-catches
longline fishing, 545, 584
conservation *see* Conservation
cost of flight, 497
distribution, 497
influence of oceanic currents, 497
diving, 565
flying *vs.*, 546–547
effects of and on fisheries, 500–501
exploitation *see* Human exploitation
feeding methods, 499–500, 499F
fish consumption, 500, 500T
fisheries interactions, 541–542, 569
effects of overfishing *see* Overfishing
foraging ecology *see* Seabird foraging ecology
foraging methods, 500

Seabird(s) (*continued*)
 hazards/threats, 555–556, 556F,
   557–558, 584
  habitat loss/modification, 557
  management *see* Conservation
  military activities, 561
  predation, 557
 importance of krill, 69
 indicators of ecosystem change, 501
 as indicators of ocean pollution, 560
  heavy metals, 560
  oil, 559
 life history, 499
 migration *see* Seabird migration
 nesting and feeding of young, 497–499
 population dynamics *see* Seabird
   population dynamics
 prey, 499–500
 reproduction *see* Seabird reproductive
   ecology
 reproductive rates, 499
 role in oceanic carbon cycling, 500
 species included, 497, 498T
 threats, 501
 vulnerabilities, 556–557, 556F
  activity patterns, 557
  adults, 557
  chicks, 556–557
  ecosystem, 556F, 557
  egg stage, 556
  juveniles, 557
 wing loadings, 546–547
  migration patterns and, 573, 573F
Seabird foraging ecology, **562–570**
 behavior, 564–565
  wind and, 566
 habitat, 562
  environmental gradients, 562–563
  fronts, 563, 563F
  sea ice, 564, 564F
  topographic features, 563–564, 563F
  water masses, 562, 562F
 prey associations, 566
 resource partitioning, 566–567
  competition, 569
  competition with fisheries, 569
  day *vs.* night feeding, 568
  habitat type, 567–568
  kleptoparasitism, 565, 568
  morphological/physical factors,
    568–569
  mutualism, 565, 568
  prey/predator size, 567, 568F
  prey selection, 566–567, 567F, 568F
 strategies, 565
  feeding flocks, 565
  maximization of search area, 565–566
  nocturnal feeding, 565
  physical features, 565
  subsurface predators, 565
Seabird migration, **571–581**
 distances, 497
 morphological adaptations, 572–574
  wing loading, 573, 573F
  wing shape, 572–573
 orientation and navigation, 571

 sensory mechanisms, 571–572
 physiological/behavioral adaptations,
   572
  post-breeding *vs.* pre-breeding, 572
  terrestrial species *vs.*, 572
 reasons for, 571
 studies, 571
 types, 574
  dispersal, 574
  partial, 574
  true, 574
Seabird population dynamics, **582–585**
 characteristics, 582
 clutch size, 583
 colony size, 582
 delayed breeding age, 583, 584
 food availability, 582, 584
  winter, 584
 hypotheses, 584
 inshore-foragers, 582, 583, 583–584
 interannual variation, 583, 583–584
 inter-colony variation, 582–583
 life spans, 584
 offshore-foragers, 582, 583–584
 philopatry, 582
  genetic implications, 582
 survival rates, 584
Seabird reproductive ecology
 reproductive rates, 499
Sea butterflies (Thecosomata), 56
Sea cow, Steller's, 431–432, 432F
 see also Dugongidae
Sea cucumbers (Holothuroidea), 58, 130,
   177F
Seagrasses
 destruction, threat to manatees, 439
 dugong 'cultivation grazing', 479
Seahorses (Syngnathidae), 300–301F
 see also Hippocampus (sea horse)
Sea ice
 seabird abundance and, 564, 564F
Sea lions *see* Otariinae (sea lions)
Seals *see* Pinnipeds (seals)
Seamount(s), 178
 seabird abundance and, 563
Sea otter (Lutrinae), **442–449**
 adaptations to life at sea, 444
  diet, 445–446
  diving, 445–446, 446F
  locomotion, 444
  reproduction, 446, 446F
  thermoregulation, 444–445
 conservation status, 384T
 decline, 442
 food/foraging, 443, 448, 448F, 482
 habitat, 442, 443, 443F
  foraging, 443
 life history, 443–444
  longevity, 444
  mortality, 444
  reproduction, 443–444, 446, 446F,
    448–449
  survival, 444, 448F, 449
 myoglobin concentration, 460T
 physical characteristics, 442–443
  body outline and skeleton, 386F

  dentition, 442–443, 444F
  fore legs, 442–443
  fur, 442–443
  hind feet, 442–443, 444
  insulation, 442–443, 445
  tail, 442–443, 444
  weight, 442–443
 range, 443, 443F
 responses to changing ecological
   conditions, 448
  diet, 448, 448F
  populations, 449
  reproduction, 448–449
  survival, 449
 role in structuring coastal marine
   communities, 446–447
  kelp forests, 446–447, 447F, 491
  soft-sediment subtidal, 447–448
 trophic level, 488, 489F
 see also Marine mammals
Sea spiders (pycnogonids), 112T, 180F
Sea trout (*Salmo* spp.), 252, 253
Sea turtles, **450–457**
 adaptations to aquatic life, 450
  lacrimal salt glands, 450, 452, 453F
  limbs, 450, 451–452, 451F, 452F
  physiologic, 450, 452, 452F
 as endangered species, 451
 evolution, 450, 451
 fossils, 451
 general biology, 451–453
 growth rates, 451
 reproduction, 451–452
  egg(s), 450
  egg incubation, 450, 451–452
  nest chamber, 450
  nesting behavior, 451–452, 452F
  temperature-dependent sex
    determination, 450, 451–452
 taxonomy, 450
  family Cheloniidae *see* Cheloniids
  family Dermochelyidae *see*
    Dermochelyids
Sea urchins (Echinoidea)
 barrens, 447F
 sea otter effects, 446–447, 447, 447F,
   491
 threat to kelp forests, 146
Seaweed(s)
 life cycles
  environmental controls, 143
  two-phase life cycle, 143
  isomorphic *vs.* heteromorphic, 143
 see also Algae; Phytobenthos
Seaweed ecology
 effects of wave action, 144
 phytobenthos, 143–144
 successional colonization, 144, 145T
  conditions favoring opportunists, 144
  replacement of opportunists, 144
 temperate species, 143
 zonation
  deep-growing plants, 144
  estuaries, 144
  physical and biotic factors, 144
  rock pools, 144

subtidal area, 144
temperate shores, 143–144
Sea whips (Gorgonacea), 177F
Selective deposit feeders, 127, 127–128, 131
Selective tidal stream transport, 271
Semaeostomae medusas, 52, 53F
Semibalanus balanoides (acorn barnacle), 203
Sensory systems
cephalopods, 80
copepods, 48
demersal fishes, 369
dolphins and porpoises, 419–420
fish hearing, 372–373
fish lateral lines, 376–377, 376F
fish schooling, 338–339
sperm and beaked whales, 408–409
see also Bioluminescence; Fish hearing and lateral lines; Fish vision
Sepiidae (cuttlefish)
buoyancy, 80
spawning, 81
Seston, 202
Sexual dimorphism, 410, 422
dolphins and porpoises, 411
eels (*Anguilla*), 265
killer whale (*Orcinus orca*), 411
mesopelagic fish(es), 241
pinnipeds (seals), 428
Rynchopidae (skimmers), 511–512
sperm whales (Physeteriidae and Kogiidae), 404–405
Shag(s)
cormorants *vs.*, 525–526
plumage, 526
rock, behavioral displays, 529F
see also Phalacrocoracidae
Shanny (*Lipophrys pholis*), 235–236, 236F
Shear stress
resistance adaptations, 203
Shearwater(s), 539
Audubon's, 539
Manx, expansion of geographical range, 542
migration, 573F, 575–577, 576T
pursuit plunging, 569
short-tailed, 542, 573
see also Procellariiformes (petrels)
Shelf ecosystems, seabird abundance and, 563, 563–564
Shepherd's beaked whale (*Tasmacetus shepherdii*), 406
Short fin pearleye (*Scopelarchus analis*), 354F
Short-tailed shearwater (*Puffinus tenuirostris*), 497, 542
Shy albatross (*Diomedea caut*), 542, 575
see also Albatrosses
Silicon (Si), 102
radiolarian skeletons, 30
Silver eels *see* Eels
Siphonophores, 54
bioluminescence, 113
Calycophorae, 54, 55F

Cystonectae, 54, 55F
fragility of colonies, 54
Physonectae, 54, 55F
Sirenians, 381, 386, 386F, **431–441**
behavior, 437
classification, 384T, 431–432, 465, 465T
family Dugongidae *see* Dugongidae
family Trichechidae *see* Manatees
conservation, 439–441
status, 384T
distribution, 431–432
ecology, 437–439
evolution, 431, 469–470
fossil history, 431
exploitation, 437–438, 439F
future outlook, 439
grinding molars, 482F
mating behavior, 436–437
migration and movement patterns, 479
morphology, 435–437, 436F
physiology, 435
population biology, 437
social organization, 437
species
extinct, 431, 470–471
living, 431–432, 465T, 470–471, 471F
threats, 437–438, 438F, 439F, 440F
see also Marine mammals
Skimmers *see* Rynchopidae (skimmers)
Skuas
migration, 579
see also Charadriiformes
Small subunit ribosomal RNA (SSUrRNA), 21, 22F
Snares penguin, 548T, 550–551
see also Eudyptes (crested penguins)
Snow petrels (*Pagodroma nivea*), 543
see also Procellariiformes (petrels)
Sockeye salmon (*Oncorhynchus nerka*), 253–254, 256, 256F, 259, 305F
SOIREE *see* Southern Ocean Iron Enrichment Experiment (SOIREE)
Sole (*Solea solea*), 281
Solea solea (sole), 281
Solenosmilia variabilis coral, 188–189
Songs, marine mammals, 484, 484F, 485F
Sooty shearwater (*Puffinus griseus*), 540F
migration routes, 575–577, 578F
see also Shearwater(s)
Sooty tern, 513F
see also Sternidae (terns)
Sotalia fluviatilis (tucuxi), 411, 418
Sousa spp. (humpback dolphins), 415
Southern bluefin tuna (*Thunnus maccoyii*), 309–310
Southern Ocean
benthic foraminifera, 150T
impact of conditions on fish, 218
isolation, 218
krill *see* Krill (Euphausiacea)
temperature variations, 218
see also Antarctic fish(es); Polar ecosystems

Southern Ocean Iron Enrichment Experiment (SOIREE), 103
Southern phocids *see* Monachinae (southern phocids)
Spawning
cephalopods *see* Cephalopods
copepods, 47
coral reef fishes, 224
cuttlefish (*Sepia* spp.), 81
deep-sea fishes, 231
demersal fishes, 369–370
intertidal fishes, 237, 237–238
mesopelagic fish(es), 240–241
Octopodidae (octopuses), 81
Sepiidae (cuttlefish), 81
see also Fish reproduction
Species flocks, Antarctic fishes, 218
Spectacled porpoise (*Phocoena dioptrica*), 416, 421
Sperm whales (Physeteriidae and Kogiidae), 382–383T, 403, 404T, 408F, **403–410**
acoustics and sound, 408–409
acoustic behavior, 408–409
sound production mechanism, 409
spermaceti organ, 404–405, 409
anatomy and morphology
asymmetry of the skull, 405
cranial features, 404–405
rostrums, 410
sexual dimorphism, 404–405
spermaceti organ, 404–405, 409
conservation, 409–410
current threats, 409
anthropogenic noise, 410
whaling industry, 409
defense from predators, 483
distribution and abundance, 405–406
global distribution, 405–406
diving characteristics, 459T
evolution, 469
foraging ecology, 406
competition between species, 406
diving ability, 406
squid diet, 406
suction feeding, 406
growth and maturation, 388F, 389
home ranges, 389
lack of knowledge, 403
migration and movement patterns, 474, 474F
myoglobin concentration, 460T
Physeter macrocephalus, 403, 408F
predation, 409
Kogia spp. defence mechanisms, 409
predators and defence mechanisms, 409
relationship to other cetaceans, 405F
social organization, 406–408
allomaternal care, 407
epimeletic behavior, 407, 410
female groups, 407
Kogia spp., 407
male groups, 406–407
sound production
mechanism, 409

Sperm whales (Physeteriidae and Kogiidae) (*continued*)
  trophic level, 489*F*
  see also Odontocetes (toothed whales)
Spheniscidae (penguins), 69
Sphenisciformes (penguins), 69, **546–554**
  adaptations to aquatic lifestyle, 546, 547*T*, 552–553
  balance of life at sea and time ashore, 553
  breeding
    chick-rearing, 553
    success, determinants of, 553
  characteristics, 546
  conservation, 553–554
  distribution, 546, 548*T*
  diving, 553
    flying *vs.*, 546–547
  fat storage, 553
  flightlessness, 546, 547, 553–554
  foraging, 553
    inshore foragers, 553
    offshore foragers, 553
  fossils *see* Fossil(s)
  insulation, 553
  migration, 574
  molting, 553
  nearest living relatives, 546, 547*F*
  porpoising, 552–553
  species, 547–549, 548*T*
  threats, 553–554
    global warming, 553
    predation, 553–554
  underwater swimming, 552–553
  see also Seabird(s)
Spheniscus, 548–549
  breeding patterns, 549
  characteristics, 548*T*, 549, 549*F*
  discovery, 548
  distribution, 548*T*, 549
  feeding patterns, 548*T*, 549
  migration, 574
  nests, 548*T*, 549
  species, 548
  see also Sphenisciformes (penguins)
Spheniscus demersus (African penguin), 548, 548*T*, 549, 553–554
Spheniscus humboldti (Humboldt penguin), 548, 548*T*, 549, 553–554
Spheniscus magellanicus (magellanic penguin), 548, 548*T*, 549, 549*F*
Sphyraena spp. (barracuda), 300–301*F*
Spinner dolphin (*Stenella longirostris*), 411–415, 420–421, 420*F*, 421–422
Spiny eel (*Notacanthus chemnitzii*), 357*F*
Spiochaetopterus oculatus polychaete worm, 204
Spio setosa polychaete worm, 204
Spirula cephalopod, buoyancy, 80
Spotted dogfish (*Scyliorhinus canicula*), 369
Sprat, 249
  description and life histories, 249
  distribution, 249
Sprat (*Clupea sprattus*), 249

Sprat (*Sprattus sprattus*), 280
Sprattus sprattus (sprat), 280
Spur dog (*Squalus acanthias*), 280
Squalus acanthias (spur dog), 280
Squid (Teuthida), 58, 81
  bioluminescence, 115
SSUrRNA, 21, 22*F*
Stargazers (*Astrocopus* and *Uranoscopus* spp.), 216
Steller's sea cow (*Hydrodamalis gigas*), 431–432, 432*F*, 437–438, 438–439
  see also Dugongidae
Stenella (dolphins), 417–418
Stenella attenuata (pantropical spotted dolphin), 420*F*, 421–422
Stenella coeruleoalba (striped dolphins), 421
Stenella longirostris (spinner dolphin), 411–415, 420–421, 420*F*, 421–422
Stenella longirostris roseiventris (dwarf spinner dolphin), 415
Steno bredanensis (rough-toothed dolphin), 419
Stercoracidae
  migration, 579
  see also Charadriiformes
Sterna fuscata (sooty tern), 513*F*
Sterna hirundo (common tern), 513*F*
Sternidae (terns), **510–521**
  body size
    colony size and, 515, 516*F*
    territory size and, 515, 516*F*, 518
  breeding ecology and behavior, 510, 515, 516*F*, 518–519
    chick-rearing, 518, 519
    clutch size, 515–516, 519
    courtship, 518
    egg incubation, 518
    'fish fights', 518–519
    mobbing, 515
    nest site selection, 518, 519
    phenology, 516
    range, 512*T*
  conservation, 519–521
    measures, 521
    status, 520, 520*T*
  daily activity patterns, 519
  distribution, 510, 512*T*, 515, 518
  foraging, 519, 568–569
  habitat, 514, 515
    nesting, 515
  migration, 515, 579, 580*T*, 581*F*
  physical appearance, 511–512, 513*F*, 514
    Laridae (gulls) *vs.*, 514
    plumage, 514
  species, 510, 512*T*
  taxonomy, 510
  threats, 520–521
  of world (types), 512*T*
  see also Seabird(s)
Stictocarbo magellanicus (rock shag)
  behavioral displays, 529*F*
  see also Shag(s)
Stingray (*Dasyatis sayi*), 351*F*

Stomiiform fishes, bioluminescence, 116–117
Stoplight loosejaws (*Malacosteus* spp.), 360*F*
Storm petrels (Oceanitidae), 539
  migration, 576*T*, 577
  see also Procellariiformes (petrels)
Storm surges
  areas affected
    North Sea *see* North Sea
  drag coefficient, sea surface, 184
  see also Drag coefficient
  see also Fish feeding and foraging; Tide(s)
Strandings, 422
Stratification
  impact on primary productivity, 101
Striped dolphins (*Stenella coeruleoalba*), 421
SubAntarctic diving petrel, 540*F*
  see also Procellariiformes (petrels)
Sublittoral zone, 126*T*
Submersibles
  cold-water coral reefs knowledge/research, 189*F*
Subtropical convergence areas, 271
Subtropical gyres, 271
Sula leucogaster (brown booby), 524*F*
  see also Sulidae (gannets/boobies)
Sulfur dioxide (SO$_2$)
  pollution, 560
Sulidae (gannets/boobies), 522, 525
  breeding patterns, 525
  characteristics, 525, 528*T*
  distribution, 525
  feeding patterns, 525, 528*T*, 569
  migration, 577–579
  species, 524*F*, 525
  see also Pelecaniformes
Sunfish (*Mola mola*), 300–301*F*
Suprabenthos (hyperbenthos), 166, 176, 199
  see also Benthic boundary layer (BBL)
Suspension feeders, 127, 127–128, 131, 166
Susus (*Platanista* spp.), 418, 419, 420, 421
Swell shark (*Cephaloscyllium ventriosum*), 353*F*
Swimbladder retia, 271
Swordfish (*Xiphias gladius*), 216
Swordfishes (Xiphiidae), 300–301*F*
Symbiotic relationships
  algae-gelatinous zooplankton, 52
  algae-radiolarians, 28–29, 32
  bacteria-deep-sea animals, 130
  bacteria-fish, 115–116
  bacteria-squid, 81, 115
  chemoautotrophic bacteria-vent animals, 178–179
  dinoflagellates-radiolarians, 51–52
Syngnathidae, 300–301*F*
Synthliboramphus, 503*T*
  reproduction, 506, 507
  see also Alcidae (auks)

T

Tanaid crustaceans, 180*F*

Tasmacetus shepherdii (Shepherd's beaked whale), 406

Taxonomy *see* Classification/taxonomy (organisms)

Teleosts, 212, 212–213, 213*F*
 acanthopterygian spp., 212–213
 body fluids and osmoregulation, 215
 buoyancy, 215
 classification difficulties, 213
 euteleosts, 212–213
 features, 213*F*
 fins, 298, 298*F*
 four main radiations, 212, 212*F*
 habitat refinements, 209
 majority of all fish, 213
 osmoregulation, 215
 radiations, 212

Temperature
 effect on copepods, 46–47, 46*F*
 effect on primary production, 106
 effect on radiolarians, 31–32
 effects on fish larvae, 291–292
 influence on fish distribution, 274
 latitudinal differences, 123–124
 vertical differences, 123–124

Temperature-dependent sex determination (TSD), sea turtles, 450, 451–452

Tench (*Tinca tinca*), 356*F*

Terns *see* Sternidae (terns)

Teuthida (squids), 58, 81
 bioluminescence, 115

Thalassocalycida ctenophores, 54–56

Thalassoma bifasciatum (cleaner wrasse), 328

Thaliacea tunicates, 9, 58
 Doliolida, 59–60
 Pyrodomida, 58–59
 Salpida, 59*F*, 60

Thecosomata (sea butterflies), 56

Theragra chalcogramma (Alaska/walleye pollock), 310–311, 311*F*, 319–320, 323*F*, 328*F*

Thick-billed murre, 503, 505*F*
 see also Alcidae (auks)

Thorson, Gunnar, 127*T*, 169
 classification of the benthos, 125
 distribution of benthic larvae, 128, 128*F*

Three-spined stickleback (*Gasterosteus aculeatus*), 283

Thunnus (tuna), 282, 298
 optimization of behavior, 282–283

Thunnus alalunga (albacore tuna), 309–310, 309*F*

Thunnus maccoyii (southern bluefin tuna), 309–310

Thunnus thynnus (Atlantic bluefin tuna), 216, 217, 282–283, 300–301*F*

Thysanoessa inermis (Euphausiidae), 65–66

Thysanoessa raschii (Euphausiidae), 65–66

Tidal cycles
 seal haul-out sites and, 476–478

Tidal streams
 effects on migration, 313–314, 313*F*, 314*F*

Tide(s)
 continental shelf *see* Continental shelf
 impact on vertical migration, 318–319

Tidepool sculpin (*Oligocottus maculosus*), 235

Tinca tinca (tench), 356*F*

Toadfish (*Opsanus tau*), 374–375

Toothed whales *see* Odontocetes (toothed whales)

Topographic features, seabird abundance and, 563–564, 563*F*

Toxins
 biotoxins, 422
 plankton viruses enabling hosts to produce, 17–18

Trace element(s)
 requirements for primary productivity, 101, 102
  see also Iron; Manganese; Molybdenum; Nutrient(s)

Trachymedusae medusas, 52, 53*F*

Trade-off
 feeding-predator avoidance, 328

Transduction
 DNA, 19

Transformation, DNA, 19

Traps
 crab fishing, 456

Trichechidae *see* Manatees

Trichechus inunguis (Amazonian manatee), 434–435, 434*F*
 see also Manatees

Trichechus manatus (West Indian manatee), 432–434, 433*F*
 see also Manatees

Trichechus manatus latirostris (Florida manatee), 432, 433*F*, 436*F*, 438*F*
 see also Manatees

Trichecus senegalensis (West African manatee), 434, 434*F*
 see also Manatees

Trophic levels
 definition, 488

Tropic birds *see* Phaethontidae (tropic birds)

Trumpet fish (*Aulostomus maculatus*), 282

Tucuxi (*Sotalia fluviatilis*), 411, 418

Tuna (*Thunnus*), 250, 282, 298
 Atlantic/Pacific fisheries, 250
 description and life histories, 250
 fisheries *see* Tuna fisheries
 principle species, 250
  see also Thunnus (tuna)

Tuna fisheries
 Mediterranean, 250

Tunicates
 bioluminescence, 112*T*, 114–115
 Larvacea/Appendicularia, 58, 59*F*, 60
 Salpida, 59*F*
 Thaliacea, 58

Doliolida, 59–60
Pyrodomida, 58–59
Salpida, 59*F*, 60

U

Ultrasonic sounds, 410

Upwelling ecosystems
 planktonic foraminifera, 36
 seabird populations and, 583
  see also Pelagic biogeography; Pelagic fish(es); Plankton

Uranascopus spp. (stargazers), 216

Urchin barrens, 446–447, 447*F*
 see also Sea urchins

Uria, 503*T*
 diet, 506
  see also Alcidae (auks)

Uria aalge, 503

Uria lomvia (Brunnich's guillemot), 503, 505*F*

Ursids, 381, 384*T*

V

Vanadium (V), 101

Vaquita (*Phocoena sinus*), 416*F*, 421

Vargula spp. ostracod, 111–113

Ventral grooves, definition, 402

Vertical migration *see* Fish vertical migration

Viperfish (*Chauliodus macouni*), 76*F*

Viruses, 13, 15, 16, 19
 abundance, 15
 attack of bacterioplankton, 27
 blooms, control by, 18
 contributing to bacteria mortality, 13, 16–17
 description, 13–14
 effects on host species composition, 18
 general properties, 13–14
 genetic transfer, 19
 life cycles, 14*F*
 lytic infections, 15–16
 plankton *see* Plankton viruses
 reproduction, 13–14, 14*F*
 resistance to, 18–19
  see also Plankton viruses

Vision
 dolphins and porpoises, 419
 fish *see* Fish vision
 marine mammals, 463–464

Visual pigments, fish vision, 355–356
 spectral absorption, 356–359, 359*F*
  bioluminescence detection, 358, 360*F*
  cone visual pigment, 358–359, 361*F*
  fresh/saltwater species differences, 357–358
  optical environment, 358, 360*F*
  optimization, 359
  range, 356–357
 spectral responses, 361
  limits, 361
  proving color vision, 361
 structure, 355–356
  chromophore and opsin, 355–356

Visual pigments, fish vision (*continued*)
  factors affecting absorption spectrum,
    356
Viviparity, 131, 333

W

Waipara (New Zealand), fossil penguins,
  546
Walleye pollock *see Theragra*
    chalcogramma (Alaska/walleye
    pollock)
Walruses *see* Odobenidae (walruses)
Wandering albatross (*Diomedea exulans*),
    539, 540F, 543, 545, 575
  see also Albatrosses
Water
  depth, seabird abundance and,
    563–564
Water mass(es)
  seabird distribution and, 562, 562F
Water pressure
  relationship to depth, 123
Waved albatross (*Phoebastria irrorata*),
    574–575, 575
  see also Albatrosses
Weddell seal (*Leptonychotes weddellii*)
  dive duration and blood lactate
    concentration, 461, 461F
  myoglobin concentration, 460T
    see also Phocidae (earless/'true' seals)

West African manatee (*Trichecus*
    senegalensis), 434, 434F
  see also Manatees
Western boundary current(s), 271
West Indian manatee (*Trichechus*
    manatus), 432–434, 433F
  see also Manatees
Whaling industry, sperm and beaked
    whales, 409
Whiptail gulper (*Saccopharynx* spp.), 76F
White-flippered penguin, 549
Wideawake (*Sterna fuscata*), 513F
  see also Sternidae (terns)
Wilson's phalarope, 531
  appearance, 531
  diet, 536
  distribution, 535
  habitat, 534–535
  migration, 536
  names, 536, 537T
  surface-tension feeding, 533
    see also Phalaropes
Wilson's storm petrel, 540F
  see also Procellariiformes (petrels)
Wind(s)
  seabird foraging behavior and,
    566
Wrasse (*Labroides* spp.), 223, 224, 282,
    300–301F
Wrasse blenny (*Hemiemblemaria*
    simulus), 328

X

Xenodermichthys copie (bluntsnout
    smooth-head), 357F
Xiphias gladius (swordfish), 216
Xiphiidae (swordfishes), 300–301F

Y

Yangtze river dolphin (*Lipotes vexillifer*),
    416F, 417, 418, 421
Yellow eels *see* Eels
Yellow-eyed penguin, 548T, 549–550
  breeding patterns, 549–550, 553–554
  characteristics, 548T, 549–550, 550F
  feeding patterns, 548T, 550, 553
  migration, 574
  nests, 548T, 549–550
    see also Sphenisciformes (penguins)
Yellow-nosed albatross (*Diomedea*
    chlororhynchos), 543, 545
  see also Albatrosses

Z

Ziphius cavirostris (Cuvier's beaked
    whale), 406, 407F
Zonation, 271
Zooplankton, 4
  gelatinous *see* Gelatinous zooplankton
  see also Copepod(s)